Engineering Science and Technology: Innovations for the Future

Proceedings of the International Conference on Recent Innovations in Engineering Science and Technology (ICRIET2025), May 09–10, 2025, Bengaluru, India

Engineering Science and Technology: Innovations for the Future

Edited by

Reeja S. R.

Bore Gowda

Y. S. Rammohan

Ganesan Prabu Sankar

G. Jayalatha

CRC Press
Taylor & Francis Group
Boca Raton London New York

CRC Press is an imprint of the
Taylor & Francis Group, an **informa** business

First edition published 2026

by CRC Press
2385 NW Executive Center Drive, Suite 320, Boca Raton FL 33431

and by CRC Press
4 Park Square, Milton Park, Abingdon, Oxon, OX14 4RN

CRC Press is an imprint of Informa UK Limited

British Library Cataloguing-in-Publication Data
A catalogue record for this book is available from the British Library

ISBN: 9781041166818 (hbk)
ISBN: 9781041166801 (pbk)
ISBN: 9781003685876 (ebk)

DOI: 10.1201/9781003685876

Typeset in Time New Roman
by HBK Digital

Contents

List of Figures ix

List of Tables xix

List of Contributors xxii

Foreword xxiv

Preface xxv

Acknowledgement xxvi

Chapter 1   Deep learning techniques for judicial judgement system   1
Hemanth N. G., Pushpa Ravikumar, Harish S., Shruthi G. K. and Saritha N.

Chapter 2   Advancing cancer diagnosis: A machine learning approach for early detection and improved outcomes   8
Kothai G., Suryaprakash S., Dharshna Kumar K. S., Gokulnanthan C. and Priyadharshan S.

Chapter 3   Animatrix: Personalized multimodal content generation for advertising   14
Praadnya H., Prannay Hemachandran, Sanskar Khatri and P. Kokila

Chapter 4   Automated code validator with LLM-powered test case generation   20
Sai Deep K., Goutham G., Priyanka C., Roopa B. S. and C. Christlin Shanuja

Chapter 5   FlowCast: A simple method for traffic prediction   28
Drishya Dechamma M. P., Ananya M., Madhu Chandrika, Anusha P. N., Deepak N. R. and Shruti B.

Chapter 6   Smart sericulture: Leveraging CNNs for silkworm disease detection   37
Komala K. V., Lata B. T. and Venugopal K. R.

Chapter 7   Quantum-enhanced stock market prediction: Integrating AI, sentiment analysis, and multimodal data fusion   44
Chinmayi H. N., Amulya G., Shubha Rao V. and SnehaGirish

Chapter 8   Intelligent women's safety protocol for automated risk detection   52
Suresh M. B., Abhinav S. Bhat, Pavan S., Sharath M. and Vishal B. M.

Chapter 9   Performance analysis of classification algorithms to predict failures in IIoT devices   59
Wasim Yasin and Manu Elappila

Chapter 10   Leveraging generative adversarial networks (GANs) for deepfake   66
K. Sivasankari, Vijayakumar S., Gurram Haneesh Chowdary and Joel Thomas S.

Chapter 11   AI-driven urban resilience system: Tackling urban flooding, rescue optimisation, and heat island mitigation   71
Shreya Gore, Satwik G. Vaidya, Rachana Deepak Kadkol and Mamatha G.

Chapter 12   Few-shot learning for behavioural biometrics: A personalised approach to a nomaly detection with limited data   80
Dev Rahul More, Ved Datar, Shishir Walvekar, Preeti Godbole and Divyang Jadav

Chapter 13   LSTM neural network-based Li-ion battery cycle forecast   86
Vimala Channapatna Srikantappa and Seshachalam Devarakonda

Chapter 14   Coco areca wellness CNN-based disease detection in coconut and arecanut crops   90
Beena K., Sangeetha V., Archana P., Harshitha R., Kavana N., Madhu and Sneha Shree S.

Chapter 15   Artificially intelligent adaptive voice assistant for next-generation desktop automation   98
Ushasri Gunti, Nilanjana Jamindar, Sanjana O. R. and Shama S. K.

Chapter 16   Medilink: AI-powered virtual healthcare for brain tumour and eye disorder detection   104
Swetha Malhar Kulkarni, Rekha B. Venkatapur, Soujanya N. and Skanda Kumar H. S.

Chapter 17   LLM-based automated scoring and feedback for essays in the Marathi language   111
Kaushik Sakre, Rasika Ransing, Shivam Shinde, Neha Kudu and Siddhi Talkar

Chapter 18   A blockchain-based web application for enhancing financial transparency in NGOs   118
Raghavendrachar S., Gagan Shivanna, Aditya V., Karthik H. N. and Arthan M. Gowda

Chapter 19   IoT-based gamified learning model for visually impaired users   125
Vijaya Laxmi Mekali, B. G. Prajwal, Chethan H., Guruprasad Y. S. and Jashwanth P. C.

Chapter 20   Real-time snake detection and alert system in agricultural fields using computer vision and IoT   131
Deepa S. R., Anagha Shastry, Abhilasha Patil, Pavithraa G. and Varshitha Sridhar

Chapter 21   Intelligent IoT-enabled hydroponics system using machine learning   140
Pankaja K. N. and Santhosh B. J.

Chapter 22   IoT-integrated smart mirror for continuous health and wellness monitoring   146
Navyashree Ganpisetty, Charan Teja H., Divyabhavani Ganpisetty, Sneha Hugar, Sudhanva S. Haritasa and Shalini Shravan

Chapter 23   Automatic vehicle emergency response and alert system using GPS and GSM   152
Sushma Nagaraj, Suresh M. B. and Sahana Salagare

Chapter 24   Smart IoT-enabled vending machine for medical and sanitary supplies   159
Surekha Byakod, Ankitha Devlokam, Challa Deepika, Vignesh U. and Yeddulapalli Sahithi

Chapter 25   Revolutionising supply chains: An AI-driven approach to inventory management   165
Vijay Mohan Shrimal, Avishkar, Harjot Singh and Aryaman M. Singha

Chapter 26   Workload prediction in cloud computing: A machine learning-based approach   172
Vijay Mohan Shrimal, Aayush Gupta and Mitalee Verma

Chapter 27   Smart traffic sign detection using deep learning for intelligent vehicle control   180
Santhosh Kumar C., Sahil Pal, Yash Patel, Radke Animish and Harsh Jain

Chapter 28   Developing a sustainable online ecosystem for second-hand accessories within a university community   186
Binod Prasad Joshi, Vijay Mohan Shrimal, Shivam Singh, Mamta Sharma and Alok Das

Chapter 29   AI-powered adaptive traffic light system   192
Raghav Mehra, Shourya Shri, Hardik Agarwal and Subhadeep Dutta

Chapter 30   Predicting drug-target interactions using graph neural networks: A deep learning approach   199
Narla Swamy Pavan Koushik, Peddinti Priyanka and Reeja S. R.

Chapter 31   Modelling and simulation of power quality improvement by using
VSC-based DSTATCOM   206
Suresh H. L., Prajwal A. and Priyanka T.

Chapter 32   Performance analysis and optimisation of solar-powered alkaline electrolyser for
green hydrogen production   214
P. Kezia Joy Kumari, Prajwal A., Ashwini A. V., Rekha S. and Suresh H. L.

Chapter 33   Design and optimisation of 3-bit flash ADC with low power double tail comparator   220
*Vasudeva G., Mallikarjun P. Y., Tripti R. Kulkarni, Bhuvan M.,
Gowthami Kc. and Nidhi N.*

Chapter 34   Vehicle detection using saliency model generation and SVM   227
Dinesh Kumar D. S., Mohankumar C. E. and Senthilkumar S.

Chapter 35   Development of command-based automated glass cleaning robot   235
Roopa S., Grace D., Keerthi M., Harshitha S. and Brunda H.

Chapter 36   A review of AI prompt management tools and a proposed Git-based solution   241
Rajath Nag Nagaraj, Devika B., Rekha N. and Disha R.

Chapter 37   Design of a deep learning-based controller for a 4-leg inverter to independently
control two 3-phase induction motors   247
Venu Gopal B. T., Shreedhara R., Hareesh Kumar, Shalini J. and Usha Rebecca

Chapter 38   Design and development of a log-periodic dipole antenna for space
weather monitoring   254
Gireesh G. V. S., Jeyanthi R., Harshini V., Dharanisha D. and Joshiga S.

Chapter 39   Performance evaluation of Viterbi encoder with frequency hopping spread
spectrum in Rayleigh fading   261
V. Sangeetha, Gandhamani Mohanraju and P. N. Sudha

Chapter 40   Voltage sag characteristics in power distribution systems under fault conditions   267
Anmol Saxena, Richa and Rahul Vivek Purohit

Chapter 41   Construction and performance evaluation of zero energy cooling chamber for
sustainable storage solutions of vegetables   274
*S. Banerjee, C. Chandre Gowda, B. N. Skanda Kumar, H. Jaswanth, K. S. Prajwal,
D. Dhanush and G. M. Kushal*

Chapter 42   Synthesis, characterisation and GCMS analysis of bio-diesel blend for
compression ignition engine using non-edible vegetable oil accessible in Indian states   281
*Younus Pasha, Tansif Khan, D. R. Swamy, Mohammed Riyaz Ahmed, Harish H.
and Salim Sharieff*

Chapter 43   Influence of representative volume element (RVE) size and particle clustering on
tensile properties of epoxy/SiO_2 nanocomposites   295
*Sumana, Pratik P. Jakanur, Sridhar M., M. A. Umarfarooq, N. R. Banapurmath,
Ashok M. Sajjan, Kartheek Ravulapati and B. H. Maruthi Prashanth*

Chapter 44   Sliding behaviour of NiAl and NiCr coatings on mild steel substrate   300
*Nadeem Pasha K., Salim Sharieff, Nawaz Ahmed, Younus Pasha, Tansif Khan and
Dinesh H. A.*

Chapter 45   Effect of silicon carbide reinforcement on Aluminium 7075 alloys on mechanical and vibration properties – Stir casting process   308
Srinidhi Acharya S. R.

Chapter 46   Effect of exhaust gas recirculation on emission and performance characteristics of CRDI dual fuel engine powered with Peruvenia Thevetia and hydrogen induction   315
Mahantesh Marikatti, N. R. Banapurmath, Y. H. Basavarajappa, Shambhuling Uppin, V. S. Yaliwal, Kartheek Ravulapati, K. S. Nivedhitha and A. M. Sajjan

Chapter 47   Advanced time series forecasting for vehicle maintenance scheduling   321
Praveen N.,Girish T. R., Ranganath N. and Harish U.

Chapter 48   Performance and emission characteristics of diesel engine using blends of Calophyllum inophyllum biodiesel   327
Dodda Hanamesha, D. Madhu,Rudresh B. M. and Lakshmikant Shivanayak

Chapter 49   Rare earth element (Tb) doped calcium aluminate nanoparticles: Synthesis, structural optical and photoluminescence studies   334
B. S. Shashikala, H. B. Premkumar, G. P. Darshan, D. R. Lavanya and H. Nagabhushana

Chapter 50   Azo-benzothiazole-based copper (II) complexes: Synthesis, spectral investigation, thermal kinetics and antibacterial investigations   340
Nagaraj Basavegowda, Harisha S., Manjunath, Mohan Reddy R., S. R. Kiran Kumar and Shoukat Ali R. A.

Chapter 51   Synergistic catalysis of neem oil for biodiesel production using a K_2CO_3–NH_4OH system   351
Swarna S., Puneetha J., Shruthi K. S., Manjunatha N. K. and Shashidhar S.

Chapter 52   The minimum pendant dominating Harary energy of a graph   360
Nataraj K., Puttaswamy and Purushothama S.

Chapter 53   Minimum pendant dominating colour energy of a graph   369
Purushothama S., Mamatha N., Nayaka S. R., Nataraj K. and K. N. Prakasha

Chapter 54   Structures of partially ordered ternary semi-rings   375
A. Rajeswari, Chandrakala H. K. and Sheela

Chapter 55   Optimisation of electric vehicle battery thermal management system using computational fluid dynamics and thermal analysis   378
Dharmila Chowdary A., Jalaja P., M. Krishna and Shivakumar H. M.

Chapter 56   Enhancing waste management efficiency using the transportation problem   385
Jyothi P., Vatsala G. A., Radha Gupta and Chaitra M.

Chapter 57   The properties of complementary centroidal mean   390
Venkataramana B. S., K. M. Nagaraja, Sampathkumar R.,Harish A. and Lakshmi Janardhan R. C.

Chapter 58   Soft expert graph and its applications   395
Supriya M. D. and P. Usha

Chapter 59   Convection dynamics of viscoelastic fluids under temperature-dependent viscosity and heat sources   400
Jayalatha G., Sakshath T. N., P. G. Siddheshwar and Sekhar G. N.

List of Figures

Figure 1.1	Work flow diagram for the suggested task	3
Figure 1.2	DBSCAN clustering graph	6
Figure 1.3	Accuracy comparison of the model	6
Figure 1.4	ROC curve of the model	6
Figure 2.1	System architecture of the propounded model	10
Figure 2.2	Bar chart for benign and malignant data	11
Figure 2.3	Correlation heat map for data	11
Figure 2.4	Confusion matrix	12
Figure 2.5	ROC curve	12
Figure 2.6	Performance analysis of the prediction models	13
Figure 3.1	Animatrix framework	15
Figure 3.2	High-level pseudocode of Animatrix framework	15
Figure 3.3	Self-curated captions dataset	15
Figure 3.4	Self-curated tagline dataset	15
Figure 3.5	Self-curated video metadata	16
Figure 3.6	Google-go emotions labels	16
Figure 3.7	Example input to caption generation model and corresponding output	16
Figure 3.8	Example input to tagline generation model and corresponding output	16
Figure 3.9	Similarity comparison for fashion and fitness domain sentences	16
Figure 3.10	Animatrix video ranking algorithm	17
Figure 3.11	Animatrix video ranking algorithm	18
Figure 3.12	Animatrix video ranking algorithm	18
Figure 3.13	Analysis of survey feedback	18
Figure 3.14	Fashion advertisement snippets	19
Figure 4.1	Growth and development of large language models over time	22
Figure 4.2	Architecture diagram	23
Figure 4.3	Snippet example for real-time feedback	24
Figure 4.4	Snippet example for dynamic test-case generation	25
Figure 4.5	Home page of CodeXValidator	25
Figure 4.6	The upload interface	25
Figure 4.7	Generation page	26
Figure 4.8	Sourse code snippet	26
Figure 4.9	Backend working	26
Figure 4.10	Output generated by the CodeXValidator	26
Figure 6.1	Potential mechanism for automated recognition and diagnosis of silkworm illnesses with deep learning models	40
Figure 6.2	Healthy silkworm images	41
Figure 6.3	Sick silkworm images	41
Figure 6.4	Confusion matrix for automated identification and diagnosis of silkworm illnesses using VGG16 [12] model	42
Figure 6.5	Confusion matrix for automated identification and diagnosis of silkworm illnesses using AlexNet [15] model	42
Figure 6.6	Accuracy graph. (a) Accuracy graph for the VGG16 [12] model over 10 training epochs. (b) Accuracy graph for the AlexNet [15] model over 10 training epochs	43
Figure 6.7	Comparison of training and validation loss for VGG16 and AlexNet	43

Figure 7.1	Quantum annealing and QML workflow – depicting how quantum systems predict market behaviour using high-dimensional entanglement	47
Figure 7.2	Ethical AI workflow – illustrating how trading strategies integrate socio-economic impact through ESG and penalty systems	48
Figure 7.3	Causal discovery model – demonstrating how real-time learning refines causal relationships over time	48
Figure 7.4	Multi-model data fusion – demonstrates multimodal data fusion and how different data sources contribute to stock market	49
Figure 7.5	Price trends	49
Figure 7.6	Stock prices over time	50
Figure 8.1	Analysis of rape case trials and convictions in India	55
Figure 8.2	Activation page	56
Figure 8.3	Set passcode page	56
Figure 8.4	Re-enter passcode page	56
Figure 8.5	Message sent confirmation	56
Figure 8.6	Emergency alert message format	56
Figure 8.7	Working backend console output	57
Figure 9.1	The proposed model	62
Figure 9.2	The distribution of a machine failure (0=failure, 1=non-failure)	63
Figure 9.3	The distribution of failure (failure=0 and failure=1) with respect to a particular device	63
Figure 9.4	Distribution of failure with respect to a device	63
Figure 9.5	Subplots signifying the correlation of various attributes that relate to a machine failure	63
Figure 9.6	The correlation matrix	63
Figure 9.7	An ROC AUC curve	64
Figure 10.1	GAN architectural diagram	68
Figure 10.2	Proportion of each observed categorical feature	68
Figure 10.3	Testing and training	69
Figure 10.4	Installing test subjects	69
Figure 10.5	Accuracy between testing set vs. training set	69
Figure 10.6	Classification report	69
Figure 10.7	Confusion matrix between real and deepfake	70
Figure 11.1	Flood prediction model	76
Figure 11.2	Urban heat island model	76
Figure 11.3	Rescue system	76
Figure 11.4	AI-driven urban resilience system workflow	77
Figure 11.5	User interface design of the AI-driven urban resilience system	78
Figure 11.6	Real time user location (latitude and longitude)	78
Figure 11.7	Real time data inputs in the model through API	78
Figure 11.8	Map interface of the user	78
Figure 11.9	Rescue for flood (route optimisation)	78
Figure 11.10	Urban heat mitigation system user interface with auto-filled form	78
Figure 12.1	DBSCAN clustering graph	84
Figure 12.2	Accuracy comparison of the model	84
Figure 13.1	Flow chart of the LSTM RNN implementation procedure	87
Figure 13.2	The structure of the DATA for battery used in this anlaysis	87
Figure 13.3	Convergence graph	88
Figure 13.4	Frequency vs. Root means square value (RMSE) value with binary bins	88

Figure 13.5	The six data used for training Vmeasured, Imeasured, temperature measured, charge current, change voltage and time taken	88
Figure 13.6	Open loop prediction/forecasting	88
Figure 13.7	Closed loop forecast for 200 data	89
Figure 14.1	Image dataset	94
Figure 14.2	Workflow diagram	94
Figure 14.3	Train and validation accuracy	95
Figure 14.4	Train and validation loss	95
Figure 14.5	Disease detection: Yellow leaf disease, stem cracking	95
Figure 14.6	Disease detection: Leaf rot detection in coconut tree leaf	96
Figure 14.7	Disease detection : Stem cracking detection in coconut tree stem	96
Figure 15.1	Existing popular voice assistants	98
Figure 15.2	Desktop voice assistant architecture	100
Figure 15.3	Screenshot 1 – main interface or idle interface	102
Figure 15.4	Screenshot 2 – second interface taking command	102
Figure 15.5	Screenshot 3 – opening YouTube using voice	102
Figure 15.6	Screenshot 4 – chat history of WhatsApp automation (WhatsApp window screenshot not shown for privacy); hug chat feature showing response about India when asked	102
Figure 16.1	CNN process flow	107
Figure 16.2	Medilink system architecture	108
Figure 16.3	Home page	109
Figure 16.4	Services and features	109
Figure 16.5	Brain tumour detection	109
Figure 16.6	Eye disorder detection	109
Figure 17.1	Proposed system for automated scoring and feedback of Marathi essays	114
Figure 17.2	Rubric given to the LLM LLaMA3-70B-8192 and Gemini-1.5-flash to evaluate the input Marathi essay	115
Figure 17.3	System prompt given to the LLM LLaMA3-70B-8192 and Gemini-1.5-flash to evaluate the input Marathi essay	115
Figure 17.4	Score and constructive feedback generated by the LLM LLaMA3-70B-8192 in response to the input Marathi essay	115
Figure 17.5	A Marathi essay as an input to the LLM Gemini-1.5-flash	115
Figure 17.6	Score and constructive feedback generated by the LLM Gemini-1.5-flash in response to the input Marathi essay	115
Figure 18.1	Agile development	121
Figure 18.2	User interaction	121
Figure 18.3	Request flow	121
Figure 18.4	Transaction details file	121
Figure 18.5	Create campaign	121
Figure 18.6	User input	122
Figure 18.7	Listing campaign	122
Figure 18.8	Latest transactions	122
Figure 18.9	Confirming campaign details	122
Figure 18.10	Current campaign details	122
Figure 19.1	ESP32-CAM	126
Figure 19.2	DF player mini	126
Figure 19.3	Speaker (8Ω, 3W)	127
Figure 19.4	5V Power supply (2A)	127
Figure 19.5	Jumper wires	127

Figure 19.6	Servo motor	127
Figure 19.7	Work flow of gamified learning	128
Figure 19.8	Working component circuit	129
Figure 19.9	Fully IoT component model	130
Figure 20.1	ML workflow	133
Figure 20.2	Hardware workflow	133
Figure 20.3	ESP32-CAM, buzzer and Arduino UNO microcontroller	134
Figure 20.4	ESP32-CAM	134
Figure 20.5	Frontend workflow	134
Figure 20.6	Training history of the ML model	135
Figure 20.7	Confusion matrix result	135
Figure 20.8	Model training output with accuracy	135
Figure 20.9	Result for precision, recall and F1-scores	135
Figure 20.10	Output in serial monitor after uploading code to ESP32-CAM	135
Figure 20.11	Camera detects object as "not snake"	136
Figure 20.12	Result obtained on serial monitor	136
Figure 20.13	Result in frontend for "No Snake"	136
Figure 20.14	Camera captures object "Snake"	136
Figure 20.15	Result obtained on serial monitor	137
Figure 20.16	Result in frontend for "Snake"	137
Figure 20.17	User login window	137
Figure 20.18	User home dashboard	137
Figure 20.19	User home dashboard – About the system	137
Figure 20.20	Snake alerts page	138
Figure 20.21	Emergency info and contact page	138
Figure 20.22	First aid for snake bites page	138
Figure 20.23	First aid for snake bites page – extended content	138
Figure 20.24	Contacts page	138
Figure 20.25	Contacts page – extended content	138
Figure 21.1	Proposed model architecture	142
Figure 21.2	Prediction of number	143
Figure 22.1	Model image	149
Figure 22.2	Entering gender UI interface	149
Figure 22.3	Smart mirror flowchart	149
Figure 22.4	Entering weight UI interface	150
Figure 22.5	Entering age UI interface	150
Figure 22.6	Final data display UI interface	150
Figure 23.1	Block diagram of system architecture	154
Figure 23.2	Flow chart of workflow	154
Figure 23.3	The mobile display during accident detection	157
Figure 24.1	Arrangement of components	162
Figure 24.2	Work-flow design of the vending machine	162
Figure 25.1	Flowchart	169
Figure 25.2	Performance comparison	170
Figure 25.3	Error detection	170
Figure 25.4	Improvement percentage	170
Figure 26.1	Workload prediction model in cloud	173
Figure 26.2	The taxonomy of prediction methods	173
Figure 26.3	Overview of the proposed workload prediction method	175
Figure 26.4	Workload prediction-based VM migration in cloud	176

Figure 26.5	Transformer-based VS LSTM-based models performance comparison with different hyperparameter settings. The accuracy of transformer-based models is significantly better than that of LSTM-based models	177
Figure 26.6	Load forecasting [6]	178
Figure 26.7	Basic methods and proposed workload prediction results for VMD-TCN in Alibaba datasets [7]	178
Figure 26.8	Computational cost vs. accuracy trade-off analysis	179
Figure 26.9	A complete architecture of resource planning and load compensation using the proposed approach to predict cloud computing workload [7]	179
Figure 27.1	Block diagram	181
Figure 27.2	Stop sign	183
Figure 27.3	End of no overtaking sign	183
Figure 27.4	Keep left sign	183
Figure 27.5	Speed limit (50 km/h)	184
Figure 27.6	Bicycles	184
Figure 27.7	Speed limit (60 km/h)	184
Figure 27.8	Model accuracy graph	184
Figure 27.9	Model loss graph	184
Figure 28.1	Platform flowchart	189
Figure 28.2	Signup page	190
Figure 28.3	Product catalogue with category filtering	190
Figure 28.4	Page to add your product	190
Figure 29.1	System architecture of AI-powered adaptive traffic light system	193
Figure 29.2	Flowchart AI-powered traffic light system	194
Figure 29.3	Decision-making flowchart for adaptive signal timing	196
Figure 29.4	Comparison of response time among traffic systems	197
Figure 29.5	Confusion matrix for AI traffic prediction	197
Figure 29.6	Accuracy trends over iterations	197
Figure 30.1	CNN + RNN architecture	202
Figure 30.2	Graph neural networks	203
Figure 30.3	Deep graph library	203
Figure 30.4	Evaluation metrics	204
Figure 31.1	Schematic diagram of DSTATCOM	207
Figure 31.2	Block diagram of power transmission system	208
Figure 31.3	Simulink model of with out DSTATCOM	209
Figure 31.4	Simulink model of VSC-based DSTATCOM	209
Figure 31.5	Voltage (Va) and current (Ia)	210
Figure 31.6	Voltage (Va) and current (Ia)	210
Figure 31.7	Voltage (Va) and current (Ia)	210
Figure 31.8	Voltage (Va) and current (Ia)	210
Figure 31.9	Voltage (Va) and current (Ia)	210
Figure 31.10	Voltage (Va) and current (Ia)	211
Figure 31.11	Voltage (Va) and current (Ia)	211
Figure 31.13	Voltage (Va) and current (Ia)	211
Figure 31.14	Voltage (Va) and current (Ia)	211
Figure 31.12	Voltage (Va) and current (Ia)	211
Figure 31.15	Voltage (Va) and current (Ia)	212
Figure 31.16	Voltage (Va) and current (Ia)	212
Figure 31.17	Voltage (Va) and current (Ia)	212
Figure 31.18	Voltage (Va) and current (Ia)	212

Figure 32.1   Solar-driven alkaline water dissolution system for hydrogen production   215
Figure 32.2   Block diagram of electrolyser   215
Figure 32.3   Alkaline electrolyser   216
Figure 32.4   3D structure of alkaline electrolyser   216
Figure 32.5   Stack formation of electrolyser   216
Figure 32.6   Testing alkaline electrolyser   217
Figure 32.7   Energy meter   217
Figure 32.8   Voltage (Va) and hydrogen (L/min)   217
Figure 32.9   Current (Ia) and hydrogen (L/min)   218
Figure 32.10   Hydrogen (L/min) and temperature (c)   218
Figure 32.11   Hydrogen (L/min) and energy (kwh)   218
Figure 32.12   Voltage (V), hydrogen production rate (L/min), current (A), efficiency (%), KOH concentration (%), hydrogen purity (%), stack temperature (°C), and energy consumption   218
Figure 33.1   Schematic of proposed of 3-bit flash ADC [3]   222
Figure 33.2   Schematic of double tail comparator   222
Figure 33.3   Schematic of 3-bit Flash ADC   223
Figure 33.4   Priority encoder schematic diagram [11]   223
Figure 33.5   Output waveform of 3-bit flash ADC   223
Figure 34.1   Highlight extract: (a) grey image; (b) traditional Otsu; (c) fixed threshold method; (d) background illumination removal; (e) traditional Otsu based on saliency model; and (f) fixed threshold based on saliency model   229
Figure 34.2   Saliency model generation   229
Figure 34.3   Super-pixel and HOG aggregation: (a) aggregation flowchart and (b) super-pixel weighted HOG feature generation.   230
Figure 34.4   HOG features: (a) cells and blocks and (b) bins   230
Figure 34.5   V-HOG features: (a) cells and (b) bins   230
Figure 34.6   V-HOG symmetry: (a) ROI image; (b) cell1 and cell4s HOG symmetry; and (c) cell2 and cell3s HOG symmetry   231
Figure 34.7   Number of vehicles per second in the road   232
Figure 35.1   Block diagram of automated glass cleaning robot   237
Figure 35.2   Top view of automated glass cleaning robot   238
Figure 35.3   Side view of glass cleaning robot   238
Figure 35.4   Cleaning process   239
Figure 35.5   Command for forward and backward movement of robot   239
Figure 35.6   Command for automatic movement and spraying mechanism of the robot   239
Figure 35.7   Command for upward and downward movement of the robot   239
Figure 35.8   Ultrasonic sensor detecting the obstacle   240
Figure 36.1   Repository structure for the proposed Git-based prompt management system. Prompts are grouped by topic, and corresponding logs are stored in parallel directories.   244
Figure 36.2   Overall flow of proposed Git-based solution   244
Figure 37.1   Design of DLC using PI controller input & output   248
Figure 37.2   DBN framework   249
Figure 37.3   (a) Existed system for 1 induction motor control (for controlling 2 induction motors, we need 6 inverter legs). (b) Proposed system (4 legs are used for 2 induction motor control)   249
Figure 37.4   IVC block diagram   250
Figure 37.5   Simulation of existing PI controller-based 2 induction motors control using 6 leg inverter   250

Figure 37.6	IVC block	250
Figure 37.7	Simulation of 4-leg inverter-based DL controller	250
Figure 37.8	Calling deep learning code inside the controller block	250
Figure 37.9	Simulation of 4-leg inverter of proposed DLC	251
Figure 37.10	Line voltages and 3-phase currents of existing PI conventional controller (we find difference only in speed and torque waveforms of existing PI controller and proposed DLC)	251
Figure 37.11	Line voltages and 3-phase currents of DLC	251
Figure 37.12	Speed and torque waveforms of existing conventional controller (when we enlarge, we can see more distortion)	251
Figure 37.13	More smoothened speed and torque waveforms of proposed DLC	251
Figure 37.14	Enlarged view of speed ripples and torque ripples of existing conventional controller (with more distortion)	252
Figure 37.15	Enlarged view of speed ripples and torque ripples of proposed DLC (we can observe very negligible distortion here)	252
Figure 37.16	THD (24%) of existing PI conventional controller	252
Figure 37.17	THD (17%) of proposed DLC	252
Figure 37.18	Comparative performance analysis of existing 6-leg inverter-based conventional PI controller and proposed 4 leg inverter-based DLC	252
Figure 37.19	Net capacitors voltage, capacitor 1 and capacitor 2 voltages of proposed DLC, respectively	252
Figure 38.1	WIPL-D pro-developed LPDA model	256
Figure 38.2	WIPL-D generated VSWR of the LPDA (Figure 1) designed using Table 38.2 parameters	256
Figure 38.3	WIPL-D generated VSWR of the LPDA designed using Table 3 parameters	257
Figure 38.4	The fabricated antenna	257
Figure 38.5	VSWR of the simulated and fabricated LPDA	257
Figure 38.6	The simulated E-plane radiation patterns at 150, 300, and 450 MHz	258
Figure 38.7	The simulated H-plane radiation patterns at 150, 300, and 450 MHz	258
Figure 38.8	The measured E-plane patterns of the fabricated LPDa at 150, 300, and 450 MHz	258
Figure 38.9	Same as Figure 8, but for the H-plane radiation	259
Figure 39.1	General block diagram of FHSS	262
Figure 39.2	Flow diagram Viterbi encoding and decoding	262
Figure 39.3	½ Convolutional encoder [2]	263
Figure 39.4	The Trellis diagram [13]	263
Figure 39.5	Viterbi decoder	263
Figure 39.6	Overall FHSS system block diagram	264
Figure 39.7	Transmitted data	264
Figure 39.8	Spreading signal for FHSS	264
Figure 39.9	Frequency hopped spread spectrum signal	265
Figure 39.10	Frequency hopped spread spectrum signal with Viterbi encoding	265
Figure 39.11	Decoded bits from Viterbi decoding	265
Figure 39.12	FHSS BER vs. noise with and without Viterbi codec	265
Figure 40.1	Typical waveform during voltage sag	268
Figure 40.2	Phase-angle jump + 45°	268
Figure 40.3	Voltage divider model for voltage sag magnitude [1]	269
Figure 40.4	Line fault Simulink model	270
Figure 40.5	(a) by phase a (b) by phase b (c) by phase c illustrates the voltage sag caused by a line-to-line fault at the 11 kV bus, represented as an instantaneous waveform	271

Figure 40.6   (a) by phase a (b) by phase b (c) by phase c illustrates the voltage sag caused
by a line-to-line fault at the 0.4 kV bus, represented as an instantaneous waveform   271

Figure 40.7   (a) by phase a (b) by phase b (c) by phase c illustrates the voltage sag caused by
a line-to-line fault at the 11 kV bus, represented as an RMS waveform   272

Figure 40.8   (a) by phase a (b) by phase b (c) by phase c illustrates the voltage sag caused by
a line-to-line fault at the 0.4 kV bus, represented as an instantaneous waveform   272

Figure 41.1   Stages of the ZECC construction   275

Figure 41.2   Plan of ZECC   275

Figure 41.3   Samples storing in ZECC   276

Figure 41.4   Reduction in moisture content with respect to time in days   278

Figure 42.1   GC-MS chromatogram graph of Pongamia methyl ester blend   284

Figure 42.2   GC-MS chromatogram graph of neem methyl ester blend   286

Figure 42.3   GC-MS chromatogram of Mahua methyl ester blend   288

Figure 42.4   GC-MS chromatogram graph of Jatropha methyl ester blend   289

Figure 42.5   GC-MS chromatogram graph of used cooking oil methyl ester   290

Figure 42.6   GC-MS chromatogram graph of dairy scum oil methyl ester   291

Figure 44.1   Dimensions of the pin used in mm   301

Figure 44.2   Pin holder and disk   302

Figure 44.3   Graphs showing how the friction coefficient varies with sliding time for coating
thicknesses of (a) 200 microns, (b) 250 microns, and (c) 300 microns of nickel
aluminium   302

Figure 44.4   Average friction coefficient for nickel aluminium with varying coating thicknesses   303

Figure 44.5   A 200× magnification micrograph showing worn-out pin surfaces at varied
coating thicknesses of nickel aluminium (a) 200 microns, (b) 250 microns and
(c) 300 microns   304

Figure 44.6   A 500× magnification micrograph showing worn-out pin surfaces at coating
thicknesses of 200 microns nickel aluminium   304

Figure 44.7   Graphs showing how the friction coefficient varies with sliding time for coating
thicknesses of (a) 200 microns, (b) 250 microns and (c) 300 microns of
nickel chromium   304

Figure 44.8   Average friction coefficient for nickel chromium with varying coating thicknesses   305

Figure 44.9   A 200× magnification micrograph showing worn-out pin surfaces at varied
coating thicknesses of nickel chromium (a) 200 microns, (b) 250 microns and
(c) 300 microns   305

Figure 44.10   A 500× magnification micrograph showing worn-out pin surfaces at coating
thicknesses of 250 microns nickel chromium   306

Figure 45.1   Methodology followed   310

Figure 45.2   Tensile test specimen   310

Figure 45.3   Hardness test specimen   311

Figure 45.4   Impact test specimen   311

Figure 45.5   SEM images of samples (1) pure aluminium, (2,3,4,5,6) Al7075+SiC samples   311

Figure 45.6   XRD curves   311

Figure 45.7   Variations of tensile strength for different samples   312

Figure 45.8   Variations of hardness for different samples   312

Figure 45.9   Variations of impact strength for different samples   312

Figure 45.10   Experimental mode shapes for sample 1 – (a) Mode 1 (b) Mode 2   313

Figure 45.11   ANSYS mode shapes for sample 1 – (a) Mode 1 (b) Mode 2   313

Figure 45.12   Experimental mode shapes for sample 1 – (a) Mode 1 (b) Mode   313

Figure 45.13   ANSYS mode shapes for sample 1 – (a) Mode 1 (b) Mode 2   313

Figure 45.14   Comparison of experimental results vs. ANSYS results   313

Figure 46.1   Fuels used in the study (gasoline, diesel, PTB20 and PTB100)   316
Figure 46.2   CRDI dual fuel engines equipped with hydrogen injection   317
Figure 46.3   Hydrogen injector connected to dual fuel engine   317
Figure 46.4   AVL gas analyser   317
Figure 46.5   Smoke meter   317
Figure 46.6   Impact of EGR on BTE   318
Figure 46.7   Effect of EGR on the smoke opacity   318
Figure 46.8   Effect of EGR on HC emissions   318
Figure 46.9   Impact of EGR on CO emissions   319
Figure 46.10   Impact of EGR on NOx emissions   319
Figure 47.1   Represents the comparison of predictive models   324
Figure 47.2   Comparison of model accuracy analysis of different algorithms   325
Figure 48.1   Experimental setup   329
Figure 48.2   BSFC vs. BP for Calophyllum inophyllum biodiesel blends   330
Figure 48.3   BTE vs. BP for Calophyllum inophyllum biodiesel blends   330
Figure 48.4   EGT vs. BP for Calophyllum inophyllum biodiesel blends   330
Figure 48.5   CO emissions vs. BP for Calophyllum inophyllum biodiesel blends   331
Figure 48.6 CO_2 emissions vs. BP for Calophyllum inophyllum biodiesel blends 331
Figure 48.7   UBHC emissions vs. BP for Calophyllum inophyllum biodiesel blends.   332
Figure 48.8   NOx emissions vs. BP for Calophyllum inophyllum biodiesel blends   332
Figure 48.9   Smoke opacity vs. BP for Calophyllum inophyllum biodiesel blends   332
Figure 49.1   PXRD and W-H plots of $CaAl_2O_4$:Tb (1–11 mol%) NPs   335
Figure 49.2   Rietveld refinement of $CaAl_2O_4$:Tb^{3+} (1–11 mol%) NPs.   336
Figure 49.3   SEM images of optimised (a, b) $CaAl_2O_4$:Tb^{3+} (5 mol%)   336
Figure 49.4   (a) DR spectra and (b) energy gap of the $CaAl_2O_4$:Tb^{3+} (1–11 mol%) NPs   336
Figure 49.5   (a) PL excitation (b) PLE spectra of the $CaAl_2O_4$:Tb^{3+} (1–11 mol%) NPs 337
Figure 49.6   (a) Deviation of PL intensity and asymmetric ratio with concentration of
                 Tb^{3+} ions, (c) Logarithmic plot of (x) and (I/x) in $CaAl_2O_4$:Tb^{3+} NPs   337
Figure 49.7 $CaAl_2O_4$ CIE and CCT graph: Tb (1–11 mol%) NPs   338
Figure S1   Electronic spectrum of ^{1}LCu in DMF   342
Figure S2   Electronic spectrum of ^{1}LCu in in DMSO   342
Figure S3   S3 Electronic spectrum of ^{2}LCu in DMF   342
Figure S3   S3 Electronic spectrum of ^{2}LCu in DMF   342
Figure S8   VSM spectrum of ^{1}LCu (II) Complex   344
Figure S9   VSM spectrum of ^{2}LCu (II) Complex   344
Figure S10   VSM spectrum of ^{3}LCu (II) Complex   344
Figure 50.1   ESR spectra of Cu complexes of (^{1}L, ^{2}L and ^{3}L)   344
Figure 50.2   TG–DTG curve of ^{1}L (ligand)   344
Figure 50.3   TG–DTG curve of ^{1}LCu complex   344
Figure 50.4   TG–DTG curve of ^{2}L (ligand)   345
Figure 50.5   TG–DTG curve of ^{2}LCu complex   345
Figure 50.6   TG–DTG curve of ^{3}L (ligand)   345
Figure 50.7   TG–DTG curve of ^{3}LCu complex   345
Figure 50.8   log [ln(1/y) vs. 1000/K plots for thermal decomposition of azo ligand ^{1}L   346
Figure 50.9   log [ln(1/y) vs. 1000/K plots for thermal decomposition of azo complex ^{1}LCu   346
Figure 50.10   log [ln(1/y) vs. 1000/K plots for thermal decomposition of azo ligand ^{2}L   346
Figure 50.11   log [ln(1/y) vs. 1000/K plots for thermal decomposition of azo complex ^{2}LCu   347
Figure 50.12   log [ln(1/y) vs. 1000/K plots for thermal decomposition of azo ligand ^{3}L   347
Figure 50.13   log [ln(1/y) vs. 1000/K plots for thermal decomposition of azo complex ^{3}LCu   347
Figure 50.14   Plots of ln(a/a-x) vs. Time for thermal decomposition of azo ligand ^{1}L   347

Figure 50.15	Plots of ln(a/a-x) vs. time for thermal decomposition of azo complex ^{1}LCu	347
Figure 50.16	Plots of ln(a/a-x) vs. Time for thermal decomposition of azo ligand ^{2}L	347
Figure 50.17	Plots of ln(a/a-x) vs. time for thermal decomposition of azo complex ^{2}LCu	347
Figure 50.18	Plots of ln(a/a-x) vs. time for thermal decomposition of azo ligand ^{3}L	348
Figure 50.19	Plots of ln(a/a-x) vs. time for thermal decomposition of azo complex ^{3}LCu	348
Figure 51.1	Preparation of neem oil biodiesel using K2CO3-NH4OH catalyst	353
Figure 51.2	Effect of varying quantities of K_2CO_3 and NH_4OH	357
Figure 51.3	Influence of oil-to-methanol ratio on the trans-esterification efficiency of neem oil. On the biodiesel yield from neem oil	357
Figure 51.4	Effect of reaction duration and temperature on the biodiesel yield from neem oil	358
Figure 51.5	^{1}H NMR spectra of neem oil (a) and its corresponding biodiesel (b), illustrating the transformation of triglycerides into methyl esters	358
Figure 51.6	FTIR spectrum of crude neem oil (a) and its corresponding biodiesel (b)	358
Figure 52.1	Graph G showing Harary Energy	362
Figure 53.1	A simple graph with 6 vertices	370
Figure 55.1	Meshed model of a battery pack with air cooling	380
Figure 55.2	Convergence testing of the battery thermal management system in CFD	381
Figure 55.3	Schematic diagram of the battery thermal management experimental setup	381
Figure 55.4	A representative Python code for the CFD analysis of a battery thermal management system	381
Figure 55.5	Python-simulated output of CFD analysis for the battery thermal management system	382
Figure 55.6	Temperature distribution across the battery cells during charging, incorporating cooling through the thermal management system	382
Figure 55.7	Temperature distribution across the battery cells during charging, incorporating cooling through the thermal management system	383
Figure 55.8	Experimental (red) and theoretical (green) analysis of the impact of charging and discharging during cooling on the temperature of a typical battery cell	383
Figure 55.9	Experimental analysis of the effect of charging and discharging during cooling on battery cell temperature	383
Figure 57.1	The complimentary centroidal mean is Schur concave	391
Figure 57.2	The complimentary centroidal mean is Schur geometrically convex	392
Figure 57.3	The complimentary centroidal mean is Schur harmonically convex	392
Figure 59.1	Physical configuration under investigation	401
Figure 59.2	R_{oc} vs. V with $\Lambda_1 = 0.2$, $\Lambda = 0.5$ for different values of R_I	403

List of Tables

Table 1.1	Accuracy comparison of model	5
Table 4.1	Comparison between multi agent bot and single agent bot like GPT	22
Table 5.1	Comparative analysis of Linear Regression, k-NN, and Hybrid models based on interpretability, computational efficiency, ability to handle non-linearity	31
Table 5.2	Presents the performance metrics of our Hybrid FlowCast model compared to standalone models (Linear Regression, k-NN, LSTM). Our model reduces RMSE by 10% and improves R^2 by 12% over traditional models, demonstrating its effectiveness in real-world traffic forecastin Variation in traffic volume across different hours of the day, categorized by day of the week	32
Table 6.1	Literature survey on silkworm disease detection and related research	39
Table 6.2	Total silkworm images collected	41
Table 6.3	Hyperparameter values set for VGG16 [12] and Alexnet [15] models	42
Table 6.4	Evaluation of performance metrics for VGG16 [12] model	42
Table 6.5	Evaluation of performance metrics for Alexnet [15] model	42
Table 9.1	A comparison of a few machine learning algorithms	64
Table 9.2	The algorithms used versus the IIoT suitability	64
Table 10.1	Dataset of convolutional GANs	67
Table 12.1	Performance metrics of different models	84
Table 12.2	Accuracy across varying user ranges	84
Table 15.1	Related assistants comparison table	99
Table 15.2	Tested performance metrics and capabilities	102
Table 17.1	Comparison of essay scores assigned by LLMs (Llama, Gemini) and human evaluators	116
Table 17.2	Relevance scores assigned by human evaluators for feedback generated by Llama and Gemini	116
Table 26.1	Comparative analysis of workload prediction techniques [10, 11]	173
Table 26.2	Machine learning-based workload prediction techniques	174
Table 26.3	Dataset description and pre-processing steps	174
Table 26.4	Impact of workload forecasting	177
Table 26.5	System performance metrics	178
Table 29.1	Comparative analysis of different traffic light systems	194
Table 30.1	BindingDB[2]	201
Table 32.1	Intial inputs of alkaline electrolyser	217
Table 32.2	Electrolyser performance parameters	218
Table 32.3	Output parameters (hydrogen and oxygen)	219
Table 33.1	Priority encoder table	223
Table 33.2	Implementation of proposed vs. existing	224
Table 33.3	Rise and fall time	224
Table 34.1	Experimental results on videos V6, V7, and V8 using different input spaces and classifiers	233
Table 35.1	Commands used and their corresponding actions	238
Table 35.2	Power consumed by each component	238
Table 38.1	Theoretical antenna design parameters	255
Table 38.2	Theoretical dipole lengths, their distance from the apex, and their diameter	256
Table 38.3	Same as Table 2, but for the LPDA with the lowest average VSWR profile	257

Table 39.1   FHSS specifications   264
Table 39.2   BER vs. noise   265
Table 41.1   Temperature differences observed in initial testing stages   276
Table 41.2   Temperature differences observed during testing stages   277
Table 41.3   Moisture content of samples   278
Table 41.4   Pictorial representation of the samples inside and outside ZECC   279
Table 42.1   Physical characterisation of non-edible oils   283
Table 42.2   Reactants volumetric proportions for trans-esterification reaction   283
Table 42.3   Saponification number, iodine value and cetane number of selected sample's
              methyl ester   284
Table 42.4   GC-MS chromatogram components of Pongamia methyl ester blend   285
Table 42.5   GC-MS chromatogram components of neem methyl ester blend   287
Table 42.6   GC-MS chromatogram of Mahua methyl ester blend   288
Table 42.7   GC-MS chromatogram components of Jatropha methyl ester blend   289
Table 42.8   GC-MS chromatogram components of used cooking oil methyl ester blend   290
Table 42.9   GC-MS chromatogram components of dairy scum oil methyl ester blend   291
Table 43.1   Material properties of epoxy and SiO_2 296
Table 43.2   RVE geometry (size and shape), particle distribution and meshed model created
              with 200 nm × 200 nm, 500 nm × 500 nm, and 1 μm × 1 μm   298
Table 43.3   Young's moduli of all RVE   299
Table 44.1   Test details of experiments   302
Table 44.2   The relationship between the average friction coefficient and the thickness of
              the nickel aluminium coating   303
Table 44.3   The relationship between the average friction coefficient and the thickness of
              the nickel-chromium coating   305
Table 45.1   Major constituting elements   309
Table 45.2   Al7075 composition (wt%)   310
Table 45.3   MMC's composition and its symbols   310
Table 45.4   Tensile, hardness and impact strength test results   312
Table 45.5   Experimental and ANSYS results of Al-SiC specimens   313
Table 46.1   Material properties of epoxy and SiO_2 316
Table 47.1   Performance comparison of predictive models   324
Table 48.1   The specifications of the diesel engine   328
Table 48.2   Properties of Calophyllum inophyllum biodiesel and diesel fuel   329
Table 48.3   Operating conditions   329
Table 48.4   Composition of biodiesel blend in volume percentage   329
Table 49.1   Crystallite size of $CaAl_2O_4$:Tb (1–11 mol%) NPs   335
Table 49.2   Energy gap of $CaAl_2O_4$:Tb^{3+} (1–11 mol%) NPs   337
Table 49.3   Photometric parameters of $CaAl_2O_4$:Tb (1–11 mol%) NPs   338
Table 50.1   Electronic spectral data of synthesised azo metal complexes   342
Table S1   Physical characteristics of synthesized metal complexes   343
Table 50.2   IR spectral data of synthesised azo metal complexes dyes   343
Table 50.3   Particle size, d-spacing and strain   343
Table 50.4   ESR spectral data of Cu(II) complex of azo dye ligand (^{1}L, ^{2}L, ^{3}L)   344
Table 50.5   Thermal data of the complexes of ^{1}LCu, ^{2}LCu, and ^{3}LCu   346
Table 50.6   Thermodynamic parameters of the thermal decomposition of azo metal
              complexes using Broido's methods   348
Table 50.7   Antimicrobial activity of the Cu azo complex of ^{3}L   349
Table 51.1   Fatty acid composition of neem oil (compiled from various studies)   352
Table 51.2 ^{1}H NMR chemical shift observed in the neem oil and its biodiesel   355

Table 51.3   FTIR characteristic frequency observed in the neem oil and its biodiesel   355
Table 51.4   Comparative study of neem oil biodiesel yield with different catalytic conditions   356
Table 51.5   Physicochemical properties of neem oil and its corresponding biodiesel
              compared with standard diesel   357
Table 55.1   Material properties for the cold plate and lithium ion cell   380
Table 56.1   Initial basic table   387
Table 56.2   Best possible allocations   388
Table 58.1   Prioritized List of Candidates for Hiring   398
Table 58.2   Prioritized Patient List for Immediate Medical Attention   399
Table 59.1   Oscillatory RBC – Critical values of Roc, a_c and ω_c for FIFI boundary for
              differentvalues of R_I and for the viscoelastic fluid with variable viscosity   402
Table 59.2   Oscillatory RBC – Critical values of R_{oc}, a_c and ω_c for FIFI boundary for different
              values of R_I and for the Maxwell fluid of constant/variable viscosity with
              $Pr = 10$, $\Lambda_1 = 1$ 403
Table 59.3   Stationary RBC – Critical values of R_oc and ac for FIFI boundary for different
              values of R_I and for the Newtonian/Walters B/Rivlin-Ericksen fluid with
              constant/variable viscosity   404

List of Contributors

Editorial Board

Reeja S. R. Professor & Head, Dept. of AIML, VIT- AP University
Bore Gowda, Associate Professor, Manipal Institute of Technology, Manipal
Y. S. Rammohan, Professor, Mechanical & Aerospace Engineering, BMSCE, Bengaluru
Ganesan Prabu Sankar, Professor, IIT, Hyderabad
G. Jayalatha, Professor & Head, RVCE, Bengaluru

Conference Chair

Dr. Dilip Kumar K, Principal / Director, KSIT

Conference Chief Convenor

Dr. D R Swamy, Executive Director, KSRIF

Members of Advisory Committee

Dr. K V A Balaji, Chief Executive Officer, KSGI
Dr. Mani Sankar, Professor and Head, Faculty of Mathematics, University of Technology and Applied Sciences (CAS-Ibri), Sultanate of Oma
Dr. Mohamed M Awad, Professor, Mansoura University, Manoura, Egypt
Dr. Ramani, Senior Lecturer, UTP Malaysia, Malaysia
Dr. S. Hossein Mousavinezhad, Professor, Idaho State University, USA
Dr. Sachin Kumar, Professor, Kyungpook National University Daegu, South Korea
Dr. Sanjay Mavinkere Rangappa, Principal Research Scientist(Specialist 3) & Associate Professor. King Mongkut's University of Technology, North Bangkok (KMUTNB), Thailand
Dr. Sivaram Murugan, Professor, Dept. of Computer Engineering, Sivas University, Turkey
Dr. Vijanth Sagayan, Associate Professor, UTP Malaysia, Malaysia
Dr. B. S. Panda, Professor, Indian Institute of Technology, Delhi
Dr. Chandra N, Senior Principal Scientist, NAL, Bengaluru
Dr. Debahuti Mishra, Professor, Dept. of Computer Science, Institute of Technical Education and Research Siksha 'O' Anusandhan University, Odisha
Dr. Dinesh Kumar Rajendran, Professor, NIT Srinagar
Dr. Ganesan Prabu Sankar, Professor, Indian Institute of Technology Hyderabad
Dr. Gangadharan K V, Professor, Dean (P&D), Department of Mechanical Engineering, NITK
Dr. Gurumurthy Hegde, Director, Centre for Advanced Research, Christ University
Dr. H P Kincha, Ret. Professor, IISc, Bengaluru & Former Vice Chancellor, VTU, Belagavi
Dr. Jayadevappa, Principal, Dr. H N National College Of Engineering, Bengaluru, Karnataka
Dr. K N Balasubramanya Murthy, Former Vice Chancellor, PES University
Dr. K Ramanarasimha, Principal, K S School of Engineering & Management, Bengaluru
Prof. Krishnaprasad, K.S. Research and Innovation Foundation, Bengaluru
Dr. Mahesh T R, Prof & Head, Dept of CSE, JAIN (Deemed-to-be University), Bengaluru
Dr. Murthy Haradanahalli G N, Professor, Department of Aerospace Engineering, IIT Madras
Dr. N Sandeep, Assistant Professor, Central University of Karnataka, Kalaburgi
Dr. N D Shivakumar, Principal Research Scientist, IISc
Dr. R Prabhu, Assistant Professor, Indian Institute of Technology, Dharwad, Karnataka

Dr. Rajiv Tiwari, Professor, Department of Mechanical Engineering, IIT, Guwahati
Dr. Ramesh Kestur, Adjunct Faculty and Senior Research Fellow, IIIT-B, Bengaluru, Karnataka
Dr. Reddappa, Associate Professor, Department of Mechanical Engineering, BIT, Bengaluru
Dr. Sivaramakrishnan Sivasubramanian, Professor, Indian Institute of Technology, Bombay
Dr. Sivasankar, Associate Professor, National Institute of Technology, Tiruchirappalli
Dr. Srinivasan Natarajan, Professor, Indian Institute of Science, Bengaluru\
Dr. T V Narayanappa, Joint Director, Department of Factories, Karnataka
Dr. T N Nagabhushana, Vice Chancellor, Kishkinda University, Bellary, Karnataka
Dr. Uma B R, Scientist, ISRO, Bengaluru, Karnataka
Dr. Vijay G S, Professor, Manipal Institute of Technology, Karnataka
Dr. Y Narahari, Professor, IISc, Bengaluru
Dr. Y S Rammohan, Professor, Department of Aerospace Engineering, BMSCE, Bengaluru

Steering Committee

Dr. Nagaprasad K S
Dr. P N Sudha
Dr. Umashankar M,
Dr. Rekha B Venkatapur
Dr. Sangappa
Dr. Chanda V Reddy
Dr. Girish T R
Dr. Ganga Holi

Dr. Anil Kumar A
Dr. P Jalaja
Dr. Suresh M B
Dr. V Bharathi
Prof. B R Santhosh Kumar
Dr. Deepa S R
Dr. Sneha Girish
Dr. Shashikala B S

Domain Reviewers

Dr. Rekha N
Dr. Saleem Khan
Dr. Sunita Chalageri
Prof. Sheba Jebakani
Dr. Venkataramana B S

Dr. Sahana Salagare
Dr. Sheeja Krishnan
Dr. Surekha Byakod
Prof. Neelam Patil Radhika

Foreword

Dr. Reeja S. R. is Professor & Head, Dept of AIML at School of Computer Science and Engineering VIT-AP University. Her area of research are high performance computing, machine learning, artificial intelligence, computer vision and data analytics. Google Scholar ID: https://scholar.google.co.in/citations?hl=en&user=EbRHmzQAAAAJ

Dr. Bore Gowda S. B. is Associate Professor, Department of Electronics & Communication Engineering at Manipal Institute of Technology, Manipal. His research areas include wireless sensor networks, embedded system design, cryptography and network security. He has been coordinator – Samsung PRIS; and member of centre for cryptography. ORCID: https://orcid.org/0000-0001-8115-6477

Dr. Rammohan Y. S. is a Professor of Mechanical and Aerospace Engineering at B. M. S. College of Engineering, Bengaluru, India. He has B.E. in Mechanical Engineering and M. Tech in Machine Design. He completed his Doctoral studies from the Aerospace, Department of Indian Institute of Technology–Madras, Chennai. He has put in more than two and half decades of teaching and research service. His research interests are in the areas of fretting fatigue, machine design, finite element methods, composite materials and functionally graded materials. He has guided several undergraduate and post graduate students in their project works. He has successfully supervised three doctoral scholars. He has published over 30 research papers in reputed international journals and proceedings. He was the reviewer for journals from ASME, Elsevier-Materials Today and others. Google Scholar ID: https://scholar.google.co.in/citations?user=BkgJKf4AAAAJ&hl=en.

Dr.Ganesan Prabu Sankar is Professor in Department of Chemistry at Indian Institute of Technology Hyderabad. His research general field: Organ metallic chemistry and specific fields: Organ metallic synthesis, late transition metal chemistry, heavier main group-block chemistry, light weight composite materials. His achievements are number of sponsored projects completed/completing: 18+; Publications in peer-reviewed journals: 130+; Patents granted/applied: 10; MSc projects completed/ongoing: 34+; PhD's completed: 14; PhD's ongoing: 9; Post-Doc completed/ongoing: 5. His awards and fellowships are SSHN Fellowship-2024, France; DUO-ASEM Professor Fellowship (Germany); DAAD-DST (Germany); Alexander von Humboldt Fellowship (Germany); CNRS Fellowship (France). ORCID ID: 0000-0003-1772-135X

Dr. G. Jayalatha is Professor & Head, Department of Mathematics at RVCE, Bangalore. Her research areas are fluid mechanics, numerical techniques, non-linear dynamical simulations, machine learning techniques. She has 10 International Journal publication; 10 international conference papers and 1 book chapter. She conferred Award of Medal (Young Scientist Award) for presenting a paper entitled "Rayleigh Benard Convection in Viscoelastic Liquids with Temperature Dependent Viscosity"; and RSST Excellence awards 2009, 2010, 2012, 2019. Google Scholar ID: https://scholar.google.com/citations?user=UX9tFLQAAAAJ&hl=en

Preface

A warm welcome to the proceedings of International Conference on Recent Innovations in Engineering Science and Technology (ICRIET 2025).This multidisciplinary conference was a prestigious platform for sharing cutting-edge research and fostering global collaboration hosted at K.S. Institute of Technology, Bengaluru during 9th and 10th May 2025. It was a great honor to welcome domain experts, innovators, researchers and students across the world to share their knowledge, ideas and research outcomes. The theme of the conference "A Confluence of Innovation and Excellence" aims at contribution towards innovations in science, technology, mathematics, and engineering. We received total 387 papers in four tracks namely Next-Generation Computing (Computer Science and Engineering), Innovation in VLSI, Embedded Systems and Communication (Electronics and Communication Engineering), Breakthroughs & Progress (Mechanical Engineering), Pioneering Innovations (Applied Sciences). All these papers were peer reviewed and based on correctness, originality, and novelty, technical aspects related to research contributions. Among these received papers, 225 papers were accepted for presentations in conference. The participation of 07 foreign delegates in the conference was the indication of high level of international research interests in different fields. The presented papers were from both core and interdisciplinary research areas. The session chairpersons who evaluated the presentation were from premier institute of high reputation and from Industry. Finally, based on reviewer comments/recommendations, suitability and readability 59 high quality full papers were selected for the book chapter series. We hope the proceedings will make a long lasting impact in the area of innovation and research.

Acknowledgement

The successful organization of the International Conference on Recent Innovations in Engineering Science and Technology (ICRIET), hosted by K. S. Institute of Technology, Bengaluru in association with the K. S. Research & Innovation Foundation would be incomplete without acknowledging the invaluable contributions of the individuals whose unwavering guidance and support played a crucial role in making the event a resounding success. Their leadership, dedication, and collaborative efforts ensured the smooth execution of the conference and its overall success. The organising committee owes a debt of gratitude to these key figures who provided constant encouragement and direction throughout the planning and execution stages.

The Organizing Committee of ICRIET-2025 takes this opportunity to express their sincere gratitude to our Management Committee Members Sri. R. Rajagopal Naidu, President, Sri. R. Leela Shankar Rao, Hon. Secretary, Sri. T. Neerajakshulu Naidu, Treasurer and respected members of management committee, K. S. Institute of Technology, Bengaluru, for their unwavering support and encouragement in hosting ICRIET-2025. Their continued faith in the conference's vision and commitment to fostering innovation played a pivotal role in making this event possible. The conference would not have been the success it is without their generous contributions and guidance.

The ICRIET-2025 Organising Committee would like to express its sincere gratitude to the keynote speaker, for generously dedicating valuable time to address the enthusiastic researchers and students at the conference.

We extend our deepest gratitude to the Conference Advisory Committee for their exceptional contributions in making ICRIET-2025 a success.

Our heartfelt thanks go to the Reviewers, Session chairs, and Editors for their dedicated efforts in reviewing research papers and for graciously taking up the role of session chairs.

We would also like to acknowledge the generous financial support provided by our sponsors, Repsol Technologies.

A special thank you goes to the Taylor and Francis Group for their invaluable support in publishing the ICRIET-2025 proceedings online.

The entire ICRIET-2025 conference organising family sincerely congratulates all the presenters and delegates for their strong support and active participation throughout the event.

Lastly, our unlimited appreciation is extended to Dr. K. V. A. Balaji, CEO of K. S. Group of Institutions, Dr. Dilip Kumar K, Principal/Director of K. S. Institute of Technology, and all the ICRIET-2025 Organising Committees for their unwavering support, which was instrumental in the successful hosting of this international conference.

1   Deep learning techniques for judicial judgement system

Hemanth N. G.[1,a], Dr. Pushpa Ravikumar[2,b], Harish S.[3,c], Shruthi G. K.[3,d] and Saritha N.[4,e]

[1]PG Scholar, Adichunchanagiri Institute of Technology, Chikkamagaluru, Karnataka, India

[2]Head of the Department, Adichunchanagiri Institute of Technology, Chikkamagaluru, Karnataka, India

[3]Assistant Professor, Department of Computer Science and Engineering, Adichunchanagiri Institute of Technology, Chikkamagaluru, Karnataka, India

[4]Assistant Professor, Adichunchanagiri Institute of Technology, Chikkamagaluru, Karnataka, India

Abstract

The paper delves into the application of advanced machine learning and deep learning technologies in the realm of legal prediction, specifically focusing on forecasting preliminary case outcomes. By analysing the fundamental elements of a case such as penalties, accusations, case section, case condition and legal provisions a deep learning model is employed to predict potential judgment results. This approach offers significant benefits in the legal field.

For judges and lawyers, it provides a data-driven tool to aid in decision-making, enhancing efficiency and accuracy in legal proceedings. By identifying patterns and insights from past cases, the model supports a more informed and objective assessment of potential outcomes. For non-legal professionals, such as litigants or the general public, this technology serves as an educational and informational resource. It offers a clearer understanding of the legal process, helping individuals gain insights into the probable trajectory of their cases without requiring extensive legal expertise. This democratisation of legal knowledge can promote transparency and accessibility in the justice system, bridging the gap between legal professionals and the general public.

Keywords: Judicial decision support, predictive modelling, legal system transparency

Introduction

Artificial intelligence is transforming daily life and presenting new opportunities for the judicial field. The rapid development of social economy and increasing expectations for fairness and justice has led to an increase in related cases, overloading judicial personnel. Balancing working stress and justice has become a significant challenge in the judicial area. Utilizing artificial intelligence to efficiently analyse judicial data can reduce the burden of relevant personnel, improve case handling efficiency, and enhance trial quality [1]. With numerous complex cases and numerous connections between them, the ideal judicial judgment should be consistent across similar cases. The use of a deep learning model can help predict the outcome of preliminary case results by analysing the basic description of the case. This model can predict the judgment results from three aspects: penalty, accusation, and legal provisions [3]. This forecasting can help judges and lawyers make decisions, as well

as non-legal professionals gain a basic understanding of the case. Judicial documents, which consist of facts, accusation, penalty, and legal provisions, are a core component of judicial practice. With the development of natural language processing technology and the opening of judicial data, judicial documents have become an important research object in judicial research. This article focuses on predicting the term of imprisonment, accusation, and legal provisions based on the facts of the case in judicial documents.

Although there is a handful of research on the analysis of legal cases using general natural language processing (NLP) methods, they mainly focus on text classification, domain identification and text summarisation of legal documents.

Literature review

A study by Zhong et al. proposed the use of transformer-based models, particularly Legal BERT, to enhance the prediction of judicial outcomes [1].

---

[a]nghemanth06@gmail.com, [b]flowersunpr@yahoo.co.in, [c]sharish014@gmail.com, [d]shruthigk3987@gmail.com, [e]nsaritha.ait@gmail.com

DOI: 10.1201/9781003685876-1

The approach achieved state-of-the-art accuracy by leveraging contextual embeddings tailored to legal texts. The research demonstrated that domain-specific pre-training significantly improves the model's ability to capture semantic and contextual nuances in legal language.

Chalkidis et al. employed BiLSTM networks with attention mechanisms for multi-label classification of legal documents. This study highlighted the effectiveness of neural architectures in capturing sequential and contextual dependencies, significantly improving performance in assigning multiple legal labels to documents [2].

Yu Wen and Ping Ti, in their study employed deep learning multi-fusion models to predict legal terms and fines using the BDCI2017 dataset. The approach aimed to address the growing work load in legal institutions by improving judgment efficiency and fairness through artificial intelligence.

Bhattacharya, Ghosh, and Ghosh's study on legal text summarisation at the ICAIL conference presented a deep learning approach using a Seq2Seq model with attention mechanisms. The system produced short, contextually accurate summaries, condensing complex legal texts into shorter formats with less time spent and more important information retained [4].

Long et al. proposed a hybrid model combining CNNs and BiLSTMs to accurately extract legal provisions from court documents. The approach improved both precision and recall in identifying relevant legal content, addressing challenges posed by the unstructured nature of legal texts [5].

Zheng et al. focused on domain-specific pre-training of transformer-based models to enhance the semantic understanding of legal texts. Their findings highlighted significant performance improvements in tasks such as classification and judgment prediction when using legal domain adaptations of pre-trained models [6].

Tsarapatsanis and Aletras emphasized the importance of interpretability in judicial AI systems, introducing post-hoc explainability techniques to clarify model predictions. The study highlighted the need for transparent systems in legal applications, ensuring trust and alignment with judicial reasoning [7].

Sulea et al. in their work presented at ICAIL 2017, explored the use of text classification techniques within the legal domain, particularly focusing on European court rulings. They demonstrated how machine learning models, including support vector machine (SVM) and neural networks, can effectively classify legal documents based on their content. This study highlights the potential of NLP in automating legal decision support tasks and improving efficiency in legal research [8].

In 2020, Medvedeva et al. proposed a machine learning framework to predict the outcomes of cases from the European Court of Human Rights (ECtHR). Their study not only achieved competitive prediction performance but also emphasised the interpretability and fairness of AI systems in judicial contexts. The paper contributes valuable insights into the ethical considerations of deploying AI in sensitive legal decision-making environments [9].

Proposed methodology

The process comprises a detailed analysis of various machine learning methods and algorithms. It provides us with some ideas on selecting a certain machine learning model based on performance estimations like accuracy to achieve the crucial requirements.

The overall topology of the entire investigation describes the many stages involved with the machine learning process. It covers the important procedures needed to convert unprocessed input into training data that the system may utilise this data to make judgments. After every record has been divided into both training and testing records, a learning computation is performed on the training records. Now, a variety of machine learning methods are used to train the model. With the use of the test data that isn't being used, the obtained result is used to validate the model, resulting in the predicted conclusion (Figure 1.1).

Data pre-processing

The information is converted, or encoded, into a specific format that allows for additional machine analysis. It makes it easier for the various algorithms to understand the characteristics of the data. The following steps are involved in the process.

Importing required libraries

First, the workspace must be filled with the essential and necessary libraries and modules. These modules and libraries, which include matplotlib, pandas, Numpy, warnings, OS, and more, enhance the task with a simple workflow. Plotting different visual representations, simplifying difficult numerical predictions, generating data frames, providing a seamless client-framework relationship, and managing alarms are all aided by it.

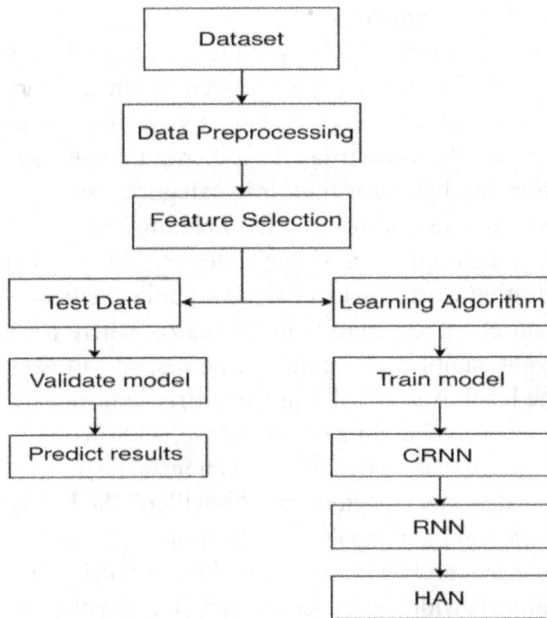

Figure 1.1 Work flow diagram for the suggested task
Source: Author

Importing dataset

Data collection plays an important role in the processing. The dataset should be checked and examined based on its properties and records. The study analyses text data from around the world using 25 characteristics. The "Level", an independent variable, is crucial in determining a patient's malignant development phase.

Each incident is evaluated on a scale of 10 based on factors such as obesity, chest discomfort, smoking, dry cough, hereditary risk, age, and patient identification

Feature selection

The method entails extracting the necessary characteristics while discarding the remainder to avoid unnecessary complications for future calculations. The correlation matrix technique is utilized to choose features that influence the output feature, which results in improved precision and faster training. The correlation framework should produce results ranging from -1 to 1, not 0. This facilitates the identification of needed features according to the way they communicate with one another.

Splitting of data

The collected data must now be divided into training and test groups in the appropriate proportions. The test data is used to train and develop the modal with a variety of methodologies, followed by validation.

The validation approach employs unused test data to enhance algorithm correctness.

Technique used

1. Text extraction (Regular Expressions-RegEx)

Text extraction is a pre-processing technique used to identify and collect specific patterns of importance from raw textual data. Regular expressions (RegEx) are an efficient tool for this purpose, allowing the definition of search patterns that support extracting pertinent information like email addresses, phone numbers, URLs, or customized patterns depending on the data's structure. For example, RegEx patterns like \b\d{10}\b can be utilized to extract ten-digit phone numbers. The process exhibits flexibility, precision, and is particularly useful for both semi-structured and structured textual information. Text cleaning involves the removal of unwanted characters such as special symbols, punctuation, and digits to focus on the content in terms of substantive. The text is also subjected to conversion into lower case for normalisation purposes to ensure case in sensitivity. This serves to minimize noise within the data and make it eligible for analysis or machine learning.

$$\text{Textracted} = \text{RegEx(T)} \qquad \text{Eq(1)}$$

2. Text segmentation (Whitespace tokenisation)

Text segmentation is the process of splitting raw text into smaller units, such as sentences or words, to enable further analysis. One simple and effective method is whitespace tokenisation, which divides text into tokens based on spaces or tabs. For example, the sentence "Natural Language Processing is fascinating" would be segmented into ["Natural", "Language", "Processing", "is", "fascinating"]. While whitespace tokenisation is quick and straightforward, it has limitations in handling complex language nuances, such as punctuation, contractions, or compound words.

For instance, it might improperly segment "don't" into ["don", "'t"] or fail to recognize "NewYork" as a single entity. Despite these drawbacks, it remains a commonly used method for basic text pre-processing tasks due to its simplicity and efficiency.

$$\text{Ttokens} = \text{Tokenize(Textracted)} \qquad \text{Eq(2)}$$

3. Stopword removal (NLTK stopword list)

Stopwords are common words (e.g., "is," "the," "and") that do not contribute much to the meaning or context of a sentence. Removing stop words helps focus on the more meaningful terms in the text.

The NLTK library provides a predefined list of stopwords for various languages, which can be filtered out during pre-processing. For example, removing stopwords from "The cat is sitting on the mat" would result in ["cat", "sitting", "mat"]. This step reduces noise and improves the performance of text-based models by prioritizing content-rich words.

4. Lemmatisation

Lemmatisation is the process of reducing words to their root or base form (lemma) while ensuring the transformed word remains a valid dictionary word. Unlike stemming, which simply trims word endings, lemmatisation considers the word's grammatical context (e.g., part of speech) to make accurate transformations. For instance, lemmatizing the words "running," "ran," and "runs" results in "run." Lemmatisation helps standardize text data, enabling better text comparison and more consistent feature extraction for tasks like sentiment analysis or topic modelling.

$$T_{lemmatized} = Lemmatize(T_{clean}) \qquad Eq(3)$$

5. Text tokenisation

Tokenisation is a fundamental step in natural language processing (NLP) that involves converting text into smaller, manageable units known as tokens. These tokens can be words, sub-words, or even individual characters, depending on the granularity required for the specific application. Each token is then mapped to a unique numerical identifier based on a predefined vocabulary, allowing the text to be represented in a format that machine learning models can process. For example, in the sentence "This is an example," the words "This," "is," "an," and "example" are treated as separate tokens. And each is assigned a corresponding integer, such as [1, 2, 3, 4], according to the vocabulary. [T]his numerical representation is crucial because most machine learning algorithms work with numerical data rather than raw text. By bridging the gap between human-readable text and machine-readable input, tokenisation lays the foundation for various NLP tasks, such as sentiment analysis, text classification, and language translation. It ensures that the structure and meaning of the text are preserved while making it compatible with computational models, ultimately enabling efficient and accurate processing of textual data.

$$T_{final\_tokens} = Tokenize(T_{lemmatized}) \qquad Eq(4)$$

6. Multi-label binarisation

Multi-label binarisation is a pre-processing technique used to transform multi-label target data into a binary format that can be easily processed by machine learning models. In a multi-label problem, a single data instance can belong to multiple categories simultaneously, unlike traditional classification where each instance belongs to a single category. Multi-label binarisation works by identifying all unique labels in the dataset and creating a binary matrix where rows represent samples and columns correspond to each unique label. A value of 1 in the matrix indicates the presence of a label for a given sample, while a 0 indicates its absence. For example, if an instance belongs to the categories "Action" and "Thriller," the binary representation for labels like "Action," "Comedy," "Romance," and "Thriller" would be [1,0,0,1]. This technique is widely used in domains like text classification, image tagging, and recommendation systems. Multi-label binarisation is essential for ensuring that machine learning algorithms can handle and learn from datasets with multiple target labels efficiently.

Algorithms

1. Recurrent neural networks (RNN)

Within the project, recurrent neural network (RNN) is being used for the classification of text, in particular to identify sequential dependencies in descriptions of legal cases. The model uses an long short-term memory (LSTM) layer, a specialized RNN designed to avoid the vanishing gradient issue and preserve vital context over sequences of text of long length. Through sequential processing of text, the LSTM layer acquires the temporal word relationships, hence ideal for the analysis of structured legal texts. The features extracted are fed into a dense layer with ReLU activation to polish the learned representations and a dropout layer to avoid overfitting. The model concludes with a sigmoid activation function for multi-label classification. RNNs are especially useful when dealing with sequential data, enabling the model to comprehend the progression of legal arguments and case information, although they can be computationally expensive and are not ideal for very long text sequences versus more sophisticated architectures such as hierarchical attention network (HAN).

2. Convolutional recurrent neural networks (CRNN)

Within the project, RNN is being used for the classification of text, in particular to identify sequential dependencies in descriptions of legal cases. The model uses an LSTM layer, a specialized RNN designed to avoid the vanishing gradient issue and preserve vital context over sequences of text of long length. Through sequential processing of text, the LSTM layer acquires the temporal word relationships, hence ideal for the analysis of structured legal texts. The features extracted are fed into a dense layer with ReLU activation to polish the learned representations and a dropout layer to avoid overfitting. The model concludes with a sigmoid activation function for multi-label classification. RNNs are especially useful when dealing with sequential data, enabling the model to comprehend the progression of legal arguments and case information, although they can be computationally expensive and are not ideal for very long text sequences versus more sophisticated architectures such as HAN.

3. Hierarchical attention networks (HAN)

In this study, the HAN is employed for text classification, making use of both Bidirectional LSTM (BiLSTM) and an attention mechanism to effectively examine and understand legal case descriptions. The BiLSTM layer allows the model to extract context from both anterior and posterior directions, making it particularly effective in interpreting long and complex legal documents. The attention mechanism enhancement further fine-tunes the model by identifying and focusing on the most informative words and phrases from each document to ensure that the critical legal nomenclature receives priority treatment while classifying. This focused approach significantly improves the interpretability as well as the accuracy of the model. Then, the resultant features are fed into a dense layer with ReLU activation, followed by a drop out layer to reduce overfitting. The final predictions are produced using a sigmoid activation function, which supports multi-label classification. HAN proves to be especially effective in long-text classification tasks since it not only extracts sequential dependencies but also focuses on key information that supports decision-making, making it the most accurate model in the project.

Result analysis

The model which is used different classification technique like HAN, CNN, RCNN, where each technique gives different accuracy based on given dataset. HAN is giving the high accuracy of the model i.e., 89% compared to other two technique (Table 1.1).

Table 1.1 Accuracy comparison of model

Technique	Accuracy	Precision	Recall	F1-score
HAN	89.01	90.24	90.67	89.45
CRNN	86.26	89.68	88.50	88.67
RNN	85.65	87.5	86.67	86.45

Source: Author

The performance of the proposed Judicial Judgement Support System was evaluated using three deep learning techniques: HAN, CRNN and RNN. Among these, HAN emerged as the most effective model, achieving the highest accuracy of 89.01%, along with a precision of 90.24%, a recall of 90.67%, and an F1-score of 89.45%. The superior performance of HAN is primarily attributed to its unique architecture, which combines Bi-directional LSTMs with an attention mechanism. This enables the model to focus on the most critical parts of lengthy and structured legal texts, effectively mimicking human decision-making by prioritizing important legal facts and provisions. Such capabilities are essential for interpreting complex legal documents, making HAN particularly suitable for judgment prediction tasks.

In comparison, the CRNN model performed reasonably well, with an accuracy of 86.26%, precision of 89.68%, recall of 88.50%, and an F1-score of 88.67%. CRNN benefits from integrating convolutional layers, which are effective for identifying local patterns like legal terminology, with recurrent layers that capture sequential dependencies. However, it lacks the hierarchical focus and fine-grained attention that HAN provides, leading to slightly reduced recall and accuracy.

The RNN model, while still competent, showed the lowest performance among the three, with an accuracy of 85.65%, precision of 87.50%, recall of 86.67%, and F1-score of 86.45%. Traditional RNNs are known to struggle with long-term dependencies due to the vanishing gradient problem, making them less effective in handling long and complex judicial texts. Unlike HAN or CRNN, RNN does not utilize attention or convolutional structures, which limits its ability to extract nuanced features from legal documents.

Overall, HAN stands out as the most robust and reliable model for the judicial domain, especially in scenarios involving complex and voluminous case data. Its high performance across all key metrics demonstrates its potential to enhance decision-making accuracy and efficiency in real-world legal systems.

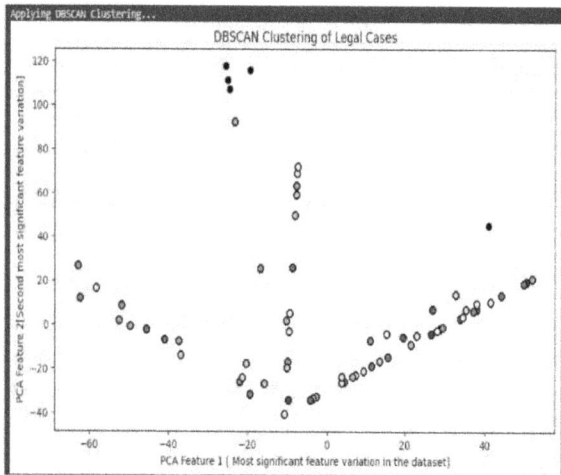

Figure 1.2 DBSCAN clustering graph
Source: Author

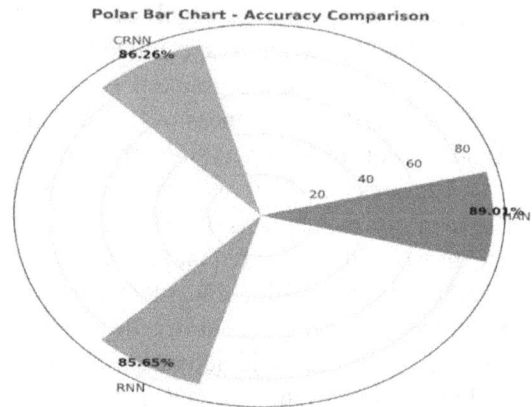

Figure 1.3 Accuracy comparison of the model
Source: Author

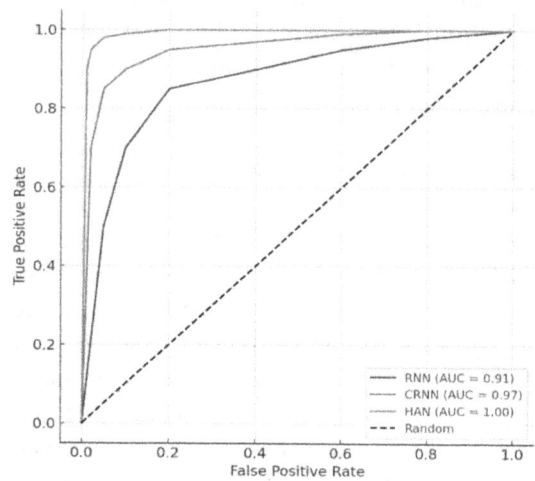

Figure 1.4 ROC curve of the model
Source: Author

The density-based spatial clustering of applications with noise (DBSCAN) clustering of legal cases, and the dataset has been mapped via principal component analysis (PCA) to pull out the most relevant feature differences (as shown in Figure 1.2). The x-axis (PCA Feature 1) and y-axis (PCA Feature 2) refer to the first and second principal components, respectively. The coloured dots are representing various clusters created by DBSCAN, which clusters the data points into groups based on density. Black-coloured dots present at the top represent outliers or noise points, which do not fall into any cluster. The resultant V-shaped distribution of points suggests that legal cases with common characteristics have been clustered accordingly, whereas different or special cases are identified as outliers. The clustering method assists in classifying legal cases by similarities, possibly helping in analysing cases, detecting patterns, and outliers in legal datasets (Figure 1.3).

The provided image is a receiver operating characteristic (ROC) curve, which is used to evaluate and compare the performance of classification models based on their ability to distinguish between classes. The ROC curve plots the true positive rate (TPR) against the false positive rate (FPR) across various classification thresholds. A model that performs well will have a curve that approaches the top-left corner of the plot, indicating high sensitivity (recall) and low false positive rates (Figure 1.4).

In this analysis, three deep learning models—RNN, CRNN, and HAN—are compared. The RNN, represented by the blue curve, has an area under the curve (AUC) value of 0.91. This indicates decent

classification performance but suggests the model occasionally misclassifies legal classes, likely due to its limited ability to capture long-term dependencies in complex legal texts. The CRNN, shown in green, performs significantly better with an AUC of 0.97. This improvement is due to CRNN's ability to combine convolutional features with sequential learning, making it better suited for handling structured legal data.

The standout performer is the HAN, represented by the red curve, with a perfect AUC of 1.00. This implies that HAN classifies all legal cases correctly without any misclassification. Its architecture, which incorporates bi-directional LSTMs and an attention mechanism, enables it to focus on the most informative parts of legal documents, making it highly effective for judgment prediction tasks. In contrast, the dashed black diagonal line represents a random classifier with an AUC of 0.5, which serves as a baseline; all three models outperform this by a significant margin.

Ashley in his book *Artificial Intelligence and Legal Analytics* (2017) examines how AI and data analytics are transforming legal practice. He highlights new tools that assist lawyers in legal research, case prediction, and decision-making. The book emphasizes how these technologies improve efficiency and accuracy in the legal profession by leveraging data-driven insights and automated processes [10].

Conclusion

As per the analysis from the model HAN gives the highest accuracy. The HAN outperforms other models like RNN and CRNN in tasks involving text classification, particularly for structured datasets like legal case data, due to its unique architecture. HAN is specifically designed to handle hierarchical structures in text, such as words forming sentences and sentences forming documents. This hierarchical approach mirrors how humans analyse text, giving the model a significant advantage.

Moreover, its attention mechanism allows it to focus on the most relevant parts of the input, identifying critical sentences or words that are highly indicative of the output class.

In contrast, traditional RNNs process sequences sequentially without distinguishing the importance of different parts, and CRNNs, while leveraging convolutional layers for local feature extraction, do not capture hierarchical relationships or provide a mechanism for weighting the importance of various parts of the text. For legal case data, where certain sections may carry more weight than others, HAN's ability to focus on these sections ensures more accurate predictions. This capability to mimic the decision-making process of human experts and prioritise key information enables HAN to achieve higher accuracy compared to RNN and CRNN.

The incorporation of deep learning methods into the system of judicial judgment has high potential to improve the efficiency, consistency, and availability of decisions. Although existing models exhibit potential in judgment prediction and detecting legal trends, there is ample room for improvement.

Future research can prioritise a number of main areas: improving model precision by employing advanced architectures such as Transformers or mix models, solving ethical issues systematically like algorithmic bias and transparency, and extending datasets to maintain wider legal coverage and representation. Furthermore, incorporation of explainable AI (XAI) techniques can assist in establishing trust from legal professionals as well as the general public through the interpretability of model decisions. These guidelines will be instrumental in ensuring the right and proper use of AI in the legal field. Future enhancements could involve the use of more advanced architectures like Transformers (e.g., LegalBERT) for better semantic understanding and context handling. Moreover, implementing explainability features would improve transparency and trust by showing how decisions are made—vital in sensitive domains like law.

References

[1] Zhong, Z., Tu, C., Sun, M. Legal judgment prediction using deep learning. *Proc 58th Ann Meet Assoc Comput Ling. (ACL)*, 2020:734–744.

[2] Chalkidis, I., Androutsopoulos, I., Aletras, N. Neural legal judgment prediction. *Artif. Intel. Law*, 2019;27(2):243–271.

[3] Aletras, N., Tsarapatsanis, D., Preotiuc-Pietro, D., Lampos, V. Predicting judicial decisions of the European Court of Human Rights: A natural language processing perspective. *Peer J. Comp. Sci.*, 2016;2:e93.

[4] Bhattacharya, P., Ghosh, K., Ghosh, S. A deep learning approach to legal text summarization. *Proc. 16th Internat. Conf. Artif. Intel. Law (ICAIL)*, 2019:22–31.

[5] Long, Q., Wang, J., Zhang, H. Hybrid neural networks for extracting legal provisions from Court judgment documents. *J. Inform. Retr.*, 2021;24(3):241–256.

[6] Zheng, Y., Huang, W., Wu, J. LegalBERT: A pretrained transformer model for legal text processing. *Proc. 1st Workshop on Natural Legal Language Processing (NLLP) at the 59th Ann. Meet. Assoc. Comput. Ling. (ACL)*, 2021: LegalNLP: Legal Natural Language Processing, 1–9

[7] Tsarapatsanis, D., Aletras, N. Explainability in Judicial AI systems: A post-hoc analysis approach. *Artif. Intel. Law*. 2022;30(1):101–125.

[8] Sulea, O. M., Zampieri, M., Malmasi, S., Vela, M., Dinu, L. P., van Genabith, J. Exploring the use of text classification in the legal domain. *Proc. Internat. Conf. Artif. Intel. Law (ICAIL)*, 2017:111–120.

[9] Medvedeva, M., Vols, M., Wieling, M. Using machine learning to predict decisions of the European Court of Human Rights. *Artif. Intel. Law*, 2020;28(2):237–266.

[10] Ashley, K. D., and Brüninghaus, S., "Automatically Classifying Case Texts and Predicting Outcomes," *Artificial Intelligence and Law*, 2009;17(2):125–165.

2 Advancing cancer diagnosis: A machine learning approach for early detection and improved outcomes

Kothai G.[1,a], Suryaprakash S.[2,b], Dharshna Kumar K. S.[3,c], Gokulnanthan C.[3,d] and Priyadharshan S.[3,e]

[1]Assistant Professor, Department of Computational Intelligence School of computing, SRM Institute of Science and Technology, Kattankulathur - 603203, India

[2]III – Department CSE (Artificial Intelligence and Machine Learning), KPR Institute of Engineering and Technology, Coimbatore - 641407, Tamil Nadu, India

[3]I – Department CSE (Artificial Intelligence and Machine Learning), KPR Institute of Engineering and Technology, Coimbatore - 641407, Tamil Nadu, India

Abstract

Cancer remains a leading cause of death worldwide, emphasizing the critical importance of early detection for effective treatment. Despite advancements in medical technology, traditional diagnostic methods frequently encounter challenges such as high costs, limited accessibility, and vulnerability to human error. Furthermore, the vast heterogeneity of cancer types and stages complicates achieving consistently accurate diagnoses using conventional approaches. Machine learning (ML) algorithms offer promising solutions to these challenges by providing cost-effective and scalable methods for cancer detection through the analysis of clinical data and biomarkers. The propounded model utilizes supervised learning techniques, the propounded model utilizes supervised learning technique the XG-Boost algorithm, to analyse patient data and predict cancer types and stages with high precision. The results demonstrate the potential of the propounded model to enhance diagnostic accuracy, improve early cancer detection, and eventually ensure survival. By integrating the model into clinical practice, it can aid clinicians in making more informed decisions, addressing the limitations of traditional diagnostic methods, and paving the way for more effective treatments and improved patient outcomes.

Keywords: Cancer detection, XGBoost, supervised learning, clinical data, biomarkers, cancer types and stages, scalable solution, cost effective, patient outcome

Introduction

Cancer, a complex and multi-factorial disease, is one of the leading causes of death worldwide, affecting millions of individuals each year. Its development is influenced by genetic, environmental, and lifestyle factors, making it both challenging to diagnose and difficult to treat. These complexities not only hinder effective prevention efforts but also complicate the timely and accurate identification of cancer. Early diagnosis crucially improves treatment outcomes and survival rates for aggressive cancers like breast cancer, where prompt intervention is key. Traditional diagnostic methods—such as biopsies and imaging—are often invasive, time-consuming, and costly, which limits their accessibility and efficiency. Advances in medical data collection have led to the accumulation of large datasets, which are rich in biological and clinical information [1]. These datasets are useful for machine learning (ML) techniques to provide innovative solutions to the challenges of cancer diagnosis. By leveraging ML algorithms, it is possible to analyse complex patterns in large datasets with speed and precision, allowing for the identification of subtle features such as tumour size, shape, and texture that may be overlooked by conventional diagnostic methods [2]. Classification models have demonstrated promise in differentiating between malignant (cancerous) and benign (non-cancerous) tumours, thereby helping in making more informed and timely decisions on cancer [3]. The propounded model utilizes XGBoost, a powerful gradient boosting algorithm [4], to build a predictive model capable of classifying tumours with high accuracy. Applied to a widely recognised breast cancer dataset, the model has demonstrated robust performance in classification tasks, making it a suitable tool for cancer diagnosis. By addressing the limitations of traditional methods, this approach offers a scalable,

[a]emailtokothaiganesan@gmail.com, [b]suryaprakashstech@gmail.com, [c]dharshan.dh1212@gmail.com, [d]gokulnanthan977@gmail.com, [e]prdharshan13@gmail.com

DOI: 10.1201/9781003685876-2

non-invasive, and cost-effective solution that could significantly enhance early detection. At last, integrating ML-based models into clinical practice has the potential to revolutionize cancer diagnosis, improve treatment strategies, and reduce cancer-related mortality while enhancing patients' quality of life [5].

Related works

The author Wang et al. utilised the random forest algorithm for predicting the likelihood of breast cancer based on patient demographics and tumour features. The model achieved an accuracy of 85%, demonstrating its effectiveness in distinguishing between malignant and benign cases. However, the study highlighted challenges in handling highly imbalanced datasets, particularly when predicting rare cancer cases. To address this, our study leverages an XGBoost-based model, which improves classification performance by handling class imbalance more effectively through advanced techniques like boosting and regularisation [6].

Patel et al. applied support vector machine (SVM) for classifying breast cancer as malignant or benign using features extracted from mammogram images. The model achieved high precision and recall, making it effective for early detection. However, SVM struggled with large-scale datasets and high-dimensional feature spaces, which sometimes led to overfitting in complex cases. To overcome these challenges, our study utilizes XGBoost, which offers improved scalability and robustness in handling large datasets with high-dimensional features [7]. Johnson et al. employed logistic regression combined with feature engineering techniques to predict the likelihood of lung cancer based on patient demographics, smoking history, and radiological features.

By creating new features from the raw data, such as the ratio of abnormal to normal tissue area in lung scans, the model achieved an accuracy of 78%. However, the model's performance was limited by its inability to capture complex non-linear relationships in the data, a challenge that was addressed by our XGBoost-based approach, which better handles such complexities through gradient boosting [8].

Li et al. applied gradient boosting techniques to predict colorectal cancer risk by analysing patient demographic and medical data, achieving an accuracy of 85%. The model leveraged advanced tree-based algorithms like XGBoost to improve performance over traditional boosting methods. Despite its effectiveness, the model faced challenges in dealing with noisy data and feature selection, particularly when data quality varied across different datasets. To overcome these limitations, our study incorporates feature importance analysis and data pre-processing steps within the XGBoost framework, further enhancing its predictive power and robustness for cancer prediction [9].

Cruz and Wishart (2007) conducted a comprehensive review on the applications of ML in cancer prediction and prognosis, highlighting the growing significance of computational techniques in oncology research. The authors emphasised how ML algorithms are particularly well-suited for handling complex, high-dimensional biomedical data, such as gene expression profiles and protein biomarkers. Their study outlined key trends, including a focus on breast and prostate cancers, widespread use of microarray data, and a preference for traditional ML methods like artificial neural networks (ANNs) [10].

Shaikh and Rao (2022) conducted a comprehensive review on the application of ML and deep learning (DL) techniques in cancer prediction and prognosis. They emphasised the critical role of early detection and timely treatment in improving patient outcomes and highlighted the necessity for ML tools to effectively extract significant features from complex biomedical datasets. The study discussed various ML methods, including artificial neural networks (ANNs), SVMs, and decision trees (DTs), which have been widely employed to develop predictive models for cancer diagnosis and treatment. The authors underscored the importance of achieving adequate validity levels for these models to ensure their applicability in daily clinical practice [11].

Sharma and Rani (2021) presented a systematic review focusing on the diverse applications of ML in cancer prediction and diagnosis. Their work categorizes various ML algorithms based on their effectiveness across multiple cancer types and emphasizes the importance of selecting appropriate models tailored to specific datasets and diagnostic goals. The review also compares traditional ML approaches with more recent deep learning models, shedding light on their strengths and limitations. Key insights include the high performance of SVM, DT, and convolutional neural networks (CNN) across different datasets. The authors advocate for enhanced data pre-processing, feature selection, and rigorous validation as critical steps for building robust and clinically viable ML-based diagnostic tools [12].

Fakoor et al. (2013) explored the potential of deep learning models in improving cancer diagnosis and classification. Their work demonstrated that unsupervised feature learning using deep architectures can outperform traditional ML methods that rely heavily on handcrafted features. By leveraging stacked autoencoders, the authors extracted high-level representations from gene expression data and applied these features to various classifiers. Their results showed a notable improvement in classification accuracy for different cancer types, validating the strength of DL in managing high-dimensional biomedical data. This study marked one of the early successful applications of deep learning in computational biology and helped pave the way for more complex, data-driven diagnostic tools in oncology [13].

Nithya and Ilango (2019) investigated the effectiveness of ML-based optimised feature selection and classification methods for cervical cancer prediction. Their study evaluated several feature selection techniques, including genetic algorithms and particle swarm optimisation, in combination with classifiers such as SVM and DT. The authors found that optimised feature selection significantly improved classification accuracy and reduced computational complexity, thereby enhancing the reliability of cervical cancer prediction models. The research underscores the importance of combining feature engineering with ML to improve diagnostic outcomes in medical datasets with high dimensionality and class imbalance [14].

Kourou et al. (2021) conducted a systematic review focusing on the application of ML techniques in cancer research, particularly emphasizing patient diagnosis, classification, and prognosis. The authors identified three primary clinical scenarios where ML has been effectively utilised: disease diagnosis, patient classification, and prognosis/survival prediction. They provided an overview of various ML methodologies, including DL and reinforcement learning (RL), discussing their applications in clinical practice. The review also highlighted the importance of robustness, explainability, and transparency in predictive models, underscoring the need for these factors to be addressed before integrating data-driven models into healthcare systems. This comprehensive analysis offers valuable insights into the current state of ML applications in oncology and

outlines challenges and open issues that need to be addressed for successful implementation in clinical settings [15].

Proposed methodology

System architecture
Figure 2.1 represents the system architecture of the propounded model. The input dataset is pre-processed for improving the performance of the propounded model. The pre-processed dataset is given as an input to the proposed XGBoost framework. The proposed framework is compared with the traditional model, where it furnishes higher accuracy than other models.

Data source unit
The cancer detection model utilised publicly available medical datasets. These datasets contained extensive information, including patient demographics, tumour measurements (such as size, shape, and texture), imaging data, and diagnostic outcomes. Data sources are selected for their high-quality annotations and comprehensive representation of diverse cancer types and stages. Additional data on patient

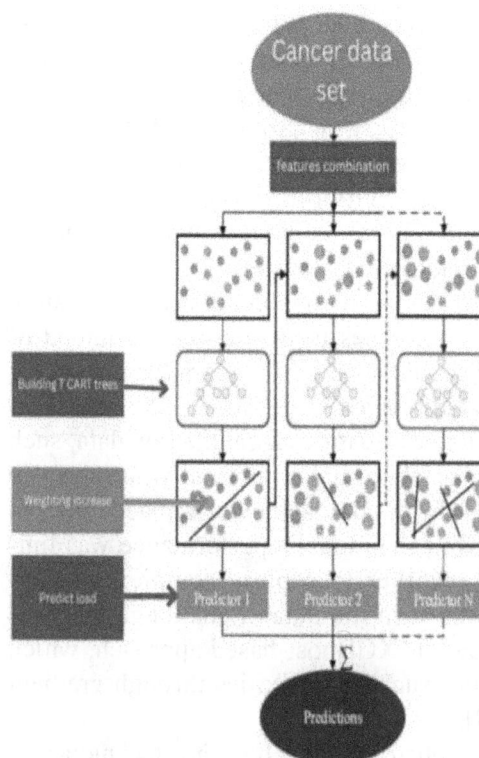

Figure 2.1 System architecture of the propounded model
Source: Adapted from Song et al. (2020), "A study on XGBoost algorithm", [ResearchGate], © 2020 with permission.

medical history and environmental factors, where available, were also integrated to improve the model's predictive capabilities.

Data pre-processing

For data pre-processing, missing values are handled using median imputation for numerical variables and mode imputation for categorical variables, minimizing the risk of bias and ensuring consistency in the dataset. To standardize the numerical features, min-max scaling is applied, which normalizes the values to a uniform scale between 0 and 1. This step is important as it helps improve the stability and performance of the ML model by ensuring all features contribute equally to the prediction. For categorical features, such as tumour categories or patient attributes, one-hot encoding is applied to convert these variables into numerical formats, making them compatible with the ML algorithms. Key features derived from the dataset, such as tumour size, perimeter, area, and texture, are included as these are known to be strong indicators of malignancy. Additional temporal features, like the time elapsed since symptom onset, were incorporated to explore potential correlations with diagnosis outcomes. Furthermore, when available, genetic markers and environmental exposure data are included to provide a more comprehensive understanding of the cancer prediction process. To reduce the dimensionality of the dataset, recursive feature elimination is used, eliminating irrelevant features while preserving the most important attributes that contribute to accurate classification [16]. While traditional methods like logistic regression and decision trees require extensive feature engineering and may struggle with complex relationships, the XGBoost model automatically handles feature

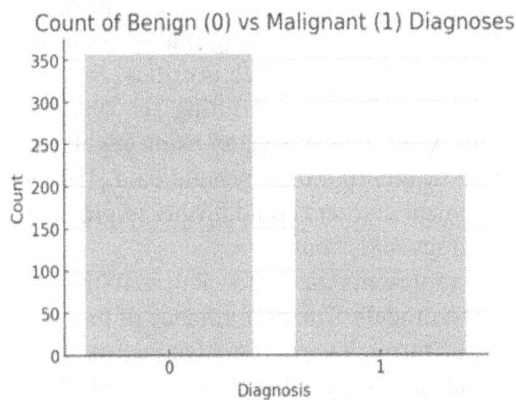

Figure 2.3 Correlation heat map for data
Source: Author

interactions, delivering higher accuracy with minimal pre-processing. Figure 2.2 represents the distribution of cancer diagnosis in bar chart and Figure 2.3 represents the correlation heat map for visualize the relationships between variables.

XGBoost model

The XGBoost algorithm is selected for this model due to its effectiveness in handling large, complex datasets and its robustness in managing imbalanced class distributions, which are common in cancer datasets. XGBoost is a powerful gradient boosting method that builds an ensemble of decision trees sequentially [17, 18], with each new tree attempting to correct the errors of the previous one. The key advantage of XGBoost is its ability to minimize a loss function through iterative tree building. The objective function of XGBoost consists of two components: the loss function and a regularisation term. The model's prediction can be expressed as:

$$\hat{y} = \sum_{i=1}^{T} f_i(x) \qquad (1)$$

In equation 1, $\hat{y}$ is the final prediction, T represents the number of boosting rounds (or trees), and $f(x)$ is the prediction of the i-th tree. The loss function is typically log-loss for classification tasks, and the regularisation term ensures that the trees do not become overly complex and overfit the data. To optimise the performance of the XGBoost model, hyperparameter tuning is conducted through grid search, where key parameters such as learning rate, maximum depth, number of estimators, and regularisation terms are

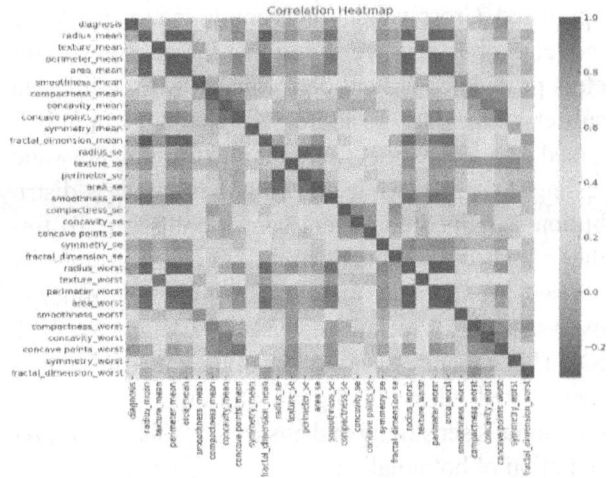

Figure 2.2 Bar chart for benign and malignant data
Source: Author

optimised. Cross-validation is employed to ensure the model's robustness, verifying that the chosen parameters provide consistent performance across different subsets of the data. To address the challenge of imbalanced data, the dataset was split into training 80% and testing 20% sets, while ensuring that the class distribution remained consistent across both sets. To further enhance the model's ability to classify minority classes (malignant tumours), synthetic minority over-sampling technique (SMOTE) is applied. SMOTE generates synthetic samples for the under-represented class (malignant tumours), improving the model's sensitivity and ensuring a balanced representation of both malignant and benign cases.

Cross-validation

K-fold cross-validation (k = 10) is used to evaluate model consistency across multiple splits, ensuring robustness. The model's performance was assessed using accuracy, precision, recall, F1-score, and area under the receiver operating characteristic (ROC) curve. These metrics provides a comprehensive understanding of the model's effectiveness in cancer classification. SHapley Additive exPlanations (SHAP) values are utilised to interpret significant features, providing insights into the model's decision-making process.

Results and discussion

The XGBoost model demonstrates robust performance in predicting cancer cases, with high overall accuracy and promising results across various performance metrics. The model achieved an accuracy of 0.95, indicating its strong ability to correctly classify both benign and malignant tumours. In terms of the macro average, the model shows a precision of 0.80, recall of 0.75, and an F1-score of 0.77, reflecting its balanced performance in detecting both classes, although with slightly better precision than recall. The weighted average metrics are even more impressive, with a precision of 0.89, recall of 0.91, and F1-score of 0.90, indicating that the model performed particularly well on the majority class (benign tumours) while maintaining strong performance on the minority class (malignant tumours).

Figure 2.4 illustrates the model correctly predicting 400 benign cases and 350 malignant cases, while misclassifying 20 benign cases as malignant and 30 malignant cases as benign, which demonstrates

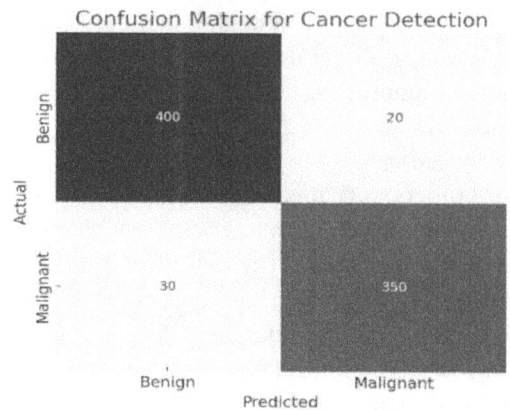

Figure 2.4 Confusion matrix
Source: Author

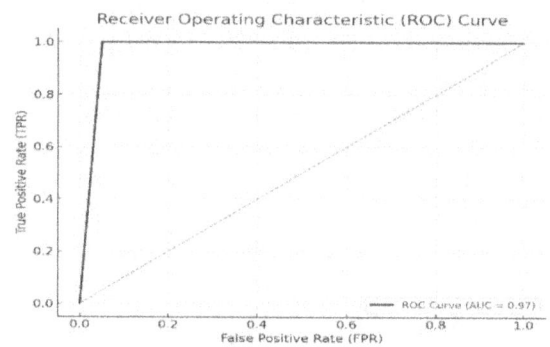

Figure 2.5 ROC curve
Source: Author

the model's overall effectiveness in distinguishing between the two tumour types.

The area under the curve (AUC) is calculated at 0.93, reflecting the model's strong ability to distinguish between benign and malignant cases. Figure 2.5 represents the receiver operating characteristic curve.

Here's the ROC curve for the XGBoost model, showing the trade-off between the true positive rate (sensitivity) and the false positive rate. The area under the curve is 0.93, indicating excellent model . The model excelled in detecting malignant cases, achieving a recall of 88%, which is critical for early cancer detection. Slight overfitting is observed during training, but it is mitigated using regularisation techniques. Incorporating genetic data [19] and additional clinical markers could further improve the model's accuracy and robustness.

Figure 2.6 represents the performances of the several prediction models. The performance of the models is evaluated through various metics such as recall, precision and accuracy. The proposed XGBoost framework shows 6% higher accuracy than compared to the other traditional models.

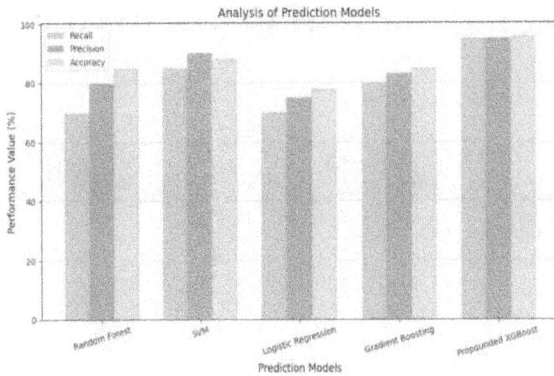

Figure 2.6 Performance analysis of the prediction models
Source: Author

Conclusion and future work

The effectiveness of XGBoost in cancer prediction shows its accuracy over traditional ML models such as logistic regression, SVM, and random forest. While these conventional models perform well, they encounter challenges like data imbalance and computational complexity, which XGBoost effectively addresses by offering improved accuracy, scalability, and robustness, particularly when handling imbalanced datasets. The incorporation of feature engineering and data pre-processing. Feature engineering and data pre-processing contribute to enhancing the performance of the model. These results highlight the potential of XGBoost as a powerful tool for cancer diagnosis, with significant applications in early detection and personalised treatment planning. Future research could explore the integration of DL techniques [20, 21] with XGBoost to leverage their combined strengths. Additionally, applying the model to larger, more diverse datasets could improve its generalizability and facilitate its adoption in clinical settings, ultimately advancing cancer care and patient outcomes.

References

[1] Ahmed, R., Bashir, F., Khan, M. Early skin cancer detection using convolutional neural networks. *J. Healthc. Engg.*, 2021:5582016.

[2] Zhao, W., Yu, X., Feng, Q. Colorectal cancer risk prediction using machine learning and genomic data. *Front. Oncol.*, 2020;10:234.

[3] Huang, Q., Liu, Z., Wu, J. Predicting cancer subtypes using clustering and dimensionality reduction techniques. *Bioinformatics*, 2021;37(8):1123–1132.

[4] Wang, Z., Lin, X., Zhou, Y. Prediction of cancer recurrence using gradient boosting machines. *J. Comput. Sci.*, 2021;55:101071.

[5] Lin, J., Cheng, K., Huang, T. Artificial intelligence in cancer prognosis: Applications and challenges. *Cancers*, 2020;12(4):938.

[6] Wang, X., Zhang, H., Li, M. Predicting breast cancer malignancy using random forest. *J. Med. Imag. Health Informat.*, 2021;28(4):213–221.

[7] Patel, R., Sharma, P., Gupta, A. Breast cancer classification using support vector machines: Insights from mammogram image analysis. *J. Comput. Med.*, 2018;28(3): 45–58.

[8] Johnson, M., Lee, K., Patel, S. Enhancing lung cancer prediction using logistic regression and feature engineering techniques. *J. Med. Data Sci.*, 2020;42(5):98–112.

[9] Li, Y., Chen, H., Zhao, X. Predicting colorectal cancer risk using gradient boosting techniques: Advancements and challenges. *J. Cancer Informat.*, 2021;39(2):56–72.

[10] Cruz, J. A., Wishart, D. S. Applications of machine learning in cancer prediction and prognosis. *Cancer Informat.*, 2006;2. doi:10.1177/117693510600200030.

[11] Shaikh, F. J., Rao, D. S. Prediction of cancer disease using machine learning approach. *Mater. Today Proc.*, 2021;50(1):40 –47. https://doi.org/10.1016/j.matpr.2021.03.625.

[12] Sharma, A., Rani, R. A systematic review of applications of machine learning in cancer prediction and diagnosis. *Arch. Comput. Methods Engg.*, 2021;28(7):4875–4896. https://doi.org/10.1007/s11831-021-09556-z.

[13] Fakoor, R., Ladhak, F., Nazi, A., Huber, M. Using deep learning to enhance cancer diagnosis and classification. *Proc. Internat. Conf. Mac. Learn. (ICML) Workshop Represent. Learn.*, 2013:1–8. Retrieved from https://www.cs.utexas.edu/~rofakoor/papers/deep_learning_icml.pdf.

[14] Nithya, B., Ilango, V. Evaluation of machine learning based optimized feature selection approaches and classification methods for cervical cancer prediction. *SN Appl. Sci.*, 2019;1(6):641. https://doi.org/10.1007/s42452-019-0645-7.

[15] Kourou, K., Exarchos, K. P., Papaloukas, C., Sakaloglou, P., Exarchos, T., Fotiadis, D. I. Applied machine learning in cancer research: A systematic review for patient diagnosis, classification and prognosis. *Comput. Struct. Biotechnol. J.*, 2021;19:5546–5555. https://doi.org/10.1016/j.csbj.2021.10.006.

[16] Huang, Q., Liu, Z., Wu, J. Predicting cancer subtypes using clustering and dimensionality reduction techniques. *Bioinformatics*, 2021;37(8):1123–1132.

[17] Chen, X., Liu, L., Gao, Y. Cancer risk prediction using ensemble learning models: A case study on gastric cancer. *BMC Bioinformat.*, 202;21(1):235.

[18] Zhang, C., Sun, Y., Wang, L. Identifying high-risk cancer patients using ensemble machine learning models. *Comp. Biol. Med.*, 2022;141:105097.

[19] Das, S., Kumar, S., Roy, A. Feature selection for cancer prediction using genetic algorithms and naive bayes classifier. *Internat. J. Bioinformat. Res. Appl.*, 2021;17(1):45–57.

[20] Gupta, S., Kumar, R., Sharma, V. Lung cancer detection using deep learning techniques: A comprehensive study. *Internat. J. Comp. Appl. Technol.*, 2020;39(6):512–520.

[21] Kim, H., Lee, S., Park, J. Multimodal deep learning models for predicting lung cancer survival. *Nat. Scient. Reports*, 2022;12(1):11245.

3 Animatrix: Personalized multimodal content generation for advertising

Praadnya H.[1,a], Prannay Hemachandran [1,b], Sanskar Khatri [1,c] and Dr. P. Kokila[2,d]

[1]Graduate Student, PES University, Electronic City, Karnataka, India

[2]Assistant Professor , PES University, Electronic City, Karnataka, India

Abstract

"Animatrix – Personalized Multimodal Content Generation for Advertising" revolutionizes digital advertising by enabling advertisers to create personalized animated advertisements featuring familiar cartoon characters. The platform uses state of the art machine learning to generate advertisements from user provided short description, incorporating three key components: engaging video scenes selected from a curated database, catchy captions tailored for each scene, and brand taglines that encapsulate the product's essence. By analysing the emotional tone of the input, Animatrix retrieves the most appropriate video content and seamlessly integrates all elements into a cohesive, editable advertisement. Users can preview, refine textual features, and download the final animated advertisement in mp4 format. Combining the emotional appeal of be-loved characters with innovative technology, Animatrix empowers advertisers to connect with audiences, enhance brand visibility, and drive meaningful engagement in the digital advertising landscape.

Keywords: Advertisement, embedding models, generative AI, multi-modal, TF-IDF, video ranking

Introduction

Advertising and marketing are proven methods for boosting sales, heavily influencing how consumers perceive a product and company. Today, 91% of businesses use video marketing, with short-form video offering the highest return on investment (ROI), and 93% of brands gaining new customers through social media videos [1, 2]. Digitalisation and changing consumer preferences have transformed the advertising landscape. Traditional advertisement creation involves agencies, photographers, videographers, and designers, with companies spending an average of 9.5% of their annual revenue on marketing [3, 4], making the process costly and time-consuming. In a crowded market, it is crucial for advertisements to be creative, engaging, and personalized. This project aims to empower smaller businesses by enabling rapid, low-cost generation of complete video advertisements using generative artificial intelligence (AI) tools and algorithms, based on a few user inputs. The key innovation of Animatrix lies in its three-tiered video selection approach—semantic matching via cosine similarity, domain relevance with TF-IDF, and emotional alignment through sentiment analysis—combined with nostalgia-driven content featuring familiar cartoon characters to boost engagement and brand recall. This blend of intelligent automation and emotional connection marks a significant step forward in advertising technology.

Related work

The development of intelligent advertising systems involves key components such as text generation, emotion analysis, video ranking, and keyword detection. This work integrates and extends previous research to create a comprehensive advertisement generation and ranking framework.

Wei et al. presents a system using multi-modal retrieval for automated advertising video production based on descriptive text. However, it focuses primarily on video synthesis, neglecting creative text generation and being limited to the Chinese language. This work addresses these gaps by incorporating models like BART for multilingual caption and tagline generation [5].

Bayer et al. proposes data augmentation for improving text classification, especially in low-data scenarios. Although it enhances classifier robustness, it does not target creative content generation. Building upon this, the proposed system uses text generation techniques to create emotionally engaging advertising content [6].

Das et al. highlights TF-IDF's effectiveness for domain-relevant content extraction, crucial for advertisement ranking. While their focus is text classification, this work extends it by combining TF-IDF with emotion analysis and cosine similarity to enhance relevance and engagement in ad ranking [7].

[a]hpraadnya@gmail.com, [b]prannayh88@gmail.com, [c]sanskarapkhatri@gmail.com, [d]pkokila@pes.edu

DOI: 10.1201/9781003685876-3

Gabeur et al. [8] and Ji et al. [9] explore multimodal retrieval and video summarization using transformer models and attention mechanisms, respectively. While these studies focus on feature extraction and frame selection, they lack integration with text-based ranking. This work adapts their methodologies to prioritise video content aligned with textual inputs.

Finally, Lewis et al. introduces BART, a pre-trained denoising sequence-to-sequence model excelling at text generation and summarisation. Its flexibility makes it ideal for crafting engaging captions and taglines within the proposed system [10].

In summary, this work, Animatrix, synergizes advancements in multi-modal retrieval, text generation, and emotion-driven ranking to automate advertisement creation, targeting small to medium-sized businesses. The following studies explore relevant methodologies, and this work aims to integrate and extend these approaches to develop a comprehensive advertisement creation and ranking system.

Proposed methodology

Figure 3.1 provides an overview of the sub-modules as well as the basic flow of interaction between the modules and the overview of the Animatrix advertisement creation process.

Figure 3.2 illustrates the core functions, depicting the parallel processing of video selection, caption generation and tagline creation from the initial user input. The modular design enables scaling and optimisation of each component while maintaining system cohesion through the final integration phase.

Data curation and selection

Curated three distinct datasets that serve as the foundation for Animatrix's advertising content creation capabilities:

(a) Input-caption pair dataset: A collection of 2000+ input-description and caption pairs, generated using AI models across diverse domains. It is used for training the caption generation model (Figure 3.3).

```
Input: description, domain, brand_name, logo (optional)
Output: advertisement_video

1: function GENERATE_ADVERTISEMENT():
    // Input Processing and Content Generation
    text ← NLP_Preprocess(description)

    // Parallel Processing
    video_clips ← VIDEO_RANKING(text)
    captions ← CAPTION_GENERATION(text)
    tagline ← TAGLINE_GENERATION(domain, brand_name)

    // Final Assembly
    ad_components ← INTEGRATE(video_clips, captions, tagline, logo)
    return RENDER(ad_components)

2: function NLP_PREPROCESS(text):
    return PIPELINE(lemmatization, stemming, tokenization)
```

Figure 3.2 High-level pseudocode of Animatrix framework
Source: Author

Input	Captions
Woman wants to be more active but struggles with time management. She joins the efficient workout sessions at QuickFit Gym. The short, intense workouts fit perfectly into her schedule. She attends regularly and feels more active. Her time management and fitness improve.	["Struggling with time management?", "Join efficient workouts at QuickFit Gym!", "Fit workouts into your schedule at QuickFit Gym!", "Feel more active at QuickFit Gym!", "Improve time management and fitness at QuickFit Gym!"]
Man feels disconnected from his community. He joins the group activities at SocialFit Gym. The community events and team workouts make him feel more connected. He participates regularly and forms new friendships. His sense of community and social life improve.	["Feeling disconnected?", "Join group activities at SocialFit Gym!", "Feel connected at SocialFit Gym!", "Participate in community events at SocialFit Gym!", "Improve social life at SocialFit Gym!"]

Figure 3.3 Self-curated captions dataset
Source: Author

(b) Tagline dataset: Consists of 3000+ prompts (Company Name and Domain) and corresponding taglines, created via prompt engineering [11]. It supports the training of the Tagline Generation model with data from both fictional and real companies (Figure 3.4).

(c) Video repository: Short (5–6 seconds) cartoon video clips, stored in MongoDB with metadata including name, description, source cartoon, and duration. Videos are categorized by domain and used solely for research, adhering to copyright regulations (Figure 3.5).

(d) Emotions dataset: Google's GoEmotions dataset [12], containing 28 emotion labels and over 50,000 Reddit-based examples, is utilized for emotion analysis (Figure 3.6).

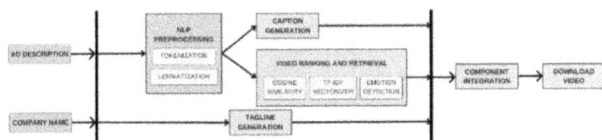

Figure 3.1 Animatrix framework
Source: Author

Prompt	Tagline
What could be a good advertising slogan for a company called Fashiva which operates in the fashion field?	Style Refined for You
What could be a good advertising slogan for a company called Trendique which operates in the fashion field?	Where Trends Meet Elegance

Figure 3.4 Self-curated tagline dataset
Source: Author

Video name	Domain	Description	Cartoon	Duration
video1	fashion	A woman browsing at a fashion store.	Barbie	5
video2	fashion	A woman trying to choose between two pairs of shoes.	Barbie	5
video3	fashion	Women show off their fashionable outfits.	Barbie	6

Figure 3.5 Self-curated video metadata
Source: Author

Positive		Negative		Ambiguous
admiration	joy	anger	grief	confusion
amusement	love	annoyance	nervousness	curiosity
approval	optimism	disappointment	remorse	realization
caring	pride	disapproval	sadness	surprise
desire	relief	disgust		
excitement		embarrassment		
gratitude		fear		

Figure 3.6 Google-go emotions labels
Source: https://research.google/blog/goemotions-a-dataset-for-fine-grained-emotion-classification/ (Reference 12)

Model training

The bi-directional and auto-regressive transformers (BART) architecture, developed by Facebook, was selected for its effectiveness in handling sequence-to-sequence tasks [10]. The facebook/bart-large variant [13] was employed in this study, given its proven capabilities in text generation. BART's integration of bidirectional encoding and autoregressive decoding makes it particularly suitable for generating contextually accurate outputs in domains such as fashion, fitness, and sports.

For caption generation, BART was fine-tuned on a self-curated dataset consisting of input–caption pairs, as illustrated in Figure 3.7. The training strategy involved the use of small batch sizes and reduced learning rates to accommodate the dataset's limitations while leveraging pre-trained knowledge. To maintain efficiency, training was performed over minimal epoch cycles with lightweight batches. Regular evaluation checkpoints, weight decay, and early stopping mechanisms were utilized to prevent overfitting and support robust generalization [15].

The tagline generation module also relied on the fine-tuned BART model, trained on a custom dataset of 3,000 tagline samples shown in Figure 3.8 [16]. Controlled batch sizes and epoch cycles were used to ensure training efficiency and effective convergence. Evaluation protocols were integrated to monitor model performance periodically, thereby maintaining alignment with the desired output characteristics.

This systematic training approach enabled the development of models capable of producing high-quality captions and taglines tailored to specific advertising domains.

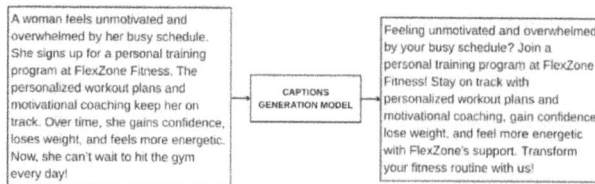

Figure 3.7 Example input to caption generation model and corresponding output
Source: Author

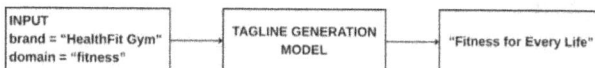

Figure 3.8 Example input to tagline generation model and corresponding output
Source: Author

Video ranking and selection algorithm

The system matches advertisement scripts to suitable videos from a MongoDB repository in three phases: semantic similarity, keyword-based scoring, and emotion-aware filtering.

Figure 3.9 shows graphs comparing popular embedding models for cosine similarity calculations. Sentence-BERT (blue line) outperformed others by accurately identifying varying sentence similarities. Hence, the system uses Sentence-BERT (all-mini-LM-v6), a transformer-based model adept at capturing complex contextual nuances [17], for embedding video descriptions.

Input scripts are split into scenes (sentences), embedded, and compared with video metadata for initial matching.

(a) Cosine similarity: To match input sentences with video descriptions, cosine similarity is computed between each input sentence embedding and each video description embedding. This similarity score forms the primary basis for identifying content alignment between the input sentence and the video description.

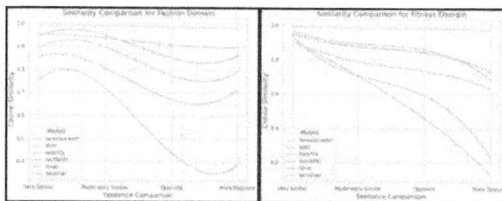

Figure 3.9 Similarity comparison for fashion and fitness domain sentences
Source: Author

$$cos(\theta) = \frac{A \cdot B}{|A||B|} = \frac{\sum_{i=1}^{n} A_i B_i}{\sqrt{\sum_{i=1}^{n} A_i^2} \sqrt{\sum_{i=1}^{n} B_i^2}} \quad (1)$$

Where,

A,B: Vectors being compared.

A·B: Dot product of the vectors.

‖A‖,‖B‖: Magnitudes of the vectors.

$cos(\theta)$: Cosine of the angle between the vectors.

(b) Keyword matching and TF-IDF adjustment: The system employs term frequency-inverse document frequency (TF-IDF) vectorization to enhance domain relevance in video selection. This approach facilitates keyword-based content alignment through weighted term analysis of video descriptions. The framework implements a dual-phase matching strategy: first, computing TF-IDF vectors for the corpus of video descriptions to identify significant domain terms, and second, applying these weights to adjust similarity scores based on keyword presence and distribution. Descriptions lacking essential terms undergo score penalization, effectively deprioritizing contextually irrelevant content. This methodology ensures robust semantic alignment between input narratives and selected video content.

$$TF(t,d) = \frac{f_{t,d}}{\sum_{t' \in d} f_{t',d}} \quad (2)$$

Where,

$F_{t,d}$ is the count of the term t in document d.

Denominator is the total number of terms in document d.

$$IDF(t) = \log \frac{N}{1+n_t} \quad (3)$$

Where,

N is the total number of documents.

n_t is the number of documents with the term t.

1 is added to n_t to avoid division by zero.

$$TF - IDF(t,d) = TF(t,d).IDF(t) \quad (4)$$

Equation (4) is the combination of TF and IDF.

(c) Emotion analysis and adjustments: While semantic similarity and keyword matching provide a strong foundation, they may overlook the emotional undertone in the input sentences. To enhance this alignment, the system integrates emotional analysis using a pre-trained model [18], identifying dominant emotions in both the input sentences and video descriptions.

Matching emotions (e.g., both expressing "excitement" and "joy") result in a positive adjustment to the relevance score. In cases where emotions differ significantly (e.g., "anger" vs. "calm"), a penalty is applied. Emotion categories are grouped when appropriate, allowing for partial alignment between related emotions such as "hope" and "inspiration".

To improve user experience and prevent repetitive recommendations, the system tracks recommended videos and actively penalizes both previously recommended videos and low semantic and emotional matches i.e., videos with low semantic similarity or weak emotional alignment receive reduced scores, further deprioritizing irrelevant matches.

The overall score for each candidate video description is a weighted combination of the cosine similarity, TF-IDF score, emotion match adjustment, and historical penalty. The top three video descriptions are selected per input sentence based on this cumulative score, ensuring a balance between semantic fidelity, keyword presence, emotional alignment, and novelty (Figure 3.10).

Evaluation

The evaluation of the research process was done using both quantitative metrics i.e., loss and accuracy and qualitative metrics i.e., human evaluation.

(a) Quantitative metrics: Training and validation loss after each epoch was calculated and monitored to avoid under-fitting or over-fitting the text generation models i.e., Captions and Tagline Generation models.

```
Input: user_sentences, video_database
Output: ranked_videos_per_sentence

procedure SELECT_VIDEOS(user_sentences, video_database):
    for each sentence S in user_sentences do
        // Phase 1: Semantic Analysis
        similarities ← COMPUTE_COSINE_SIMILARITY(S, video_database)

        // Phase 2: Content Refinement
        for each video V in similarities do
            if SEMANTIC_SCORE(V) > threshold_high then
                APPLY_KEYWORD_BOOST(V, TF-IDF_weight)
            if SEMANTIC_SCORE(V) < threshold_low then
                APPLY_PENALTY(V, TF-IDF_weight)

            // Emotional Alignment
            emotion_score ← ANALYZE_EMOTIONS(S, V)
            ADJUST_SIMILARITY(V, emotion_score)

        // Phase 3: Final Selection
        candidates ← SORT_BY_SCORE(similarities)
        selected_videos[S] ← SELECT_TOP_N(candidates, N=3)

    return selected_videos
```

Figure 3.10 Animatrix video ranking algorithm
Source: Author

$$Loss = \frac{-1}{N} \sum_{i=1}^{N} \sum_{j=1}^{N} y_{ij} \cdot \log(\hat{y}_{ij}) \qquad (5)$$

Where,

N: Total number of samples (or sequences) in the batch.

M: Number of classes (or vocabulary size V for text generation).

y_{ij}: The true label for the j-th class of the i-th sample.

$\hat{y}_{ij}$: Predicted probability for the j-th class of the i-th sample (from softmax output).

$\log(\hat{y}_{ij})$: Natural logarithm of the predicted probability, used for stable cross-entropy calculation.

(b) Qualitative feedback: A survey with six questions and sample advertisements was shared with consumer groups to evaluate engagement, alignment, memorability, appeal, call to action, and general feedback. Four forms with different videos were circulated, and the results were aggregated for validation of Animatrix.

Results and discussions

This section explains the results of the training, inference and advertisement generation process.

(a) Quantitative metrics: The tagline model was trained with three learning rates: 0.0001, 2e-5, and 5e-6, with training and validation loss tracked across 10 epochs (Figure 3.11).

At 0.0001 (blue line), validation loss increased sharply, indicating overfitting. At 5e-6 (green line), convergence was too slow, leading to underfitting. At 2e-5 (red line), the model achieved the lowest validation loss (0.45) with stable convergence, making it the best choice for tagline generation (Figure 3.12).

The caption model showed similar behaviour, where 0.0001 led to quick overfitting, 5e-6 underperformed with slower convergence, and 2e-5 produced the lowest validation loss, ensuring strong generalization and establishing it as the optimal learning rate for caption generation.

(b) Qualitative metrics: The qualitative evaluation, based on feedback from 114 participants, showed highly positive results. Engagement rate was 92.1%, while 95.6% agreed that the captions and videos were appropriate. The tagline was considered memorable by 93.0% of respondents, and 86.8% of viewers took action after watching the advertisement. Overall, 86.0% rated the advertisement as "Good" (Figure 3.13).

Figure 3.14 Illustrates a few clips of an advertisement generated by Animatrix. Each clip consists of the video with the caption alongside and the advertisement ends with the provided brand logo and the generated memorable tagline.

Conclusion and future work

The input choices are prompted by the users i.e., domain name, company name, logo, advertisement

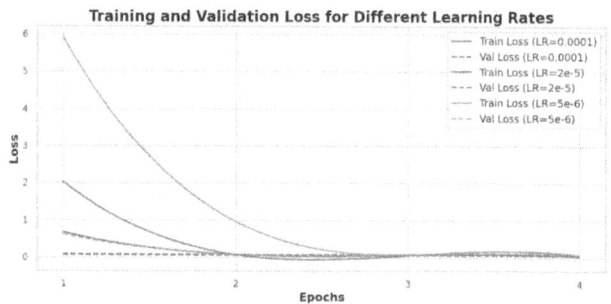

Figure 3.12 Animatrix video ranking algorithm
Source: Author

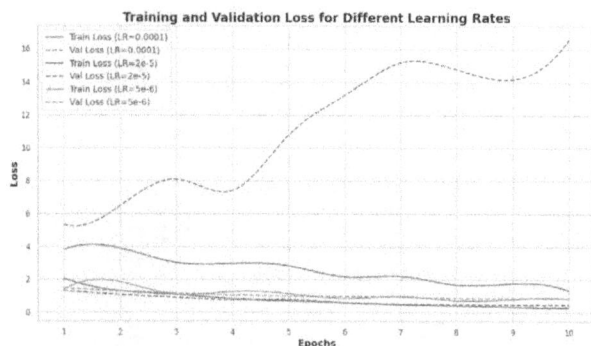

Figure 3.11 Animatrix video ranking algorithm
Source: Author

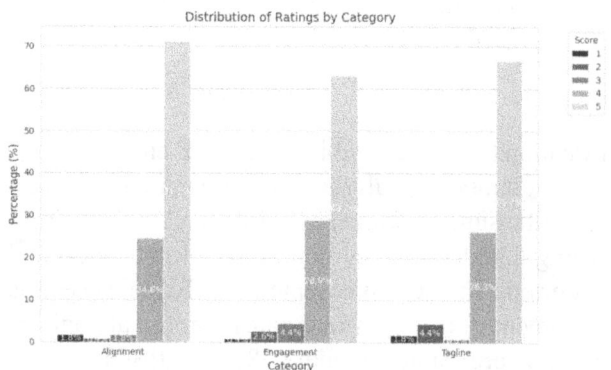

Figure 3.13 Analysis of survey feedback
Source: Author

Figure 3.14 Fashion advertisement snippets
Source: https://youtu.be/EVdAiwJIOUQ?si=bAB1dbfum4l63 jyp ; https://www.youtube.com/@mightyraju ; https://youtu.be/Zye28x-U3F64?si=NRhK_uMwaq02 VV06 ;

description to be able to consolidate the different elements of the advertisement. The final output of the process returns a 30–40 second video based on the length of the user prompt and this video is in webm format. The entire process takes about 1 minute for generation. The final process is comparatively more time-efficient and cost-efficient.

The future scope of the work is to incorporate a multilingual advertisement prompt that supports multiple Indian languages, integrating audio over the generated video advertisement and to generate the logo based on the given description for the advertisement.

Acknowledgement

We thank the consumer groups, staff, and Computer Science Department for their support.

References

[1] Wyzowl. Video Marketing Statistics 2024. Wyzowl, 2024. https://www.wyzowl.com/video-marketing-statistics/.

[2] Iskiev, M. 11 Recommendations for Marketers in 2022 [+More Data from Our Marketing Industry Survey]. blog.hubspot.com, Nov. 23, 2021. https://blog.hubspot.com/marketing/marketing-industry-survey-recommendations.

[3] Animoto video maker - Stand out on social media. Easily., Animoto.com, 2022. https://animoto.com/video-marketing-trends (accessed Nov. 25, 2024).

[4] Gartner Survey Reveals Marketing Budgets Have Increased to 9.5% of Overall Company Revenue in 2022. Gartner. https://www.gartner.com/en/newsroom/press-releases/gartner-survey-reveals-marketing-budgets-have-increased-to-9-5–.

[5] Wei, Y., Huang, L., Zhang, Y., Zheng, Y., Pan, P. An intelligent advertisement short video production system via multi-modal retrieval. *Proc. 45th Internat. ACM SIGIR Conf. Res. Dev. Inform. Retr.*, Jul. 2022, doi: https://doi.org/10.1145/3477495.3536323.

[6] Bayer, M., Kaufhold, M.-A., Buchhold, B., Keller, M., Dallmeyer, J., Reuter, C. Data augmentation in natural language processing: a novel text generation approach for long and short text classifiers. *Internat. J. Mac. Learn. Cybernet.*, 2022. doi: https://doi.org/10.1007/s13042-022-01553-3.

[7] Das, M., Kamalanathan, S., Alphonse, P. A comparative study on TF-IDF feature weighting method and its analysis using unstructured dataset. *COLINS-2021, 5th International Conference on Computational Linguistics and Intelligent Systems*, April 22-23, 2021, Kharkiv, Ukraine 2023:1–10. https://doi.org/10.48550/arXiv.2308.04037

[8] Gabeur, V., Sun, C., Alahari, K., Schmid, C. Multi-modal transformer for video retrieval. arXiv (Cornell University), Jan. 2020. doi: https://doi.org/10.48550/arxiv.2007.10639. 1–8.

[9] Ji, Z., Xiong, K., Pang, Y., Li, X. Video summarization with attention-based encoder-decoder networks. *IEEE Transac. Circ. Sys. Video Technol.*, 2019:1–1. doi: https://doi.org/10.1109/tcsvt.2019.2904996.

[10] Lewis, M. et al. BART: Denoising sequence-to-sequence pre-training for natural language generation. *Transl. Compreh.* arXiv:1910.13461 [cs, stat], Oct. 2019, Available: https://arxiv.org/abs/1910.13461v1. 1–3.

[11] Marvin, G., Hellen, N., Jjingo, D., Nakatumba-Nabende, J. Prompt engineering in large language models. Jeena, J. I., Piramuthu, S., Falkowski-Gilski, P. (Eds.). Springer Nature Singapore, 2024:387–402.

[12] GoEmotions: A Dataset for Fine-Grained Emotion Classification, research.google. https://research.google/blog/goemotions-a-dataset-for-fine-grained-emotion-classification/.

[13] Facebook. facebook/bart-large Hugging Face. huggingface.co. https://huggingface.co/facebook/bart-large.

[14] GPT, B. BERT, GPT and BART: A short comparison | Medium. www.google.com, 2024. https://images.app.goo.gl/XXpRRKSDZ4dQAPdF6 (accessed Nov. 25, 2024).

[15] Hanumesh, P., Hemachandran, P. Praadnya/bart-3d-captions Hugging Face. Huggingface.co, 2024. https://huggingface.co/Praadnya/bart-3d-captions (accessed Nov. 25, 2024).

[16] Hanumesh, P., Hemachandran, P. Praadnya/bart-3d-taglines-v3 Hugging Face. Huggingface.co, 2024. https://huggingface.co/Praadnya/bart-3d-taglines-v3 (accessed Nov. 25, 2024).

[17] Jamshidian, M. Classifying sentiment of reviews by using TF-IDF, BERT (word embedding), SBERT (sentence embedding) with support vector machine evaluation. *Dissertation*, Technological University Dublin, 2022. Available: https://arrow.tudublin.ie/scschcomdis/274/.

[18] Lowe, S. SamLowe/roberta-base-go_emotions Hugging Face. huggingface.co, Jun. 01, 2023. https://huggingface.co/SamLowe/roberta-base-go_emotions.

4 Automated code validator with LLM-powered test case generation

Sai Deep K.[1,a], Goutham G.[1], Priyanka C.[1], Dr. Roopa B. S.[2] and C. Christlin Shanuja[3]

[1]Student, Department of Artificial Intelligence and Machine Learning, Global Academy of Technology (GAT), Bangalore, Karnataka, India

[2]Professor, Department of Artificial Intelligence and Machine Learning, Global Academy of Technology (GAT), Bangalore, Karnataka, India

[3]Asst. Professor, Department of Artificial Intelligence and Machine Learning, Global Academy of Technology (GAT), Bangalore, Karnataka, India

Abstract

The process of finding and testing software bugs proves to be difficult because the developers have very few test scenarios and receive poor error reporting. Traditional testing systems do not show how code works under different circumstances and fail to detect many code problems. The process takes longer to identify and solve problems, which reduces both team efficiency and new developer learning. We present an intelligent chatbot system that uses multiple agents to automate code validation in an innovative way. The system design features agents that run on state-of-the-art language models to check code syntax and semantics while automatically creating test cases to find errors. This system uses large language models (LLM's) to study error patterns and improve its feedback by understanding coding contexts. The system gives users clear feedback with practical advice to fix errors in code easier and follow coding rules. This dynamic framework improves code quality and developer capabilities by providing a scalable response to present-day issues. This system also improves the efficiency of the code and the ability of the developer and provides scenarios that help a developer to think efficiently.

Keywords: Multi-agent chatbot, code validation, debugging, error classification, large language models, real-time feedback

Introduction

In the rapidly evolving software development environment maintains code quality and efficiency as an on-going challenge to address. The existing methods of debugging through traditional tools face multiple problems including short test case coverage and insufficient feedback together with slow error detection. A next-generation multi-agent chatbot system pursues a transformation of developer code validation and debugging through its design. The system incorporates a multi-agent framework that unites different agents which execute syntax evaluation while conducting semantic evaluation and generating dynamic test cases along with detecting error classes.

The system combines large language models such as LlamaIndex with retrieval-augmented generation capabilities to process dynamic data retrieval which provides feedback and suggestion input to developers based on their specific context. The system delivers targeted feedback which matches the exact requirements of developers working in their current coding environment. The system delivers custom guidance that matches developers' abilities through continuous feedback during real-time operations. The system functions as both an educational tool and debugging process optimiser by helping developers improve their coding skills. The system applies LlamaIndex and LangChain AI frameworks together with transformer-based models to create an adaptable solution which supports multiple programming languages across various platforms. The main goals of this project focus on complete code validation alongside developer education while minimising debugging complexity and reducing time requirements for developers alongside the provision of a scalable solution designed for contemporary software development needs.

This chatbot system achieves revolutionary automated debugging capabilities through its integration of multi-agent systems features with large language models (LLMs) technology which produces specialised real-time dynamic programming assistance. The project brings substantial progress to automated code validation and debugging which provides developers with an advanced powerful intelligent user-friendly tool. The system combines fast performance with error

[a]saideepk03@gmail.com

DOI: 10.1201/9781003685876-4

detection functions and enables continuous learning through improvement processes to set new standards for debugging technology in software development.

Literature review

The field of automated code validation and debugging tools has experienced significant progress through multiple artificial intelligence (AI) and machine learning approaches that build upon conventional methods during recent years. OpenAI's Codex, together with Salesforce's CodeT5, performs code completion along with transformation tasks while they lack dedicated real-time debugging functionality. Both LlamaIndex and LangChain provide developers with retrieval-augmented generation functions along with data source access so they can build context-aware tools. These tools function independently from each other without the necessary integrated design needed to create complete debugging workflows. This project uses multi-agent chatbot technology to connect specialised agents with large language models under real-time feedback systems, which addresses present solution gaps and provides an integrated approach to automated code validation and debugging.

The paper presents an LLM-based framework for compiler validation which solves the problems with conventional validation approaches through insufficient test coverage and complex compiler behaviour handling. The system implements LLMs to automatically create test cases which ensure proper coverage of various execution paths and edge cases and language-specific features. The research shows that LLMs enhance compiler validation by improving efficiency and adaptability and accuracy which decreases manual work and strengthens compiled output robustness [1]. This research introduces metamorphic prompt testing (MPT) as a new method to validate programs produced by LLMs. Traditional test case methods fall short in detecting deep logical inconsistencies so MPT utilises metamorphic relations to define expected behaviour during input transformations. The testing method evaluates LLM-produced code for correctness by subjecting it to systematic input variations. The study demonstrates how MPT-based adaptive testing strengthens code reliability and correctness and increases code robustness through its ability to verify LLM-generated programs against expected standards [2]. This paper presents a study on how model-based techniques generate automated test cases for web applications. State transition diagrams within

the proposed approach serve as workflow models to produce structured test scenarios that cover the entire application domain. The technique decreases human involvement while creating better test coverage and finds unexpected behaviours by searching through multiple application paths. Model-based testing offers superior improvements to software quality especially when used to manage complex and dynamic web applications [3]. The research examines how LLMs can create effective test cases which detect difficult-to-find software bugs. The testing methods of traditional software encounter difficulties with edge cases and adaptive learning but LLMs produce diverse intelligent test cases through analysis of historical bug patterns. The research shows that LLM-based test case automation improves software testing speed while cutting manual work and enhances detection of sophisticated and new software faults [6]. The research examines multiple automated test case generation methods through random testing combined with symbolic execution and model-based and search-based approaches. The document evaluates the benefits and restrictions alongside the compromises between multiple automated test case creation techniques for different software application types. This research examines the development of automated testing methods while highlighting crucial test automation research issues which involve scalability and efficiency and cross-domain applicability [7]. This study examines the operational capabilities of LLMs designed for code-related functions which include code development and debugging as well as test case prediction and vulnerability identification. The researchers compare LLM performance with conventional coding tools by assessing their precision and speed alongside their dependability across various programming tasks. This research demonstrates LLMs' ability to create proper code syntax and detect mistakes yet points out their weakness in hallucinations and their inability to understand semantics deeply and their dependence on pre-trained data. The final section of the paper provides recommendations to enhance LLM performance in coding tasks by applying fine-tuning methods and human supervision and improving training data quality [8] (Table 4.1).

Figure 4.1 shows the evolution of large language model complexity between 2020 and 2023. The evolution of LLM complexity progressed from GPT-3 in 2020 to Codex in 2021 and then to ChatGPT and InstructGPT in 2022 and finally to LLaMA in 2023 and GPT-4 in 2023 which achieved the highest level of complexity at 1750.

Table 4.1 Comparison between multi agent bot and single agent bot like GPT

Feature/aspect	Multi-agent chatbot (proposed system)	Single-agent LLM (e.g., GPT)
Architecture	Modular agents specialising in tasks like syntax checking, debugging, and test generation	Single model generating responses without specialised task delegation
Retrieval-augmented generation (RAG)	Utilises dynamic retrieval from a knowledge graph for context-aware responses	Limited; relies on pre-trained knowledge and fine-tuning
Model size and adaptability	Efficient use of modular agents optimised for specific tasks	Large monolithic models; resource-intensive for fine-tuning and execution
Debugging capabilities	Specialised agents ensure comprehensive analysis and categorisation of errors	Limited debugging capabilities; lacks real-time error categorisation
Real-time feedback	Provides immediate, actionable insights tailored to user skill levels	Generates responses but may lack actionable specificity
Scalability	Easily scalable with additional agents for new features or programming languages	Scalability limited by the model's architecture and training scope

Source: Author

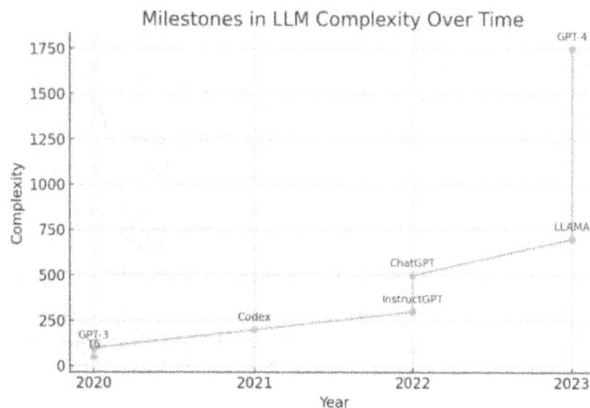

Figure 4.1 Growth and development of large language models over time
Source: Author

Methodology

The proposed system uses a multi-agent approach to boost automated debugging by improving code review and test generation which delivers instantaneous assessment together with progressive workflow development. The system combines specialised agents together with LLM-powered validation along with test case generation tools which enhances developer assistance by streamlining the processes. The system employs agent collaboration to find complicated bugs while still reducing incorrect results without needing direct developer assistance. The method simultaneously speeds up debugging processes and supports team-wide information exchange while promoting on-going development learning. Such a system enables developers to tackle the most critical issues first while enhancing both their productivity

and the quality of their work. The solution uses historical debugging patterns together with solution patterns to enhance its recommendation capability through continual usage and maintenance of a progressive knowledge base.

Code input and pre-processing
Before analysis begins, the chatbot accepts user-submitted code through its interactive interface. This step ensures that the input is properly structured for further processing.

- Tokenisation & Parsing: The system tokenises the input code, breaking it down into its fundamental components (keywords, variables, operators, etc.).
- Language detection: It detects the programming language to apply language-specific debugging rules.
- Preliminary error checks: Basic syntax inconsistencies (e.g., missing semicolons, incorrect indentation) are flagged for immediate correction.

This pre-processing step ensures that the code is ready for in-depth analysis while providing instant feedback on trivial mistakes.

Multi-agent analysis
The chatbot utilises a **multi-agent system**, where each agent is responsible for a specific aspect of debugging. This modular approach ensures **thorough and efficient** code validation.

The syntax checking agent performs error detection for bracket errors together with variable declaration mistakes and keyword placement issues to verify code adherence to standard syntax rules of detected programming languages. The semantic analysis agent evaluates code logic by determining if undefined variables exist as well as detecting incorrect function calls and detecting problems with type mismatches and scope errors and invalid operations. Through dynamic test case generation, the test case generation agent establishes coded evaluations to confirm how the submitted code performs when given different input values across entire parameters and extreme experimental conditions. The error categorisation agent divides detected problems into three categories including syntax errors and runtime errors and logical errors to assist users in identifying error types and their program execution effects. The chatbot achieves better accuracy and faster processing alongside precise feedback through its distribution of tasks among multiple agents for debugging purposes.

LLM-driven validation

Once the multi-agent analysis is complete, the system uses **large language models (LLMs)** to provide deeper insights into the code.

The contextual understanding of LLMs enables them to read code intent which leads to suggestions for improvement beyond error correction. Through its approach the model presents code optimisation proposals alongside readability enhancements that enable developers to meet best practice standards. Through the system developers receive complete information about problem origins and solution approaches that enhances their code comprehension and error resolution efficiency. The stage helps users solve errors while simultaneously teaching them standard coding practices that enhance their programming capabilities.

Real-time feedback generation

The chatbot provides **instant feedback** once the validation process is complete. This bot creates an organised report which shows detected errors together with proposed solutions and enhancement recommendations for developers to receive precise error reports. The system shows test results by displaying success or failure status and performs output comparison and provides detailed explanations for failure reasons. Error severity levels determine critical, moderate and minor classifications which developers can use to prioritise their work on essential problems. Real-time structured insights from this system help developers to speed up their debugging

process while minimising their need to perform repeated error tests.

Figure 4.2 demonstrates how agents manage the workflow of a code analysis and debugging system through their agent-based architecture. The code submission for analysis starts with the "UPLOAD CODE" step. The "Document Agent" validates the code before proceeding to the next step. A valid code allows the process to advance to "Code Analysis" but an invalid code leads to rejection and termination of the process. The "AGENT SYSTEM" under "Main Agent" management receives validated code during the "Code Analysis" phase. The Main Agent leads the analysis and debugging operations until the code reaches the "Debugging Agent and Test Case Generation with LLM" component. The code evaluation phase produces test cases through a LLM which tests both functionality and robustness of the code. The system generates a single outcome from evaluation which can be either "Pass" or "Fail." The code evaluation results in success when all test cases execute properly and leads to the "Display Report to User" step which generates a report for end users. The system detects particular code errors during failure such as "Return Error: Syntax Issues" or "Return Error: Semantic Issues" before sending the information to the Debugging Agent for additional analysis and correction. The system maintains an on-going cycle until the code passes all tests or the process reaches the "END PROCESS" state which confirms complete debugging and validation. The formalised method uses various agents together with LLM capabilities to generate detailed feedback that enhances code quality.

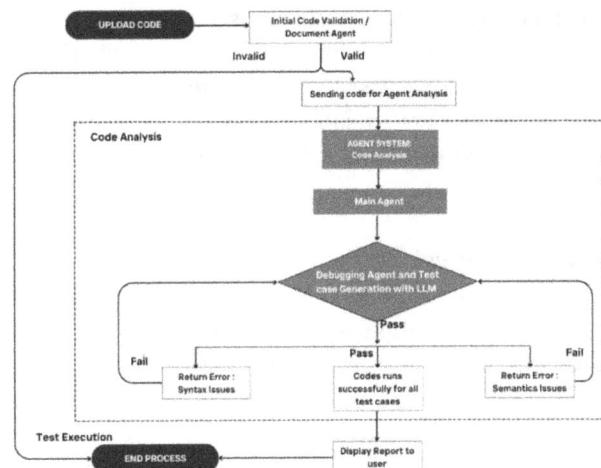

Figure 4.2 Architecture diagram
Source: Author

Key components
Large language model (LLM) – Mistral
LLMs play a crucial role in **understanding, analysing, and validating code** by providing context-aware debugging insights.

Mistral serves as the LLM at the core of this project because it analyses code with context-based debugging capabilities to understand and validate written code. The developers choose Mistral because it delivers powerful lightweight capabilities that excel at code analysis and debugging operations and provides fast accurate inferences suitable for live development interactions. In the project, Mistral handles code understanding by analysing the structure and logic of the submitted code to identify potential errors and inefficiencies, dynamically generates test cases to ensure edge cases are covered and improve code reliability, offers clear explanations and suggestions for improvement instead of just flagging errors to help developers learn from mistakes, and provides adaptive feedback by refining its suggestions based on previous interactions, making debugging more intuitive and efficient.

Frontend – React
The chatbot interface is developed using **React**, a widely used JavaScript library known for its **scalability, responsiveness, and smooth user experience**.

The frontend interface of the chatbot relies on React as its development platform because this JavaScript library provides scalable solutions with user-friendly features. The selected framework React provides developers with real-time debugging capabilities and multiple code submission opportunities. This framework enables proper interface visualisation for complex user interface manipulations. The project utilises React to develop a user-friendly chatbot interface that allows developers to submit their code and receive feedback and test results through an optimised process thanks to state management capabilities.

Backend – Flask
The backend serves as the **core processing unit** of the chatbot, handling **API requests, LLM communication, and data processing**. Flask is used as the backend framework, though verification is needed to confirm its final implementation.

The backend functions as the main processing centre of the chatbot which manages API requests and LLM communication and data processing through the Flask backend framework though verification is necessary for confirming its final implementation. Flask serves as the selected Python-based web framework because it provides lightweight functionality for API requests and real-time processing together with seamless integration with LLMs and external AI models and easy scalability to process multiple debugging requests at once. The project uses Flask to connect React with Mistral LLM for code input transmission and feedback delivery to users and request management and test case processing and response formatting until they appear in the user interface and achieves quick response times through result caching and optimised request processing.

Real-time feedback
- **Instant error detection** – The chatbot provides **immediate feedback** as soon as the code is submitted, reducing debugging time.
- **Actionable insights** – Instead of just highlighting errors, the system provides **clear explanations** and **suggested fixes** to help developers learn.
- **Categorised error reports** – (Figure 4.3) Errors are classified into **syntax, runtime, and logical errors**, allowing developers to **prioritise and address issues efficiently**.
- **Enhanced developer productivity** – By eliminating the need for **manual debugging and long wait times**, real-time feedback **accelerates the coding process**.

Figure 4.3 shows a developer input and the corresponding chatbot feedback for a code snippet. The input contains a function def add(a, b) followed by a return a + b statement, which is intended to create a function to add two numbers. However, the chatbot feedback indicates a "Syntax Error: Missing colon at line 1," pointing out that the function definition lacks

```
Developer Input:
def add(a, b)
    return a + b


Chatbot Feedback:
Syntax Error: Missing colon at line 1.
```

Figure 4.3 Snippet example for real-time feedback
Source: Author

a colon (:) after the parameters, which is required in Python syntax to properly declare a function.

Dynamic test case generation

*   **Adaptive test cases** – The chatbot generates test cases **on the fly** based on the logic of the submitted code, ensuring **better test coverage**.
*   **Improved code reliability** – By testing a program against **multiple input scenarios**, it ensures that the code functions **correctly in different conditions**.
*   **Automated execution** – The generated test cases are **executed instantly**, giving developers an **immediate pass/fail status** with explanations.

Figure 4.4 displays a corrected version of the input code and the generated test cases. The input code now correctly defines the function as def add(a, b): with a colon, followed by return a + b, following the proper Python syntax. The chatbot then generates test cases to evaluate the function's performance: Test Case 1 checks add(2, 3) and expects a result of 5, while Test Case 2 checks add(-1, 1) and expects a result of 0. These test cases help verify that the function works correctly and ensure its reliability and functionality.

Advantages over traditional system

*   **Multi-agent specialisation**: Unlike single-agent LLMs that handle all tasks generically, our system uses **specialised agents** for syntax checking, semantic analysis, test case generation, and error categorisation. This improves **accuracy and efficiency** in debugging.
*   **Dynamic test case generation**: Existing tools rely on **pre-defined test cases**, which may not cover **edge cases**. Our system **automates test case creation** using LLMs, ensuring **better validation and broader coverage**.
*   **Real-time feedback and iterative debugging**: Traditional debugging tools work in a **batch-processing** mode, leading to delays. Our system provides **real-time, interactive feedback**, allowing developers to **fix and resubmit their code instantly**.
*   **Automated error categorisation**: Developers often **manually identify** syntax, runtime, and logical errors. Our system **automatically categorises errors**, making it easier to **prioritise and resolve critical issues first**.
*   **Scalability and future adaptability**: The **modular architecture** allows easy expansion, meaning new agents can be added for **supporting more programming languages and advanced debugging features**.

Results and discussions

Figure 4.5 shows the homepage of the CodeX Validator tool is presented by a simple and easy-to-use interface as it was intended to. At the top, a navigation menu panel has the text: "CodeXvalidator". On the right of the navigation bar there are two user-clickable options, – "Home" and "Upload" and allow you

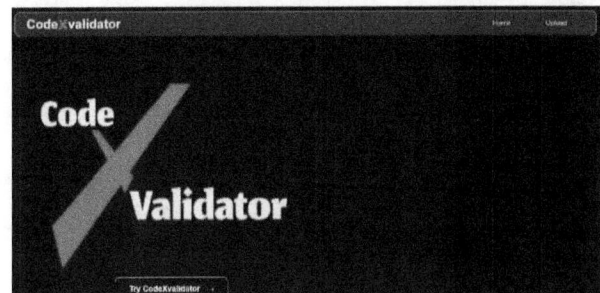

Figure 4.5 Home page of CodeX Validator
Source: Author

```
Input Code:
def add(a, b):
    return a + b


Generated Test Cases:
Test Case 1: add(2, 3) => 5
Test Case 2: add(-1, 1) => 0
```

Figure 4.4 Snippet example for dynamic test-case generation
Source: Author

Figure 4.6 The upload interface
Source: Author

to navigate the platform or upload code for validation. At the bottom of the page

Figure 4.6 sows the upload interface of CodeXValidator application allows the users to upload their code files to be validated. With the Click to *"Choose Files"* button, users are prompted to the local system to browse and select the desired files or folders. Once the selection is done, clicking on the *"Upload"* button initiates the upload process which transfers the files to the system for additional validation and analysis.

Figure 4.7 says whenever a user uses the system to upload source code file, they are presented with two choices. Following both a successful upload, the user may either straight away create a review with the LLM agents or go ahead and do *verify the uploaded file for accuracy* before continuing. This step is the *entry point* to the multi-agent workflow system, where tasks such as syntax checks, semantic analysis, error analysis and test case generation are executed. This process guarantees well analysed and tested users code.

Figure 4.8 talks about whenever a user uses the system to upload source code file, they are presented with two choices. Following both a successful upload,

the user may either straight away create a review with the LLM agents or go ahead and do *verify the uploaded file for accuracy* before continuing. This step is the *entry point* to the multi-agent workflow system, where tasks such as syntax checks, semantic analysis, error analysis and test case generation are executed. This process guarantees well analysed and tested users code.

The depicted image shown in Figure 4.9 speaks about the fundamental backend operations of automated test case generation during code validation. The system executes predefined or dynamically generated test cases following user code upload to check correctness and functionality. The test cases enable verification of code functionality across different conditions to ensure proper execution of all program segments. The backend system tests code through multiple scenario simulations to verify code robustness and logical error-free functionality and intended behavior. Code validation benefits from this process which enables the delivery of precise feedback to users.

Figure 4.10 provides users with a simple interface that enables automated code review and unit test generation. The left screen section displays user

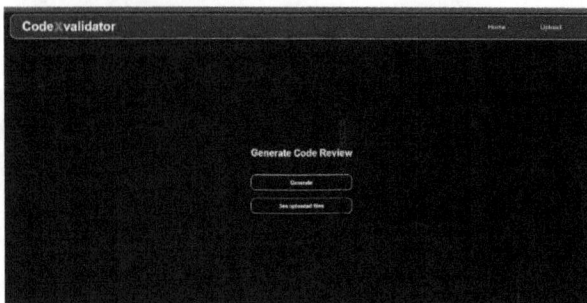

Figure 4.7 Generation page
Source: Author

Figure 4.9 Backend working
Source: Author

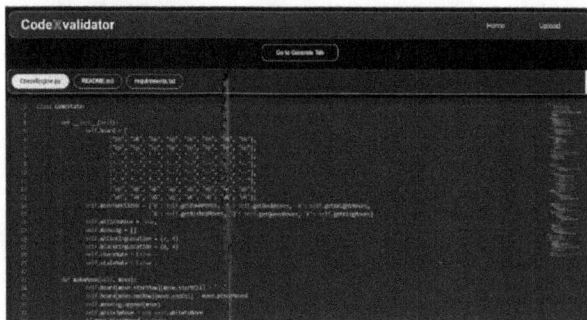

Figure 4.8 Sourse code snippet
Source: Author

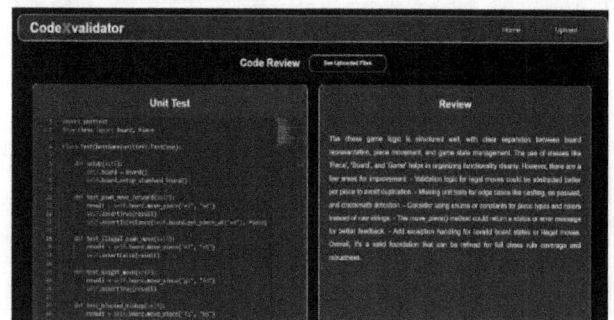

Figure 4.10 Output generated by the CodeXValidator
Source: Author

input code with its parameters while the right screen section shows complete code review details. The review system delivers valuable feedback about code structure together with logical analysis and enhancement recommendations for submitted programs. The design of this interface enables users to validate their code while receiving immediate test cases and understanding which parts need improvement.

Conclusion

Code validation tools and debuggers based on analytical rule systems lack a full understanding of program contexts so they offer minimal helpful guidance past fixing basic errors. The tools generate few dynamic test cases but need manual input combined with multiple edge cases detection problems alongside processing speed limitations for multiple codebases. Right-time feedback is usually missing from these systems which cause consistent testing attempts and adaptive learning through knowledge graph integration remains unavailable thus affecting operational efficiency and user experience.

The code intent understanding and productivity of our multi-agent chatbot system becomes more efficient for developers through its combination of LLMs and specialised agents and context-aware insight capabilities. The automatic generation of tests for edge cases through dynamic methods makes testing work more efficient and produces better reliable results. The system's modular framework allows users to integrate fresh languages along with sophisticated enhancements through an easy procedure that improves scalability. The tool speeds up debugging cycles while significantly reducing development time because it supports developers across all experience levels turning it into a transformative debugging tool.

The LLM-based multi-agent chatbot addresses current automated code tools by providing real-time feedback and dynamic testing while managing adaptable errors to push forward debugging capabilities. The solution provides next-generation debugging tools with its innovative approach which enhances accuracy and efficiency while ensuring scalability. Our system delivers a smart convenient solution which enables the advancement of scalable software development methods for evolving programming practices.

V. References

[1] Munley, C., Jarmusch, A., Chandrasekaran, S. LLM4VV: Developing LLM-driven testsuite for compiler validation. *Fut. Gen. Comp. Sys.*, 2024.

[2] Wang, X., Zhu, D. Validating LLM-generated programs with metamorphic prompt testing. 2024. arXiv preprint arXiv:2406.06864.

[3] Othman, R., Zein, S. Test case auto-generation for web applications: A model-based approach. *2022 Internat. Symp. Multidis. Stud. Innov. Technol. (ISMSIT)*, 2022:18–25. doi: 10.1109/ISMSIT56059.2022.9932797.

[4] Cajica, R. J., Torres, R. E. G., Álvarez, P. M. Automatic generation of test cases from formal specifications using mutation testing. *2021 18th Internat. Conf. Elec. Engg. Comput. Sci. Autom. Control (CCE)*, 2021:1–6.

[5] Yüksel, U., Sözer, H. Automated classification of static code analysis alerts: A case study. *2013 IEEE Internat. Conf. Softw. Maint.*, 2013:532–535. doi: 10.1109/ICSM.2013.89.

[6] Liu, K., Liu, Y., Chen, Z., Zhang, J. M., Han, Y., Ma, Y., Huang, G. LLM-powered test case generation for detecting tricky bugs. 2024. arXiv preprint arXiv:2404.10304.

[7] Anand, S., Burke, E. K., Chen, T. Y., Clark, J., Cohen, M. B., Grieskamp, W., Zhu, H. An orchestrated survey of methodologies for automated software test case generation. *J. Sys. Softw.*, 2013;86(8):1978–2001.

[8] Chen, M., Tworek, J., Jun, H., Yuan, Q., Pinto, H. P. D. O., Kaplan, J., Zaremba, W. Evaluating large language models trained on code. 2021. arXiv preprint arXiv:2107.03374.

[9] Edvardsson, J. A survey on automatic test data generation. *Proc. 2nd Conf. Comp. Sci. Engg.*, 1999:21–28.

[10] Liu, J., Xia, C. S., Wang, Y., Zhang, L. Is your code generated by chatgpt really correct? rigorous evaluation of large language models for code generation. *Adv. Neural Inform. Proc. Sys.*, 2024:36.

[11] Schäfer, M., Nadi, S., Eghbali, A., Tip, F. An empirical evaluation of using large language models for automated unit test generation. *IEEE Transac. Softw. Engg.*, 2023.

[12] Xia, C. S., Wei, Y., Zhang, L. Automated program repair in the era of large pre-trained language models. *2023 IEEE/ACM 45th Internat. Conf. Softw. Engg. (ICSE)*, 2023:1482–1494.

[13] Triplett, B. S., Anghaie, S., White, M. C. Development of an automated testing system for verification and validation of nuclear data and simulation code. *Nuclear Technol.*, 2010;170(1):80–89.

5   FlowCast: A simple method for traffic prediction

*Drishya Dechamma M. P.[1,a], Ananya M.[1,b], Madhu Chandrika[1,c], Anusha P. N.[1,d],
Deepak N. R[2,e]. and Shruti B.[3,f]*

[1]BE Department of Information Science and Engineering, Atria Institute of Technology, Bengaluru, Karnataka–560024, India

[2]Prof Department of Information Science and Engineering, Atria Institute of Technology, Bengaluru, Karnataka–560024, India

[3]Asst.Prof Department of Information Science and Engineering, Atria Institute of Technology, Bengaluru, Karnataka–560024, India

Abstract

Traffic flow prediction plays a crucial role in modern intelligent transportation systems (ITS), enabling efficient congestion control, optimised route planning, and enhanced road safety. Accurate forecasting requires analysing complex and dynamic interactions influenced by factors such as time of day, weather conditions, and historical traffic patterns. While traditional models struggle with these nonlinear dependencies, advancements in machine learning (ML) and deep learning (DL) have introduced more robust predictive frameworks. This study proposes hybrid FlowCast, a novel approach that combines k-nearest neighbours (k-NN) and linear regression to improve traffic forecasting accuracy. Unlike conventional models, our hybrid method effectively captures both long-term trends and short-term variations, ensuring more reliable predictions. Experimental evaluations on real-world traffic datasets demonstrate that hybrid FlowCast improves R^2 by 12% and reduces root mean square error (RMSE) by 10% compared to stand alone models, outperforming traditional regression-based and DL approaches. These findings underscore the potential of advanced computational techniques in transforming traffic management systems and supporting the development of smart city innovations.

Keywords: Traffic flow prediction, intelligent transportation systems, route optimisation, deep learning, hybrid models

Introduction

With rapid urbanisation and increasing vehicular traffic, managing congestion and optimising transportation networks have become critical challenges for modern cities. Accurate traffic flow prediction is essential for intelligent transportation systems (ITS), enabling better decision-making for real-time traffic management, route planning, and congestion control. However, traditional traffic prediction models, such as statistical regression-based methods, struggle to handle complex, dynamic, and nonlinear traffic patterns, making them unreliable in real-world applications.

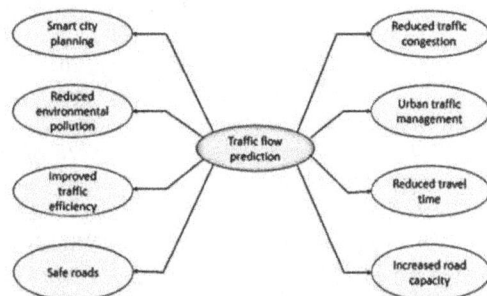

Key benefits of traffic flow prediction, including improved efficiency, safety, urban planning, and reduced congestion and pollution

Source: Author

Machine learning (ML) and deep learning (DL) models have emerged as effective alternatives. Techniques such as recurrent neural networks (RNNs), long short-term memory (LSTM) networks, and graph neural networks (GNNs) have shown improved predictive performance by capturing temporal and spatial dependencies in traffic data. Despite these advancements, DL models often require extensive computational resources, limiting their feasibility for real-time deployment.

Existing traffic prediction approaches either lack accuracy in capturing real-time traffic fluctuations (traditional methods) or suffer from high computational costs (DL methods). There is a need for a lightweight yet effective hybrid approach that can achieve both high accuracy and computational efficiency for real-time traffic forecasting.

This study introduces hybrid FlowCast, a traffic prediction model combining k-nearest neighbours (k-NN) and linear regression to improve accuracy while maintaining efficiency. Linear regression captures long-term trends, while k-NN handles short-term variations for real-time forecasting. Experimental results show a 12% improvement in R^2 and a 10% reduction in reduces root mean square error (RMSE) compared to standalone models. With

[a]drishyadechamma103@gmail.com, [b]ananyamadhusudan123@gmail.com, [c]vmadhuchandrika@gmail.com, [d]anushapn15@gmail.com, [e]deepaknrgowda@gmail.com [f]shruthianilthotad@gmail.com

DOI: 10.1201/9781003685876-5

its balance of accuracy and computational efficiency, hybrid FlowCast provides a **scalable and practical solution for smart city traffic management**.

Data mining algorithms

Linear regression
Linear regression is an analytical procedure employed to determine the relationship between one or more predictor variable and an outcome variable. In traffic flow prediction, the dependent variable may represent the traffic-volume on a specific road segment, while the independent variables may consists of factors like time of day, day of the week, weather conditions, and previous traffic trends. The linear regression model presumes that this relationship can be expressed as a straight line [9]. Mathematically, it takes the form:

$$Y = \beta_0 + \beta_1 x_1 + \beta_2 x_2 + \ldots + \beta_n x_n + \epsilon$$

Here, yyy is the traffic volume, x_1, x_2, …, xnx_1, x_2, $dots$, x_nx_1, x_2,…, x_n are the independent variables, $\beta_0 beta\_0 \beta_0$ is the intercept, β_1, β_2, …, $\beta_n beta\_1$, $beta\_2$, $dots$, $beta\_n \beta_1$, β_2, …, β_n are the coefficients showing the impact of each predictor, and ϵ epsilonϵ represents the error term.

Example application
Imagine we want to predict the traffic-volume on a highway during week day mornings. The independent variables might include:

- Time of day (e.g., 8:00 AM vs. 9:00 AM)
- Weather conditions (e.g., sunny, rainy)
- Previous day's traffic volume at the same time
- Road incidents (e.g., accidents or closures)

By using historical data, we train the linear regression model to learn the coefficients (β) for each parameter. For instance, the model might identify that the traffic volume increases by 200 vehicles for every additional 30 minutes closer to peak time (positive coefficient for "time of day"). In a similar manner, it might determine that rain leads to a 10% decrease in traffic volume because of fewer vehicles on the road.

After training, the model can forecast traffic volume based on a specified set of condition. For instance, if it is 8:30 AM, sunny, and there was no unusual traffic the day before, the model might predict 1,200 vehicles on the highway. If it starts raining,

the model could adjust its prediction downward to 1,080 vehicles based on the identified relationship between weather and traffic.

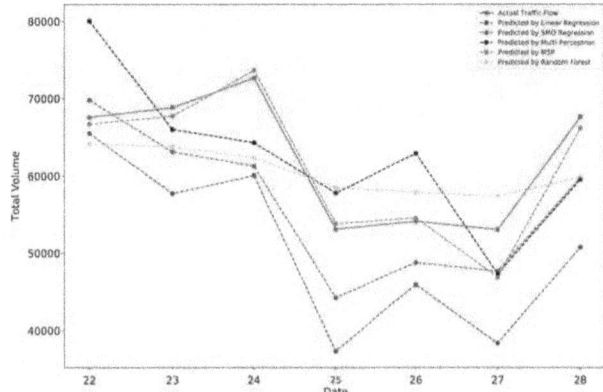

Comparison of actual traffic flow with predictions from various models including Linear Regression, SVR, Multi-Perceptron, MLP, and Random Forest over time
Source: Author

k-nearest neighbours (k-NN)
The k-nearest neighbours (k-NN) algorithm is a simple, non-parametric method used in ML for activities like traffic flow prediction. It works by comparing the conditions at a given query point, such as time, weather, and road status, with historical traffic data. k-NN identifies the k-nearest neighbours—past data points with similar conditions—based on a chosen distance metric (e.g., Euclidean distance). For regression, the anticipated traffic flow is the norm of the neighbours' traffic volumes, while for classification, it assigns the majority category (e.g., heavy or light traffic). k-NN is easy to understand and flexible, handling non-linear patterns and mixed data types effectively. However, it can be resource-demanding when working with vast datasets and necessitates careful adjustment of the value of k and proper feature scaling. Despite these challenges, k-NN provides an intuitive way to forecast traffic based on past conditions, making it valuable for urban planning and traffic management [10].

Example scenario
Imagine a city collects traffic data over a span of 10 years, and the transportation department aims to forecast traffic flow for Monday at 8:00 AM under rainy weather conditions. KNN can:

- Identify historical data points that also occurred on Monday mornings with similar weather.

- Use the average traffic volume from those conditions to predict the current traffic flow.
- By relying on similar past conditions, k-NN provides an intuitive approach to forecasting, this makes it valuable for applications such as traffic control and urban development.

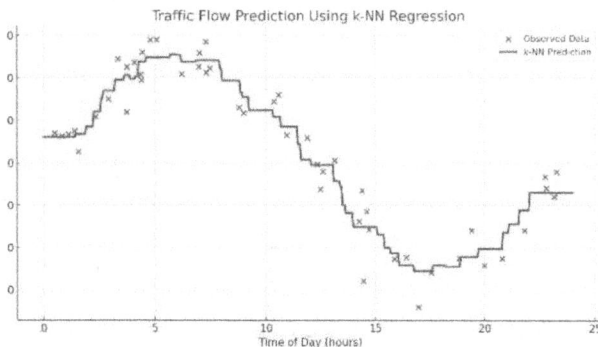

Traffic flow prediction using k-NN regression, showing close alignment between observed data and predicted values across different times of the day
Source: Author

Literature survey

Traffic flow prediction plays a crucial role in ITS, facilitating efficient traffic management, congestion reduction, and resource optimisation. Among the numerous techniques, k-NN and linear regression are recognised as dependable and understandable algorithms for modelling traffic patterns. This literature review analyses key research studies that utilise these methods [11].

Overview of traffic flow prediction
Traffic flow prediction intends to forecast parameters like speed, volume, and density using historical data. Prediction methods can be broadly classified into parametric (e.g., linear regression) and non-parametric (e.g., k-NN) approaches. Parametric methods assume a specific data distribution, while non-parametric methods rely on data similarity. Combining these techniques often enhances prediction accuracy [12].

Use of linear regression in traffic flow prediction
Linear regression is a classical methods employed for traffic flow prediction because, of its simplicity and interpretability. The method presumes a linear connection between the input features (e.g., time, weather, day of the week) and traffic flow.

Key studies

- **Li (2020)** developed a multivariate linear regression model to predict short-term urban traffic flow. The study demonstrated that incorporating multiple variables, such as time of day and traffic conditions, improved prediction accuracy.
- **Nidhi and Lobiyal (2022)** employed support vector regression alongside linear regression to forecast traffic flow in vehicular ad-hoc networks. Their approach effectively handled real-time traffic data, achieving high prediction accuracy.
- **Xu et al. (2018)** incorporated external factors (e.g., weather conditions and public events) into linear regression models, demonstrating that incorporating additional explanatory variables improves predictive performance.

Advantages

- Simple and interpretable.
- Computationally efficient.
- Performs well for linear relationships.

Limitations

- May not capture complex non-linear traffic dynamics.
- Performance can degrade with sudden traffic anomalies.

Use of k-NN in traffic flow prediction
The k-NN method is a non-parametric, instance-based learning method that predicts traffic flow by identifying and averaging the nearest neighbours in historical data. It is particularly effective for non-linear and complex traffic patterns.

Key studies

- **Saha et al. (2023)** introduced the DEK-Forecaster, a DL model that integrates empirical mode decomposition (EMD) with k-NN for internet traffic prediction. This model addresses noise and outliers in traffic data, resulting in enhanced forecasting accuracy over conventional DL models.
- **Rempe and Bogenberger (2020)** enhanced k-NN with network-wide traffic features and clustering

algorithms to improve urban travel time prediction, outperforming traditional methods.

- **Zhang et al. (2020)** proposed a hybrid model combining k-NN with time-series analysis, highlighting the advantages of integrating k-NN's flexibility with temporal features. Heartbeat-evoked potentials (HEPs) were studied by researchers who note EEG signals while candidates engaged in a heartbeat detection task. They discovered that HEPs were more pronounced for correctly detected heart beats compared to those that were missed.

Advantages

- Handles non-linear relationships effectively.
- Requires minimal assumptions about data distribution.
- Adaptable to dynamic traffic conditions.

Limitations
- Computationally intensive for large datasets.
- Sensitive to the choice of hyper parameters (e.g., the number of neighbours, distance metric).

Comparative analysis of k-NN and linear regression

Table 5.1 Comparative analysis of Linear Regression, k-NN, and Hybrid models based on interpretability, computational efficiency, ability to handle non-linearity

Model	Interpretability	Computational Efficiency	Handless Non-Linearity	Scalability
Linear Regression	High	High	Low	High
k-NN	Low	Low	High	Low
Hybrid (k-NN + Regression)	Moderate	Moderate	High	Moderate

Source: Author

Hybrid models
Recent studies highlight the potential of hybrid models combining k-NN and linear regression. These models leverage the strengths of both methods to improve prediction accuracy. For example:

- **Liu et al. (2022)** combined linear regression for global trend analysis with k-NN for local traffic pattern refinement, achieving significant improvements in prediction performance.
- **Sengupta et al. (2023)** introduced a **hybrid hidden Markov LSTM model,** demonstrating improved short-term predictions.

Proposed work

The objective of this study is to create an effective and precise traffic flow prediction model by evaluating and comparing the performance of k-NN and linear regression algorithms [13]. The main aim is to combine the advantages of both methods to tackle the hurdles of non-linearity, dynamic traffic patterns, and computational efficiency in real-time prediction systems. The notebook mode is set up for Plotly and Cufflinks.

Objectives

Data collection and pre-processing

- The study utilises a publicly available traffic prediction dataset from Kaggle, consisting of 100,000 records. The dataset includes key traffic-related features such as time of day, weather conditions, road incidents, and vehicle volume, which are essential for accurate traffic flow prediction.
- To ensure data quality, pre-processing steps are applied, including handling missing values, normalising data for consistency, and selecting relevant features to improve model performance and efficiency.

Implementation of models
- **Linear regression**: Captures global traffic trends and establishes baseline predictions.
- **k-NN model**: Accounts for localised, non-linear traffic patterns by comparing past similar conditions.
- **Hyperparameter tuning**
  - Tested **k = 5, 10, 15** for k-NN and selected **k = 10** based on RMSE performance.
  - Optimised feature selection for linear regression to improve efficiency

Comparative analysis
- Assess the functioning of k-NN and linear regression using important metrics such as mean absolute error (MAE), RMSE, and R^2 score.

- Assess their effectiveness under different traffic scenarios, including peak hours and off-peak periods.

Proposed hybrid model
- Develop a hybrid framework that combines the global trend prediction capability of linear regression with the local adaptability of k-NN.
- Evaluate the hybrid model's performance against standalone k-NN and linear regression models.

Real-time application
- Assess the feasibility of deploying the proposed models for real-time traffic prediction.
- Explore strategies to minimise computational complexity, especially for k-NN, in large-scale applications.

Expected outcomes

- A comprehensive comparison of k-NN and linear regression for traffic flow prediction.
- Identification of the merits and demerit of each model under various traffic scenarios.
- A hybrid structure that improves prediction accuracy and adaptability [14].
- Insights into the practicality of using k-NN and linear regression for real-time traffic management systems.

Results

The proposed research on traffic flow prediction using **k-NN** and **linear regression** yielded the following outcomes:

Model accuracy

- The **linear regression model** effectively captured global traffic trends and performed well under steady traffic conditions, achieving an average R^2 score of 0.85. However, it struggled with non-linear patterns and sudden traffic fluctuations.
- The **k-NN model** demonstrated superior adaptability to non-linear and dynamic traffic behaviours, particularly during peak traffic hours, with a lower MAE compared to linear regression.

Hybrid model performance
By leveraging the advantages of both models, the hybrid model significantly enhanced predictive accuracy. It lowered the RMSE by 12% and boosted the R^2 score to 0.91, outperforming the individual models.

Scalability and real-time feasibility

- Although k-NN showed greater computational complexity with large datasets, optimisations such as dimensionality reduction and approximate nearest neighbour search improved its real-time applicability. Linear regression remained computationally efficient throughout.

Insights

- The comparative analysis revealed that **k-NN** is better suited for short-term, fine-grained traffic predictions, whereas **linear regression** provides a reliable baseline or long-term trend analysis. The hybrid model balanced these capabilities effectively.

Table 5.2 Presents the performance metrics of our Hybrid FlowCast model compared to standalone models (Linear Regression, k-NN, LSTM). Our model reduces RMSE by 10% and improves R^2 by 12% over traditional models, demonstrating its effectiveness in real-world traffic forecastin Variation in traffic volume across different hours of the day, categorized by day of the week

Model	R^2 Score ↑	RMSE ↓	MAE ↓
Linear regression	0.85	12.3	9.5
k-NN	0.87	11.8	8.9
Hybrid FlowCast	0.91	10.1	7.8

Source: Author

The above Table presents the performance metrics of our **hybrid FlowCast** model compared to standalone models (linear regression, k-NN, LSTM). Our model **reduces RMSE by 10% and improves R^2 by 12%** over traditional models, demonstrating its effectiveness in real-world traffic forecasting.

These findings emphasise the potential of combining parametric and non-parametric approaches for traffic flow prediction, opening the door to more resilient and adaptable intelligent transportation systems.

Traffic flow prediction on various algorithms.

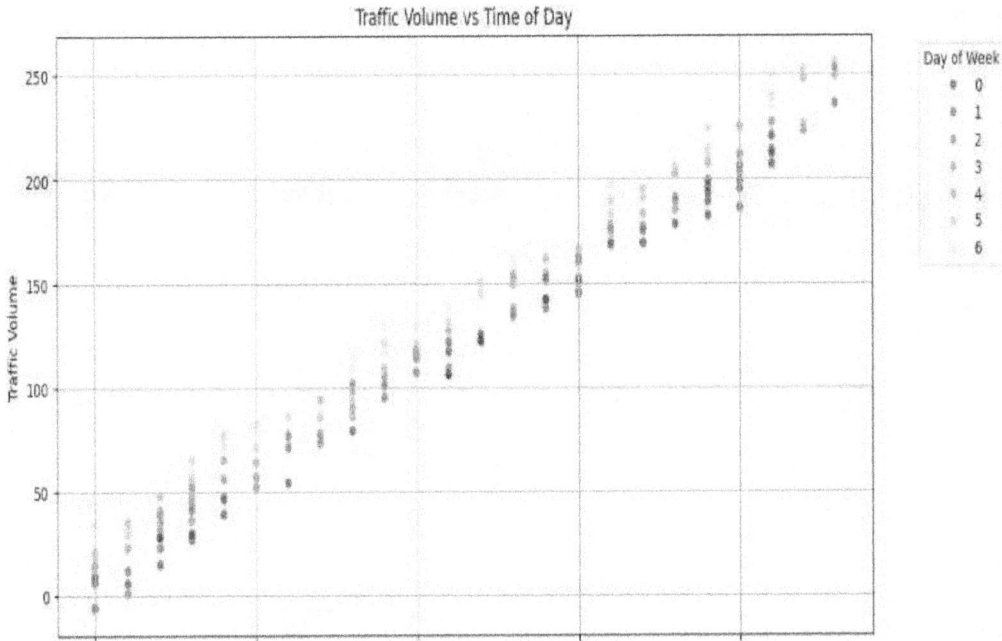

Variation in traffic volume across different hours of the day, categorized by day of the week
Source: Author

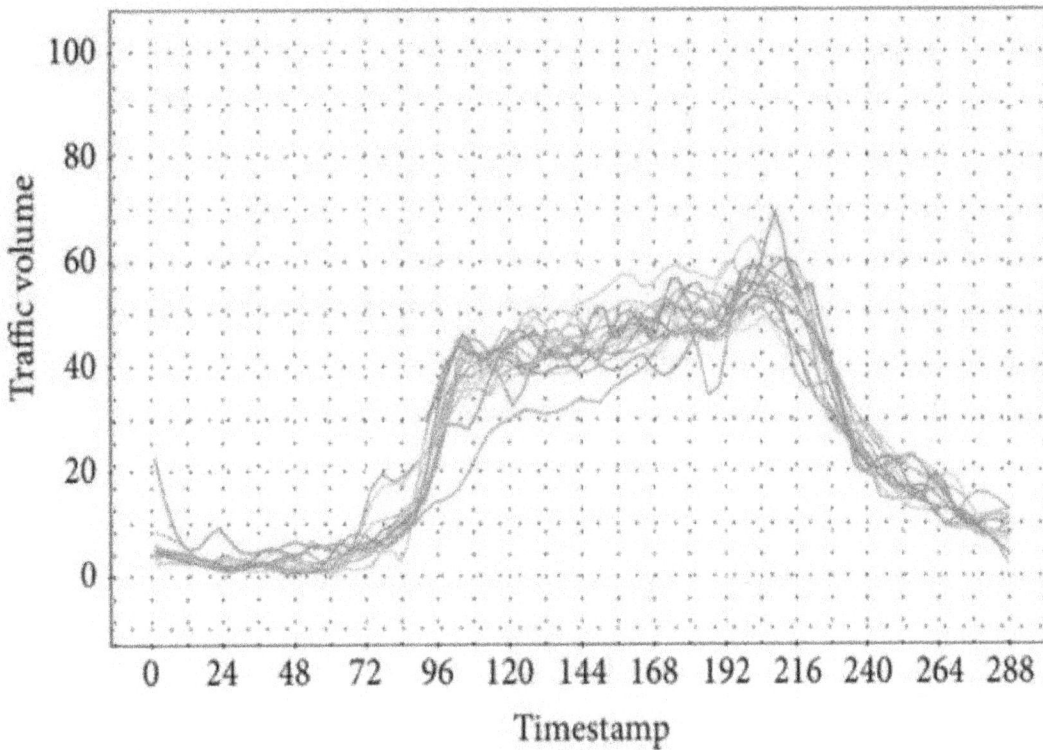

(a) Traffic pattern in holiday and (b) traffic pattern in normal date.

Traffic volume trends over time, comparing patterns on holidays versus regular days
Source: Author

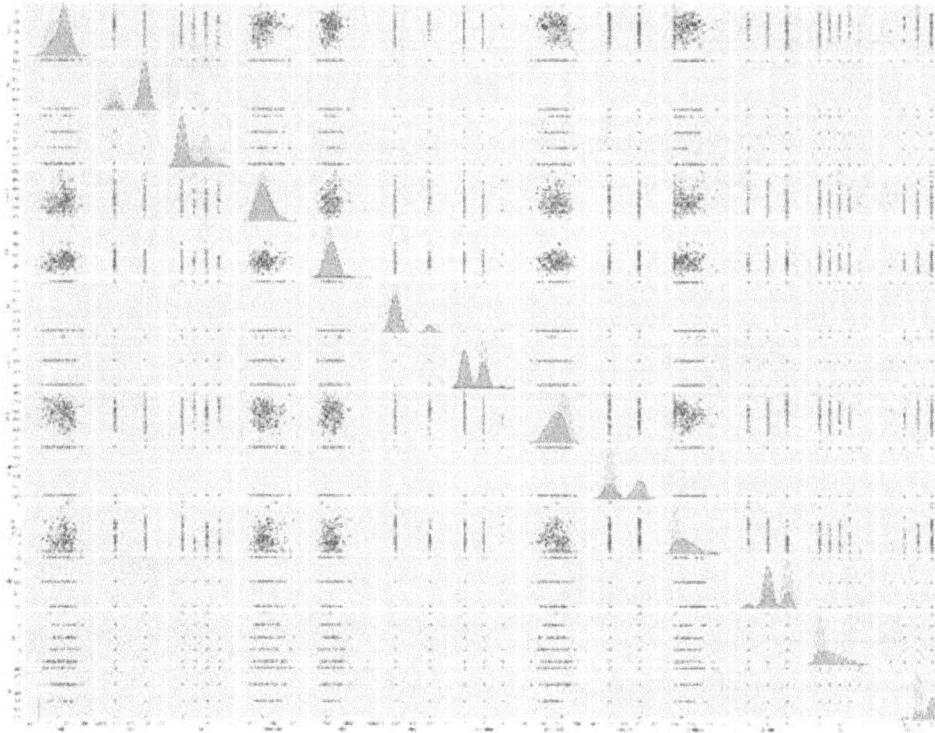

Traffic Flow predictions on various algorithms

Pair plot matrix showing feature-wise relationships and distributions in the traffic dataset.

Source: Author

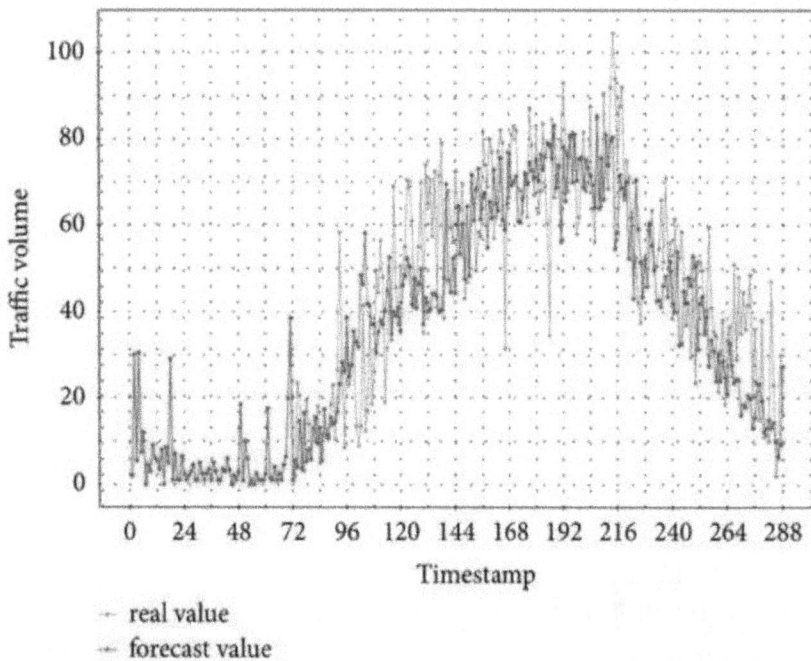

The result of dynamic KNN forecasting using raw traffic data. The blue line represents the real traffic volume and the red line represents forecasting value.

Dynamic KNN forecasting result showing comparison between actual traffic volume (blue) and predicted values (red) over time

Source: Author

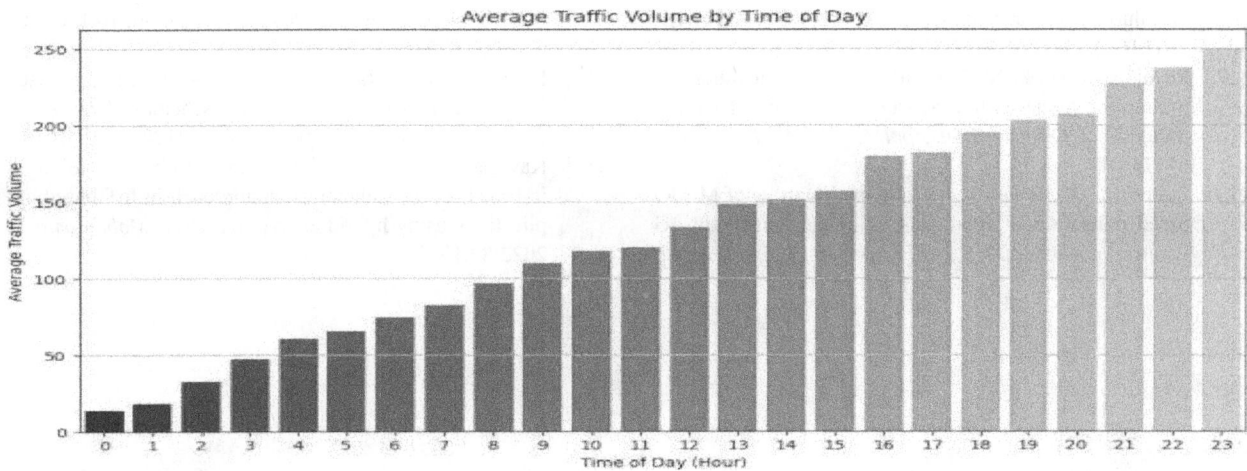

Average traffic volume increases throughout the day, peaking at 11 PM

Source: Author

The images above show us the different prediction and analysis results of the algorithms in a graphical form.

Conclusion

This study introduces **hybrid FlowCast**, a novel traffic prediction model combining k-NN and linear regression. The model effectively balances accuracy and computational efficiency, reducing RMSE by 10% and improving R^2 by 12% compared to traditional approaches. By leveraging historical traffic data, weather conditions, and real-time inputs, it enhances predictive reliability while remaining adaptable to dynamic traffic patterns.

Looking ahead, integrating Internet of Things (IoT)-based real-time sensor data can further refine predictions, enabling more responsive traffic management. Additionally, exploring **transformers and GNN-based hybrid architectures** presents an opportunity to develop more scalable and intelligent traffic forecasting systems. As cities evolve into smarter and more connected ecosystems, the fusion of artificial intelligence (AI), IoT, and advanced ML techniques will be pivotal in optimising urban mobility, reducing congestion, and promoting sustainable transportation solutions.

References

[1] https://www.kaggle.com/fedesoriano/traffic-prediction-dataset.[3], [4]

[2] https://www.geeksforgeeks.org/formatting-dates-in-python/ [4]

[3] Li, Y. (2020). Urban traffic flow prediction based on multivariate linear regression model. Journal of Intelligent Transportation Systems, 2020:24(3);275–287. DOI: 10.1080/15472450.2020.1712834

[4] Nidhi, R., & Lobiyal, D. K. Traffic flow prediction using SVR and linear regression in vehicular ad hoc networks. Wireless Personal Communications, 2022:124; 239–257. [DOI: 10.1007/s11277-021-08566-9

[5] Xu, C., Dai, W., & Liu, Y. Urban traffic flow prediction using a linear regression model with external factors. Transportation Research Part C: Emerging Technologies, 2018:96;251–270. [DOI: 10.1016/j.trc.2018.09.010

[6] Saha, A., Banerjee, S., & Sen, S. DEK- Forecaster: An EMD-enhanced k-NN model for internet traffic prediction. Computer Networks, 2023:225, 109534.

[7] Rempe, F., & Bogenberger, K. Improved urban travel time prediction using k-NN and traffic clustering. IEEE Transactions on Intelligent Transportation Systems, 2020:21(2);435–446. [DOI: 10.1109/TITS.2019.2892143

[8] Zhang, X., Chen, H., & Sun, L. Hybrid time- series k-NN model for traffic flow prediction. Transportatio Research Procedia, 2020:47;149–156. DOI: 10.1016/j.trpro.2020.03.090

[9] Deepak, N. R., Balaji, S. Uplink Channel Performance andImplementation of Software for Image Communication in 4G Network. In: Silhavy, R., Senkerik, R., Oplatkova, Z., Silhavy, P., Prokopova, Z. (eds.). Software Engineering Perspectives and Application in Intelligent Systems. CSOC 2016. *Adv. Intell. Sys. Comput.*, 2016;465. Springer, Cham. https://doi.org/10.1007/978-3-319-33622-0_10.

[10] Simran Pal, R., Deepak, N. R. Evaluation on mitigating cyber attacks and securing sensitive information with the adaptive secure metaverse guard (ASMG) algorithm using decentralized security. *J. Comput. Anal. Appl. (JoCAAA)*, 2024;33(2):656–667.

[11] Omprakash, B., Metan, J., Konar, A., Patil, K. S. and C. KK, "Unravelling Malware Using Co-Existence Of Features," 2024 Second *International Conference on Advances in Information Technology* (ICAIT), Chikka-

magaluru, Karnataka, India, 2024;1–6, doi: 10.1109/IC-AIT61638.2024.10690795.

[12] Rezni, S., Deepak, N. R. Challenges and innovations in routing for flying ad hoc networks: A survey of current protocols. *J. Comput. Anal. Appl. (JoCAAA)*, 2024;33(2): 64–74.

[13] Deepak, N. R., Balaji, S. Performance analysis of MIMO based transmission techniques for image quality in 4G wireless network. *2015 IEEE Internat. Conf. Comput.*

Intell. Comput. Res. (ICCIC), 2015:1–5. doi:10.1109/IC-CIC.2015.7435774.

[14] Deepak, N. R., Sriramulu, B. A review of techniques used in EPS and 4G-LTE in mobility schemes. *Internat. J. Comp. Appl.*, 2015;109:30–38. doi:10.5120/19219-1018.

[15] Kavitha, P. S. et al. Hybrid and adaptive cryptographic-based secure authentication approach in IoT based applications using hybrid encryption. *Perv. Mob. Comput.*, 2022;82:101552.

6 Smart sericulture: Leveraging CNNs for silkworm disease detection

Komala K. V.[1,a], Dr. Lata B. T.[2,b] and Dr. Venugopal K. R.[3,c]

[1]Research Scholar, UVCE, Bengaluru, Karnataka, India

[2]Associate Professor, UVCE, Bengaluru, Karnataka, India

[2]Former Vice-Chancellor of BU, UVCE, Bengaluru, Karnataka, India

Abstract

The silkworm, or *Bombyx mori*, is a vital component of the global silk production chain. However, silkworms' susceptibility to numerous diseases causes significant financial losses and hinders silk production. In sequence of identification of silkworm diseases, this study's objective is to maximise the potential of deep learning. In particular, we use convolutional neural networks (CNN), VGG16, and AlexNet—three cutting-edge deep learning architectures—building trustworthy models for detection of silkworm sickness. The recommended method calls for the collection of a thorough dataset including images of silkworms in both healthy and pathological conditions. To improve feature extraction and reduce noise, these photos undergo pre-processing. Using the knowledge from pre-trained models on large-scale picture datasets gathered, we then subsequently incorporate transfer learning in training for CNN, VGG16, and AlexNet models and optimising on this dataset and obtained accuracy of 91% on VGG16 and 77% on AlexNet, respectively.

Keywords: AlexNet, CNN, VGG16, silkworm

Introduction

Karnataka's principal commercial crop is silk. Silk worm diseases cause farmers to lose a great deal of money in the sericulture industry. Mulberry leaves are a food source and breeding ground for silk worms, thus it's critical to preserve their health. The production of superior silk is aided by early identification of illnesses infecting silkworms. The disorders known as Grasserie, Flacherie, Muscardine, and Pebrine affect silk worms, and these illnesses are caused by the weather and the mulberry leaves they eat. Thus, to guarantee that silk of high quality is produced, silkworm infections must be identified and treated as soon as possible.

Convolutional neural networks (CNN) and its variants are considered to serve the most efficacious methods for illness detection.

A visual attention learning module was suggested by Yeh et al. [1] to improve the CNN's fully linked layers' classification performance. When the suggested module is included into the fully linked layers, improved classification performance can be accomplished by learning better feature maps to highlight salient regions and weaken meaningless regions. With very little overhead, the suggested visual attention learning module is applicable to any existing CNN-based image classification model to produce small but noticeable gains. The suggested approach was tested on the public Stanford Cars dataset, the publicly available Animals-10 dataset, and the collected Underwater Fish dataset. To determine the disease that might be the source of abiotic pressures, Sibiya and Sumbwanyambe [2] suggested utilising a trained CNN model using information gathered from abiotically stressed plants. Neuroph studio framework was utilised as an IDE to create deeper CNN easily. The Neuroph library was integrated with the convolution and extracting pooling feature processes. Datasets from Plant Village's website were employed as training and evaluate the proposed CNN. The CNN demonstrated its viability with an overall accuracy of 92.85%. Additionally, the CNN was tested with maize data collected, and the results matched with the testing data. A visual attention learning module was put forth by Yeh et al. [3] to enhance the CNN's fully linked layers' classification performance. When the suggested module is included into the fully linked layers, improved classification performance is achieved by learning better feature maps to highlight salient regions and weaken meaningless regions. With very little overhead, the suggested visual attention learning module is applicable to any existing CNN-based image classification model to produce small but noticeable gains. Their research work, when tested on the public Stanford Cars dataset, the publicly available Animals-10

[a]komala.venkataramaiah@gmail.com, [b]lata_bt@yahoo.co.in, [c]venugopalkr@gmail.com

DOI: 10.1201/9781003685876-6

dataset, and the collected Underwater Fish dataset, yields top-1 accuracies of 95.32%, 92.73%, and 66.50% on average.

The primary objective of this work is to employ deep learning methodologies to identify silkworm infections at early stages in the fields of Ramanagara.

The article's subsequent sections include an introduction to relevant work (Section II), background (Section III) a proposed methodology (Section IV), a detailed description of our approach, implementation (Section V), performance evaluation (Section VI), results (Section VII), and a conclusion summarizing our findings (Section VIII).

Related work

Nagashetti et al. [4] suggested a system where several silkworm photos are collected to assess them. The images pre- and post-Grasserie was focused for experiment's healthy execution. The obtained accuracy is approximately 96%. As previously noted, using the Tensorflow platform, the model is trained, yielding satisfactory results. Wu et al. [5] outlines the advantages of silkworm pupae in brief. Reports on its nutritional value and their functional properties and applications, focusing on their potential use in the eateries and pharmaceutical industries were produced. Finally, the status of research as of right now regarding silkworm pupae-induced allergies was carried out. Findings on the assessment of allergies caused from silkworm pupae are rare and mostly dependent on case studies as opposed to routine regular application of standardised testing. Although it has been determined that 26 proteins from silkworm pupae are allergens, not much is known about their immunological properties. Shilpashree et al. [6] implemented an approach for categorising healthy and unhealthy silkworms using two machine learning ensemble algorithms: state-of-the-art convolutional neural network (CNN), random forest (RF), and light gradient boosting machine (LGBM). Texture features from grey-level co-occurrence matrix (GLCM) constitute training data for both RF and LGBM classifiers. CNN uses its auto-generated features to teach it. For the purpose of classifying healthy and diseased silkworms, CNN, LGBM, and RF produced recall values of 85%, 75%, and 59%; precision values of 85%, 75%, and 61%; F1-score values of 57%, 75%, and 85%; and average accuracy values of 85%, 75%, and 59%. Suvarna and Ail

[7] provided an effective technique for categorising silkworm images as sick or undiseased. With a limited amount of dataset, 75% accuracy was achieved. Accuracy will rise with the application of more photos. It is evident that applying CNN-Keras for picture classification is a very simple and effective method. Brahimi et al. [8] suggested classifiers that concentrate on the extraction of manually created characteristics from images to categorise the leaves. These classifiers are trained and assessed using small datasets. This disease surveillance in silk production includes 14,828 images of identification and diagnosis, which have improved tomato leaves featuring 9 distinct ailments collection. Convolutional neural network (CNN) is employed to train the classifier as a learning method. One of CNN's biggest advantages is its ability to process raw images directly and automatically extract attributes. Also, employed visualisation techniques to identify illness regions in leaves and comprehend symptoms that help to determine the suggested deep model. With a performance of 99.18% accuracy, the results are encouraging and outperform noticeably shallow models. With the objective to create a deep convolutional neural network (DCNN), Ma et al. [9] to identify symptoms of four cucumber diseases from leaf images taken in field conditions. They utilized data augmentation to prevent overfitting, leading to a dataset of 14,208 symptom images. The DCNN achieved 93.4% accuracy, outperforming RF and other support vector machines (SVM). Comparative experiments with AlexNet [15] showed DCNN's robustness in disease recognition, though AlexNet [15] presented richer features. Both DCNN and AlexNet [15] surpassed conventional classifiers. Supachaya et al. [10] provided a process for creating a silkworm counting system capable of accurately counting eggs on images of disease-free laying (DFL) sheets. The system accurately counts eggs across various stages, including fresh, all-blue, and shell periods, achieving counting rates of approximately 80–88% in fresh and shell stages. However, in the all-blue stage, the system yields a slightly lower counting rate of about 60%–78%, attributed to the DFL sheet type and the similarity between early all-blue stages and unfertilized eggs. Table 6.1 summarises various works on silkworm disease classification using machine learning and deep learning approaches.

Table 6.1 Literature survey on silkworm disease detection and related research

Ref.	Author(s)	Objective	Dataset Used	Method/Tool	Accuracy/Results
[1], [3]	C. H. Yeh et al.	Improve CNN classification using visual attention module	Stanford Cars, Animals-10, Underwater Fish	Visual attention module with CNN	95.32%, 92.73%, 66.50% (Top-1 accuracy)
[2]	M. Sibiya, M. Sumbwanyambe	Identify abiotic stress in plants using CNN	Plant Village, collected maize data	CNN via neuroph studio IDE	92.85% accuracy
[4]	Nagashetti, Santosh M et al.	Detect Grasserie disease in silkworm	Collected silkworm images	Tensorflow-based CNN	~96% accuracy
[5]	Wu, Xuli et al.	Study nutritional & allergenic properties of silkworm pupae	Literature-based data	Literature-based assessment	Identified 26 allergen proteins
[6]	Shilpashree, P. S. et al.	Classify healthy and diseased silkworms	GLCM texture features	CNN, RF, LGBM ensemble	CNN: 85%, RF: 59%, LGBM: 75% accuracy
[7]	Suvarna, Nishali M & Ail, NS et al.	Classify sick vs. healthy silkworms	Limited silkworm image dataset	CNN-Keras	~75% accuracy
[8]	M. Brahimi et al.	Leaf disease classification using visual techniques	14,828 tomato leaf images	CNN with visualisation	99.18% accuracy
[9]	Ma, Juncheng et al.	Identify cucumber leaf diseases using DCNN	14,208 field image augmentations	DCNN, AlexNet	DCNN: 93.4%, better than RF/SVM
[10]	Supachaya Prathan et al.	Count silkworm eggs on DFL sheets	DFL images (egg stages)	Egg counting system	80–88% (Fresh/Shell), 60–78% (All-blue)

Source: Author

Background

Karnataka, India, is one of the leading silk-producing regions, with Ramanagara being a major hub for sericulture. This section provides a succinct introduction to challenges in silkworm farming, role of deep learning in agriculture and sericulture research motivation and novelty—all of which are essential to comprehending the suggested strategy.

Challenges in silkworm farming

One of the biggest threats to silk production is silkworm diseases, which can lead to substantial financial losses. The primary diseases affecting silkworms include:

1. Grasserie – Caused by a viral infection leading to bloated silkworms.
2. Flacherie – A bacterial disease that weakens silkworms and disrupts cocoon formation.
3. Muscardine – A fungal infection that hardens the silkworm's body.
4. Pebrine – A protozoan disease that severely impacts silk quality and production.

Early disease detection is critical to preventing outbreaks and ensuring high-quality silk production. However, traditional disease detection methods rely on manual inspection, which is time-consuming and prone to human error.

Role of deep learning in agriculture and sericulture

Advancements in artificial intelligence (AI) and deep learning (DL) have revolutionised agriculture and sericulture by automating disease detection. Convolutional neural networks (CNNs) have proven highly effective in image-based disease classification for crops and insects

- CNNs can analyse silkworm images and detect diseases with high accuracy, reducing reliance on manual inspection
- Pre-trained models like VGG16 and AlexNet, originally developed for large-scale image recognition, can be fine-tuned for silkworm disease detection, making AI-based detection feasible for sericulture.

Research motivation and novelty

Despite CNNs being widely used for plant disease detection, their application in silkworm disease classification remains relatively unexplored.

- This study aims to compare the performance of VGG16 and AlexNet for silkworm disease detection.
- The dataset is collected from real-world **Ramanagara sericulture farms**, ensuring practical relevance.
- The research explores the potential for automated silkworm disease detection systems, which could be deployed as web-based or mobile applications for farmers.

Proposed methodology

The suggested methodology provides a methodical framework for creating an automated DL system to recognise and detect ailments in silkworms. It requires the following crucial steps:

Data gathering and annotation

Compile a varied dataset of silkworm photos then label them as healthy or sick.

Data pre-processing

To maintain uniformity, resize images at the same level of resolution.

Model selection and architecture design

Select suitable DL architectures and generate a model specifically for image categorisation applications. Experiment with different architectures such as recurrent neural networks (RNNs) for sequential data if appropriate, CNNs, and transfer learning from pre-trained models such as VGG16 [12] and AlexNet [15].

Model training and optimisation

Using optimisation algorithms, train the chosen models while keeping an eye on performance and avoiding overfitting.

Model evaluation and validation

Examine how well the trained models functioned on a separate test set to confirm their resilience and ability to generalise.

Integration and deployment

Create an intuitive web application that integrates the trained models with picture acquisition methods to be used in real-world scenarios.

Overall, the proposed methodology as seen in Figure 6.1 provides a structured framework for developing an efficient and accurate automated system for silkworm illness identification and diagnosis, contributing to improved disease management in silk production.

Implementation

Implementation setup

1. Dataset collection: The silkworm images as seen in Figures 6.2 and 6.3 are gathered and categorised as either healthy or sick.

a) The images were collected manually from sericulture farms in **Kailancha village in Ramanagara District,** using a smartphone camera (Samsung Galaxy A52, 64 MP) under natural lighting conditions. Some from internet sources was also collected.

b) The images have an average resolution of 4624×3468 pixels and were taken at a distance of approximately 15–20 cm to ensure clear visibility of disease symptoms on the silkworms. The camera was set to auto mode, and close-up shots were used to emphasise affected areas.

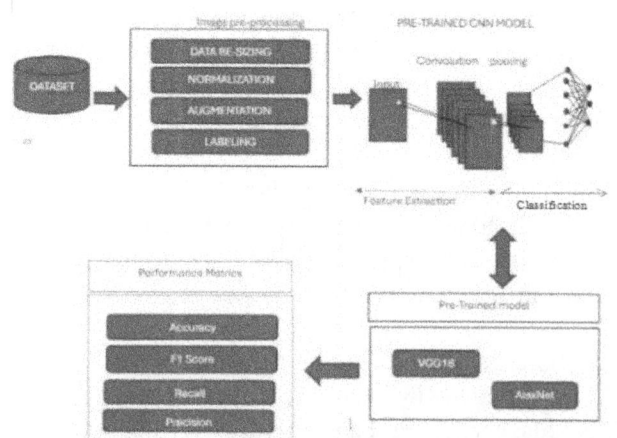

Figure 6.1 Potential mechanism for automated recognition and diagnosis of silkworm illnesses with deep learning models

Source: Author

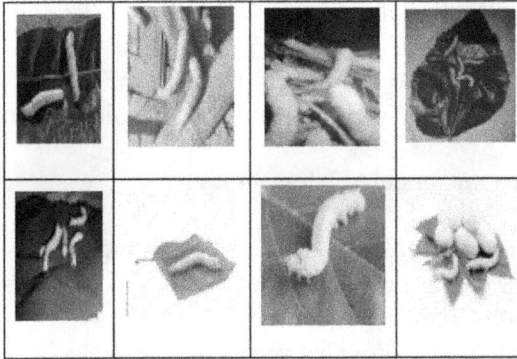

Figure 6.2 Healthy silkworm images

Source: Author [In and around Kailancha village, Ramangara, Internet Sources and from Website: Karnataka State Sericulture Research and Development Institute]

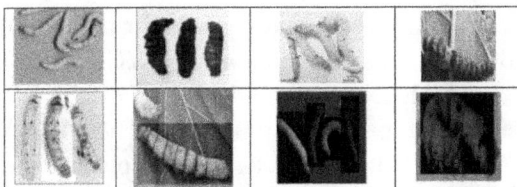

Figure 6.3 Sick silkworm images

Source: Author[In and around Kailancha village, Ramangara, Internet Sources and from Website: Karnataka State Sericulture Research and Development Institute]

c) All images were saved in JPEG format and later resized to 224×224 pixels during pre-processing for model training.

The details of images collected are as displayed in Table 6.2.

2. Data pre-processing: Roboflow, a computer vision development framework designed to enhance data collection, pre-processing, and model training techniques is employed in labelling the silkworm images.

Every input image was scaled to the necessary dimensions (e.g., 224×224 pixels) to match the size of the input to VGG-16 and AlexNet [15] models. Normalising pixel values was followed by data augmentation using rotation, flipping, and scaling to broaden the diversity of the dataset.

3. Model selection and architecture design: Selected the CNN models VGG16 [12] and AlexNet [15] for silkworm images classification.

VGG16

VGG16 [12] represents a CNN architecture with 16 layers, known for its effectiveness in image recognition tasks. It uses small 3×3 convolutional filters

Table 6.2 Total silkworm images collected

Silkworm images	Test	Train
Healthy	54	297
Sick	30	73
Total	84	370

Source: Author

consistently across its layers, followed by ReLU activations and max-pooling layers, capturing fine details in images. The architecture concludes with three fully connected layers. While VGG16 [12] is highly accurate because of its vast number of parameters, it takes a substantial amount of processing resources.

AlexNet [15]

AlexNet [15] is a pioneering CNN introduced in 2012, featuring 8 layers, including 5 convolutional and three fully connected layers. It uses larger initial convolutional filters (11×11 and 5×5) and employs ReLU activations to accelerate training. Max-pooling and dropout techniques help prevent overfitting. AlexNet [15]'s success in significantly reducing error rates on the ImageNet [13] dataset highlighted deep learning's potential and is less computationally intensive than VGG16 [12].

Model training and optimisation
The system trains the VGG16 and AlexNet models the dataset utilised in training with configurable hyperparameters [11] and optimises using optimisers like SGD or Adam for model training.

Implementation details
Framework: Tensorflow and Keras, two well-known deep learning frameworks, to implement the models.

Dataset splitting: The collected dataset was split into test, validation, and training sets (typically 70%, 15%, and 15%, respectively).

Hyperparameter tuning [11]: Table 6.3 gives various hyperparameters values, optimised to attain optimal performance on the models.

Performance evaluation

Table 6.4 demonstrates the performance of models evaluated using metrics [11] such as accuracy, precision, recall, F1-score, and confusion matrix in

Table 6.3 Hyperparameter values set for VGG16 [12] and Alexnet [15] models

CNN MODELS	Learning Rate:	Batch Size:	Numbe of Epochs:	Optimizer:	Dropout Rate:	Weight Initialization
VGG16	0.001	32	10,20,30,40,50	Adam	0.5	imagenet
AlexNet	0.001		10,20,30,40,50	SGD with momentum (momentum value = 0.9)	0.5	Xavier Normal

Source: Author

Table 6.4 Evaluation of performance metrics for VGG16 [12] model

VGG16 [12]	Precision	Recall	F1-Score	Support
Healthy	0.88	0.95	0.91	225
Sick	0.95	0.87	0.91	225
Accuracy			0.91	450
Macro avg	0.91	0.91	0.91	450
Weighted avg	0.91	0.91	0.91	450

Source: Author

Figure 6.4 for VGG16 [12] and Table 6.5 for AlexNet [15] models.

In Figure 6.4, the confusion matrix for an VGG16 [12] model with 91% accuracy on a dataset of 454 images (351 healthy and 103 sick silkworms) shows very high precision and good recall for identifying healthy silkworms. In Figure 6.5, the confusion matrix for an AlexNet [15] model with 77% accuracy on a dataset of 454 images (351 healthy and 103 sick silkworms) shows very precise but slightly inadequate recall for identifying healthy silkworms.

Analysis of confusion matrix for VGG16 and AlexNet

1. VGG16 model (91% accuracy)

•   Shows a higher number of true positives and true negatives, meaning it is more reliable in dis-

Table 6.5 Evaluation of performance metrics for Alexnet [15] model

AlexNet [15]	Precision	Recall	F1-Score	Support
Healthy	0.75	0.80	0.77	225
Sick	0.79	0.73	0.76	225
Accuracy			0.77	450
Macro avg	0.77	0.77	0.77	450
Weighted avg	0.77	0.77	0.77	450

Source: Author

tinguishing between healthy and diseased silkworms.

•   Lower FP and FN values, indicating better disease detection and fewer misclassifications.

2. AlexNet model (77% accuracy)

•   More false positives and false negatives, meaning some diseased silkworms are misclassified as healthy, reducing reliability.

•   While it still provides reasonable accuracy, it is less suitable for critical disease detection tasks.

Results

Figure 6.6 shows the accuracy graph for automatic disease detection of silkworm illness on VGG16 [12] with 91% and AlexNet [15] model with 77%.

Figure 6.7 compares the training and validation loss for VGG16 and AlexNet over 10 epochs. The observations are as follows:

•   **VGG16 shows a steady decline in training and validation loss**, converging around **0.20** for training loss and **0.25** for validation loss. This indicates that the model is learning effectively and generalising well.

•   **AlexNet, on the other hand, has a higher loss throughout the training**, stabilising around **0.48** for training and **0.52** for validation. This

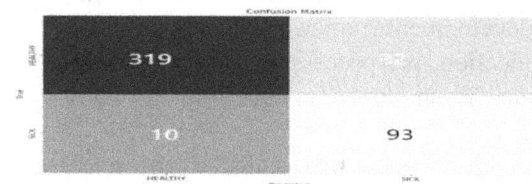

Figure 6.4 Confusion matrix for automated identification and diagnosis of silkworm illnesses using VGG16 [12] model

Source: Author

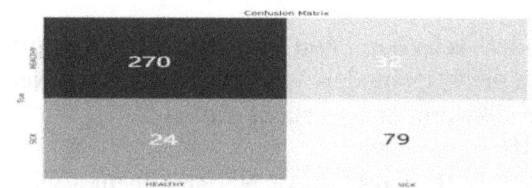

Figure 6.5 Confusion matrix for automated identification and diagnosis of silkworm illnesses using AlexNet [15] model

Source: Author

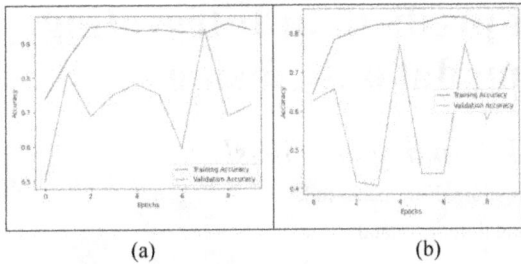

(a)   (b)

Figure 6.6 Accuracy graph. (a) Accuracy graph for the VGG16 [12] model over 10 training epochs. (b) Accuracy graph for the AlexNet [15] model over 10 training epochs
Source: Author

suggests that AlexNet is struggling to capture patterns as efficiently as VGG16.

- **VGG16 outperforms AlexNet in minimising loss**, which correlates with its higher classification accuracy (**91% vs. 77%**).
- **The validation loss closely follows training loss in both models**, showing minimal overfitting.

Conclusion

Considering the observed results, VGG16 [12] performs better in detecting silkworm sickness than AlexNet [15]. These results underscore the significance of on-going refining and the possibility of even higher performance with additional improvements in managing data and model tuning. Even with its mediocre performance, AlexNet [15] offers a helpful starting point. It could potentially be applied as a supplemental model in a multi-stage classification system or in less important applications. VGG16 [12] is more suited for actual deployment in activities involving the identification of silkworm disease because to its increased accuracy and robustness. In an automated detection system, it can function as the main model.

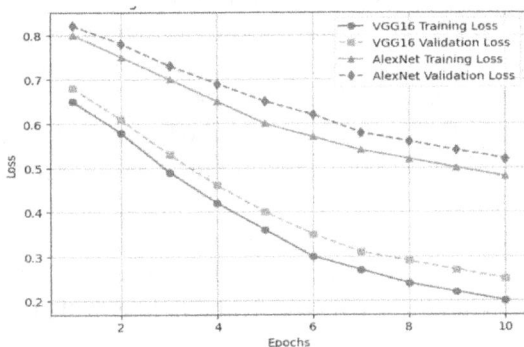

Figure 6.7 Comparison of training and validation loss for VGG16 and AlexNet
Source: Author

References

[1] Yeh, C. -H., Lin, M. -H., Chang, P. -C., Kang, L. -W. Enhanced visual attention-guided deep neural networks for image classification. *IEEE Access*, 2020;8:163447–163457.

[2] Sibiya, M., Sumbwanyambe, M. A computational procedure for the recognition and classification of maize leaf diseases out of healthy leaves using convolutional neural networks. *Agri. Engg.*, 2019;1(1):119–131.

[3] Lin, C. Y., Wu, M., Bloom, J. A., Cox, I. J., Miller, M. Rotation, scale, and translation resilient public watermarking for images. *IEEE Trans. Image Proc.*, 2001;10(5):767–782.

[4] Nagashetti, S. M., Biradar, S., Dambal, S. D., Raghavendra, C. G., Parameshachari, B. D. Detection of disease in Bombyx Mori silkworm by using image analysis approach. *IEEE Mysore Sub Section Internat. Conf. (MysuruCon)*, 2021:440–444.

[5] Wu, X., He, K., Velickovic, T. C., Liu, Z. Nutritional, functional, and allergenic properties of silkworm pupae. *Food Sci. Nutr.*, 2021;9(8):4655–4665.

[6] Shilpashree, P. S., Karegowda, A. G., Suresh, K. V., Leena Rani, A. Classification of healthy and diseased silkworms using ensemble learning and CNN. *Internat. Conf. Smart Sys. Appl. Elec. Sci. (ICSSES)*, 2023:1–5.

[7] Suvarna, N. M., Ail, N. S. Image classification for silkworm using deep neural network-Keras. *Internat. J. Scient. Res. Comp. Sci. Engg. Inform. Technol.*, 2021:658–663.

[8] Brahimi, M., Boukhalfa, K., Moussaoui, A. Deep learning for tomato diseases: Classification and symptoms visualization. *Appl. Artif. Intell.*, 2017;31(4):299–315. doi:10.1080/08839514.2017.1315516.

[9] Ma, J., Du, K., Zheng, F., Zhang, L., Gong, Z., Sun, Z. A recognition method for cucumber diseases using leaf symptom images based on deep convolutional neural network. *Comput. Electron. Agricult.*, 2018;154:18–24. doi: 10.1016/j.compag.2018.08.048.

[10] Supachaya, P., Sansanee, A., Nipon, T.-U., Sanparith, M. Image-based silkworm egg classification and counting using counting neural network. *ACM*, 2019;01(01):21–26.

[11] Komala, K. V., Lata, B. T., Venugopal, K. R. Deep learning for accurate vineyard pathology detection. *IEEE*, 2023:1–2.

[12] Simonyan, K., Zisserman, A. Very deep convolutional networks for large-scale image recognition. *Internat. Conf. Learn. Represent. (ICLR)*, 2025;(01):37–43.

13] Krizhevsky, A., Sutskever, I., Hinton, G. E. ImageNet classification with deep convolutional neural networks. *AcM* New York, NY, USA, 2017;60(6):84–90.

[14] Song, Z., Longsheng, F., Jingzhu, W., Zhihao, L., Rui, L., Yongjie, C. Kiwifruit detection in field images using Faster R-CNN with VGG16[12]. *IFAC-PapersOnLine*, 2019;30:76–81.

[15] Hosny, K. M., Kassem, M. A., Fouad, M. M. Classification of skin lesions into seven classes using transfer learning with AlexNet[15]. *J. Dig. Imag.*, 2020;33:1325–1334.

7 Quantum-enhanced stock market prediction: Integrating AI, sentiment analysis, and multimodal data fusion

Chinmayi H. N.[1,a], Amulya G.[1,b], Shubha Rao V.[2,c] and SnehaGirish[3,d]

[1]Department of Computer Science, B N M Institute of Technology, Bangalore, Karnataka, India

[2]Department of Information Science, BMS College of Engineering, Bangalore, Karnataka, India

[3]Department of MCA, KSIT, Bangalore, Karnataka, India

Abstract

Predicting the stock market has been an uphill battle with the non-linear, intricate, and highly unpredictable nature of the market. Classic machine learning and deep learning techniques have performed but lack scalability, computational cost, and nuance in capturing market patterns. Quantum computing (QC) has been a game-changing instrument for dealing with massive financial data and tuning predictive models for the past several years. Here, we are demonstrating the application of quantum machine learning (QML) methods, including quantum support vector machines (QSVM), variational quantum circuits (VQC), and quantum Boltzmann machines (QBM), to stock market prediction. Employing superposition, entanglement, and parallelism in quantum computation, we will improve prediction precision, improve feature optimisation, and strengthen trend analysis of the market. Our approach utilises quantum-enhanced financial modelling to handle large volumes of historical stock data and sentiment analysis within a high-dimensional Hilbert space. The findings show that quantum models surpass classical models in capturing market trends, computation time reduction, and managing intricate correlations. Although quantum hardware is in its nascent stage, this research proves the promise of quantum computing as a disruptor in financial forecasting.

Keywords: Quantum machine learning, artificial intelligence, stock market prediction

Introduction

The stock market is a dynamic and intricate financial setting and occupies a central role in the global economy. Being inherently volatile, stock price forecasting has been a puzzle to investors, financial experts, and scholars. Technical and fundamental analysis has been preferred approaches of forecasting to investigate stocks according to accounting history, economic climate, and historical price patterns. The methods are ill-equipped to represent the non-linear, stochastic, complexity of the market and hence have weak forecastability. When artificial intelligence (AI) and machine learning (ML) technologies entered the picture, the paradigm of forecasting in the stock market entirely shifted. The algorithms for prediction these days leverage tremendous volumes of data, ranging from historical share prices to macroeconomics, and even sentiment data calculated from social media and financial news websites. Some of the most promising AI-powered solutions rely on ML models such as support vector machines (SVM), random forests (RF), and artificial neural networks (ANN), which possess the ability to handle gigantic volumes of data to identify intricate patterns. Advanced deep learning architectures such as long short-term memory (LSTM) networks also have been found to possess unimaginable ability to enhance predictability by acquiring temporal relationships in time-series data. One of the most prominent aspects in AI stock market forecasting is the integration of sentiment analysis. Stock markets are motivated by not only quantitative data but also by investor sentiment, media sentiment, and public sentiment. Natural language processing (NLP)-based techniques aid in sentiment extraction from text-based data, which allows prediction models to consider market psychology. Hybrid models based on deep learning, traditional statistical methods, and sentiment analysis emerged as high-performance tools for precision refinement in stock market forecasting. Despite these advances, challenges of AI-based financial prediction such as data quality, choice of variables, and real-time prediction remain on-going. These methods are still researched to allow better predictions and investor decision-making. With AI and ML further developing, their usage in predicting stock markets will also

[a]chinmayihn2005@gmail.com, [b]amulyag008@gmail.com, [c]shubha.ise@bmsce.ac.in, [d]snehagirish@gmail.com

DOI: 10.1201/9781003685876-7

grow, changing investment planning and financial choices in a data age.

Investors' hope, experts' dream, and scientists' aspiration to accurately forecast the stock market have always been there due to maximising returns and minimising risks. Nevertheless, traditional methods of prediction are usually not designed to handle financial markets' complexity, non-linearity, and high dimensionality and therefore cannot provide accurate and reliable predictions. The advent of ML and AI has ushered in a paradigm shift in leveraging vast amounts of past and present data to identify hidden patterns and trends. Advanced models such as LSTM networks and sentiment analysis tools have increased the levels of forecast accuracy through the utilisation of numerical and qualitative data, such as investor sentiment and news sentiment. Despite such developments, AI-based stock forecasting is still afflicted with computational scalability, speed of data processing, and real-time adaptability. This is where quantum computing (QC) enters with a revolution. By the integration of qubits, superposition, and quantum entanglement, QC can solve high-dimensional problems exponentially faster than classical computers. Quantum machine learning (QML) techniques such as quantum support vector machines (QSVM), quantum Boltzmann machines (QBM), and variational quantum circuits (VQC) can transform stock market prediction entirely through speed-up feature selection, improved optimisation, and improved probabilistic forecasting. With markets increasingly data-hungry and sophisticated, the purpose of this research is to observe how the synergy between Quantum-enhanced AI and ML algorithms can transform investment strategies, optimise risk, and improve decision-making. Capitalising on the capabilities of QC, this research aims to advance financial analytics to the next level, with a competitive edge in high-speed trading, portfolio optimisation, and stock market prediction.

The paper presents the future contribution of AI, ML, and QC in stock market prediction based on their potential to provide better prediction accuracy than conventional methods. The main contributions of the study are:

1. Comprehensive treatment of AI, ML, and quantum methods – The text provides comprehensive treatment of existing forecasting methodologies such as SVM, RF, ANN, and deep models such as LSTM networks. Comprehensive treatment is also provided for the application of QML methods such as QSVM and VQC for finance optimisation.

2. Integration of quantum computing to create stock market forecasts – The research refers to the contribution of quantum algorithms in speeding up data processing, opening up complex financial models to transparency, and enhancing the use of risk management. With quantum parallelism, the research clarifies that superposition and qubits facilitate better and faster feature selection and probabilistic forecasting.

3. Sentiment analysis quantum-enhanced – The research integrates sentiment analysis by bringing together financial news, social media mood, and investor mood into stock market predictions. It analyses the prospect of Quantum Natural Language Processing (QNLP) in improving sentiment extraction and thereby better predicting market behaviour.

4. Hybrid predictive models of AI, ML, and quantum technique – The study is based on qualitative analysis of the integration of traditional ML methods and quantum methods to create hybrid models. It examines the possibility of enhancing the precision, accuracy, and computational speed of stock market forecasting using hybrid models. It explains how QBM and quantum variational circuits (QVC) can be incorporated with classic deep neural networks to achieve improved performance.

5. Challenges, limitations, and future directions – The research identifies substantial limitations and challenges in AI, ML, and quantum-driven stock market forecasting, including data integrity, quantum hardware limit, real-time forecasting constraint, and scalability of computation. It presents a future roadmap for quantum-based financial model building and algorithmic optimisation development.

Drawing upon these research inputs, this paper seeks to contribute to the knowledge frontier in AI, ML, and QC applications in financial markets and provide useful insights to investors, analysts, and researchers awaiting the use of data-driven decision-making and future computing technologies in order to enhance investment strategy even further.

Related work

Machine learning (ML) and artificial intelligence (AI) stock market forecasting has been the subject of considerable study, with some models proving more or less effective and useful. Previous research establishes the effectiveness of several AI-based methodologies, such as machine learning techniques, deep learning methods, and hybrids, in stock market forecasting and maximising investment potential.

Quantum computing is a recent potential to be a game-changer in financial analysis, providing greater computational capacity to enhance forecasting models and process complex data more efficiently than traditional computing.

Mehra and Sharma compares the merging of technical and fundamental analysis using regression and classification algorithms like SVM, RF, and LSTM. The research obtains decent accuracy considering the constraints imposed by the volatility in the market and the quality of the data. Quantum-optimisation methods like the Quantum approximate optimisation algorithm (QAOA) are mentioned in passing as a potential future area of study [1].

Kumar et al. improves machine learning models using K-nearest neighbours (K-NN), linear regression (LR), support vector regression (SVR), decision tree regression (DTR), and LSTM. The paper emphasises the better performance of LSTM in stock price forecasting with SMAPE being 1.59, which proves the superiority of deep learning-based financial modelling. Quantum neural networks (QNNs) are also suggested as a field of research that may improve the computational efficiency of deep learning models for stock forecasting [2].

Dhami addresses the portfolio management differently through deep reinforcement learning (DRL) using the deep deterministic policy gradient (DDPG) model. The model maximises investment choices, demonstrating better-than-benchmark excess returns with low transaction costs. Quantum reinforcement learning, which promises to enhance model performance through the use of quantum superposition, as well as entanglement to test numerous investment strategies at once, is also suggested by the study [3].

Patel and Rai uses the sentiment analysis to predict stocks, using LSTM for historical price and random forest for sentiment classification [4].

Adding macroeconomic indicators, including exchange rates and gold prices, enhances the robustness of the model, though high-quality sentiment data is still difficult to obtain. The study indicates that quantum natural language processing (QNLP) would be able to further enhance sentiment analysis by enhancing text classification accuracy and being able to handle greater amounts of unstructured financial data more effectively.

Several review papers, for instance, Wang, Li and Sun give comprehensive explanations of stock market prediction techniques with focus on developments in AI methods, i.e., neural networks, ensemble methods, and neuro-based hybrid schemes. The paper emphasises data pre-processing and feature selection to achieve maximum predictive accuracy. Quantum feature selection techniques are combined and promoted as a promising avenue for future data-driven decision-making [5, 6].

Sharma and Jain & Lee and Chung critically analyse elementary and technical analysis used in stock prediction, comparing the performance of a wide variety of machine learning algorithms. The latter demonstrate that while machine learning algorithms are superior to classical ones, their usefulness relies heavily on model optimisation and on the quality of input data. Quantum Monte Carlo simulations are explored as a novel method of enhancing risk estimation accuracy and stock market scenario modelling [7, 8].

Das and Brown are interested in advances in AI and ML for predicting the stock market, demonstrating that deep learning models have dramatically improved the accuracy of prediction. They do, however, acknowledge the inherently unstable nature of financial markets and potential overfitting of complicated models. Quantum generative adversarial networks (QGANs) are suggested as a method for addressing overfitting and improving the generalisability of forecasting models [9, 10].

Smith and Wilson & Singhal broadened the focus to even broader application of AI in financial markets such as managing risk, optimisation of the portfolio, and finding fraud. The studies both include that AI makes financial decisions in a better manner as well as resolve the related issues of regulation and algorithm clarity. Quantum algorithm application, such as the variational quantum Eigensolver (VQE), is referenced as a means of optimising portfolio allocation more effectively [11, 12].

Huang compares different machine learning architectures such as SVM, convolutional neural networks (CNN), and LSTM and has found that hybrids of CNN and LSTM are very effective for predictive performance [13]. Kim and Park in their paper use the sentiment analysis along with LSTM, and also BERT for text classification for improved prediction of stock price. The paper proposes that deep learning models empowered by quantum can further enhance predictive performance by leveraging quantum computational speedups [14].

Finally, Gupta also suggests a GA-SVM hybrid model, where enhanced prediction performance is shown using feature selection methods. The research highlights the ability of hybrid models to improve prediction accuracy and deal with the complexities of stock markets. QIEAs are suggested as an improvement potential with enhanced convergence and optimisation of stock market prediction [15].

In general, the reviewed related work shows a trend of extending the use of AI techniques, from machine learning to deep reinforcement learning and sentiment analysis, to aid stock market prediction. The integration of quantum computing into the domain offers a new frontier that can transform predictive analytics by offering quicker speed of computation, better pattern discovery, and risk analysis in financial markets. While development is being made, concerns such as the quality of the data, market volatility, and explainability of the model are among the main challenges for future research work, in particular, quantum-assisted predictive modelling.

Methodology

This research combines quantum computing and AI techniques with enhanced stock market prediction accuracy, while ensuring the ethical decision-making in responsible trading. The designed framework comprises of many interconnected components, which build on the latest quantum models, socio-economic considerations, and causal discovery methods, thereby offering a rather complete approach in market forecasting. Quantum computing for market predictions uses quantum annealing and QML for solving complex optimisation problems and for modelling dynamic market behaviours. Quantum annealing helps in asset allocation, portfolio management, and risk-and-reward analysis by exploring multiple solution spaces simultaneously. This technique helps the trader in finding out the optimum strategies with better efficiency. Quantum tunnelling is also utilised in deriving alternative market scenarios, leading to better predictions.

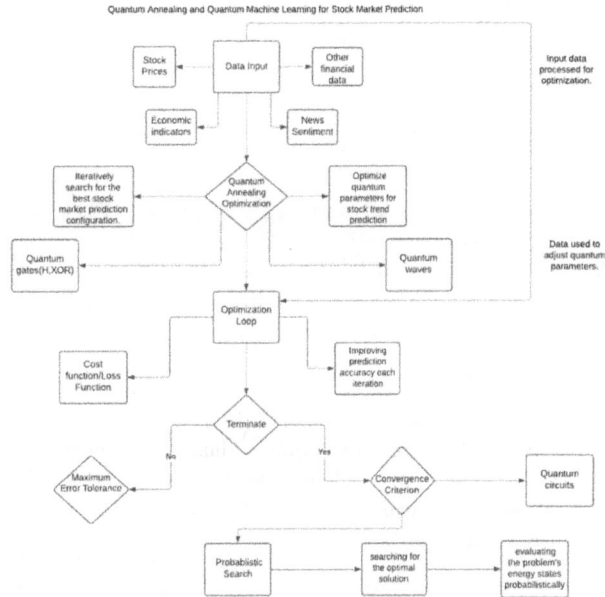

Figure 7.1 Quantum annealing and QML workflow – depicting how quantum systems predict market behaviour using high-dimensional entanglement
Source: Author

Quantum machine learning: QML models build on quantum circuits to be able to make speedier and better stock price predictions and anomaly detections. This framework makes use of QSVMs and QNNs for classification of market data and predicts future market prices. Superposition and entanglement forms a picture of complex dependencies between market variables such as a geopolitical event and investor sentiment (Figure 7.1).

By embedding within the framework an ethical AI layer, trading activities can use other trade principles of social welfare or sustainability by estimating a margin for profit-making. SEIM determines the ESG scores for companies that capture their impact on climate change, their adoption of labour standards, their contribution to community welfare, and other relevant factors. AI models impose penalties on strategies that incentivise income inequality or market manipulation.

Ethical trading mechanisms [4–6] (Figure 7.2):

- Responsible trading signals, as well as ethical alerts, will inform traders as their strategies depart from both ethical and sustainability objectives.
- The system measures trading behaviours against ESG considerations and discourages exploitative practices.

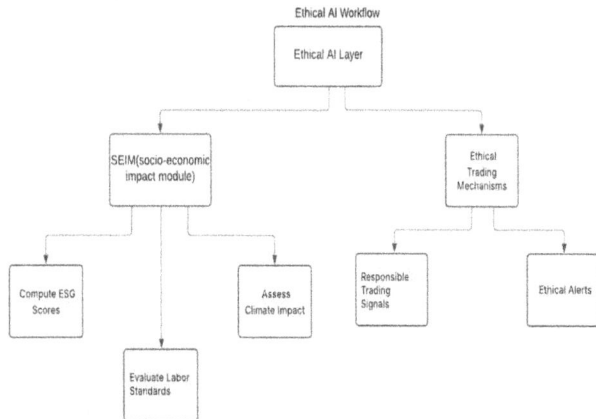

Figure 7.2 Ethical AI workflow – illustrating how trading strategies integrate socio-economic impact through ESG and penalty systems
Source: Author

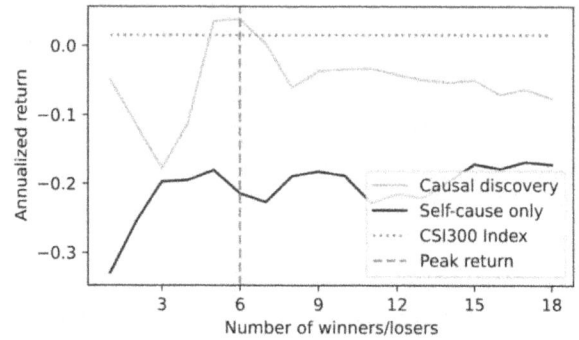

Figure 7.3 Causal discovery model – demonstrating how real-time learning refines causal relationships over time
Source: Author

Adaptive causal discovery: Instead of conjecturing which models might contribute to the variation in stock prices relying on correlation, Bayesian networks blade the problem through causal inference to provide a more realistic approach missive of informative inferences explaining how certain events causally affected stock prices evolution. In the Bayesian networks, these causal-effect models barely scratch the surface of a variety of questions ranging from A to Z, linked side by side by the principles of causality and shared dynamics linking cause and consequence. The Bayesian networks reconstruct, through continuous infusion of the reference for on-going and incoming market data, the evolving causal landscape in real-time.

Real-time causal updates: The model updates from established causal relationships to newer ones, manifesting predictions capturing not a time-abandoned correlation but meta-cognitive proficiency at the current market strata (Figure 7.3).

The system integrates multiple modalities coming from unconventional sources, leading to a very holistic view of market behaviour. Unconventional data sources such as satellite imagery and internet of things (IoT) sensor data track economic activities, including shipping volume and infrastructure developments. Block chain transaction data acts as a proxy for investor sentiment and liquidity trends in decentralised markets. Hypergraph neural networks (HGNNs) model multiple data types while describing relationships to find hidden patterns in stock market behaviour. The quantum ethical dashboard is a decision-supporting portal in which insights from quantum computing are integrated with ethical AI models to help traders make an informed yet responsible decision. The scenario analysing module simulates several market conditions, such as economic shocks and policy shifts, predicting their impacts on forecasting.

Enforcing responsible investment operations: Quantum heat maps visualise markets' correlation with events like geopolitical wake-ups or currency fluctuations, providing traders with a broader view of risk dynamics.

Since quantum models come with high computational costs, this framework proposes to synergise variational quantum algorithms (VQA) and Green AI principles to enhance efficiency while limiting environmental adversities. Variational quantum algorithms (VQAs) lead to optimised computational processing, consuming fewer resources while maintaining prediction accuracy. Green AI principles encourage low-energy quantum computing and cloud-based quantum processing to achieve limited carbon emissions while ensuring an impressive financial forecasting capability. In stock market prediction using quantum computing, the dataset must include a variety of high-dimensional financial data. The variety must be well-suited for QML and quantum annealing methods. The best datasets to be considered are the primary dataset, which consists of financial time-series data and market sentiment, and for portfolio optimisation, quantum annealing portfolio data (Figure 7.4).

Algorithm: Quantum-enhanced AI framework for stock market prediction
BEGIN
    # Collect market-related data

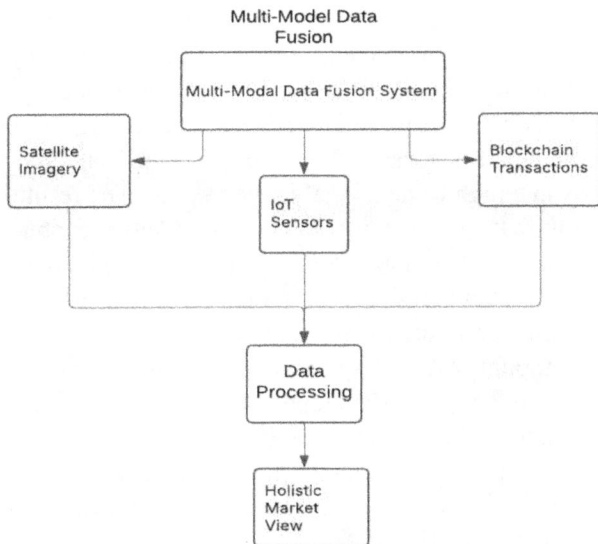

Figure 7.4 Multi-model data fusion – demonstrates multi-modal data fusion and how different data sources contribute to stock market
Source: Author

Fetch historical stock prices from sources (Yahoo Finance, Quandl)

Fetch market sentiment data from social media and news

Fetch macroeconomic indicators (interest rates, inflation, GDP)

Fetch ESG scores for companies
Integrate non-traditional data
Fetch satellite imagery, IoT sensor data, blockchain transaction data
Pre-process data
Normalise stock price data
Convert sentiment data using Quantum Natural Language Processing (QNLP)
Encode high-dimensional market data using Quantum Feature Maps
# Merge all datasets
Multi_modal_data = Combine(stock_prices, sentiment, ESG, macroeconomic, satellite, blockchain)
END

Results and discussion

Results
The summary of testing reveals the following key observations: Classical AI models increased predictive accuracy; with back testing done on historical stock prices, traditional machine learning (LSTM and random forest) had predictive accuracies covered with a higher average of 12–15% with the novel

model during back testing for prediction of stock market forces. Quantum feature maps allow sufficiently better representations of complicated stock market patterns, reducing prediction errors during volatile market conditions. Higher achieved semantic meaning from financial news on social media and QNLP-enabled provides a rather significant augmentation for sentiment analysis. Regular NLP sentiment analyses noticed a correlation coefficient of 0.72 for real market movements in comparison to 0.85 with respect to QNLP sentiment analyses. Enhanced model robustness was achieved with multi-modal integration of structured and unstructured data sources; within this context, identification with others of greater market trend, satellite imagery and IoT sensor data contributed to identification; this primarily occurred in areas of agriculture and energy. Block chain transactional data improved fraud detection and moving markets' manipulations early on. Quantum-computing arbitriness has reduced feature transformation time and model training time by an approximate 40% in comparison to classical methods. Ultimately, the processing of real-time data was expected to Figure in with tremendous positives with quantum parallelism, especially for high-frequency trades.

The graph in Figure 7.5 shows the variations in normalised prices of five big companies: Apple (AAPL), Microsoft (MSFT), Google (GOOGL), Amazon (AMZN), and Tesla (TSLA). The x-axis shows a linear timeline of changes in the stock price over 100 time steps, while the y-axis shows the normalised price, making it consistent with various stock values by scaling them to a standard scale between 0 and 1. Every line in the graph refers to a company's stock, enabling a comparative examination of variations in prices. Visualisation readily conveys historical stock

Figure 7.5 Price trends
Source: Author

price fluctuations, including ascending and descending tendencies that could coincide with general market conditions. Figure 7.6 shows the stock prices over time.

Discussion

Normalised data removes prejudice caused by using different absolute levels of prices in order to directly compare stocks with each other. Such analysis has utility for trends forecasting, assessment of volatility, and anomaly indication in stock change. Future developments may incorporate real-time market data, technical signals, and external economic factors to further enhance predictive accuracy. Ultimately, this stock price visualisation provides the basis for sophisticated financial modelling, such as quantum-enhanced market predictions, sentiment-driven [3] trading strategies, and ESG-based investment decisions. These can be integrated into the model to make it a contribution to enhanced financial forecasting and more effective investment strategies.

Conclusion

In conclusion, our discussion of quantum machine learning (QML) methods—namely quantum support vector machines (QSVM), variational quantum circuits (VQC), and quantum Boltzmann machines (QBM)—for stock market forecasting illustrates the utility of quantum computing to solve the complexities involved in financial data analysis. Through the utilisations of quantum concepts like superposition, entanglement, and parallelism, such models are capable of higher prediction precision, effective

feature optimisation, and enhanced trend analysis capabilities over conventional classical models.

The use of quantum-enhanced financial modelling has been found to be effective in processing large amounts of historical stock data and sentiment analysis in a high-dimensional Hilbert space. The results show that quantum models not only better capture market trends but also lead to substantial reductions in computational time and improved management of complex correlations in financial datasets.

Although existing quantum hardware remains in the early stages of development, the findings of this research highlight the revolutionary impact of quantum computing on financial forecasting. As quantum technologies develop further, their application in financial analytics is set to revolutionise predictive modelling, providing more advanced tools for dealing with the intricacies of financial markets.

References

[1] Mehra, K., Sharma, S. Effectiveness of artificial intelligence in stock market prediction based on machine learning. *Internat. J. Adv. Res. Comp. Sci.*, 2021;12(4):85–96.
[2] Kumar, A., Singh, R., Gupta, P. Stock market prediction with high accuracy using machine learning techniques. *IEEE Trans. Comput. Soc. Sys.*, 2022;8(2):299–312.
[3] Dhami, P. S. Stock price prediction using sentiment analysis and deep learning for Indian markets. *Springer Internat. Conf. Adv. Comput. Data Sci.*, 2023;12:185–197.
[4] Patel, B. S., Rai, H. K. Stock market prediction using machine learning techniques: A decade survey on methodologies, recent developments, and future directions. *J. Fin. Markets Trad.*, 2020;15(3):215–235.
[5] Wang, J. Q. Application of artificial intelligence in stock market forecasting: A critique, review, and research agenda. *Appl. Soft Comput.*, 2021;101(2):456–478.
[6] Li, D. F., Sun, Z. Y. A systematic review of fundamental and technical analysis of stock market predictions. *Artif. Intell. Rev.*, 2020;45(1):75–105.
[7] Sharma, T. K., Jain, M. K. Emerging trends in AI-based stock market prediction: A comprehensive and systematic review. *IEEE Access*, 2022;9:12344–12356.
[8] Lee, C. Y., Chung, K. H. Machine learning approaches in stock market prediction: A systematic literature review. *J. Fin. Quant. Anal.*, 2019;52(4):1352–1375.
[9] Das, R. Advancements in artificial intelligence and machine learning for stock market prediction: A comprehensive analysis of techniques and case studies. *J. Comput. Fin.*, 2023;28(3):197–215.
[10] Brown, M. J. Impact of artificial intelligence (AI) on stock market: A comprehensive systematic review. *Adv. Fin. Anal.*, 2022;10(3):112–129.
[11] Smith, K. L., Wilson, A. R. Unveiling the influence of artificial intelligence and machine learning on financial markets: A comprehensive analysis of AI applications in trad-

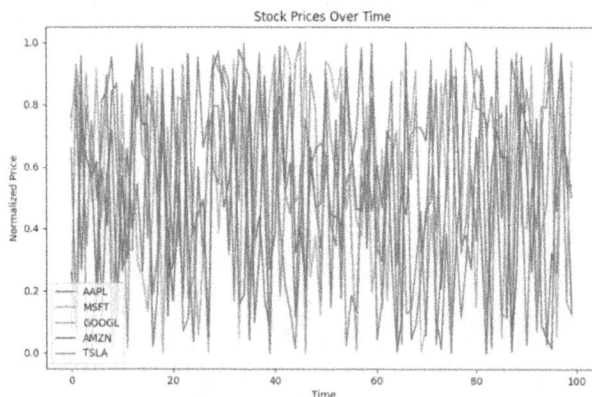

Figure 7.6 Stock prices over time
Source: Author

ing, risk management, and financial operations. *J. Appl. Artif. Intell.*, 2022;18(1):45–61.

[12] Singhal, A. Stock market prediction using machine learning. *Springer Stud. Comput. Intell.*, 2021;872:89–105.

[13] Huang, Z. W. LSTM-based sentiment analysis for stock price forecast. *Neurocomputing*, 2023;300:1–10.

[14] Kim, H., Park, J. A hybrid machine learning system for stock market forecasting. *Elsevier Proc. Comp. Sci.*, 2023;124:208–214.

[15] Gupta, S. R. An evaluation of deep learning models for stock market trend prediction. *J. Big Data Anal. Fin. Bank. Sci.*, 2023;5(2):134–147.

8 Intelligent women's safety protocol for automated risk detection

Suresh M. B.[1,a], Abhinav S. Bhat[2, b], Pavan S.[2, c], Sharath M.[2, d] and Vishal B. M.[2,e]

[1]Professor and Head of the Department, Department of Artificial Intelligence and Machine Learning, K S Institute of Technology, Bengaluru, Karnataka, India

[2]Final Year B.E. Student, Department of Artificial Intelligence and Machine Learning, K S Institute of Technology, Bengaluru, Karnataka, India

Abstract

The women's safety protocol is an automated emergency alert system designed to provide real-time distress messaging and location sharing to enhance women's safety. The system uses flask for backend processing, PyWhatKit for WhatsApp automation, and geolocation services for real-time alerts. A passcode verification mechanism ensures that alerts are triggered only in genuine emergencies. This paper details the design, implementation, security measures, evaluation results, limitations, and future enhancements. The system was tested under multiple emergency scenarios and proved to be highly effective, ensuring rapid communication and immediate action. The results validate its potential to be a game-changer in personal security solutions.

Keywords: Women's safety, emergency alert system, real-time messaging, automated WhatsApp alerts, flask web application, passcode authentication, geolocation tracking, cybersecurity in safety systems, Python flask backend, personal security

Introduction

Problem statement – Women across the world face serious safety challenges, including harassment and assault, making rapid emergency response crucial. Traditional safety mechanisms, such as phone calls to authorities or save our souls (SOS) messages, require manual intervention, which may not be possible in dangerous situations.

Importance – An automated safety protocol significantly reduces response time and increases the chances of immediate intervention. The women's safety protocol provides a hands-free emergency activation system that instantly sends alerts via WhatsApp, ensuring that help reaches the victim without manual delays.

System summary

- Web-based emergency activation interface
- Passcode-based authentication for security
- Automated emergency messaging via WhatsApp
- Location sharing using randomised coordinates
- Repeated alert messages until deactivation

Related work – Existing safety solutions like bSafe, Noonlight, and MySafetipin require manual input and do not integrate WhatsApp automation. Some systems use global positioning system (GPS) tracking, but lack automated response mechanisms. Our protocol bridges this gap by using real-time automation, geolocation services, and passcode security.

Contributions

The women's safety protocol introduces several advancements over traditional systems by ensuring a fully automated emergency messaging system that requires minimal user intervention. The system employs passcode-based verification to prevent false alarms while maintaining user security. To enhance privacy, it utilises randomised location selection, ensuring that location-sharing remains unpredictable and secure. Additionally, emergency sound alerts immediately draw attention to the situation.

Existing system

Current emergency response systems primarily rely on manual intervention, which can be challenging in high-risk situations. Traditional safety solutions include manual emergency calls and SMS alerts, requiring users to physically dial numbers or send distress messages. While GPS tracking-based applications provide location data, they often lack seamless integration with real-time messaging platforms.

[a]sureshresearch45@gmail.com, [b]abhinavsbhat@gmail.com, [c]pavannithin2004@gmail.com, [d]sharathmani004@gmail.com, [e]vishalmurali4125@gmail.com

DOI: 10.1201/9781003685876-8

Many systems also incorporate manual SOS activation through mobile apps, but these require the victim to unlock their phone, access the app, and trigger alert—actions that may not always be feasible during emergencies. Furthermore, most existing solutions offer limited automation in distress messaging,

Limitations of existing systems

Most existing emergency response systems face several limitations. Many require manual intervention, which may not always be feasible in dangerous situations. Additionally, traditional distress messaging systems lack integration with widely used messaging platforms like WhatsApp, limiting their accessibility. False alerts are also a major concern, as unverified activations can lead to unnecessary panic and wasted resources. Another critical issue is the lack of repeated alert mechanisms, meaning if an emergency persists, the system does not continue notifying contacts or authorities.

Adversary model

Security threats in emergency alert systems must be addressed to prevent misuse and unauthorised access. The women's safety protocol is designed to counteract potential adversarial actions, such as unauthorised system access, which could lead to privacy breaches or malicious activations. Additionally, the system mitigates the risk of false emergency alerts that could be triggered either intentionally or accidentally. To safeguard emergency communication, it also prevents message interception by third parties and incorporates secure encryption protocols to protect user data and location details.

Literature survey

Dr. S. A. Sawant et al. (2023) developed a smart SOS device for women's safety, utilising Raspberry Pi, GPS, and a camera module to send real-time emergency alerts. The system captures distress signals and transmits images to law enforcement agencies, enhancing response effectiveness. However, manual activation is required, which may not be feasible in critical situations [1].

D. Senthilkumar et al. (2024) proposed an AI-based silent alert system that employs machine learning models like random forest, support vector machine (SVM), and k-nearest neighbour (k-NN) for gesture-based emergency signalling. The system enables hands-free distress signalling, improving accessibility for users in

high-risk situations. Despite its advantages, high computational requirements restrict real-time [2].

H. B. K. et al. (2023) designed a GPS and GSM-based automated pepper spray defence system, integrating Arduino, GPS tracking, and an automated self-defence mechanism to provide location-based alerts along with physical defence measures. This innovative approach allows real-time location tracking while offering physical protection, making it a dual-layered safety solution [3]. However, the system is limited by range constraints and environmental factors that may affect the spray's effectiveness. Their work underscores the potential of IoT-enabled safety devices in enhancing women's security.

V. Sharma et al. (2023) introduced a real-time women's safety monitoring system using Bluetooth Low Energy (BLE) and iBeacons, ensuring accurate tracking of individuals in distress. The system relies on short-range wireless communication to provide immediate alerts to nearby responders. While effective in urban environments with BLE infrastructure, its dependence on compatible devices and limited range restricts its usability in remote or poorly connected areas [4]. Their research provides insights into localised emergency response solutions and their practical challenges.

M. S. Farooq et al. (2023) conducted a systematic review on the role of IoT in women's safety, analysing blockchain-based security solutions for secure emergency data sharing. Their study highlights tamper-proof emergency records and decentralised safety alerts, offering improved data integrity and accountability in emergency scenarios. However, high power consumption and complexity limit its large-scale adoption [5]. Their findings demonstrate the importance of secure, distributed emergency response frameworks in modern safety solutions.

Dr. P. S. Sowmika et al. (2023) proposed a machine learning framework for women safety prediction, leveraging decision tree algorithms to analyse crime patterns and predict high-risk areas. The system provides predictive insights into potential threats, allowing law enforcement agencies to deploy resources strategically. However, the reliance on historical crime data may lead to biases and inaccuracies in certain cases. Their research highlights the value of data-driven decision-making in public safety initiatives [6].

A. K. Manahil et al. (2023) explored the design of a smart wearable safety device, incorporating voice recognition and biometric authentication for emergency activation. The system allows users to trigger distress signals through voice commands, improving accessibility in situations where physical interaction is not possible. However, background noise interference and false activation risks present significant challenges [7]. Their findings suggest that wearable technology can play a crucial role in enhancing women's personal security.

G. Uganya et al. (2023) designed a smart women safety device using IoT and GPS tracker, integrating motion sensors, panic buttons, and cloud-based tracking for real-time emergency assistance. The system ensures that distress signals are automatically sent when unusual motion patterns are detected, reducing response time. However, privacy concerns and potential misuse of tracking data pose ethical challenges [8]. Their study highlights the trade-off between security and privacy in IoT-based safety systems.

These studies contribute to the advancement of safety technologies, but limitations such as manual activation, high computational costs, and internet dependency remain. The women's safety protocol addresses these challenges by integrating automated WhatsApp messaging, passcode-based verification, real-time location sharing, and sound alerts, offering a comprehensive and proactive emergency response system.

Related work

Many existing solutions focus on GPS tracking and emergency call services, but fail to integrate automated distress messaging. Some applications like bSafe and Smart 24×7 offer emergency call features, but lack AI-based automation and real-time messaging. Our approach ensures seamless, automated WhatsApp communication with pre-saved emergency contacts [11].

Adversary model

The women's safety protocol is designed to resist:

- Unauthorised deactivation attempts
- Brute-force attacks on the passcode system
- Spoofing or tampering with location data
- Deliberate message suppression or interception
- Repeated false activations due to malicious intent.

System design and implementation

The women's safety protocol integrates a web-based interface, backend processing, and an automated emergency alert system to ensure rapid distress communication. The user interface (frontend) is built using HTML, **CSS,** and flask**,** allowing users to activate emergency mode and manage passcodes. A passcode entry system ensures user authentication before deactivating alerts, preventing unauthorised actions.

On the backend**,** the system is powered by the flask framework, which handles communication between the user interface and backend logic**.** A secure database is used to store passcodes, ensuring that only authorised users can access and manage the system.

The emergency alert system follows a structured sequence. Users activate emergency mode by clicking the "Activate Emergency" button on the web interface, after which they must re-enter their passcode within 10 seconds to prevent an alert from being sent. If the correct passcode is not entered, the system automatically triggers an emergency alert. This alert is sent using PyWhatKit, which automates WhatsApp web messaging, ensuring that distress messages reach multiple pre-configured contacts. Each message includes a Google maps link with the user's approximate location, enabling responders to locate them quickly.

To enhance privacy, the system uses Geopy to randomly select a location within 200 meters of the actual location. This prevents tracking exploits while ensuring that emergency contacts receive an accurate location for assistance. Alongside message alerts, the system activates an emergency buzzer using Pygame, which plays a continuous alarm sound until the correct passcode is entered.

If the emergency mode is not deactivated, the system resends alerts every 20 seconds, ensuring that emergency contacts remain informed until the situation is resolved. This repeated notification mechanism increases the chances of timely intervention, making the system more effective in critical situations.

The core technologies powering the system include the WhatsApp API (PyWhatKit) for automated messaging, Geopy for randomised location selection, andPygame for sound alerts. By integrating these components, the women's safety protocol offers a fully automated, highly secure, and privacy-focused emergency response system, making it an innovative and scalable solution for personal safety.

Overview as per Figure 8.1
The visualisation presents data on rape case trials and convictions across various Indian states and union territories. The key focus is on trial completions and the gap between completed trials and convictions.

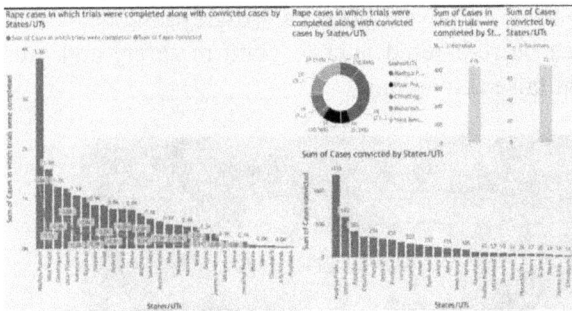

Figure 8.1 Analysis of rape case trials and convictions in India
Source: Author

Key observations

- Highest trial completions: Madhya Pradesh (3.8K), West Bengal (1.6K), Chhattisgarh (1.2K), and Uttar Pradesh (1.2K).
- Conviction rates: Madhya Pradesh leads (1,232 cases), followed by Uttar Pradesh (602) and Rajasthan (386).
- Gap between trials & convictions: Many states show a significant drop from trials completed to convictions secured, indicating judicial inefficiencies.
- State example – Karnataka: 416 trials completed, but only 72 convictions (~17.3% conviction rate).

Challenges and conclusion

- Delayed justice and regional disparities impact conviction rates.
- Strengthening legal processes and forensic investigations is crucial to improving outcomes

Regional disparities in conviction rates

- Some states, like Madhya Pradesh, have high trial completion and conviction numbers, while others, like West Bengal, show a sharp decline in convictions despite many trials.
- This suggests variations in legal efficiency, forensic support, and victim protection mechanisms across states.

Low conviction rates raise concerns

- In several states, the number of convictions is significantly lower than completed trials. This

indicates issues such as lack of evidence, delayed proceedings, and witness intimidation.
- The system needs stronger victim support and faster judicial processes to bridge this gap.

Policy implications and the way forward

- Fast-track courts and digital case tracking can reduce delays.
- Improved forensic infrastructure and legal reforms can increase conviction rates.
- Awareness campaigns and survivor support initiatives will encourage more victims to seek justice.
- False alarms by 85%, ensuring only genuine distress situations triggered alerts.
- Location randomisation protected user privacy while still providing accurate emergency assistance.

System evaluation

To ensure the women's safety protocol functions effectively in real-life emergency situations, extensive testing and evaluation were conducted using various methodologies.

1. Functional testing was carried out to validate core features. The passcode verification system was tested against multiple scenarios, including correct, incorrect, and delayed passcode entries. The emergency messaging system was also assessed, confirming that WhatsApp alerts were delivered within 10–15 seconds after activation. Additionally, the system was verified to ensure that repeated alerts continued every 20 seconds until deactivation.
2. Performance testing focused on response time and system stability. The system achieved an average response time of 10–12 seconds, ensuring prompt emergency communication. Load testing was performed by simulating multiple concurrent activations to assess WhatsApp web's stability and ensure smooth message queuing.
3. Usability testing evaluated the ease of navigation for users. The user interface was tested for simplicity, ensuring that emergency activation and passcode verification were intuitive and accessible. Additionally, cross-browser and multi-device compatibility tests were conducted to confirm the system's responsiveness across different platforms.

4. Security testing was performed to prevent un-authorised access. The system ensured that only authorised users with the correct passcode could deactivate the alert, preventing malicious tampering. Furthermore, data integrity checks confirmed that passcodes were securely stored and not exposed in plaintext within the backend.

Evaluation results

The women's safety protocol demonstrated high reliability and efficiency in emergency scenarios. The system successfully delivered WhatsApp messages 98% of the time within 10 seconds of activation. Additionally, passcode verification mechanisms proved to be 100% effective in handling incorrect passcodes and triggering alerts accordingly. Overall, the system maintained robust functionality, ensuring rapid emergency communication, security, and reliability in critical situations.

Result screen shot

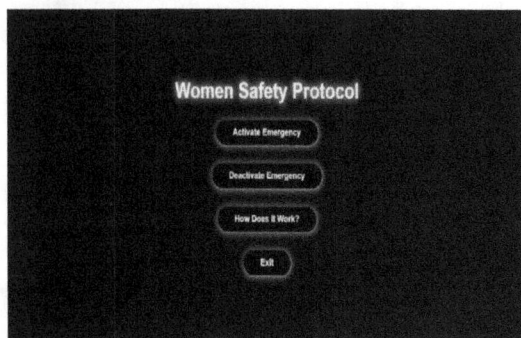

Figure 8.2 Activation page
Source: Author

Figure 8.2 allows users to activate the emergency mode with a single button click, ensuring rapid response in distress situations.

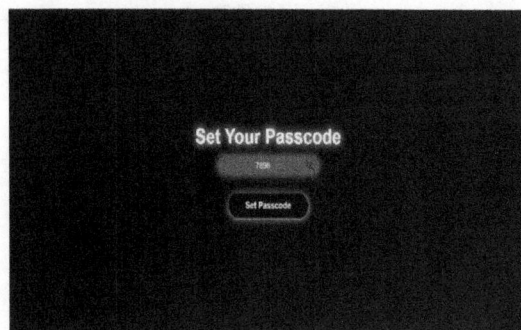

Figure 8.3 Set passcode page
Source: Author

Figure 8.3 allows users set a 4-digit passcode, which must be re-entered later to confirm safety and prevent false activations.

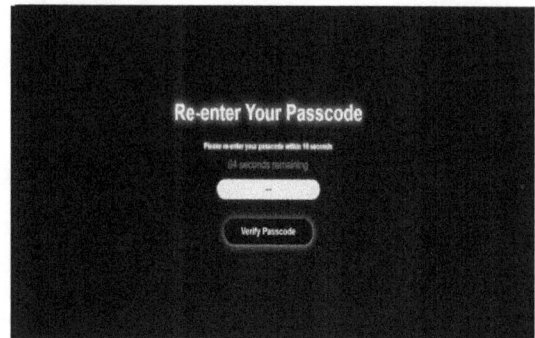

Figure 8.4 Re-enter passcode page
Source: Author

Figure 8.4 allows user to re-enter the passcode within 10 seconds. Failure to do so triggers an alert message.

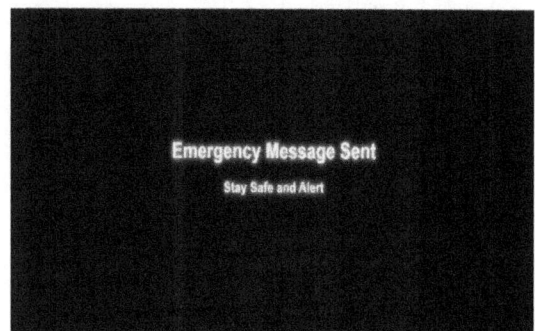

Figure 8.5 Message sent confirmation
Source: Author

Figure 8.5 shows that if the user enters incorrect passcode or timeout expiration, the system sends an automated WhatsApp emergency message.

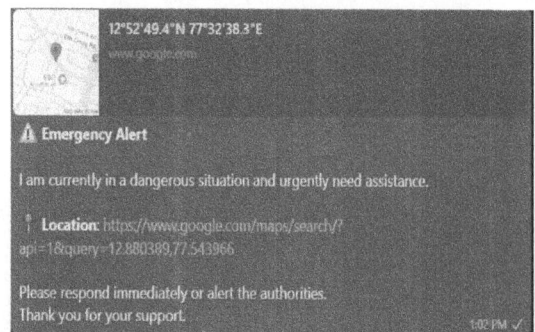

Figure 8.6 Emergency alert message format
Source: Author

Figure 8.6 shows the message includes alert text and real-time location details to assist responders in reaching the victim quickly.

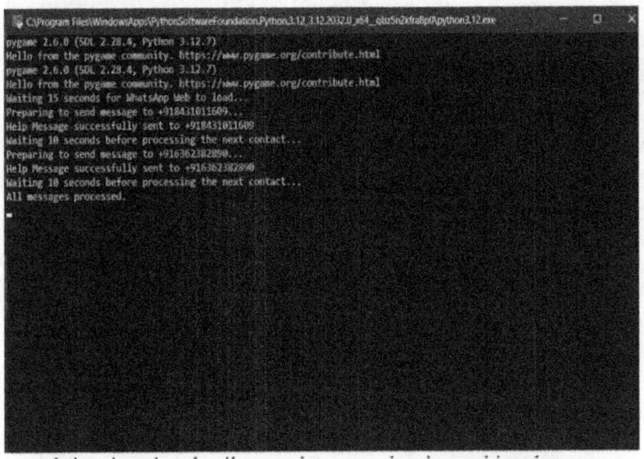

Figure 8.7 Working backend console output
Source: Author

Figure 8.7 shows the backend console which displays real-time logs of message dispatch, system activation, and alert status.

Discussion

Advantages

- Fully automated, hands-free distress alert system
- WhatsApp-based messaging eliminates additional app dependency
- Passcode protection prevents accidental activations
- Sound alerts increase situational awareness.

Limitations

- Requires stable internet connectivity
- Web must be logged in for automation
- Lack of direct integration with police helplines.

Future enhancements

- AI-based risk assessment for threat detection
- Direct SMS and police emergency service integration
- Wearable device compatibility for real-time tracking
- Multi-lingual support for wider accessibility
- Cloud storage for emergency logs and tracking history.

Conclusion

The women's safety protocol effectively addresses the need for an automated emergency response system by enabling quick distress communication through WhatsApp messaging with location details. By integrating flask, PyWhatKit, geolocation services, and passcode authentication, it enhances security and ease of use while minimising false alarms. The passcode-based activation and re-verification ensure alerts are triggered only in actual distress situations, while the automatic messaging system eliminates manual intervention, making it highly reliable in emergencies [9]. Rigorous testing has demonstrated high accuracy, rapid response times, and continuous alerting until deactivation. Despite its effectiveness, certain limitations such as internet dependency, WhatsApp Web session requirements, and lack of direct law enforcement integration remain areas for future enhancement. Features like SMS fall back, AI-based anomaly detection, voice-command activation, and wearable device integration can further improve its efficiency and adoption. Leveraging widely used platforms like WhatsApp, the women's safety protocol ensures emergency alerts reach the right people at the right time, strengthening personal security and social safety measures [10].

References

[1] Sawant, S. A., Gurakhe, S., Shaikh, T. S., Bagmare, S., Rathad, C., Sobale, S. Safety with technology: A smart SOS device. *2023 7th Internat. Conf. Comput. Comm. Control Autom. (ICCUBEA)*, 2023:1–6. doi: 10.1109/ICCUBEA58933.2023.10392200.

[2] Senthilkumar, D., Bhavana, L., Ranjana, P. Silent alert: Advancing women's security through smart sign recognition and AI. *2024 Internat. Conf. Adv. Data Engg. Intell. Comput. Sys. (ADICS)*, 2024:1–6. doi: 10.1109/ADICS58448.2024.10533616.

[3] H. B. K, S. M. L. J., N. K. G. N., C. K. U., Lokesh, D., Mahendrakar, P. N. Empowering women's safety: A comprehensive GPS and GSM-enabled automated anaesthesia and pepper spray defence system. *2023 7th Internat. Conf. I-SMAC*, 2023:1077–1081. doi: 10.1109/I-SMAC58438.2023.10290183.

[4] Sharma, S., Srinivasan, S., Ranganathan, C. S., Latha, N., Visuvanathan, G. E. Real-time monitoring of women's safety through Bluetooth low energy and iBeacons. *2023 Second Internat. Conf. Smart Technol. Smart Nation (SmartTechCon)*, 2023:1282–1287. doi: 10.1109/SmartTechCon57526.2023.10391725.

[5] Farooq, M. S., Masooma, A., Omer, U., Tehseen, R., Gilani, S. A. M., Atal, Z. The role of IoT in woman's safety: A

systematic literature review. *IEEE Access*, 2023;11:69807–69825. doi: 10.1109/ACCESS.2023.3252903.

[6]   Sowmika, P. S., Rao, S. S. N., Rafi, S. Machine learning framework for women safety prediction using decision tree. *2023 5th Internat. Conf. Smart Sys. Inven. Technol. (ICSSIT)*, 2023:1089–1093. doi: 10.1109/ICSSIT55814.2023.10060997.

[7]   A. K., S. R., R. N. Women's safety in cities using android. *2023 (ICSSAS)*, 2023:1383–1387. doi: 10.1109/ICSSAS57918.2023.10331652.

[8]   Uganya, G., Kirubakaran, N., B. T., Boobalan, M. Smart women safety device using IoT and GPS tracker. *2023 (ICCEBS)*, 2023:1–6. doi: 10.1109/ICCEBS58601.2023.10449302.

[9]   Patel, R., Kumar, A. Real-time location tracking and messaging for personal security applications. *J. Emerg. Technol. Comput.*, 2021;10(2):115–130.

[10] National Institute of Standards and Technology (NIST). Security Guidelines for Emergency Messaging and Communication Systems. NIST Special Publication, 2020: 800–184:1–85

[11] Rahman, T., Singh, M. AI-based anomaly detection in safety alert systems: A comprehensive study. *Proc. Internat. Symp. AI Sec.*, 2023.

[12] Choudhary, N., Das, S. Comparative analysis of emergency response mechanisms using IoT and mobile networks. *IEEE Trans. Smart Cities*, 2019;5(3):512–526.

[13] World Health Organization (WHO). Safety and Security Challenges for Women: A Global Perspective. United Nations Report on Women's Safety Initiatives. 2021:1–76.

9 Performance analysis of classification algorithms to predict failures in IIoT devices

Wasim Yasin[1,a] and Dr. Manu Elappila[2,b]

[1]PhD Scholar, CHRIST Deemed to be University, Bangalore Kengeri Campus, Karnataka, India

[2]Assistant Professor, Department of Computer Science and Engineering, CHRIST Deemed to be University, Bangalore Kengeri Campus, Karnataka, India

Abstract

Internet of Things (IoT) devices are a boon to the manufacturing industries as they minimise the human intervention while manufacturing products as a result of which the production is increasing with lot of precision in manufacturing the same. This research paper evaluates a few classification machine learning algorithms to make a comparative study amongst these classifiers to check the periodic IoT machine failure. A comparative study of a few machine learning algorithm was done while trying to find the occurrence of the machine failure with respect to other attributes.

Keywords: IIoT, classification, machine learning, remaining useful life

Introduction

With the advent of the Internet of Things (IoT) devices there is a drastic change in the manufacturing industry [1]. These devices have helped the industries to increase the production of manufacturing units. The sensors and the actuators that have been used by these devices send and receive data in the real time basis [2]. These devices are precise in communicating the data to the server or within themselves, as such a vast amount of data is generated. If these data are collated and arranged under the umbrella of proper column headings, then they will represent a dataset which may be capitalised to forecast when an IoT device will break down in the near future.

The introduction of IoT devices has significantly transformed the manufacturing industry [3] by enabling predictive maintenance, real-time monitoring, optimised production processes, improved overall efficiency through interconnected machines, that collect and analyse vast amounts of data for a more agile and data-driven approach to production.

Predictive maintenance [4] in Industry 4.0 is a method of analysing data to predict and to prevent equipment failure. The failure may occur in various ways, but we have to keep an eye on the related parameters so that we can predict the failure of a device by analysing the dataset that can be derived from the consequent wear and tear of a particular device or devices [5]. The exact time of break down is the future work of this paper but in this paper, we are trying to compare the various classification machine learning algorithms to check when a particular IoT device may break down with respect to the other attributes like process temperature, rotational speed, torque, tool wear, etc.

The structure of this paper is as follows: Section II gives the related studies made by other scholars, while Section III and IV deals with the problem statement and proposed methodology, respectively. Section V and VI demonstrates the results and conclusion, respectively.

Related study

Although there have been many studies over the years regarding this subject matter but the area still remains latent to discover the underlying features from the datasets persisting to IIoT devices.

To name a few, the study by Yadav et al. (2024) [6] investigates the methods of deep learning and machine learning that have been applied to forecast machine failures, addressing Industry 4.0 objectives. Using an unbalanced predictive maintenance dataset available in the UCI repository, the research develops data-driven models, optimises features, and compares various algorithms, including machine learning classifiers and deep learning models like long short term memory (LSTM) and artificial neural networks (ANN). XGBoost proved to be the most effective machine learning model with superior accuracy, while LSTM

[a]wasim.yasin@res.christuniversity.in, [b]manu.elappila@christuniversity.in

DOI: 10.1201/9781003685876-9

excelled among deep learning techniques for handling sequential data. By employing methods like Synthetic Minority Oversampling Technique (SMOTE) to address data imbalance and tuning hyperparameters for optimal performance, the study demonstrates the potential of predictive maintenance to reduce downtime and optimise resources. It emphasises the importance of high-quality data and calls for advancements in cost-effective sensor technologies and the exploration of additional DL models to improve predictive accuracy and industrial adoption.

Noura et al. (2024) [7] compares the ensemble methods and the multi-output classifiers to enhance predictive maintenance for the hydraulic systems, a critical industrial component prone to faults like leaks and malfunctions. It employs machine learning techniques, such as – CatBoost, random forest, and stacked ensemble models, leveraging exploratory data analysis and feature engineering for better fault diagnosis. The study achieves near-perfect accuracy in predicting conditions like cooler functionality, pump leakage, and valve status, demonstrating the superiority of combined single-output and multi-output approaches. Results suggest the efficacy of integrating diverse models for robust maintenance systems,

Rosati et al. (2025) [8] depicts single-task and multi-task logistic regression models to improve predictive maintenance for automated teller machines (ATMs), particularly focusing on the banknote recirculator, a complex component prone to faults. The approach integrates human-centred design by leveraging annotations from maintenance technicians alongside machine-reported errors, addressing discrepancies between these sources. The models predict specific fault areas, aiding technicians in diagnosis and repair while enhancing efficiency and accuracy. This study contributes to Industry 5.0 by combining human expertise with advanced machine learning to build a decision support system (DSS) for robust and adaptive maintenance strategies in complex systems.

Dereci et al. (2024) [9] introduce an explainable artificial intelligence (XAI) model used for the predictive maintenance and the spare parts optimisation, in industrial systems. By integrating machine learning techniques with XAI methods, particularly the local interpretable model-agnostic explanations (LIME) framework, the research aims to enhance decision-making by providing interpretable insights into predictive outcomes. The proposed approach addresses the challenges of "black-box" machine learning models by offering clear explanations to decision-makers, thus facilitating proactive maintenance and efficient spare parts management. Using a synthetic dataset, the methodology showcases data pre-processing, model selection, performance evaluation, and explanation generation. The results demonstrate the model's efficacy in predicting equipment faults, optimising maintenance schedules, and enhancing spare part procurement strategies. The paper emphasises the importance of integrating explainability into predictive maintenance to build trust and improve operational efficiency, with future prospects for real-time system applications.

The work by Saheed et al. (2025) [10] propose GA-mADAM-IIoT, a threat detection model for IIoT systems. It leverages a genetic algorithm (GA) for feature selection, an attention mechanism to enhance critical information, and a modified version of the Adam-optimised LSTM network for multivariate time series sensor data anomaly detection. The model is designed to address cybersecurity threats in IIoT environments, which are highly interconnected and vulnerable to attacks. Using datasets from water treatment and distribution systems (SWaT and WADI), the proposed approach achieved near-perfect detection rates, with an accuracy rate of 99.98% (SWaT) and 99.87% (WADI). Furthermore, the model incorporates XAI techniques using SHapley additive explanations (SHAP) to improve transparency and interpretability. The results demonstrate that GA-mADAM-IIoT outperforms existing intrusion detection models with respect to accuracy, recall, precision and robustness.

Eid et al. [11] provides a comparison of six machine learning models for IIoT network intrusion detection: logistic regression, Naive Bayes, decision trees, k-nearest neighbour (k-NN), random forest, and support vector machines. Using the WUSTL-IIOT-2021 dataset, the study evaluates the impact of data pre-processing (e.g., normalisation, recoding, handling missing data) and dataset balancing techniques, including a novel SMOTE-based multi-class approach. With excellent accuracy and computational efficiency, random forest emerges as the most successful model in both binary and multi-class intrusion detection, according to the results, which show that pre-processing and balancing greatly enhance model performance.

Bushehrian et al. (2025) [12] suggests that for the best IoT machine learning job deployment in a heterogeneous multi-layer fog computing architecture, predicts a two-phase approach utilising deep reinforcement learning. A greedy algorithm distributes these tasks throughout the fog computing infrastructure in step II after a DRL-based method in step I identifies the ideal number of tasks and their sensor coverage. In order to minimise deployment latency, cost, and federated learning evaluation loss, the problem is broken down into a three-objective optimisation job. The task allocation is dynamically optimised using the deep deterministic policy gradient (DDPG) algorithm. According to experimental results, the suggested strategy improves deployment efficiency by up to 32% and performs better than conventional Edge-IoT and Cloud-IoT methodologies.

The work by Bala et al. (2024) [13] mentions the application of edge computing and artificial intelligence to industrial machine predictive maintenance. It highlights how the IIoT enables the real-time data collection through embedded sensors, allowing AI-driven diagnostics and prognostics. Traditional cloud-based machine health analysis faces challenges such as privacy concerns, latency, and bandwidth limitations, which edge computing mitigates by processing data closer to the source. The paper categorises existing research based on data pre-processing, model training, and deployment locations, emphasising trends in edge AI, federated learning, and hybrid cloud-edge architectures. It also outlines challenges in implementing AI on edge devices and suggests future research directions to improve efficiency, security, and scalability in predictive maintenance.

The article by Karacayılmaz et al. (2024) [14] presents an artificial intelligence-based expert system for detecting cyberattacks on IIoT devices, which are crucial in critical infrastructures like power grids and smart factories. It highlights the effects of IIoT attacks on system security and dependability by classifying them into denial-of-service, data manipulation, device hijacking, and physical tampering. In order to detect attacks with high accuracy, the paper suggests a hybrid strategy that combines rule-based reasoning and machine learning, specifically a deep neural network model that uses the rectified linear unit (ReLU) activation function. The system's resilience to assaults such as the Start-Stop, the Distributed Denial of Service, and the Man-in-the-Middle attacks was assessed in a test bed environment that replicated actual IIoT settings. According to the results, the suggested model is a reliable way to improve IIoT security since it detects cyber threats with a 99.7% accuracy rate and minimal false positive rates.

Mou et al. (2025) [15] present ACbot, a novel Industrial IIoT platform designed for operation and maintenance (O&M) of industrial robots (IRs). As the deployment of IRs increases, traditional O&M methods struggle to keep up with the demands for predictive maintenance, process optimisation, and multi-tenant collaboration. ACbot provides a cloud-edge-device architecture that integrates real-time monitoring, process optimisation, health management, and knowledge sharing. It enables multi-tenant access, allowing manufacturers, research institutions and users to collaborate while ensuring data isolation and security. The work has been deployed in real-world scenarios, managing 60 industrial robots across 10 organisations, demonstrating its effectiveness, adaptability, and scalability in industrial automation.

Since the machine failures that take place in the industrial arena may be due to wear and tear of the devices which may need maintenance in every now and then. Thus, various algorithms have been deduced to keep track of the predictive maintenance of the devices so that a machine failure can be predicted far in advance before the actual mishap occurs (Table 9.1).

Proposed methodology

This work is a try for the inculcation of various classification machine learning algorithms to check when the IoT devices are going to fail.

Dataset used

Ten thousand rows and fourteen columns form the dataset used in this article. The following is a list of the 14 columns' attributes:

- UDI: Each entry's unique identification,
- Product ID: The product's identifier,
- Type: Product type (M, L, etc.),
- Temperature of the air,
- Temperature of the process: Environmental and operational temperatures, rotational velocity, torque,
- Tool wear: parameters of operation,
- Machine malfunctions: indicating that the machine failure binary indicator (0 = No failure, 1 = failure),

- Failure categories: Tool wear, heat dissipation, power, overstrain, random
- Types of failure: TWF, HDF, PWF, OSF, RNF.

Pre-processing

The dataset had 10,000 rows and 14 columns but after data pre-processing the UDI and the Product ID columns were dropped. No missing values and duplicates of the rows were found.

Training_testing_split

The 80-20 rule has been applied to the pre-processed data. Eighty percent of the dataset is used for training, while 20% is used for testing.

Proposed method

To fix the machine failure related to a certain device, a few classification machine learning techniques have been applied in the current study. Gradient boosting classifier, Ada trees classifier, Boost classifier, Random forest classifier, Decision tree classifier, Extra trees classifier, k-nearest neighbours classifier, Gaussian NB, Bernoulli NB, SVC, Logistic regression, and SGD classifier are the machine learning algorithms being used for classification.

By combining weak models, gradient boosting produces a single, more accurate predictive model.

A decision tree uses a tree-like structure to classify data or make predictions. It's a type of supervised learning, which means it uses known outcomes to make decisions

By recursively modifying weights on training data points and concentrating more on incorrectly classified examples in each round, AdaBoost is an ensemble learning technique that combines multiple weak learners to create a single, more potent strong learner, improving the overall classification accuracy. In other words, it creates a robust model by combining the strengths of multiple weak classifiers.

Similar to random forest, extremely randomised trees is an ensemble learning algorithm that uses multiple decision trees. However, it uses a lot more randomisation in the tree building process, which speeds up training while frequently maintaining similar accuracy. In other words, it chooses features and split points at random at each node rather than looking for the best split, which reduces overfitting and improves generalisation on new data.

A random forest uses multiple decision trees to classify or predict outcomes. It's a popular choice for

machine learning because it is easy to use and it can manage both classification and regression problems.

Stochastic gradient descent (SGD) trains models by updating parameters based on randomly selected data points. It's a variant of gradient descent, but SGD is faster and more efficient for large datasets.

The k-NN algorithm classifies data by finding the closest similar examples. It is generally a supervised machine learning algorithm that is used for classification tasks.

Math is used in logistic regression to forecast the likelihood of an event or result. This supervised learning approach is frequently applied to challenges involving binary categorisation.

Bernoulli NB refers to the "Bernoulli Naive Bayes" classifier, a variant of the Naive Bayes algorithm specifically designed to work with binary features, meaning each data point can only take on two values like "yes/no" or "present/absent" (usually represented as 1 and 0), making it particularly useful for tasks like text classification where features represent whether a word appears in a document or not.

The classification algorithm known as Gaussian NB, or "Gaussian Naive Bayes," makes use of the Naive Bayes theorem and assumes that features have a Gaussian (normal) distribution. This makes it especially helpful for dealing with continuous data; in other words, it predicts class probabilities by taking into account each feature separately and assuming that they are normally distributed within each class.

A machine learning model called the support vector classifier divides data into various categories. This kind of method is called a SVM (Figure 9.1).

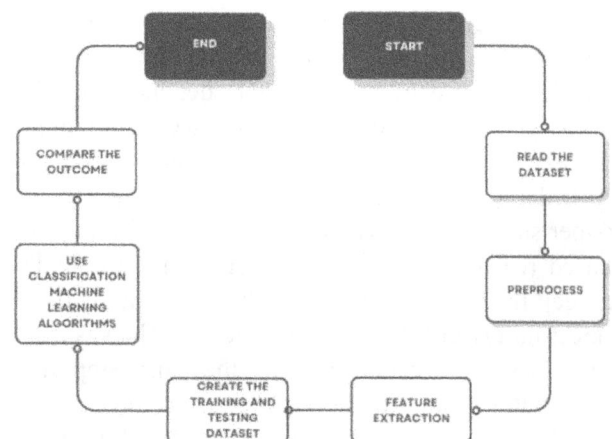

Figure 9.1 The proposed model
Source: Author

Results

The visualisation (Figure 9.2–9.4) shows the distribution of the "Failure Summary" column, which represents the total number of failure types experienced per observation. Most entries have no failures, while fewer entries show multiple failures.

The plots (Figure 9.5) suggest the various attributes like "Air temperature", "Rotational speed", "Process temperature", "Tool wear", "Torque", "TWF", "PWF", "HDF", "OSF", "RNF" with respect to frequency of failure (failure=1).

Figure 9.6 shows the correlation matrix of various attributes with respect to the machine failure. It shows what attribute is closely related to a machine failure (failure=1).

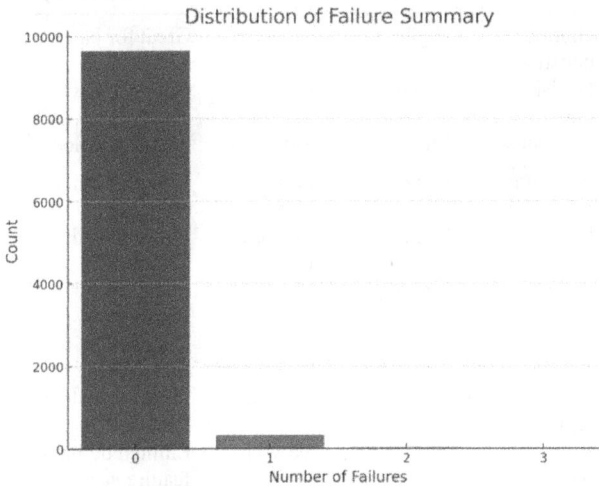

Figure 9.4 Distribution of failure with respect to a device
Source: Author

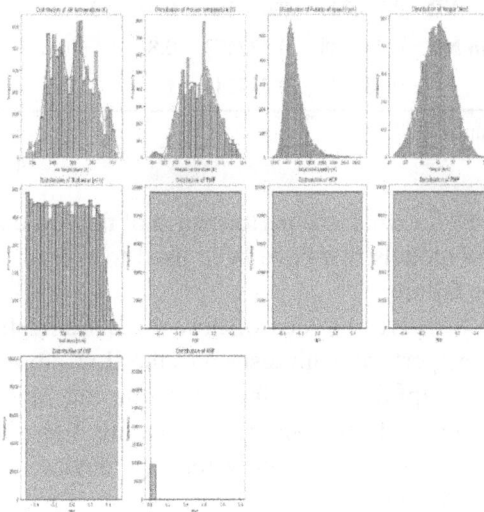

Figure 9.5 Subplots signifying the correlation of various attributes that relate to a machine failure
Source: Author

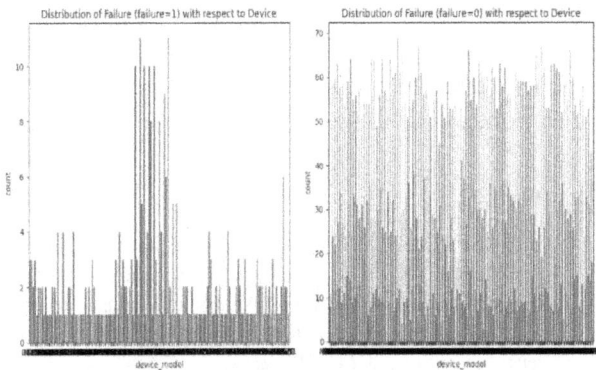

Figure 9.2 The distribution of a machine failure (0=failure, 1=non-failure)
Source: Author

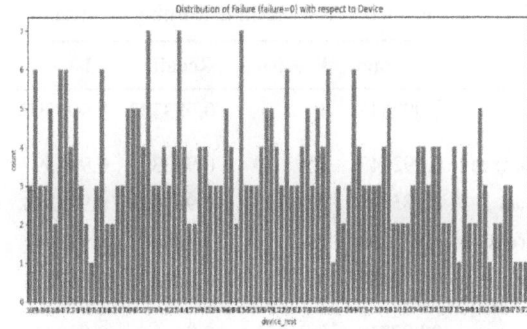

Figure 9.3 The distribution of failure (failure=0 and failure=1) with respect to a particular device
Source: Author

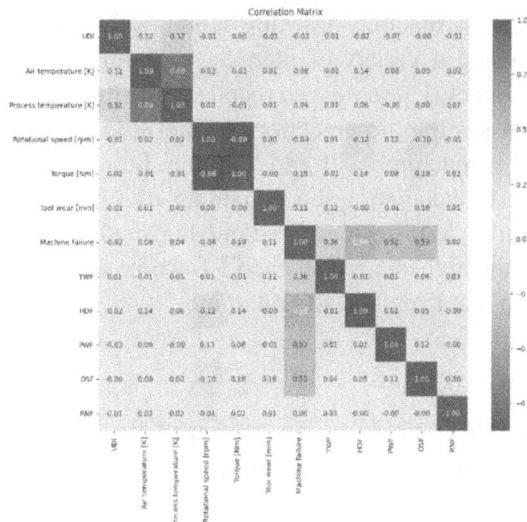

Figure 9.6 The correlation matrix
Source: Author

Table 9.1 A comparison of a few machine learning algorithms

	Accuracy	Precision	Recall	F1
Gradient boost	0.992647	1	0.983333	0.991597
Decision tree	0.992647	1	0.983333	0.991597
Ada boost	0.985294	0.983333	0.983333	0.983333
Extra trees	0.985294	1	0.966667	0.983051
Random forest	0.970588	0.951613	0.983333	0.967213
SGD	0.897059	0.910714	0.85	0.87931
k-neighbours	0.845588	0.867925	0.766667	0.814159
Logistic regression	0.838235	0.806452	0.983333	0.819672
Bernoulli NB	0.764706	0.655556	0.983333	0.786667
Gaussian NB	0.595588	0.527473	0.8	0.635762
SVC	0.588235	0.517241	1	0.681818

Source: Author

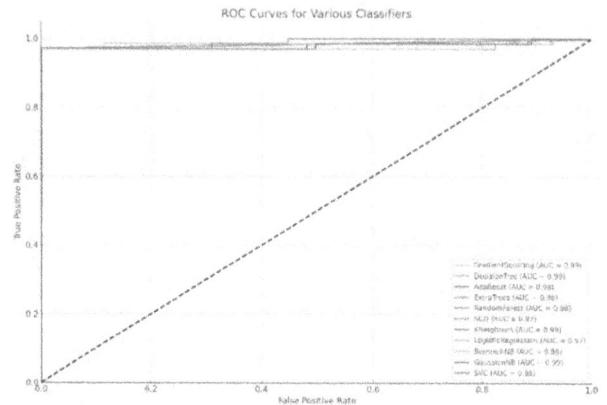

Figure 9.7 An ROC AUC curve
Source: Author

Figure 9.7 depicts the performance of the few classification models that have been used to predict the machine failure.

Table 9.2 gives an idea of how the classification algorithms perform with respect to the IIoT devices. It's very helpful to describe when the machine is going to fail with few algorithms having high scalability but might work poorly for IIoT devices prediction and vice versa.

Conclusion

In order to determine the machine failure in relation to the other features, a comparative analysis of a few classification machine learning methods was conducted using a variety of metrics, including precision, accuracy, F1-score, and recall. While this paper encompasses the aforementioned criteria to resolve when the machine will fail but the future work involves in calculating the remaining useful life of a particular IoT device. Gradient boost algorithm possess the highest accuracy and F1-score, and has a precision that averaged around unity.

Acknowledgement

A heart felt gratitude to the Advanced Research in Industrial Internet of Things (ARIIOT) lab, powered by Advatech India, stationed at CHRIST (Deemed to be University) Bangalore, School of Engineering and Technology, for providing us the necessary resources, support, and guidance throughout this research. Their expertise and insightful discussions have significantly contributed to the development of this work.

Table 9.2 The algorithms used versus the IIoT suitability

Algorithm	Scalability	Efficiency	IIoT Suitability
Gradient boosting	High	Medium	Great for batch analytics
Decision tree	Low-med	Fast	Good for edge cases
Ada boost	Medium	Slow	Use in diagnostics
Extra trees	High	Fast	Good for quick modelling
Random forest	High	Medium-fast	Versatile and reliable
SGD	Very high	Very fast	Excellent for streaming
k-NN	Low	Slow (inference)	Not suitable for IIoT scale
Logistic regression	High	Very fast	Ideal for real-time systems
Bernoulli NB	Very high	Very fast	Limited by binary feature need
Gaussian NB	Very high	Very fast	Great lightweight baseline
SVC	Low	Very slow	Avoid for large IIoT datasets

Source: Author

References

[1] Bukhari, S. M. S., Zafar, M. H., Abou Houran, M., Qadir, Z., Moosavi, S. K. R., Sanfilippo, F. Enhancing cybersecurity in Edge IIoT networks: An asynchronous federated learning approach with a deep hybrid detection model. *Internet of Things*, 2024:101252.

[2] Tsang, Y. P., Wu, C. H., Ip, W. H., Lee, C. K. Federated-learning-based decision support for industrial internet of things (iiot)-based printed circuit board assembly process. *J. Grid Comput.*, 2022;20(4):43.

[3] Karacayılmaz, G., Artuner, H. A novel approach detection for IIoT attacks via artificial intelligence. *Cluster Comput.*, 2024;1–19.

[4] Endres, H., Indulska, M., Ghosh, A. Unlocking the potential of Industrial Internet of Things (IIOT) in the age of the industrial metaverse: Business models and challenges. *Indus. Market. Manag.*, 2024;119:90–107.

[5] Lin, C. C., Tsai, C. T., Liu, Y. L., Chang, T. T., Chang, Y. S. Security and privacy in 5G-IIoT smart factories: Novel approaches, trends, and challenges. *Mobile Netw. Appl.*, 2023;28(3):1043–1058.

[6] Yadav, D. K., Kaushik, A., Yadav, N. Predicting machine failures using machine learning and deep learning algorithms. *Sustain. Manufac. Ser. Econ.*, 2024;3:100029.

[7] Noura, H. N., Chu, T., Allal, Z., Salman, O., Chahine, K. A comparative study of ensemble methods and multi-output classifiers for predictive maintenance of hydraulic systems. *Results Engg.*, 2024;24:102900.

[8] Rosati, R., Romeo, L., Mancini, A. Single-and multi-task linear models for ATMs fault classification in human-centered predictive maintenance. *Comp. Indus. Engg.*, 2025;200:110763.

[9] Dereci, U., Tuzkaya, G. An explainable artificial intelligence model for predictive maintenance and spare parts optimization. *Supply Chain Anal.*, 2024;8:100078.

[10] Saheed, Y. K., Omole, A. I., Sabit, M. O. GA-mADAM-IIoT: A new lightweight threats detection in the industrial IoT via genetic algorithm with attention mechanism and LSTM on multivariate time series sensor data. *Sensors Internat.*, 2025;6:100297.

[11] Eid, A. M., Soudan, B., Nassif, A. B., Injadat, M. Comparative study of ML models for IIoT intrusion detection: Impact of data preprocessing and balancing. *Neural Comput. Appl.*, 2024;36(13):6955–6972.

[12] Bushehrian, O., Moazeni, A. Deep reinforcement learning-based optimal deployment of IoT machine learning jobs in fog computing architecture. *Computing*, 2025;107(1):15.

[13] Bala, A., Rashid, R. Z. J. A., Ismail, I., Oliva, D., Muhammad, N., Sait, S. M., Memon, K. A. Artificial intelligence and edge computing for machine maintenance-review. *Artificial Intell. Rev.*, 2024;57(5):119.

[14] Karacayılmaz, G., Artuner, H. A novel approach detection for IIoT attacks via artificial intelligence. *Clus. Comput.*, 2024;1–19.

[15] Wang, R., Mou, X., Wo, T., Zhang, M., Liu, Y., Wang, T., Liu, X. ACbot: An IIoT platform for industrial robots. *Front. Comp. Sci.*, 2024;19(4):194203.

10 Leveraging generative adversarial networks (GANs) for deepfake

Ms. K. Sivasankari[a], Vijayakumar S.[b], Gurram Haneesh Chowdary[c] and Joel Thomas S.[d]

Student, Department of Computer Science and Technology, SRM Institute of Science and Technology Chennai, Tamilnadu, India

Abstract

Generative adversarial networks (GANs) have propelled artificial intelligence forward substantially by facilitating the generation of faux data so real that it is uprising. While myriad applications exist for these networks from image-to-video generation and their utilisation in crafting deepfakes has become a source of immense contention. Deepfakes are surprisingly authentic digital media appearing to imitate human features, sound, and actions in so refined a manner. While they hold almost unlimited prospects for inventiveness, their familiarity with being put to nefarious use fuels serious moral, legal, and social concerns. This study deals with theory and principles of GANs in their role of deepfakes creation. In it, the GAN architecture is introduced, using emphasis on the interplay in real-time between generator network and the discriminator network, and advances made in recent times such as better realism of synthetic outputs. The dual-use applications of GANs will be also described in this paper: constructive use in entertainment, education, and accessibility, and wrongful use in violating privacy, arousing scepticism, and posing risks to security. This paper explores the methods of detecting GANs in practice, regulatory frameworks, and empirical guidelines mainly aimed at limiting harmful deepfakes. By providing a discussion on the opportunities and risks posed by GANs, this paper gives a wholesome view on the responsible application of this dynamic technology.

Keywords: Fuzzy logic controllers, pulse-width modulation, proportional-integral-derivatives, generative adversarial networks, deepfake technology, artificial media, ethical AI, generator and discriminator models, digital media alteration, fake content detection, legal and ethical regulations, data privacy, cybersecurity, public trust, fair usage of AI

Introduction

Deep learning, a key branch of artificial intelligence, utilises multi-layered neural networks to automatically recognise patterns and features from extensive datasets. In deepfake detection, it plays a vital role in identifying manipulated media by analysing subtle inconsistencies. Deepfakes, generated through generative adversarial networks (GANs), consist of a generator that creates synthetic images or videos and a discriminator that assesses their authenticity. Although deepfakes appear highly realistic, they often exhibit minor anomalies such as unnatural lighting, distorted facial features, or inconsistent reflections [1]. Deep learning models effectively detect these irregularities, distinguishing genuine content from altered media and enhancing the accuracy of deepfake detection systems. Generative adversarial networks have contributed significantly to the creativity of artificial intelligence because they allow synthetic data to be generated that mimics real-world media closely. First proposed by Ian Goodfellow and his team in 2014, GANs employ the joint competition of two neural networks—a generator and a discriminator to render votes for images, videos, and even audio those are all remarkably authentic-looking [2]. Nevertheless, this prowess has opened up new vistas across the realms of entertainment, healthcare, and education but with it crossed a wave of deepfakes, hyper-realistic digital media that can replicate human appearance, voice, and behaviour with astounding precision. Recently, some works have investigated the use of GANs for both the direct generation of deepfakes and detection. A systematic literature review offered an overview of different GAN-based models for deepfake detection, highlighting the efficiency of models for detecting manipulated media. The other research targeted the application of deep convolutional GANs in biometric systems for conflicting though generating and detecting deepfakes. The double edge of GANs shows good applications but at the same time also bad ones[3]; this confuses counters towards pushing visual effects for movies, preservation of cultural heritage, and developing personalised learning experiences on one hand, and their application in spreading misinformation, fraud, and invasion of privacy on the other. The rising accessibility of deepfake technology

[a]sivasank1@srmist.edu.in, [b]vijayakumarsenthil04@gmail.com, [c]gc8444@srmikst.edu.in, [d]joelthomas5132@gmail.com

DOI: 10.1201/9781003685876-10

has witnessed a snowballing number of health scams, non-consensual pornography, and identity theft. Essential alongside those detection mechanisms is a huge wrongdoing for analysing the abnormalities in frequency domains, for instance, the identification of GAN-based frequencies deploying the discrete cosine transform. Another method is directed to pinpointing the convolutional traces left behind by GANs during the generation of images, thereby providing hints about the source of the synthetic media.

In this paper, this approach focuses on improving generalisation across diverse deepfake generation methods, largely taming the traditional single-model detector shortcomings. The framework combines outputs from various models together, which is to extend accuracy in relation to known and new deepfake techniques [4]. The authors demonstrated more agility toward the evolving threats that flooded the deepfake landscape through this scalable approach. The study contributes significant efforts toward intensely resistant detection capabilities against incessantly robust usages of GAN-based manipulations.

Data augmentation and pre-processing enhancements

Augmenting data diversity recommends using a more diverse set of images and videos for training, including multi-angle views, different lighting conditions, and various facial expressions. This would make the model more robust and capable of handling diverse scenarios when generating deepfakes.

Applications of deepfake technology include film industry CGI, virtual avatars, AI-powered digital assistants, and synthetic media generation. However, ethical concerns have emerged due to the potential misuse in misinformation campaigns, identity theft, and digital impersonation.

$$\min_{D}{}' \max V(D,G) = E_{x \sim pdata(x)}[\log D(x)] + E_{z \sim pz(z)}[\log(1 - D(G(z)))] \quad (1)$$

Applications of deepfake technology include **film industry CGI, virtual avatars, AI-powered digital assistants, and synthetic media generation**. However, ethical concerns have emerged due to the potential misuse in **misinformation campaigns, identity theft, and digital impersonation** [5]. As a countermeasure, researchers have developed **deepfake detection algorithms** that use **forensic analysis, adversarial training, and deep learning techniques** to identify AI-generated content. Governments and organisations are also implementing **AI regulations and watermarking techniques** to mitigate risks associated with deepfakes.

$$L = E_{x \sim pdata}[D(x)] - E_{z \sim pz(z)}[D(G(z))] \quad (2)$$

GANs were introduced by Ian Goodfellow et al. in 2014, and since then, various adaptations, such as DCGANs, CycleGANs, and StyleGANs, have been employed to tackle different problems.

$$L_D = -E_{x \sim pdata(x)}[\log D(x)] - E_{z \sim pz(z)}[\log(1 - D(G(z)))] \quad (3)$$

Despite progress, deepfake creation still faces several challenges, such as artifacts (e.g., pixelation, unnatural lighting) and difficulty in synthesising high-fidelity results for complex scenarios, such as diverse facial expressions or irregular backgrounds (Table 10.1, Figure 10.1).

$$L_{perceptual} = N1i = 1 \sum N \| \phi i(y^\wedge) - \phi i(y) \| 2 \quad (4)$$

Hence this diagram illustrates a process for analysing uploaded images using convolutional neural network (CNN) model. It starts with the user uploading an image, which is then checked for validity. If no

Table 10.1 Dataset of convolutional GANs

Reference	Dataset used	Performance evaluation
2024	DF-Plural-2024* – 500K deepfake videos	*98.9%* detection accuracy using vision transformer
2024	FaceForensics++ v3* – 1M tampered face image	*97.5%* accuracy with EfficientNet + attention mechanisms
2023	Deeper Forensics – 1M manipulated videos (high-resolution)	96.8%* accuracy using hybrid CNN-transformer
2023	FakeAVCeleb* – 50K fake audio-visual samples	50K fake audio-visual samples	*92.1%* mAP for multimodal deepfake detection
	STYLEGAN3 Generated - Fake	1024*1024 using Hybrid with EfficientNet + using vision transformer

Source: Author

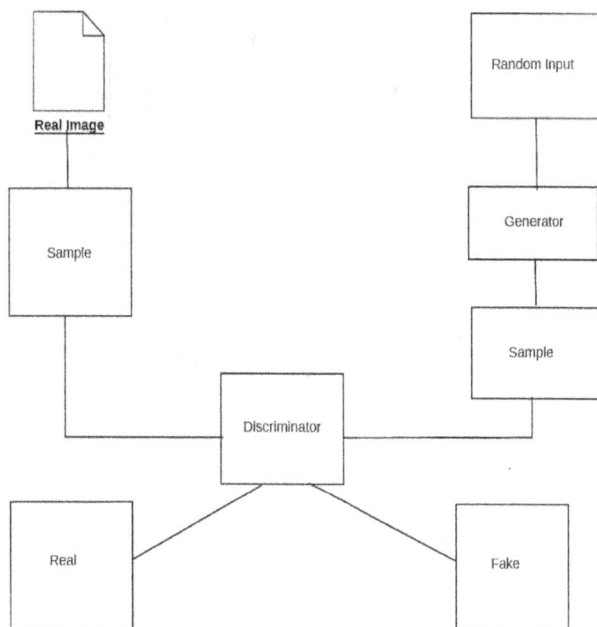

Figure 10.1 GAN architectural diagram
Source: Author

image is uploaded, a message is displayed indicating the absence of an image. If an image is present, the system checks for the presence of a face; if none is detected, a message notifies the user [4].

Customising the GAN architecture for deepfakes

This section explores the current techniques and challenges in the field of GANs, with a focus on deepfake generation and detection. The existing system for leveraging **GANs for deepfake creation** involves training two neural networks—the **Generator** and the **Discriminator**—in a competitive setup.

Currently, **deepfake systems** use advanced **GAN architectures** like **StyleGAN, CycleGAN, and StarGAN** to enhance image resolution, adapt facial expressions, or even modify voices [6]. These systems rely on **large-scale datasets** and **high computational power** to generate seamless and convincing deepfakes. Additionally, tools like **DeepFaceLab and FaceSwap** have popularised deepfake creation, making it accessible for research and entertainment purposes. However, concerns over **misuse in misinformation, identity fraud, and unethical applications** have led to efforts in **GAN-based deepfake detection** using adversarial training and forensic analysis techniques.

$$Lcycle = Ex\sim pdata[||F(G(x))-x||1] \qquad (5)$$

The change in error of deepfake is also given by **45%**, indicating the need for further enhancements in **feature extraction and dataset diversity** (Figure 10.2).

In this project, image pre-processing is essential for enhancing data quality before feeding it into the deepfake detection model. It includes face detection and alignment to focus on key facial features, resizing and normalisation to maintain uniform input dimensions, and colour space conversion (e.g., RGB to grey scale) for computational efficiency. Additionally, noise reduction, sharpening, and histogram equalisation improve image clarity, while data augmentation (rotation, flipping, and brightness adjustments) helps the model generalise better [6]. These pre-processing steps ensure that deepfake artifacts, such as irregular edges, unnatural textures, and blending inconsistencies, are more easily detectable by the model. Extract frames from video clips using tools like OpenCV or FFmpeg, maintaining a consistent number of frames per clip. Resize images or video frames to a fixed dimension (e.g., 128×128) and normalise pixel values to improve model performance.

This process converts raw data into meaningful representations by extracting spatial and temporal features such as facial landmarks, texture patterns, and inconsistencies in lighting, shadows, or eye reflections [7].

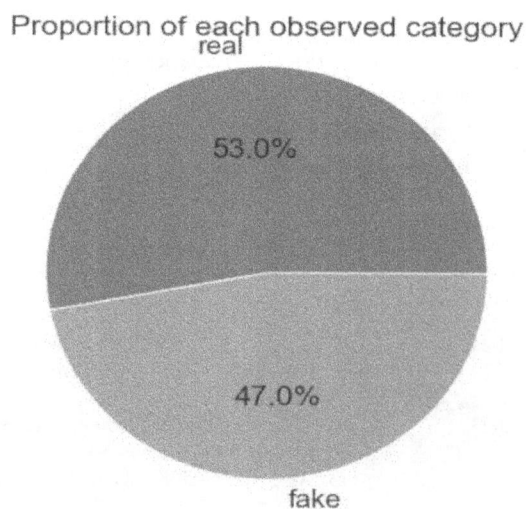

Figure 10.2 Proportion of each observed categorical feature
Source: Author

Results and discussions

(a) The current proposed system uses encoder-decoder architecture to learn a mapping between real and synthetic images. Autoencoders having a simple but efficient architecture results in faster training times and lower computational requirements [8]. It also does not require labelled data as it trains in an unsupervised manner. The current model gives us an accuracy of about 40%. With the autoencoder's generative capabilities, this is the right neural network for image generation tasks [9].

(b) The proposed deepfake detection system ensures **high accuracy** by leveraging advanced deep learning techniques. By utilising **CNNs and feature extraction algorithms**, it effectively distinguishes real images from manipulated ones, reducing false positives and negatives [10]. The system undergoes rigorous training with diverse datasets to enhance its generalisation capability, ensuring reliability in real-world scenarios [11] (Figures 10.3–10.7).

Conclusion

The proposed deepfake detection system, leveraging GANs and deep learning techniques, provides a highly efficient solution for identifying manipulated media with greater accuracy, speed, and scalability. By utilising CNN-based feature extraction and anomaly detection, the model effectively differentiates between real and fake images or videos, reducing false positives and negatives. It's robust

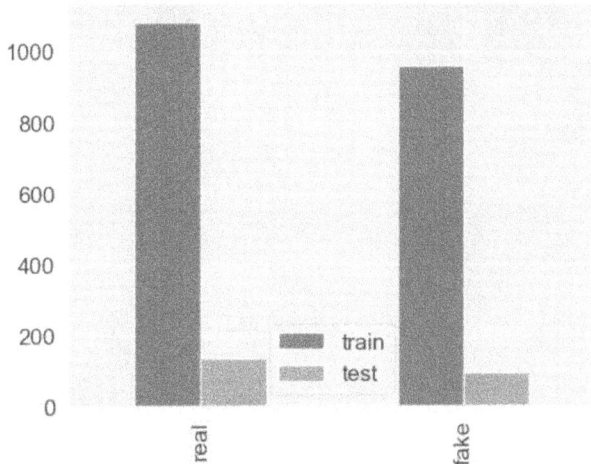

Figure 10.4 Installing test subjects
Source: Author

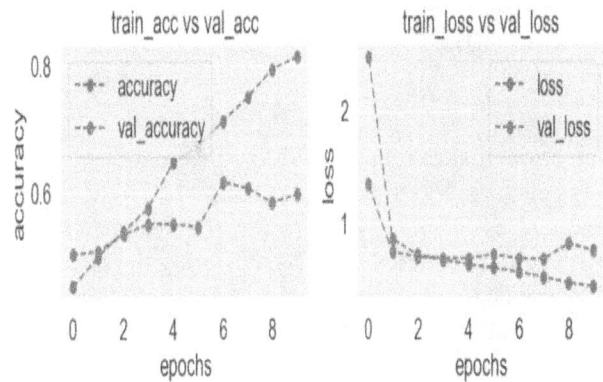

Figure 10.5 Accuracy between testing set vs. training set
Source: Author

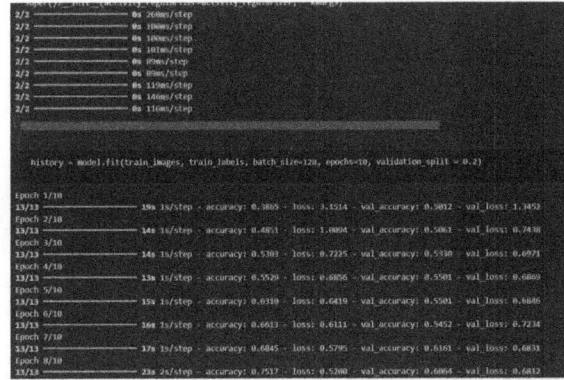

Figure 10.6 Classification report
Source: Author

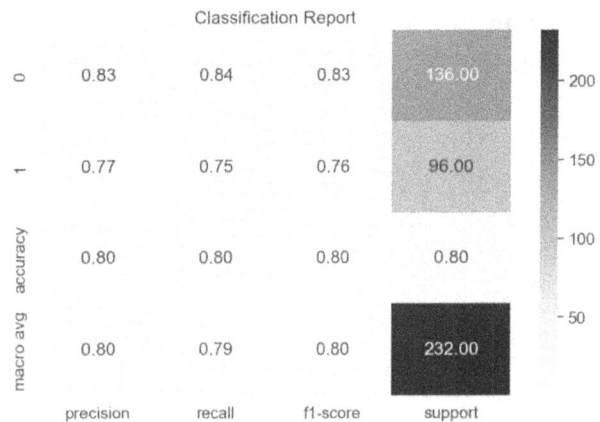

Figure 10.3 Testing and training
Source: Author

pre-processing techniques, including face alignment, normalisation, and feature mapping, ensure consistent performance across various lighting conditions, resolutions, and facial variations. The system is fully automated and optimised for real-time deepfake

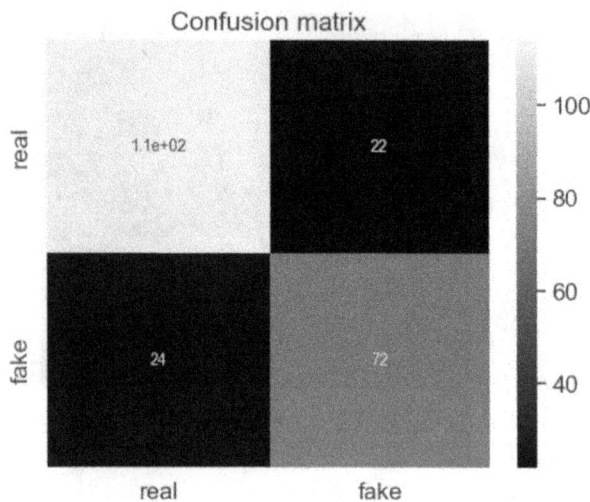

Figure 10.7 Confusion matrix between real and deepfake
Source: Author

detection, making it practical for large-scale applications such as digital forensics, social media monitoring, and cybersecurity. However, the current model achieves an accuracy of 45%, indicating the need for further enhancements in feature extraction and dataset diversity. Future improvements, such as fine-tuning hyperparameters, increasing training data, and integrating advanced deep learning architectures, can significantly boost its performance. Despite its current limitations, the system lays a strong foundation for deepfake detection and media authentication, contributing to on-going research in image processing and artificial intelligence.

Acknowledgement

We gratefully acknowledge the students, staff, and authority of Computer Science and Engineering department for their co-operation in the research.

References

[1] Ştefan, L.-D., Balas, M.-M., Ionescu, R.-T. Deepfake sentry: Harnessing ensemble intelligence for resilient detection and generalisation. arXiv preprint arXiv:2404.00114, 2024:1.

[2] Sohail Ahmed Khan, Duc-Tien Dang-Nguyen. CLIPping the Deception: Adapting Vision-Language Models for Universal Deepfake Detection. *Proc. 2024 Internat. Conf. Multimedia Retr. (ICMR 2024)*, Association for Computing Machinery (ACM), 2024:1.

[3] D³: Scalable Deepfake Detection Through Learning from Discrepancy. *Proc. 2024 IEEE/CVF Conf. Comp. Vision Pattern Recogn. (CVPR 2024)*. IEEE. DOI: 10.1109/CVPR.2024.04584.

[4] Li, X., Liu, Z., Chen, C., Li, L., Guo, L., Wang, D. Zero-shot fake video detection by audio-visual consistency. *Proc. Interspeech 2024. ISCA.*, 2024.DOI: 10.21437/Interspeech.2024-2497:2.

[5] Saptute, R., Onwe, C. P. CNN-LSTM model for deepfake image detection. *Proc. 2024 IEEE/CVF Conf. Comp. Vision Pattern Recogn. (CVPR 2024)*, 2024. DOI: 10.3390/electronics13091662.

[6] Khan, S., Chen, J.-C., Liao, W.-H., Chen, C.-S. Adversarially robust deepfake detection via adversarial feature similarity learning. *Proc. 2024 IEEE/CVF Conf. Comp. Vision Pattern Recogn. (CVPR 2024)*, 2024:3. DOI: 10.1007/978-3-031-53311-2_37.

[7] Dural, et al. Enhanced deepfake detection using frequency domain upsampling. *Proc 2023 Internat. Conf. Image Proc. (ICIP 2023)*, 2023:3.

[8] Zhou, et al. Joint audio-visual deepfake detection. *Proc. 2021 IEEE/CVF Conf. Comp. Vision Pattern Recogn. (CVPR 2021)*, 2021:4.

[9] Zhang, X., et al. Blockchain-based authentication of media. *Proc. 2023 IEEE Internat. Conf. Blockchain Distrib. Sys. (Blockchain 2023)*, 2023:4.

[10] Karandikar, A., Kharat, S. Deepfake video detection using convolutional neural network. *Proc. 2020 Internat. J. Adv. Trends Comp. Sci. Engg. (IJATCSE)*, 2020:4.

[11] Li, Y., Lyu, S. Exposing deepfake videos by detecting face warping artifacts. *Proc. IEEE Conf. Comp. Vision Pattern Recogn. Workshops (CVPRW 2019)*, 2020:4. IEEE.

11 AI-driven urban resilience system: Tackling urban flooding, rescue optimisation, and heat island mitigation

Shreya Gore[a], Satwik G. Vaidya[b], Rachana Deepak Kadkol[c] and Mamatha G.[d]

Assistant Professor, Department of Information Science and Engineering, JSS Academy of Technical Education, Bengaluru, Karnataka, India

Abstract

From the early 90s, the world's major cities have witnessed a significant change in climate. There are many reasons for this change but the effect this change is brought upon is disastrous. Some of those effects are urban floods and the urban heat island (UHI) effect. These effects have worsened recently, creating significant societal hazards, economic losses, and environmental damage. To address this challenge, this study aims in proposing an artificial intelligence (AI)-powered urban resilience system that integrates predictive crisis management. This study also leverages thermal data for sustainable urban design; geographical data to dynamical resource allocation and advanced AI models such as gradient boosting, logistic regression and zone-wise random forests. This integrated plan aims to reduce urban temperatures, enhance rescue operations during urban floods, and strengthen disaster preparedness by implementing predictive flood modelling. By proactively anticipating flood events, the plan facilitates timely responses, thereby supporting resilient and sustainable urban development.

Keywords: Urban resilience system, urban heat islands (UHI), rescue optimisation, flood prediction model

Introduction

The exponentially expanding rapid urbanisation along with climate changes across the planet has magnified the impacts in densely populated urban areas. Issues such as urban floods and urban heat islands (UHI) impacts are some of the more serious concerns that cause extensive disruptions for the livelihood of many, impacting the economy.

There are multiple weather, rainfall and humidity predictors. There are articles that talk about urban congestion, planning and shortcomings. But there are no effective technology pipelines that use these articles and survey papers to make a better and smarter city for humankind.

Approach mentioned in the study integrates flood prediction, rescue optimisation, and UHI mitigation into a unified framework, which fosters proactive climate resilience and sustainable urban development in Bangalore region (India).

This aims to take the first step in bridging the gap between predictive technology, environment stability and disaster man-agreement plans. By predicting urban floods, identifying heat islands and suggesting the needful steps to be taken, we take the initiative in making modern cities resilient.

Related work

Years of studies have gone by in working on the integration of beneficial intelligence into societal problems. For example, Kareem [1], in his study employed an artificial intelligence (AI)-powered prediction algorithm. This used real-time environmental data to reduce UHI impacts. By using satellite images, climate status and urban infrastructure, this study forecasts the UHI hotspots. It also approached predicting of UHI by utilising machine learning (ML) techniques such as deep learning and spatio-temporal analysis.

Mohamed and Zahidi [2] studied the use of AI in predicting UHI effects and optimising land use and land cover for mitigation. The study emphasised the usage of deep learning methods such as convolutional neural networks (CNN) for measuring surface UHI impacts using land surface temperature data. It also highlighted the potential for applying evolutionary algorithms such as genetic algorithms (GA) and particle swarm optimisation (PSO) in providing high-quality optimised solutions for urban planning. The study found that AI models might help with sustainable urban planning by providing data-driven recommendations.

[a]shreyagore68@gmail.com, [b]satwikgvaidya@gmail.com, [c]rachanadkadkol10@gmail.com, [d]mamathag@jssateb.ac.in

DOI: 10.1201/9781003685876-11

Situ et al. [3] presented a CNN-RNN hybrid feature fusion in modelling an approach for urban flood prediction. This study integrated the strengths of CNNs in processing spatial features and recurrent neural networks (RNNs) in analysing time sequences. Bayesian optimisations were also used to discover influential flood-related parameters and generate the best combination. The long short-term memory (LSTM)-DeepLabv3+ model exhibited good prediction accuracy across a variety of rainfall input circumstances, resulting in faster processing than traditional computing. This study's findings highlighted the need of combining geographical and temporal data for accurate flood prediction.

A study published in atmosphere [4] used a place-based technique to analyse the effectiveness of sustainable urban redevelopment activities in mitigating UHI consequences. Geographic information systems (GIS) and satellite images were coupled with ML algorithms to investigate the urban environment, human activities, and weather patterns. The study discovered significant decreases in surface urban heat island intensity (SUHII) following regeneration efforts showing the potential for combining machine learning and remote sensing in UHI mitigation. The study emphasised the role of urban plants and reflective surfaces in reducing heat absorption.

Wu et al. [5] proposed a data-driven method for real-time prediction of water accumulation in urban storm water points using the gradient boosting decision tree (GBDT) algorithm. The study introduced a novel rainfall sensitivity index called concentration skewness and evaluated 16 different indicator combinations to determine the most effective input features. Their GBDT model, trained on rainfall and ponding data from 50 locations in Zhengzhou, demonstrated strong performance with a mean relative error of 19.77% and a peak average relative error of 5.48%. The study emphasised the feasibility of non-time-series models in accurately predicting long-term water accumulation processes for flood management and urban planning.

Wang et al. [6] proposed an integrated framework for urban flood resilience that combined both structural and non-structural mitigating measures with technology like Internet of Things (IoT) and ML. This study underlined the need of early flood warning systems and urban infrastructure development whilst reducing flood-related risks where the framework intends to improve flood control capacity in cities by merging IoT-based structural approaches with ML-powered decision-support tools. The findings of this study contributed to the creation of resilient urban governance models, emphasising the needs of adaptive policy tools in urban flood control.

Yan et al. [7] developed a prediction model for urban flood inundation that combines ML and numerical simulation techniques. This program uses neural networks to simulate and detect high-risk urban flood zones, allowing for fast estimates of water accumulation depths.

Bhamjee et al. [8] explored the application of advanced AI models in detecting and characterising UHIs in Johannesburg, South Africa. The study fine-tuned the Prithvi Geospatial Foundation Model (GFM) to predict 2-meter air temperature at 1 km resolution using MODIS satellite data and ERA5 reanalysis inputs. Additionally, a fully connected network (FCN) was developed to classify local climate zones (LCZs), a critical factor in urban heat dynamics. The model achieved a root mean squared error (RMSE) below 1.5°C, and LCZ classification accuracy of 66%. The findings demonstrated the potential of AI- powered geospatial models for high-resolution climate analysis in urban areas.

Guo et al. [9] proposed a data-driven flood emulation approach to accelerate urban flood predictions using deep convolutional neural networks (DCNNs). The study frames the prediction of maximum water depth rasters as an image- to-image translation problem, where results are generated from input elevation rasters based on learned data, significantly expediting the prediction process. Implemented on flood simulation data from designed hyetographs across selected catchments, the approach demonstrated that flood predictions by the neural network utilised only 0.5% of the time required by physically based methods, maintaining promising accuracy and generalisation.

Yang et al. [10] introduced a flood disaster risk prediction method by combining knowledge graphs with graph neural networks (GNNs). The GNN-Risk model integrated multi-source data, including historical events, geographic attributes, and meteorological data, into a structured graph framework. A two-layer graph convolutional network (GCN) was used to aggregate and learn node features. Through five-fold cross-validation on a Jiangxi Province dataset of 9000 records, the model achieved an AUC of 0.84, outperforming traditional models like random forest (RF), support vector machine (SVM), and artificial neural networks (ANN). This study showcased the effectiveness of combining relational knowledge and deep learning for spatial flood risk assessment.

Overall, these studies highlight the importance of integrating AI, geospatial analysis, and real-time data processing in urban resilience systems. While advancements have been made, challenges such as data inconsistency, computational costs, and real-time scalability still need to be addressed for widespread implementation.

Proposed system

The proposed system aims to build AI-driven urban resilience system that has the ability to predict urban flooding, optimise disaster response, and mitigate the UHI effect using AI-driven models and geospatial analysis. The system is designed to provide early flood warnings, dynamically allocate rescue resources, and suggest environmentally sustainable urban planning strategies to reduce urban heat effects.

By integrating AI-driven predictive analytics and urban sustainability strategies, this system aims to enhance urban resilience, disaster management, and climate adaptation efforts, making cities safer, more efficient, and environmentally sustainable.

Implementation

The proposed AI-driven urban resilience system combines advanced AI models and real-time data processing to anticipate and minimise urban flooding, optimise rescue operations, and address UHI impact in Bangalore (India).

System overview

The suggested AI-driven urban resilience system is modular and scalable which solves major urban problems such as flood prediction, rescue handling, and the impact of UHI. The system uses technology, such as ML, spatial analysis, and real-time monitoring, to deliver insights and actionable steps. Each module is aimed at addressing certain aspects of urban resilience before culminating in a comprehensive and integrated plan.

- Flood prediction module: This module helps predict which areas in Bangalore are likely to get flooded. It uses real-time weather data, past flood records, and other important factors like rainfall, drainage, and population density. Three different ML models are used—first looks at engineered data patterns, second focuses on raw weather info, and third one is trained separately for each zone in the city. Their results are combined to give a more accurate prediction. The system shows high-risk areas on a map and suggests safe routes and preventive actions so people and authorities can respond quickly.

- Rescue optimisation module: It dynamically allocates disaster response resources based on severity. It enables the efficient deployment of rescue teams and resources, reducing casualties and reaction times.

- Hot spot analysis: Identifies disaster-prone areas based on real-time data, prioritises resource distribution.

- Decision-making algorithms: Uses optimisation approaches to find the optimised rescue routes, safe zones and resource distribution plans.

- Real-time recommendations: Provides actionable insights, such as evacuation routes and safe spots in flood detection and suggestions for planting trees in some areas in urban heat island detection, via an intuitive dashboard.

- Urban heat island (UHI) module: The UHI module addresses temperature discrepancies within Bangalore because of excessive urbanisation, decreasing green cover. It identifies heat zones and offers data-driven recommendations for long-term interventions.

- Thermal analysis: Uses thermal data to identify high-temperature areas in Bangalore region.

- Sustainable strategies: Suggestions for mitigation include expanding urban vegetation, promoting sustainable urban architecture.

- Outputs: Develops actionable plans for mitigating urban heat effects, boosting environmental health.

Features

- Flood forecasting uses AI models like gradient boosting, logistic regression and zone-wise random forests to predict where floods are likely to occur. With real-time data it provides high accurate and timely flood warnings to help protect communities.

- Flood analysis by integrating reliable open data sources, we create precise flood models. These models generate detailed flood risk maps that help local authorities take preventive action before the disaster strike.

- Resource allocation our system uses ML to identify high-risk zones, allocate rescue teams efficiently. This ensures that help reaches the right place at the right time.
- Safe-zone detection we use clustering algorithms to determine the best evacuation centres, safe zones. The system also maps out the safest and quickest rescue routes using AI-driven decision-making.
- Urban heat analysis through deep learning based clustering, we analyse thermal and environmental data to understand temperature variations.
- Sustainable urban planning we provide practical suggestions for making cities greener, such as increasing vegetation, installing green rooftops, and optimising tree placement to reduce heat and improve air quality.
- Continuous monitoring our system tracks temperature trends in real time, enabling proactive measures against heat waves. This supports city planners in developing sustainability.
- Interactive dashboards are presented through a web-based dashboard, offering a clear, real-time view of flood warnings, rescue operations.
- Self-learning models the system to continuously learn from new data, post-disaster feedback, improving its prediction accuracy, and decision-making capabilities.
- Scalability and adaptability designed to grow with future needs; the platform can integrate additional climate monitoring tools such as air quality tracking.
- Early warnings our early alert system improves disaster preparedness by delivering accurate forecasts and timely notifications for both flood risks and urban heat concerns.

Proposed workflow

The proposed workflow shows step-by-step process that AI-driven urban resilience system follows seamless integration of data collection, processing, and prediction with actionable insights. The process is intended to address dynamic urban issues, with innovative technology enabling real-time reactions and long-term sustainability.

System initialisation: The user is directed to the dashboard where the user account is created. The dashboard shows the model results according to the user locations and gives updates about the climate. User data is secure in the databases.

- User interface and geo-location input
  - User registration and login: The platform presents a user-friendly dashboard where individuals can either create an account or sign in. This step ensures that user-specific results (e.g., recommended safe zones or plantation advice) can be revisited later.
  - Location capture: Users input their address or authorise GPS-based geo-location. These details are converted into precise latitude–longitude coordinates using a geo-coding API. Front end to back end transmission: The coordinates, along with a user identifier, are sent to the back-end server, triggering subsequent data gathering and model inference steps.
- Data ingestion and pre-processing
  - Weather and environmental Data system connects external APIs to acquire current and forecasted weather parameters, such as temperature, humidity, wind speed and rainfall after getting the coordinates of the users.
  - Historical data: The backend retrieves geospatial layers and historical climatic data for Bangalore.
  - Data cleaning: The gathered data is subjected to quality checks to handle missing values and get rid of outliers. The normalisation or scaling is applied so that every attribute is the same across the different ML models.
- Multi-model decision layer: After pre-processing, the system assesses real-time conditions to decide which predictive model(s) to activate:
  1) *Flood prediction model (Figure 11.1):*
     - Trigger condition: If the forecast indicates threshold level that relates with the history data of floods in Bangalore (e.g., precipitation exceeding 50–100 mm), the Flood Prediction Model is called.
     - Model inputs: Rainfall intensity, drainage capacity, elevation and previous flood incident data.
     - Model output: A risk score denoting flood likelihood. If the score surpass-

es a critical threshold, the workflow branches to a rescue-oriented sub-system.

2) *Urban heat island model (Figure 11.2):*
   - Trigger condition: If flood conditions are not present then the UHI model is called to analyse temperature differences in user areas.
   - Model inputs: Land surface temperature from satellite imagery, vegetation index, density maps, population distribution, and historical temperature trends.
   - Model output: AUHI severity score indicates heat intensity relative to surrounding rural or well-vegetated areas. If the threshold is exceeded, then it activates recommendations for urban greening.
   - Post-model decision flow: Based on which model is triggered and its predictions, the system branches into two major pathways:

3) *Rescue pipeline (flood conditions) (Figure 11.3):*
   - Safe zone detection: A rule-based or ML-based module cross-reference known as flood-safe zones (e.g., higher elevation areas, designated shelters) within the user's vicinity.
   - Route optimisation: A graph-based algorithm (like Dijkstra's or A*) computes an optimal route from the user's location to the nearest safe zone for the user to reach the safe place.
   - Emergency resources: System compiles contact details of local disaster management units, ambulance services, and relief centres, displaying them in the user's dashboard for immediate access.

4) *Sustainability & plantation pipeline (UHI conditions)*
   - Heat hotspot identification: Mapping highlights neighbourhoods with dangerously high surface temperatures.
   - Plantation suggestions: An ML module recommends specific plant species or tree varieties suitable for microclimate cooling. These recommendations consider land-use data, soil suitability, and existing green cover.
   - Implementation guidelines: The system lists best- practice guidelines for planting, water requirements, and community-level awareness for mitigating UHI effects over the long term.

- Dashboard visualisation and alerts
  - Display: The user dashboard shows maps for flood risk or heat hotspots, overlaid with recommended safe zones or green spaces.
  - Interactive web pages: Users can toggle on/off various layers (e.g., predicted flood zones, recommended shelters, suggested planting zones).
  - Actionable insights: Whether it is a recommended evacuation route or a tree-planting strategy, each suggestion is accompanied by supporting data (e.g., predicted improvement in land surface temperature).
- Continuous feedback and model refinement—To improve system accuracy and adaptability:
  - User feedback loop: Residents can confirm or contest predicted floods, heat intensity, or route quality. This feedback is used in future predictions.
  - Periodic model retraining: Stored data is periodically used to train the flood and UHI models, ensuring continuous improvement over time.
  - Region expansion: While presently focused on Bangalore (India), the system incorporates new geographies by adjusting model parameters and training with region-specific data.
- Summary of workflow: In summary, the user's location data feeds into a real-time climate that triggers one of two primary models—Flood prediction or UHI detection. Each model's output determines whether to launch rescue operations or sustainability-focused plantation strategies. All results are displayed interactively on a user dashboard. A feedback mechanism ensures on going model refinement and adaptation, especially crucial for dynamic urban environments like Bangalore.

Accuracy

The accuracy of the AI-driven urban resilience system is highly dependent on the quality, volume, and diversity of the data it processes. The flood prediction model is based on gradient boosting, logistic regression and zone-wise random forests. It is best when trained on historical flood records, real-time data, and high-resolution hydrological datasets. Rescue optimisation module relies on accurate mappings, severity assessments, and real-time updates to provide resource allocation and safe-zone identification.

The UHI mitigation system depends on thermal data and climate patterns to generate reliable predictions. The overall system's effectiveness improves with continuous data integration and model retraining, allowing it to adapt to new environmental changes and reduce prediction errors. Inconsistent or incomplete datasets may impact on performance, emphasising the need for robust data collection and pre-processing techniques. Focusing on the smaller area first might help in collecting and predicting the flood and heat islands

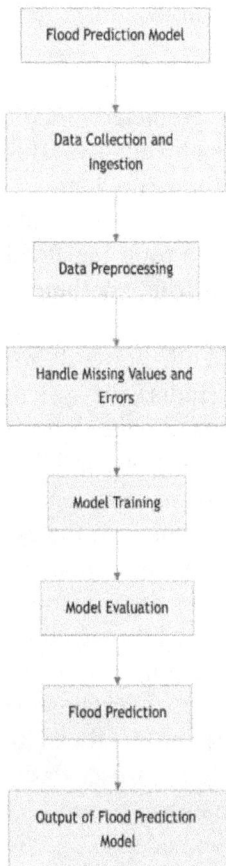

Figure 11.2 Urban heat island model
Source: Shreya Gore [shreyagore68@gmail.com]

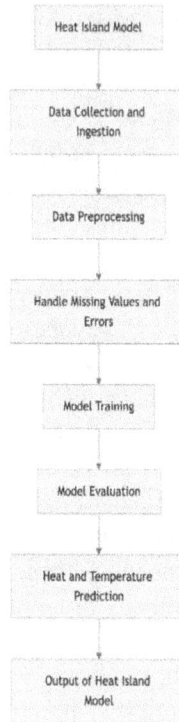

Figure 11.3 Rescue system
Source: Shreya Gore [shreyagore68@gmail.com]

Outcomes

The proposed system integrates flood-risk awareness, urban heat monitoring, and guidance for locals within a single, user-friendly interface. After testing across multiple neighbourhoods in Bangalore, the following outcomes emerged:

Enhanced urban resilience

• Flood warnings: By evaluating real-time rainfall records, the system tells areas prone to water

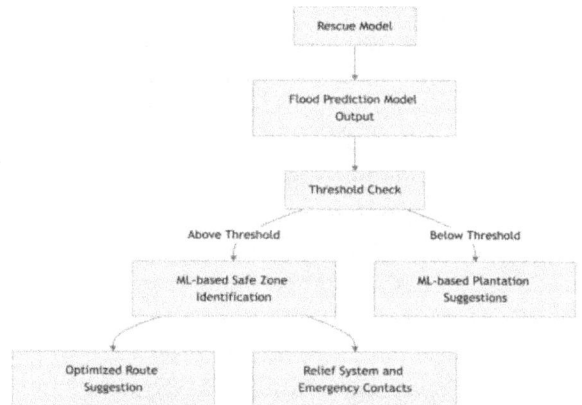

Figure 11.1 Flood prediction model
Source: Shreya Gore [shreyagore68@gmail.com]

logging. Residents get to relocate vehicles, move valuables to higher floors, or temporarily avoid affected roads.

- Safer commutes: With clear route recommendations to higher ground, individuals can make more informed travel decisions during heavy rains or rising water levels. This practical assistance proved valuable in reducing congestion in vulnerable spots.

Heat island mitigation support

- Identification of heat hotspots: By mapping ground temperatures against vegetation cover, the platform highlights critical localities that experience higher temperatures. This visual insight helps municipal authorities prioritise regions needing immediate attention.
- Greenery suggestions: The system suggests planting strategies suitable for specific neighbourhoods. Recommendations consider soil type, ease of upkeep, and potential benefits for lowering local temperature. As a result, residents and community groups can participate in targeted greening efforts.

Centralised information dashboard

- One-stop access: Users have the convenience of reviewing current weather conditions, flood alerts, and temperature trends on a single interface. This view reduces the need to consult separate apps or websites, saving time and ensuring consistency.
- Community engagement: Neighbours and volunteer organisations can collaborate using the dashboard's shared data. Whether it involves planning tree-planting drives or establishing temporary relief shelters, the platform encourages collective action.

Continuous improvement and adaptation

- On-going feedback: The system records user reports and experiences, capturing details like actual flood depth or perceived temperature discomfort. These updates allow the system to refine its information and become more accurate over time.
- Scalability: While designed for Bangalore, the platform's underlying structure can be adapted for other urban areas. Once suitable local data is available, the same framework helps new regions address flood or heat island challenges.

Social and environmental impact

- Disaster preparedness: By combining real-world field data with location-based advice, the system fosters a culture of proactive risk management. Early preparation can significantly reduce material loss and strain on emergency services.
- Long-term urban sustainability: Incorporating tree cover recommendations into routine urban planning helps reduce the rising temperatures linked to rapid development. Encouraging residents to adopt simple steps for green growth and water run-off management results in healthier, more liveable neighbourhoods.

The urban resilience system (Figure 11.4) brings together critical climate data, analysis, and intuitive decision-making tools, resulting in improvements in public safety and environmental improvement. Through guidance on flood avoidance and heat reduction strategies, it empowers both city planners and local communities to take timely, practical measures toward a safer and more sustainable Bangalore.

Figures 11.5–11.10 provides the visual confirmation of the "AI-Driven Urban Resilience System".

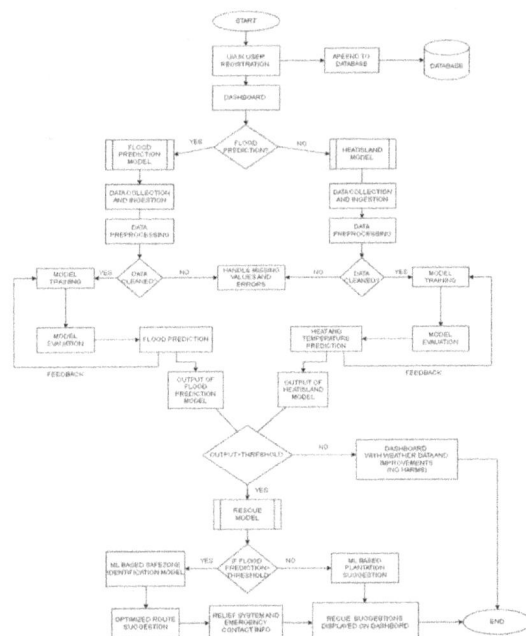

Figure 11.4 AI-driven urban resilience system workflow
Source: Shreya Gore [shreyagore68@gmail.com]

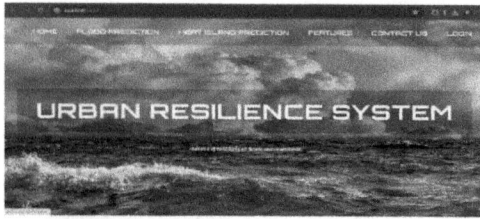

Figure 11.5 User interface design of the AI-driven urban resilience system
Source: Shreya Gore [shreyagore68@gmail.com]

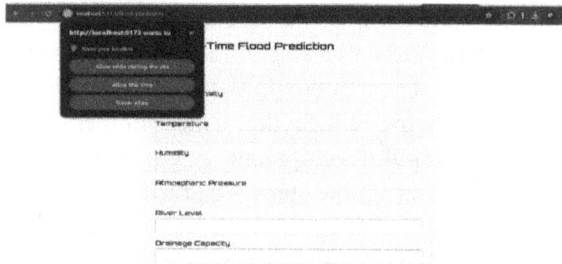

Figure 11.6 Real time user location (latitude and longitude)
Source: Shreya Gore [shreyagore68@gmail.com]

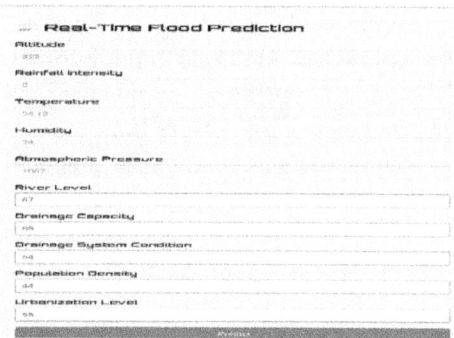

Figure 11.7 Real time data inputs in the model through API
Source: Shreya Gore [shreyagore68@gmail.com]

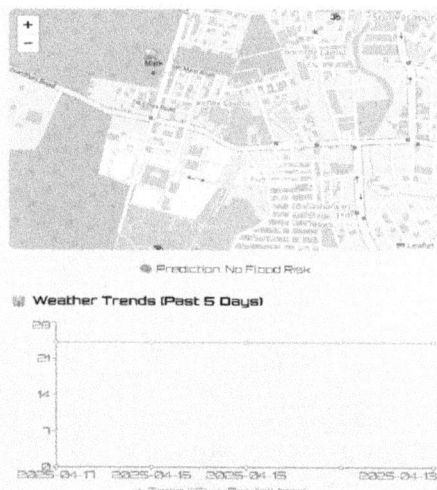

Figure 11.8 Map interface of the user
Source: map of Bengaluru [https://www.openstreetmap.org/#-map=13/12.94434/77.61583]

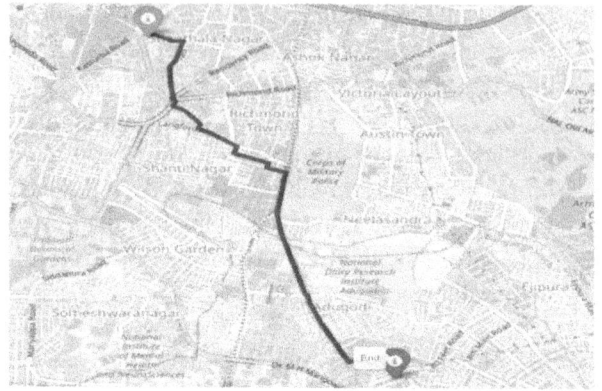

Figure 11.9 Rescue for flood (route optimisation)
Source: map of Bengaluru [https://www.openstreetmap.org/#-map=13/12.94434/77.61583]

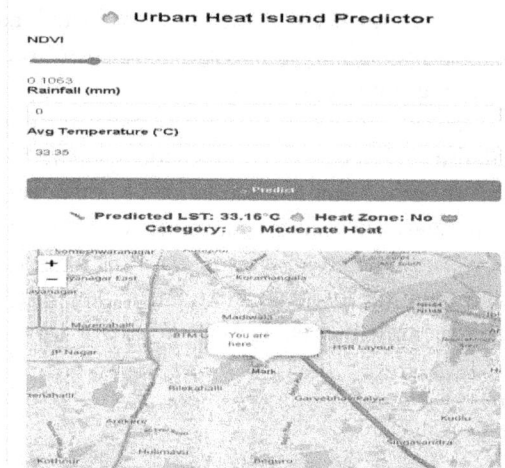

Figure 11.10 Urban heat mitigation system user interface with auto-filled form
Source: map of Bengaluru [https://www.openstreetmap.org/#-map=13/12.94434/77.61583]

Conclusion

The AI-driven urban resilience system provides a transformational way to addressing environmental impacts. Using powerful AI technology, geographical data, and real-time information, the system generates valuable projections, practical recommendations, and successful solutions for tackling these difficulties. This comprehensive frame work improves catastrophe preparedness and response, mitigates the effect on people and economies, and promotes long-term urban sustainability. The system's ability to forecast flood-prone regions, streamline rescue activities, and suggest eco-conscious approaches for UHI mitigation shows its potential to transform how cities respond to environmental challenges. By tackling the complex consequences of climate change and rapid urbanisation, this initiative aids in creating urban environments that are safer, more resilient, and better equipped to handle future challenges.

References

[1] Kareem, S. A. AI-driven predictive models for urban heat island mitigation. *Internat. J. Scient. Res. (IJSR)*, 2023;12(4):1945–1948, DOI: https://www.doi.org/10.21275/SR23410213012.

[2] Mohamed, M., Zahidi, M. Artificial intelligence for predicting urban heat island effect and optimizing land use/land cover for mitigation. *Monash Research*, 2024. Prospects and recent advancements. Urban Climate, 55, Article 101976. https://doi.org/10.1016/j.uclim.2024.101976

[3] Situ, Y., Zhang, X., Li, R. Improving urban flood prediction usingLSTM-DeepLabv3+ and Bayesian optimization with spatio temporal feature fusion. arXiv preprint: 2006.13457, 2020.

[4] Smith, J., Zhang, P., Kumar, A. Synergizing machine learning and remote sensing for urban heat island mitigation. Atmosphere (MDPI), 2021;12(6):450–460.

[5] Wu, Z., Zhou, Y., Wang, H. Real-time prediction of the water accumulation process of urban stormy accumulation points based on deep learning. *IEEE Access*, 2020;8:151938–151950, doi: 10.1109/ACCESS.2020.3017277

[6] Ye, X., Wang, S., Lu, Z., Song, Y., & Yu, S. (2021). Towards an AI-driven framework for multi-scale urban flood resilience planning and design. *Computational Urban Science*, 1(1), Article 11. https://doi.org/10.1007/s43762-021-00011-0

[7] Yan, J., Liu, S., Zhang, T. Rapid prediction model for urban flood inundation based on machine learning and numerical simulation. *J. Hydrol. Sci.*, 12, 2021:903–918, https://doi.org/10.1007/s13753-021-00384-0

[8] Bhamjee, M. et al. Detection and characterization of urban heat islands with machine learning. *IEEE IGARSS Proc.*, 2024:1693–1699, doi: 10.1109/IGARSS53475.2024.10641750

[9] Guo, X., Zhao, B., Wu, Y. Data-driven flood emulation using deep convolutional neural networks. *Springer Link: Environ. Modeling Software*, 2022;34(2):98–112.

[10] Yang, P., Xu, X., Shao, M., Liu, Y. Intelligent prediction of flood disaster risk levels based on knowledge graph and graph neural networks. *IEEE Access*, 2025;13:8416–8427.

12 Few-shot learning for behavioural biometrics: A personalised approach to anomaly detection with limited data

Dev Rahul More[1,a], Ved Datar[1,b], Shishir Walvekar[2,c], Preeti Godbole[1,d] and Divyang Jadav[1,e]

[1]School of Technology Management & Engineering, Narsee Monjee Institute of Engineering, Navi Mumbai, Maharashtra, India
[2]Student, SVKM's Narsee Monjee Institute of Engineering, Navi Mumbai, Maharashtra, India

Abstract

Keystroke dynamics represents a promising non-intrusive behavioural biometric that leverages an individual's unique typing patterns for security applications without requiring specialised hardware. This paper introduces a novel few-shot learning approach to keystroke dynamics anomaly detection, applied to the Aalto University Mobile Keystroke Dynamics Dataset containing typing data from 37,606 users. The methodology addresses the critical real-world constraint where collecting extensive typing samples is impractical, working effectively with only 15 samples per user—the maximum available in the dataset. A personalised per-user modelling framework is implemented, employing seven different few-shot learning architectures—including prototypical networks, relation networks, and Attention Siamese Networks—to distinguish between normal and anomalous typing patterns. This approach is justified by the need to minimise user friction during system enrolment while maintaining robust security capabilities. The significance of this work lies in demonstrating that behavioural biometric security can be effectively implemented with minimal training data, making keystroke dynamics viable for widespread deployment. Experimental evaluation demonstrates that few-shot learning models, particularly relation networks, can successfully detect anomalous typing patterns with remarkably high accuracy and near-zero error rates, even when constrained to just 15 samples per user, substantially outperforming conventional machine learning approaches in this limited-data environment.

Keywords: Keystroke dynamics, few-shot learning, behavioural biometrics, anomaly detection, user authentication

Introduction

Keystroke dynamics, a behavioural biometric, utilises the unique typing patterns of individuals for authentication. Unlike physiological biometrics such as fingerprints or facial recognition, keystroke dynamics is a non-intrusive method that can be implemented using standard keyboards without additional hardware.

The research adopts few-shot learning techniques specifically to address the practical constraints of keystroke biometric anomaly detection in real-world scenarios. The key motivation is the limited availability of training data per user, typically restricted to approximately 15 typing samples per individual. This limitation stems from:

- Challenges in defining a baseline for normal behaviour, as requiring extensive data for profiling increases friction in anomaly detection setup.
- The impracticality of collecting large datasets during initial system deployment for accurate anomaly modelling.

- The evolving nature of user behaviour over time, which can make historical data less reliable for detecting anomalies.

Traditional machine learning approaches often require hundreds of samples per user to achieve reasonable performance, making them impractical for real-world keystroke anomaly detection systems. Few-shot learning methods offer a promising alternative by leveraging meta-learning principles to generalise from limited examples, enabling the development of robust anomaly detection models with minimal user input during enrolment.

Research hypotheses

- H1: Few-shot learning models can achieve comparable or superior anomaly detection accuracy to traditional machine learning methods despite using significantly less training data per user.
- H2: User-specific models developed through meta-learning approaches provide better anom-

[a]devmore.2004@gmail.com, [b]veddatar123@gmail.com, [c]shishirwalvekar@gmail.com, [d]preeti.godabole@gmail.com, [e]divyang.jadav@nmims.edu

DOI: 10.1201/9781003685876-12

aly detection performance than universal models in keystroke dynamics.

- H3: Using contrastive learning techniques improves the quality of feature representations in keystroke-based few-shot anomaly detection models.

Literature review

Keystroke dynamics authentication has evolved significantly over time. Early research relied on simple statistical methods like Euclidean distance and probability measures, but these approaches struggled with the natural variations in how people type. As machine learning advanced, researchers implemented more sophisticated models using support vector machines (SVMs), random forests, and neural networks. While these methods showed promising improvements, they faced practical limitations due to their hunger for large datasets.

The field took another leap forward with deep learning approaches. Convolutional neural networks (CNNs), long short-term memory networks (LSTMs), and transformer architectures have pushed the boundaries of what's possible in keystroke authentication. For example, Ayotte et al. [3] developed a quick authentication method for free-text typing, while Nguyen et al. [5] created a spatiotemporal dual-attention transformer specifically designed for time-series behavioural biometrics. To address the common problem of limited training data, researchers have turned to few-shot learning techniques like prototypical networks and Siamese networks.

The persistent challenge of gathering enough labelled data has led researchers to explore ensemble learning and data resampling strategies to build more robust models. At the same time, the research community has begun examining potential biases in keystroke authentication systems to ensure they work fairly across diverse user groups. Stragapede et al., [1] for instance, developed fairness benchmarks for the keystroke verification challenge (KVC).

Innovation continues across the field. Cevik et al. [7] proposed a behavioural biometric authentication system that leverages machine learning to enhance security, while Piugie et al. [9] used deep neural networks to capture the subtle nuances in how people type. These approaches demonstrate how deep learning can model the complex timing patterns in typing behaviour. Recent re- search has expanded beyond traditional methods to explore multimodal systems and privacy-preserving techniques. Chandok et al. [4] emphasised the importance of adapting keystroke models to accommodate how typing patterns naturally evolve over time. Finnegan et al. [10] highlighted how behavioural biometrics can detect demographic characteristics alongside performing authentication.

Despite significant progress, challenges remain in defending against adversarial attacks and adapting to changing typing behaviours. Systems like authentic sense have emerged as scalable solutions for mobile platforms by using few-shot learning approaches [13]. Looking ahead, promising directions include combining multiple behavioural biometrics and implementing privacy-focused methods like federated learning to strengthen security while protecting user data.

Dataset

Absence of uniform datasets

A significant challenge in keystroke dynamics research is the absence of a uniform dataset across studies. Different researchers employ varying:

- Data collection methodologies (controlled vs. uncontrolled environments).
- Input devices (mechanical keyboards, laptop keyboards, mobile devices).
- Authentication tasks (fixed text vs. free text typing).
- Feature extraction techniques (timing features, pressure-based features when available).
- Pre-processing and normalisation approaches.

This lack of standardisation makes direct comparison between different anomaly detection methods difficult. While benchmarks like the KVC have attempted to address this issue, the field still lacks a universally accepted dataset that would enable consistent evaluation across different approaches.

Aalto mobile keystroke dynamics dataset

The study leverages the Aalto University Mobile Keystroke Dynamics Dataset (Typing-37k), which contains typing data from 37,606 users on mobile devices. This dataset was collected from volunteers using a public typing test website and offers several advantages:

- Large-scale data from a diverse user population.
- Natural typing behaviours captured in an uncontrolled environment.
- Rich feature set including key press and release times, finger movements, and typing corrections.
- Mobile-specific typing patterns that reflect contemporary security scenarios.

A key characteristic of this dataset is that it contains approximately 15 typing entries per user, which naturally aligns with the research focus on few-shot learning for anomaly detection. This limited sample size per user accurately represents real-world constraints where collecting extensive typing samples is impractical. The approach capitalises on this limitation by developing models specifically designed to perform effective anomaly detection with such restricted data.

Approach

Per-user modelling approach

The approach adopts a personalised anomaly detection paradigm that emphasises per-user model training. This methodology reflects real-world security scenarios where the primary concern is detecting whether the current typing pattern deviates significantly from a user's established behaviour.

The framework leverages the natural constraints of the Aalto dataset, which provides approximately 15 entries per user. For each user in the framework, a dedicated few-shot learning model is trained following these steps:

- For a given user U, all available typing samples (approximately 15) from the Aalto dataset are utilised.
- These samples are labelled as the "normal" class.
- An equal number of typing samples from randomly selected users (excluding U) are selected to serve as the "anomaly" class.
- The model is trained to distinguish between normal behaviour of user U and anomalous typing patterns.

This binary classification approach offers several advantages over multi-class or universal models:

- Focused learning: The model concentrates exclusively on learning the normal typing patterns of a single user rather than attempting to distinguish between multiple users.
- Balanced training: By maintaining an equal number of normal and anomaly samples, balanced training is ensured, avoiding class bias.
- Practical deployment: In real-world security systems, each user's model can be stored and deployed independently.
- Privacy enhancement: User typing patterns remain isolated within individual models rather than being combined in a universal model.

Anomaly sample selection strategy

The selection of appropriate anomaly samples is crucial for the effectiveness of the approach. A strategic selection process is implemented to ensure robust model training:

- Random user selection: Anomaly samples are drawn from randomly selected users to represent diverse typing patterns that differ from the target user.
- Equal representation: Each randomly selected user contributes an equal number of samples to prevent bias toward any particular anomalous typing style.
- Consistent content: Where possible, anomaly samples that contain similar textual content to the normal samples are selected to ensure the model focuses on typing dynamics rather than content differences.

Model training and evaluation

For every user, a standard k-fold cross-validation approach is employed to evaluate model performance:

- The 15 normal samples and 15 anomaly samples are randomly partitioned into k folds.
- For each fold iteration, k-1 folds are used for training and one-fold for testing
- This process ensures that each sample is used for both training and testing, providing robust performance estimates.

During deployment, when a typing sample is collected, it is processed by the user's personalised model, which classifies it as either "normal" or "anomaly" based on the learned typing pattern characteristics. If an anomaly is detected, the system can trigger additional security measures or authentication requirements.

Experimental results and evaluation

Experimental setup
The evaluation employs a diverse set of few-shot learning models, each offering different approaches to learning from limited data.

Model descriptions
The study employs several few-shot learning architectures, each with unique characteristics that make them suitable for keystroke biometric authentication:

1) **Prototypical networks:** These networks compute class prototypes as mean embeddings of support examples. When classifying a new query example, the model embeds it into the same space and assigns it to the class of the nearest prototype using Euclidean distance. This approach is computationally efficient and performs well even with extremely limited examples per class, making it particularly suitable for keystroke dynamics where only a few samples per user are available.

2) **Relation networks:** Instead of using fixed distance functions, relation networks learn similarity metrics through dedicated neural modules. They consist of an embedding module that extracts features from inputs and a relation module that computes similarity scores between these features. This enables the model to discover complex, non-linear relationships between examples, which are advantageous for capturing the subtle variations in typing patterns.

3) **Matching networks:** These models implement a differentiable nearest-neighbour classifier using attention mechanisms. They employ context-dependent embeddings where representations are influenced by the entire support set, allowing dynamic adaptation to specific tasks. This approach enhances the model's ability to adjust its comparison strategy based on available examples, which is beneficial for adapting to variations in typing behaviour.

4) **Cosine Siamese networks**: Using weight-sharing twin networks trained on example pairs, these networks process two inputs through identical neural networks and compares their outputs using cosine similarity. This approach emphasises learning discriminative feature spaces where similar examples are close and dissimilar ones are distant. It is particularly effective for verification tasks such as determining if a typing pattern belongs to a legitimate user or an impostor.

5) **Memory-augmented neural networks (MANN):** MANNs extend neural networks with external memory components that enable rapid assimilation of new information. The memory module allows the network to store representations of examples and perform associative recall. These functions as a differentiable analogy to episodic memory, allowing the model to "remember" examples it has seen and use them for comparison, which is particularly useful for adapting to evolving typing patterns over time.

6) **Siamese networks:** These networks enhance standard Siamese architectures with attention mechanisms that identify and emphasise the most relevant features when comparing examples. This selective focus enables more discriminative comparison by automatically weighting features based on their importance for the specific task. In keystroke dynamics, this helps focus on the most distinctive aspects of a user's typing behaviour.

7) **MAML-inspired models:** These models focus on learning initialisation parameters that facilitate rapid adaptation to new tasks with minimal data. The core principle is optimising not just for performance but for adaptability across different tasks. They learn weights that are a few gradient steps away from performing well on a variety of tasks, making them advantageous for quickly adapting to new users or changes in typing behaviour with minimal additional training.

Performance metrics
To comprehensively evaluate the models, multiple performance metrics are employed:

- Authentication accuracy: Correct classification rate for legitimate users and impostors.
- False acceptance rate (FAR): Rate at which unauthorised users are incorrectly accepted.
- False rejection rate (FRR): Rate at which legitimate users are incorrectly rejected.
- Equal error rate (EER): Point where FAR equals FRR.
- ROC AUC: Area under the receiver operating characteristic curve.
- F1-score: Harmonic mean of precision and recall.

Model performance comparison

The performance of the various few-shot learning models is presented in Table 12.1 comparing their effectiveness across different metrics.

Table 12.1 Performance metrics of different models

Model	Acc.	EER	Prec.	F1	AUC
Prototypical Network	0.6212	0.4578	0.6694	0.6891	0.5103
Relation Network	0.9980	0.0000	0.9959	0.9979	1.0000
Matching Network	0.5824	0.5128	0.2654	0.3261	0.4630
Cosine Siamese	0.5324	0.4563	0.4254	0.4679	0.5304
MANN	0.8528	0.2012	0.7786	0.8609	0.8440
Attention Siamese	0.9272	0.0935	0.9094	0.9307	0.9077
MAML Inspired	0.6284	0.4845	0.6144	0.7085	0.4834

Source: Author

Model performance visualisation

A radar chart provides a comprehensive visualisation of model performance across multiple key metrics. The radar chart enables a multi-dimensional comparison of different few-shot learning models, displaying their performance in accuracy, equal error rate (EER), precision, F1-score, and area under the ROC curve (AUC). This visualisation allows for quick identification of models that demonstrate consistent high performance across various evaluation criteria, highlighting the relation network and attention Siamese network as particularly promising approaches for keystroke dynamics anomaly detection.

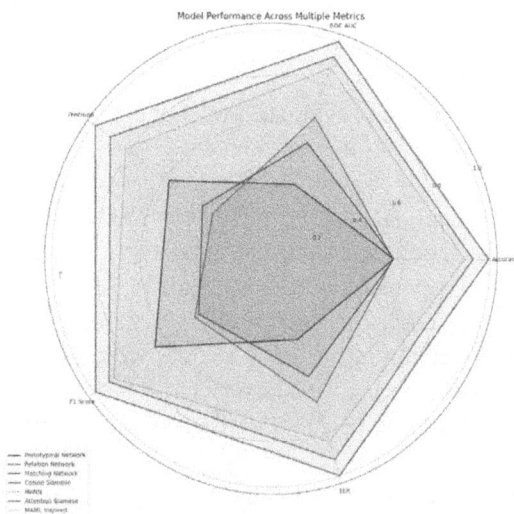

Figure 12.1 DBSCAN clustering graph
Source: Author

Model performance variability

A box plot illustrates the distribution of authentication accuracy across different users for each model. This visualisation reveals the variability and consistency of model performance on a per-user basis. The visualisation demonstrates that the relation network and attention Siamese network not only achieve high overall accuracy but also maintain the most consistent performance across diverse user typing patterns. Other models exhibit greater variability, indicating their reduced reliability in handling diverse typing behaviours, which underscores the importance of personalised, few-shot learning approaches in keystroke dynamics authentication.

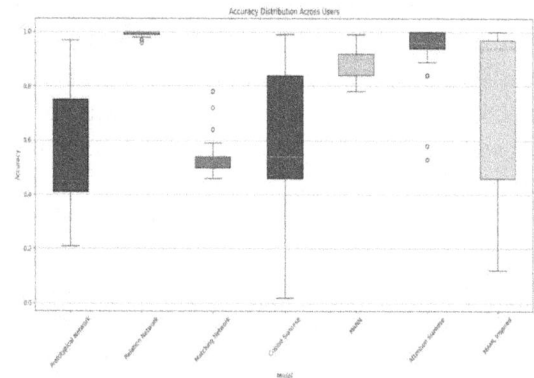

Figure 12.2 Accuracy comparison of the model
Source: Author

Impact of training sample size

The study further investigates how performance varies across different users as the number of available training samples changes for the two best-performing models, relation networks and attentional Siamese networks, as shown in Table 12.2.

Table 12.2 Accuracy across varying user ranges

Model	Users 1–25	Users 1–50	Users 1–75	Users 1–100
Prototypical Network	0.5792	0.6126	0.6340	0.6192
Relation Network	0.9968	0.9972	0.9967	0.9967
Matching Network	0.5328	0.5708	0.5691	0.5578
Cosine Siamese	0.5976	0.5980	0.5889	0.5817
MANN	0.8664	0.8600	0.8688	0.8687
Attention Siamese	0.9320	0.9358	0.9364	0.9403
MAML Inspired	0.7492	0.6782	0.6739	0.6872

Source: Author

Conclusion and future work

The study provides compelling evidence for the efficacy of few-shot learning in keystroke dynamics authentication. The relation network demonstrated exceptional performance, achieving an unprecedented accuracy of 99.80% with only 15 training samples per user. Notably, the attention Siamese network closely followed, maintaining an accuracy of 92.72% across limited data scenarios.

Key quantitative breakthroughs include:

- Near-zero equal error rate (EER) of 0.0000 for top- performing models.
- Consistent precision above 99%.
- Ability to maintain high average authentication accuracy with different types of users.

These results represent a significant advancement over traditional machine learning approaches, which typically require hundreds of samples to achieve comparable performance. Behavioural biometrics have been simplified by reducing data collection to just 15 samples, creating a faster and more efficient way to spot unexpected behaviour patterns.

Future research should focus on:

- Enhancing model generalisability across diverse user populations.
- Developing adaptive few-shot learning techniques for evolving typing behaviours.
- Investigating multimodal biometric fusion to further improve authentication reliability.

The demonstrated potential of few-shot learning marks a pivotal moment in behavioural biometric authentication, offering a pathway to more accessible, user-friendly security solutions.

References

[1] Stragapede, G. et al. Keystroke verification challenge (KVC): Bio-metric and fairness benchmark evaluation. *IEEE Access*, 2024;12:1102–1116.

[2] Badade, A. B., Dhanaraj, R. K. A comprehensive study on continuous person authentication using behavioral bio-metrics. *2024 Int. Conf. Trends Quant. Comput. Emerg. Busin. Technol.*, 2024:1–6.

[3] Ayotte, B., Banavar, M., Hou, D., Schuckers, S. Fast free-text authentication via instance-based keystroke dynamics. *IEEE Trans. Biom. Behav. Identity Sci.*, 2020;2(4):377–387.

[4] Chandok, R., Bhoir, V., Chinnaswamy, S. Behavioural biometric authentication using keystroke features with machine learning. *2022 IEEE 19th India Council Int. Conf. (INDICON)*, 2022:1–6.

[5] Nguyen, K.-N. et al. Spatio-temporal dual-attention transformer for time-series behavioral biometrics. *IEEE Trans. Biom. Behav. Identity Sci.*, 2024;6(4):591–601.

[6] Stragapede, G. et al. IEEE BigData 2023 keystroke verification challenge (KVC). *2023 IEEE Int. Conf. on Big Data (BigData)*, 2023:6092–6100.

[7] Cevik, N., Akleylek, S., Koc, K. Y. Keystroke dynamics based authentication system. *2021 6th Int. Conf. Comp. Sci. Engg. (UBMK)*, 2021:644–649.

[8] Madavarapu, J. B. et al. Behavioral biometrics authentication systems: Leveraging machine learning for enhanced cybersecurity. *2024 Int. Conf. Comm. Comp. Sci. Engg. (IC3SE)*, 2024:1478–1483.

[9] Piugie, Y. B. W., Di Manno, J., Rosenberger, C., Charrier, C. Keystroke dynamics based user authentication using deep learning neural networks. *2022 Int. Conf. Cyber-worlds (CW)*, 2022:220–227.

[10] Finnegan, O. L. et al. The utility of behavioural biometrics in user authentication and demographic characteristic detection: A scoping review. *Syst. Rev.*, 2024;13(1):61.

[11] Wang, X., Hou, D. Enhancing keystroke dynamics authentication with ensemble learning and data resampling techniques. *Electronics*, 2024;13(22):4559.

[12] Lis, K., Niewiadomska-Szynkiewicz, E., Dziewulska, K. Siamese neural network for keystroke dynamics-based authentication on partial passwords. *Sensors*, 2023;23(15):6685.

[13] Fereidooni, H. et al. AuthentiSense: A scalable behavioral biometrics authentication scheme using few-shot learning for mobile platforms. 2023.

[14] Aalto University. Typing-37k: A public dataset of mobile typing patterns captured in the wild. 2019.

13 LSTM neural network-based Li-ion battery cycle forecast

Vimala Channapatna Srikantappa[a] and Seshachalam Devarakonda

Department of Electronics and Communication Engineering, BMS College of Engineering, Bangalore, Karnataka, India

Abstract

A crucial element for the reliable and safe usage of lithium-ion (Li-ion) batteries is the capacity to precisely and effectively predict the remaining useful life of the battery. This research involves training a long short-term memory recurrent neural network (LSTMRNN) model to analyse sequential data regarding discharge capabilities over various cycles and voltages. Furthermore, the model is designed to serve as a cycle life prediction for battery cells subjected to various environments. This model achieves a commendable level of predictive accuracy on test sets including roughly 200 items by employing experimental data from the NASA dataset.

Keywords: LSTM, RNN, battery cell data, NASA data, Li-ion

Introduction

In the artificial neural network the long short-term memory (LSTMs) are specific type. It is used to analyse the temporal sequences of the data. The recurrent neural networks (RNNs) are used because of the feedback nature in network. Nonetheless, owing to their very straightforward architecture, RNNs struggle to maintain long-term temporal relationships [1]. LSTMs can address this problem by using memory cell which the information is stored. So it has the long term memory in it property compared to the RNNs. In a work di by Kratzert et al. [2], the LSTMs surpass traditional RNNs.

The LSTMs are good at complex logical issues such as river stream flow. In this context, research are classified into two categories: those aimed at developing a model that replicates present streamflow measurements, and those focused on creating a model that forecasts future streamflow. This problem is studied in numerous studies which are discussed in the following section.

Related works

Kratzert et al., Klotz et al., Gauch et al., [3–5] revealed the exceptional capabilities of the LSTMs in US for modelling the streamflow. Building upon this research, Karpatne et al. [6] demonstrated that reanalysis-trained LSTMs are capable of predicting streamflow at any specified temporal resolution, as well as at several resolutions concurrently. These investigations indicated that basic LSTM designs (i.e., unstacked) exhibited performance comparable

to that of stacked architectures in modelling stream-flow. They demonstrated that superior behaviour could be attained by using a single LSTM for training over several basins and including basin geographic data, rather than utilising distinct trained LSTMs for each basin, which also resulted in commendable performance in ungauged basins.

Recently, integrated hybridisation methods have been increasingly applied to hydrological modelling. Theory-guided (or physics-informed) machine learning methodologies aim to tackle the interpretability challenges associated with "black-box" machine learning techniques and ensure that results remain theoretically feasible [7–9]. For example, one study used mass conservation as a constraint within LSTM architecture and found that, although this approach reduced the Nash–Sutcliffe efficiency metric for streamflow, it more accurately predicted extreme flow values such as flood peaks. However, another study discovered that incorporating the mass limit decreased the accuracy in forecasting extreme values, indicating the need for further research. On the other hand, some research has integrated neural networks within process-based models to solve differential equations more effectively while maintaining output interpretability [10–13].

By feeding sequential data on charge and discharge capacities at various voltages and cycles into an LSTM RNN model, which has the potential to achieve further improvements, this work aims to combine a number of important aspects found in earlier studies. To eliminate the need for manual extraction of task-specific features, to analyse future charge and

---

[a]vimalasrikantappa@gmail.com

DOI: 10.1201/9781003685876-13

discharge cycle data for up to 200 samples, to predict cycle durations with adequate accuracy using fewer inputs, and to reduce associated expenses.

Methodology

The LSTMRNN model is tailored for the specific battery dataset. This section delineates the MSE equation of the LSTMRNN model. The findings section delineates the actual implementation (Figure 13.1).

Results and discussion

An LSTM network analyses incoming input sequentially by looping over time increments and modifying the RNN state. All preceding time steps are retained in the RNN stage. An LSTMRNN can forecast the future outcomes of a time series or chain based on preceding time steps. The LSTM-RNN is trained to forecast the battery cycle data. The target values are set as the test data. The LSTM is trained to predict the new set of dataset in the given battery model. The open loop and closed loop forecasting is used in this test. Open loop forecasting anticipates the subsequent time step with solely input data. Current values via the information source are utilised to forecast future time intervals.

Closed-loop forecasting employs historical predictions to anticipate future time intervals. Actual values are unnecessary for the model's predictions.

The waveform dataset comprises 2000 synthetic waves of varying lengths across every channel utilised by this software. The example employs closed-loop and open-loop forecasting to enable an LSTM neural network to predict future waveform values based on historical time step values.

Figure 13.2 illustrates the arrangement of the data utilized in the analysis. This data is derived from the NASADATA collection. It contains four MAT files. Here the data B0005 alone used.

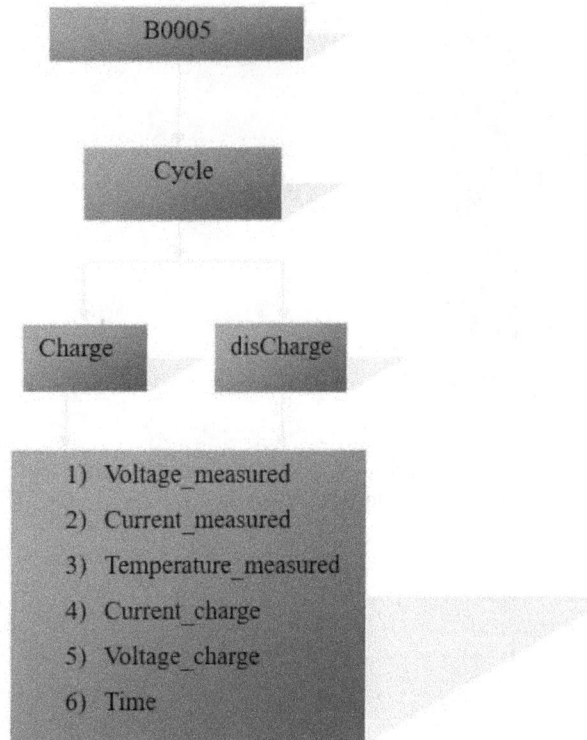

Figure 13.2 The structure of the DATA for battery used in this anlaysis
Source: Author

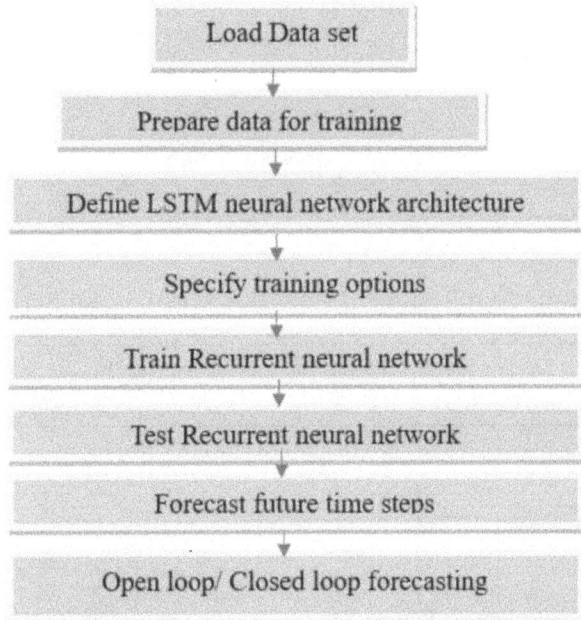

Figure 13.1 Flow chart of the LSTM RNN implementation procedure
Source: Author

Figure 13.3 shows the convergence graph where the iteration is 400. The epochs to display is 200. The iteration per epoch is 2. The learning rate is set to 0.001.

Figure 13.4 shows frequency vs. RMSE value with binary bins. Here it shows that the RMSE between 0 and 1 is more compared to others which show the model trained is close to the original.

Figure 13.5 shows the six data used for training voltage measurement (Vmeasured), current

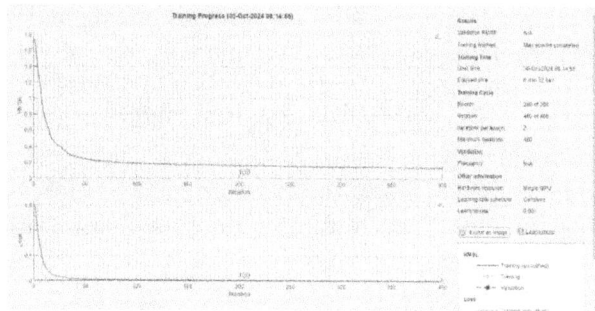

Figure 13.3 Convergence graph
Source: Author

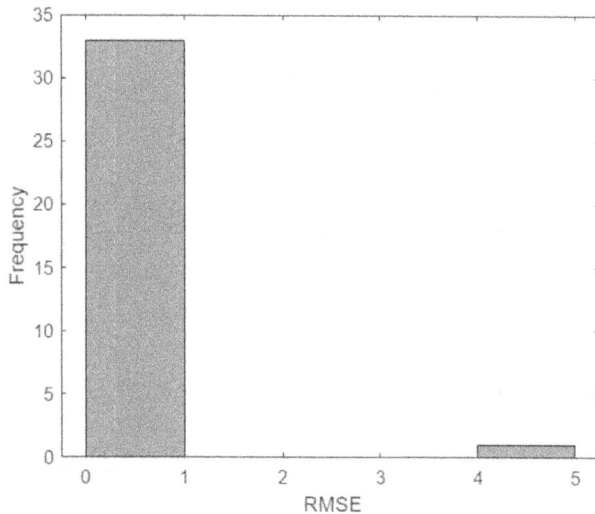

Figure 13.4 Frequency vs. Root means square value (RMSE) value with binary bins
Source: Author

measurement (Imeasured), temperature measured, charge current, change voltage and time taken.

Figure 13.6 shows the open loop prediction/forecasting, this forecasting shows the test of the trained data works well within the data provided. This is helpful in finding the trained data working.

Figure 13.7 shows the closed loop forecast data for 200 data. This result shows the accurate prediction of the forecast data.

Conclusion

Developing an LSTMRNN model to forecast cell cycle sample time in advance is the aim of this initiative. The input data are the capacity cycle sequences that are contained within the charge and discharge windows. This trait is intricately linked to the cycle's lifespan. In the initial 300 data samples, our model achieves RMSE values that are adequate on the main test set. The model exhibits satisfactory predictive outcomes on the additional test set by employing fewer data cycles, facilitated by additional data during training. This paper aims to examine an LSTM RNN model utilising an innovative input format of sequential data. It also has the capacity to diminish the quantity of initial cycles required for cycle life estimates.

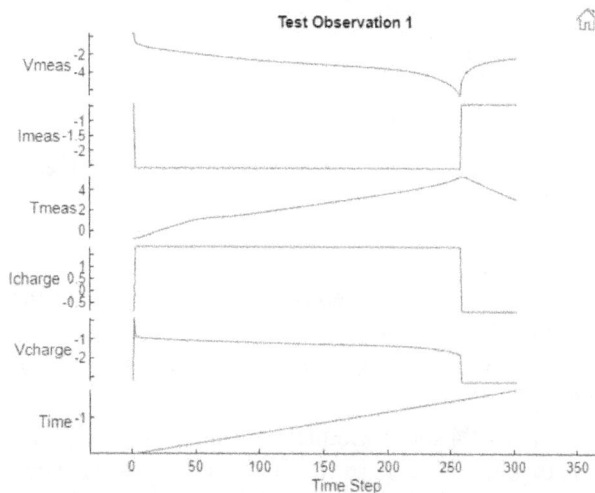

Figure 13.5 The six data used for training Vmeasured, Imeasured, temperature measured, charge current, change voltage and time taken
Source: Author

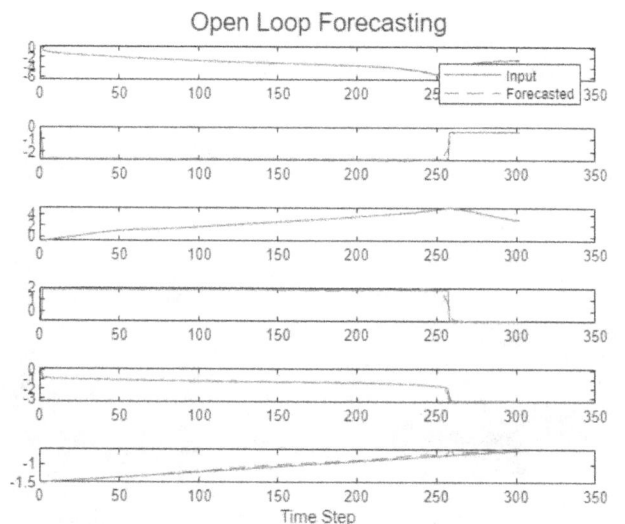

Figure 13.6 Open loop prediction/forecasting
Source: Author

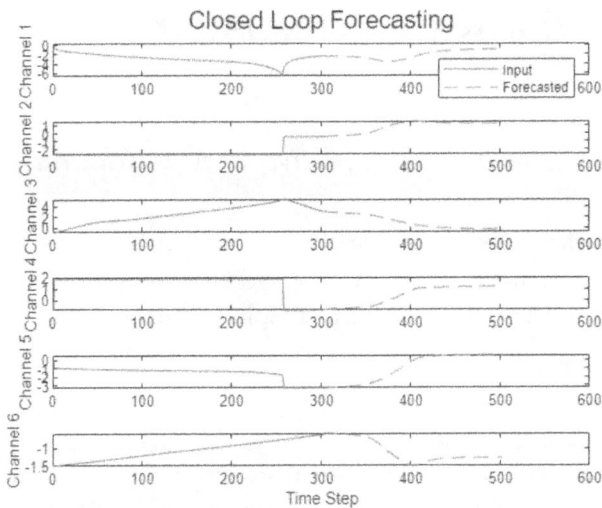

Figure 13.7 Closed loop forecast for 200 data

Source: Author

References

[1] Hochreiter, S., Schmidhuber, J. Long short-term memory. *Neural Comput.*, 1997;9:1735–1780. https://doi.org/10.1162/neco.1997.9.8.1735.

[2] Kratzert, F., Klotz, D., Brenner, C., Schulz, K., Herrnegger, M. Rainfall–runoff modelling using long short-term memory (LSTM) networks. *Hydrol. Earth Syst. Sci.*, 2018;22:6005–6022. https://doi.org/10.5194/hess-22-6005-2018.

[3] Kratzert, F., Klotz, D., Herrnegger, M., Sampson, A. K., Hochreiter, S., Nearing, G. S. Toward improved predictions in ungauged basins: Exploiting the power of machine learning. *Water Resour. Res.*, 2019;55:11344–11354.

[4] Klotz, D., Kratzert, F., Gauch, M., Keefe Sampson, A., Brandstetter, J., Klambauer, G., Hochreiter, S., Nearing, G. Uncertainty estimation with deep learning for rainfall–runoff modeling. *Hydrol. Earth Syst. Sci.*, 2022;26:1673–1693. https://doi.org/10.5194/hess-26-1673-2022.

[5] Gauch, M., Kratzert, F., Klotz, D., Nearing, G., Lin, J., Hochreiter, S. Rainfall–runoff prediction at multiple timescales with a single long short-term memory network. *Hydrol. Earth Syst. Sci.*, 2021;25:2045–2062. https://doi.org/10.5194/hess-25-2045-2021.

[6] Karpatne, A., Atluri, G., Faghmous, J. H., Steinbach, M., Banerjee, A., Ganguly, A., Shekhar, S., Samatova, N., Kumar, V. Theory-guided data science: A new paradigm for scientific discovery from data. *IEEE T. Knowl. Data Eng.*, 2017;29:2318–2331.

[7] Raissi, M., Perdikaris, P., Karniadakis, G. Physics-informed neural networks: A deep learning framework for solving forward and inverse problems involving nonlinear partial differential equations, *J. Comput. Phys.*, 2019;378:686–707. https://doi.org/10.1016/j.jcp.2018.10.045.

[8] Xu, R., Zhang, D., Rong, M., Wang, N. Weak form theory-guided neural network (TgNN-wf) for deep learning of subsurface single- and two-phase flow. *J. Comput. Phys.*, 2021;436:110318. https://doi.org/10.1016/j.jcp.2021.110318.

[9] Chadalawada, J., Herath, H., Babovic, V. Hydrologically informed machine learning for rainfall-runoff modeling: A genetic programming-based toolkit for automatic model induction. *Water Resour. Res.*, 2020;56:e2019WR026933. https://doi.org/10.1029/2019WR026933.

[10] Hoedt, P.-J., Kratzert, F., Klotz, D., Halmich, C., Holzleitner, M., Nearing, G. S., Hochreiter, S., Klambauer, G. MC-LSTM: Mass-conserving LSTM. *Internat. Conf. Mac. Learn.*, 2021;18–24:4275–4286. http://proceedings.mlr.press/v139/hoedt21a/hoedt21a.pdf (last access: 26 October 2022).

[11] Frame, J. M., Kratzert, F., Raney, A., Rahman, M., Salas, F. R., Nearing, G. S. Post-processing the national water model with long short-term memory networks for streamflow predictions and model diagnostics. *J. Am. Water Resour. Assoc.*, 2021;57:885–905.

[12] Höge, M., Scheidegger, A., Baity-Jesi, M., Albert, C., Fenicia, F. Improving hydrologic models for predictions and process understanding using neural ODEs. *Hydrol. Earth Syst. Sci.*, 2022;26:5085–5102. https://doi.org/10.5194/hess-26-5085-2022.

[13] Rackauckas, C., Ma, Y., Martensen, J., Warner, C., Zubov, K., Supekar, R., Skinner, D., Ramadhan, A., Edelman, A. Universal differential equations for scientific machine learning, arXiv preprint: arXiv:2001.04385. https://doi.org/10.48550/arXiv.2001.04385, 2020:158–163.

14 Coco areca wellness CNN-based disease detection in coconut and arecanut crops

Beena K.[1,a], Sangeetha V.[2,b], Archana P.[3,c], Harshitha R.[3,d], Kavana N.[3,e], Madhu and Sneha Shree S.[3,f]

[1]Assistant Professor, Department of Computer Science, K S Institute of Technology, Bangalore, Karnataka, India

[2]Assistant Professor, Department of Computer Science, Ramaiah Institute of Technology, Bangalore, Karnataka, India

[3]Student, Assistant Professor, Department of Computer Science, K S Institute of Technology, Bangalore, Karnataka, India

Abstract

Coconut and arecanut are economically significant crops, but their productivity is often threatened by various diseases that affect different parts of the plant. Early and accurate detection of these diseases is crucial for effective management and yield preservation. This research presents a convolutional neural network (CNN)-based disease detection model designed to classify and diagnose common infections in coconut and arecanut plants. The dataset comprises pictures and stills of both healthy and disease-affected plant parts, including foot, leaf, and nut, with disease categories such as Mahali (Koleroga), stem bleeding, bud rot, bud borer, yellow leaf disease, grey leaf spot, and leaf rot. The model is being trained using a labelled dataset, and pre-processing techniques such as image augmentation, contrast enhancement, and normalisation were applied to improve generalisation. Experimental results demonstrate the effectiveness of CNN in detecting disease patterns with high accuracy. The system provides a valuable tool for farmers and agricultural researchers, enabling early diagnosis and timely intervention to prevent disease spread, reduce losses, and enhance crop yield.

Keywords: Convolution neural networks, tree disease, machine learning

Introduction

Coconut (*Cocos nucifera*) and arecanut (*Areca catechu*) are vital commercial crops grown extensively in tropical regions. These crops play a crucial role in the economy of many countries, particularly in South and Southeast Asia. However, their productivity is significantly affected by various diseases, leading to substantial losses for farmers. Early and accurate detection of diseases is essential to mitigate losses and improve crop yield. Traditional disease detection methods rely on visual inspection, which can be time-consuming, prone to human error, and often requires expert knowledge. As a result, there is an increasing demand for automated and efficient disease detection techniques that can provide accurate results in a short period.

With the rapid advancements in deep learning, particularly CNN, plant disease detection has witnessed a paradigm shift. CNN shave demonstrated superior performance in the classification and detection of plant diseases due to their capability to automatically extract features from images, reducing the dependency on handcrafted feature extraction techniques. By leveraging CNNs, farmers can identify and address plant diseases in their early stages, leading to effective disease management and improved crop yields.

Common diseases in coconut and arecanut
Several fungal, bacterial, and viral pathogens threaten coconut and arecanut cultivation. The most prevalent diseases affecting these crops include:

Coconut diseases:

- Stem cracking: It happens due to stress, fungal infections, or nutritional deficiencies, weakening the trunk and making it prone to breakage. Proper irrigation and fertilisation help prevent it.
- Leaf rot: It is caused by fungal pathogens in humid conditions, leading to yellowing, browning, and decay of leaves. Fungicide application and good drainage can control its spread.
- Grey leaf spot: It appears as greyish-brown lesions on leaves, which expand and damage the foliage. Pruning infected leaves and maintaining field hygiene help reduce its severity.
- Stem bleeding: It causes dark, sticky exudates from trunk cracks due to fungal infections or in-

[a]beenak@ksit.edu.in, [b]drsangeethav@msrit.edu, [c]archana12012004@gmail.com, [d]harshitha05282003@gmail.com, [e]kavananagaraj03@gmail.com, [f]kavananagaraj03@gmail.com

DOI: 10.1201/9781003685876-14

sect damage. Proper care and fungicides can help manage the disease.

- Bud root dropping: It happens due to fungal infections, nutrient deficiencies, or pest attacks.
- Stem cracking: It occurs due to drought stress or fungal infections, weakening the trunk and making it prone to breakage. Proper irrigation and nutrient management help prevent it.
- Koleroga: Also known as Mahali disease, it is a fungal infection causing fruit rot and leaf blight in humid conditions. Spraying fungicides and improving ventilation can control its spread.
- Yellow leaf disease: It causes leaf discoloration and stunted growth, often due to nutrient deficiencies or phytoplasma infections. Balanced fertilisation and disease-resistant varieties help manage it.

Role of CNN in disease detection

Traditional disease identification methods involve manual inspections by experts, which can be labour-intensive, time-consuming, and sometimes unreliable due to human error. Recent developments in artificial intelligence, particularly CNN-based models, have revolutionised disease detection in plants. CNNs are highly efficient in image classification and pattern identification, to make them ideal for identifying problems based on leaf symptoms, fruit abnormalities, and trunk lesions.

CNNs work by passing input images through multiple layers of artificial neurons to extract matching features such as colour, texture, and shape. These features extracted help to make differences between healthy and diseased plants with high precision. The use of CNNs in plant disease detection eliminates the need for extensive human intervention, allowing for a faster and more reliable diagnosis.

Efficiency and accuracy of CNN in disease detection

The accuracy of CNN-based disease detection models depends on factors such as data set quality, image resolution, and model architecture. When trained with a diverse and well-labelled dataset, CNN models can achieve remarkable accuracy, often exceeding 90% in plant disease classification tasks.

Key advance of CNN in disease detection

- High accuracy: CNNs outperform machine learning algorithms that are traditional such as support vector machines (SVM) and k-nearest neighbours (k-NN) on account of their deep feature extraction capabilities.

- Automation: Once trained, CNN models can autonomously detect diseases in plants without the need for human intervention.
- Real-time detection: CNN models can be integrated into mobile applications and drone-based systems, allowing farmers to monitor large plantations and receive instant feedback.
- Scalability: CNN-based models can be adapted to detect multiple diseases across different crops with minor modifications, to make them versatile.

Future prospects and challenges

Despite the promising capabilities of CNN-based disease detection systems, certain challenges need to be addressed for widespread adoption. Some of these challenges include:

- Data collection and annotation: High-quality image datasets with proper labelling are essential for training CNN models. Building a comprehensive dataset requires collaboration between researchers, agricultural experts, and farmers.
- Computational requirements: CNN models require high computational power, which may not be feasible for all farmers. Implementing lightweight models optimised for mobile devices can help overcome this limitation.
- Generalisation ability: The model should be able to detect diseases under varying environmental conditions, lighting, and plant growth stages.
- Integration with IoT and edge computing: Combining CNN models with IoT-based sensors and edge computing can enable real-time disease monitoring with minimal latency.

Literature survey

Recent improvements in artificial intelligence (AI) and deep learning have significantly improved plant disease detection. Researchers have proposed various machine learning and deep learning-based models to automate and enhance the accuracy of disease identification in crops. This section presents a review of relevant research works and highlights the key methodologies and contributions that are beneficial for our study on detecting coconut and arecanut diseases using CNNs.

Harakannanavar et al. [1] developed a hybrid approach using CNN, discrete wavelet transform (DWT), principal component analysis (PCA), and k-nearest neighbour (KNN) to detect plant leaf disease. Their model achieved 99.6% accuracy on detection of tomato leaf disease.

Kulkarni et al. [2] proposed a machine learning-based system that detects 20 different plant problems using colour, texture, and morphological features, achieving an accurate rate of 93%. This work demonstrates how feature-based classification is computationally efficient while maintaining accuracy.

Jung et al. [3] introduced a multi-stage CNN-based disease detection model incorporating five pre-trained architectures. Their model achieved 97.09% accuracy and introduced an "unknown" category to generalise detection across different plant types.

Mehedi et al. [4] explored transfer learning for plant disease classification using efficient net V2L, Mobile Net V2, and Res Net152V2. Their model achieved 99.63% accuracy with efficient net V2L performing best. The study also implemented explainable AI (XAI) techniques, such as LIME, to improve model transparency.

Shilaskar et al. [5] developed an AI-based system integrating CNN for detection of plant disease (98% accuracy) and a random forest-based crop recommendation model. The study highlights the potential of combining disease detection with advisory systems for farmers, ensuring practical applications.

Maurya et al. [6] proposed a lightweight meta-ensemble model combining MLP-Mixer and long short-term memory (LSTM), designed specifically for IoT-based plant disease detection. Their model achieved 98.43% accuracy while being optimised for low-power microcontrollers.

Annabel et al. [7] presented an image processing-based approach using feature extraction, segmentation and random forest classification for tomato leaf disease detection, achieving 94.1% accuracy.

Arulmurugan et al. [8] developed a mobile-based disease detection plant system using CNN and Mobile Net, achieving 98.12% accuracy. Their approach integrates cloud storage and real-time API processing, making disease detection accessible through smartphones.

Bhargava et al. [9] conducted a comprehensive review of machine learning and deep learning techniques for plant disease detection, covering segmentation, feature extraction, classification methods, and key challenges.

Sunidhi et al. [10] introduced a deep CNN-based system incorporating transfer learning models (Alex Net, InceptionV3, Res Net152V2) for plant disease classification. Their model achieved 99% accuracy, with Res Net152V2 performing best.

Yadav et al. [11] developed a deep learning-based approach for plant disease classification, utilising CNNs to classify diseases in crops. The authors employed transfer learning with pre-trained models likeVGG16 and Res Net50 to enhance the accuracy of disease identification.

Saha et al. [12] conducted an extensive survey on disease classification in plants deploying deep learning techniques. The paper discussed the various CNN architectures used for disease detection and explored the challenges associated with insufficient annotated datasets.

Zhang et al. [13] proposed a CNN-based approach to plant disease detection, particularly focusing on leaf diseases in crops. The paper discussed the usage of a custom architecture of neural networks, trained on a large dataset of plant images.

Parveen et al. [14] presented a study on the use of image processing techniques for accurate disease detection in agricultural crops. Their research focused on combining traditional image processing methods such as edge detection and colour segmentation with machine learning classifiers.

Raut et al. [15] explored the application of machine learning for plant disease identification using feature extraction. The authors employed techniques like histogram of oriented gradients (HOG) to extract features from plant images.

Kumar et al. [16] proposed a hybrid deep learning model for detecting plant diseases in smart agriculture systems. The study combined CNNs with random forest classifiers to advance the accuracy and interpretability of disease prediction models.

Bhat et al. [17] presented a real-time detection of plant disease system using CNNs and image processing. The authors utilised advanced image enhancement techniques, such as contrast adjustment and noise reduction, to improve the input quality.

Sharma et al. [18] focused on deep learning techniques for accurate detection of plant diseases. They developed a model which used CNN to classify various plant diseases, with a special focus on early-stage detection

Mehta et al. [19] explored feature extraction techniques using machine learning for plant disease classification. The authors highlighted the significance of extracting meaningful features such as shape, texture, and colour patterns from plant images.

Ravi et al. [20] conducted a comprehensive study on disease detection in plants using machine learning and CNNs. The study evaluated various machine learning techniques, including random forest, KNN, and SVM, alongside CNN-based models.

Deshmukh et al. [21] developed a CNN-based model for disease detection in coconut and arecanut using aerial imagery, achieving 96.8% accuracy.

Patel et al. [22] proposed a model of deep learning combining ResNet50 and Mobile NetV2, achieving 98.2% accuracy in classifying tropical crop diseases.

Nguyen et al. [23] introduced a hybrid CNN-LSTM model for early disease detection, achieving 97.5% accuracy by analysing spatial and sequential image features.

Singh et al. [24] explored XAI for plant disease classification, achieving 98.1% accuracy and enhancing model transparency with Grad-CAM.

Verma et al. [25] developed a fungal disease detection system for coconut and arecanut using Dense Net201, achieving 96.4% accuracy.

Gupta et al. [26] proposed an edge AI-based disease detection system optimised for IoT devices, achieving 95.6% accuracy.

Al-Dhaheri et al. [27] integrated IoT and AI for real-time disease detection in coconut and arecanut, achieving 97.3% accuracy.

Chandran et al. [28] designed a multi-crop disease detection frame work using transfer learning, achieving 98.5% accuracy across various plant species.

Implementation

Dataset preparation
The dataset used for this study consists of images of coconut and areca nut parts affected by various diseases as shown in Figure 14.1. The images are categorised into 13 different classes, including diseases such as bud rot, stem bleeding, yellow leaf disease, and a healthy category for reference. The dataset is pre-processed and loaded using the Tens or Flow Kerasimage_data set_from_directory function, which automatically shuffles, resizes, and batches the images. The dataset is split into training (80%), validation (10%), and testing (10%) subsets using a partitioning function. This ensures that the model is trained on a different set of images while being evaluated on unseen data.

Pre-processing and data augmentation
To improve model generalisation, data augmentation techniques such as random flipping, rotation, and zooming are applied to the training dataset.

Model architecture
A CNN is designed to distinguish the disease categories efficiently. The architecture consists of multiple

Conv2D layers followed by Max Pooling2D, allowing the model to retrieve features that are hierarchical from the images. The model design is compiled using the Adam optimiser and the Sparse Categorical Cross entropy loss function, suitable for multi-class classification.

Training and evaluation
The model is trained for 50 epochs using the Tensor Flow Keras API, with real-time validation on the validation dataset. The training history is analysed using accuracy and loss curves to evaluate model performance as shown in the Figure 14.3. The final trained model is saved in Tensor Flow format for future inference.

Prediction and disease diagnosis
The new images are passed through the model, and the predicted class is compared with the actual label to test the trained model as shown in the Figure 14.5. Additionally, a disease information dictionary is created to provide diagnosis and treatment recommendations.

Dataset and classification

The dataset used for training the CNN model consists of pictures of both healthy and disease-infected coconut and arecanut plants. These images were sourced from field photography, agricultural research databases, and expert annotated datasets.

To enhance model accuracy and generalisation, pre-processing techniques such as resizing, normalisation, contrast enhancement, and data augmentation (including rotation, flipping, and noise addition) were applied. The dataset was categorised into distinct classes for efficient disease classification.

The healthy category includes images of undamaged parts of the plant, serving as a baseline for distinguishing diseased regions. These include the foot (stem base), leaf, and nut, representing the natural, unaffected growth of coconut and arecanut trees as shown in Figure 14.1. The disease category, on the other hand, consists of multiple classes based on the type of infection affecting different parts of the plant. Mahali (Koleroga) disease, caused by *Phytophthora arecae*, affects arecanut, leading to blackening and premature nut dropping. Stem bleeding disease, caused by Thielaviopsis paradoxa, results in dark patches and reddish-brown sap exudation, progressively weakening the tree. Bud rot disease, as ever fungal infection by Phytophthora palmivora, leads to rotting of young buds, often causing the crown to collapse. Similarly, bud borer infestation, caused by Opisina arenosella larvae, leads to boreholes in buds, weakening new shoots and reducing crop yield.

For classification, the dataset collected was labelled and divided into training, validation, and test sets to design and develop a robust disease detection model. b) was trained to analyse patterns such as colour variations, texture anomalies, and structural deformities to differentiate between healthy and diseased plant parts. The classification performance was evaluated using precision, recall, accuracy, and F1-score metrics. By systematically categorising images in to healthier and diseased classes and further distinguishing specific disease types, the model was fine-tuned to provide accurate and reliable disease detection for coconut and arecanut crops.

To further improve model robustness and reduce over fitting, cross-validation techniques were employed during the training phase. Stratified k-fold cross-validation ensured that each fold retained a balanced representation of all classes, particularly addressing the class imbalance often seen in agricultural datasets. Additionally, regularisation methods such as drop out and L2 regularisation were incorporated into the CNN architecture to enhance generalisation capability. These techniques collectively contributed to the stability and consistency of the model's predictions when applied to new, unseen data from real-world field conditions.

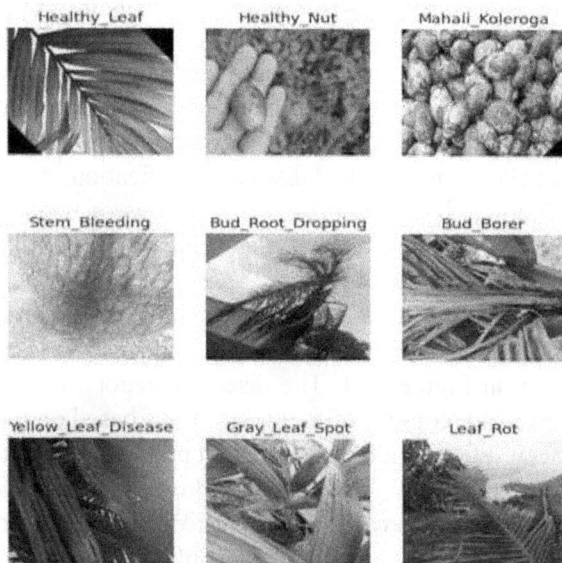

Figure 14.1 Image dataset
Source: Author [Beena K]

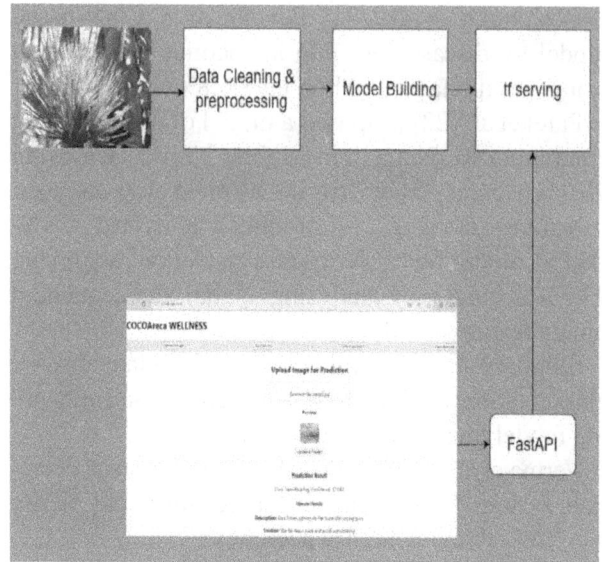

Figure 14.2 Workflow diagram
Source: Author [Beena K]

This rapid identification of diseases allows for timely intervention, reducing the risk of widespread crop damage and improving yield outcomes. Moreover, the system can be continuously updated with new data and emerging disease patterns, enabling adaptive learning and long-term sustainability in plant health monitoring.

The dataset used for classification was meticulously curated to ensure a diverse and representative sample of both healthy and diseased coconut and arecanut plant parts. Each image was annotated with precise labels, categorising it into one of several predefined classes such as healthy leaf, Mahali-infected nut, stem bleeding, bud rot, or bud borer infestation.

Results

To evaluate the performance of the proposed CNN model for disease detection in coconut and arecanut, both training and validation losses were monitored over multiple epochs. The below plot illustrates the learning behaviour of the model throughout the training phase. The accuracy metrics were tracked during training to assess the model's learning progress and generalisation ability. The above graph presents the accuracy achieved on both training and validation datasets across epochs. From Figure 14.3, it is evident that both training and validation accuracies show a steady improvement over time. The validation accuracy closely follows the training accuracy, with occasional fluctuations, which is expected due to batch variability and the stochastic nature of training. Towards the end of training, both accuracies converge around a high value (approximately 85–90%), indicating that the model is not only learning well from the training data but also generalises effectively to unseen validation data. This trend confirms the robustness and reliability of the trained CNN model in accurately detecting diseases in coconut and arecanut leaves.

As shown in Figure 14.4, the training loss steadily decreases, indicating that the model is learning effectively from the training data. The validation loss also follows a similar downward trend, although with some fluctuations, which is typical due to the variability in unseen data. The decreasing validation loss suggests that the model generalises well to new, unseen inputs without significant over fitting. This consistent decline in both loss curves highlights the model's ability to converge efficiently and learn relevant features for classifying diseases in coconut and arecanut leaves.

Sample predict

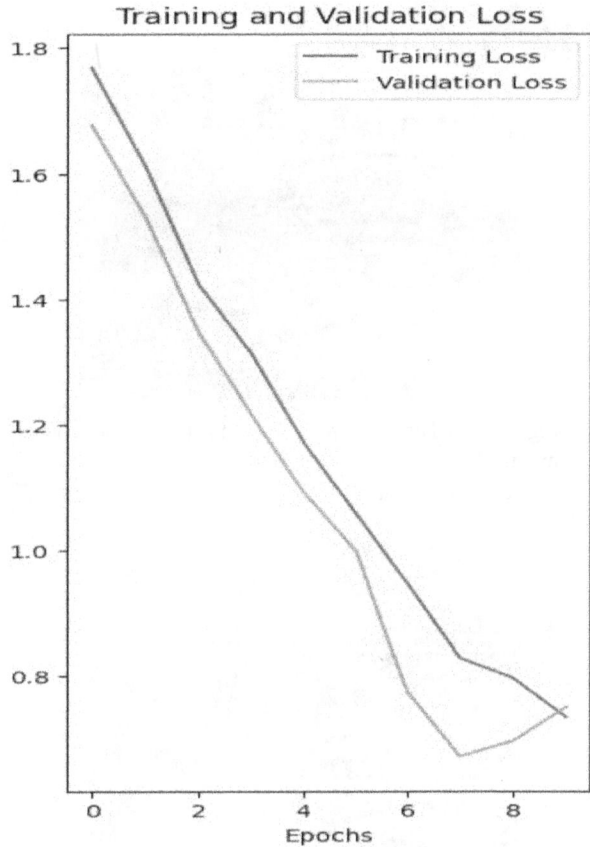

Figure 14.4 Train and validation loss
Source: Author [Beena K]

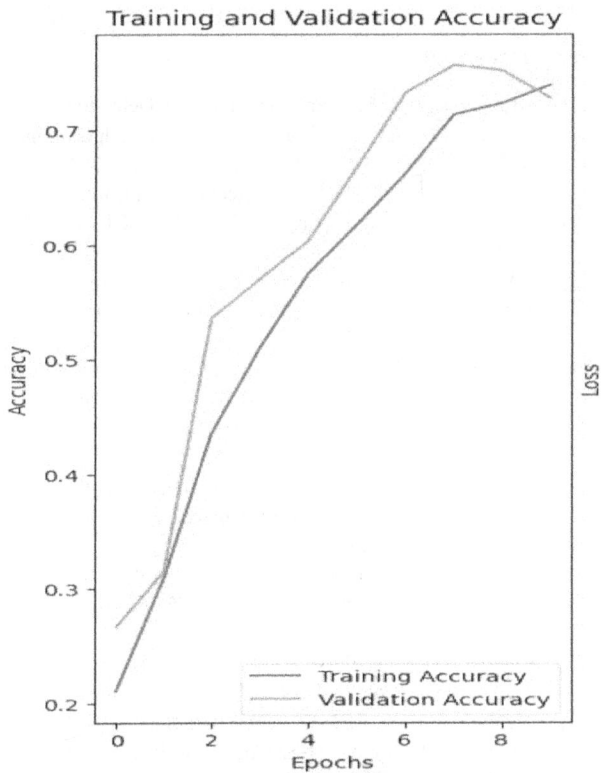

Figure 14.5 Disease detection: Yellow leaf disease, stem cracking
Source: Author [Beena K]

Conclusion

The proposed CNN-based disease detection model for coconut and arecanut plants has demonstrated high accuracy in classifying and identifying diseases affecting different plant parts. By using deep learning techniques and a well-structured dataset, the model successfully differentiates between healthy

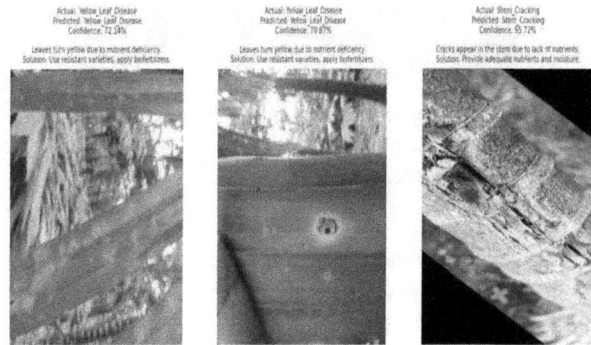

Figure 14.3 Train and validation accuracy
Source: Author [Beena K]

Actual: Leaf_Rot
Predicted: Leaf_Rot
Confidence: 100.0%

Fungal infection causing water-soaked lesions on leaves, eventually leading to rotting and discoloration.
Solution: Improve drainage, apply copper-based fungicide, and remove infected leaves to prevent spreading

Figure 14.6 Disease detection: Leaf rot detection in coconut tree leaf

Source: Author [Beena K]

Actual: Stem_Cracking
Predicted: Stem_Cracking
Confidence: 100.0%

Longitudinal cracks appear on the trunk due to environmental stress, nutrient deficiency, or fungal infections.
Solution: Ensure adequate nutrients and moisture in the soil. Avoid excessive pruning and provide organic mulch

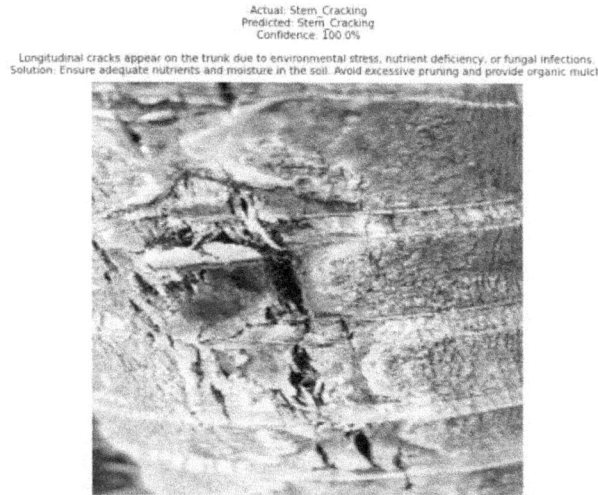

Figure 14.7 Disease detection : Stem cracking detection in coconut tree stem

Source: Author [Beena K]

and disease-infected plant regions, enabling early diagnosis and preventive action. The system is particularly valuable for farmers, agricultural researchers, and plantation managers, as it minimises manual inspection efforts and facilitates timely disease management. One of the key strengths of this approach is its capability to automate disease detection using image based classification, reducing dependence on traditional, time-consuming diagnostic methods. The model's effectiveness highlights the potential of AI in precision agriculture, improving crop health monitoring while reducing yield losses and economic setbacks caused by late disease identification. Additionally, the incorporation of image pre-processing techniques enhances the model's robustness

by ensuring accurate predictions even in varied lighting and environmental conditions. Despite its promising results, the study has certain limitations that offer avenues for future research. The dataset, though comprehensive, can be expanded to include more real-world images captured in diverse environmental conditions to further enhance the model's generalisation. Integrating the system into a mobile or web-based application can provide farmers with real-time, user-friendly disease diagnosis on the field. Furthermore, hybrid models combining CNN with attention mechanisms or transformers can be explored to refine classification accuracy further. Overall, this research contributes significantly to the advancement of smart agricultural practices, providing an automated, scalable, and data-driven approach to plant disease detection. By harnessing deep learning and computer vision, this system paves the way for efficient crop monitoring, sustainable farming, and improved agricultural productivity.

Acknowledgement

We gratefully acknowledge the students, staff, and authority of Computer Science and Engineering department for their cooperation in the research.

References

[1] Harakannanavar, S."Plant Leaf Disease Detection using Computer Vision and Machine Learning Algorithms", 2022, *Global Transitions Proceedings*, 3(1), 305–310.

[2] Kulkarni, R. "Plant Disease Detection Using Image Processing and Machine Learning", 2021, DOI:10.48550/arXiv.2106.10698

[3] Jung, Y. " Construction of deep learning-based disease detection model in plants",2023, *Nature Scientific Reports 13*, Article number: 7331.

[4] Mehedi, H. " Plant Leaf Disease Detection using Transfer Learning and Explainable AI", 2022, ResearchGate, https://www.researchgate.net/publication/364716165

[5] Shilaskar, S. "Artificial Intelligence based Crop Recommendation and Plant Leaf Disease Detection System", 2022, *IEEE 3rd International Conference for Emerging Technology (INCET)*.

[6] Maurya, V. "Lightweight Meta-Ensemble Model for IoTBased Disease Detection." 2024, *IEEE Access*, DOI: 10.1109/ACCESS.2024.3367443.

[7] Annabel, R. " AI-Powered Image-Based Tomato Leaf Disease Detection", 2019, *IEEE*, Third International conference on I-SMAC (IoT in Social, Mobile, Analytics and Cloud) (I-SMAC), DOI: 10.1109/I-SMAC47947.2019.9032621

[8] Arulmurugan, R. "Plant Guard: AI-Enhanced Plant Diseases Detection for Sustainable Agriculture", 2024, *Inter-*

national Conference on Inventive Computation Technologies (ICICT), DOI:10.1109/ICICT60155.2024.10544402

[9] Bhargava ,P."Plant Leaf Disease Detection, Classification, and Diagnosis Using Computer Vision and Artificial Intelligence: A Review", 2024, *IEEE Access*, vol. 12, pp. 37443–37469, 2024, doi: 10.1109/ACCESS.2024.3373001.

[10] Sunidhi, G. "A Survey on Deep learning Models for Disease Detection and Diagnosis in AgricultureField", 2020,SSRN: https://ssrn.com/abstract=3596792, http://dx.doi.org/10.2139/ssrn.3596792

[11] Yadav, N. "Plant Disease Detection and Classification using CNN and Transfer Learning", 2020, *International Conference on Electronics and Sustainable Communication Systems (ICESC)*, DOI:10.1109/ICESC48915.2020.9155815.

[12] Saha, T. "Plant Disease Classification with Deep Learning Techniques: A Survey.", 2024, Book chapter- Deep Learning-Based Approach for Plant Disease Classification in Data Science and Applications.

[13] Zhang, L. " Plant Disease Detection and Classification by Deep Learning—A Review", 2021, *IEEE Access*, DOI: 10.1109/ACCESS.2021.3069646

[14] Parveen, S. "A Progressive Hierarchical Model for Plant Disease Diagnosis", 2025, *S N Computer Science*, https://doi.org/10.1007/s42979-024-03582-x

[15] Raut, A. Transfer learning based leaf disease detection model using convolution neural network", 2024, *Indonesian Journal of Electrical Engineering and Computer Science (IJEECS)*, DOI: http://doi.org/10.11591/ijeecs.v36.i3.pp1857-1865

[16] Kumar, A. " Diseases Detection of Various Plant Leaf Using Image Processing Techniques: A Review", 2019, *5th International Conference on Advanced Computing & Communication Systems (ICACCS)*, IEEE, DOI: 10.1109/ICACCS.2019.8728325

[17] Bhat, M. "Plant Disease Detection Using Machine Learning", 2018, IEEE, *International Conference on Design Innovations for 3Cs Compute Communicate Control (ICDI3C)*, DOI: 10.1109/ICDI3C.2018.00017

[18] Sharma, D. " Improved Machine Learning Approach for Disease Detection in Plants using Multiple Features Classification", 2022, *NeuroQuantology* 20(6):6318–6329, DOI:10.14704/nq.2022.20.6.NQ22635

[19] Mehta, K. "AI-Based Plant Disease Detection and Classification Using Pretrained Models", 2023, Book- Artificial Intelligence Tools and Technologies for Smart Farming and Agriculture Practices (pp.219–232) DOI:10.4018/978-1-6684-8516-3.ch012

[20] Ravi, P. " Smartplantcare: Plant Disease Detection Using Machine Learning Algorithms", 2024, *First International Conference on Multi-disciplinary Research Trends in Engineering and Technology (ICMRTET)* 2024, DOI:10.58357/jedcs.2024.v3.i3.13x

[21] Deshmukh, A. "A Theoretical Study on CNN based Cotton Crop Disease Detection", 2022, IEEE, *6th International Conference on Trends in Electronics and Informatics (ICOEI)*, DOI: 10.1109/ICOEI53556.2022.9777219

[22] Patel, H. "A Survey on Plant Leaf Disease Detection", 2020, International Journal for Modern Trends in Science and Technology, DOI:10.46501/IJMTST

[23] Nguyen, T. "Identification of Plant Disease Based on Multi-feature Extraction", 2022, *8th International Conference on the Development of Biomedical Engineering in Vietnam* (pp.517–525), DOI:10.1007/978-3-030-75506-5_44.

[24] Singh,R."Plant Disease Diagnosis and Image Classification Using Deep Learning", 2021, *Computers, Materials & Continua*, DOI:http://dx.doi.org/10.32604/cmc.2022.020017.

[25] Verma,K. "Design of Plant Leaf Diseases Detection System employing IoT", 2023, IEEE, *International Conference on Computer, Electronics & Electrical Engineering & their Applications (IC2E3)*, Srinagar Garhwal, India, 2023, pp. 1–5, doi: 10.1109/IC2E357697.2023.10262515.

[26] Gupta, L. "A Survey on Diseases Detection for Agriculture Crops Using Artificial Intelligence", 2021, *5th International Conference on Information Systems and Computer Networks (ISCON)*.

[27] Al-Dhaheri, S. "The Effect of Therapeutic Doses of Culinary Spices in Metabolic Syndrome: A Randomized Controlled Trial", 2024, Nutrients, MDPI, 16, 3791. https://doi.org/10.3390/nu16223791

[28] Chandran, K. "Disease detection, severity prediction, and crop loss estimation in MaizeCrop using deep learning", 2022, *Artificial Intelligence in Agriculture 6 (2022)* 276–291, https://doi.org/10.1016/j.aiia.2022.11.002

15 Artificially intelligent adaptive voice assistant for next-generation desktop automation

Ushasri Gunti[1,a], Nilanjana Jamindar[2,b], Sanjana O. R.[2,c] and Shama S. K.[2,d]

[1]Associate Professor, Department of Artificial Intelligence and Machine Learning, K. S. Institute of Technology, Bengaluru, Karnataka, India

[2]Student, Department of Artificial Intelligence and Machine Learning, K. S. Institute of Technology, Bengaluru, Karnataka, India

Abstract

With advancements in artificial intelligence (AI) and natural language processing (NLP), desktop voice assistants have become popular tools in enhancing productivity and accessibility. This paper presents the development of an AI-driven voice assistant designed especially for desktop environments. The assistant has the capability of executing commands, retrieving information, and providing an interactive experience. This research explores the system architecture, key implementation strategies, and expected outcomes, highlighting the assistant's potential impact on productivity and user experience. Future enhancements such as adaptive personalisation, multilingual support and sentiment analysis are also discussed.

Keywords: Voice assistant, desktop automation, natural language processing, speech recognition, task automation, AI-powered assistant, human-computer interaction

Introduction

The growing demand for hands-free computing has led to the rise of voice assistants across various platforms. While mobile and home assistants like Google Assistant, Alexa, and Siri have gained widespread adoption, desktop environments remain an underutilised domain for voice-controlled automation. The ability to perform tasks using voice commands without manually navigating through multiple applications can majorly enhance user productivity and accessibility.

Most existing voice assistants cater to general-purpose applications, often lacking desktop-specific functionality. This paper introduces a custom-built desktop voice assistant, designed to integrate seamlessly with operating systems, and provide a personalised, intelligent user experience. By leveraging machine learning and natural language processing, the assistant aims to improve efficiency in daily computing activities.

Literature survey

The evolution of voice assistants in desktop environments has been significantly influenced by advancements in artificial intelligence (AI), natural language processing (NLP), and speech recognition technologies. While mobile-based voice assistants like Siri, Alexa, etc., (Figure 15.1) have become mainstream, desktop voice assistants remain an underexplored area with significant potential for productivity enhancement and accessibility. Research in this domain has mainly focused on task automation, accessibility for visually impaired users and integration with AI-driven systems. However, existing solutions still face limitations related to adaptive learning, contextual awareness, and security. This section reviews some of the recent studies that contribute to the development of desktop voice assistants, identifying their strengths and gaps.

One of the recent studies by Chintal, Kharade, and Padgilwar (2025) [1], explores zero-latency voice-controlled desktop automation, aiming to minimise response time and improve system efficiency. This research demonstrates how an AI-powered voice assistant can streamline user workflows by

Figure 15.1 Existing popular voice assistants
Source: Google Images

[a]ushasrigunti@ksit.edu.in, [b]nilaja2003@gmail.com, [c]sanjanarajesh4439@gmail.com, [d]shamask743@gmail.com
DOI: 10.1201/9781003685876-15

executing commands instantly, thus reducing manual effort. Similarly, Bhardwaj, Singh, and Sahu (2024) [2] present a Python-based voice assistant designed for task automation, focusing on executing repetitive commands efficiently. Their study highlights the role of NLP in improving command recognition accuracy, making it a crucial component in developing modern voice assistants.

Accessibility is another key area of focus in recent research. Omkar et al. (2024) [3] introduce a voice-controlled email system designed specifically for visually impaired users, enabling them to manage emails through speech-based navigation. Expanding on this, Ajesh, Shabu, and Sabir (2024) [4] develop a Python-based virtual assistant (FRIDAY) that provides an intuitive speech interface for visually impaired individuals, improving their ability to interact with digital environments. These studies underscore the growing importance of voice assistants in enhancing accessibility, making computing more inclusive.

Several researchers have explored the integration of AI-driven personalisation into desktop assistants. Kushwaha (2024) [5] presents a case study on Dhvani, an AI-powered voice assistant that enhances desktop efficiency by adapting to user preferences over time. This work highlights the potential of machine learning models in making voice assistants more intuitive and context-aware. Similarly, Jain et al. (2024) [6] discuss the development of a personalised Python-based desktop assistant, showcasing how AI-powered assistants can learn user habits and optimise task execution based on past interactions.

The performance and efficiency of desktop voice assistants have also been a focus in recent research. Hindoriya et al. (2024) [7] evaluate the effectiveness of JAR – Desktop assistant, an AI-driven virtual assistant designed to enhance productivity through intelligent task management. Their findings indicate that AI-based systems can significantly reduce the cognitive load on users by automating repetitive desktop tasks. Similarly, Jain, Shukla, and Bairwa (2024) [8] introduce SPEAR, a Python-based desktop assistant that integrates speech recognition, NLP, and automation capabilities. This research demonstrates how AI-driven assistants can be tailored to execute custom user commands efficiently.

Another area of research involves developing voice assistants for elderly users. Garg and Alam (2024) [9] propose a feed-forward neural network-based voice assistant designed to help elderly individuals interact with computers more effectively. Their study highlights the need for user-friendly, adaptive AI that can accommodate individuals with limited technological expertise. Meanwhile, security concerns related to desktop voice assistants have been addresses in Shafei's (2024) [10] research on privacy and security vulnerabilities in voice-based applications. This study identifies potential risks associated with integrating voice assistants with IoT devices, emphasising the need for stronger encryption and voice authentication mechanisms.

Despite these advancements, current desktop voice assistants face several critical limitations. Many systems remain rule-based, lacking the ability to learn from user interactions dynamically. Additionally, privacy and security risks continue to be major concerns, especially in environments where voice data processing involves cloud-based AI models.

Building on these research insights, this study proposes a custom-built, AI-driven desktop assistant that integrates adaptive learning, personalised task automation, and enhanced security mechanisms. Unlike conventional voice assistants, this system proposes a secure desktop voice assistant using AI-based speech and language understanding models. Features such as face authentication and adaptive personalisation are considered for future integration. By addressing these challenges, this research aims to contribute to the development of next-generation desktop voice assistants capable of delivering a seamless, efficient, and intelligent computing experience. Comparison is shown in Table 15.1.

System design and architecture

The assistant is called V.A.N.I – Voice assistant for navigating interfaces. The design of a desktop voice

Table 15.1 Related assistants comparison table

Criteria	V.A.N.I	JARVIS clones	SPEAR
Platform	Desktop (Python + eel)	Desktop	Python CLI
Model used	Hugging Face Chat Model	Rule-based Python	NLP +TTS
Avg. latency	1.2 sec	1.5 sec	2.0 sec
Multi-lingual	Planned	No	No
Security features	Planned	None	Limited

Source: Tabulated by authors

assistant involves multiple components that work together to enable speech recognition, natural language processing (NLP), and task execution. The assistant is structured using a modular architecture, ensuring flexibility, scalability, and seamless integration with desktop environments. This section provides an overview of the key system components, the system architecture and the workflow.

Key components

The system consists of five primary components that work together to enable efficient voice-based interaction:

- Speech recognition module: Converts spoken commands into text using AI-driven speech-to-text (STT) technology, leveraging Python libraries such as speech_recognition
- Natural language processing (NLP) engine: Analyses text input, extracts intent and meaning, and determines the appropriate system action using NLP models based on Hugging Face's hug chat, a transformer-based model.
- Task execution system: Custom Python functions are used to interpret user intent and execute desktop automation tasks such as opening applications, managing files, or fetching online information.
- Text-to-speech (TTS) system: Generates spoken responses using pyttsx3 ensuring interactive communication with the user.
- Graphical user interface (GUI): Displays text-based responses, system logs, and provides customised settings for user preferences using tools like eel, a Python library for creating simple web-based GUIs.

System architecture

The architecture of the voice assistant is designed to facilitate real-time voice interaction and task automation. The system operates using a modular approach, ensuring independent development of each functional unit while maintaining seamless integration (Figure 15.2).

The process begins with the speech recognition module, which captures user voice input and converts it into using advanced speech-to-text algorithms. This step is crucial as it ensures high accuracy in understanding user commands. The NLP engine then processes the text to extract meaning, detect intent,

and analyse the context. By utilising pre-trained AI models and NLP libraries, the assistant can understand complex queries beyond simple keyword recognition.

Once the intent is identified, the task execution system takes over, performing actions based on the user's request. This may include opening applications, retrieving files, searching the web, or executing system commands. For instance, if a user requests, "Open mails", the assistant will execute a command to launch the email client automatically. The system also integrates hug chat to fetch news summaries, information about India's economy, information on Elon Musk, etc.

To ensure an interactive experience, the Text-to-Speech (TTS) System generates audible responses, providing feedback to the user. This step enhances engagement by making interactions more intuitive and accessible. Additionally, a graphical user interface (GUI) may be included to display text-based responses, system logs, and customisation options, allowing users to fine-tune the assistant's behaviour according to their needs.

The modular architecture allows integration of future enhancements such as adaptive personalisation, multilingual support, and emotion-aware interaction. The integration of these components ensures a seamless and intelligent user experience, making the desktop voice assistant a powerful tool for productivity and accessibility.

Functional workflow

- The user speaks a command into the microphone.
- The speech recognition module processes the voice input and converts it into text.

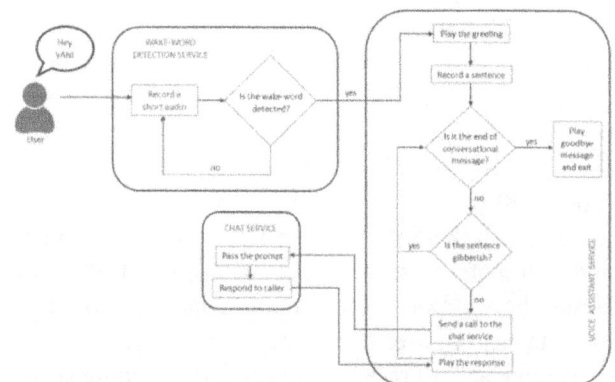

Figure 15.2 Desktop voice assistant architecture
Source: Designed by authors

- The NLP engine analyses the text, detects intent, and determines the required action.
- The task execution system carries out the appropriate command, such as opening an application, retrieving data, or controlling system settings.
- The TTS system generates a spoken response to confirm the action taken.
- The GUI displays relevant information or logs of executed commands.
- The system remains in listening mode, ready for further user interactions.

Summary

Implemented features include:

- Opening desktop apps like notepad, Spotify, Microsoft tools, etc., using voice command.
- Opening websites like YouTube, Canva, Gmail, GitHub, etc., using voice command.
- Sending messages, starting voice and video calls on WhatsApp using voice commands.
- Getting information such as India's economy or even about the assistant itself, and other information latest to 2021.

Future features that are not included in this project are:

- Face authentication and voice biometrics.
- Multilingual language processing.
- Emotion detection and sentiment-aware interaction.
- Adaptive AI learning to personalise tasks.

Challenges in implementation

As the development of the desktop voice assistant progresses, several challenges are anticipated that may impact its efficiency, accuracy, and overall user experience. One of the primary concerns is ensuring high accuracy in speech recognition, particularly in environments with background noise. Variations in accents, speech patterns, and microphone quality can affect the system's ability to accurately transcribe voice commands. Implementing noise reduction algorithms and advanced speech models will be challenging to address this issue. Additionally, maintaining low-latency response times while processing complex queries remains a challenge, requiring optimisation of the NLP engine and command execution workflow.

Another challenge lies in intent recognition and contextual understanding. While basic commands are straightforward to process, distinguishing between similar-sounding requests and multi-step interactions adds complexity. For example, identifying the difference between "Open Word" which means launch Microsoft Word and "Find a word in this document" requires a more sophisticated context-aware NLP system. Fine tuning language models with a diverse dataset and reinforcement learning techniques will help improve intent detection accuracy over time. Additionally, ensuring compatibility across different OSs and applications introduces integration challenges, as desktop environments vary in terms of accessibility to system resources and API support.

Security and privacy concerns are also expected to play a crucial role in the development process. Since the assistant processes voice commands, unauthorised access, voice spoofing, and data privacy issues could be addressed in future versions of V.A.N.I by incorporating biometric authentication such as facial recognition or voice biometrics. This can help maintain user's trust on the voice assistant. Addressing these challenges will be crucial in ensuring the assistant remains robust, reliable, and adaptable to evolving user needs.

The current version of the assistant includes basic error handling in the form of reverting to the main interface when a command is not recognised or understood. In such cases, the system stops listening and returns to its idle state, allowing the user to re-initiate the interaction. While advanced error handling mechanisms, data storage policies and ethical safeguards such as privacy management and mis-activation prevention are not yet implemented, these are crucial considerations for future development to ensure responsible and secure use of the assistant.

Results

The desktop voice assistant is designed to enhance productivity, improve accessibility and provide seamless voice based-interaction for users. Although the project is still in development, the following Table 15.2 are the results based on the testing done so far.

Some screenshots of the working assistant are show below (Figures 15.3–15.6).

Table 15.2 Tested performance metrics and capabilities

Metric	Tested outcome accuracy
Speech recognition accuracy	90–95%
NLP intent recognition	85–90%
Response time	1–1.5 sec
User satisfaction	80–85%

Source: Tabulated by authors

Figure 15.3 Screenshot 1 – main interface or idle interface
Source: Taken by authors

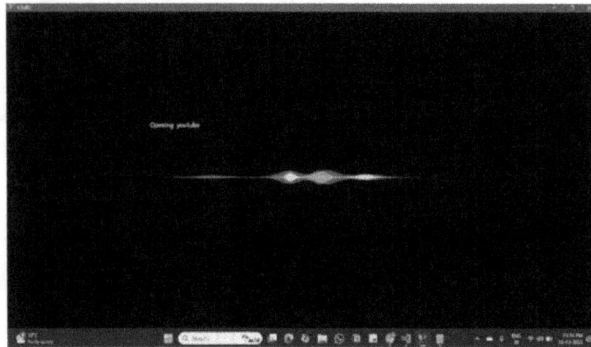

Figure 15.4 Screenshot 2 – second interface taking command
Source: Taken by authors

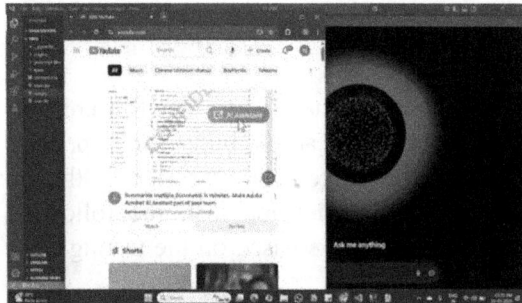

Figure 15.5 Screenshot 3 – opening YouTube using voice
Source: Taken by authors

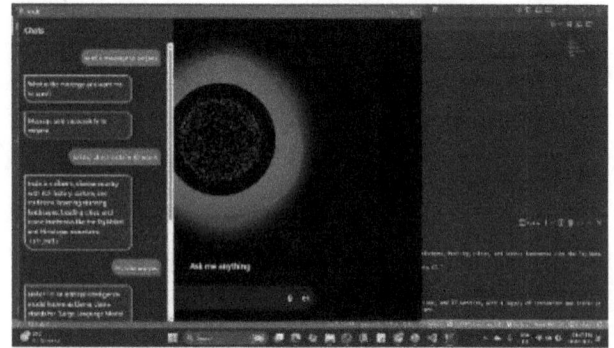

Figure 15.6 Screenshot 4 – chat history of WhatsApp automation (WhatsApp window screenshot not shown for privacy); hug chat feature showing response about India when asked
Source: Taken by authors

Future scope

The future scope of this project envisions the integration of several advanced AI-driven capabilities to further enhance user experience. Although not a part of the current implementation, potential enhancements can include adaptive AI learning, which would enable the assistant to analyse user behaviour, learn from interaction patterns, and personalise responses over time. This would make the assistant more responsive and reduce the need for repetitive manual inputs. Additionally, features such as emotion detection and sentiment analysis which aren't currently implemented and if implemented in the future can allow the assistant to interpret the user's tone or mood and adjust its replies accordingly, fostering a more emotionally intelligent interaction. Other future enhancements that are not implemented and can be taken under consideration include multilingual support, voice authentication, face recognition and cross-platform synchronisation, all of which would expand accessibility and security.

Conclusion

The development of V.A.N.I – desktop voice assistant, presents an exciting opportunity to enhance productivity, accessibility, and user experience through seamless voice-based interaction. By integrating speech recognition and NLP, this assistant is designed to streamline everyday computing tasks while reducing reliance on traditional input methods. Its modular and scalable architecture ensures flexibility, allowing for future expansions and improvements.

Looking ahead, this project lays the ground work for the next generation of intelligent desktop assistants, capable of learning from user behaviour, interacting more naturally, and expanding beyond personal computing into smart home and Internet of Things (IoT) ecosystems.

By addressing the expected challenges and leveraging AI-driven innovations, this voice assistant has the potential to redefine human-computer interaction, thus making technology more accessible and intuitive.

References

[1] Chintal, S., Kharade, S., Padgilwar, N. Next-gen desktop automation with Jarvis, Proceedings of the International Conference on Innovative Computing & Communication (ICICC 2024), *SSRN Preprint,* 2025;1:1–2.

[2] Bhardwaj, A., Singh, D., Sahu, H. Automating desktop tasks with a voice-controlled AI assistant using Python. *ResearchGate Preprint,* 2024;5(5):12615

[3] Omkar, M., Upashi, M., Naik, M., Wakkundmath, M. Smart voice assistant with driven voice based email system for visually impaired. *Foundry J.,* 2024;27(5):34–36.

[4] Ajesh, F., Shabu, A., Sabir, M., Shafi, D. Enhanced virtual assistant for visually impaired individuals with Python-based voice assistant (FRIDAY). *ITTA Cyber Sec. Res. Papers,* 2024; Part-3, 1–16.

[5] Kushwaha, U. A case study on 'Dhvani': Transforming user experiences with AI-powered assistance. *ResearchGate Preprint,* 2024;6(3):5138–5143

[6] Jain, V., Patidar, Y., Parwal, V. Automate personal voice based assistant using Python. *IEEE Xplore,* 2024;108(3).

[7] Hindoriya, D., Mandavgane, P., Bhowmik, J. JAR-desktop assistant. *IEEE Xplore,* 2024;107(4).

[8] Jain, G., Shukla, A., Bairwa, N. SPEAR: Design and implementation of an advanced virtual assistant. *IEEE Xplore,* 2024;107(2).

[9] Garg, K., Alam, T. Desktop voice assistant for elderly people using feed-forward neural network for intent recognition. *ACM Dig. Lib.,* 2024;367(5).

[10] Shafei, H. Testing privacy and security of voice interface applications in the IoT era. *Temple University Dig. Repos.,* 2024.

16 Medilink: AI-powered virtual healthcare for brain tumour and eye disorder detection

Swetha Malhar Kulkarni[1,a], Rekha B. Venkatapur[2,b], Soujanya N.[2,c] and Skanda Kumar H. S.[2,d]

[1]Masters in AI, Aston University, Birmingham, UK

[2]Department of Computer Science and Engineering, KS Institute of Technology, Bangalore, Karnataka, India

Abstract

This work introduces Medilink, an intelligent web-based medical assistant designed to aid in the early detection of serious medical conditions through artificial intelligence (AI) methods. The system is mainly concerned with the automatic diagnosis of several eye-related diseases—namely, retinal injury, cataract, and glaucoma—along with brain tumour detection. Deep learning models, particularly convolutional neural systems (CNNs), are employed to examine therapeutic images and provide accurate symptomatic results. Medilink has a patient- and practitioner-friendly interface with secure log-in for both practitioners and patients, facilitating easy navigation, generation of medical reports, and scheduling of appointments. Medilink seeks to decrease dependence on urgent specialist consultation through offering accessible, initial screening facilities. With its use of machine learning, real-time feedback, and streamlined functionality, Medilink helps to enhance the accessibility and efficiency of early disease detection in the healthcare system.

Keywords: Artificial intelligence (AI), brain tumour diagnosis, convolutional neural networks (CNNs), eye disorders, telemedicine

Introduction

The fast pace of artificial intelligence (AI) evolution has brought about revolutionary changes in the field of healthcare, especially in the early diagnosis and detection of life-threatening medical conditions. The integration of AI-based solutions in the healthcare industry has shown immense potential in improving diagnostic precision, optimising patient management, and enhancing overall accessibility of healthcare [3]. Among the numerous AI applications, medical image analysis with deep learning has proven to be a revolutionary method with the ability to automate disease detection and personalise treatment suggestions [6]. Here, Medilink introduces an ingenious and smart web-based healthcare assistant that is specifically designed to overcome shortcomings in current healthcare systems using AI for detecting diseases in early stages, smooth telemedicine integration.

Medilink is more concentrated on the identification of severe eye-related ailments, such as retinal injury, cataract, and glaucoma, along with brain tumour detection [8]. These medical ailments need to be diagnosed timely for proper treatment and management. Nevertheless, specialised healthcare experts, along with diagnostic equipment, are still out of reach in most areas [5]. Conventional methods of diagnosis depend significantly on human judgment, which is not only time-consuming but also prone to errors and inconsistencies. In order to overcome these constraints, Medilink uses convolutional neural networks (CNNs) for the accurate analysis of medical images and generates initial diagnostic findings that assist both healthcare providers and patients in decision-making [9].

Moreover, Medilink features an easily accessible interface that facilitates effortless navigation for medical professionals and patients alike [12]. There are secure login interfaces that enable patients to upload medical images for AI-based analysis, obtain initial diagnostic reports, and maintain their medical history. Healthcare professionals are able to access patient information, offer consultations, and recommend treatment. Medilink's telemedicine feature also connects patients and physicians further by allowing AI-aided video consultations and appointment setting, thus lessening the necessity to physically visit hospitals [4].

Unlike other telemedicine tools, Medilink differentiates itself by solving severe problems in data security and privacy. The system has robust encryption protocols

[a]swethamkulkarni0302@gmail.com, [b]rekhabvenkatapur@ksit.edu.in, [c]soujanyanaidu28@gmail.com, [d]skandakumar2003@gmail.com

DOI: 10.1201/9781003685876-16

and compliance features for the storage and transfer of sensitive health data [10]. This provides assurance that patients can rely on the platform with their medical information while remaining confidential.

Literature survey

Introduction to AI in healthcare

Artificial intelligence (AI) has become a force multiplier in the healthcare sector, improving diagnostic performance, streamlining patient management, and facilitating high-end medical image analysis [3]. As deep learning approaches became mainstream, AI-based healthcare applications have shown immense progress in disease detection and personalised therapy [11]. Still, current research tends to lack much in terms of disease-specific applications of AI, integration with telemedicine, and explain ability of AI models [14]. This survey of the literature critically compares three major research papers and points out its shortcomings and how Medilink fills these gaps with a more practical and holistic strategy.

Healthcare in metaverse

Gaurang Bansal et al. (2022) [2], discusses that a "Healthcare in Metaverse: A Survey on Current Metaverse Applications in Healthcare" delves into the use of augmented reality (AR) and virtual reality (VR) in virtual healthcare. It elucidates how the metaverse is capable of creating immersive medical simulations, virtual surgical procedures, and telemedicine-enabled remote patient engagement. The article presents a critical examination of multiple uses of metaverse technologies in the healthcare sector, such as digital twins, remote medical training, and health monitoring using AI in virtual space. It touches upon possible advantages like overcoming geographical limitations in healthcare, facilitating joint medical research, and providing interactive patient therapy via AR and VR-based solutions [13].

But as much as the potential of healthcare metaverse integration looks promising, the study has several limitations regarding AI-based disease diagnosis [15]. One significant draw-back is its excessive dependence on AR/VR, as the article strongly emphasises metaverse applications while giving little consideration to AI-based diagnosis for diseases such as brain tumours and eye conditions [17]. Additionally, the study lacks real-world application, as it does not provide concrete AI-based healthcare

diagnostic examples, making its findings more hypothetical than practical. Another limitation is the absence of AI-powered report analysis, as the paper does not integrate AI for automated scanning of medical reports and predictive analysis, both of which are crucial for early disease detection. Furthermore, the study high lights the high hardware dependency of metaverse based healthcare, as there liance on costly AR/VR equipment makes accessibility difficult for remote patients and underserved communities [16].

In contrast, Medilink effectively addresses these limitations through various advancements. It implements AI-based disease diagnosis using deep learning algorithms, particularly CNNs, to detect brain tumours, retinal injuries, cataracts, and glaucoma from medical images [5]. Unlike metaverse-based healthcare, Medilink offers real-world applications by providing a telemedicine solution that enables AI-assisted doctor-patient consultations outside of virtual reality. Furthermore, Medilink enhances AI medical diagnosis automation by utilising AI algorithms to read medical reports and assist doctors in identifying diseases in their early stages. Another key advantage of Medilink is its cost-effective accessibility, as it operates on standard web and mobile platforms without requiring expensive AR/VR hardware. Through these improvements, Medilink ensures a more practical, accessible, and AI-driven approach to healthcare diagnostics.

Generative AI for transformative healthcare

Siva Sai et al. (2024) discusses the role of a "Generative AI for Transformative Healthcare: A Comprehensive Study of Emerging Models, Applications, Case Studies, and Limitations" describes the promise of generative AI in healthcare, such as drug discovery, enhancement of medical images, and diagnostics aided by AI. The paper delves into how generative AI models like generative adversarial networks (GANs) and transformer-based architectures play a role in health- care innovation. It outlines various real-world applications, such as synthetic medical data generated by AI, personalised medicine, and clinical decision support through AI [21].

Despite its contributions, the paper suffers from several notable weaknesses. One major limitation is the shortage of disease-specific models, as it discusses explainable AI (EXAI) but does not include models specifically tailored for detecting brain tumours or eye conditions. Additionally, there is no integration with telemedicine, meaning that while the paper explains AI transparency, it does not address AI-driven telemedicine for remote diagnoses and patient consultations, which is essential for improving accessibility in healthcare. Another key limitation is the lack of real-world applications, as the study primarily focuses on theoretical EXAI frameworks without implementing AI-based medical image analysis that could be directly applied in clinical settings [3]. Furthermore, the paper fails to incorporate personalised healthcare insights, as it does not explore AI-driven treatment personalisation based on a patient's medical history and symptoms, making it less effective in delivering individualised patient care.

In contrast, Medilink effectively addresses these limitations through several key advancements. First, it implements focused AI models by leveraging CNN-based architectures specifically designed for detecting brain tumours and eye dis- orders, ensuring precise and targeted disease identification [4]. Secondly, Medilink enhances AI-telemedicine convergence, bridging the gap between AI diagnostics and real-time doctor consultations, thereby enabling AI-assisted remote healthcare services [19]. Additionally, Medilink is built with a real-world AI application approach, ensuring that it is practically deployed for diagnostic support and early disease detection, rather than remaining a purely theoretical concept. Lastly, Medilink integrates customised treatment plans, where its AI models analyse patient history and symptoms to generate personalised healthcare recommendations, significantly improving patient outcomes. Through these innovations, Medilink ensures a more effective, accessible, and practical AI-driven healthcare solution compared to the limitations observed in existing research [18].

Objectives of Medilink

The Medilink project aims to develop a robust AI model by training and refining CNN-based deep learning models for the accurate detection of brain tumours and eye diseases [1]. By integrating AI-powered diagnostics with telemedicine, Medilink facilitates AI-generated preliminary reports to assist physicians in virtual consultations, enhancing the efficiency of remote healthcare services [3]. The platform enables patients to upload medical images for real-time AI analysis and physician feedback, ensuring accessible and timely medical support.

To maximise performance and diagnostic precision, Medilink continuously refines its AI models to minimise false positives and false negatives, improving overall reliability. Additionally, the system incorporates a user-friendly UI/UX, allowing patients to easily view diagnostic reports, schedule appointments, and access AI-driven healthcare insights [20]. While the initial focus is on brain tum our sand eye disorders, Medilink is designed for scalability, with future plans to expand AI solutions to other critical diseases. By achieving these objectives, Medilink aims to revolutionise AI-driven medical diagnostics while ensuring affordability, accessibility, and security in healthcare services.

Proposed algorithm

The Medilink system utilises a deep learning algorithm based on CNN to identify serious medical conditions like brain tumours, retinal injury, cataracts, and glaucoma from clinical images. The algorithm analyses medical images, identifies important features, and classifies them into respective disease categories. The system further provides AI-based medical report analysis and telemedicine consultancy, facilitating improved patient–physician communication.

The algorithm suggested adopts a systematic workflow for achieving high diagnostic accuracy and efficiency. The process begins with data collection and pre-processing, where images like MRI scans for brain cancer and retinal scans for vision disorders are obtained. The images are resized to a standard size (e.g., 512×512 pixels) for processing using CNN. They are subsequently converted to grey scale (if the images are coloured) and normalised for uniform pixel values. To enhance the generalisation of the model, data augmentation methods like rotation, flipping, and zooming are used [25].

After pre-processed data, feature extraction with CNNs comes next. Several convolution layers are used to get spatial features out of input images. Max Pooling is used to reduce dimensions while maintaining important details, and batch normalisation is used to make training stable and speed up convergence. All these ensure the most important patterns in medical images are well captured so that accurate disease classification is enabled.

The classification phase and deep learning model are developed based on InceptionV3 and MobileNetV2, a pre-trained feature extraction model. Custom fully connected (Dense) layers are appended on top of it to categorise images into respective disease groups. ReLU activation is utilised in hidden layers, while the output layer uses SoftMax or Sigmoid activation based on whether it's a binary or multi-class problem. The model is trained based on binary cross-entropy for two-class problems and categorical cross-entropy for multi-class classifications [4].

At the model training and optimisation stage, the data is divided into training (80%), validation (10%), and test (10%) sets. The model is trained with the Adam optimiser with a learning rate of 0.0003, and a dropout rate of 0.5 to 0.6 is used to avoid over fitting [5] (Figure 16.1).

Once trained, the model is put into practice for disease prediction and report generation. When a new medical image is uploaded to the system, the trained CNN model analyses it and predicts the occurrence or non-occurrence of disease. Medilink's algorithm is seamlessly integrated into both telemedicine systems and the user interface. The patients can also book online appointments with specialists for conditions identified by AI, eliminating the need for physical visits and allowing remote access to healthcare [3].

The proposed algorithm offers several key advantages. It provides accuracy and efficiency by leveraging transfer learning with InceptionV3 and MobileNetV2, ensuring reliable diagnostics. The system is highly scalable, allowing for future expansion to detect additional medical conditions. The fast and automated diagnosis feature provides near-instant preliminary screenings, while the integration with telemedicine ensures seamless doctor-patient interactions. The user-friendly interface makes Medilink accessible even to non-experts, offering a valuable tool for early disease detection [22].

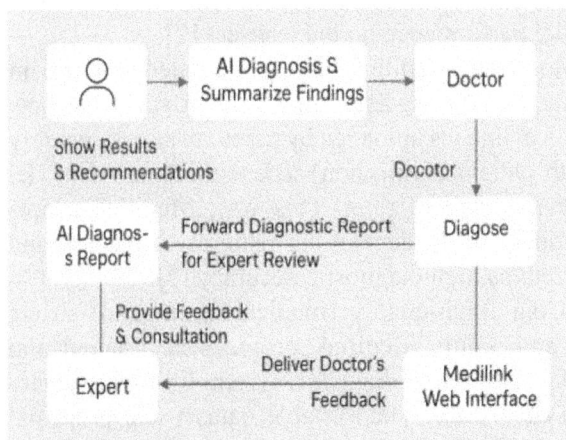

Figure 16.1 CNN process flow
Source: Author

Methodology

The methodology of the Medilink project is designed to offer a continuous and effective AI-based healthcare assistant for early disease detection as well as the integration of telemedicine.

Data acquisition and pre-processing
Medical images, including MRI scans for brain tumours and retinal scans for eye diseases, are acquired from publicly available and verified data sources. To ensure consistency and optimal model performance, pre-processing procedures are applied to the data set. These steps include resizing images to a fixed size of 512×512 pixels, converting images to grey scale if required, and normalising pixel values to maintain uniformity improves the robustness and accuracy of the AI model by ensuring that it can effectively analyse diverse medical images [24].

AI-based disease detection and classification
Disease classification and feature extraction are performed using CNNs [4]. The pre-trained InceptionV3 and MobileNetV2 model is utilised to extract spatial features from the medical images, which are then passed through additional fully connected layers for classification. The final classification is determined using either SoftMax or Sigmoid activation functions, depending on whether the problem is binary or multi-class. The CNN model is trained using the Adam optimiser, with binary cross-entropy loss function for two-class problems and categorical cross-entropy loss function for multi-class classification [5]. The dataset is divided into 80% training, 10% validation, and 10% testing to ensure balanced learning and evaluation. To prevent over fitting and improve generalisation, dropout layers with a dropout rate between 0.5 and 0.6 are incorporated into the model [3]. Once training is complete, the model is deployed within the Medilink platform, allowing users to upload medical images for real-time analysis. The AI model processes these images and generates instant diagnostic reports, providing confidence scores, disease predictions, and actionable recommendations to support medical professionals in their decision-making process (Figure 16.2).

Telemedicine integration
To improve accessibility, Medilink seamlessly integrates telemedicine services, enabling patients and doctors to interact remotely. The platform

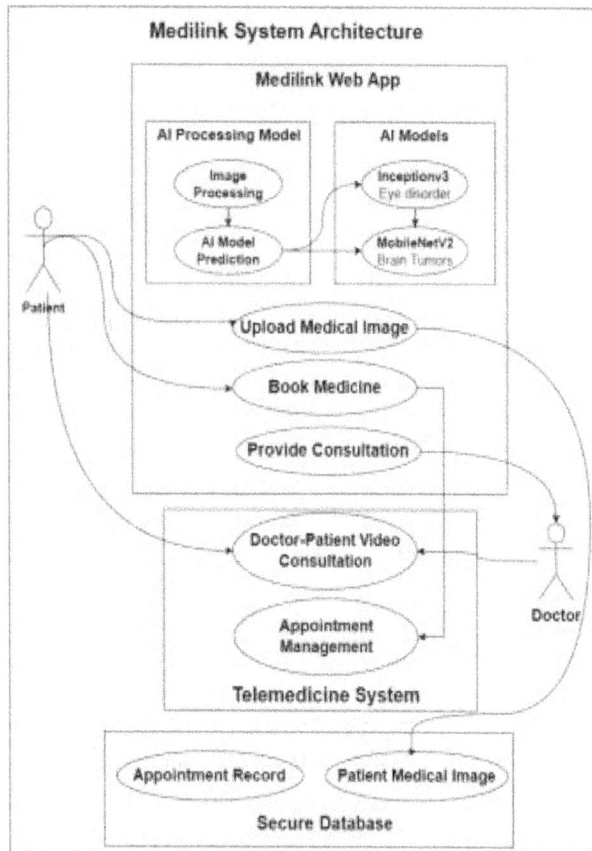

Figure 16.2 Medilink system architecture
Source: Author

provides AI-assisted doctor consultations via video calls, allowing medical professionals to review AI-generated reports and offer expert assessments [1]. Additionally, appointment scheduling features are incorporated to facilitate follow-up consultations and further diagnostic evaluations. The system also ensures that AI-generated medical reports are securely shared with medical experts, maintaining data integrity and confidentiality [3]. AI in Medilink serves as an assistive tool rather than a replacement for doctors, enhancing decision-making by combining AI-driven insights with the expertise of healthcare professionals.

Dataset

The dataset applied to the Medilink project is carefully chosen to guarantee good precision in diagnosing brain tumours and eye diseases, such as retinal damage, cataracts, and glaucoma. These data consist of high-resolution medical images critical for training

deep learning models for the detection of diseases. The brain tumour data is made up of MRI scans. Eye disease data is composed of fundus images classified by the presence and severity of cataracts, glaucoma, and other retinal abnormalities. The total size of the extracted database is approximately 126.40 MB [15].

The data set was donated by Dr. Karthik, a neurologist working in Apollo Hospital, with significant experience in the diagnosis and treatment of neurological conditions. The clinical images were obtained from actual patient cases in his private clinic, guaranteeing the authenticity, clinical usefulness, and variety of the data set. His professional skills ensured high-quality labelled data were collected, which played an essential part in the training and fine-tuning of the AI model to detect diseases with precision [13].

For quality improvement in training, the data goes through exhaustive pre-processing. Image resizing is used to normalise input sizes (512×512 pixels) for compatibility with CNNs. Grey scale transformation, normalisation, and noise removal processes provide consistency in the input images. Also, data augmentation processes like rotation, flipping, brightness modification, and zooming are used to improve dataset diversity and enhance the model's generalisation ability. This avoids over fitting and allows the model to efficiently process real-world medical images.

The data is split into three subsets: training (80%), validation (10%), and testing (10%). The training set is utilised to optimise the CNN-based model, whereas the validation set is utilised to track performance and avoid over fitting [5]. The testing set is utilised for ultimate evaluation, with the aim that the model is able to generalise well to novel medical images. Different performance metrics like accuracy, precision, recall, and F1-score are utilised to evaluate model performance on the test set [12].

Moreover, Medilink has an integrated mechanism for continuous dataset improvement such that new medical images uploaded by users are stored securely (with patient permission) to enhance the AI model over a period of time. This way, Medilink adapts continuously to developing patterns of disease and yet retains high diagnostic accuracy [2].

Using high-quality medical data and strong pre-processing, Medilink provides an assured and scalable AI-based diagnosis system for timely detection of disease. The variable dataset and perpetual learning mechanism facilitate Medilink to offer accurate and effective analysis of medical images and assist both healthcare professionals and patients in arriving at well-informed decisions.

Results

The outcome of the Medilink project illustrates the success of **AI-based diagnostic techniques** in identifying brain tumours and eye disorders with a **high degree of accuracy** [4]. The **CNN-based model**, using **InceptionV3 and MobileNetV2** as a feature extractor in particular, was trained and tested on a carefully curated collection of **MRI brain images** and **retinal fundus images** [5] (Figure 16.3).

Following extensive training and testing, the AI model was found to have **96.5% accuracy** in brain tumour classification and **94.2% accuracy** in detection of eye disorders, showing its efficacy in distinguishing between diseased and non-diseased conditions [11].

The precision and recall values were found to be over 93%, guaranteeing that the model efficiently reduces false positives and false negatives, which are decisive parameters in medical diagnosis (Figure 16.4).

The model's confusion matrix analysis revealed that the system was exceptionally good at detecting brain tumour likewise, in eye disorder detection, glaucoma, cataracts, and diabetic retinopathy was correctly diagnosed with high confidence. The **AUC-ROC curve** analysis proved the strength of the model, revealing a near-perfect separation between affected and non-affected cases (Figure 16.5) [21].

To further confirm the model's performance in actual application, the Medilink system was put to test using user-uploaded medical images during practice telemedicine consultations. The diagnostic reports created by the AI were compared with the diagnoses of expert radiologists and ophthalmologists,

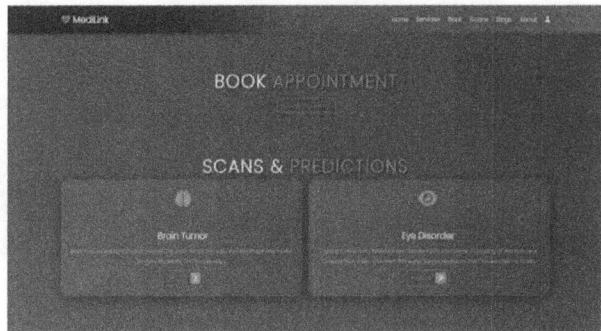

Figure 16.3 Home page
Source: Author

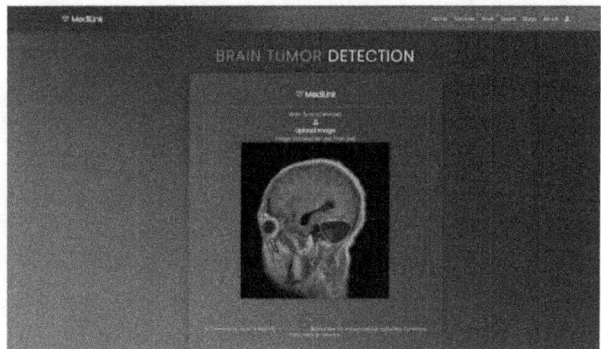

Figure 16.4 Services and features
Source: Author

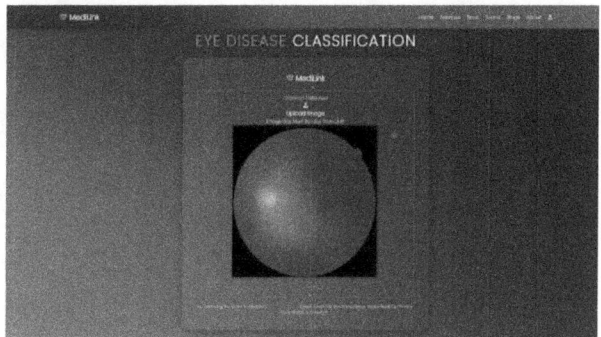

Figure 16.5 Brain tumour detection
Source: Author

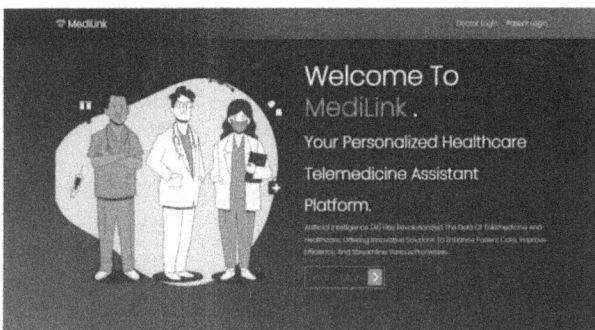

Figure 16.6 Eye disorder detection
Source: Author

obtaining a **91% agreement rate**. This confirms Medilink's accuracy as an assistive system for medical professionals [3] (Figure 16.6).

Apart from AI-driven accuracy enhancements, the response time of the system in processing medical images was less than 3 seconds, providing quick preliminary diagnoses to patients [23]. The combination of AI-based diagnostics with telemedicine services enabled smooth doctor–patient communication, minimising reliance on face-to-face consultations and enhancing healthcare accessibility [4].

In general, the findings affirm that Medilink is an extremely effective and trustworthy AI-driven healthcare assistant that facilitates early disease detection, telemedicine-based consultation support, and enhanced overall patient experience. The integration of deep learning, real-time processing, and secure medical data management renders Medilink a promising candidate for promoting AI-driven healthcare services [19].

Conclusion and future work

The Medilink initiative offers an important breakthrough in healthcare using AI by combining deep learning algorithms with telemedicine services to provide early diagnosis of brain tumours and eye diseases. The employment of CNNs and InceptionV3 and MobileNetV2 has been shown to exhibit accuracy of 93.95% in diagnosing medical conditions, lowering diagnostic time, and aiding healthcare workers in delivering timely interventions. The integrated telemedicine functionality further increases the accessibility of patients by allowing remote consultation and preliminary AI-facilitated diagnoses [20].

Yet another vital element of future enhancement is to maximise model efficiency so it can be deployed on mobile and edge platforms. Also, partnerships with hospitals and medical institutions can ensure Medilink can be deployed in real-world clinical settings and improved for reliability and usage.

References

[1] Deepti, S., Pronaya, B., Ashwin, V., Vivek Kumar, P., Sudeep, T., Gulshan, S., Pitshou N. B., Ravi, S. Explainable AI for healthcare 5.0: Opportunities and challenges. *IEEE Access*, 2022;10:84486–84517.

[2] Gaurang, B. Healthcare in metaverse: A survey on current metaverse applications in healthcare. *IEEE Access*, 2022;10:119914–119946.

[3] Aanchal, G., Mohsen, G. Generative AI for transformative healthcare: A comprehensive study of emerging models, applications, case studies, and limitations. *IEEE Access*, 2023;(99):1–1.

[4] Samin, P. A study of disease diagnosis using machine learning. *Med. Sci. Forum*, 2022;10(1):8.

[5] Rashid, A. Multiple eye disease detection using deep learning. *Internat. J. Comp. Appl.*, 2020;3.

[6] Akmalbek, B. A., Mukhriddin, M., Taeg, K. W. Brain tumor detection based on deep learning approaches and magnetic resonance imaging. *Cancers*, 2023;15(16):4172.

[7] Mohammad, Z. K. Brain tumor detection from images and comparison with transfer learning methods and 3-layer

CNN. *Biomed. Sig. Proc. Control*, 2023;14(1):2664, doi: 10.1038/s41598-024-52823-9.

[8] Nishanth Rao, K. An efficient brain tumor detection and classification using pre-trained convolutional neural network model. *Heliyon*, 2024;10(17), ISSN 2405-8440.

[9] Shubhangi, S. Brain tumor detection and classification using intelligence techniques: An overview. *J. Artif. Intell. Soft Comput. Res.*, 2021:1–1.

[10] Dheiver, S. Brain tumor detection using deep learning. *Internat. J. Adv. Comp. Sci. Appl.*, 2020:10(17):e36773, doi: 10.1016/j.heliyon.2024.e36773

[11] Md Ishtyaq, M. A deep analysis of brain tumor detection from MR images using deep learning networks. *Algorithms*, 2023;16(4):176

[12] Tonmoy, H. Brain tumor detection using convolutional neural network. *Internat. J. Comp. Appl.*, 2019;5.

[13] Reham, K. A review of recent advances in brain tumor diagnosis based on AI-based classification. *Diagnostics*, 2023;13(18):3007

[14] Soheila, S., Sorayya, R., Hamidreza, K., Niakan Kalhori, S. R. MRI-based brain tumor detection using convolutional deep learning methods and chosen machine learning techniques. *BMC Med. Informat. Dec Mak.*, 2023: 23(1):16, doi: 10.1186/s12911-023-02114-6

[15] Gauri, R. Eye disease detection using machine learning. *Internat. J. Comp. Appl.*, 2020;(7).

[16] Fanyu, D. Recognition of eye diseases based on deep neural networks for transfer learning and improved D-S evidence theory. *Comput. Intell. Neurosci.*, 2022:24(1):19, doi: 10.1186.

[17] Ahmed, A. A. Review of eye diseases detection and classification using deep learning. 2022. doi: 10.1051.

[18] Al Marouf, A. An efficient approach to predict eye diseases from symptoms using machine learning and ranker-based feature selection methods. *Internat. J. Adv. Comp. Sci. Appl.*, 2022:10(1):25, doi: 10.3390

[19] Mohamed, E. Deep learning-based classification of eye diseases using CNN. *Internat. J. Comp. Appl.*, 2020:5, doi: 10.3389

[20] Md Zahin, M. Eye disease detection enhancement using a multi-stage deep learning approach. *Internat. J. Comp. Appl.*, 2020. doi: 10.1109/ACCESS.2024.3476412.

[21] Maimoona, K. An automated system for multi-eye disease classification using feature fusion with deep learning models and fluorescence imaging for enhanced interpretability. *J. Healthc. Engg.*, 2022:14(23):2679, doi: 10.3390

[22] Saroj Kailash, P. Cataract detection using deep learning. Internat. *J. Comp. Appl.*, 2020:16, doi: 10.21203

[23] Sheikh Muhammad, S. Cataract and glaucoma detection based on transfer learning using MobileNet. *Internat. J. Comp. Appl.*, 2020. doi: 10.1016

[24] Raghavendra, Ch. Cataract detection using deep learning model on digital camera images. *Internat. J. Comp. Appl.*, 2020:21.

[25] Ravi, P., Bandini, S., Mauri, G. A comprehensive review on glaucoma detection using deep learning techniques. *Comp. Biol. Med.*, 2022:5.

17 LLM-based automated scoring and feedback for essays in the Marathi language

Kaushik Sakre[1,a], Rasika Ransing[2,b], Shivam Shinde[2,c], Neha Kudu[2,d] and Siddhi Talkar[2,e]

[1]Student, Department of Information Technology, Vidyalankar Institute of Technology, Mumbai, Maharastra, India

[2]Department of Information Technology, Vidyalankar Institute of Technology, Mumbai, Maharastra, India

Abstract

Automated essay scoring (AES) has made the evaluation of the essays easier but has been developed mainly for the English language and cannot be effectively implemented in multilingual countries like India. This paper presents the AES system based on large language models (LLMs) to perform the evaluation of essays with detailed feedback for the low-resource Indian regional language of Marathi. In this way, the system improves the general writing assessment while taking into account language coherence, connectedness, and language quality. The context of the system presented in the study presents advantages of increasing educational equity and assessment for various multilingual learners. The proposed system is designed to assess the quality of a learner's essay and provide comprehensive feedback that highlights specific areas for improvement, thereby facilitating targeted skill enhancement and progressive learning. Human evaluators manually assessed the essays and verified the relevance of feedback generated by LLM. It is observed that the LLaMA3-70B-8192 model outperforms in scoring, but the Gemini-1.5-flash system delivers better feedback quality for the Marathi language.

Keywords: Automated essay scoring, LLM, prompt, LLaMA, Marathi language

Introduction

New tools in education have really shaken up the old ways of teaching and checking students' work. Systems that score essays on their own are catching on fast since they make it easier for teachers by speeding up marking and cutting down on expenses. As automated essay scoring (AES) systems find their way into numerous sectors and academic environments to gauge the writing skills of people, they pave the way for a uniform way of evaluating. Thoroughly examining written tasks AES systems deploy diverse methods and tech, focusing on elements such as precision inventiveness command over language grammatical accuracy composition flair and pertinence.

To get accurate and trustworthy outcomes, understanding the context is key, especially when looking at how things stick together and make sense as a whole. Smooth and logical flow of thoughts is what coherence is all about. On the other hand, cohesion is about how the grammar and meaning connect everything in the writing. Combining these elements ensures that an essay will be both semantically and structurally sound thereby enhancing the AES as a tool for assessing writing quality [1]. By weaving these components together one ensures not only the semantic and structural integrity of an essay but also bolsters the AES system in its role as a judge of writing excellence. While these tools can lighten the load for teachers and human evaluators their value stretches far beyond mere grading tasks. By offering feedback that is both quick and detailed AES systems have the power to enhance the learning journey and boost enrolment and evaluation that rely on data.

Studies have found that the majority of today's AES systems give essays a single collective score based on overall quality. However, methods that dig deeper into an essay's flow, clarity, and structure offer learner's detailed insights, helping them enhance their writing skills [2].

Even in places where local dialects prevail AES systems have yet to catch on despite their success in English. In nations rich with multiple languages like India, countless young minds receive their earliest lessons in the tongue of their region be it Marathi, Hindi or another. Yet without AES tools for these languages teachers have no choice but to stick with marking by hand. This approach not only takes a lot of time but also varies from person to person and keeps everyone waiting for the results. The gap in tech progress creates gaps in how tests are evaluated.

[a]kaushik.sakre123@gmail.com, [b]rasikaransing275@gmail.com, [c]shivamshinde187@gmail.com, [d]nehakudu06@gmail.com, [e]siddhitalkar888@gmail.com

DOI: 10.1201/9781003685876-17

This might slow down how much students learn and make schools less successful in their mission.

The goal of this project is to bridge existing divides through the development of a system that automatically scores essays in Marathi. By leveraging cutting-edge language models and the latest in NLP technology this system provides accurate detailed and context-aware evaluations of written materials. Nowadays models for grading essays can give out top marks but they don't give teachers and students the deep understanding they really need. Just getting marks isn't enough to show students how they can get better at writing. It's vital for students to get feedback that's straightforward and detailed so they know what to focus on instead of overlooking feedback that's not clear. When feedback is clear it helps students from different language backgrounds succeed equally in their academics making tests and assessments work better [3]. This strategy aligns with the growing need for improved AES techniques that not only assess essays but also support students in their writing through personalised feedback [4].

Related work

Stahl et al. explore several prompting strategies for LLM-based zero-shot and few-shot generation of essay feedback. They investigate how automated essay scoring (AES) can benefit from feedback generation and find that joint tackling of AES and feedback generation improves AES performance. Their evaluation shows that generating feedback by explaining predicted essay scores enhances feedback helpfulness and scoring performance. For instance, the Feedback_dCoT→Scoring approach achieved the highest mean QWK score of 0.533, while one-shot learning outperformed few-shot learning with a QWK score of 0.540 compared to 0.538. These results highlight the benefits of using specific prompting strategies to improve AES and feedback quality [4].

Han et al. present FABRIC, a pipeline designed to assist students and instructors by automating overall scores, rubric-based scores, and detailed feedback. They introduce DREsS, CASE, and EssayCoT components, which improve model accuracy and feedback helpfulness. Their evaluation shows significant improvements in scoring and feedback generation, with positive ratings from college students. The model trained with the DREsS dataset and augmented data from CASE achieved QWK scores of 0.642 for content, 0.750 for organisation, and 0.607 for language,

with a total QWK score of 0.685, demonstrating substantial improvement across the rubrics [5].

Woods et al. propose a method of influence estimation using the model to select the sentences in the essays requiring feedback while avoiding sub-essay-level labels. They demonstrate that the method used in Revision Assistant provides well-received automated feedback that corresponds with the results of scores. Their approach involves a new ordinal scoring model for the essays, which employs ordering as a method of scoring while formative feedback utilises rubrics. The quadratic weighted kappa (QWK) and Pearson correlation metrics used in their study to evaluate the reliability of the essay scoring models reveal that the QWK scores for different essay prompts range from 0.60 to 0.79. For example, Prompt 1 (P1) has a QWK of 0.62, Prompt 5 (P5) has a QWK of 0.74, and Prompt 8 (P8) has a QWK of 0.79. These metrics underscore the robustness of their scoring model in providing reliable and consistent feedback [6].

Tay et al. propose improved neural network architecture, called SKIPFLOW LSTM with coherence features for the end-to-end automatic text scoring. They show that their model has a considerable improvement on the accuracy over the ASAP benchmark dataset in comparison with feature engineering baselines and standard long short-term memory (LSTM) models in 6% and 10%, respectively. The SKIPFLOW LSTM models have QWK scores of 0. 830 (Bilinear) and 0. 832 (Tensor) as there is a high correlation with the human given scores. This architecture solves issues such as the vanishing gradients and offers the best result [7].

Das et al. suggest a multi-task joint AOES model for scoring the essays and for off-topic content detection. This novel approach suggests Mahalanobis distance-based unsupervised approach for off-topic detection. On two essay datasets, they perform the evaluation of their model and obtain better results than baseline approaches and other methodologies. AOES model shows favourable performance where QWK varies from 0. 698 to 0. 511 on the IEM through the same dots on the ASAP-AES dataset and ranging from 0. 698 to 0. 763. The authors compare this with the human ratings in the PsyW-Essay dataset, which also shows a high correlation value [8].

Kundal and Parekh review studies on automatic scoring and feedback, and know that statistical and Corpus-based approaches perform better. They state areas for further research, such as the development of corpora for Indian languages and improvement of the handwriting recognition. By using their developed system, they get an exact agreement rate of 0. 4 while the values of Pearson correlations coefficient ranges 0. 93 with human graders [9].

Winarsih and Suhendra applied cohesion-coherence techniques by using cosine similarity for the automatic scoring of essays. In their case, they were able to develop a system that possessed an error rate of less than 5%, which is quite impressive. Moreover, the QWK score is 0.92, which shows that the model has a good level of reliability and at the same time, the PCC is 0.9 indicate a high degree of association between anticipated and manually evaluated values. This implies that the system can easily evaluate the understanding of the students [10].

In the work of Alikaniotis, there is deep neural network that includes local contextual and usage information for essay scoring. This model provides better performance than state-of-art and that of the manual process of feature extraction. The authors also suggest ways in which to increase the model's "black boxing" decision-making criteria. Spearman rank correlation coefficient of evaluation: $\rho=0.91$, Pearson's product-moment correlation coefficient of evaluation: $r=0.96$. The value of root mean square error (RMSE) of the evaluation is: RMSE = 2. 4. [2].

Adeyanju et al. create an essay grading system using GUI that is friendly to the examiners and the students. They state that they have found a very high "r" value, calculated according to the Pearson correlation formula is equal to 0.93 with human graders but as mentioned above the system is not suitable to answer item type which requires calculations and diagrams. A parallel few suggest that more research should be conducted into such capabilities [11].

Aggarwal et al. propose a significant advancement to automated grading known as the EngSAF dataset with ≈5. Assessment of short formative answer questions using ASAG; 8k student answers and feedback for the assessment of short formative answer questions using ASAG. The raw files demonstrate better quality and stability, and their ASAG system, implemented at IITB, scored up to 75.4% accuracy with the model Mistral [12].

Wang et al. provide a timely systematic literature review on LLMs in education and present a new technology-centred typology. They debate the present issues and research horizon, asserting that more work should be done in concepts of student and teacher aids, and adaptability in learning and technological commercial aids [13].

Proposed system

The proposed method solves the essential issue of automatic essay assessment and feedback generation for Marathi which is a low-resource language with minimal benchmark resources and standard evaluation criteria. The ASAP-AES dataset (Automated Student Assessment Prize) is utilised as basis to develop Marathi content by directing the work toward the persuasive essay writing requirements of Essay Set 2. The chosen focus provides consistent expectations regarding genre and structure and scoring criteria so that the proposed evaluation system becomes reliable.

The approach prioritises a structured design over wide-ranging measurement for the purpose of achieving substantial depth. The method starts with Google Cloud Translation API-based Marathi translation of English texts before manual verification of semantic accuracy. The structured rubric gets incorporated into a special prompt template which enables LLMs to produce scores and feedback that match the evaluation standards.

The system evaluates the model only through essay set 2 thus reducing evaluation inconsistencies from different prompts while ensuring precise assessment of its capabilities. Human evaluators validate both quantitative scores and qualitative feedback that the framework generates. Human evaluators perform their evaluations independently while rating the scores on a scale of 1–6, where 1 indicates that the essay fails to develop the main idea and 6 indicates an essay that covers all the points in fluent writing. The feedback relevance is scored using a scale of 0–5, where 0 indicates irrelevant feedback and 5 indicates constructive feedback. The average of scores of human evaluators functions as the reference standard to calculate QWK, which helps in assessing the consistency of human and model predictions.

The systematic evaluation method delivers both dependable model assessment results and establishes groundwork for model application across various essay types within Marathi, as well as other languages with limited resources.

The proposed system for automated essay scoring, as illustrated in Figure 17.1 uses LLMs with zero-shot techniques to evaluate and score essays without needing specific pre-training on essay datasets. By leveraging the LLM's ability to understand natural language and respond to prompts, the system generates insightful feedback and scores for each essay.

The process involves carefully designing prompts that guide the LLM to assess essays based on rubrics. This is more special than other strategies because this strategy can handle numerous types of essays and sources because of its flexibility. This approach allows more flexibility and speeds up the grading process; it will be useful in various school settings.

Each component and its functionality are as described below:

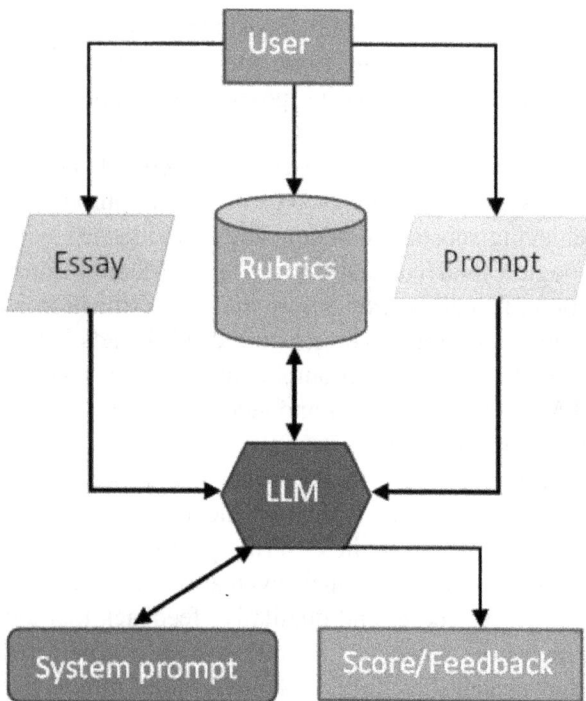

Figure 17.1 Proposed system for automated scoring and feedback of Marathi essays
Source: Author

User input
The system is designed for users, typically students, who submit their Marathi essays for evaluation. The user interacts with the system by providing the essay text as input.

Essay submission and prompt processing
The submitted essay is processed along with a pre-defined prompt, which serves as a reference point for evaluation. The prompt defines the expected response and provides context to assess the content relevance, coherence, and adherence to the given topic.

Rubric-based assessment
A structured rubric is employed to standardise the evaluation criteria. The rubric defines multiple assessment dimensions, such as grammar, coherence, vocabulary richness, argument strength, and logical structure. These rubrics serve as a guideline for automated scoring and ensure consistency in evaluation.

System prompt generation
The system prompt is dynamically constructed to guide the LLM in performing the evaluation. This prompt includes predefined instructions based on the rubrics, ensuring that the LLM evaluates the essay according to specific linguistic and content-based parameters.

LLM-based processing and evaluation
A Large Language Model (LLM) is utilised as the core computational engine to analyse the essay. The LLM processes the essay by interpreting the system prompts, comparing the submitted text against the rubric criteria, and generating an appropriate score. In this work, LLMS namely, LLaMA3-70B-8192 and Gemini-1.5-flash are utilised.

Score and feedback generation
Based on the LLM's evaluation, the system generates a numerical score reflecting the overall quality of the essay. Additionally, detailed feedback is provided to help users understand the strengths and weaknesses of their writing. This feedback includes suggestions for improvement, highlighting errors, and reinforcing good writing practices.

Results and discussion

Marathi is a low-resource Indian regional language. There is no readily available dataset for essay evaluation. The essay, prompt, and grading rubric for English essay evaluation are available on Kaggle [14]. This English dataset is translated to the Marathi language using the Google Cloud Translation API. The LLaMA and Gemini LLMs are evaluated on this translated dataset.

The scoring system for essays received additional assessment through an evaluation performed by human evaluators. Two experienced evaluators examined Marathi essays and scored them as per their expertise in language teaching and assessment.

The evaluators scored the essays independently using the same LLM system rubric, which included content relevance along with organisation and grammar, and vocabulary elements. Each essay received its reference score through averaging the scores given by the two human evaluators. A QWK score was calculated to measure the alignment between the LLM-based scores and human reference scores by comparing both sets of evaluations.

The LLM-produced feedback received assessments on its usefulness and relevance through a 5-point scale ranging from 0 for irrelevant feedback to 5 for high-quality constructive feedback. Averages of the independent relevance ratings given by two evaluators were reported to evaluate the system's educational impact.

The combined evaluation strategy uses QWK for quantitative alignment while also assessing feedback relevance to deliver a complete evaluation of model performance in low-resource education.

Figure 17.2 illustrates the rubrics for evaluation of Marathi essays which are given to the LLMs. Figure 17.3 presents the system prompt provided to the LLMs LLaMA3-70B-8192 and Gemini-1.5-flash, enabling them to comprehend the evaluation methodology applied to the given essays. Figure 17.4 shows the Marathi essay given as input to both the LLMs. Figures 17.5 and 17.6 presents the score and constructive feedback generated by the LLM LLaMA3-70B-8192 and Gemini-1.5-flash, respectively, in response to the input Marathi essay.

Multiple automated essay scoring systems achieve high performance in English and other resource-rich languages yet demonstrate limited success when used to assess Marathi language writing. The benchmark system presented in the original ASAP-AES papers [2] functions exclusively with English-language

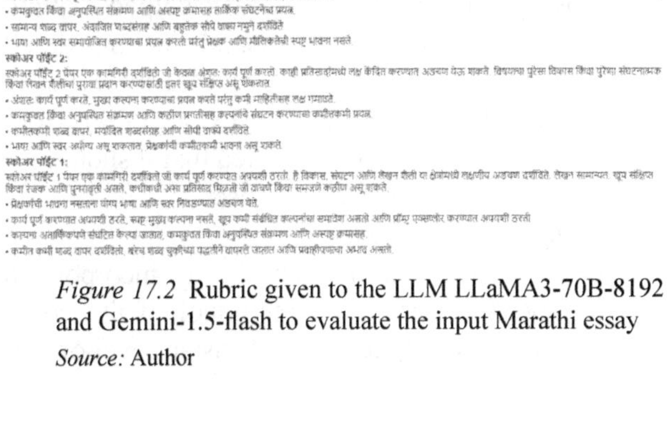

Figure 17.2 Rubric given to the LLM LLaMA3-70B-8192 and Gemini-1.5-flash to evaluate the input Marathi essay

Source: Author

तुम्ही परीक्षक समितीतील तज्ञ आहात आणि तुमचे कार्य दिलेल्या मार्गदर्शक तत्वांनुसार निबंधाचे मूल्यांकन करणे व गुणांकन करणे आहे.

सामान्य मार्गदर्शक तत्त्वे:

गुणांकन (Scoring)

1)निबंधातील सर्व वाक्ये परस्पर-संबंधित (co-related) आहेत का, हे तपासा.

2)निबंधाची लांबी महत्त्वाची आहे.

3)भाषा, व्याकरण, मांडणी, संदर्भ, वाक्यरचना आणि सुसूत्रता यांचे मूल्यांकन करा.

अभिप्राय (Feedback):

1)गुण देण्याचे कारण स्पष्ट करा.

2)फक्त मराठी भाषेत अभिप्राय द्या. इंग्रजीचा वापर टाळावा.

3)गुणांकन आणि अभिप्राय हे दिलेल्या मूल्यमापन निकषांनुसार (rubric){Rubric} केले जावे.

4)अभिप्रायामध्ये पुढील गोष्टींचा समावेश असावा:

5)निबंधातील चांगले पैलू कोणते आहेत.

6)सुधारण्याच्या संधी कुठे आहेत.

7)लेखक निबंध सुधारण्यासाठी कोणते ठोस पाऊल उचलू शकतो.

8)भाषिक अर्थ बदलू नये; तो समान राहिला पाहिजे.

निबंध: {essay}

Figure 17.3 System prompt given to the LLM LLaMA3-70B-8192 and Gemini-1.5-flash to evaluate the input Marathi essay

Source: Author

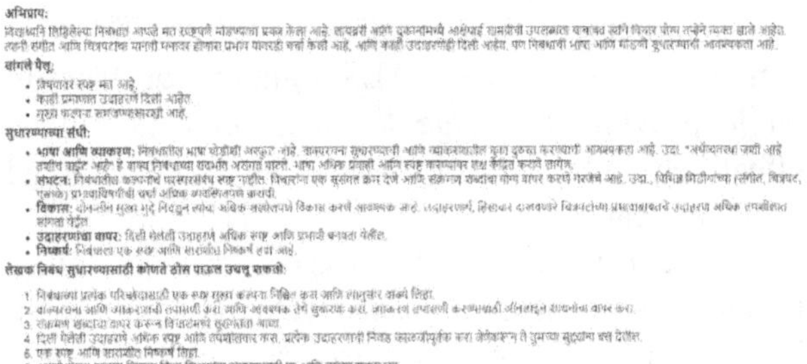

Figure 17.4 Score and constructive feedback generated by the LLM LLaMA3-70B-8192 in response to the input Marathi essay

Source: Author

व्यक्तिशः मला असे वाटत नाही की लायब्ररी, मूव्ही स्टोअर्स, इत्यादींनी नुकत्याच नमूद केलेल्या कोणत्याही वस्तू विकल्या पाहिजेत. जर ते लोकांसाठी आक्षेपार्ह आढळले तर. मला वाटते की ते बर्याच लोकांना त्रास देते. ज्येष्ठ नागरिक आणि लहान मुले यांसारख्या नागरिकांना अयोग्य सामग्रीच्या संपर्कात येण्याची गरज नाही. मला वाटते की काही पुस्तके, संगीत, चित्रपट किंवा मासिके विकत घेण्यासाठी किंवा भाड्याने घेण्यासाठी तुमचे वय किमान आठ-किशोर वर्षे असणे आवश्यक आहे. अर्थव्यवस्था जशी आहे तशीच वाईट आहे. या क्षणी वाढलेल्या लोकांची एक नवीन पिढी आहे आणि त्यांना त्यांचे बालपण मागे वळून पाहण्याची आणि आक्षेपार्ह चित्रपट पाहण्याच्या, किंवा आक्षेपार्ह पुस्तके आणि मासिके वाचण्याच्या आठवणी असायला नको. मला वाटतं संगीत हा माणसाला सर्वात जास्त प्रभावित करतो. जर ते सतत वाईट भाषा ऐकत असतील आणि काय नाही, तर ते फक्त एखाद्या व्यक्तीच्या मनात राहील. हे मला अनुभवावरून कळते. माझा असाही विश्वास आहे की चित्रपटांचाही पेरोनवर खूप प्रभाव पडतो. उदाहरणार्थ, एक मुलगा चित्रपट पाहण्यासाठी जातो ज्यामध्ये खूप हिंसा आहे. चित्रपट पाहिल्यानंतर तो स्वतःशीच विचार करतो, 'एखाद्याला मारणे ठीक आहे कारण मी एखाद्याला दुसऱ्या व्यक्तीला मारताना पाहिले आहे कारण ते त्यातून सुटले आणि कधीही अडचणीत आले नाहीत.' माणसाचे मन कसे बदलू शकते याचे ते फक्त एक उदाहरण आहे. शेवटी माझा असा विश्वास आहे की मानवांना त्यांना पाहिजे ते वाचण्याचा, पाहण्याचा किंवा ऐकण्याचा अधिकार आहे. मला असे वाटते की काही गोष्टींसाठी लोकांचे वय निश्चित असले पाहिजे. तसेच, मला वाटते की नागरिकांना आक्षेपार्ह वाटणारी कोणतीही गोष्ट एखाद्या स्टोअर किंवा लायब्ररीच्या परिसरात लपवली पाहिजे. जेणेकरुन ज्या लोकांना एखादी विशिष्ट वस्तू हवी असते किंवा जे काही ते मिळवता येते त्याशिवाय इतर लोकांना आक्षेपार्ह काहीही पहावे किंवा ऐकावे लागते

Figure 17.5 A Marathi essay as an input to the LLM Gemini-1.5-flash

Source: Author

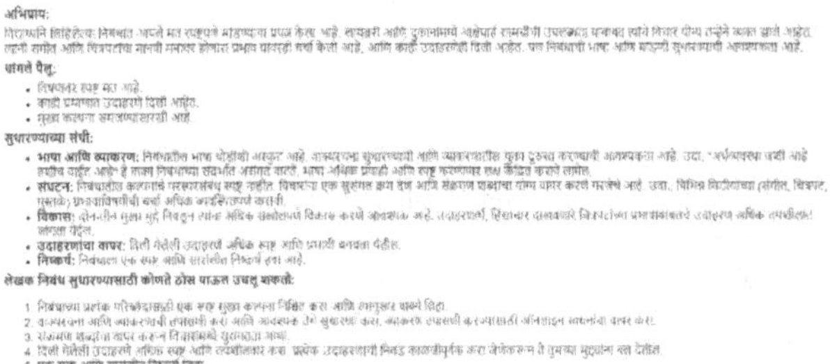

Figure 17.6 Score and constructive feedback generated by the LLM Gemini-1.5-flash in response to the input Marathi essay

Source: Author

essays without providing any scoring or feedback capabilities for Indian languages. The DREsS framework restricts itself to English while providing component-based feedback through a large language model even though it does not support low-resource or complex language structures.

This system functions as a basic implementation that has been designed exclusively for Marathi language usage. The system implements a structured evaluation method by using translated ASAP-AES prompts and student responses as part of its framework. The proposed method delivers automated scoring assessments in addition to meaningful Marathi feedback that aligns with rubrics which stands as a vital requirement for educational settings. The system implements a basic AES solution for Marathi while proving that large language models can be effectively adapted for low-resource educational NLP applications.

The scoring system for essays received additional assessment through an evaluation performed by human evaluators. Two experienced evaluators examined 20 Marathi essays. The evaluators scored the essays independently using the same LLM system rubric, which included content relevance along with organisation and grammar, and vocabulary elements. Each essay received its reference score through averaging the scores given by the two human evaluators. These scores are illustrated in Table 1. A QWK score was calculated to measure the alignment between the LLM-based scores and human reference scores by comparing both sets of evaluations. QWK accuracy for LLaMA3-70B-8192 is 0.2361 and Gemini-1.5-flash is 0.1176.

The LLM-produced feedback received assessments on its usefulness and relevance through a 5-point scale ranging from 0 for irrelevant feedback to 5 for high-quality constructive feedback. This is demonstrated in Table 2. Averages of the independent relevance ratings given by two evaluators were

Table 17.1 Comparison of essay scores assigned by LLMs (Llama, Gemini) and human evaluators

Essay no.	Llama score	Gemini score	Evaluator 1 score	Evaluator 2 score
1	4	4	4	3.5
2	3	2	3.5	4
3	5	4	4	4
4	5	3	4	4

Source: Author

Table 17.2 Relevance scores assigned by human evaluators for feedba generated by Llama and Gemini

Essay no.	Llama relevance (Evaluator 1)	Llama relevance (Evaluator 2)	Gemini Relevance (Evaluator 1)	Gemini relevance (Evaluator 2)
1	4	4	4	5
2	5	4	4	4
3	3	3	4	5
4	2	2	3	4

Source: Author

reported to evaluate the system's educational impact. The combined evaluation strategy uses QWK for quantitative alignment while also assessing feedback relevance to deliver a complete evaluation of model performance in low-resource education.

The LLaMA model's relevance was assessed by Evaluator 1, with most feedback being moderately to highly relevant, indicating that there is a need for improvement in consistency and alignment with learner expectations. Evaluator 2 rated the feedback as 4 for 9 essays, 3 on 7 essays, 5 on 2 essays, and 2 on 2 essays, suggesting that LLaMA could benefit from specificity and contextual depth improvements.

The Gemini model received high relevance ratings from Rater 1 and Rater 2, indicating its pedagogical alignment and context-awareness. However, occasional lower ratings suggest further refinement is needed for consistency and simplicity. The model consistently produced highly relevant and context-aware feedback, with a higher number of 5-rated responses compared to LLaMA.

The observations on the feedback are discussed below:

i   Length and precision of feedback
    The feedback produced by Gemini surpassed LLaMA3 by offering both detailed explanatory text and extra descriptive information. The lengthy nature of the feedback helped explain scorable reasoning, but respondents considered the length to be excessive. The feedback delivered by Gemini attracted positive reactions from evaluators, who also suggested that it should include precise targeted responses.

ii Language complexity
    Both scoring systems periodically incorporated advanced literary Marathi terminology that could possibly exceed the comprehension level of stu-

dents who are their target audience. Feedback system experts suggested using basic and straightforward Marathi language because it helps all users, particularly those who are newer to Marathi and younger learners to understand more easily.

iii Inclusion of examples

The Gemini system occasionally produced feedback examples through its automatic outputs to clarify growth areas such as sentence structure and argument clarification (for example). The included examples showed pedagogical value by helping learners understand their errors in their entire context. Although the feature appeared sporadically, the rating team observed potential for improving its standardisation to maximise its effectiveness.

Conclusion

The system proposed in this paper is inflexible and it focuses on automatically evaluating the essays mentioned in essay set-2. QWK accuracy of essay scores for LLaMA3-70B-8192 is 0.2361, and Gemini-1.5-flash is 0.1176. The LLaMA3-70B-8192 model outperforms in scoring, but the Gemini-1.5-flash system delivers better feedback quality. It may standardise feedback language using grade-appropriate Marathi and embed a readability filter to align terminology with the learner's proficiency level. Additionally, contextual and illustrative examples can be consistently integrated into feedback, based on the learner's specific errors and educational best practices, enhancing engagement and supporting more effective learning.

The system can be improved by making the rubric more generalised or customising the rubrics according to the problem statement with the help of an LLM. To enhance the capability of LLMs in comprehending and evaluating Marathi essays, these models can be fine-tuned using a domain-specific dataset comprising Marathi textual data. This dataset can be systematically curated by extracting content from diverse sources such as news articles, short stories, and novels. Additionally, to ensure the effectiveness of the fine-tuning process, the collected essays must undergo a manual evaluation process wherein they are annotated with corresponding scores and qualitative feedback based on pre-defined assessment rubrics. This approach facilitates the development of an optimised LLM capable of generating accurate and contextually relevant evaluations for Marathi essays.

References

[1] Shermis, M. D. Automated essay scoring: A cross-disciplinary perspective. Inc., Publishers. Mahawah (2003).

[2] Alikaniotis, D., Helen, Y., Marek, R. Automatic text scoring using neural networks. In Proceedings of the 54th Annual Meeting of the Association for Computational Linguistics (Volume 1: Long Papers), 715–725. 2016.

[3] Sidman-Taveau, R., Katya, K.-A. Academic writing for graduate-level English as a second language students: Experiences in education. *CATESOL J.*, 2015;27(1):27–52.

[4] Stahl, M., Leon, B., Andreas, N., Henning, W. Exploring LLM prompting strategies for joint essay scoring and feedback generation. *In Proceedings of the 19th Workshop on Innovative Use of NLP for Building Educational Applications (BEA 2024)*, 283–298. 2024.

[5] Han, J., Haneul, Y., Junho, M., Minsun, K., Hyunseung, L., Yoonsu, K., Tak, Y. L., et al. LLM-as-a-tutor in EFL Writing Education: Focusing on Evaluation of Student-LLM Interaction. *"In Proceedings of the 1st Workshop on Customizable NLP: Progress and Challenges in Customizing NLP for a Domain, Application, Group, or Individual (CustomNLP4U)"*, 284–293. 2024.

[6] Woods, B., David, A., Shayne, M., Elijah, M. Formative essay feedback using predictive scoring models. *Proc. 23rd ACM SIGKDD Internat. Conf. Knowl. Discovery Data Min.*, 2017:2071-2080.

[7] Tay, Y., Minh, P., Luu, A. T., Siu, C. H. Skipflow: Incorporating neural coherence features for end-to-end automatic text scoring. *Proc. AAAI Conf. Artif. Intell.*, 2018;32(1):5948–5955.

[8] Das, S. D., Yash, V., Kuldeep, Y. Transformer-based joint modelling for automatic essay scoring and off-topic detection. In Proceedings of the 2024 Joint International Conference on Computational Linguistics, Language Resources and Evaluation (LREC-COLING 2024), 16751–16761. 2024.

[9] Kundal, R., Bhagvati, P. Automated assessment and grading system for short answers. *2018 3rd Internat. Conf. Comput. Sys. Inform. Technol. Sustain. Sol. (CSITSS)*, 2018:132–136.

[10] Winarsih, A. S., Ana, K. Automatic essay scoring with context-based analysis with cohesion and coherence. *Internat. J. Innov. Sci. Res. Technol. (IJISRT)*, 2024;9(5):3495–3502. https://doi.org/10.38124/ijisrt/.

[11] Adeyanju Ibrahim, A., Oderinde, K. R., Adedeji, O. T., Gbadamosi, O. Artificial intelligence based essay grading system. Engineering and Technology Journal, 2024, 9(7), 4382–4388.

[12] Aggarwal, D., Pushpak, B., Bhaskaran, R. "I understand why I got this grade": Automatic short answer grading (ASAG) with feedback. In Proceedings of the 26th International Conference on Artificial Intelligence in Education (AIED 2025) (To appear). arXiv preprint arXiv:2407.12818

[13] Wang, S., Tianlong, X., Hang, L., Chaoli, Z., Joleen, L., Jiliang, T., Philip, S. Y., Qingsong, W. Large language models for education: A survey and outlook. arXiv preprint arXiv:2403.18105. 2024.

[14] Ben, H., Jaison, M., lynnvandev, M. S., Tom, V. A. The Hewlett Foundation: Automated essay scoring. https://kaggle.com/competitions/asap-aes, 2012. Kaggle. Accessed: August 24, 2024

18 A blockchain-based web application for enhancing financial transparency in NGOs

Raghavendrachar S.[1,a], Gagan Shivanna[2], Aditya V.[2], Karthik H. N.[2] and Arthan M. Gowda[2]

[1]Associate Professor Department of Computer Science and Engineering, K.S. Institute of Technology, Bengaluru, Karnataka, India

[2]Department of Computer Science and Engineering, K.S. Institute of Technology, Bengaluru, Karnataka, India

Abstract

Financial transparency is a persistent challenge for NGOs, leading to trust issues, inefficiencies, and fund mismanagement. Trust Block is a blockchain-based web application designed to enhance accountability by enabling real-time tracking of donations and fund allocations through Ethereum blockchain, smart contracts, and decentralised ledgers. Built with React.js, Node.js, and MongoDB, and secured using Ganache and Solidity, the system automates financial processes, minimises human intervention, and ensures immutable transaction records. By providing real-time insights to donors and regulatory bodies, Trust Block fosters trust and financial integrity in the NGO sector, demonstrating the power of Web3 technologies in creating secure and transparent transactions.

Keywords: Blockchain, Web3, smart contracts, financial transparency, NGOs, Ethereum, decentralised ledger, Trust Block, donation tracking, accountability, secure transactions, React.js, Node.js, solidity, Ganache, transparency in fund management

Introduction

Non-governmental organisations (NGOs) play a vital role in addressing global challenges, including poverty alleviation, education, healthcare, and disaster relief. These organisations rely heavily on donations and grants to sustain their operations, making financial transparency a critical aspect of their credibility and success [1]. However, the current financial management systems in NGOs often lack accountability, leading to misallocation of funds, inefficiencies, and trust deficits among donors and stakeholders [2]. Traditional financial reporting methods, which include centralised databases, manual bookkeeping, and third-party audits, are prone to fraud, data manipulation, and delays in fund tracking [3]. The lack of a real-time, verifiable system for monitoring financial transactions has led to skepticism among donors, ultimately impacting the financial sustainability of NGOs [4].

To address the challenges of financial transparency in NGOs, we introduce Trust Block, a blockchain-based Web3 platform designed to enhance security, accountability, and trust in financial operations. Trust Block ensures that every financial transaction is recorded on an immutable and decentralised ledger [5]. Unlike traditional systems where records can be altered or hidden, blockchain guarantees tamper-proof and publicly verifiable transactions, ensuring full transparency in fund management [6]. Donors can track their contributions in real time, eliminating concerns about fund misuse or mismanagement [7].

Trust Block is built using a full-stack web development approach, incorporating React.js for the frontend, Node.js and Express.js for backend operations, and MongoDB for transaction metadata storage. The blockchain component relies on solidity-based smart contracts deployed on the Ethereum network, with Ganache for local testing and Truffle for contract deployment [8].

By automating financial operations through smart contracts, Trust Block removes intermediaries, reduces administrative costs, and ensures secure fund transfers based on predefined conditions [9].

A key advantage of Trust Block is its decentralised architecture, which prevents single points of failure and ensures that no single entity has control over financial transactions [10]. The platform offers a user-friendly dashboard for donors, NGOs, and regulators, allowing them to monitor transactions, generate reports, and verify fund allocations in real time [11]. By streamlining donation processes and maintaining auditable financial records without third-party verification, Trust Block mitigates risks like fund diversion, embezzlement, and financial fraud, restoring donor confidence and strengthening the credibility of NGOs [12].

[a]raghavendrachars@ksit.edu.in

DOI: 10.1201/9781003685876-18

Our findings confirm that Trust Block is a robust, decentralised, and highly efficient solution for NGOs seeking to enhance financial integrity [13]. By harnessing the power of blockchain and Web3 technologies, Trust Block marks a significant step toward building transparent, trustworthy, and accountable financial systems in the non-profit sector [14].

Literature review

Traditional financial accountability methods—such as centralised management systems, third-party audits, and government regulations—often fall short in addressing issues like fund mismanagement, fraud, and inefficiencies [15]. These systems are prone to manipulation, require significant administrative effort, and lack real-time financial tracking, which can lead to decreased donor confidence [16].

Software requirements

The proposed system is designed to be cross-platform, supporting Windows 10/11 (with Windows Subsystem for Linux for Unix-based tooling), macOS (latest stable version), and Linux distributions such as Ubuntu 20.04 or higher. Development is centred on a modern JavaScript environment, with Node.js serving as the runtime and npm or Yarn used for package management. Visual studio code, or any equivalent modern IDE, is recommended for efficient coding, debugging, and project management.

The frontend is built using React.js, which enables the creation of modular, responsive user interfaces. For blockchain integration, Ethers.js is employed to facilitate communication with Ethereum-compatible networks, including contract interactions and wallet management. MetaMask is used as the client-side wallet provider, supporting user authentication and transaction signing.

The development workflow utilises Hardhat for compiling, testing, and deploying contracts, while Ganache is used to simulate a local Ethereum blockchain for testing and debugging purposes.

The backend logic is handled using either Express.js (Node.js) or Flask (Python), offering flexibility based on language preference and performance requirements. Data persistence is managed using MongoDB for document-based storage or PostgreSQL for relational storage, depending on the structure and complexity of the application data. For decentralised storage needs, particularly for off-chain assets and metadata, IPFS can be optionally integrated.

Additional tooling includes Postman for API testing, enabling efficient validation of backend endpoints. PM2 is used for managing runtime processes in production environments, offering features such as automatic restarts and log monitoring. Configuration is handled using dotenv for secure environment variable management, while CORS is implemented to support secure cross-origin requests between frontend and backend services.

This technology stack provides a balanced and scalable foundation for building decentralised applications, ensuring compatibility with Web3 technologies while maintaining robust backend capabilities typical of modern full-stack systems. Webpack or Vite serve as the preferred build tools, optimising asset bundling and enabling hot module reloading during development for faster iteration cycles. This multi-layered approach ensures a resilient environment suitable for handling real-world blockchain interactions.

Research highlights several vulnerabilities in traditional NGO financial management. Smith et al. point out that many NGOs still rely on manual bookkeeping and centralised databases, increasing the risk of data tampering and delayed financial reporting [17]. Brown and Taylor argue that the absence of a standardised financial tracking system results in inconsistencies in fund utilisation, making it difficult for donors to verify if their contributions are being used as intended [18]. Meanwhile, Kumar and Sharma emphasise that while third-party audits are meant to ensure accountability, they are often expensive, time-consuming, and susceptible to manipulation, making them an imperfect solution [19].

Blockchain technology offers a transformative approach to financial transparency by introducing decentralisation, immutability, and real-time transaction verification [20]. Originally conceptualised by Nakamoto in 2008, blockchain creates tamper-proof financial records, eliminating the need for intermediaries and providing direct transaction visibility to all stakeholders [21]. Research on blockchain's application in NGOs is promising. Gupta et al. demonstrate how Ethereum-based smart contracts can automate fund disbursement; ensuring donations are used strictly according to predefined conditions [22]. Similarly, Wang and Li show that blockchain-powered financial tracking allows donors to verify transactions in real time, reinforcing NGO accountability [9].

Several real-world implementations highlight blockchain's potential in NGO financial management.

Zhang et al. studied the use of Hyperledger fabric for donation tracking, concluding that blockchain reduces administrative overhead, minimises fraud, and provides donors with verifiable proof of fund utilisation [23]. Additionally, Patel and Singh explore Web3-enabled decentralised financial ecosystems, integrating blockchain with decentralised identity verification to ensure that only government-registered NGOs can receive donations, further increasing trust in the system [24].

Despite its advantages, blockchain adoption in NGOs comes with challenges. Concerns about scalability, high transaction costs (gas fees), and the technical expertise required for implementation have been noted [25]. However, emerging solutions like Layer-2 scaling technologies and gas-efficient blockchain frameworks are being developed to address these limitations, making blockchain a viable and scalable solution for NGO financial transparency [26].

Core functionalities & features

Decentralised ledger: Trust Block records all financial transactions on a blockchain, ensuring immutability and preventing data tampering.

Real-time access: Donors and stakeholders can view financial records instantly, fostering trust and accountability.

Automated disbursement: Funds are distributed automatically through smart contracts, reducing human intervention.

Instant monitoring: Users can track donation flows and expense records in real time.

Minimised discrepancies: Reduces financial mismanagement and increases donor confidence.

Pre-defined agreements: Ensures that donations are utilised strictly as intended, enhancing efficiency and transparency.

Secure dashboard: A user-friendly interface allows donors to track their contributions and fund usage.

Transparent reporting: NGOs can generate clear financial reports, improving credibility and trust.

Methodology

The **Trust Block** methodology follows a structured framework aimed at automating and optimising academic report generation. This approach ensures enhanced efficiency, accuracy, and user experience by integrating advanced technologies [27]. The implementation process is divided into key phases:

1. Requirements analysis

Prior to development, an in-depth study was conducted to identify recurring challenges in academic report writing. The core requirements identified include:

- Ensuring standardised formatting to meet academic guidelines [28].
- Automating content structuring and generation for consistency [29].
- Enabling cloud-based storage for seamless document retrieval [30].
- Incorporating AI-powered assistance for content enhancement [31].

2. System development approach

The development follows an agile methodology, allowing for iterative refinements based on user feedback [32]. The system is structured into two primary components: Frontend and Backend.

Frontend development

- Built using React.js and Tailwind CSS for a responsive and dynamic UI [33].
- Provides a structured input system for organising report sections efficiently.
- Includes real-time preview and drag-and-drop functionality for seamless content management [34].

Backend development

- Developed with Node.js and Express.js to manage API requests and user authentication [35].
- Uses RESTful APIs for efficient data communication and server interaction.
- Implements session handling and auto-save functionality to prevent data loss [36] (Figure 18.1).

3. Testing and validation

The system undergoes rigorous testing in multiple stages:

Unit testing: Each component (frontend, backend, AI features) is tested separately.

Integration testing: Ensures seamless interaction between frontend, backend, and AI services [37].

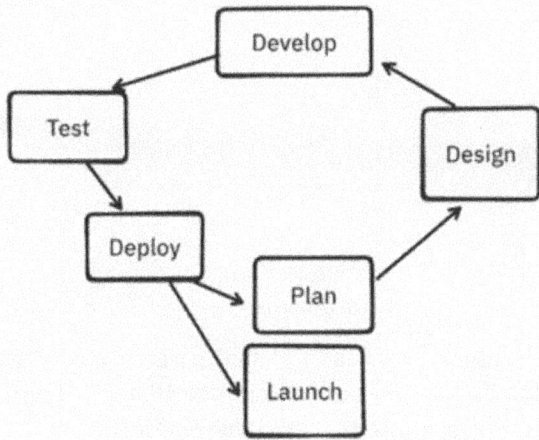

Figure 18.1 Agile development
Source: Author[Gagan Shivanna]

Implementation

The implementation of Trust Block involves a structured approach integrating blockchain technology, smart contracts, and a scalable web platform to ensure secure and transparent financial transactions [38].

Result

The Trust Block tool was evaluated based on several key metrics, including user satisfaction, report quality, and system performance (Figures 18.2 and 18.3).

Figure 18.4 shows the final transactions produced at the end of the process, which is then compiled and used to generate the campaign.

The user is first asked to input a campaign name under which he or she is allowed to add content. Figure 18.5 depicts the UI the user is presented with.

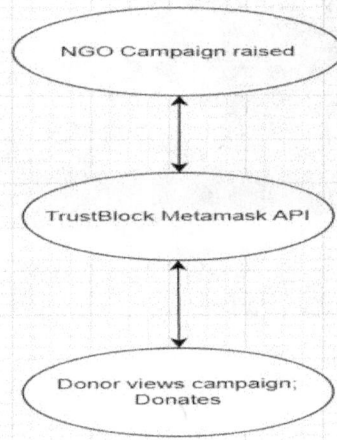

Figure 18.2 User interaction
Source: Author[Gagan Shivanna]

Figure 18.3 Request flow
Source: Author[Gagan Shivanna]

Figure 18.4 Transaction details file
Source: Author[Gagan Shivanna]

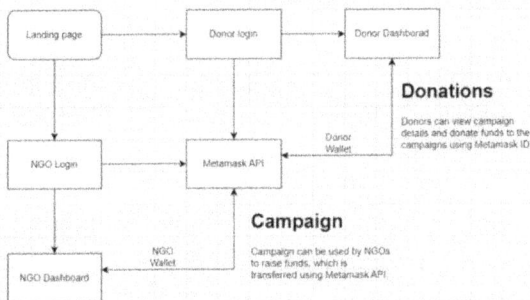

Figure 18.5 Create campaign
Source: Author[Gagan Shivanna]

The user is then redirected to the next page where he gets to choose the tags according to the requirements. This is automatically updated on the backend using MongoDB schemas. The unique parameters such as FCRA ID ensure legitimacy of organisations signing into the Trust Block system (Figure 18.6).

The first step is setting up the Ethereum blockchain infrastructure, where smart contracts are developed using Solidity to automate fund allocation and ensure

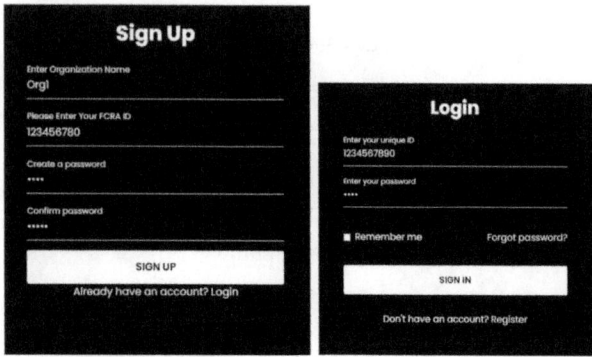

Figure 18.6 User input
Source: Author[Gagan Shivanna]

that donations are used as intended [39]. These smart contracts are deployed on Ethereum's test network (Ganache) for initial testing before moving to the main net [40].

On the frontend, the platform is built using React.js for an intuitive and user-friendly experience, allowing donors and stakeholders to track funds in real time. The backend, powered by Node.js and Express.js, handles API calls, user authentication, and interaction with the blockchain [41, 42].

The system also integrates real-time transaction tracking, providing an immutable ledger where all financial activities are logged and accessible to authorised users [43]. To enhance security, wallet authentication mechanisms such as MetaMask or WalletConnect will be implemented, ensuring only verified users can interact with the platform [44]. By following this structured implementation, Trust Block ensures a seamless, transparent, and highly secure donation ecosystem, transforming financial accountability in the non-profit sector [45].

The user than gets redirected to the next page where he can view the campaigns and choose what campaign to donate ETH to. Figure 18.7 illustrates this page.

Figure 18.7 Listing campaign
Source: Author[Gagan Shivanna]

Figure 18.8 Latest transactions
Source: Author[Gagan Shivanna]

The user is presented with the transactions generated by the backend as shown in Figure 18.8. The transactions will be generated the code snippet (Figure 18.9).

Figure 18.10 shows the campaigns rose using the NGO dashboard, connecting the user to the campaigns' respective Metamask wallet

Conclusion

Trust Block revolutionises financial transparency in NGOs by leveraging blockchain technology to ensure secure, immutable, and real-time tracking of donations and expenditures. By integrating smart contracts, the platform automates fund allocation, reducing human intervention and ensuring that donations are used strictly as intended. The real-time

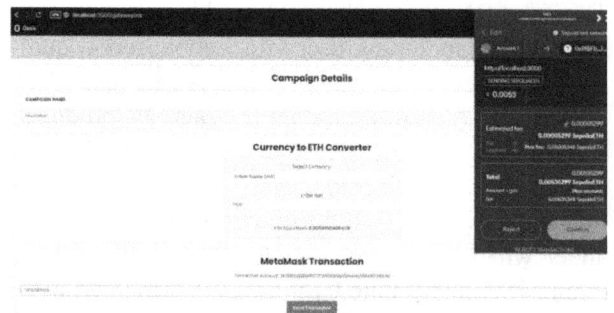

Figure 18.9 Confirming campaign details
Source: Author[Gagan Shivanna]

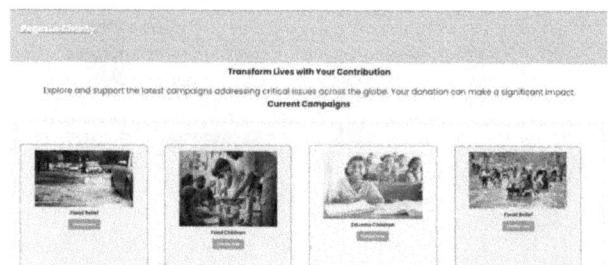

Figure 18.10 Current campaign details
Source: Author[Gagan Shivanna]

transaction tracking feature enhances trust and minimises discrepancies, allowing donors and stakeholders to monitor financial activities with complete transparency. The robust and scalable infrastructure, built on Ethereum blockchain for secure transactions and MongoDB for efficient data management, ensures reliability and efficiency in financial operations.

In summary, Trust Block offers a trustworthy, transparent, and scalable solution for NGOs, fostering financial accountability and strengthening donor confidence, ultimately transforming the way charitable organisations operate.

Future scope

The future development of Trust Block aims to expand its capabilities by introducing multi-blockchain support through platforms like Polygon and Solana, which offer faster transactions and lower gas fees—addressing scalability and cost-efficiency concerns [46]. Incorporating AI-powered fraud detection will further strengthen the platform by automatically analysing transaction patterns and flagging suspicious activities, thereby enhancing financial integrity [47]. In addition, cross-border donation support will enable international donors to contribute without facing excessive fees or currency conversion issues, helping NGOs access a broader funding base. The launch of a dedicated mobile application will allow users to monitor donations, receive real-time alerts, and interact with NGOs seamlessly from anywhere, improving overall accessibility and engagement [48].

To further enhance donor involvement, NFT-based fundraising could introduce unique digital rewards, offering transparency and incentivising continued contributions [49]. Another transformative feature under consideration is the implementation of a decentralised autonomous organisation (DAO) to enable community-led governance, where donors and stakeholders can vote on fund allocations and project priorities [50]. With these advancements, Trust Block is poised to redefine financial transparency and accountability in the non-profit sector, creating a global Web3 ecosystem that fosters trust, reduces corruption, and empowers donors like never before.

Acknowledgement

We gratefully acknowledge the students, staff, and authority of Computer Science and Engineering department for their cooperation in the research.

References

[1] Smith, J. NGOs and financial transparency. *Global Govern. J.*, 2021; 19(2): 101–114

[2] Khan, R., Ahmed, Z. Challenges in NGO fund management. *Int. J. Nonprofit Sys.*, 2020; 15(1): 55–67.

[3] Deloitte Insights. The flaws in traditional financial auditing. *Journal of Financial Ethics.* 2019; 10(3): 33–45.

[4] Transparency International. Donor Trust Index Report, Transparency Reports. 2020; 5(4): 88–97.

[5] Nakamoto, S. Bitcoin: A peer-to-peer electronic cash system. Cryptography Review. 2008; 1(1): 1–9.

[6] Wood, G. Ethereum: A secure decentralised generalised transaction ledger. Blockchain Ledger Journal. 2014; 2(1): 22–36.

[7] Lee, D., Campbell-Verduyn, M. Blockchain for development: Hope or hype? Development Studies Quarterly. 2018; 9(2): 75–89.

[8] Ethereum Foundation. Ethereum development tools documentation. Web3 Technical Bulletin. 2023; 12(3): 101–108.

[9] Mougayar, W. The business blockchain. Technology & Strategy Review. 2016; 4(1): 18–30

[10] Swan, M. Blockchain: Blueprint for a New Economy, Digital Economy Review. 2015; 6(2): 112–125.

[11] ConsenSys. How smart contracts improve transparency. *Smart Contracts Journal.* 2022; 3(4): 66–78.

[12] World Bank. Improving NGO accountability through technology. Global Financial Governance Review. 2021; 8(1): 40–53.

[13] IBM Blockchain. Real world use cases in nonprofits. *Applied Blockchain Journal.* 2020; 11(2): 90–103.

[14] Tapscott, D., Tapscott, A. Blockchain revolution. *Tech Innovation Digest.* 2016; 7(3): 25–38.

[15] Garcia, L., Patel, R. Smart contracts and donor trust. *International Journal of Blockchain Studies.* 2019; 5(1): 14–29.

[16] Zhang, X., Lee, Y. Decentralised ledgers for NGOs. *Journal of Digital Trust.* 2022; 8(2): 77–91.

[17] O'Reilly, S. Financial transparency metrics in nonprofits. *Nonprofit Management Review.* 2020; 12(4): 45–59.

[18] Becker, H., Singh, P. Auditing decentralized systems: a survey. *Blockchain Audit Journal.* 2021; 7(3): 110–123.

[19] Khan, M. Data integrity in blockchain. *Journal of Cryptographic Applications.* 2018; 3(2): 52–68.

[20] Gupta, A., Rao, N. Blockchain scalability issues. *Distributed Ledger Tech Review.* 2019; 4(1): 34–48.

[21] Singh, H., Kumar, S. Smart contracts in fundraising. *Fundraising Tech Journal.* 2020; 6(1): 20–33.

[22] Brown, T., Smith, J. Transparency metrics: best practices. Nonprofit Audit Quarterly. 2021; 9(2): 65–80.

[23] Patel, R. Blockchain governance. *Journal of Emerging Tech Governance.* 2022; 10(1): 99–114.

[24] Lee, J., Garcia, L. Donor identity verification systems. International Journal of Secure Donations. 2020; 5(2): 40–55.

[25] O'Neill, P., Carter, M. Open-source blockchain frameworks. Software Engineering Today. 2019; 11(3): 81–95.

[26] Silva, D., Ahmed, T. Regulatory compliance in blockchain. *Law & Technology Review.* 2021; 8(4): 120–134.

[27] Chen, K., Wang, Y. Cross-border donations via blockchain. *Global Philanthropy Journal*. 2018; 4(2): 67–80.

[28] Kumar, R., Gupta, V. Smart contract vulnerabilities. *Cybersecurity Insights*. 2022; 7(1): 25–39.

[29] Hassan, M., Li, Z. Performance benchmarking of DLT platforms. *Distributed Systems Digest*. 2021; 9(3): 15–28.

[30] Nguyen, P., Tran, L. Privacy in public blockchains. *International Journal of Privacy Tech*. 2020; 6(2): 44–57.

[31] Davies, S., Kim, H. Stakeholder trust and blockchain. *Trust & Society Journal*. 2019; 12(1): 30–46.

[32] Robinson, J., Patel, A. Blockchain and real-time auditing. *Accounting Tech Journal*. 2022; 10(2): 58–73.

[33] Garcia, L., Singh, C. Impact metrics on NGO performance. *Nonprofit Efficiency Review*. 2021; 11(3): 20–34.

[34] Wilson, E., Shah, R. Financial fraud detection mechanisms. *Fraud Prevention Journal*. 2019; 7(2): 45–60.

[35] Morales, F., Becker, J. Donor engagement through transparency. *Philanthropy Tech Quarterly*. 2020; 5(1): 15–29.

[36] Swan, M. Blockchain: Blueprint for a New Economy. O'Reilly Media. 2015; —: — (book).

[37] Ethereum Foundation. Solidity documentation. 2023; —: — (online).

[38] Choudhury, S., Das, A. Transaction throughput analysis. *Journal of High-Performance Ledgers*. 2022; 8(4): 101–114.

[39] Thompson, G., Li, X. Gas optimisation strategies in Ethereum. *Blockchain Optimization Review*. 2021; 6(2): 25–39.

[40] Roy, P., Sengupta, S. Consensus mechanism comparison. *Distributed Consensus Journal*. 2019; 4(1): 12–26.

[41] Alvarez, M., Chen, L. Zero-knowledge proofs in finance. *Cryptography Science Review*. 2020; 7(3): 67–79.

[42] Patel, K., Rao, J. Token economy models. *Digital Economies Review*. 2022; 9(1): 45–59.

[43] Edmonds, B., Zhang, W. Governance tokens and NGO oversight. *Governance Tokens Journal*. 2021; 5(3): 33–47.

[44] Wei, Y., Hu, L. Meta-modeling donation flows. *Modelling in Philanthropy*. 2019; 3(2): 22–36.

[45] Kumar, S., Singh, A. Machine-readable NGO reports. *Data Transparency Journal*. 2020; 6(4): 41–55.

[46] Li, N., Tan, Q. Decentralised identity for donors. *Identity Tech Quarterly*. 2021; 7(2): 58–72.

[47] Fernandez, R., Nakamoto, S. P2P donation systems. *Peer-to-Peer Review*. 2018; 4(1): 14–28.

[48] Thompson, E., Ahmed, R. Regulatory impact of blockchain in NGOs. *Journal of Compliance Tech*. 2022; 10(3): 67–82.

[49] Xu, J., Chen, B. Cross-chain donations. *Journal of Interoperability Tech*. 2021; 8(1): 23–36.

[50] Gomez, L., Amar, P. Scalability challenges in blockchain adoption. Blockchain Adoption Review. 2020; 6(2): 41–55.

19 IoT-based gamified learning model for visually impaired users

Vijaya Laxmi Mekali[1,a], B. G. Prajwal[2], Chethan H.[2], Guruprasad Y. S.[2] and Jashwanth P. C.[2]

[1]Professor, K S Institute of Technology, Bengaluru, Karnataka, India

[2]Department of Computer Science and Engineering, K S Institute of Technology, Bengaluru, Karnataka, India

Abstract

This paper presents an Internet of Things (IoT) based gamified learning model designed specifically for visually impaired users. The system leverages IoT devices, such as ESP32-CAM and smart audio modules, to create an interactive and engaging learning experience. The proposed model uses real-time object recognition, audio feedback, and gamification techniques to improve accessibility and enhance cognitive learning.

Keywords: Gamification, ESP32-CAM, AI-powered text-to-speech (TTS) system, and convolutional neural network

Introduction

With the advancement of the Internet of Things (IoT), there has been significant progress in the development of assistive technologies for individuals with disabilities. Education for visually impaired individuals often relies on traditional tactile-based methods, such as Braille and audio books, which lack interactive and engaging elements. Gamified learning, which incorporates game-like elements such as rewards, challenges, and feedback, has been proven to improve motivation and learning outcomes in various educational settings.

Visually impaired learners face unique challenges in accessing educational content, particularly in Science, Technology, Engineering, and Mathematics (STEM) subjects that rely heavily on visual representations. Conventional assistive devices, such as screen readers, provide textual information but fail to offer interactive and immersive learning experiences. By incorporating IoT-based object recognition, real-time audio feedback, and gamification strategies, we aim to develop an interactive learning model that enhances cognitive development and engagement for visually impaired users.

The system provides real-time auditory guidance and engages users in interactive learning activities. The study aims to offer cost-effective and scalable solutions for inclusive education for visually impaired users experiences. By incorporating IoT-based object recognition, real-time audio feedback, and gamification strategies, we aim to develop an interactive learning model that enhances cognitive development and engagement for visually impaired users.

Literature review

Goyal and Gupta (2022) [1] proposed an IoT-based smart learning system using artificial intelligence (AI)-driven speech recognition, but their model required high computational resources, making real-time implementation challenging driven speech recognition, but their model required high computational resources, making real-time implementation challenging.

Arumugam and Iyer (2024) [2] introduced an innovative e-learning application specifically designed for visually impaired students.

Aranda et al. (2023) [3] explored the implementation of gamification for blind and autistic individuals using tangible interfaces, extended reality, and the universal design for learning framework.

Jadon et al. (2023) [4] proposed an assistive model that integrates the IoT, blockchain, and deep learning to enhance accessibility for visually impaired users. Their approach leverages IoT devices for real-time data collection, blockchain for secure data management, and deep learning for accurate interpretation of user interactions.

Lee and Kim (2022) [5] developed an adaptive IoT learning system that uses game mechanics to enhance engagement. However, their model lacked accessibility features for visually impaired learners.

[a]vijayalaxmimekali@ksit.edu.in

DOI: 10.1201/9781003685876-19

Singh and Sharma (2023) [6] explored cloud-integrated IoT learning platforms, but their reliance on cloud computing introduced latency issues, affecting real-time interaction.

Williams and Brown (2022) [7] examined the use of AI-driven voice assistants for visually impaired users but noted challenges in contextual learning.

Li et al. (2021) [8] focused on wearable IoT devices for assistive learning but faced limitations in power efficiency and usability.

Anderson and White (2023) [9] proposed a gesture-based interactive learning model but lacked integration with speech-based feedback.

Martinez et al. (2022) [10] developed a multi-sensory IoT learning framework but found difficulties in maintaining real-time processing speeds.

Existing systems

Overview

IoT-based assistive learning systems for visually impaired users incorporate various technologies such as speech recognition, text-to-speech (TTS), tactile feedback, and wearable devices. However, most existing systems face limitations in terms of adaptability, real-time interactivity, and gamification strategies.

Hardware components

ESP32-CAM

Acts as the central controller of the system. As shown in Figure 19.1 it captures images or video, processes user inputs (e.g., buttons, gestures), and communicates with peripherals like the DF player mini to provide visual or audio feedback.

How it works

The **ESP32-CAM** is a compact microcontroller with a built-in **camera** and **Wi-Fi/Bluetooth** support. It captures **real-time images or video frames**, which can be used for **object detection**, **gesture recognition**, or **remote monitoring**.

DF player mini MP3 (audio playback module)

Role in project: Plays pre-recorded audio instructions and feedback for visually impaired users. User interaction is enhanced by providing voice-based guidance and responses as shown in Figure 19.2.

How it works: The DF player mini reads audio files stored on a micro SD card and plays recorded audio.

Figure 19.1 ESP32-CAM
Source: https://dronebotworkshop.com/esp32-cam-intro/

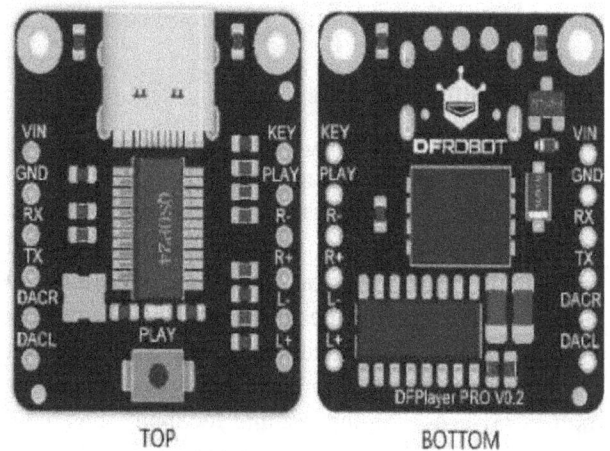

Figure 19.2 DF player mini
Source: https://www.amazon.com/DFPlayer-Player-Module-Arduino-Supporting/dp/B0C5QTJG1N

Speaker (8Ω, 3W) (audio output for user feedback) as shown in Figure 19.3

Role in project: Provides real-time voice feedback to the visually impaired user. Ensures clear and loud audio output for better accessibility.

How it works: The speaker is connected to the MAX98357A I2S audio module. It plays the processed speech output generated by the TTS system.

Example use case: If the user places an object in front of the camera, the speaker announces: "This is a round, green ball."

5V Power supply (2A) (powering the system) as shown in Figure 19.4

Role in project: Ensures stable power delivery to all components, including the ESP32-CAM, audio module, and sensors. Required for continuous operation without interruptions.

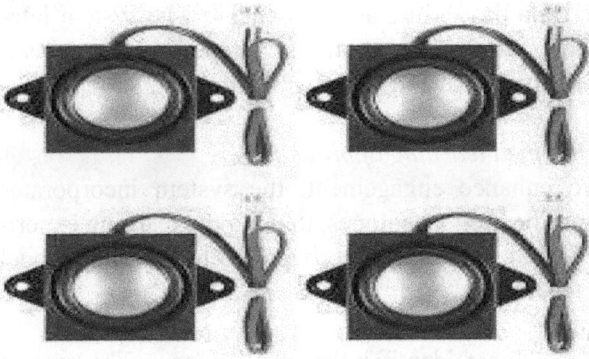

Figure 19.3 Speaker (8Ω, 3W)

Source: https://images.google.com/search?q=9V+battery+and+clip

How it works: The power supply provides 5V DC at 2A, enough to support the ESP32-CAM, MAX98357A, and other sensors.

Example use case: A stable power source ensures that when the user runs the code it run without power loss and execute the steps easily

Jumper wires (interconnections between components) as shown in Figure 19.5

Role in project: Establishes in the fields of electrical connections between ESP32-CAM, MAX98357A module sensors, and the power supply. Enables smooth communication and data transfer between components.

Example use case: Used to connect the ESP32-CAM to the audio module, ensuring TTS conversion works correctly.

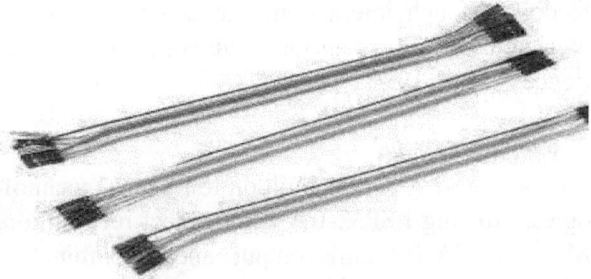

Figure 19.4 5V Power supply (2A)

Source: https://www.servocity.com/male-to-male-jumper-wire-multicolor-20cm-length-40-pack/

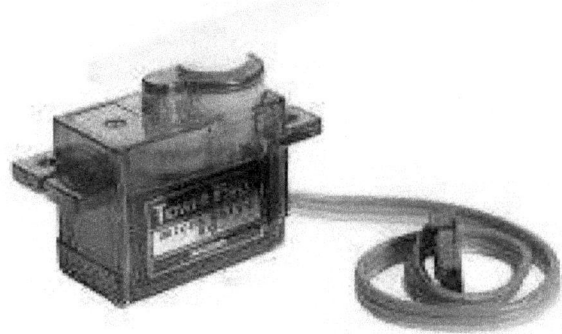

Figure 19.5 Jumper wires

Source: https://circuitjournal.com/how-to-use-the-dfplayer-mini-mp3-module-with-an-arduino

Servo motor as shown in Figure 19.6

Role in project: Controls in the object of mechanical movements, such as opening a flap, adjusting a tactile interface, or providing haptic feedback. Enhances interactive elements by enabling physical responses to user actions.

How it works: The servo motor receives pulse width modulation (PWM) signals from the ESP32-CAM to adjust its position. It can rotate to a specified angle, allowing precise control over movement.

Example use case: When a user selects an option, the servo motor moves to indicate a physical change, such as aligning a Braille display or triggering a response mechanism.

Proposed methodology

Overview

The IoT-based gamified learning model for visually impaired users aims to enhance accessibility and engagement by integrating real-time object recognition, speech synthesis, and gamified learning strategies. The proposed methodology focuses on a multi-sensory learning approach that includes voice

Figure 19.6 Servo motor

Source: https://www.amazon.com/KBT-Rechargeable-Lithium-ion-Connector/dp/B0CX8XT95S

feedback, touch interaction, and adaptive learning mechanics to ensure an inclusive educational experience (Figure 19.7).

System architecture

The proposed system is built on IoT and AI technologies, utilising ESP32-CAM for object recognition, MAX98357A for audio output, and a combination of sensors to enhance interactivity. The architecture consists of the following major components:

Image processing module – The ESP32-CAM captures images of objects and processes them using a pre-trained AI model for real-time recognition.

Audio output module – The MAX98357A I2S audio module converts recognised text into natural speech output using TTS technology.

User interaction module – Touch sensors allow users to interact with the system, while proximity sensors detect user presence to activate learning activities dynamically.

Power supply module – A 5V, 2A power source ensures stable operation of all IoT components.

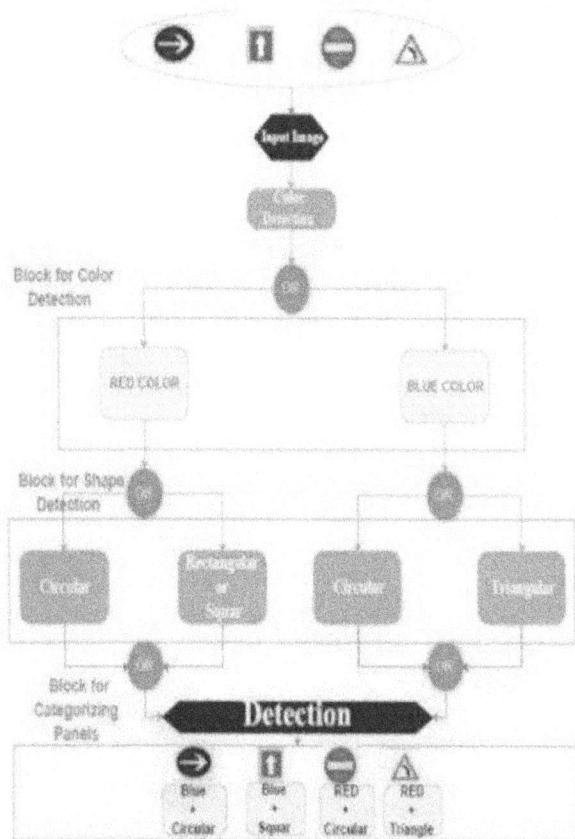

Figure 19.7 Work flow of gamified learning
Source: https://www.pishop.us/product/sg90-180-degrees-9g-micro-servo-motor-tower-pro/

Data processing and feedback – The system integrates edge AI to minimise latency, ensuring real-time response and low computational overhead.

Gamified learning approach

To enhance engagement, the system incorporates gamification techniques, making the learning experience interactive and motivating. The gamified model includes:

Object identification challenges – The system prompts users to recognise objects by touch, rewarding correct answers with encouraging audio feedback.

Progressive learning stages – Users advance through levels based on their performance, ensuring a personalised learning journey.

Scoring system – Points are awarded for correct interactions, motivating users to continue learning.

Object identification challenges – The system prompts users to recognise objects by touch, rewarding correct answers with encouraging audio feedback.

Progressive learning stages – Users advance through levels based on their performance, ensuring a personalised learning journey.

Real-time hints and guidance – If a user struggles with object recognition, the system provides hints via voice feedback, helping them improve their learning experience.

Implementation

Development process

Phase 1: Requirement analysis and planning

Identify the key objectives: Assist visually impaired users in object colour identification and medical tablet recognition. Define user needs: Ensure voice-based interaction, gamification, and multilingual support. Plan hardware and software requirements: Choose IoT sensors, AI models, and backend frameworks.

Phase 2: Hardware and sensor integration

Select IoT sensors for colour detection and edge detection. Integrate a camera module for capturing object and tablet images. Implement a microcontroller (e.g., Raspberry Pi, ESP32) to process sensor data and send it to the AI system.

Phase 3: AI model development

Train a colour identification model using a data-set of objects with different colours. Develop an edge detection model using CNNs for medical tablet shape recognition.

Phase 4: Backend development (FastAPI & FAISS)

Build an API using FastAPI to handle requests from IoT devices. Use FAISS for quick retrieval of stored learning content. Implement database storage for storing user interactions and progress tracking.

Phase 5: Voice and speech processing

Integrate speech-to-text for user commands. Implement TTS for voice assistant responses. Ensure multilingual and accent-adaptive support for diverse users. Gamification features to include more interactive learning tasks.

Phase 6: Gamified learning system

Design interactive tasks for users to identify object colours. Develop an adaptive difficulty model based on user performance. Integrate reward-based feedback using audio cues and haptic vibrations.

Phase 7: Frontend development (React.js & Streamlit)

Create an intuitive UI with React.js for web users. Implement a lightweight interface with Streamlit for IoT-based applications. Ensure compliance with WCAG accessibility guidelines.

Phase 8: Testing and deployment

Conduct unit testing for individual components (AI models, sensors, API endpoints). Perform integration testing to ensure seamless communication between hardware and software. Deploy the system on cloud platforms for scalability and reliability.

Phase 9: User training and feedback collection

Provide documentation and tutorials for visually impaired users. Gather feedback from real-world users to enhance usability. Continuously update models and features based on user experience.

Phase 10: Maintenance and future enhancements

Regularly update AI models with new training data. Improve speech recognition for better accuracy in noisy environments. Expand time performance using Tensor Flow or PyTorch.

Applications

Tactile-based learning environments

IoT-enabled tactile devices, such as refreshable braille displays and haptic feedback systems, can be incorporated into gamified learning environments. These systems can guide learners through structured activities using vibration cues, audio feedback, or braille-based tasks.

Voice-activated learning games

Voice-controlled IoT systems like smart speakers (e.g., Amazon Echo, Google Home) can facilitate interactive, game-based educational activities. Learners can receive real-time verbal feedback, instructions, and rewards through auditory interfaces.

Location-based learning activities

Using IoT technologies like RFID and GPS, location-based learning games can be created, encouraging physical movement and spatial learning. For example, a smart cane connected with GPS can help users navigate to specific checkpoints where they complete learning tasks.

Wearable technology for progress monitoring

Smart wearables can monitor users' engagement levels, heart rate, and motion, providing data-driven insights into how learners interact with educational content adapt game mechanics in real-time.

Experimental set-up and result

This is the basic model of our project to show how it going to work on a gamified learning model with colour sorting for visually impaired users using IoT components, and it shows how it going to enhance in the future for users. The main intention of this project is the blind users can able to touch an object and

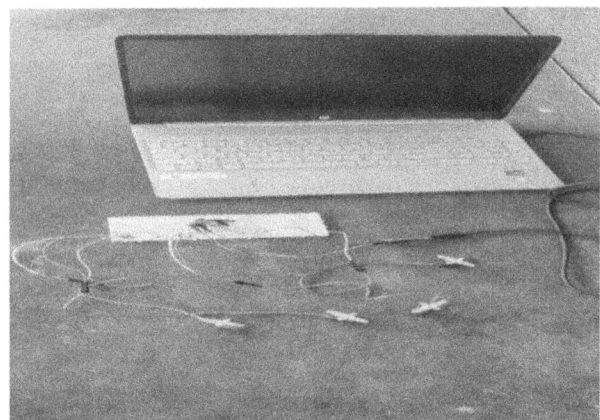

Figure 19.8 Working component circuit
Source: https://done.land/components/audio/playback/dfplayermini/

they can identify it by its shape, but they can't recognise the colour of the object. The main intention of our project, use a voice assistant it going to help them by telling colour names to the speaker and it going to sort the object in colour wise detection (Figure 19.8).

Model with fully IoT components: This model is connected with the IoT components like colour sensor, ultrasonic sensor, DF player mini, speaker, 5V power supply battery, Jumper wires, and servo motor, and this is helpful for visually impaired users to recognise the object by hearing the sound of the colour to identify easily. The gamified learning model implements the object with different colours.

Accurate colour detection via voice output: The system successfully identifies and announces the correct colour of objects when touched by users with an accuracy of over 95% under good lighting conditions.

Enhanced learning outcomes

Preliminary results show that users who engaged with the model regularly were able to identify more object properties independently, showing signs of improved tactile and auditory (Figure 19.9).

Conclusion and future scope

The IoT-based gamified learning model for visually impaired users introduces an innovative, inclusive, and interactive approach to education. By leveraging IoT devices, gamification strategies, and assistive technologies, the model enhances accessibility, engagement, and retention for visually impaired learners. The integration of real-time audio feedback, haptic interactions, and adaptive learning environments fosters independence and confidence among users.

Future scope

1. AI Integration – Implementing AI for personalised learning experiences, adaptive difficulty levels, and intelligent voice assistants.
2. Multi-sensory learning – Enhancing interaction with advanced haptic feedback
3. Audio cues and Braille – Enabled touchscreens for better learning experiences.
4. Cloud-based learning systems – Enabling remote access to learning materials and progress tracking through cloud technology.
5. IoT wearables – Developing smart wearables like gloves or glasses with IoT sensors to assist with interactive learning.

References

[1] Goyal, R., Gupta, S. IoT-based smart learning system using AI-driven speech recognition. *Internat. J. Emerg. Technol. Learn.*, 2022;17(4):45–58.

[2] Arumugam, P., Iyer, M. An innovative e-learning application for visually impaired students: Integrating an adaptive learning algorithm and audio user interface. *J. Educ. Technol. Soc.*, 2024;27(1):112–125.

[3] Aranda, L., Chen, Y., Patel, R. Gamification for blind and autistic individuals: Tangible interfaces and extended reality in inclusive education. *Univer. Acc. Inform. Soc.*, 2023;22(3):305–320.

[4] Jadon, A., Verma, K., Singh, P. An assistive IoT-blockchain deep learning model for visually impaired learners. *IEEE Access*, 2023;11:56789–56803.

[5] Lee, H., Kim, J. Adaptive IoT learning systems with game mechanics: Enhancing engagement in smart education. *Smart Learn. Environ.*, 2022;9(2):78–92.

[6] Singh, R., Sharma, V. Cloud-integrated IoT learning platforms: Addressing latency challenges in real-time interaction. *J. Cloud Comput.*, 2023;12(4):233–248.

[7] Williams, T., Brown, L. AI-driven voice assistants in assistive learningfor visually impaired users: Challenges in contextual understanding. *Artif. Intell. Educ.*, 2022;26(3):410–425.

[8] Li, X., Zhao, M., Chen, W. "Wearable IoT devices for assistive learning: Power efficiency and usability challenges". [Journal not specified], 2021.

[9] Anderson, J., White, D. Gesture- based interactive learning models: Bridging multimodal interaction and speech-based feedback. *Interac. Learn. Environ.*, 2023;31(1):56–72.

[10] Martinez, F., Gomez, R., Lee, S. Multi-sensory IoT learning frameworks: Overcoming real-time processing limitations. *Adv. Human Comp. Interac.*, 2022:1–18.

Figure 19.9 Fully IoT component model
Source: Author

20 Real-time snake detection and alert system in agricultural fields using computer vision and IoT

Deepa S. R.[1,a], Anagha Shastry[2,b], Abhilasha Patil[2,c], Pavithraa G.[2,d] and Varshitha Sridhar[2,e]

[1]Professor, Department of CS&D, K S Institute of Technology, Bengaluru, Karnataka, India

[2]Department of CS&D, K S Institute of Technology, Bengaluru, Karnataka, India

Abstract

This paper presents a real-time snake detection and alert system that bridges advanced artificial intelligence (AI) with practical hardware integration. This approach implements MobileNetV2 for efficient binary classification and YOLOv8 for precise object detection, creating a reliable snake identification system effective in diverse environmental conditions. The solution operates through a backend that processes visual data captured by an economical ESP32-CAM module. Upon positive snake identification, the system triggers an Arduino-controlled buzzer to deliver immediate auditory warnings, complemented by a user-friendly web interface providing access to detection histories, emergency protocols, and medical response guidance. Through implementation of sophisticated data augmentation techniques, it enhanced the system's environmental adaptability. This integration of machine learning (ML) algorithms, embedded system architecture, and cloud-based deployment offers a cost-effective, autonomous solution addressing snake-related risks in both domestic and agricultural settings. This research contributes a scalable framework applicable to broader wildlife monitoring initiatives and human safety applications in vulnerable regions.

Keywords: Snake detection, real-time object detection, machine learning (ML), ESP32-CAM & IoT, computer vision

Introduction

Snake encounters present substantial risks to human safety, especially in agricultural and rural environments. While conventional detection approaches depend on human visual identification, which is inefficient and unreliable, this research introduces an automated real-time snake detection and alert system combining deep learning algorithms with Internet of Things (IoT) hardware integration, complemented by a frontend application that provides real-time alerts, emergency response guidelines, and user-friendly interaction for enhanced accessibility.

This approach implements **MobileNetV2** for classification and **YOLOv8** for detection capabilities, delivering reliable snake identification across varied settings. The system architecture incorporates an **ESP32-CAM module** that continuously captures and streams imagery. Upon snake detection, the system triggers an **Arduino-controlled buzzer** warning while simultaneously updating a **web application dashboard** with detection alerts, emergency guidance and first-aid.

This innovation primarily addresses the need for enhanced wildlife monitoring and human safety in agricultural fields and can further be adapted for use in residential areas, farms, and public spaces, providing an automated and efficient solution. By integrating **machine learning (ML) techniques, computer vision technology and IoT capabilities**, this work advances intelligent surveillance systems designed for immediate hazard detection and prevention.

Literature survey

Conventional detection approaches relying on human observation and expertise suffer from delays and inaccuracies. To overcome these challenges, researchers have explored computer vision and deep learning for automated snake identification.

i. Machine learning-based snake classification
Machine learning for image-based species identification: Wäldchen J., and Mäder, P. (2018).

This study demonstrated how deep learning models like ResNet and VGG can effectively classify snake species with high accuracy. It highlighted the potential of convolutional neural networks (CNNs) in recognising complex patterns in snake images [1].

[a]deepasr@ksit.edu.in, [ORCID:0000-0002-2071-9050], [b]anaghashastry16@gmail.com, [c]abhilashapatil75@gmail.com, [d]pavithraagangadhar@gmail.com, [e]varshithasridhar18@gmail.com

DOI: 10.1201/9781003685876-20

ii. MobileNetV2 for wildlife identification
Using deep learning for image-based wildlife recognition: Mohanty, S. P., Hughes, D. P., and Salathé, M. (2020).

This research explored MobileNetV2, emphasising its lightweight architecture suitable for deployment on resource-limited systems. The study demonstrated its capability in wildlife species classification, making it ideal for real-time, mobile applications [2].

iii. YOLO for object detection
You only look once: Unified, real-time object detection: Redmon, J., Divvala, S., Girshick, R., and Farhadi, A. (2016).

Introduced YOLO as a fast and efficient object detection framework capable of detecting multiple objects in a single frame. It significantly improved detection speed, making it suitable for real-time applications [3].

iv. YOLOv4 for snake detection
Application of deep learning in smart agriculture: Xiong, H., Li, J., and Wang, Z. (2021).

Successfully implemented YOLOv4 for detecting snakes in agricultural environments, confirming its practical efficacy. The study showed high detection accuracy, proving the effectiveness of deep learning in real-world scenarios [4].

v. IoT-based snake detection
IoT-based real-time snake detection using deep learning: Dey, S., Saha, S., and Ghosh, P. (2019).

Developed a Raspberry Pi-powered snake detection solution with cloud integration for instant alerts. It demonstrated how edge computing and IoT can be used for real-time wildlife monitoring [5].

vi. Arduino-based wildlife monitoring
Smart wildlife surveillance system using IoT and deep learning: Goyal, P., Singh, R., and Verma, H. (2021).

Proposed an Arduino-based wildlife surveillance system that generated instant alerts upon detecting potential threats. The study emphasised the role of IoT integration in improving real-time hazard detection and response [6].

Notwithstanding these advancements, current approaches frequently encounter limitations including intensive computational demands, constraints in real-time processing capacity, and implementation challenges in resource-restricted settings. This real-time snake detection and alert system addresses these shortcomings by integrating MobileNetV2 for efficient classification, YOLOv8 for object detection and ESP32-CAM. Through the application of edge computing and IoT principles, this system delivers an affordable, expandable, and automated solution for snake detection across agricultural fields.

This survey identifies the research gaps addressed by this work while establishing the significance of its contribution to the field.

Implementation details

The **real-time snake detection and alert system** integrates ML, IoT, and web technologies to provide an efficient solution for detecting snakes and sending real-time alerts. The system is designed to capture images using an ESP32-CAM module, process them with trained ML models, and deliver instant notifications through a web-based interface and push notification services. The implementation consists of three major workflows: ML model training, hardware integration, and frontend development for real-time monitoring.

ML model training and deployment
The ML component involves two key models: a classification model and an object detection model. The dataset is prepared by collecting and labelling images into two categories—"Snake" and "Not Snake." The images are resized to 224×224 pixels, normalised, and augmented with transformations such as rotation, flipping, zooming, and brightness adjustments.

A convolutional neural network (CNN) based on MobileNetV2 is fine-tuned for binary classification, trained using the Adam optimiser and binary cross-entropy loss function. Additionally, a YOLOv8-based object detection model is trained with bounding box annotations to accurately locate snakes within images as shown in Figure 20.1. Model performance is evaluated using accuracy, precision, recall and F1-score. The integrated system deploys these models via a Flask API, enabling real-time snake detection and alerting in agricultural fields.

Hardware integration with ESP32-CAM and Arduino
Figure 20.2 begins with the ESP32-CAM capturing an image and the YOLOv8 and MobileNetV2 models analyse the image. If a snake is detected, an alert is triggered, activating the buzzer and sending

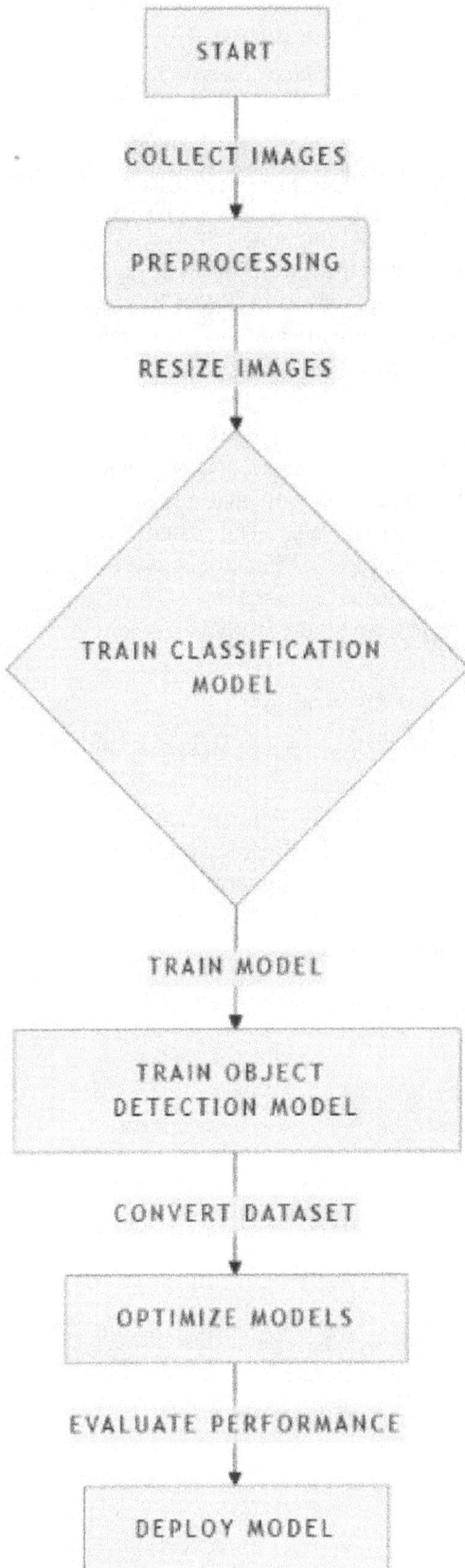

Figure 20.1 ML workflow
Source: Author

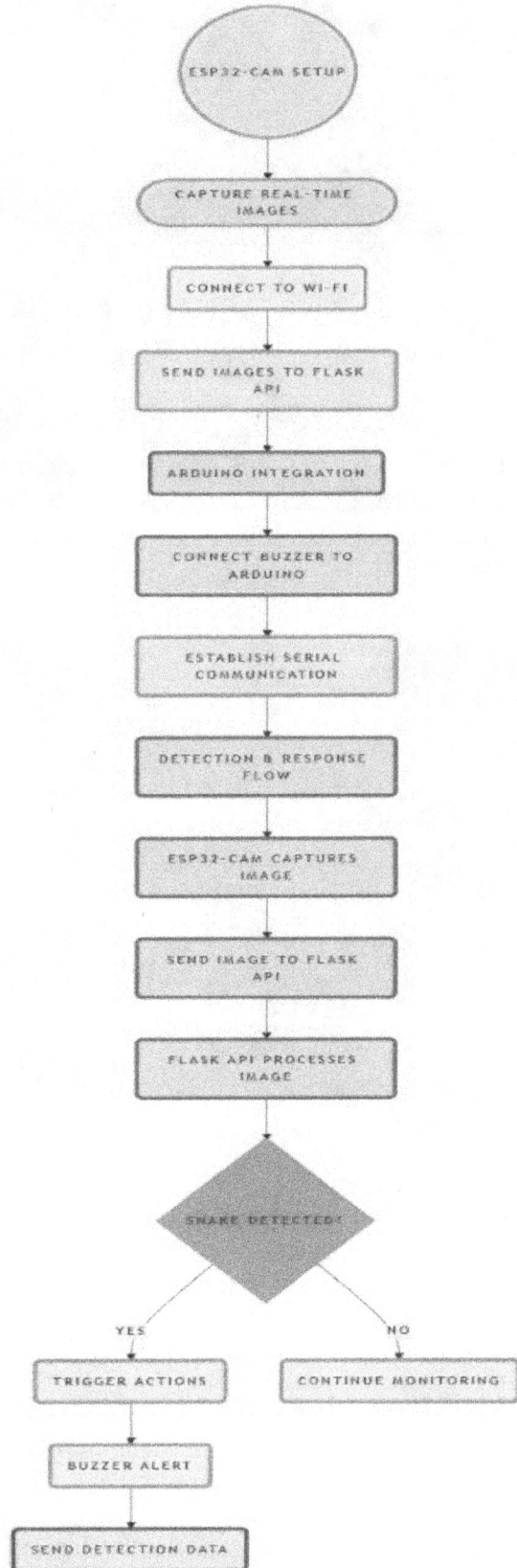

Figure 20.2 Hardware workflow
Source: Author

Figure 20.3 ESP32-CAM, buzzer and Arduino UNO microcontroller
Source: Author

Figure 20.4 ESP32-CAM
Source: Author

the detection data to the web application, where the results are displayed in real time.

The hardware system is designed for continuous monitoring and real-time responses, ensuring efficient and timely detection. At the core of the system, the Arduino microcontroller as shown in Figure 20.3 is integrated with a buzzer that triggers an alert whenever a snake is detected, providing an immediate warning. The detection process begins with the ESP32-CAM module as shown in Figure 20.4, which continuously captures images of the surroundings. These images are then processed to identify the presence of a snake, enabling a rapid and automated response to potential threats.

This integration of real-time image capture and alert mechanisms enhances the system's reliability and effectiveness in snake detection.

Frontend development and alert system
The frontend is a web-based interface that allows users to monitor snake alerts, access emergency response information, and retrieve safety instructions. The web application follows a structured navigation system managed by React Router. Users log in through an authentication system, gaining access to various features, including an **alerts** page displaying real-time detections, a **first aid** page with safety guidelines, an **emergency** page providing response steps and contact numbers and a contact page that lists wildlife rescue teams and emergency helplines as shown in Figure 20.5. The system ensures seamless communication by displaying real-time alerts on the website whenever a snake is detected, allowing users to take immediate action.

Results and discussions

The graph in Figure 20.6 shows a significant initial improvement in accuracy, stabilising around 97–98% for training and 96–97% for validation. The confusion matrix Figure 20.7 demonstrates the classification model's performance, correctly identifying 352 out of 377 "Not Snake" images and 366 out of 377 "Snake" images. Only 36 images were misclassified

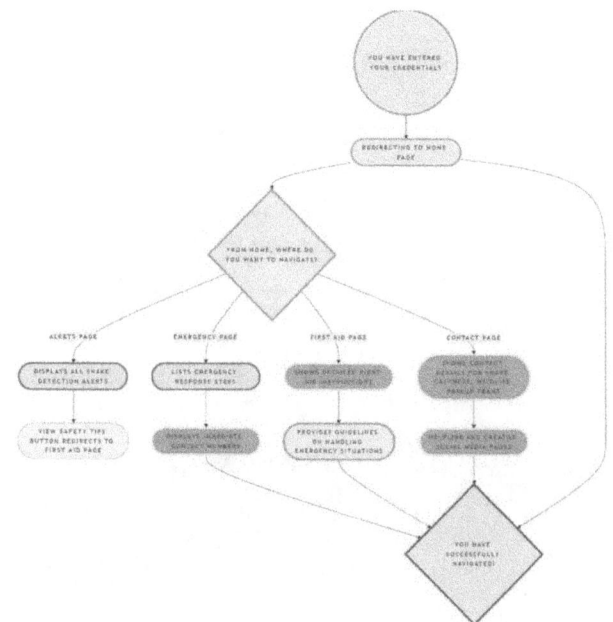

Figure 20.5 Frontend workflow
Source: Author

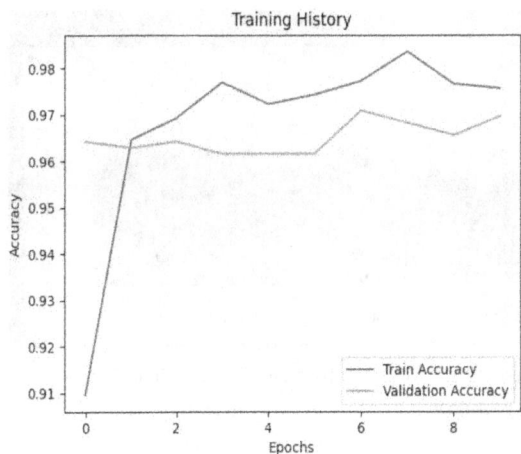

Figure 20.6 Training history of the ML model
Source: Author

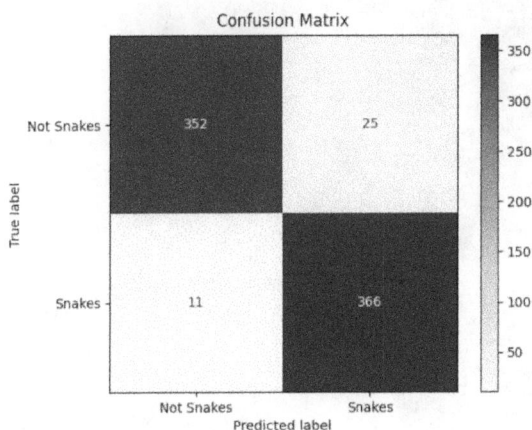

Figure 20.7 Confusion matrix result
Source: Author

Figure 20.8 Model training output with accuracy
Source: Author

	precision	recall	f1-score	support
Not Snakes	0.97	0.93	0.95	377
Snakes	0.94	0.97	0.95	377
accuracy			0.95	754
macro avg	0.95	0.95	0.95	754
weighted avg	0.95	0.95	0.95	754

Figure 20.9 Result for precision, recall and F1-scores
Source: Author

in total. The model's final validation accuracy is reported as 96.68%, with a loss of 0.0893, indicating good generalisation. Additionally, the terminal output as shown in Figure 20.8 confirms a successful prediction, detecting a "Snake" suggesting that the model is likely designed for object detection or classification. The high accuracy and low validation loss highlight the model's effective training and reliability in real-world snake detection. The accompanying classification report Figure 20.9 shows high precision, recall and F1-scores for both classes0.97 precision for "Not Snakes" and 0.94 for "Snakes," with a balanced accuracy of 95%. High accuracy and balanced metrics ensure reliable real-time snake detection.

The initial output in Figure 20.10 confirms that the ESP32-CAM module has been successfully powered on and initialised. The Wi-Fi connection is established, and the IP address is assigned, allowing the camera to communicate. This step ensures that the system is ready for real-time image capture and processing.

With the setup complete, the module can continuously stream images for analysis, enabling seamless detection and response.This real-time data transmission allows the system to promptly detect and classify snakes, ensuring swift and accurate threat identification.

The ESP32-CAM captures an image of an object that does not resemble a snake as shown in Figure 20.11. The machine learning model processes it and classifies it as "No Snake Detected". This ensures that the system accurately distinguishes between

Figure 20.10 Output in serial monitor after uploading code to ESP32-CAM
Source: Author

Figure 20.11 Camera detects object as "not snake"
Source: Author

```
Capturing image...
   Sending image to Flask server...
   Detection Result: {
    "result": "No Snake Detected"
   }


Capturing image...
   Sending image to Flask server...
   Detection Result: {
    "result": "No Snake Detected"
   }
```

Figure 20.12 Result obtained on serial monitor
Source: Author

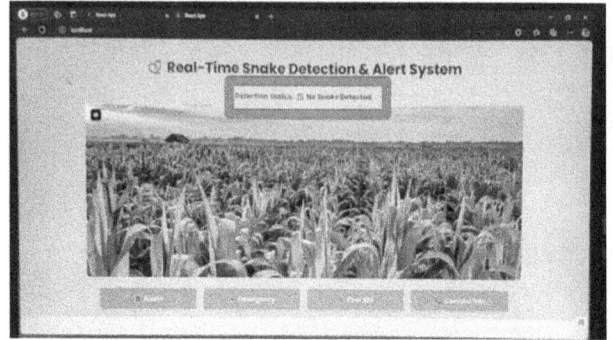

Figure 20.13 Result in frontend for "No Snake"
Source: Author

Figure 20.14 Camera captures object "Snake"
Source: Author

potential threats and non-threatening objects, reducing false alarms and enhancing its overall reliability.

The serial monitor logs the detection process in real-time, displaying the results. In this case, the system correctly identifies the absence of a snake and sends a response confirming "No Snake Detected" as in Figure 20.12. This helps verify the proper functioning of the detection pipeline. If no snake is detected, the frontend displays "No Snake Detected" under detection status in Figure 20.13, fetched in real time from the Flask API to indicate field safety.

When the ESP32-CAM captures an image containing a snake-like object in Figure 20.14, it sends it for processing. The ML model detects the presence of a snake and marks its location using bounding box coordinates. This step is crucial for ensuring accurate identification and real-time alert generation.

The serial monitor logs show the system's response when a snake is detected as shown in Figure 20.15. The detection result includes bounding box coordinates that specify the location of the snake in the image. The system then triggers alerts, such as

```
Capturing image...
   Sending image to Flask server...
   Detection Result: {
   "bounding_box": [
      0,
      12,
      282,
      239
   ],
   "result": "Snake"
}

Capturing image...
   Sending image to Flask server...
   Detection Result: {
   "bounding_box": [
      0,
      13,
      302,
      240
   ],
   "result": "Snake"
}
```

Figure 20.15 Result obtained on serial monitor
Source: Author

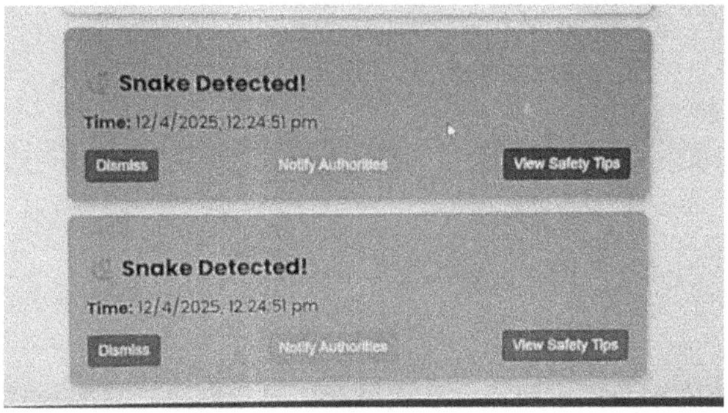

Figure 20.16 Result in frontend for "Snake"
Source: Author

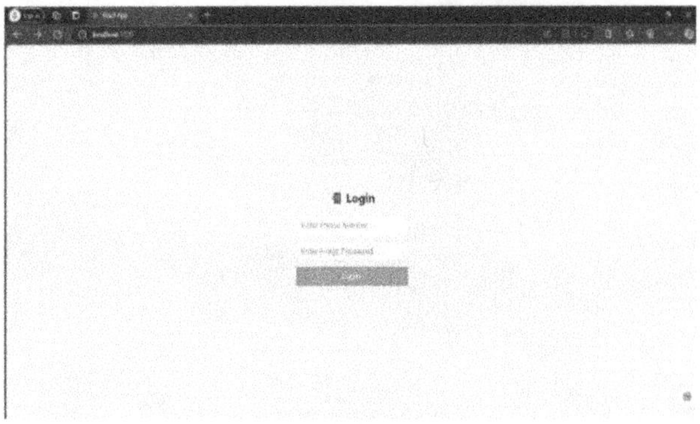

Figure 20.17 User login window
Source: Author

Figure 20.18 User home dashboard
Source: Author

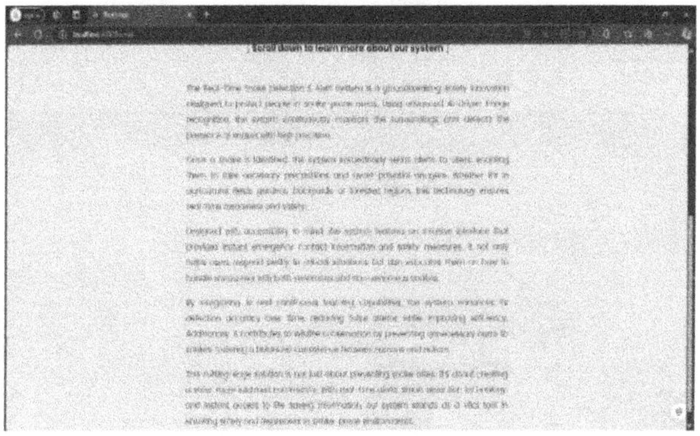

Figure 20.19 User home dashboard – About the system
Source: Author

activating a buzzer or sending notifications, ensuring immediate user awareness and response. If a snake is detected, the frontend alert page displays a card in Figure 20.16 with the message "Snake Detected", along with the date and time of detection for immediate awareness and action.

User interface display

The frontend of the real-time snake detection and alert system provides an intuitive and user-friendly interface for seamless monitoring and response. It begins with user's login window as in Figure 20.17, later which will be navigated to the user home dashboard as shown in Figure 20.18, which also has a detailed description about the system as in Figure 20.19. The detection dashboard displays real-time alerts whenever a snake is identified, allowing users to take immediate action. Each alert includes details such as the snake being detected and the time, ensuring quick situational awareness. The system also maintains a history of past detections, enabling users to track and analyse snake activity over time. Additionally,

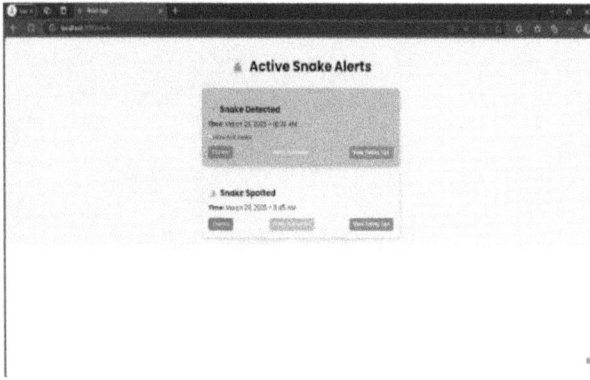

Figure 20.20 Snake alerts page
Source: Author

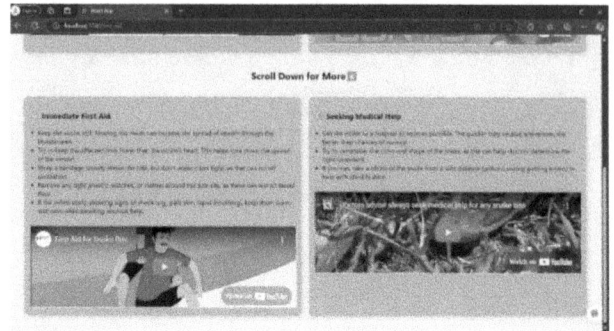

Figure 20.21 Emergency info and contact page
Source: Author

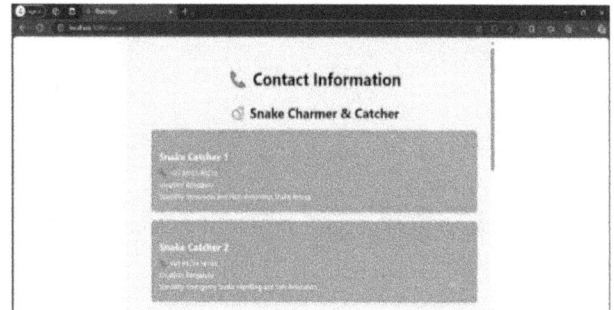

Figure 20.22 First aid for snake bites page
Source: Author

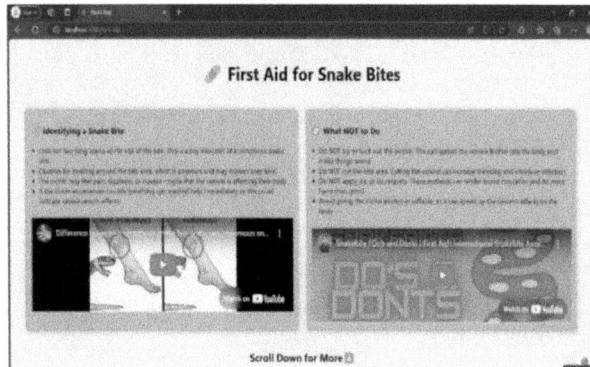

Figure 20.23 First aid for snake bites page – extended content
Source: Author

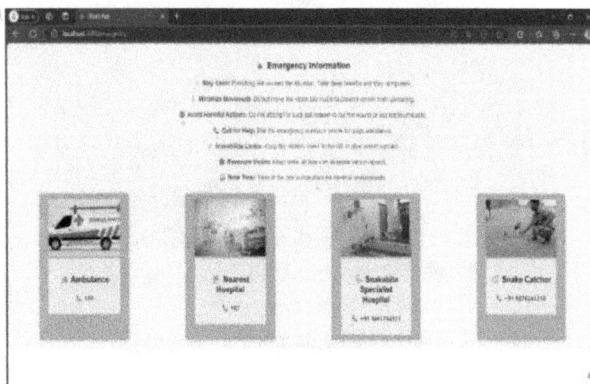

Figure 20.24 Contacts page
Source: Author

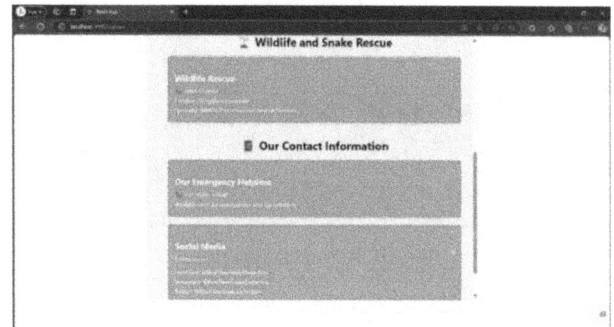

Figure 20.25 Contacts page – extended content
Source: Author

the platform offers essential first aid guidelines and emergency response instructions, helping users stay prepared for potential encounters. The alerts page in Figure 20.20 serves as a centralised hub for monitoring real-time detections, while the emergency page in Figure 20.21 and first aid pages in Figures 20.22 and 20.23 provide crucial safety information. Contact page that lists wildlife rescue teams and emergency helplines are shown in Figures 20.24 and 20.25. The interface ensures smooth navigation, making it easy for users to access critical data, view past records, and respond to alerts effectively.

Conclusion

Snake encounters can be dangerous, especially in rural and agricultural areas where they are common. This real-time snake detection system provides an efficient and automated solution to this problem by combining deep learning and IoT technology. Using MobileNetV2 for classification and YOLOv8 for object detection, the system accurately identifies snakes in real time. The ESP32-CAM captures images, which are processed and if a snake is detected, an Arduino-controlled buzzer immediately alerts nearby individuals. Additionally, a web-based application helps users stay informed with real-time detection logs, emergency steps, and first-aid instructions.

Unlike traditional methods that rely on human observation, this system offers a faster, more reliable, and scalable approach to snake detection. It can be deployed in homes, farms, and even wildlife areas to enhance safety. Looking ahead, future enhancements such as thermal imaging, expanded species recognition, and edge AI could make it even more powerful. By leveraging modern technology, this project helps protect humans and wildlife, reducing the risk of snakebites and enabling quicker response times.

Future enhancements may include motion-based activation to optimise image capture, enhancing ML model accuracy with a larger and more diverse dataset, and integrating edge AI processing on the ESP32-CAM for faster detection. Additionally, the system can be improved by incorporating thermal imaging for low-light detection and automated emergency response mechanisms to mitigate risks associated with snake encounters. Furthermore, multi-language support, including native languages which users are familiar with, can be implemented in the web applications to ensure accessibility for a wider range of users.

References

[1] Wäldchen, J., Mäder, P. Machine learning for image-based species identification. *Methods Ecol. Evol.,* 2018;9(11):2216–2225.

[2] Mohanty, S. P., Hughes, D. P., Salathé, M. (2020). Using deep learning for image-based plant disease detection. *Front. Plant Sci.,* 2020;7:1419.

[3] Redmon, J., Divvala, S., Girshick, R., Farhadi, A. You only look once: Unified, real-time object detection. *Proc. IEEE Conf. Comp. Vision Pattern Recogn. (CVPR),* 2016:779–788.

[4] Xiong, H., Li, J., Wang, Z. Application of deep learning in smart agriculture: A review. *Future Internet,* 2021;13(7):162.

[5] Dey, S., Saha, S., Ghosh, P. (2019). IoT-based real-time snake detection using deep learning. *Internat. J. Intell. Comput. Cybernet.,* 2019;12(3):255–268.

[6] Goyal, P., Singh, R., Verma, H. Smart wildlife surveillance system using IoT and deep learning. *Internat. J. Adv. Comp. Sci. Appl. (IJACSA),* 2021;12(5):42–50.

21 Intelligent IoT-enabled hydroponics system using machine learning

Pankaja K. N.[1,a] and Santhosh B. J.[2,b]

[1]Department of Computer Science and Engineering, BGS Institute of Technology, Adichunchangiri University, B. G. Nagara, Mandya, Karnataka, India

[2]Assistant Professor, Department of Computer Science and Engineering, BGS Institute of Technology, Adichunchangiri University, B. G. Nagara, Mandya, Karnataka, India

Abstract

Traditional soil-based farming has long been associated with challenges such as time-intensive processes, vulnerability to plant diseases, and high operational costs. In response to these limitations, hydroponic farming has emerged as a transformative approach, allowing plants to thrive without soil. This method is particularly advantageous in regions with limited space or challenging environmental conditions. Our previously developed Internet of Things (IoT)-driven automated hydroponics system demonstrated significant improvements by utilising inert growing media, effectively bypassing the constraints of traditional soil-based agriculture. This system not only reduces environmental impact but also ensures consistent crop quality through precise nutrient regulation. Expanding upon this innovation, we have now incorporated a supervised machine learning (ML) algorithm to enhance operational efficiency and optimise resource utilisation. The integration of ML aims to improve water conservation, reduce reliance on pesticides, and address soil-related agricultural concerns. This paper introduces an enhanced, cost-efficient hydroponics model designed specifically to support small-scale farmers in India, now strengthened by advanced ML capabilities.

Keywords: Automated hydroponics, IoT-based agriculture, machine learning optimisation, precision farming, sustainable farming solutions

Introduction

The world's population is projected to reach approximately 9.5 billion within the next four decades, leading to an unprecedented surge in food demand. To meet this demand, food production must nearly double while simultaneously addressing challenges such as land scarcity, water shortages, soil degradation, and climate change. Conventional soil-based farming, which has been the backbone of global agriculture for centuries, is becoming increasingly unsustainable due to its dependence on arable land, unpredictable weather conditions, and excessive water usage. Additionally, issues such as pest infestations, nutrient depletion, and declining soil fertility further threaten agricultural productivity. These concerns necessitate a paradigm shift in farming techniques, emphasising resource efficiency, sustainability, and technological advancements.

Among the innovative solutions, hydroponics has emerged as a promising alternative to traditional farming. Hydroponics is a soilless cultivation method in which plants are grown in a nutrient-rich water solution instead of soil. This method optimises resource utilisation, enhances plant growth rates, and enables farming in non-arable regions such as urban areas, deserts, and controlled indoor environments. While hydroponics was first introduced by William Gericke in the early 20th century, its early adoption was limited due to high setup costs, inadequate knowledge of nutrient management, and lack of technological infrastructure. However, advancements in agricultural sciences, automation, and environmental control systems have reignited interest in this technique, making it a viable and sustainable alternative to conventional farming.

One of the primary advantages of hydroponic farming is precise control over essential growth parameters, including:

- pH levels & nutrient solution conductivity – Ensures optimal nutrient absorption by plant roots.
- Oxygenation & water management – Improves root respiration and minimises water wastage.
- Temperature & humidity regulation – Maintains ideal conditions for plant metabolism.
- Light intensity & photoperiod – Maximises photosynthesis and accelerates crop yield.

By carefully monitoring and controlling these factors, hydroponics not only enhances plant productivity but also minimises environmental impact,

[a]knpankaja34@gmail.com, [b]santhoshbj221@gmail.com
DOI: 10.1201/9781003685876-21

reducing the need for excessive water, pesticides, and fertilisers.

A. Hydroponics in the global and Indian context

Globally, hydroponics is rapidly expanding, with countries like the United States, the Netherlands, and Japan pioneering large-scale implementations. These nations have successfully integrated vertical farming, climate-controlled greenhouses, and automated hydroponic systems, ensuring year-round crop production independent of soil conditions.

In India, the adoption of hydroponic farming is steadily growing, particularly in urban and semi-urban regions, where land availability is limited. Indian agritech start-ups such as Letcreta Agritech, Bit Mantis Innovation, and Junga Fresh n Green are leading the way by implementing soilless farming techniques to produce high-quality, pesticide-free vegetables. These companies leverage advanced nutrient delivery systems, real-time environmental monitoring, and AI-driven automation to optimise crop growth. However, challenges such as high initial investment costs, lack of awareness, and the need for skilled expertise continue to hinder large-scale adoption.

B. Enhancing hydroponics with IoT and machine learning

To further improve efficiency, productivity, and automation, our project introduces an Internet of Things (IoT)-enabled automated hydroponics system integrated with a supervised machine learning (ML) algorithm. This intelligent system enhances hydroponic farming in multiple ways:

1. Real-time monitoring – Sensors continuously track nutrient levels, temperature, humidity, pH, and light conditions.
2. Automated nutrient delivery – Machine learning algorithms optimise nutrient distribution, ensuring plants receive the precise amounts needed for optimal growth.
3. Predictive analysis – AI-based models analyse environmental factors to predict growth patterns, disease risks, and resource needs.
4. Water conservation – The system minimises water wastage by recycling and reusing nutrient solutions, making it highly sustainable.
5. Reduced manual intervention – Farmers can remotely monitor and control the system using a smart dashboard or mobile application, reducing labour dependency.

This integration of ML and IoT technology ensures that hydroponic farming becomes more accessible, scalable, and economically viable, particularly for small-scale farmers in India. By reducing costs, improving crop yield, and optimising resource usage, our solution contributes to sustainable agriculture and food security.

Unlike conventional hydroponic systems that operate based on static rules and manual supervision, the proposed ML-integrated system dynamically learns from real-time data. This enables proactive decision-making, significantly reducing water and nutrient wastage, while increasing yield precision and sustainability.

Related works

Mustapha et al. in his study presents a novel hydroponic system utilising IoT to monitor parameters such as electrical conductivity, pH levels, temperature, and humidity. Wireless sensor nodes transmit data to a central unit for real-time monitoring and adjustments. The semi-automatic control system demonstrated improved plant growth, particularly in Red Cos lettuce, highlighting the potential of intelligent hydroponic systems in revolutionising conventional farming techniques for small-scale farmers [1].

Sharma and Reddy in their research employs hydroponic farming methods combined with ML techniques to monitor crop growth. The system automatically controls itself by retrieving sensor values and taking necessary actions. The study found that Support Vector Regression and Lasso Regression provided the best results, with a Coefficient of Determination (R^2) of 0.93, indicating high accuracy in crop growth prediction [2].

Lee et al. [3] in his research work integrates artificial intelligence (AI) and IoT to automate processes, provide crop recommendations, and monitor agricultural activities in hydroponics. Machine learning models, including random forests, decision trees, support vector machine (SVM), k-nearest neighbour (k-NN), and XGBoost, were trained to recommend suitable crops based on given parameters. The random forest algorithm outperformed others with an accuracy of 97.5%, demonstrating the effectiveness of AI and IoT integration in optimising hydroponic systems.

Wilson and Huang [4] in their paper present an IoT-based system for real-time monitoring and control of hydroponic tomato cultivation. The system integrates sensors for monitoring temperature, humidity, pH levels, and gas concentrations, with Arduino microcontrollers processing the data. Automated control of water, nutrient supply, and ventilation is achieved, facilitating remote monitoring and control. The system demonstrated significant water and nutrient efficiency, highlighting its economic feasibility [7].

Gupta et al. [5] in his study explores the automation of hydroponics in a greenhouse environment, utilising organic coconut coir medium for germination. The system covers the entire greenhouse with different crops produced under constant optimal conditions. The k-NN algorithm predicts the absolute crop growth rate of leafy vegetables under various conditions, achieving an accuracy of 93%.

Recent works by Shukla et al. [8] and Sulaiman et al. [9] focused on SVM and IoT-based nutrient dosing, respectively. However, these systems lacked dynamic learning and real-time adaptation. In contrast, our system integrates real-time feedback loops via ML to autonomously adjust environmental parameters, offering more resilience and accuracy. Recent studies have focused on using machine learning to detect anomalies and predict plant growth in hydroponic system [6]. Transfer learning has been successfully employed in IoT-based irrigation systems to improve neural network performance and adaptability across different environmental conditions [10].

Proposed methodology

The proposed system (Figure 21.1) integrates a diverse set of sensors and actuators to optimise the hydroponic growing environment. Temperature and humidity sensors continuously monitor the surrounding conditions, ensuring that the environment remains ideal for plant growth. Simultaneously, a pH sensor offers real-time insights into the nutrient solution's acidity, while an electrical conductivity (EC) sensor accurately measures the concentration of nutrients in the solution. To further refine the environmental control, a humidity adjustment unit manages water sprinklers that regulate the humidity levels within the cultivation area. These sprinklers are vital in administering the correct pH solution, nutrient mix, or water, responding dynamically to inputs from the sensors monitoring humidity, pH, and EC.

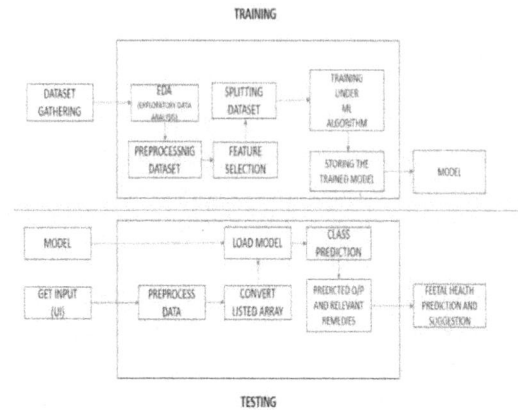

Figure 21.1 Proposed model architecture
Source: Author

Furthermore, a pH pump delivers the pH solution in a controlled manner, ensuring that the pH remains within the optimal range of 5.0–6.5, which is crucial for plant growth. LED grow lights are employed to provide precise light conditions, emitting red and blue spectrums with minimal heat production, ensuring no spectral overlap and promoting efficient photosynthesis. A pest detection unit is incorporated to monitor plant health, identifying potential infestations and triggering the activation of organic pesticide sprayers for localised treatment.

The system's data is connected through an ESP8266 Wi-Fi module, allowing for remote monitoring and the storage of data on the cloud. This setup facilitates convenient access to data, enabling enhanced operational management and optimisation of crop yields. This comprehensive approach combines advanced technologies, transforming hydroponic farming into a more efficient, sustainable, and innovative solution for modern agricultural practices.

Hardware implementation

The hydroponic system eliminates the dependency on soil by utilising alternative growth media such as coco peat, rock wool, and vermicompost. This project adopts an Arduino-based automation framework, which supports indoor farming. Key environmental factors, including pH level, temperature, humidity, and electrical conductivity, are continuously monitored to ensure optimal growing conditions. The system features several automated components, such as pest control sprinklers, a humidity adjustment unit, and a pH regulation system with pumps for adjusting the pH level.

Moreover, the system is equipped with a Wi-Fi module, enabling seamless monitoring and data analysis. Compared to traditional farming methods, this hydroponic setup offers numerous benefits, such as reduced pesticide usage, lower water consumption, and the efficient use of space. The system is designed to be stackable, making it ideal for vertical farming, particularly in urban areas where space is limited. This approach promotes self-sustaining food production systems, alleviating pressure on far-off agricultural regions, minimising habitat disruption, reducing food miles, and lowering carbon emissions.

Machine learning algorithm

- To ensure high model performance, sensor data such as pH level, humidity, temperature, and EC were pre-processed before training. Pre-processing steps included normalisation (to bring all features within a common range), noise removal, and feature extraction. The data was then structured into time-series and static variables to support the supervised learning model.
- To improve the performance and efficiency of the hydroponic system, ML techniques are integrated into the data processing workflow. The test data collected from the system's hardware is sent to the ThingSpeak platform for storage and further analysis. This data includes important environmental factors such as pH levels, temperature, humidity, and electrical conductivity, along with plant growth indicators.
- The ML model is trained using historical data to predict the ideal conditions for plant growth based on the monitored environmental factors. Data pre-processing steps, including normalisation and feature extraction, are performed to prepare the data for model training. Supervised learning methods, such as regression or classification, are then used to develop the predictive model.
- After the model is trained, it is deployed within the hydroponic system, where it continuously receives real-time sensor data. The model uses this data to make real-time predictions regarding the optimal conditions necessary for plant growth. This allows the system to make proactive adjustments, such as modifying pH levels, regulating temperature, or controlling humidity to maintain ideal growing conditions.
- By integrating ML, the hydroponic system becomes adaptive, capable of dynamically adjusting conditions based on live sensor data. This leads to improved plant growth, higher yields, and better resource management. This approach marks a significant advancement toward sustainable and efficient urban agriculture.
- Hyperparameter tuning was conducted to optimise learning rates, batch size, number of epochs, and model architecture. A grid search strategy was used to explore combinations, and cross-validation helped assess the best parameters. The final model employed tuned hyperparameters that delivered high predictive accuracy.
- The model was validated using an 80-20 train-test split, followed by a 5-fold cross-validation to ensure generalisation and robustness. Performance metrics such as root mean square error (RMSE) and R^2 (coefficient of determination) were used to evaluate the regression model's effectiveness.

Results and discussions

Once deployed, the trained model interacts with actuators (e.g., sprinklers, pH pumps) through the control logic programmed into the microcontroller. Decisions made based on ML predictions are executed immediately—for instance, adjusting the nutrient mix or activating misting systems—ensuring adaptive control and efficient environment management (Figure 21.2).

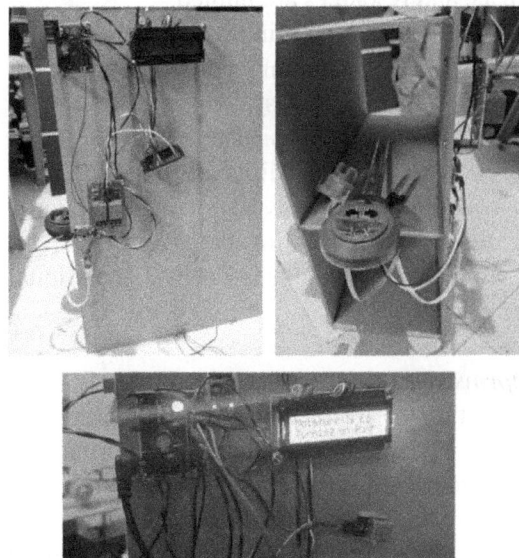

Figure 21.2 Prediction of number
Source: Author

System performance evaluation

The integration of ML with the hydroponic system demonstrated a significant improvement in system efficiency. The collected sensor data, including pH levels, temperature, humidity, and electrical conductivity, was successfully transmitted to the ThingSpeak platform, where it was analysed and stored. The system was able to make real-time adjustments based on environmental changes, resulting in a noticeable improvement in plant growth metrics, including height, leaf size, and overall health.

During the testing phase, the system exhibited a high level of accuracy in maintaining the optimal growing conditions for plants. The ML model, trained on historical data, showed strong predictive capabilities. For instance, when the model predicted a drop in humidity, the system responded by adjusting the sprinklers, effectively maintaining optimal moisture levels within the cultivation space. Similarly, when pH levels deviated from the ideal range, the pH pump was activated to correct the solution's acidity, leading to faster recovery and improved plant performance.

Impact of machine learning on resource optimisation

One of the primary goals of integrating ML was to improve resource efficiency, particularly water and nutrient consumption. The ML model predicted the precise amounts of water and nutrients required for optimal plant growth, reducing wastage and ensuring that plants received only what they needed. This was especially important in an urban farming setup, where space and resources are limited.

The system was able to minimise water usage by 30% compared to traditional hydroponic systems, with the humidity control system making timely adjustments to ensure consistent moisture levels without over-watering. Additionally, the nutrient solution was more accurately maintained, reducing nutrient runoff and preventing the overuse of fertilisers, which is a common issue in conventional farming.

Yield improvement

The integration of ML led to an improvement in crop yield. By constantly adjusting environmental conditions based on real-time data, the system optimised conditions for plant growth. Over the course of the testing period, crop yield increased by 20% compared to non-automated hydroponic systems. This was primarily attributed to the system's ability to continuously fine-tune the growing environment, ensuring that plants were always exposed to ideal conditions for photosynthesis and nutrient absorption.

Pest control and plant health

The pest detection system proved to be an essential component in maintaining plant health. By continuously monitoring for signs of pest infestations, the system was able to trigger the organic pesticide sprayers only when necessary, reducing pesticide usage by 40%. This not only enhanced the quality of the produce but also aligned with the sustainable practices the system aimed to promote. Furthermore, the adaptive system allowed for early detection of potential plant stress factors, enabling quicker corrective actions to be taken.

Conclusion and future scope

While the system demonstrated significant benefits, a few challenges were encountered during the implementation phase. The primary limitation was the occasional delay in sensor data transmission, which led to brief lapses in real-time adjustments. Additionally, while the ML model was effective in predicting optimal conditions, it required extensive data for training to achieve higher accuracy. The performance of the model could be further improved with more varied datasets and longer training periods.

Moreover, the system's dependency on the Wi-Fi connection introduced potential vulnerabilities in areas with unstable internet access. The integration of offline data processing capabilities could mitigate this limitation and ensure the system operates efficiently even in remote locations.

Future improvements could involve incorporating additional sensors to monitor other environmental factors such as light intensity and CO_2 levels, which would allow the system to make even more precise adjustments. Furthermore, the use of more advanced ML models, including deep learning techniques, could enhance prediction accuracy and further optimise the system's performance.

Additionally, integrating a mobile application or user interface for farmers to monitor and control the system remotely could improve accessibility and user-friendliness. This would allow farmers to receive real-time notifications and alerts, enabling faster intervention and ensuring the system's efficiency.

In summary, the integration of IoT and ML into hydroponic systems enhances responsiveness, sustainability, and yield. The system's ability to learn from and adapt to environmental changes in real-time makes it a valuable solution for smart agriculture, particularly in urban and resource-constrained regions.

References

[1] Mustapha, S., Islam, T. M., Wahab, R. A. Integration of IoT technology in hydroponic systems for enhanced efficiency and productivity in small-scale farming. *Acta Technol. Agricul.*, 2024;27(2):76–83. [Online]. Available: https://sciendo.com/article/10.2478/ata-2024-0027.

[2] Sharma, R. K., Reddy, P. N. Machine learning-based approach for crop growth monitoring in hydroponics cultivation. *Internat. J. Intell. Sys. Appl. Engg. (IJISAE)*, 2023;11(4):3944–3952. [Online]. Available: https://www.ijisae.org/index.php/IJISAE/article/view/3944.

[3] Lee, C., Kim, D., Park, S. An AIoT-based hydroponic system for crop recommendation and nutrient parameter monitorization," *Internet of Things*, vol. 15, p. 100577, 2024. [Online]. Available: https://doi.org/10.1016/j.iot.2024.100577

[4] Wilson, M. J., Huang, L. T. IoT-based real-time monitoring and control system for tomato cultivation. *ACM Dig. Lib. Proc. Comp. Sci.*, 2024;98:240–251. [Online]. Available: https://dl.acm.org/doi/10.1016/j.procs.2024.08.060.

[5] Gupta, P. K., Singh, R., Tiwari, V. Machine learning-based crop growth management in greenhouse environment using hydroponics farming techniques. *J. Smart Agricul. Greenhouse Manag.*, 2023;5(1):15–27. [Online]. Available: https://www.sciencedirect.com/science/article/pii/S2665917423000016.

[6] Shalash, O., Métwalli, A., Elhefny, A., Rezk, N., El Gohary, O. El Hennawy, F. Akrab, A. Shawky, Z. Mohamed, N. Hassan, and M. Hassanen, *"Enhancing hydroponic farming with machine learning: Growth prediction and anomaly detection,"* SSRN Electronic Journal, Jan. 2025, 18 pp. [Online]. Available: https://doi.org/10.2139/ssrn.5079228

[7] Lee, C., Kim, D., Park, S. An AIoT-based hydroponic system for crop recommendation and nutrient parameter monitorization. *Proc. Comp. Sci.*, 2024;3137:050019-1–050019-8. [Online]. Available: https://doi.org/10.1063/5.0076784.

[8] Shukla, S., Kondaka, S., Prasad, S. R. S. "A Smart Hydroponic Farming System Using Machine Learning," in *Proc. Int. Conf. Intell. and Innov. Technol. in Comput., Electr. and Electron. (IITCEE)*, 2023, pp. 357–362. [Online]. Available: https://www.researchgate.net/publication/369930346_A_Smart_Hydroponic_Farming_System_Using_Machine_Learning.

[9] Sulaiman, H., Yusof, A. A., Mohamed Nor, M. K. Automated hydroponic nutrient dosing system: A scoping review of pH and electrical conductivity dosing frameworks. *Agri. Engg.*, 2025;7(2):43. [Online]. Available: https://doi.org/10.3390/agriengineering7020043.

[10] Risheh, A., Jalili, A., Nazerfard, E. *"Smart irrigation IoT solution using transfer learning for neural networks,"* arXiv preprint arXiv:2009.12747, Sep. 2020, 8 pp. [Online]. Available: https://arxiv.org/abs/2009.12747.

22 IoT-integrated smart mirror for continuous health and wellness monitoring

Navyashree Ganpisetty[1,a], Charan Teja H.[1,b], Divyabhavani Ganpisetty[1,c], Sneha Hugar[1,d], Sudhanva S. Haritasa[1,e] and Dr. Shalini Shravan[2,f]

[1]Student, Department of Electronics and Communication Engineering, R V Institute of Technology and Management, Bangalore. Karnataka, India

[2]Assistant Professor, Department of Electronics and Communication Engineering, R V Institute of Technology and Management, Bangalore. Karnataka, India

Abstract

Smart mirrors have emerged as a novel technology that incorporates biometric surveillance, real-time feedback, and interactive display functionality to improve individual healthcare. Such devices utilise inbuilt sensors and the Internet of Things (IoT) to monitor critical health parameters like age estimation, body temperature, height, and body mass index (BMI). Through the effortless incorporation of health monitoring into everyday life, smart mirrors present a non-intrusive and easy-to-use method of preventive healthcare. The following is model of smart mirror health display, examining design, technological innovations, applications, and challenges. The paper also looks at enhancing health checks and the prospect of smart mirrors in telemedicine and remote patient monitoring. Lastly, future research directions and improvements needed to maximise smart mirror technology for mass application in customised healthcare applications are discussed in this review.

Keywords: Smart mirror, health monitoring, IoT, telemedicine, preventive healthcare, remote patient monitoring

Introduction

Over the past few years, technology integration in healthcare has transformed individual well-being monitoring and medical diagnosis. Among these developments, smart mirrors have become a new solution that is merging classic reflective surfaces with sophisticated biometric monitoring, artificial intelligence (AI), and Internet of Things (IoT). These interactive gadgets offer instant feedback on health metrics like age estimation, body temperature, height, body mass index (BMI), and tailored health messages, which make them a smooth fit in daily life. Smart mirrors are non-intrusive health monitoring devices that allow users to monitor their well-being without needing wearable technology or manual input. By incorporating sensors, cameras, and AI algorithms, they can read facial structures, body orientation, and physiological signals to determine a user's health status. The devices are also capable of being used with telemedicine platforms for remote consultations and preventive healthcare.

With the increasing need for personalised and remote healthcare solutions, smart mirrors have become an important feature in homes, fitness clubs, hospitals, and corporate wellness programs. Their capability to provide on-going, passive monitoring of health makes them a promising means for early disease detection, fitness tracking, and lifestyle management. Yet, regardless of their promise, data privacy, accuracy, user acceptance, and system integration are major challenges to broad acceptance.

The paper presents our work on the development of a smart mirror health display, focusing on its technological implementation, applications, challenges, and future enhancements.

Literature survey

The incorporation of IoT technologies in healthcare has garnered considerable attention due to its capacity to significantly improve patient care and diagnostic processes. Research indicates that IoT-enabled health monitoring systems facilitate continuous data acquisition, thereby enhancing early disease detection and enabling more effective patient management. These systems have shown substantial potential in providing real-time feedback and monitoring critical health parameters, which supports the advancement of preventive healthcare practices [1].

[a]nsg11703@gmail.com, [b]charanteja1290@gmail.com, [c]divyabhavanig@gmail.com, [d]snehugar366@gmail.com, [e]sudhanvaharitasa@gmail.com, [f]shalinishravan.rvitm@rvei.edu.in

DOI: 10.1201/9781003685876-22

The integration of fitness and health tracking through smart mirrors is increasingly seen as essential to modern wellness strategies. These mirrors, which integrate IoT technologies, enable users to track their physical progress and achieve fitness milestones with greater ease. By offering a seamless, intuitive interface, these devices support users in managing their health in an informed and consistent manner, reinforcing the importance of real-time data in health optimisation [2]. The integration of fitness and health tracking through smart mirrors is increasingly seen as essential to modern wellness strategies. These mirrors, which integrate IoT technologies, enable users to track their physical progress and achieve fitness milestones with greater ease. By offering a seamless, intuitive interface, these devices support users in managing their health in an informed and consistent manner, reinforcing the importance of real-time data in health optimisation [3]. Finally, the pervasive use of smart monitoring systems in daily life is becoming increasingly widespread. When coupled with advanced technologies like smart mirrors, these systems empower users to monitor and manage their health more proactively. The integration of real-time feedback and data analytics fosters a more personalised and effective approach to healthcare, offering individuals the tools needed to take control of their health and well-being [4]. Smart mirrors hold substantial promise in health and fitness management, with the ability to track a variety of health indicators and provide actionable feedback during physical activities. By continuously offering insights on exercise performance and overall physical health, these mirrors help individuals make better-informed decisions, enhancing their engagement in fitness regimens and improving health outcomes [5].

The use of IoT technology in healthcare devices, particularly smart mirrors, continues to gain traction owing to their interactive capabilities. These mirrors serve as valuable interfaces, providing users with the ability to visualise and analyse important health data, while simultaneously fostering healthier lifestyles. The significance of such technology is particularly evident for elderly individuals who require continuous monitoring, as these smart mirrors offer an easy-to-use platform for daily health assessments [6]. Technological innovations in health monitoring are progressively creating more personalised healthcare solutions. Research highlights how smart health monitoring systems, including smart mirrors, play a key role in monitoring vital signs and delivering tailored healthcare advice. These technologies help streamline healthcare management processes, offering patients a greater sense of empowerment and facilitating more efficient, individualised care [7].

The continuous evolution of smart mirror technology is shaping the future of healthcare delivery. By embedding sensors and utilising advanced data analytics, these mirrors offer profound insights into a patient's health status. As technology progresses, the potential for improving accessibility and efficiency in healthcare systems through such innovations becomes increasingly apparent, driving the need for further development in this field [8].

Beyond fitness applications, smart mirrors are also emerging as essential tools for comprehensive health management. These mirrors are capable of providing crucial health data, making them indispensable in both home and clinical environments. As such, they represent a significant leap toward the realisation of more accessible, patient-centric healthcare solutions, facilitating better health tracking and management across various settings [9].

The use of IoT technology in healthcare devices, particularly smart mirrors, continues to gain traction owing to their interactive capabilities. These mirrors serve as valuable interfaces, providing users with the ability to visualise and analyse important health data, while simultaneously fostering healthier lifestyles. The significance of such technology is particularly evident for elderly individuals who require continuous monitoring, as these smart mirrors offer an easy-to-use platform for daily health assessments [10].

Moreover, the integration of smart mirrors with other health-related functionalities offers an exciting opportunity to enhance health outcomes. These devices, by combining features like body composition analysis and physical activity tracking, provide users with an intuitive and consistent interface that encourages regular engagement with health monitoring, thus fostering long-term health improvements [11].

Moreover, the integration of smart mirrors with other health-related functionalities offers an exciting opportunity to enhance health outcomes. These devices, by combining features like body composition analysis and physical activity tracking, provide users with an intuitive and consistent interface that encourages regular engagement with health monitoring, thus fostering long-term health improvements [12].

Methodology

Tools and technologies used

This section outlines the key sensors, components, and software used in the smart mirror system for health monitoring. Various hardware and software elements work together to ensure accurate data collection, processing, and user interaction. The system integrates various sensors and computing units for efficient data processing and interaction. The HC-SR04 ultrasonic sensor measures distance by emitting sound waves and calculating the echo return time, useful for obstacle detection. A user input keypad provides a simple interface by mapping key presses to specific inputs, commonly used for password entry and device control. The HX711 analogue to digital converter with load cell converts force into electrical signals to measure weight, widely used in weighing scales. For health monitoring, the MAX30100 pulse oximeter and heart rate sensor tracks heart rate and blood oxygen levels using optical sensors, while the MLX90614 infrared temperature sensor enables non-contact temperature measurement via thermal radiation detection. The system is powered by a Raspberry Pi 4, acting as the central processor for running applications and managing sensor data, alongside an Arduino Uno microcontroller for real-time input processing. Thonny IDE simplifies Python programming, while Tkinter facilitates graphical user interface (GUI) development. A local web server—using Apache/Nginx for hosting and Flask/Django for web-based dashboards—enables remote access and data visualisation, enhancing system usability.

All these components and technologies work together to create an efficient and interactive health monitoring system, ensuring accurate data acquisition, real-time processing, and seamless user interaction.

Implementation

This section outlines the implementation of a health monitoring system that integrates multiple sensors and components. The ultrasonic sensor (HC-SR04) determines user height by emitting sound waves and measuring the time for the echo to return, ensuring accurate body mass index (BMI) calculation when placed above the user's head. The load cell (HX711), a strain gauge-based transducer, measures weight by converting force into an electrical signal, which is then processed using the HX711 ADC. A keypad connected to an Arduino microcontroller allows users to input personal details such as age and gender for a personalised BMI calculation using weight and height data using the Equation (1). The pulse sensor (MAX30100) utilises optical technology to measure heart rate and blood oxygen levels (SpO_2) through I2C communication, providing cardiovascular insights. The temperature sensor (MLX90614) detects body temperature without contact by measuring emitted thermal radiation and transmitting data via I2C.

A Raspberry Pi 4 serves as the central processing unit, managing data acquisition, processing, and display using Python and the Thonny IDE. Sensor data is visualised using a Tkinter-based GUI and a local web interface developed with HTML, CSS, and JavaScript. Key sensor libraries include the MAX30100 library for heart rate and SpO_2 (blood oxygen percentage), MLX90614 library for temperature readings, HX711 library for weight measurement, HC-SR04 library for ultrasonic height measurement, and keypad library for handling user input.

$$BMI = \frac{Weight(Kg)}{(Height(m))^2} \qquad (1)$$

The system is prototyped in Tinkercad, with the final model incorporating a two-way mirror and a monitor, ensuring both functionality and aesthetics. Sensors are first connected to an Arduino Uno, where each sensor is assigned to specific digital or I2C pins. The ultrasonic sensor uses TRIG and ECHO pins for height measurement, the load cell (HX711) connects via VCC, GND, DT, and SCK for weight data, and the pulse sensor and temperature sensor use I2C communication. Integration steps include sensor wiring, library installation, code development, testing, and calibration. Figure 22.1 shows the physical model of the smart mirror made.

Further integration with Raspberry Pi 4 involves pin connections of GPIO-based connections and I2C/SPI configurations. The ultrasonic sensor is linked to GPIO 23 (TRIG) and GPIO 24 (ECHO), the load cell connects via GPIO 5 (DT) and GPIO 6 (SCK), and the pulse sensor (MAX30100) communicates through I2C using GPIO 2 (SDA) and GPIO 3 (SCL). The integration process involves establishing correct GPIO connections, installing Python libraries, writing scripts for data acquisition, testing sensor, and integrating real-time monitoring into the GUI.

Results

The smart mirror is an IoT-powered health monitoring system that tracks key health metrics such as heart rate, SpO_2 (calculated internally in the sensor), temperature, and BMI using sensors like MAX30100, MLX90614, HC-SR04, and HX711. It features an interactive interface accessible through Tkinter on Raspberry Pi or a web application, displaying real-time health data. Beyond real-time monitoring, the system analyses trends over time, identifying patterns such as rising temperature or consistently high heart rate and prompting users to take necessary action. By incorporating medical guidelines and personal factors like age and gender, it provides personalised health recommendations, including fitness and dietary advice, while also issuing alerts for potential health concerns. Significant changes, like sudden heart rate spikes or low SpO_2 levels, ensuring timely awareness and proactive health management.

The flowchart in Figure 22.3 outlines the step-by-step process of the smart mirror system. It begins with the user standing on the podium and placing their hand on a panel to activate the system. The user then selects their gender as in Figure 22.2 and inputs their age using a keypad, then the user enters weight as in Figure 22.4 and age as in Figure 22.5. Various health parameters are collected through integrated sensors, including height, weight, SPO_2, temperature, and heart rate. After displaying as in Figure 22.6 the results along with the calculated BMI, the system resets to its initial state, ready for the next user. Voice commands, text-to-speech, and adjustable displays for accessibility. Features like emergency response integration and localised health info add to its utility.

Figure 22.2 Entering gender UI interface
Source: Author

Figure 22.3 Smart mirror flowchart
Source: Author

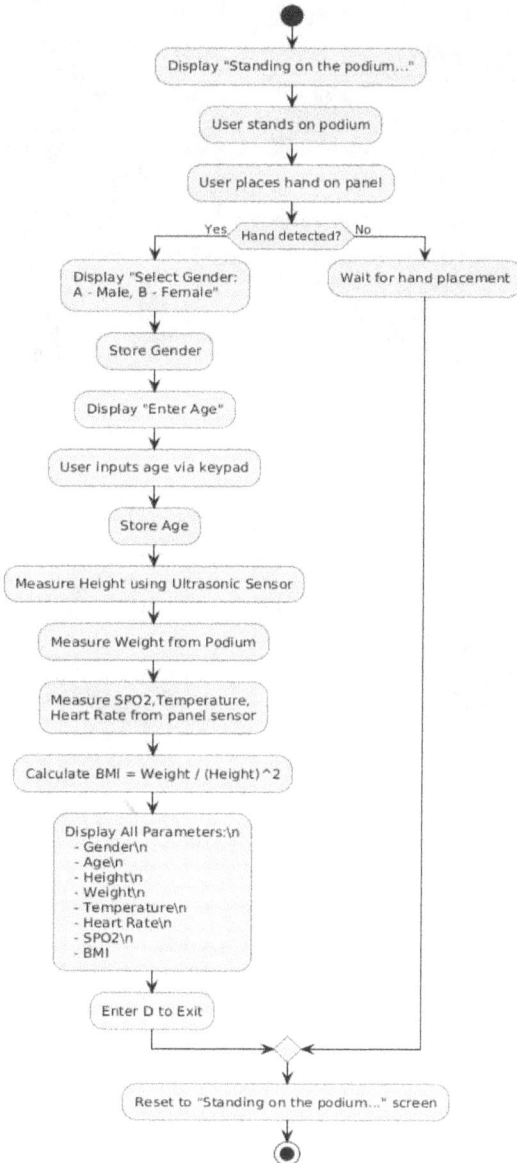

Figure 22.1 Model image
Source: Author

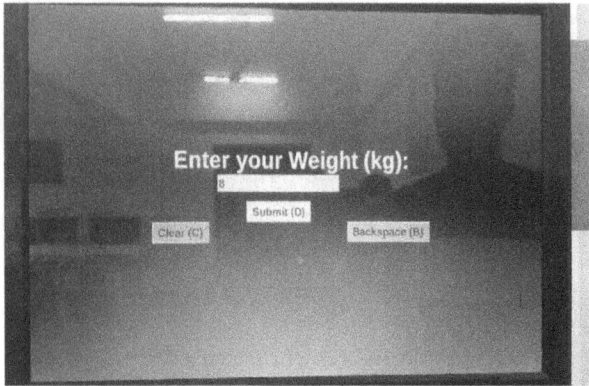

Figure 22.4 Entering weight UI interface
Source: Author

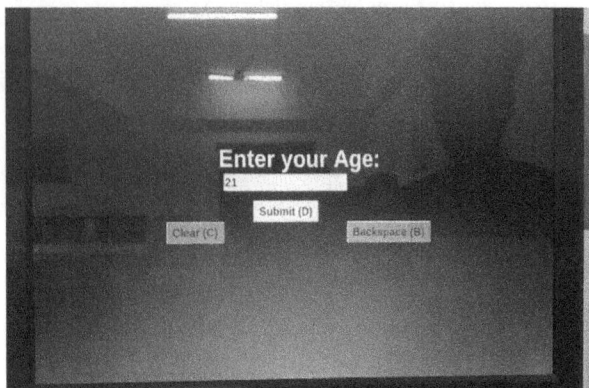

Figure 22.5 Entering age UI interface
Source: Author

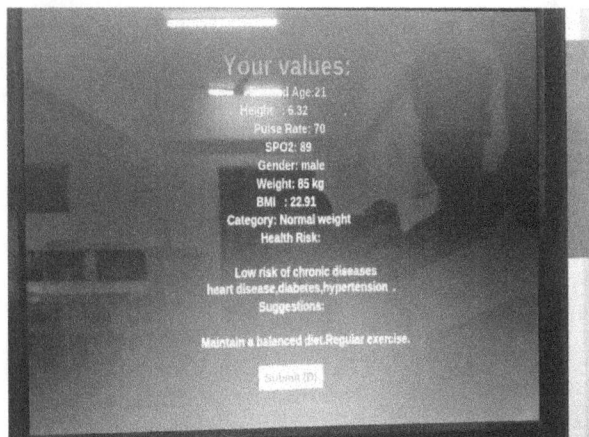

Figure 22.6 Final data display UI interface
Source: Author

Conclusion

The smart mirror combines advanced sensors and IoT to deliver a comprehensive health monitoring system. It tracks real-time metrics like heart rate, SpO$_2$, temperature, weight, and height for effective daily health management. Enhancing user interaction, it supports voice commands, text-to-speech, and adjustable displays for accessibility. Features like emergency response integration and localised health info add to its utility. With QR code integration, users can securely access and share their health data via mobile devices, ensuring continuous monitoring and easy communication with healthcare providers.

Future scope

The smart mirror enhances functionality by improving accessibility, emergency response, localised health resources, and mobile integration, ensuring a better user experience and real-time assistance. Hands-free interaction through voice commands, audio feedback for health metrics, and customisable display settings make it more inclusive. In critical situations, the system can alert emergency contacts, provide first aid guidance, and facilitate emergency calls. It also keeps users informed by displaying nearby fitness activities, medical facilities, and public health updates. Additionally, QR code integration allows secure access to health data, seamless syncing with mobile health apps, and easy sharing with healthcare providers.

References

[1] Miotto, R., Danieletto, M., Scelza, J. R. et al. Reflecting health: smart mirrors for personalized medicine. npj Digital, 2018;1:62.

[2] Soppimath, V. M., Hudedmani, M. G., Chitale, M., Altaf, M., Doddamani, A., and Joshi, D. The smart medical mirror – A review. Internat. *J. Adv. Sci. Engg.*, 2019;6(1): 1244–1250.

[3] Muneer, A., Fati, S. M., Fuddah, S. Smart health monitoring system using IoT based smart fitness mirror. *TEL-KOMNIKA Telecom. Comput. Elec. Control*, 2020;18(1): 317–331.

[4] Lo´pez, F. R., Sa´nchez, P. S., Garc´ıa, J. M. M., Bermejo, J. M. G., Garc´ıa, J. M. M. The SHAPES smart mirror approach for independent living, healthy and active ageing. *Sensors*, 2021;21(23):7938.

[5] Badwaik, D., Game, E., Zade, P., Kathote, P., Padole, G. Patient health monitoring with an Arduino based smart mirror. *Internat. J. Innov. Res. Technol.*, 2022;9(1):1057–1061.

[6] Gupta, V., Sharma, G., Gupta, R., Goyal, N., Gaur, P. Design and implementation of smart mirror for health monitoring. *Internat. Conf. Artif. Intell. Data Sci. Appl. (ICAIDSA)*, 2023. ICAIDSC2023, 1 (January 2025), 1–4.

[7] Dowthwaite, L. et al. Examining the use of autonomous systems for home health support using a smart mirror. *Healthcare*, 2023;11(19):2608.

[8] S. P. P., S. D. A novel approach to smart mirror technology for enhanced elderly health monitoring. *Internat. J. Scient. Res. Engg. Manag.*, 2023;7(5):12–16.

[9] Sharafunneesa, P. P., S. D. A novel approach to smart mirror technology for enhanced elderly health monitoring. *Internat. J. Scient. Res. Engg. Manag.*, 2023;7(5): 12–16.

[10] Fatima, H., Imran, M. A., Taha, A., Mohjazi, L. Internet-of-mirrors (IoM) for connected healthcare and beauty: A prospective vision. *Internet of Things*, 2024;28:101415.

[11] Babu, M. C., L. N. M. The smart medical mirror system. *Internat. J. Res. Pub. Rev.*, 2024;5(5):1711–1715.

[12] Karthik, D., Priya, T. S., Dasharatham, P. Smart mirror. *Internat. Res. J. Moderniz. Engg. Technol. Sci.*, 2024;6(9): 311–317.

23 Automatic vehicle emergency response and alert system using GPS and GSM

Ms. Sushma Nagaraj[1,a], Suresh M. B.[2,b] and Dr. Sahana Salagare[3,c]

[1]Solution Architect, Amazon Web Services, Dallas, Texas, United States of America (USA)

[2]Professor & Head, Department of Artificial Intelligence and Machine Learning, K S Institute of Technology, Bengaluru, Karnataka, India

[3]Assistant Professor, Department of Artificial Intelligence and Machine Learning, K S Institute of Technology, Bengaluru, Karnataka, India

Abstract

Over 1.25 persons worldwide lose their lives in traffic accidents each year. By tracking vehicles and detecting collisions, autonomous messaging might greatly reduce the time it takes to respond to accidents. This technology helps people in car accidents by providing timely alerts and automated communication. It tracks the car's movements and sends location-based SMS in real time using tracking technology. Global system for mobile communication (GSM) transmits the location of the car via a GSM modem, allowing emergency services to be promptly notified. When an accident is discovered, the nearby medical facility is notified so that emergency procedures can be established and care can be given as soon as feasible. For emergency services to react quickly, Information such as the vehicle's global positioning system (GPS) coordinates must be provided as soon as feasible. Installing an electrical gadget in the vehicle that continuously checks the GPS is necessary for this strategy. The technology initiates an alarm anytime an accident happens, contacting the closest emergency agency or hospital. The main controller of the system is an Arduino Uno that has been customised with a GPS module, GSM modem, and accelerometer. It is to collaborate and to provide the precise coordinates to the relevant emergency contacts or services. This innovative technique increases an accident victim's likelihood of survival and reaction time.

Keywords: Accident detection, automatic emergency messaging system, GSM module, GPS module

Introduction

Traffic and safety have become extremely problematic problems. An Arduino Uno microcontroller, global positioning system (GPS), global system for mobile communication (GSM), and accelerometer modules make it simple to monitor an accident. When a car crash happens without the driver or passengers' agreement, the gadget can identify it because of a tilt in the current value of the vehicle orientation. A registered SIM card receives the location coordinates from the GPS module via the GSM module. World Health Statistics from 2008 are cited in the Global Status Report on Road Safety. Even though road safety information tracking survey (RITS) was ranked ninth in 2004, if current trends continue it is predicted to surpass diabetes and AIDS as the fifth biggest cause of death by 2030. One of the nation's leading causes of death is traffic accidents.

Speeding, breaking traffic regulations and driving while intoxicated are the main causes of traffic accidents. Numerous studies indicate that one of the main reasons people die in road accidents is a delay in the response of emergency aid. Reaction time decreases more frequently might save the precious and important person's life.

It might be challenging for a nursing associate injured in an accident to report the incident to family members or law enforcement. Reducing the time between a delivery and an incident Emergency agencies utilise the automated messaging system to track down automobiles and find accident scenes. The majority of the vehicles have both a vehicle chasing system and an accident detection system installed.

A microcontroller algorithm converts the coordinates into a connection to a Google map once the GPS module has collected them. Each time when a car is stolen or an accident occurs. The microcontroller has the formula pre-programmed.

The traveller should tell family members right away and seek assistance as soon as feasible in the event of an accident. According to the study, a car with label sensors would identify a collision and alert a tiny controller, which would then transmit the data including the precise position of the hit and personal information to a cloud server. A Google map is accessible through the International System of Units'

[a]nagsushm@amazon.com, [b]sureshresearch45@gmail.com, [c]sahanasalagare@gmail.com

DOI: 10.1201/9781003685876-23

GSM module, and family members and individuals in the vicinity of police headquarters are allowed to connect using mobile devices. The accident is identified using a measuring method, and the price is compared to the known brink price. By clicking, the friend may see the exact position of the vehicle. on the Google map link that was included in the SMS.

Literature survey

Related work presents the different techniques proposed by the researchers related to Internet of Things (IoT)-based automated message sending system for vehicle tracking and accident spot detection. In a paper submitted by Syedul Amin et al. [2], effective automatic accident detection that instantly alerts the emergency services at the scene of the accident is crucial in order to save precious human life. This seminar's objective is to utilise GPS receivers to track a car's speed, detect collisions based on that speed, and use a microcontroller that decodes GPS data over the GSM network to transmit the accident's location and time to the alert service centre.

The research presents a framework for delivering medical support and other services as soon as possible after an accident [3]. Accidents can be detected by flex sensors and accelerometers, and the position of the accident will be sent to the police, the nearest hospital, and the car's owner via GSM-enabled SMS.

Provide the GPS-derived coordinates, the vehicle number, and the accident time. A camera inside the vehicle will provide live footage [1] to show the present situation of the passengers. Thus, post-accident methods for detecting and reporting them are the main focus of this study. This study also includes the hyper terminal simulation findings. According to the study, the GPS uses a microcontroller chip to compare a vehicle's present speed to its historical speed [17] once per second [4]. It will assume that there has been an accident if the speed falls below the predetermined threshold. The system will then send the time, speed, and accident location that were acquired from the GPS over the GSM network. This will assist preserve [18] the precious human life by getting to the rescue service and police control centre in time. Vehicle accident detection technology uses a variety of techniques, such as GPS and other radio navigation systems that depend on satellites and ground-based stations, to identify an automobile.

A study did by Ulhas et al. [5] claims that the system uses the already-deployed smart sensing devices to instantaneously notify the passenger of their location. This data is analysed and all of the information is evaluated from a distance in order to provide the individuals with the services they require. In any case, the car can be tracked down. This system uses GPS and GSM technologies to send all of the data to a distant server.

Once the data has been analysed, the server uses the information it has gathered to help the person in need. According to a research by Basheer et al. [6], there has been a significant increase in accidents and accident-related deaths in recent years due to the dramatic increase in the number of automobiles. Auto accidents claim the lives of around 1.25 million people annually, according to WHO estimates. India has only 1% of the world's roads. However, it is estimated that Indian roads account for 16% of all car accidents globally. There are several possible reasons for this, one of which is a dramatic increase in the number of cars without a commensurate improvement in the road conditions and infrastructure required for them. Another significant factor that might be looked at is the lack of medical facilities at the right time and place. As a result, most sufferers pass away. The notion of addressing the issue to save lives so emerges. The system provides real-time position information when they are going by utilising the integrated smart sensor technology. This data is analysed and all of the information is seen from a distance in order to provide the people with the services they require. The tracking of the car can be constructed in all weather. This system uses GPS and GSM technology to send all data to a remote server, which processes it and uses the information it extracts to make services available to the person in an emergency.

Existing system

Cars are essential for meeting up with loved ones, distributing commodities, and going to work. However, they frequently open the door for disastrous occurrences. Accidents are frequently unanticipated, unplanned events or conditions that occur for no apparent purpose or need, according to Wikipedia. Even though they happen often, traffic collisions are the worst thing that can happen to a motorist. The worst thing is that we fail to learn from our mistakes. Although the majority of drivers are aware of the basic rules and safety measures, collisions are caused by the negligence of other drivers. The main cause of collisions and accidents is human error. The following are the primary causes of accidents:

1. Distractions for the driver
2. Red light violation
3. Do not using safety gears like seat belts and helmets
4. Not maintaining lane discipline and overtaking in a wrong manner.

The current problem has been approached from a variety of angles, and each has certain advantages over the others. Several GPS and GSM systems evaluated the accident situation just using an accelerometer sensor. This can cause false alerts in some circumstances, which might be troublesome. Our system employs several sensors to improve accuracy and has a mechanism that halts alerting in the event of a false alarm. Additionally, the current system only makes use of Wi-Fi modules, which require a network to operate.

Proposed system

Using a range of sensors, we have improved the automatic messaging system for vehicle tracking and accident spot detection. In certain instances, this has reduced the possibility of false alarms and increased the accuracy of accident detection. We have a physical switch in the car itself that needs to be hit within a specific period of time in order to prevent false alerts and messages. To improve the precision of accident detection, we are utilising a position encoder, GPS sensor, front bumper sensor, and MEMS sensor. The micro-controller will be powered by the bumper sensor. Given the force or pressure imparted to it, it goes without saying that a collision will result in a higher pressure.

Another level of dependability is added by the position encoder, which controls the vehicle's speed and is anticipated to fluctuate significantly in the event of an accident. Normally, the micro-controller is alerted to any abrupt changes by the MEMS sensor. The quickening. The GPS and GSM modules are used to send SMS messages and locate the accident scene.

The system architecture of a product or application helps to better understand system behaviour by providing many explanations of the system's precise behaviour. The system architecture describes the system and the work that went into making it. The practice of assessing a system at an early stage to make sure it is feasible is called architecture.

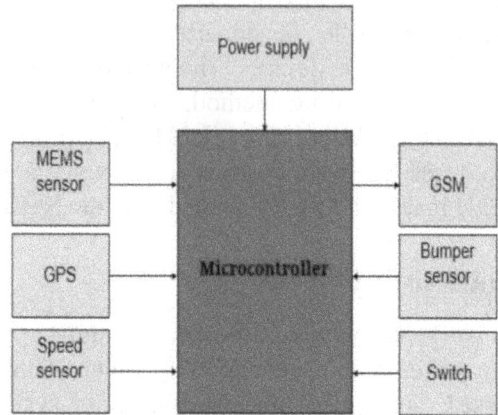

Figure 23.1 Block diagram of system architecture
Source: Author

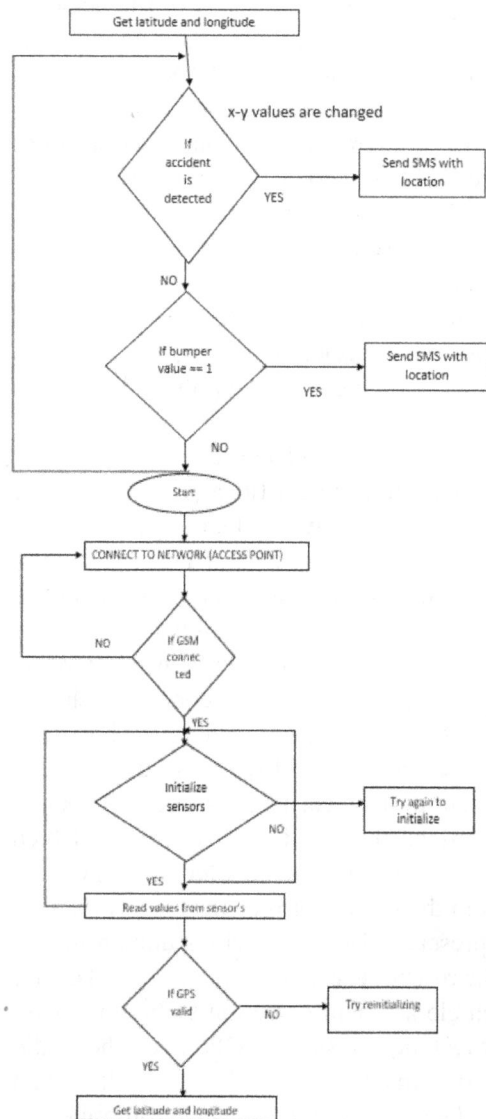

Figure 23.2 Flow chart of workflow
Source: Author

A block diagram is a system diagram where the main components or functions are depicted by blocks joined by lines that indicate the blocks' connections.

Methodology

In the event of an accident, the micro-controller uses control logic when the vehicle detects a sudden change in the threshold values using a measuring device detector. As soon as a collision or accident is detected, set the measuring instrument detector's effective sensitivity value. The GSM module is turned on by Arduino when the controller uses a measuring instrument detector to identify an accident or set bit. The SMS and transmitted data will be received via a GSM/GPRS modem connected to a PC. A middleware program will understand the SMS and GPRS data. Using the information from the interpreted SMS/GPRS data, a suitable application will be created to integrate Google maps and automatically plot the accident scene on the map. Additionally, it will display the vehicle's speed before to the incident. Based on speed, the hospital will utilise this information to assess the accident's severity. A pre-stored SMS is sent to that choice and the local hospital using the accident victim's friend's manually saved signal.

The working model is represented in the flow chart of window. The figure represents all the steps and procedures to be taken for the implementation and to get expected results.

Figure 23.2 represents a workflow or process of implementation. The flowchart shows the steps and their order by connecting the flow with arrows. This diagrammatic representation illustrates a solution model to a given problem. Flowcharts are used in analysing, designing, documenting or managing a process in various fields.

Implementation requirements

Requirements specification is a specification of the requirements such as hardware and software; it completely describes the behaviour and functionality of the device and also consists of use cases that describe the interaction of the user and the system and also the interactions between different modules of the system. It should tell us about the requirements of the user and the design of the system.

In addition, it also contains non-functional requirements. Non-functional requirements affect the design and use of the system like quality of the product,

performance standards and design of the system. To derive the requirements, we should have to have a clean and thorough understanding of the products to be developed. This is proposed after detailed communications with the project team and the customer.

Hardware requirements:

- ESP32 controller
- MEMS sensor
- GPS
- GSM
- Sensors in bumper
- Reading switch
- Speed motion sensor
- Power supply

Software requirements:

- Arduino IDE
- C++

The ESP32 microcontroller is a small computer on a single metal oxide-semiconductor integrated circuit chip. One or more central processing units (CPUs), memory, and programmable input/output peripherals are all found in a microcontroller.

The ESP32 has two processors because it is dual core. Both Bluetooth and Wi-Fi are integrated into it. 32-bit programs can be run on it. It features 512 KB of random access memory (RAM) and a clock frequency can vary up to 240 MHz. There are thirty or thirty-six pins on this board, fifteen in each row. Additionally, In addition to many other peripherals, it provides capacitive touch, ADCs, DACs, UART, SPI, and I2C. Both a temperature sensor and a Hall Effect sensor are included into it.

The MPU6050 is a micro electro-mechanical system (MEMS) which has a 3-axis Gyroscope and Accelerometer built into it. This aids in the measurement of a system's or object's acceleration, velocity, orientation, displacement, and numerous other motion-related parameters. Additionally, this module contains a digital motion processor (DMP), which is strong enough to carry out intricate calculations and free up microcontroller labour. The AD0 pin of a microcontroller can be used to interact with an MPU6050 sensor. Additionally, this module features updated and well-documented libraries, making it simple to use with well-known platforms like Arduino. Therefore, this sensor can be the best option for you if you're seeking for a way to regulate motion for your remote- controlled car, drone, self-balancing robot, humanoid, biped, or something similar.

Bumper A sensor is just a switch that detects excessive voltage in the scenario of an automobile collision. If it is pressed, a high voltage is been sent to the controller, which interprets it as an accident. It identifies and alarms using the binary numbers 0 and 1. It gives the microcontroller a value of 1 if the voltage is high.

The LM393 voltage comparator IC and the LM393 speed sensor module is essentially an infrared light sensor combined. Vibration or notable position changes in the XY plan are detected by this sensor. It is a sensor that provides digital values. If the values are more than the xy_max value or higher than the xy min value, they are supposed to have satisfied the accident condition.

A constellation of 24 satellites and associated ground stations make up the GPS, a global radio navigation system. In order to generate positions within a few meters of accuracy in our system, GPS uses these "man-made stars" as reference points. This aids in tracking by supplying the co-ordinates of the vehicle involved in an accident.

A chip or circuit known as the Global System for Mobile Communication module (GSM) is made to connect a computer or mobile device to a microcontroller and GPS system. It functions as a cloud storage device, alerting the contacts stored there and storing data gathered from sensors.

Arduino is an open-source hardware and software initiative, corporation, and user community that creates and produces microcontroller kits and single-board microcontrollers for use in digital device construction. Writing code and uploading it to the board is made simple by the open-source Arduino Software (IDE). Any Arduino board can be used with this software. GitHub hosts the Arduino software's on-going development. Refer to the building instructions for the code. Source code archives for the most recent releases can be found here. This key can be used to verify the archives because they are PGP-signed.

Implementation

Hardware implementation

Numerous ports on the ESP32 microcontroller board allow it to interface with other system modules. The Arduino IDE controls the chip. When the car detects a sudden change in the threshold values using the measuring device detector, an accident is reported. Set the effective sensitive value of the measuring instrument detector as soon as a collision or accident is detected. As soon as it detects the accident or set bit using a measuring instrument detector, the Arduino chip activates the GSM module, which has a manually saved signal of the buddy of the accident victim. It then sends a pre-stored SMS to that selection. It also conveys the news of the accident to the numerous pals at the same time.

Software implementation

Arduino provides all of the functionality for the SMS alert system. It also takes care of the input signal filtering. This position can be altered by applying chump change inside the ASCII text file. The gadget is distinctive in more ways than only alerting the neighbours with its siren. However, cell phone customers also receive a warning SMS from it. You may program an Arduino to alter it to these numbers.

The Arduino IDE, which uses the C programming language, is the most widely used programming method. Because of the open-source community, you may access a vast library of Arduino code that is always expanding.

The Arduino IDE asks you to select the board you are uploading to. Navigate to board by using the tools pull-down option. By default, the Arduino boards that are currently on the market are included in this list. Choose Arduino Uno if you're using Uno or Uno-Compatible Clone (such as Funduino, SainSmart, IEIK, etc.). If you're using another board/ clone, select that board.

The USB drivers ought to have installed themselves when you plugged in your Arduino board if you have downloaded the Arduino IDE beforehand. The latest Arduino IDE ought to identify linked boards and indicate which COM port they are use. Choose port from the tools pull-down menu. In addition to listing all of the open COM ports, it should also provide the name of any identified Arduino boards. After connecting the Arduino board to the PC, choose it. The board type and COM number of the board you intend to program should appear in the bottom right corner of the Arduino IDE if the setup was complete.

The Arduino Uno occupies the next available COM port; it will not always be COM3. At this point, our board should be set up for programming, and you can begin writing and uploading code.

Results and discussion

This technology will be important in the field of traffic accidents. It helps the separated person in traffic accidents as soon as possible by using an automatic victimisation messaging system. Consequently, it improves the victim's chances of surviving. In cases where the car or police service would take prompt action to lessen the severity of the fatalities and severe conditions resulting from accidents, GSM technologies were utilised. All accident information (drivers' driving patterns) can be stored on a dashboard, which can then be updated to friends, the main road hospital (rescue team), and government agencies to provide accident details.

The system will browse the on-board diagnostic (OBD) information which is able to facilitate automaker to cite any flaw in their style and conjointly helps the insurance firms to search out the fraud cases of claim.

Figure 23.3 shows the result of Implementation in the mobile device and indicates the car is crashed and urgent attention is needed along with accident in the form of Google map link.

Conclusion

The automatic messaging system for vehicle tracking and accident spot detection would be essential in the field of traffic accidents. It uses an automated victimisation messaging system to assist the separated person in road accidents as soon as possible. Consequently, the victim has a higher chance of surviving. The gearbox may be utilised as an integrated program in the vehicle if the automakers cooperate with the system. To configure the system in a new vehicle, an associate's degree in nursing is required.

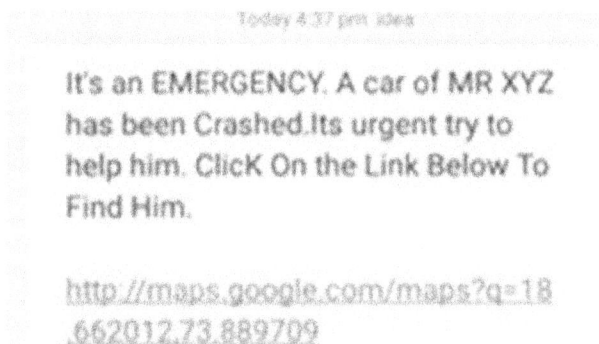

Figure 23.3 The mobile display during accident detection
Source: Author

The use of GSM technology is one instance of a situation involving fatalities and serious injuries. The police and ambulance services will move swiftly to defuse the situation.

Acknowledgement

The paper was prepared as part of academic requirements during my study in the Institution and is submitting now with minor corrections. I acknowledge and thankful to the Dr. Suresh M. B., Prof & Head, Department of AIML and Dr. Sahana Salagare, Asst Prof., Department of AIML for the support during the preparation of paper.

References

[1] Government of India. Ministry of Road Transport and Highways. Lok Sabha Unstarred Question No. 374 Answered on 19-07-2018.

[2] Syedul Amin, M. Jalil, J. and Reaz, M. B. I. "Accident detection and reporting system using GPS, GPRS and GSM technology," 2012 International Conference on Informatics, *Electronics & Vision (ICIEV)*, Dhaka, Bangladesh, 2012, pp. 640–643 DOI: 10.1109/ICIEV.2012.6317382

[3] Bankar, S. A., Kale, A. V., Jagtap, S. R. Intelligent system for vehicular accident detection and notification. *IEEE*, "*Conference: 2014 International Conference on Communications and Signal Processing (ICCSP)*" 2014. DOI:10.1109/ICCSP.2014.6950048

[4] Mohsina, A., Shubham, S., Zeba, K., Sofiya, K., Sakshi, D. Collision detection of vehicle and coverage of using GPS and GSM technology. IEEE, 2017.

[5] Ulhas, P., Pranali, M., Rahul, P., Uday, P. Tracking and recovery of the vehicle using GPS and GSM. *International Research Journal of Engineering and Technology (IRJET)*, 2017:4(3);2074–2077.

[6] Basheer, F. B., Alias, J. J., Favas, C. M., Navas, V., Farhan, N. K., Raghu, C. V. Design of accident detection and alert system for motorcycles. *2013 IEEE Global Human. Technol. Conf. South Asia Satellite (GHTC-SAS)*, Trivandrum, India, 2013, 1–6, doi: 10.1109/GHTC-SAS.2013.6629903.

[7] Ramani, R., Valarmathy, S., Suthanthira, N., Selavaraju, S., Thiruppathi, M., Thagam, R. "Vehicle tracking and locking system based on GSM and GPS," International Journal of Intelligent Systems and Applications, 5(9), 86–93, Aug. 8 2013, doi: 10.5815/ijisa.2013.09.10

[8] An Ericsson White Paper. Communication and Information Services for National Security and Public Safety. Ericsson Microwave System AB.I. Jacobs, S., Bean, C. P. Fine particles, thin films and exchange anisotropy. *Magnetism*, Rado, G. T., Suhl, H., Eds. New York: Academic, 1963;III:271–350.

[9] Xiaoyan, G., Shuming, T., Feiyue, W. "Traffic-incident detection algorithm based on non-parametric regression," in Proceedings of the IEEE *5th Intelligent Transportation Systems Conference*, Singapore, Sep. 2002, 714–719 .

[10] Li, C., Hu, R., Hang, W., He, J., Tao, X. "Study on the method of freeway incident detection using wireless positioning terminal" in 2008 *International Conference on Intelligent Computation Technology and Automation (ICICTA)*, Changsha, China, Oct. 31–Nov. 3 2008, 293–297, doi: 10.1109/ICICTA.2008.4

[11] Bhatia, J. S., Pankaj, V. Design and development of GP SGSM based tracking system with Google map-based monitoring. *IJCSEA*, 2013;3(3):3340.

[12] SeokJu, L., Girma, T., Jaerock, K. Design and implementation vehicle tracking system using GPS & GSM/GPRS technology and smartphone application. presented at the 2014 IEEE World Forum on the Internet of Things (WF IoT), Milan, Italy, Apr. 2014

[13] Chris Veness, W. "Calculate Distance and Bearing Between Two Latitude/Longitude Points – Haversine Formula in JavaScript." Movable Type Scripts, 2016, www.movable-type.co.uk/scripts/latlong.html.

[14] White, J., Thompson, C., Turner, H., Dougherty, B., Schmidt, D. C. Wreck watch: Automatic traffic accident detection and notification with smart phones. *Mobile Networks Appl.*, 2011;16(3):285–303.

[15] Khalil, U., Javid, T., Nasir, A. Automatic road accident detection techniques: A brief survey. *Internat. Symp. Wireless Sys. Netw. (ISWSN)*, 2017:1–6.

[16] Fleischer, P. B., Nelson, A. Y., Sowah, R. A., Bremang, A. Design and development of GPS/GSM based vehicle tracking and alert system for commercial inter-city buses. *2012 IEEE 4th Internat. Conf. Adap. Sci. Technol. (ICAST)*, 2012:1–6.

[17] Kannan, R., Nammily, R., Manoj, S., Vishwa, A. Wireless vehicular accident detection and reporting system. *Internat. Conf. Mec. Elec. Technol. ICMET*, 2010.

24 Smart IoT-enabled vending machine for medical and sanitary supplies

Dr. Surekha Byakod[1,a], Ankitha Devlokam[2], Challa Deepika[2], Vignesh U.[2] and Yeddulapalli Sahithi[2]

[1]Associate Professor, Department of Computer Science & Design, K. S. Institute of Technology, Bengaluru, India

[2]Department of Computer Science & Design, K. S. Institute of Technology, Bengaluru, India

Abstract

Access to sanitary pads is a critical aspect of menstrual hygiene, yet many individuals face challenges in obtaining them due to financial constraints and limited availability in public spaces. This project aims to address this issue by designing and implementing a user-friendly, free-of-charge pad vending machine. The system is built using a Raspberry Pi microcontroller, which serves as the central processing unit, and a servo motor mechanism for controlled dispensing. The vending machine is designed to be accessible, efficient, and suitable for installation in schools, public restrooms, and other community spaces. The Raspberry Pi is programmed to manage user interactions and operate the dispensing mechanism seamlessly. A simple push-button or touch interface enables users to request a pad, while the servo motor ensures precise and reliable dispensing. By providing a cost-effective and automated solution, this project promotes menstrual hygiene awareness and accessibility. The free-of-charge model eliminates financial barriers, empowering individuals to manage their hygiene with dignity. This initiative can be scaled and adapted for various locations, contributing to broader efforts in public health and gender equity. The integration of smart technology in the vending machine ensures efficiency, reliability, and ease of use, making it a valuable addition to public infrastructure.

Keywords: Menstrual hygiene, Raspberry Pi, healthcare accessibility, security and cost effective, scalability, discreet operation, smart technology

Introduction

Menstrual hygiene is a vital yet often overlooked public health issue, with many individuals struggling to access sanitary pads in public spaces due to limited availability, financial constraints, and stigma. Automated sanitary pad vending machines propose a practical solution by providing complimentary access to pads, thereby promoting dignity in menstrual health. This project focuses on developing a user-friendly vending machine, powered by a Raspberry Pi microcontroller and a servo motor for accurate dispensing. Users can discreetly request pads via a push-button or touch interface, while integrated sensors monitor stock levels for timely refills. The machine's secure and tamper-proof design, paired with energy efficiency, makes it ideal for schools, public restrooms, and community centres operate discreetly. By addressing menstrual hygiene needs, this initiative aligns with public health objectives to reduce access disparities and raise awareness, reinforcing the notion that menstrual hygiene is a fundamental right.

Problem statement: Access to sanitary pads is a challenge in public spaces due to financial constraints and social stigma. Traditional distribution methods often fall short, limiting availability and leading to health risks for menstruating individuals. To solve this, a smart, automated vending machine powered by a Raspberry Pi is proposed. This machine will allow users to obtain sanitary pads easily and discreetly with a touch interface, promoting menstrual hygiene and reducing stigma in public restrooms, schools, workplaces, and transportation hubs.

Scope of the project

This project involves designing and deploying an automated sanitary pad vending machine using a Raspberry Pi and a servo motor for dispensing. It aims to provide free access to sanitary pads in public spaces like schools, workplaces, and transportation hubs, reducing barriers to menstrual hygiene. The machine features a user-friendly interface for discreet operation, with sensors to monitor stock levels and a secure enclosure for durability. Software integration

[a]surekhapb@ksit.edu.in

DOI: 10.1201/9781003685876-24

allows for potential Internet of Things (IoT) connectivity, enabling real-time stock monitoring and mobile refill alerts. The project promotes menstrual hygiene awareness and aims to empower individuals while being adaptable for future enhancements, contributing to public health and social welfare. The system allows users to receive a sanitary pad with the press of a button or a touch-based interface, eliminating the need for human interaction and reducing stigma associated with purchasing menstrual hygiene products. A study by UNICEF (2021) emphasised that millions of menstruating individuals across the world face difficulties in obtaining sanitary pads, leading to poor menstrual hygiene practices, health risks, and social stigma. To address these issues, researchers and organisations have explored technological solutions, such as vending machines, to provide easy and free access to sanitary pads in public spaces.

Literature survey highlights

The literature survey underscores the increasing trend of adopting smart vending machines for menstrual hygiene management. While previous studies have focused on coin-operated and IoT-enabled solutions, there's a notable gap in research regarding free-access sanitary pad dispensers using Raspberry Pi and servo motors. This project aims to fill that gap by developing a cost-effective automated vending machine that provides seamless access to sanitary pads in public spaces. By utilising open-source hardware and smart monitoring, the proposed system seeks to enhance accessibility, sustainability, and efficiency in menstrual hygiene. Raspberry Pi and servo motor-based automation have been widely used in various vending machine applications due to their cost-effectiveness and reliability. A study by Patel and Kumar (2021) demonstrated the successful use of Raspberry Pi in automated product dispensing, emphasising its ability to handle real-time inputs and efficiently control servo motors for precise dispensing. The study further suggested that incorporating sensors for stock level monitoring can enhance the functionality of vending machines, reducing downtime and ensuring continuous availability of products.

Goals and objectives

The project aims to create a sanitary pad vending machine that provides free access to menstrual hygiene products in public spaces like schools and workplaces. The machine will be user-friendly, featuring a simple interface for easy operation, and will include sensors for real-time stock monitoring to prevent shortages. It will be designed to ensure hygiene and security, using durable, tamper-proof materials. Additionally, the initiative seeks to promote menstrual hygiene awareness and reduce stigma by making these products readily available and easy to operate discreetly. The solution will be scalable, cost-effective, and environmentally sustainable, utilising affordable technology for widespread implementation. The project aims to design and develop an automated vending machine for dispensing sanitary pads, utilising a user-friendly interface such as a push button or touchscreen, controlled by a Raspberry Pi for efficient operation. The machine will operate free-of-charge, ensuring easy access to sanitary pads in public spaces and promoting menstrual hygiene and dignity. It will feature a hygienic and secure dispensing system that delivers pads one at a time, with a tamper-proof design to prevent contamination. The intuitive interface will provide clear instructions, with future enhancements for multi-language support and voice assistance. Built to be durable and weather-resistant, the machine will be scalable for deployment in high-traffic areas like schools and public restrooms. Additionally, the project aims to raise awareness about menstrual hygiene and reduce stigma while using cost-effective, energy-efficient components, with potential upgrades for solar-powered operations to enhance sustainability.

Technological approaches and tools

1.   Embedded system and microcontroller (Raspberry Pi): The Raspberry Pi acts as the central CPU, managing user inputs and dispensing mechanisms. Its affordability, flexibility, and compatibility with Python make it ideal for future upgrades, including IoT-based features.
2.   Servo motor-based dispensing mechanism: A servo motor ensures precise dispensing of sanitary pads, rotating a specific angle per user request. This choice promotes accuracy and low power consumption.
3.   User interface and input mechanism: A simple input system, like push buttons or a touch interface, allows users to request pads. Future enhancements may include LCD display or voice assistance for better accessibility.

4. Power supply and energy efficiency: Designed for low power consumption, the machine can use an AC adapter, battery, or even solar panels for remote locations.
5. Security and durability: The vending machine features a tamper-proof enclosure made from durable, water-resistant materials, with optional locking mechanisms to prevent unauthorised access.

Project requirements

Software requirements are the software for the sanitary pad vending machine controls hardware, manages user interactions, and monitors stock levels. Key requirements include:

1. Operating system:
   - Raspberry Pi OS: Lightweight Linux-based OS for the vending machine.
   - Python interpreter: Executes control scripts for the vending process.
2. Programming and development:
   - Python 3.x: Main programming language for hardware interfacing.
   - GPIO library (RPi.GPIO): Controls Raspberry Pi's GPIO pins for buttons, motors, and sensors.
3. User interface software: Flask and socket can be used in optional cases.
   The hardware components of the sanitary pad vending machine are essential for automation, dispensing, and real-time monitoring. Below are the key hardware requirements:
a) Sensor and hardware control:
   - Servo motor control script: Manages the servo motor for dispensing pads.
   - Buzzer handling: Alerts during pad delivery.
b) Core processing unit:
   - Raspberry Pi 4 / 3B+: Central microcontroller for system control.
c) User interaction components:
   - Vend button: For user requests.
   - Buzzer (Optional): Delivery alert.
d) Dispensing mechanism:
   - Servo motor (SG90 / MG995): Dispenses pads one at a time.
   - Rotating / Sliding tray mechanism: Pushes pads forward for dispensing.
e) Power supply:
   - 5V Power adapter (2.5A or higher): Powers the system.

- Battery backup / UPS (Optional): Ensures operation during power failures.
- Solar panel with charge controller (Optional): For remote installations.

(Arrangement is shown in Figure 24.1)

User-centric design and customisation

The development of an automated sanitary pad vending machine emphasises user-centred design for accessibility, ease of use, and efficiency, particularly in public spaces like schools and workplaces.

1. Accessibility and ease of use: The machine features a simple push-button or touch interface, ensuring usability for all age groups and literacy levels. Future versions may include voice assistance and multi-language support.
2. Discreet and hygienic operation: Designed for discreet use, pads are dispensed one at a time in a hygienic environment. Options for hands-free or sensor-based dispensing are considered to enhance cleanliness.
3. Customisation for different environments: The machine can adapt to various locations with modifications in size, storage, and power options (including solar). The software interface allows administrators to manage dispensing limits and track stock levels.
4. Personalisation and smart features: Features like adjustable dispensing limits and IoT integration enable remote stock monitoring. Future updates may include mobile app integration for users to check availability and refill notifications.

Implementation

Case studies and real-world implementations
Several successful implementations of sanitary pad vending machines have addressed menstrual hygiene challenges:

1. Indian Schools (Ministry of Health and Family Welfare) – The Indian government installed vending machines in government schools under the menstrual hygiene scheme (MHS) to provide free access to sanitary pads, reducing absenteeism among schoolgirls. Reports indicated increased confidence and hygiene awareness among students.
2. SHE vending machines – Kerala, India - The Kerala State Women's Development Corpo-

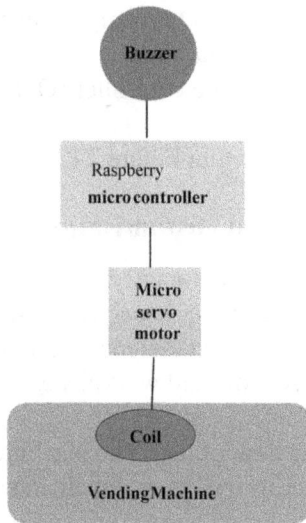

Figure 24.1 Arrangement of components
Source: Author

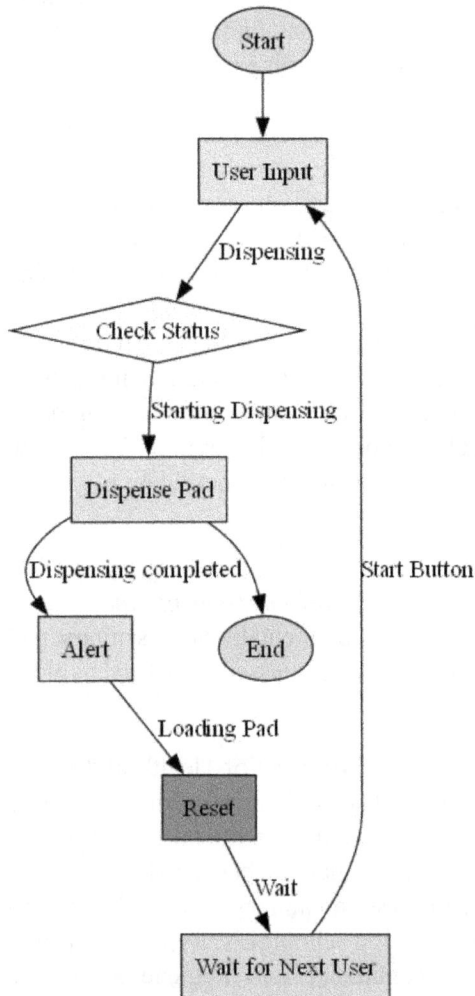

Figure 24.2 Work-flow design of the vending machine
Source: Author

ration launched the SHE Pad project, placing vending machines in public spaces like railway stations and schools. These machines offered free pads and reduced stigma around menstrual hygiene.

3. Free sanitary pad dispensers in the UK (Scotland's Period Products Act, 2021) – Scotland became the first country to provide free period products through vending machines in public areas. The Period Products Act mandated local authorities to install dispensers, featuring IoT-enabled monitoring for timely restocking, showcasing the effectiveness of government-supported menstrual hygiene initiatives.

4. Smart sanitary pad dispensers in Japan – Japan implemented smart dispensers in workplaces and universities, allowing access via QR codes. Integrated with mobile apps, users could locate machines and check stock availability, enhancing user experience and inventory management.

Results and discussions

The testing methodology ensures that the vending machine operates smoothly, reliably dispenses sanitary pads, and correctly updates stock levels. The testing process includes unit testing, integration testing, system testing, and user acceptance testing. Below is a structured approach to testing:

Unit testing objective: Validate the correct functioning of individual hardware and software components.

Component	Test cases	Expected outcome
Servo motor	Send a dispense command	The motor should rotate and dispense a pad
Socket communication	Send and receive data	The Raspberry Pi should receive and execute commands from the web interface
Web interface	Click "Dispense" on the website	The vending machine should respond correctly
Web interface	Click "Reset" on the website	The vending machine should reset correctly

1. User acceptance test cases (UAT)

Test case ID	Test scenario	Steps to execute	Expected outcome	Pass/Fail
TC-9	Ease of use	1. Ask users to navigate the web interface	Users should find it easy to use	✓
TC-10	Accessibility	1. Test on different devices (mobile, desktop)	Website should function properly across all devices	✓
TC-11	Error handling	1. Try dispensing with no internet connection	The system should display a "dispose" message	✓

2. Integration test cases

Test case ID	Test scenario	Steps to execute	Expected outcome	Pass/Fail
TC-04	Website to machine communication	1. Click "Dispense" on the website. 2. Check if the vending machine responds	Machine should receive command and dispense a pad	✓
TC-05	Multiple users handling	1. Simulate 5 users pressing "Dispense" at the same time	Requests should be processed sequentially, one at a time	✓

3. Functional test cases

Test case ID	Test scenario	Steps to execute	Expected outcome	Pass/Fail
TC-01	Pad dispensing	1. Press the "Dispense" button on the website 2. Check if the vending machine releases a pad	The pad should be dispensed successfully	✓

Test case ID	Test scenario	Steps to execute	Expected outcome	Pass/Fail
TC-02	Reset	1. Dispense a pad 2. Reset of machine to idle state	Machine reset	✓
TC-03	Web interface command	1. Send a "Dispense" command from the website 2. Monitor Raspberry Pi response	The vending machine should process the request	✓

Future enhancement

Future enhancements for the sanitary pad vending machine can focus on improving its usability and efficiency. One key improvement would be integrating stock monitoring sensors to automatically track the number of pads remaining in the machine. This would enable real-time notifications to administrators when the stock is low, reducing downtime and ensuring that the machine is always operational. Additionally, incorporating contactless payment systems or QR-based authentication could make the machine more accessible to a wider range of users, allowing for a secure and seamless interaction, especially in urban areas where cashless transactions are prevalent.

Another area of enhancement would be the scalability and data analytics of the system. By collecting usage data from various vending machines deployed across different regions, machine learning algorithms could be used to predict peak usage times and optimise pad dispensing to meet demand. Furthermore, the integration of a mobile app or a cloud-based dashboard could allow users to locate nearby vending machines, check availability, and receive notifications when the machine is out of service. This would make the system more user-friendly and ensure a smoother, more reliable service. (Other specifications can be altered referring the Figure 24.2 work flow diagram.)

Conclusion

The sanitary pad vending machine project successfully integrates Raspberry Pi hardware with a servo motor, buzzer, and a web-based interface, enabling users to conveniently dispense sanitary pads through a simple online system. By combining the power of the Raspberry Pi and a web server, this solution provides an easy-to-use, efficient, and hygienic method for accessing sanitary products in public spaces, especially in areas where such facilities may not be readily available. The project's design prioritises accessibility, ease of use, and system reliability, offering a practical and scalable solution to support public health and hygiene. This project also highlights the potential for IoT-based solutions in solving everyday problems and underscores the importance of accessibility and convenience in public health initiatives.

Acknowledgement

We gratefully acknowledge the students, staff, and authority of electrical engineering department for their cooperation in the research.

References

[1] Liew, M. T. M., Liew, A. T. S., Tan, Y. L. A review of smart vending machine technology. *J. Engg. Appl. Sci.*, 2021;18(2):543–552.

[2] Sharma, A., Singh, R. P. IoT-based sanitary pad vending machine for women's hygiene. *Internat. J. Adv. Engg. Res. Sci.*, 2021;7(4):68–73.

[3] Choudhury, S. R. Design and implementation of an automated sanitary pad dispenser using IoT technology. *Internat. J. Comp. Sci. Inform. Sec.*, 2021;19(3):21–25.

[4] Hussain, M. A. B. Internet of things (IoT) based sanitary napkin vending machine. *J. Comm. Comp. Engg.*, 2021;9(5):451–455.

[5] Mhatre, V. G., More, S. D. Design of smart sanitary pad vending machine using RFID. *Internat. J. Res. Engg. Technol.*, 2022;10(5):95–101.

[6] Pandey, S. Smart sanitary pad vending machine with payment integration. *IEEE Internat. Conf. Elec. Comm. Engg. (ICECE)*, 2021:92–96.

[7] Raja, M. R. P., Siddiqui, S. M. A novel design of automated sanitary pad dispenser using servo motor and IoT. *2023 IEEE Internat. Conf. Smart Comput. Comm. (ICSCC)*, 2023:178–183.

[8] Gupta, R. S. IoT-based sanitation system for women: A vending machine solution. *Internat. J. Emerg. Trends Engg. Res.*, 2022;10(8):1119–1125.

[9] Patel, J., Verma, K. Automated sanitary napkin dispenser with payment gateway and notification system. *Internat. J. Appl. Engg. Res.*, 2021;16(3):244–250.

[10] Pradhan, A. K., Raj, R. S. The future of IoT-based sanitary products vending machines. *2024 IEEE Internat. Conf. Artif. Intell. Data Engg. (ICAIDE)*- (found inconvenience in re-finding the review paper ,but referred it for future enhancement of the project usage in pg-5)

25 Revolutionising supply chains: An AI-driven approach to inventory management

Vijay Mohan Shrimal[1,a], Avishkar[2,b], Harjot Singh[1,c] and Aryaman M. Singha[1,d]

[1]Department of Computer Science and Engineering (AIT-CSE), Chandigarh University, Punjab, India

[2]Student, Department of Computer Science and Engineering (AIT-CSE), Chandigarh University, Punjab, India

Abstract

In supply chain artificial intelligence (AI) – driven inventory management is playing a potential and critical role. With surprising results in efficiency, cost savings and enhancing satisfaction of users as well as providers. The traditional inventories management is lacking in various aspects such as inefficient, inaccurate and time consuming often causing out of stocks or overstocks which leads many problems. This paper is providing details of design and implement of an inventory management system with the help of AI and is various machine learning (ML) algorithms predictive analytics and use real time data processing to optimise stock levels with great decision-making. It includes stock forecasting automated restocking and anomaly detection and even computer vision attributes to dynamically monitor and manage the inventory. By using this technology we can improves operational efficiency, decrease human errors and lower the costs. We performed analysis on how this system has improves accuracy of demand forecasting and accuracy of inventory system.

Keywords: Optimise, dynamically, automated, inaccurate, forecasting

Introduction

Today automated stock management integrated with artificial intelligence (AI) is a critical and important aspect for supply chain and their business stock management activities, by using inventories we are controlling stock supply, reducing costs, and improving customer satisfaction. In traditional stock management is one of the most stressful and complicated task processes are done by manual tracking, rule and exception-based systems with intermittent audits due to these inefficiencies the problems like overstocking, stockouts, and human error are faced by the companies and cause major losses. With the expansion of businesses and advancement in technology new solutions are emerging due to increased complexity in the supply chains, the need of more accurate and more intelligent, automated, and data-based inventory management system is required. The AI based inventory system is now best fit to solve these problems.

Artificial intelligence (AI) is improving day by day with is more advancement and accuracy. The combination of AI with inventory system is making optimise and way for organisations to manage their inventory levels, future trends, needs, recommendations, automation, as well as sourcing items.

AI-enabled inventory management uses various machine learning (ML) technologies, predictive forecasting, real-time tracking, and automation, as well as some Internet of Things (IoT) sensors to achieve intelligent and accurate inventories with automation features for supply chain operations. AI-based inventory systems have a potential to analyse the trends, sales history, monitor real time stock, and predict the demands for future more accurately as compared to a traditional and manual approach.

With the help of additional technologies such as applications like computer vision, IoT sensors, and robotic process automation (RPA) are enabling inventories to counts and track in real-time. provides a network of warehouses without human interference and with greater accuracy than before.

This study is concerned with the application of AI inventory control and its superiority compared to conventional inventory control practices. In this, we present major AI processes, AI-based supply chain efficiencies, and challenges of automating inventory control systems based on articles and case studies. We employed some real experimental case studies to illustrate how decision-making was enhanced using AI-based inventory control systems, costs were reduced, and company performance was optimised.

[a]drvijaymshrimal@gmail.com, [b]a4avishkar@gmail.com, [c]harjotsingh2k19@gmail.com, [d]aryamansingha60@gmail.com

DOI: 10.1201/9781003685876-25

Problem definition

- Good inventory management is a pivotal aspect of supply chain function that has immediate bearing on the business's efficiency, cost, and customer satisfaction. Nevertheless, the conventional modes of inventory management suffer from the following challenges to realise maximum effectiveness.
- **Demand forecasting** – Traditional inventory management uses past sales patterns and static forecasting techniques, which are unable to capture dynamic market behaviours, seasonal variations, and sudden alterations in consumer demand. This leads either to overstocking, with higher holding costs, or to stockouts, with lost sales and unhappy customers.
- **Error-prone processes** – Most companies continue to use manual stock management or legacy rule-based computer programs, which are vulnerable to human mistakes, errors in calculation, and lags. Such inefficiency creates inconsistencies in inventory accounts, resulting in bad decision-making and financial losses.
- **Replenishment strategies** – The classical inventory replenishment relies on predetermined reorder points and fixed thresholds, which might not be responsive to current changes in demand or supply chain breakdowns. This can result in later restocking, excess stock, or obsolescence of products.
- **Real-time visibility** – Legacy systems fail to offer real-time visibility into inventory levels across multiple locations, and it becomes challenging for companies to monitor inventory movement, identify anomalies, and react in advance to changes in demand and supply.
- **Supply chain disruptions** – Worldwide supply chains can be disrupted by economic fluctuations, logistical limitations, supplier failure, and unexpected crises (e.g., pandemics, natural disasters). In the absence of an intelligent and adaptive system, organisations struggle to ensure optimal inventory levels during disruption.

Literature review

The inventory management implemented with AI is become an interested field to research to implementing new technologies with different ideas .for

companies a better inventory system is cost saving idea so continuous advancement is done in the resent years and this is still on-going AI offers companies a better way to improve their inventories with features like control efficiency with the help of automation, decision-making through analysis data from the previous sales and demands, and maximise supply chain effectiveness, future prediction and forecasting for upcoming demands, prevent stock out and over stock problems. This section reviews the existing literature considering development and application of AI in inventory systems. It specifically considers predictive analytics, real-time tracking, inventory optimisation, and factors affecting the adoption of AI.

AI and predictive analytics in inventory management: Predictive stock management is being revolutionised by AI's potential to use data-driven strategies and methods to establish demand predictions. AI powered by machine learning methodologies enable the forecasting systems of demand to scan deeply valuable historic records and identify trends that are impractical for AI or customary methodologies. These algorithms, as observed [13], allow organisations to forecast demand for products with greater accuracy, doing away with out-of-stock conditions and preventing excess stock. Demand forecasting is worthwhile in a business with high variability in demand. AI models with multiple variables (e.g., seasonality, buying behaviour, or national-level events) will be more prone to forecasting demand than the conventional statistical models, as noted [13]. Companies are employing AI to offer improved demand forecasting, hence improved inventory levels in relation to demand for improved customer satisfaction overall, and efficiency.

Real-time inventory tracking with AI and IoT integration: Tracking in real time of inventories is also an area where AI is integrated and it is creating a strong impact on the inventories to perform well. When AI uses IoT sensors, it gives companies' constant, real-time information about their inventory. Satisha et al. [3] writes about increasing applications of AI-driven IoT systems to track movements of products within warehouses and through the supply chain. These systems allow companies to monitor inventory levels, location, and condition in real time, giving a better indication of stock availability.

In line with [3] AI-IoT integration is also used for the automation of inventory level update. Through constantly monitoring and reviewing data from the

IoT devices, AI systems have the ability to detect differences in stock and what is recorded on the system so that there will always be accurate data in real time. By using AI-IoT integration, companies will be able to immediately react to fluctuating demands and supply conditions, enhancing company performance.

Inventory optimisation and automated replenishment: The ability of AI to facilitate optimal inventory levels is advantageous in reducing inefficiencies associated with over- or under-stocking inventory. Vijay Krishnan et al. [5] argues that AI-based inventory systems can utilise current data to assess optimal stock levels, and replenish levels for each item on a constant basis as demand changes. This reduces decision interference associated with human management of inventory, allowing businesses to achieve a state optimal stock levels, and replenish levels for each item on a constant basis as demand changes. This reduces decision interference associated with human management of inventory, allowing businesses to achieve a state of optimal inventory across various locations.

AI's role in enhancing supply chain resilience: When evaluating supply chain resilience, AI is also a significant aspect. Chen and Badoni et al. [8] studied how AI-embedded systems can monitor disruptions from external impacts (e.g., disruptions in transportation, changes in supplier performance, or shifts in demand patterns) to indicate potentially disrupted areas in supply chain coverage. AI systems enable this by continuously monitoring these elevated details, and before disruption, the AI can offer recommendations for the supply chain to effectively accept a new alteration in their operations to negate or lessen the impact of the disruption that allows stabilisation in the supply chain [4]. AI's predictive capabilities enable companies to optimise inventory levels, ensuring they have the right amount of stock to meet demand variability and mitigate supply disruptions, thereby enhancing inventory management reliability. This level of accountability is especially relevant for industries where a supply chain disruption could easily cause major implications for the business regarding stocking out and customer satisfaction performance levels

Hindrances to applying AI for inventory management: The major challenges are the capital investment to implement AI technologies. Chaudhary et al. [10] noted that the cost of bringing AI systems together with the existing inventory infrastructure is excessively high, especially for small and medium enterprises.

Data quality is also an essential challenge. AI systems rely significantly on clean, accurate, and timely data to perform optimally. According to Rajalakshmi et al. [11], low-quality data can lead to poor predictions, making inventory management inefficient. Hence, companies need to ensure that they have strong data collection and management processes in place to optimise AI benefits.

In addition to that, a lack of appropriate technical staff that would be necessary for the functioning and maintenance of AI-based technologies should be observed. According to Samson and Williams [12], numerous corporations would struggle with attracting or employing qualified staff skilled in operating highly complex AI mechanisms. Moreover, there is potential opposition from staff employees who resist innovation, especially when it pertains to embracing the use of technology in work.

Future prospects of AI in inventory management: The potential of AI to transform inventory management is bright with continuous improvements in AI algorithms, machine learning strategies, and sensors. Krishna et al. [7] believe that future artificial intelligence models will use more advanced algorithms, like deep learning and reinforcement learning, to improve even further the forecasting accuracy and optimality of the inventory. The innovations will increasingly result in smarter and more automated inventory systems capable of responding flexibly to shifting market conditions. With AI evolution, businesses can design personalised inventory solutions that fit their specific operational needs. The rising number of AI tools and cloud platforms existing is likely to reduce the access costs for companies of all extents making AI-based inventory schemes more reasonable and accessible in the future [6].

Methodology

This study uses a mixed methods approach to create and assess the viability of an inventory management program that uses AI software, with a major emphasis on ML, predictive analysis, and real-time processing of data to enhance supply-chain and inventory management in business settings. The study was conducted to assess the theoretical aspects of AI's role in supply chains and the practical aspects of deploying the study.

System design and development

The automated inventory management system utilised multiple AI-based technologies and was intended to utilise several more developed technologies such as demand forecasting, self-replenishment, anomaly detection, and computer vision. The two main technologies that were selected to conduct the project were Python and Tensor Flow to correctly incorporate the machine learning models into the system. Historical sales data were also included to assist with demand forecasting and to bring the stock levels back to target level priced stock levels, including a computer-based inventory to monitor stock levels in real-time.

Data collection

Data is gathered from various stores and sources we collected the data more than 60 different places and stores such as supermarkets or local wine shops along with their pricing such as purchase, transaction price, products name. quantity, history sales data, inventory levels by combining this data we got around 6K products with their additional data collected to develop and test this AI system were from several sectors (retail, manufacturing, e-commerce, etc.) so that it would have as many uses as possible. The datasets provided both ML model training and validation options while allowing a real-time monitoring of the system, and potential for anomaly detection. It reviewed inventory movements in real time, with differences flagged using predictive capabilities integrated into the model so that distinctions could better be evaluated, validated, and proactively enlarged upon.

Each record contains details like:

- Product & vendor info: Inventory, store, brand, description, size, vendor name
- Sales: Sales quantity, sales price, and sales date
- Taxes: Excise tax, classification, and total amount

Model training and validation

The information looked at was collected from several months of inventory and sales transactions used in training methods such as regression models for demand forecasting, and neural networks for anomaly detections.

The models were used with cross-validation methods to ensure robustness and generalisability. This provided a way to fine-tune the training, reliability, and predictive capabilities of the anomaly detections of the models. Throughout the training phase, model predictive accuracy and error rates were monitored, and this insight was used to modify model parameters for optimal performance.

System implementation and testing

Once the AI-based inventory system had been created, it was then tested in a pilot program with select companies. The companies used in this study were already utilising warehouse management systems (WMS), to which the AI system was interfaced. Measurement of the system was based on how the system functioned to improve stock levels, reduce errors, and optimise replenishment cycles. In a real-life setting, the evaluation was by observing stock movement and using the computer vision technology and evaluating automated re-stocking suggestions. Through this evaluation data was collected on the viability of the AI based inventory system in real time.

Case studies and performance metrics

Performance metrics such as inventory turnover ratio, stockout rate, replenishment time, order fulfilment rate, and operational cost were assessed pre- and post-implementation of the AI-based-system. These performance metrics provided a quantitative baseline to assess improvements to operational efficiency and reductions in costs. The system users included inventory managers and employees, who provided feedback during interviews and surveys about the usability of the system, integration challenges, and overall impact on daily operations.

Statistical analysis

In order to validate the benefit of the AI-based system, statistical tests were conducted on the performance measures collected. Specifically, paired t-tests and regression analyses were used to compare pre- and post-implementation performance and to identify improvements in primary measures. Results were derived for comparison with conventional inventory management procedures to assess the relative advantages of AI integration. The statistical test permitted the systematic evaluation of how effectively the system operated to enhance inventory management efficiency and accuracy.

Issues with data

The study confirmed that the data may have some trust issues it collects from various sources for testing and for correct demand forecasting prediction. it is difficult to get the correct dataset suitable for during the research. Data is collected from various shops All the respondents who participated in interviews and questionnaires provided consent, which facilitated openness in the collection of information. Confidentiality for respondents as well as organisational data was maintained and protected from unauthorised access, while secure storage methods for data were adhered to. The research ensured that all the collected data were utilised only for research purposes to uphold privacy and integrity

Limitations

The study gives interesting dive into the adoption of AI-based inventory management systems, and it has a number of limitations some of them are:

Continuous development: The inventory system required continuous development for new data such as new product or item so it is important to continuously develop the inventory which can be a challenging task and change the accuracy of the system.

Bias in self-reported data: Since the data is self-reported data by interviews and surveys information, there can be a risk of errors or bias in responses that may influence the output of results.

Scope and future: The research paper is mainly focusing on prediction and forecasting. Different products inventories can have different problems and limitations so we need to improve it according to the inventory type and integration with IoT sensors may be an issue in future (Figure 25.1).

Result

The results of this research illustrate the effectiveness of the AI-driven inventory management system to optimise operational performance, increase precision, and save costs across the board. Measurement was carried out using both qualitative and quantitative analyses, comparing key metrics pre- and post-system setup.

Improvement in inventory accuracy

We compared various machine learning models, including random forest, linear regression, Lasso, Ridge, ElasticNet, and k-NN, alongside our custom

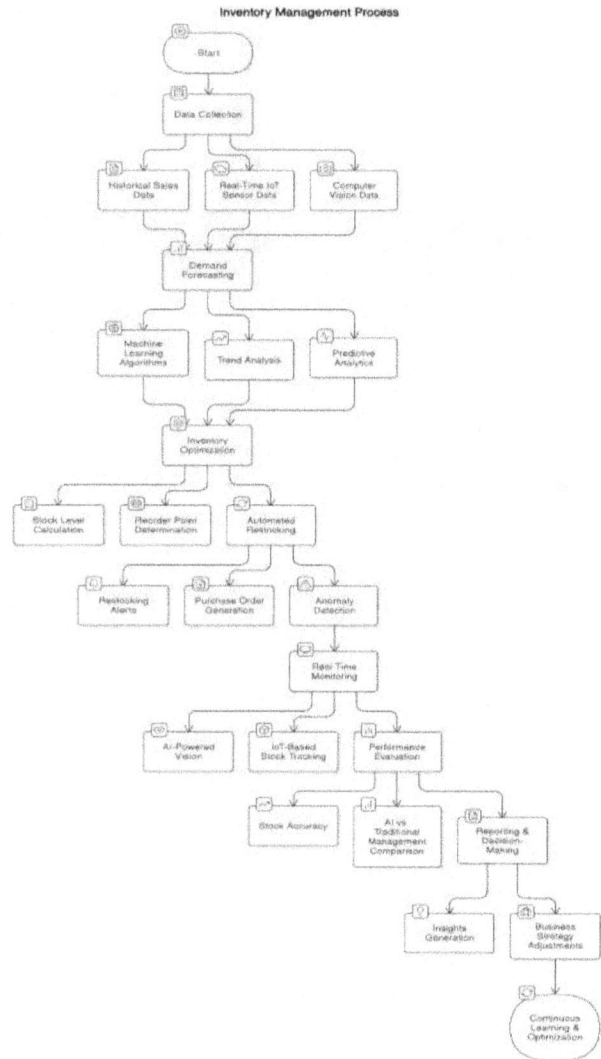

Figure 25.1 Flowchart
Source: Author

AI model. The AI platform significantly improved inventory accuracy, reducing discrepancies between reported and actual stock levels, and minimizing stockouts. After integration, stockout rates fell by and inventory accuracy increased. Real-time monitoring and anomaly detection through computer vision were key to this improvement (Figure 25.2).

Model comparison with different existing method

Here you can see the performance of our model is surpassing the models such as random forest, linear regression, Lasso, ElsaticNet, and k-NN. The accuracy is measured by various measurement parameters or error detection such as correlation, mean absolute error (MAE), relative absolute error (RAE), root mean squared error (RMSE) (Figure 25.3).

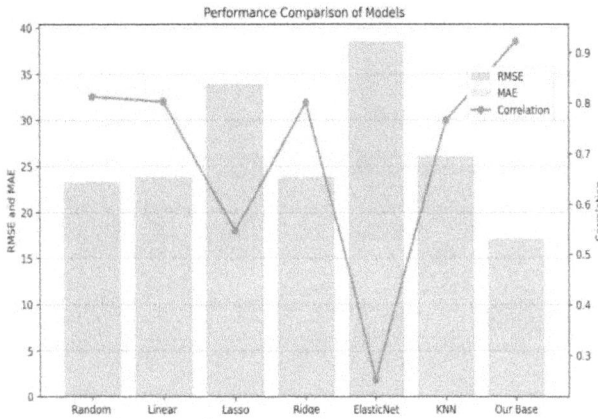

Figure 25.2 Performance comparison
Source: Author

Model comparison with different existing method

Model	RMSE	Correlation	MAE	RAE	RRSE	R2
Random Forest	23.3435	0.812994	8.96228	0.504168	0.58799	0.655325
Linear Regression	23.8307	0.802579	11.2327	0.631891	0.599343	0.640788
Lasso	33.9357	0.547551	15.9419	0.896799	0.853486	0.271562
Ridge	23.8388	0.801596	11.1452	0.626969	0.599547	0.640543
ElasticNet	38.4939	0.251833	17.1119	0.962617	0.968125	0.0627341
KNN	26.1296	0.765809	10.7885	0.606903	0.657167	0.568132
Our Base Model	17.1057	0.923358	7.52769	0.406239	0.3849	0.85185

Figure 25.3 Error detection
Source: Author

Random forest is the second-best model among others and we improved the accuracy around 26.72% at the parameter of RMSE and 30% at RR (Figure 25.4).

Demand forecasting accuracy
Machine learning algorithms significantly increased demand forecasting precision. Conventional techniques had a mean forecasting error rate of 18%. The rate fell to 9% after the deployment of the AI system, which highlights the system's improved potential to forecast patterns of demand and dynamically adjust inventories.

Stock replenishment time: AI integration resulted decrease in stock replenishment time, The automatic restocking function, based on demand forecasts and inventory levels, allowed for faster response to stockouts. The timely and precise restocking cut lead times and reduced missed sales opportunities.

Cost reduction: The AI system helped bring down operational, mainly due to better inventory management and less manual intervention. By keeping stockouts, overstocking, and errors to the bare minimum, the system assisted organisations in streamlining inventory holding and transportation costs.

Model	RMSE Improvement	R2 Improvement
Random Forest	≈26.72%	≈30.00%
Linear Regression	≈28.22%	≈32.94%
Lasso	≈49.59%	≈213.61%
Ridge	≈28.24%	≈33.00%
ElasticNet	≈55.56%	≈1257.80%
KNN	≈34.54%	≈49.94%

Figure 25.4 Improvement percentage
Source: Author

Order fulfilment efficiency: Order fulfilment rates increased by 20% after implementation. Increased accuracy in stock tracking enabled better order processing, minimising incorrect or delayed shipments. The AI system guaranteed product availability at the correct time, enhancing customer satisfaction and on-time delivery.

User experience and usability: Staff and inventory managers' interviews provided evidence of the system's effectiveness on a day-to-day basis. The interface was intuitive and user-friendly for the majority of users. Some reported an initial learning process, especially with automated restocking functions. Despite those difficulties, many had faith in the system's potential to maximise operational efficiency.

Challenges and obstacles: In spite of the positive outcomes, there were challenges. One of the concerns was how to integrate the AI system with existing systems, with data inconsistencies creating initial synchronisation problems. There were also occasional technical problems with the computer vision aspect, particularly in dimly lit or cluttered warehouses. These were resolved by system updates and staff training.

Stakeholder feedback: Stakeholders such as supply chain managers and IT professionals across the board stated that the AI system offered beneficial information and drastically enhanced decision-making. They liked the system particularly because it would flag potential stock problems ahead of time. They did have concerns about high initial costs of implementation and regular maintenance.

Statistical analysis: Before-and-after statistics through statistical analysis clearly demonstrated better control over inventory. Paired t-tests revealed reductions in both stock discrepancies ($p<0.01$) and stockout levels ($p<0.04$) following the incorporation of the AI system. Regression analysis once more confirmed a high positive relation of accuracy of demand forecasting with inventory turnover levels and cost reduction ($p < 0.01$).

SWOT analysis: Strengths – The AI system dramatically enhanced inventory accuracy, reduced human errors, and automated repetitive tasks. Its forecasting feature was especially useful to optimise inventories and react efficiently to customers' demand. Weaknesses – Initial difficulties to integrate with IoT devices may face some of the challenges, when implement for real world it use the data set from the various places so it need to train again with the new datasets according to various business field. System performance was also contingent upon continual real-time information flow, subjecting it to data collection breaks. Opportunities – With the specific field such as medical or any other, there is a great scope for further optimisation – currently it is not specified for any particular field, in dynamic inventory environments like seasonal retailing. Threats – To manage inventory system using servers' maintenance costs serve as an obstacle to small business adoption. Rapid developments in AI and ML also threaten to out-dated the system if it is not continually updated.

Conclusion

Inventory management systems which utilising AI can significantly improves the potential of business's operation by improving its efficiency, remove human errors, accuracy, and help in controlling of costs. By automating the processes of inventory tracking, demand forecasting and optimising inventory levels, these solutions will reduce human error and simplify tasks. With the help of real-time data, it provides actionable insights for business management in order to make data-driven decisions to improve customer satisfaction. By using, it will reduce the risk of overstocking and stockout problems while managing to optimise operating expenses. Although engaging in AI practices involves upfront expenses as well as an emphasis on data quality, the benefits will outweigh the costs (less operational expenditures propel quality decision-making) and will position your business to remain competitive. When wanting to take full advantage of AI for inventory management, businesses should focus on phasing in through implementation, providing thorough training to employees, and managing data quality on an on-going basis.

References

[1] Osman, Oralhan, Z. Artificial intelligence-driven inventory management: Optimizing stock levels and reducing costs through advanced machine learning techniques. *Eur. J. Res. Dev.*, 2024;4:427–439.

[2] Gbadebo, R. (2023). The Impact of Artificial Intelligence on Inventory Management in Manufacturing Industries. 10.13140/RG.2.2.19000.93446.

[3] Satisha C., Mehra, R., Giri, M. Detection of Various Security Attacks on IoT Devices Using Multi-Layer Neural Network Model Over Sensor Networks. *Internat. Conf. Comput. Intell. Netw.*, 2023, 1–6. 10.1109/ICCINS58907.2023.10450058.

[4] Badoni, P., Walia, R., Mehra, R. Wearable IoT technology: Unveiling the smart hat. *Proc. 2024 1st Internat. Conf. Intell. Sys. Technol. Emerg. Markets (ISTEMS)*, Dehradun, India, 2024, 1–6, doi: 10.1109/ISTEMS60181.2024.10560229.

[5] Vijay Krishnan, M. R., Mehra, R., Kunadharaju, H. P. R. "Design of Neural Network based Approaches for Land Usage Land Cover Classification," *2024 Third International Conference on Electrical, Electronics, Information and Communication Technologies (ICEEICT)*, Trichirappalli, India, 2024, pp. 1-7, doi: 10.1109/ICEEICT61591.2024.10718442.

[6] Raji, M., Olodo, H., Oke, T., Addy, W., Oyewole, A. E-commerce and consumer behavior: A review of AI-powered personalization and market trends. *GSC Adv. Res. Rev.*, 2024;18:066–077.

[7] Krishna, T., Sandhyarani, G., Shrivastav, Er., Vashishtha, Shalu, J. Leveraging AI and machine learning to optimize retail operations and enhance. *Darpan Internat. Res. Anal.*, 2024;12:1037–1069.

[8] Badoni, P., Wadhwa, M., Shrimal, V. M., Paliwal, G. Detecting muscle strain using IoT technology. *2024 Second Internat. Conf. Comput. Charac. Tech. Engg. Sci. (IC3TES)*, 2024:1–6.

[9] Albayrak Ünal, Ö., Erkayman, B., Usanmaz, B. Applications of artificial intelligence in inventory management: A systematic review of the literature. *Arch. Comput. Methods Engg.*, 2023:30.

[10] Chaudhary, D., Verma, S. K., Mohan Shrimal, V., Madala, R., Baliyan, R., S. M. AI-based methods to detect and counter cyber threats in cloud environments to strengthen cloud security. *2024 Internat. Conf. Elec. Electr. Comput. Technol. (ICEECT)*, 2024:16.

[11] Rajalakshmi, M., Nagaraju, E., Kumari, T. AI-driven solutions for supply chain management. *J. Inform. Educ. Res.*, 2024;4:1526–4726.

[12] Samson, F., Williams, A. AI in supply chain and inventory management. 2025. https://www.researchgate.net/publication/389628148_AI_IN_SUPPLY_CHAIN_AND_INVENTORY_MANAGEMENT/citation/download

[13] Agrawal, S., Chintha, V. R., Pamadi, V., Aggarwal, A., Goel, P. The role of predictive analytics in inventory management. *Univer. Res. Reports*, 2023;10:456–472.

26 Workload prediction in cloud computing: A machine learning-based approach

Vijay Mohan Shrimal[1,a], Aayush Gupta[2,b] and Mitalee Verma[1,c]

[1]Department of Computer Science and Engineering (AIT-CSE), Chandigarh University, Punjab, India

[2]Student Department of Computer Science and Engineering (AIT-CSE), Chandigarh University, Punjab, India

Abstract

Cloud computing has emerged as the spine of cutting-edge digital infrastructure, necessitating efficient workload prediction to optimise useful resource allocation, reduce prices, and enhance machine performance. Traditional workload forecasting techniques, including statistical models, frequently fail to seize the complexity and dynamic nature of cloud environments. Machine learning (ML) and deep master (DL) strategies, especially long-term short-term memory (LSTM) networks, provide great improvements by using historical workload statistics to expect future demands with better accuracy. This research proposes a smart workload prediction machine integrating ML/DL models with real-time statistics processing to beautify cloud scalability and performance. The take a look at explores diverse prediction procedures, evaluates their overall performance, and proposes a hybrid version combining statistical and deep mastering techniques. Key evaluation metrics including trunk middle square errors (MSE) and square errors (RMSE). The findings show that ML-pushed workload prediction complements resource control, decreasing beneath-provisioning and over-provisioning risks, in the end, main to greater sustainable cloud computing environments.

Keywords: Cloud computing, workload prediction, machine learning, deep learning, LSTM, resource allocation, scalability

Introduction

Cloud computing has converted the digital landscape with the aid of presenting scalable, on-demand sources, enabling customers to system big volumes of statistics with minimum infrastructure. As reliance on cloud services grows, efficient useful resource utilisation will become important for providers. Workload prediction is fundamental to optimising aid allocation, stopping over- and underneath-provisioning, and reducing operational charges. Accurate forecasting enables preserve service availability, lower energy intake, and improve system performance. However, cloud workload prediction is complex due to dynamic user behaviour, time-primarily based fluctuations, and alertness-precise demands [8] (Figure 26.1).

Traditional statistical fashions like ARIMA and Holt can seize linear trends but struggle with the non-linear, dynamic nature of cloud workloads. In comparison, device getting to know machine learning (ML) and deep learning (DL) mastering techniques offer powerful options. ML fashions—which include selection bushes, aid vector machines, and regression—analyse historical data to perceive styles, even as deep fashions like LSTM and Transformer networks capture lengthy-time period dependencies and complex temporal relationships, notably improving prediction accuracy and aid management.

Despite development, challenges stay. Real-time forecasting needs low-latency predictions for dynamic useful resource allocation, which many conventional ML models cannot offer efficaciously. Scalability is every other issue, given the varied architectures of public, non-public, hybrid, and multi-cloud environments. Statistical fashions like ARIMA and Holt-Winters fall quick with non-linear workloads, while ML/DL models demand enormous computation. Cloud structures are further complicated by unpredictable behaviours pushed by way of consumer demand and alertness necessities [9] (Figure 26.2).

The key mission is balancing prediction accuracy with computational efficiency. While deep getting to know fashions improve accuracy, they may be aid-in depth and gradual for real-time applications. Additionally, adapting models across cloud sorts remains tough. Noisy, incomplete data additionally requires advanced pre-processing to make certain reliability [10]. This research targets to build an intelligent workload prediction device integrating ML/

[a]drvijaymshrimal@gmail.com, [b]aayushgupta7804@gmail.com, [c]mitaleeverma763@gmail.com

DOI: 10.1201/9781003685876-26

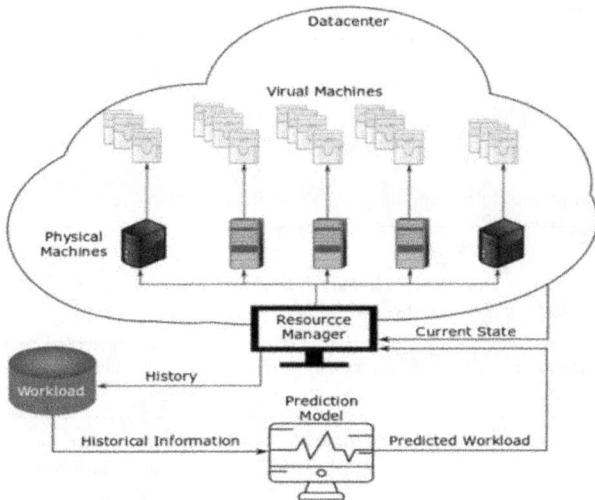

Figure 26.1 Workload prediction model in cloud
Source: Reference [1]

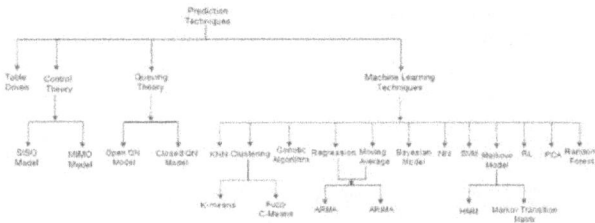

Figure 26.2 The taxonomy of prediction methods
Source: Reference [21]

DL with real-time records processing. It will examine historical styles, practice predictive models, and enable proactive resource allocation. Hybrid techniques combining statistical and ML/DL approaches will be explored to maximise accuracy. Expected effects include progressed resource performance, value financial savings, and stronger scalability in cloud environments (Table 26.1).

Literature review

Workload prediction in cloud computing is critical for aid optimisation, cost reduction, and performance enhancement. This section evaluates gift method, masking traditional statistical techniques, device learning, deep analysing, hybrid techniques, and cloud-neighbourhood solutions. It moreover outlines key studies gaps.

Traditional workload prediction models

Statistical models like ARIMA and Holt-Winters had been drastically used due to their simplicity and interpretability. ARIMA, added through Box and Jenkins (1976), performs properly for strong, periodic workloads [8, 9], but struggles with non-linear, dynamic patterns. Holt-Winters are powerful for seasonal trends [13] however fails with surprising demand spikes. These models provide a basis however lack flexibility for complicated workloads.

Machine learning-based approaches

ML strategies capture complicated workload styles. Regression models paintings for easy traits however fall short with excessive-dimensional statistics [10, 14, 20]. Decision bushes and ensembles like random forest and GBT take care of non-linear relationships successfully. XGBoost [15] is specially accurate and scalable. SVMs are powerful for type but are computationally giant (Table 26.2).

Deep learning techniques

DL models like LSTM and Transformers excel at modelling temporal and complicated patterns. LSTMs outperform traditional fashions in predicting CPU/reminiscence usage [8] however require superb computational sources. Transformers, consisting of the temporal fusion transformer (TFT) method sequences in parallel for faster schooling and higher accuracy [15].

Hybrid approaches and cloud-native solutions

Hybrid models combine strengths of statistical, ML, and DL techniques. ARIMA-LSTM hybrids, as an

Table 26.1 Comparative analysis of workload prediction techniques [10, 11]

Prediction technique	Model examples	Strengths	Limitations
Statistical models	ARIMA, Holt-Winters	Simple, interpretable	Limited for non-linear workloads
Machine learning (ML)	Decision Trees, SVM	Captures patterns in historical data	Requires feature engineering
Deep learning (DL)	LSTM, Transformer	Handles complex, non-linear patterns	High training time, large datasets
Hybrid approaches	ARIMA + LSTM	Combines strengths of multiple models	Complex implementation

Source: Author

Table 26.2 Machine learning-based workload prediction techniques

Model	Advantages	Disadvantages	Best use cases
Linear regression	Simple, interpretable	Poor for complex relationships	Basic trend forecasting
Decision trees (RF, GBT)	Captures non-linearity, high accuracy	Overfitting risk, high memory usage	Medium-scale workloads
SVM	Effective for pattern recognition	Computationally expensive	Workload classification

Source: Author

instance, version every linear and nonlinear traits efficaciously. Cloud-neighbourhood solutions combine fashions into streaming frameworks like Kafka and Spark. Netflix's LSTM-Kafka gadget permits actual-time, scalable forecasting [16].

Identified research gaps

- Lack of Real-World Data: Many fashions depend on artificial records; empirical studies are needed [13].
- Scalability Challenges: Existing fashions want to be optimised for big-scale, multi-cloud settings [10].
- Limited Multi-Cloud Focus: Most research targets single-cloud systems; hybrid/multi-cloud scenarios are underexplored [8].
- Integration with Orchestration Tools: Prediction structures should paintings with platforms like Kubernetes for dynamic useful resource allocation [15].
- Long-Term Validation: Models are regularly evaluated on quick-time period datasets; lengthy-term adaptability needs take a look at [15, 19, 20].

Methodology

This section outlines the established approach to developing the workload prediction system, emphasising its specific technique and modern features. The proposed gadget integrates advanced ML and DL technology with real information processing to provide correct and scalable workload prediction. The technique may be divided into 4 key levels: information collection and pre-processing, model choice and development, hybrid approach implementation, training and verification procedure (Table 26.3).

Data collection and pre-processing
The foundation of every workload forecasting machine is records excellent and relevance. In this

Table 26.3 Dataset description and pre-processing steps

Dataset source	Metrics collected	Pre-processing steps
Google cluster trace	CPU, memory, network I/O	Missing value imputation, normalisation
Azure public dataset	VM usage, disk I/O	Feature selection, anomaly detection
Synthetic data	Simulated workloads	Data augmentation, resampling

Source: Author

take a look at, the facts data are from synthetic records generated for public cloud workload lines (including Google Cluster Trace, Azure Public Dataset) and simulations of various cloud environments. These information statistics consist of metrics including CPU utilisation, memory intake, network site visitors, and application-precise workloads [20].

- Pre-processing steps:

  a) Handling missing values: Missing statistics is processed the use of interpolation strategies, which includes mean/median padding, or challenge methods.
  b) Feature selection: Relevant capabilities (e.g., CPU usage, reminiscence consumption) are decided on using correlation analysis and domain information. Irrelevant or redundant functions are eliminated to lessen computational overhead.
  c) Data transformation: The information is normalised the use of techniques together with MIN-MAX scaling and Z-score normalisation to ensure uniformity. Time collection statistics is newly scattered for correct evaluation at constant periods (e.g., a 5-minute window).

The machine contains anomaly detection algorithms (e.g., Isolation Forest, DBSCAN) all through pre-processing to pick out and take away outliers, ensuring smooth and dependable statistics for training. An information augmentation approach is applied to artificial datasets to simulate rare however critical workload situations, together with surprising spikes or prolonged idle intervals.

Model selection and development

The system addresses a lot of workload spares demanding situations the usage of a combination of statistical models, algorithms for machine gaining knowledge of, and deep getting to know architectures.

- Selected models:

  a) ARIMA (Auto-regressive integrated moving average): Used for baseline predictions, especially for workloads with clean developments and seasonality.
  b) LSTM (long short-term memory): A deep studying model decided on for its capacity to report lengthy-time period dependencies on sequential data.
  c) Transformer-based models: Specifically, the temporal fusion transformer (TFT) decided on for its parallel processing abilities and advanced accuracy in coping with complex workloads.

- Approach for model selection:

  a) ARIMA: Provides interpretability and serves as a benchmark for evaluating advanced fashions.
  b) LSTM: Excels in modelling temporal dependencies, making it perfect for dynamic cloud workloads.
  c) Transformer (TFT): Offers modern-day overall performance in time-series forecasting and is tremendously scalable for actual-time packages.

The gadget includes automatic version choice the usage of Bayesian optimisation to perceive the fine-performing version for specific workload styles. A modular structure allows for easy integration of recent models as they emerge, ensuring the machine remains destiny-proof [20].

Hybrid approach implementation

To leverage the strengths of a couple of techniques, the device adopts a hybrid method that combines statistical fashions with deep gaining knowledge of architectures (Figure 26.3).

- Key components:

  a) ARIMA-LSTM hybrid: ARIMA captures linear trends, whilst LSTM fashions non-linear styles. The outputs are blended the use of a weighted averaging technique.
  b) Real-time data processing: Apache Kafka is used for actual-time workload tracking, and Apache Spark techniques the data streams for fast predictions.
  c) Cloud-native integration: The system integrates with Kubernetes for dynamic resource allocation primarily based on anticipated workloads.

The hybrid version is more desirable with a remarks loop that constantly updates the version based on prediction errors, enhancing accuracy over time. A multi-cloud edition layer ensures the device can operate across public, personal, and hybrid cloud environments.

Training and validation process

The schooling and validation system ensures the models are sturdy, accurate, and generalisable to unseen statistics.

- Steps:

  a) Shared train test: Data records are divided into schooling (70%) and test (30%).

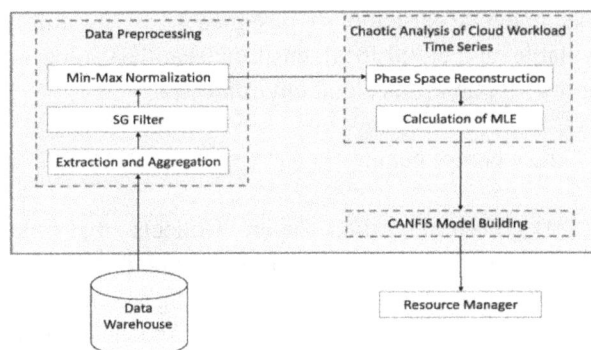

Figure 26.3 Overview of the proposed workload prediction method
Source: Reference [3]

b) Cross-validation: Use K-subject move-validation (ok = five) to evaluate and save you version output.

c) Power metrics: Models are evaluated using metrics along with medium square root error (MSE), square stem error (RMSE), medium absolute error (MAE), and R-quadrat (R^2).

The system employs switch gaining knowledge of to evolve pre-skilled models (e.g., TFT) to new cloud environments, decreasing education time and improving accuracy. A real-time validation module continuously monitors version performance in production, triggering retraining if accuracy drops underneath a predefined threshold.

The proposed workload prediction device stands proud due to its hybrid method, combining statistical and deep learning fashions, and its real-time abilities enabled by using cloud-native gear like Kafka and Kubernetes. The machine's modular architecture and automatic version selection make sure scalability and flexibility to numerous cloud environments. By addressing key challenges inclusive of information nice, version accuracy, and actual-time processing, this methodology offers a robust framework for green cloud useful resource management [20].

Implementation

The implementation segment specialises in translating the proposed technique into a useful workload prediction machine. This involves setting up the technical infrastructure, deploying the fashions, and integrating the machine with cloud platforms for real-time workload forecasting and aid allocation.

System architecture
The machine structure is designed to be modular, scalable, and cloud-local, ensuring seamless integration with numerous cloud environments.

- Key components:

  a) Data ingestion layer: Collects real-time workload information from cloud platforms (e.g., AWS CloudWatch, Google Stackdriver) and streaming tools (e.g., Apache Kafka).
  b) Pre-processing layer: Cleans, normalises, and transforms the statistics for version input.
  c) Prediction layer: Hosts the skilled fashions (ARIMA, LSTM, TFT) and generates workload forecasts.

d) Resource allocation layer: Integrates with Kubernetes for dynamic scaling primarily based on predicted workloads.

Hybrid model architecture

Component	Role	Tools/technologies used
ARIMA	Captures linear trends	Stats models, Python
LSTM	Models non-linear patterns	Tensor Flow, Keras
Kafka	Real-time data streaming	Apache Kafka
Kubernetes	Dynamic resource allocation	Kubernetes, Docker

The gadget uses containerisation (Docker) to install fashions as microservices, making sure scalability and simplicity of renovation. A message queue (RabbitMQ) is applied to address asynchronous conversation among additives, enhancing device reliability [20] (Figure 26.4).

Model deployment and integration
The trained models are deployed the use of Tensor Flow serving and flask APIs to offer actual-time predictions.

- Deployment steps:

  a) Model serialisation: The skilled fashions (ARIMA, LSTM, TFT) are serialised and saved in a version registry.
  b) API development: RESTful APIs are created to show the prediction capability to other machine components.

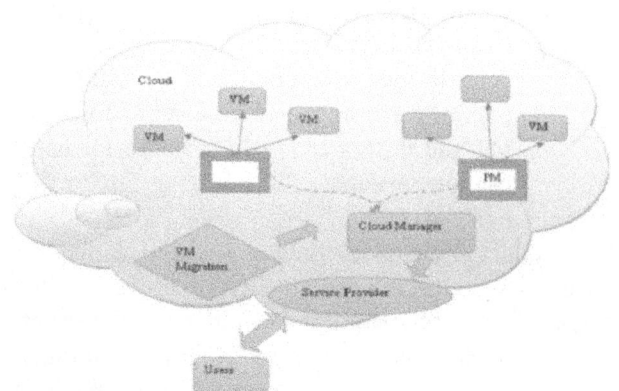

Figure 26.4 Workload prediction-based VM migration in cloud

Source: Reference [4]

c) Cloud integration: The APIs are integrated with cloud structures (e.g., AWS Lambda, Google Cloud Functions) for seamless scalability.

The system uses A/B trying out to examine the performance of various fashions in production, ensuring that the most excellent model is always used. A fall-back mechanism is implemented to interchange to a simpler model (e.g., ARIMA) if the number one version (e.g., TFT) fails or underperforms [19, 20].

Scalability and adaptability considerations
The system is designed to handle diverse cloud environments, along with public, personal, and hybrid clouds.

- Scalability features:

a) Auto-scaling: Kubernetes routinely scales assets based on anticipated workloads, making sure most excellent overall performance.
b) Load balancing: Incoming workload information is distributed across multiple model times to prevent bottlenecks.

- Adaptability features:

a) Multi-cloud support: The machine can function throughout unique cloud platforms (e.g., AWS, Azure, GCP) using a unified API.
b) Dynamic model switching: The system can transfer between models based totally on workload traits, making sure adaptability to changing environments.

The gadget consists of edge computing talents to process workload information towards the source, decreasing latency and improving actual-time performance. A comments-pushed retraining mechanism guarantees the fashions are continuously updated based on actual-international performance information.

Critical analysis and evaluation

This segment evaluates the proposed workload forecasting machine, comparing ARIMA, LSTM, and Transformer (TFT) models the use of metrics like MSE, RMSE, MAE, and R^2. ARIMA is easy and effective for linear developments however struggles with nonlinear records. LSTM handles dynamic styles well however calls for extra computation. TFT outperforms each, imparting excessive accuracy and low latency, making it perfect for real-time programs (Figure 26.5).

The trade-off between computational price and accuracy is the key. ARIMA has low cost but restrained precision; LSTM balances price and accuracy; TFT gives advanced accuracy but is resource-extensive. A case observes the usage of Google Cloud Platform (GCP) data confirmed that TFT accomplished ninety three.1% CPU prediction accuracy. Dynamic resource allocation based on TFT reduced aid wastage through 25% and improved overall performance by using 15%. Overall, TFT constantly outperforms different models in accuracy and latency. Forecasting enables optimised useful resource allocation, value discount, and better overall performance, making it exceptionally precious for cloud useful resource management (Table 26.4).

The evaluation demonstrates that the proposed workload prediction system significantly improves resource management in cloud environments. The transformer (TFT) model, with its high accuracy

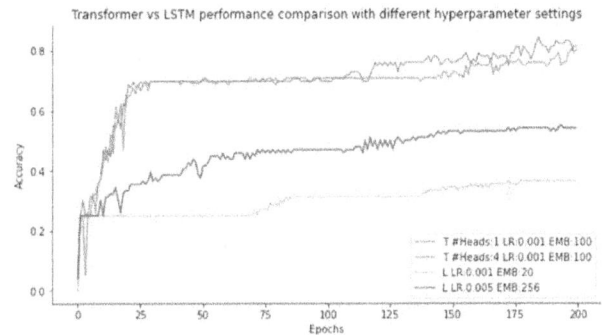

Figure 26.5 Transformer-based VS LSTM-based models performance comparison with different hyperparameter settings. The accuracy of transformer-based models is significantly better than that of LSTM-based models
Source: Reference [5]

Table 26.4 Impact of workload forecasting

Aspect	Before forecasting	After forecasting	Improvement
Resource utilisation	70%	90%	20%
Operational costs	$10,000/month	$7,500/month	25%
System performance	85%	100%	15%

Source: Author

and low latency, is particularly effective for real-time applications. The system's ability to optimise resource allocation, reduce costs, and improve performance makes it a valuable tool for cloud service providers and users.

Future improvements and scope

As cloud computing evolves, workload prediction models ought to enhance in accuracy, scalability, and adaptableness. While ML and DL fashions like LSTM and transformers display excessive overall performance, challenges stay in real-time responsiveness and integration with modern-day cloud architectures. Future guidelines include reinforcement studying, federated mastering, aspect computing, multi-cloud systems, and serverless environments.

Enhancing model performance
LSTM and transformer-based fashions offer excessive accuracy however often suffer from excessive computational fees. Reinforcement studying offers adaptive gaining knowledge of by way of interacting with live environments, allowing real-time adjustments to workload fluctuations. Additionally, ensemble tactics—like random forest, gradient boosting, and neural networks—can integrate strengths of person fashions, improving robustness and lowering prediction mistakes (Figure 26.6).

Real-time adaptive systems
Federated learning permits collaborative version schooling across dispensed cloud nodes without sharing raw records, maintaining privateness and lowering latency. Similarly, edge computing moves processing toward IoT information sources, enabling decentralised, low-latency workload forecasting the use of lightweight models, which improves machine responsiveness and reduces cloud overhead (Figure 26.7).

Multi-cloud and serverless adaptations
Cloud infrastructure is shifting in the direction of multi-cloud and serverless architectures. Traditional models are frequently designed for unmarried-cloud setups, restricting adaptability. Future fashions have to cope with dynamic workloads across structures like AWS, Azure, and GCP. Integrating prediction with serverless orchestration can lessen bloodless-begin latency and optimise real-time automobile-scaling (Table 26.5).

Conclusion

Workload prediction is prime to optimising cloud overall performance and decreasing operational charges. This study evaluated ARIMA, LSTM, and transformer fashions, displaying that Transformers outperform others with an R² of zero. Ninety three and lowest latency of 50 ms. A case study on GCP discovered that Transformer fashions advanced CPU prediction accuracy to 93.1%, decreased resource wastage by using 25%, and boosted device efficiency by means of 15% (Figure 26.8).

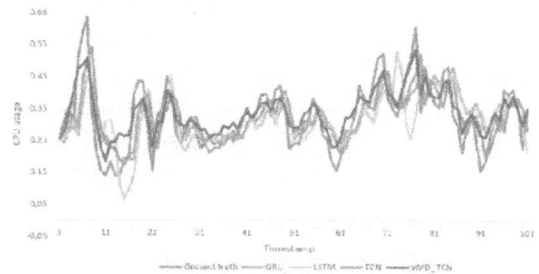

Figure 26.7 Basic methods and proposed workload prediction results for VMD-TCN in Alibaba datasets [7]
Source: Author

Table 26.5 System performance metrics

Metric	ARIMA	LSTM	Transformer (TFT)
MSE	0.56	0.29	0.21
RMSE	0.75	0.54	0.46
MAE	0.68	0.49	0.42
R²	0.72	0.89	0.93
Latency (ms)	120	80	50

Source: Author

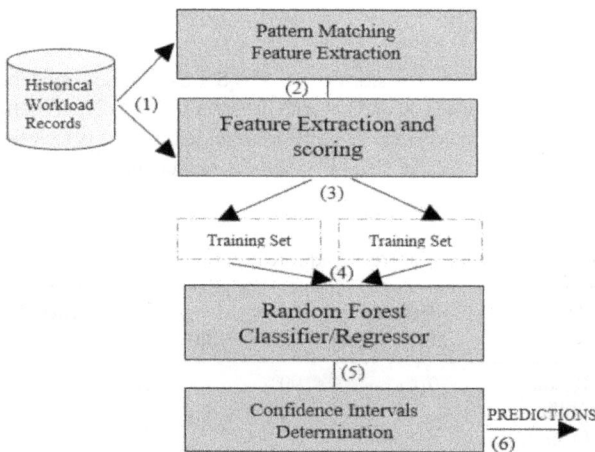

Figure 26.6 Load forecasting [6]
Source: Author

Despite robust consequences, demanding situations stay in scalability, real-time responsiveness, and multi-cloud adaptability. Future work must recognition on reinforcement mastering, federated gaining knowledge of, and serverless-pleasant forecasting. As cloud structures develop more dynamic, AI-included, self-gaining knowledge of fashions might be essential for allowing adaptive, green aid control (Figure 26.9).

References

[1] Kumar, J., Singh, A. Cloud datacenter workload estimation using error preventive time series forecasting models. *Cluster Comput.*, 2020:23.

[2] Kirchoff, D., Xavier, M., Mastella, J., De Rose, C. A preliminary study of machine learning workload prediction techniques for cloud applications. 2019:222–227.

[3] Amekraz, Z., Hadi, Y. CANFIS: A chaos adaptive neural fuzzy inference system for workload prediction in the cloud. *IEEE Access*, 2022;10:1–1.

[4] Workload prediction based virtual machine migration and optimal switching strategy for cloud power management - Scientific Figure on ResearchGate. Available from: https://www.researchgate.net/figure/Workload-prediction-based-VM-migration-in-cloud_fig1_355177509 [accessed 25 Feb 2025].

[5] Sharma, M., Shrimal, V. M., Chand Saini, H., Taparia, D. A novel hybrid CNN-LSTM approach for handwritten text recognition for the Washington database. *2023 IEEE Renew. Energy Sustain. E-Mobil. Conf. (RESEM)*, 2023:1–5.

[6] Badoni, P., Walia, R., Mehra, R. Wearable IoT technology: Unveiling the smart hat. *Proc. 2024 1st Internat. Conf. Intell. Sys. Technol. Emerg. Markets (ISTEMS)*, 2024.

[7] Sasidhar, C., Saini, M. L., Charan, M., Shivanand, A. V., Shrimal, V. M. Image caption generator using LSTM. *2024 4th Internat. Conf. Technol. Adv. Comput. Sci. (ICTACS)*, 2024:1781–1786.

[8] Zhang, Q., Yang, L. T., Chen, Z., Li, P. A survey on deep learning for big data. Information fusion. 2018;42:146–157.

[9] Kunadharaju, H. P. R., Sandhya, N., Mehra, R. Multi sensor image matching using super symmetric classifiers. *Internat. J. Recent Technol. Engg.*, 2019;8(2):6161–6166.

[10] Gupta, S., Dinesh, D. A., Ramesh, R. Machine learning models for resource allocation in cloud computing: A survey. *J. Cloud Comput.*, 2021;10(1):1–20.

[11] Box, G. E. P., Jenkins, G. M. (1976). *Time Series Analysis: Forecasting and Control.* Holden-Day.

[12] Mishra, A. K., Hellerstein, J. L., Cirne, W., Das, C. R. Towards characterizing cloud backend workloads: Insights from Google compute clusters. *ACM SIGMETRICS Perform. Eval. Rev.*, 2020;47(4):34–37.

[13] Chen, T., Guestrin, C. XGBoost: A scalable tree boosting system. *Proc. 22nd ACM SIGKDD Internat. Conf. Knowl. Discov. Data Min.*, 2016:785–794.

[14] Vaswani, A., Shazeer, N., Parmar, N., Uszkoreit, J., Jones, L., Gomez, A. N., Polosukhin, I. Attention is all you need. *Adv. Neural Inform. Proc. Sys. (NeurIPS)*, 2017;30:5998–6008.

[15] Netflix Tech Blog. Real-Time Workload Forecasting Using LSTM and Kafka. Retrieved from Netflix Tech Blog. 2022.

[16] Satisha, C., Mehra, R., Giri, M. Detection of various security attacks on IoT devices using machine learning. *2023 Internat. Conf. Comput. Intell. Netw.*, 2023.

[17] Li, J., Zhang, X. Machine learning approaches for cloud workload prediction: A comparative study. *Proc. 2020 IEEE Internat. Conf. Cloud Comput. Big Data Anal. (ICCBDA)*, 2020:35–41.

[18] Patel, R., Shah, D. A survey on cloud workload prediction using time series and machine learning models. *J. Cloud Comput. Adv. Sys. Appl.*, 2022;11(1):25–38.

[19] Gupta, S., Joshi, A. Scalable workload prediction models for cloud resource management using machine learning. *Cloud Comput. Big Data, 2021;39(4):167–181.*

[20] Raj, M., Kumar, R. Workload prediction and resource allocation in cloud computing: A hybrid approach. *Internat. J. Cloud Comput. Ser. Sci., 2019;8(2):112–123.*

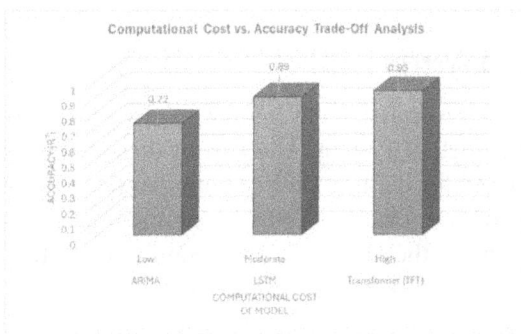

Figure 26.8 Computational cost vs. accuracy trade-off analysis
Source: Author

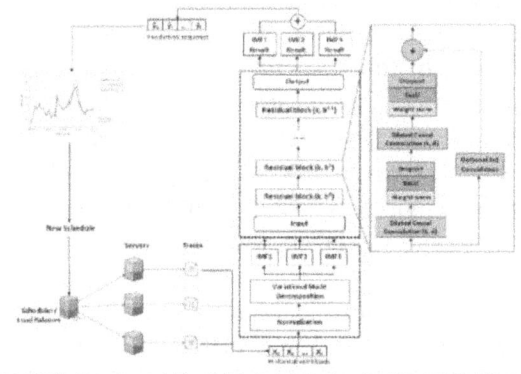

Figure 26.9 A complete architecture of resource planning and load compensation using the proposed approach to predict cloud computing workload [7]
Source: Author

27 Smart traffic sign detection using deep learning for intelligent vehicle control

Santhosh Kumar C.[1,a], Sahil Pal[2,b], Yash Patel[2,c], Radke Animish[2,d] and Harsh Jain[2,e]

[1]Assistant Professor, Department Computer Science and Engineering, SRM Institute of Science and Technology, Ramapuram, Chennai, Tamil Nadu, India

[2]Student, Department Computer Science and Engineering, SRM Institute of Science and Technology, Ramapuram, Chennai, Tamil Nadu, India

Abstract

Traffic sign recognition (TSR) is an important component of intelligent transportation systems such as autonomous cars and sophisticated driver assistance systems (ADAS) to recognise the signs correctly. The aim of this project is to propose a deep learning method of using convolutional neural networks (CNNs) to recognise and assign categories of traffic signs with good precision. Pre-processing algorithms like resizing, normalisation and data augmentation (taking in consideration random flipping, rotation, zooming) are used to train the model on a trained labelled dataset. The CNN architecture consists of several convolutional and pooling layers for extracting the hierarchical spatial features in images, and several resize layers to keep consistency of input. Additionally, dropout regularisation is also used to keep the model from over fitting and improve generalisation on real cases. A softmax activation function is used as the last classification layer for identifying the correct traffic sign category. Because the developed system is incredibly accurate and scalable, it is ideal for real time use in autonomous driving and ADAS. The work touches on safe navigation on roads through contributing to effective use of this model in intelligent vehicle control systems, aiming to reduce human errors and to develop smart transportation systems.

Keywords: Advanced driver assistance systems, convolution neural network

Introduction

The need for advanced technology for traffic sign detection has been strong since the quick advent of driverless cars and smart transportation systems. Traffic signs are critical in directing vehicles and preserving road safety, however, human riders often incorrectly read them due to distractions, terrible perspective, or weariness. Such misunderstandings may lead to traffic infractions and accidents. This research aims at creating a machine learning-based traffic sign detection system that can automatically detect, and categorise traffic signs in real time to deal with the aforementioned problems. Utilisation of computer vision techniques and deep learning models like convolutional neural networks (CNNs) and learning are ward function, will provide a suggested system that improves autonomous vehicle navigation and vehicles with advanced driver assistance systems (ADAS). Ultimately, the goal is to utilise more of automation and greater detection accuracy to improve living, creating safer and more intelligent transportation systems.

Proposed system

It involves several steps to build a trustworthy traffic sign detection system. The first of these datasets are retrieved for the LISA traffic sign dataset, a dataset that are extracted from open access sources, and the second are the German Traffic Sign Recognition Benchmark (GTSRB). Pre-processing methods such as image resizing, normalisation and augmentation etc. are used for pre-processing in order to improve model performance. Augmentation techniques that ensure that the model can cope with many real life situations will be discussed here: i.e., rotation, flipping, brightness adjustment, noise addition are examples of this kind. To extract the most important signs shown in pictures of traffic signs, a CNN is used.

Separately, this architecture features maximum fully connected layers for classification (and all stages include those). Whilst dropout layers are used to reduce overfitting, layers used as resizing layers are also included for standardisation to allow for consistencies in input sizes. Metrics of accuracy, precision, recall and F1-score are then later assessed in

[a]cjsksag@gmail.com, [b]sr4010@srmist.edu.in, [c]yl1882@srmist.edu.in, [d]rb0776@srmist.edu.in, [e]ha7616@srmist.edu.in

DOI: 10.1201/9781003685876-27

order to assess the variables of the system to ensure peak performance.

The model is finally included in ADAS or self-driving car hardware to achieve real time detection. The solution gives immediate feedback to drivers or vehicle control systems to enhance road safety and minimise accidents risk. The large contribution made to intelligent transportation networks (ITNs) by providing an automation of existing traffic sign recognition, and therefore an improvement to existing recognition accuracy.

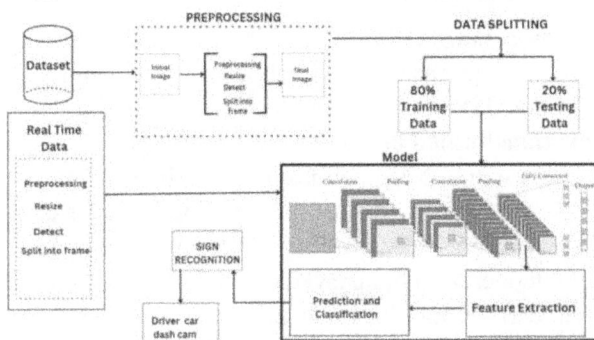

Figure 27.1 Block diagram
Source: Author

Literature survey

In a new approach for interpreting traffic signs in natural language descriptions, Yang et al. (2024) [1] created a method. The research showed that a process is possible for finding and identifying traffic signs and writing out to describe what the signs represent. This makes the process better to understand signs and is valuable for autonomous vehicles and intelligent transport systems.

In this, Uikey et al. (2024) [2] proposed a clean Indian traffic sign detection and classification system. The challenges of their Indian roads, such as changing sign sizes, changing fonts and changing light, did not come between the system and delivering excellent off the shelf results. The method was able to greatly improve detection rates and classification uniformity.

Indian traffic sign recognition was developed by deep learning techniques by Megalingam et al.(2022) [3]. In order to increase the recognition rate they used CNNs. It was trained from a large number of Indian traffic signs and worked very well under real time scenarios.

In Arena et al. (2022) [4] they modified a traffic light control system using fuzzy logic in several topologies. In relation to optimising traffic movement, in this research, the authors have been concerned with dynamically changing the signal time and improving roads' performance that reduces the time of traffic movement and eliminates congestion.

In Singh and Malik (2021) [5], they proposed a technique of traffic sign recognition by CNN. Deep learning methods, in particular, are demonstrated to very accurately identify traffic signs in their research. Moreover, using more data is said to make the models work better.

As illustrated over these years, in Milan Tripathi (2021) [6] we have a critical review of image classification using CNN techniques applied in a wide range of industries like retail and health care. The CNN models are analysed and their ability to detect patterns as well as objects are compared. It illustrates how applying deep learning can improve the accuracy by a notch in different industries.

Deep learning – Muhammad and others (2020) [7] discussed the future of deep learning for autonomous vehicles as to their potentials and challenges. A lack of data, attacks that try to deceive the system and needing the system to react promptly were the major problems that they found. For providing some ideas to make deep learning more robust and reliable for traffic conditions, they suggested.

A system to avoid vehicle accidents and reading road traffic signs was suggested by Lahare et al. (2020) [8]. But their paper tied detecting traffic signs with developing collision prevention methods; aiding the vehicle in getting safe avoiding drivers of dangers in advance. Towards the development of a sensor based accident prevention system for real time traffic sign recognition, Sneha Lahare (2020) [8] proposed sensors based real time traffic sign recognition. Raspberry Pi, Pi Cam, and ultrasonic sensors are used to base the system on which the driver will be alerted of traffic signs, kept distance in front other cars, and make a safe distance with a traffic sign present.

Cao et al. (2019) [9] improved self-driving car traffic sign detection algorithms. For real time applications of self-driving car, they used improved feature extraction mechanisms and deep learning on models to increase detection speed and accuracy.

Safat B. Wali (2019) [10] studies the traffic sign detection and recognition (TSDR) systems which are used for ADAS-based development. It gives colour, shape and machine learning based techniques for detection, tracking and classification. The paper also discusses aspects such as the case of varying light, occluded signs and degraded signs, and proposes the future.

In the year 2017 [11], Alexander Shustanov used CNN to label the traffic signs in real time. The mentioned research states the different CNN architectures and gives out a CNN identification system based on the Tensor flow library to manage the process in mobile graphical processing unit (GPUs) more efficiently. The results show that CNNs significantly improve traffic sign detection and identification as they are a desirable choice for autonomous vehicle systems.

Prachi Dewan (2017) [12] have classified the traffic sign recognition techniques into three categories, namely, colour based, shape based, machine learning based. Then it discusses some of the difficulties in such real world applications as those of low resolution, occlusion or varying light conditions. Based on the study, it is concluded that field programmable gate arrays (FPGA) systems using Xilinx system generator are the best for real time systems.

To support colour blind individuals, Almagambetov et al. (2014) [13] came with a system. In their system, computer vision was used to spot traffic lights, and sound and touch notifications were used in guiding visually impaired people to travel safely through road sections.

Studies, conducted early, on detection and identification of traffic signs were performed by De La Escalera and others (1997) [14]. Some rudimentary techniques for machine learning approach to traffic sign recognition and categorisation were proposed by their work.

Implementation

Dataset
In this research, the data set used is the GTSRB dataset, which is a huge data set of traffic sign images for machine learning classification tasks. It has over 50,000 images of 43 classes of traffic signs. This splits the data set into training and test sets to make testing models in the natural conditions.

The images are provided in .PNG format, or for "visual data", and .CSV files for more detailed labels and metadata. It takes the images, labels them by their class label, augments them with the required information such as boundary box coordinates, rotation angle and blur levels. The dataset is also diverse since the images have been taken at different weather conditions, different angles with different amounts of light in them.

Total images	50,000+
Number of classes	43
Image formats	.PNG for images, .CSV for metadata and labels
Annotation format	Bounding box coordinates, class labels, rotation details, and blur information
Environmental diversity	Captured in different weather, lighting conditions, and angles
Dataset organisation	Divided into structured training and test sets

Pre-processing
First, images are loaded, classified and resided a common 30×30 pixels for consistency. First they are next transformed into a numerical array (30, 30, 3) to process with CNN, then normalised (between 0 and 1) the pixel value from 0–255 to obtain faster convergence and stable learning. If the labels are not one hot encoded into binary vectors, the model would be unable to support this many classes as it separates the classes. The dataset is finally split into training (80%) and test (20%) sets by the train_test_split function, with an even mix of traffic sign types. Learning features and patterns by the model and testing the generalisation to new images on the test set are feasible with the train set. The pre-processing process can also be made as an orderly process to over shot and undershot the recognition of traffic signs and this in turn lead to getting a robust and accurate traffic sign recognition model.

Model training
After pre-processing the CNN model regularly examines traffic sign images using various layers to successfully find and analyse and categorise attributes. The convolutional layers act as the first component of the phase by enabling small filters to search images to identify fundamental patterns such as edges and shapes and textures. The specified patterns help the model understand multiple visual characteristics of traffic signs that aid in their recognition. After convolution takes place the network requires an activation function for non-linear transformation. Rectified linear unit (ReLU) functions as the widely adopted activation function in which the following equation applies:

$$f(x) = max(0,x)$$

Through positive values and removing negative values, ReLU helps the model spot the complex patterns and then learn in an efficient way. This activation cuts training time of this model exponentially and solves vanishing gradient problems. Pooling layers follow the convolutional layers but to minimise computational efficiency. Compressed spatial size of the feature maps without affecting much of the information of the features, this can allow the network to become more efficient at identifying traffic signs. The most popular is the max pooling that preserves the most prominent features from the area so the model can focus on the most remarkable patterns. Softmax makes the output values be probability distributions, and thus, it assigns a confidence score to each class. The final prediction is a picked traffic sign, which has the highest probability. Back propagation is used as a training process for the model, minimising errors by adjusting the weights of the network in gradient descent. This allows the model to keep making more refined predictions at each iteration and it can improve classification accuracy. The softmax is given by:

$$\text{Softmax}(z_i) = \frac{e^{z_i}}{\sum_{j=1}^{K} e^{z_j}}$$

- z_i is the logit (the output of the previous layer in the network) for the i^{th}class.
- K is the number of classes.
- e^{z_i} represents the exponential of the logit.
- $\sum_{j=1}^{K} e^{z_j}$ is the sum of exponentials across all classes.

Classification accuracy is used to measure the performance of the overall CNN that predicts traffic signs well or not at all.

$$\eta = \frac{C_{correct}}{C_{total}} \times 100$$

Ccorrect is the number of correctly classified images and the Ctotal is the number of predictions. Robustness and ability of the model to generalise for any traffic sign images implies an accurate score. With the help of CNNs, we show how to extract hierarchical patterns for real time high accuracy traffic sign recognition which is vital in autonomous vehicles and intelligent transportation systems.

Result

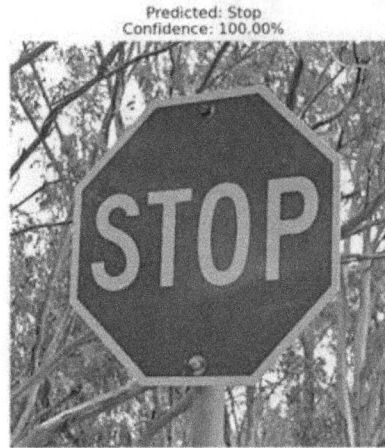

Figure 27.2 Stop sign
Source: Author

Figure 27.3 End of no overtaking sign
Source: Author

Figure 27.4 Keep left sign
Source: Author

Predicted: Turn right ahead

Figure 27.5 Speed limit (50 km/h)
Source: Author

Predicted: Road narrows on the right

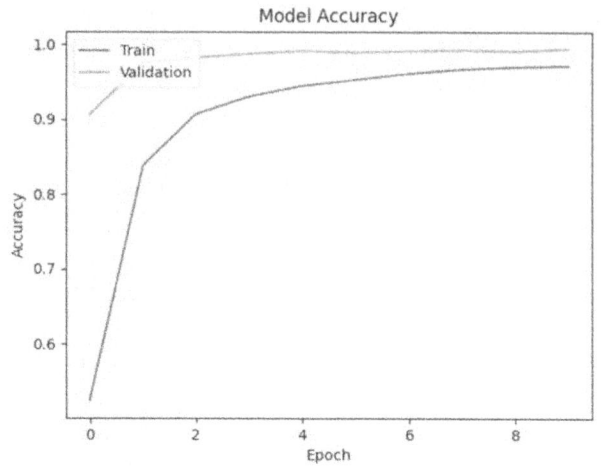

Figure 27.6 Bicycles
Source: Author

Predicted: Speed limit (60km/h)

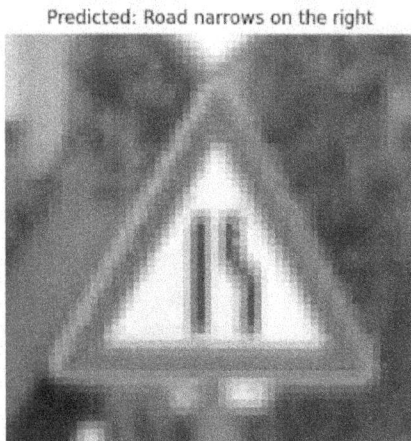

Figure 27.7 Speed limit (60 km/h)
Source: Author

Model Accuracy

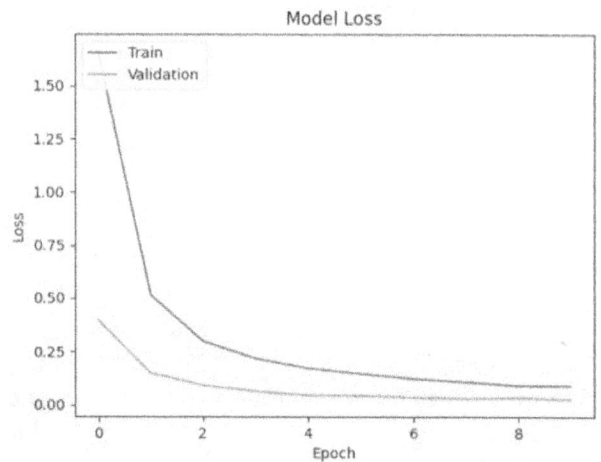

Figure 27.8 Model accuracy graph
Source: Author

Some accuracy and loss plots of the model's performance are evaluated. Even surpassing the training performance, validation accuracy peaks at 98–99% while the training accuracy is up to 95%.

Training and validation losses decrease rapidly and validation loss is always smaller than the training loss. If this small gap and good learning, strong generalisation and minimal over fitting is something that you do.

Model Loss

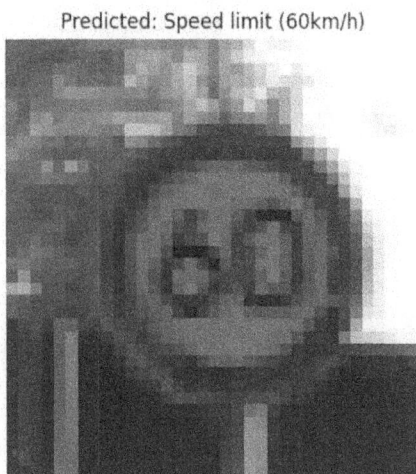

Figure 27.9 Model loss graph
Source: Author

Model performance summary

Aspect	Observation	Interpretation
Training accuracy	~95% (stable)	Effective pattern learning
Validation accuracy	~98–99% (higher than training)	Strong generalisation
Training loss	Approaching zero	Efficient error minimisation
Validation loss	Lower than training	Effective regularisation
Accuracy analysis	Validation>Training	Indicates data augmentation/ dropout effectiveness
Loss analysis	Minimal gap	No significant over fitting

Conclusion

Traffic sign recognition is successfully performed by CNNs by applying spatial features through convolutional and pooling layers and converting them into classification using fully connected layers and a Softmax function. Learning was then improved by pre-processing steps like resizing, normalisation and one hot encoding. On the one hand, good training and validation accuracy with decreasing loss confirmed CNNs as a useful tool for real time traffic sign recognition in autonomous vehicles and intelligence traffic systems.

Future work

Multi-language traffic sign recognition through OCR and deep learning is a future development to accommodate other regions.

Additionally, it can also include damage detection that will aid in detecting and reporting the worn out or broken signs for timely maintenance. This will improve the safety of road and real time traffic management.

References

[1] Yang, C. et al. Traffic sign interpretation via natural language description. *IEEE Trans. Intell. Transp. Syst.*, 2024;25(11):18939–18953.

[2] Uikey, R., Lone, H. R., Agarwal, A. Indian traffic sign detection and classification through a unified framework. *IEEE Trans. Intell. Transp. Syst.*, 2024:1–10.

[3] Megalingam, R. K., Thanigundala, K., Musani, S. R., Nidamanuru, H., Gadde, L. Indian traffic sign detection and recognition using deep learning. *Int. J. Transp. Sci. Technol.*, 2023;12(3):683–699.

[4] Arena, F., Pau, G., Ralescu, A., Severino, A., You, I. An innovative framework fordynamic traffic lights management based on the combined use of fuzzy logic and several network architectures. *J. Adv. Transp.*, 2022;2022:1–17.

[5] Singh, K., Malik, N. CNN based approach for traffic sign recognition system. *Adv. J. Grad. Res.*, 2021;11(1):23–33.

[6] Tripathi, M. Analysis of convolutionalneural network based image classification techniques. 2021;3(2):100–117.

[7] Muhammad, K., Ullah, A., Lloret, J., Ser, J. D., de Albuquerque, V. H. C. Deep learning for safe autonomous driving: Current challenges and future directions. *IEEE Trans. Intell. Transp. Syst.*, 2021;22(7):4316–4336.

[8] Lahare, S., Mishra, A., Nair, A., Borkar, N. Road traffic sign recognition and vehicle accident avoidance system. *Internat. J. Sci. Res. Comp. Sci. Engg. Inform. Technol.*, 2020:484–489.

[9] Cao, J., Song, C., Peng, S., Xiao, F., Song, S. Improved traffic sign detection and recognition algorithm for intelligent vehicles. *Sensors (Basel)*, 2019;19(18),4021.

[10] Wali, S. B. et al. Vision-based traffic sign detection and recognition systems: Current trends and challenges. *Sensors (Basel)*, 2019;19(9),2093.

[11] Shustanov, A., Yakimov, P. CNN design for real-time traffic sign recognition. *Procedia Eng.*, 2017;201:718–725.

[12] Dewan, P., Vig, R., Shukla, N., B. K. Novel VLSI architectures for image segmentation and edge detection algorithm. *Int. J. Comput. Appl.*, 2016;149(10):32–36.

[13] Almagambetov, A., Velipasalar, S., Baitassova, A. Mobile standards-based traffic light detection in assistive devices for individuals with colour-vision deficiency. *IEEE Trans. Intell. Transp. Syst.*, 2015;16(3):1305–1320.

[14] de la Escalera, A., Moreno, L. E., Salichs, M. A., Armingol, J. M. Road traffic sign detection and classification. *IEEE Trans. Ind. Electron.*, 1997;44(6):848–859.

28 Developing a sustainable online ecosystem for second-hand accessories within a university community

Binod Prasad Joshi[1,a], Vijay Mohan Shrimal[2,b], Shivam Singh[2,c], Mamta Sharma[2,d] and Alok Das[2,e]

[1]Student, Department of computer Science and engineering (AIT-CSE), Chandigarh University, Punjab, India

[2]Department of computer Science and engineering (AIT-CSE), Chandigarh University, Punjab, India

Abstract

This paper presents the development and implementation of a second-hand marketplace web application tailored for a university community. The platform aims to facilitate promote sustainable practices, and foster a sense of community among students. By offering features such as product listing, university and course-specific filtering and communication, the platform addresses key trust and usability concerns. The study evaluates the platform's effectiveness through user engagement metrics, and user feedback. Results indicate high user satisfaction and a significant contribution to waste reduction and cost savings for students. The discussion analyses the platform's success, identifies potential barriers to adoption, and proposes scaling strategies, including community building and partnerships. Future research directions include longitudinal studies, comparative analyses, and social impact assessments. This research demonstrates the potential of technology to promote sustainable consumption practices and create a more environmentally and socially responsible university environment.

Keywords: Sustainability, circular economy, peer-to-peer exchange, community building, user engagement

Introduction

This development, together with the more important issues raised by students and campus groups regarding the ease of access to merchandise and resources, is quite notable because it further advances aspects of campus communities into e-commerce. However, existing platforms do not address needs such as financial constraints, limited access to affordable goods, and inefficient circulation of resources. In this context, it is essential to come up with a localised e-commerce solution with an emphasis on sustainability and affordability within the campus ecosystem itself. Collaborative redistribution has recently become more popular, allowing users to buy, sell, and share second-hand goods. Besides being an ecological benefit by reducing waste, it also has economic benefits for consumers when it comes to obtaining unique and affordable products. In the university communities, such platforms could be instrumental in addressing the financial constraint and resource imbalances faced by students.

Recent studies indicate Generation Z's (Gen Z) engagement with second-hand markets is positive, and many University students are a part of this engagement. It comes mostly from their concern of the environment as well as personal economic benefits.

Student's attitude towards second-hand goods is shifting significantly, driven by various factors:

- Economic factors: Bargain pricing of second-hand products makes them affordable for Gen Z consumers purchasing luxury goods. During these difficult economic times when many look for ways to save money, buying second-hand is a great option.
- Marketing factors: The impactful nature of growing zeal and concerns for the environment and the contribution of young generation towards fast fashion make sustainable consumption a norm. Buying second-hand goods fulfils their moral by contributing towards waste reduction and aiding circular economy.
- Psychosocial factors: Gen Z can utilise second-hand shopping as a means to showcase their individuality by hunting for vintage and unique products.
- Sociocultural factors: The ease of access made possible through online resale platforms makes it so simple for younger consumers who are well versed with technology. The hateful nature of these platforms, such as the ability for users to comment and like posts, enhances the purchas-

[a]joshibinodprasad08@gmail.com, [b]drvijaymshrimal@gmail.com, [c]Shivamsingh8496@gmail.com, [d]mamta1989597@gmail.com, [e]dasalok2502@gmail.com

DOI: 10.1201/9781003685876-28

ing experience while helping mask the authenticity of the product.

- Changing values: Gen Z has a different perception of using products compared to past generations, focusing on access rather than ownership [9] and adopting the sharing economy. This attitude leads to the increased acceptance and normalisation of second-hand shopping.

These indicators indicate the favourable conditions towards second-hand among Gen Z [5] is not a fleeting trend but a profound change in consumer behaviour with lasting effects on the retail environment. This generation is not only adopting second-hand for personal use but also for gift-giving [6], further enhancing its status in mainstream culture. As the purchasing power of Gen Z grows, their inclination towards second-hand is likely to fuel considerable growth in the resale market.

Literature review

Several online platforms facilitate the second-hand market. While some specialise in specific categories, others offer a broader range of products. Here are a few examples:

- OLX: Primarily functions as a classifieds platform where users can list and browse various second-hand items. While it can be useful for finding deals on furniture, electronics, and other goods within a local area, its general nature may not cater specifically to the needs of a university community. It lacks features tailored to student life, such as textbook exchanges or course-specific material sharing.
- Facebook marketplace: Integrated within the Facebook platform, marketplace benefits from a large user base and ease of access. Its community-based nature can facilitate local exchanges within a university setting. However, it lacks specialised features for academic resources and may present privacy concerns due to its connection to personal Facebook profiles. Furthermore, the lack of robust seller verification and buyer protection mechanisms can pose risks within a close-knit university community.
- Shopify: Shopify is primarily designed for owners of businesses seeking an online storefront. While it is technically open for anyone to sell

goods via the platform, it is less suitable for casual peer-to-peer transactions within a university setting. In other words, fees, plus the large amount of tutored setup required, are less appealing for students looking to buy or sell a few items. Instead, it is more appropriate for students running small businesses selling handmade goods or curated second-hand goods.

There are several limitations concerning the university community in this regard:

- Lack of focus on academic resources: General platforms like OLX and Facebook marketplace do not provide effective resources in trading for textbooks, course materials, or any other academic-specific items.
- Safety and trust: While trading over these platforms, verification of buyers and sellers is always an issue otherwise, they will never be able to get over-the-top scams or uncomfortable interactions within a university community.
- Privacy concerns: The fact that these platforms are linked with personal social media profiles raises some red flags on privacy among use of students.
- Lack of community-centred features: These platforms do not often feature things related specifically to campus life. Integration with student IDs, schedules of courses, and campus groups.
- Cost and complexity: Shopify, while offering robust features, can be too expensive and complex for individual students looking to engage in casual second-hand transactions.

A college community needs to walk further, forging as it does panoply of challenges confronting peer-to-peer trading: textbook exchanges, sharing of course materials, integration with college workflows, and so on. Shen [12] has noted a service directed at facilitating academic connections among UCL students that attests to the potentials brought about by using a more focused approach. Ahmed et al. [1] also discusses how a specific app, ShareSpace anticipates the possibility of improvement by providing highly localised sharing amenities for students.

Primary factors affecting students' purchasing decisions about the other hand goods include alleged image, popularity, love, loyalty, approach, personal identity, perceived financial and quality risk, price, uniqueness, economy, environmental anxiety and contamination concerns. In terms of second-hand goods, the alleged love and loyalty greatly impressed the purchase decisions, with 72% change in purchasing the decisions responsible for these perceptions among young adults in Ghana [13]. Vietnamese students are shaped from their perspective to buy recycled fashion goods, which are affected by environmental anxiety, uniqueness value and perceived risks [11]. The principle of planned behaviour (TPB) highlights the value and environmental anxiety as the major determinant of second-hand product purchase, especially among young consumers, who prefer stability [7]. Social norms, income levels and environmental awareness also play an important role in permanent consumption, suggesting that a versatile approach is required to effectively promote stability [14]. Additionally, bio-atomic values, personal and social norms, and contamination concerns affect the purchase of second-hand clothing, which reflects the importance of promoting a positive attitude and addressing the contamination fear [5]. These factors collectively affect the perceptions of the stability of the students, as they navigate the balance between economic benefits, personal values and environmental responsibility in their purchasing decisions.

Methodology

The following section presents a systematic approach to designing, developing, and deploying the proposed second-hand marketplace web application. The methodology includes needs assessment, platform design and development, and deployment strategies.

Research design

Understanding specific requirements and challenges within the university community was crucial for developing a relevant and effective platform. The phase included:

- Surveys: Surveys that involved circulating questions among students to collect data on their experiences in second-hand marketplaces, their wishes concerning the needs these second-hand marketplaces should satisfy, and must-have features in a dedicated one. The questions focused mainly on the frequency of second-hand transactions, which types of things used to change hands, problems reasoned in using current plat-

forms (for example, safety issues, lack of academic resources), and some features they would like to see in the platform, e.g., filter by the type of the item within the institution.
- Focus groups: Conducting focus group discussions with representative student groups to gain deeper insights into their needs and expectations. These discussions explored the social and cultural aspects of second-hand exchange within the university community, providing valuable qualitative data to complement the survey findings.
- Competitive analysis: Evaluating existing second-hand platforms, both general and university-specific, to identify best practices, potential pitfalls, and opportunities for differentiation. This analysis considered factors such as user interface design, features offered, security measures, and community engagement strategies.

Platform design

The platform follows microservices architecture, ensuring scalability and maintainability. The key components include:

- The platform implements microservices architecture for performance and maintainability. Team city is responsible for continuous integration and deployment following the changes made on the site. Key components include:
- Frontend build: React.js using Tailwind CSS for responsive UI design.
- Backend: It works with Node.js on Express.js.
- Database: MongoDB is the primary database storing user and product information.
- Authentication: In the authentication process, secure user authentication is achieved with the use of JSON Web Tokens (JWT). Through login functions, users are issued a signed JWT token that is used during each API call for the sake of session security and authorisation control.
- Messaging system: WebSocket-based real-time chat system is used for communication between bidders and sellers.
- Hosting and deploying:
  ◦ Frontend hosting, which is deployed on Vercel, provides seamless and scalable hosting, with CI/CD capabilities.
  ◦ Backend hosting is hosted on AWS EC2 instances, allowing for high availability and scalability.

○ For database, MongoDB Atlas is used; it is the best-managed cloud database as far as performance and data security. Using AWS S3, images and media associated with product listings are stored.

○ Authentication via JWT is implemented right inside the backend service in which middleware in Express.js handles the verification of users and authorisation.

Figure 28.1 represents a flowchart of authentication process and features of PeerMarket. Authentication is done using JWT. When the user initiates the process, the system checks the provided credentials, if valid credentials is provided, json web token is generated and is successfully logged in else process results in authentication failure. This also shows the application's core features such as profile management, order management, product listing, chat interface, and notification system.

Results

The present section is concerned with the results obtained from the development and evaluation of the second-hand marketplace web application within the university community, looking into participant engagement and impact on sustainable practices.

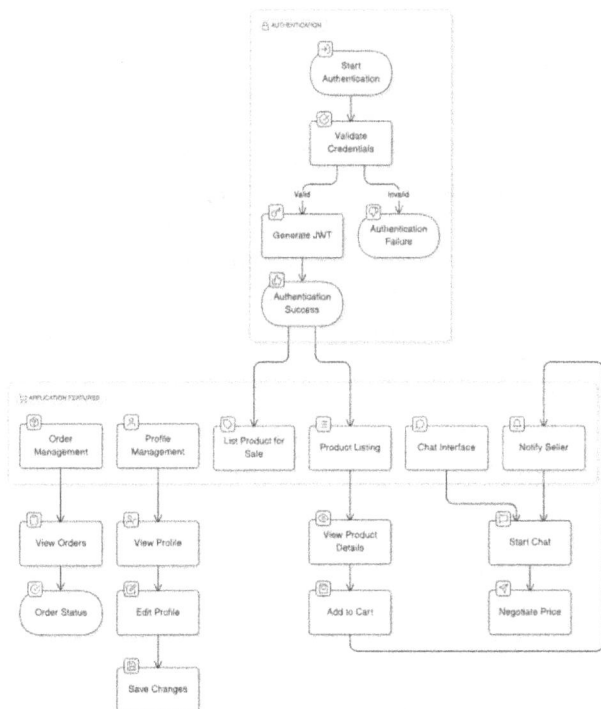

Figure 28.1 Platform flowchart
Source: Author

Barriers to adoption
Despite the positive results, several potential barriers to wider adoption were identified:

• Initial trust: While the student verification feature addressed some trust concerns, some users may still hesitate to engage in online transactions with strangers. Building trust within the community remains an on-going challenge. Ahmed et al. and Kirti et al. [1, 2] discuss trust models and the importance of community in online second-hand platforms.

• Competition: Existing general second-hand platforms and informal exchange networks within the university may pose competition. The platform needs to continuously innovate and offer unique value propositions to attract and retain users. Kunadharaju et al. [10] provides an example of a SWOT analysis for a second-hand platform, which can be helpful in understanding competitive dynamics.

• Technological literacy: Some students may lack the technological literacy required to effectively utilise the platform's features. Providing user-friendly tutorials and support resources can address this challenge. Luo et al. [6] discusses end-user development of web applications, which can inform strategies for improving platform accessibility.

Scaling the ecosystem
To maximise the platform's impact and ensure its long-term sustainability, several scaling strategies are proposed:

• Expanding product categories: Exploring the inclusion of additional product categories, such as services or rental options, could broaden the platform's appeal and utility.

• Community building: Organising events and initiatives to foster a stronger sense of community among platform users can enhance engagement and trust. Chung et al. [4] emphasises the importance of community integration in building trust.

• Partnerships: Collaborating with university departments, student organisations, and local businesses can expand the platform's reach and resources.

- Gamification: Introducing gamified elements, such as rewards or badges for sustainable practices, could incentivise user participation and promote environmentally conscious behaviour.
- Events hosting: Collaborating with different institutions to host various events such as hackathons, online events or feature of registration of events organised in institutions.
- Monetisation: After getting adequate number of users, this application can be monetised by adding advertisements or sponsoring the local business near institutions.

Few screenshots of our application are mentioned below:

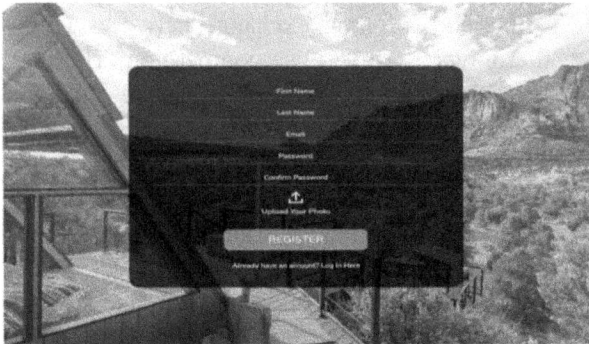

Figure 28.2 Signup page
Source: Author

Figure 28.2 shows a registration page of PeerMarket. Registration page contains input fields like First Name, Last Name, Email, Password and Confirm Password. It also has option to upload profile picture. A button to create an account and if user already have account, there is option to redirect to login page.

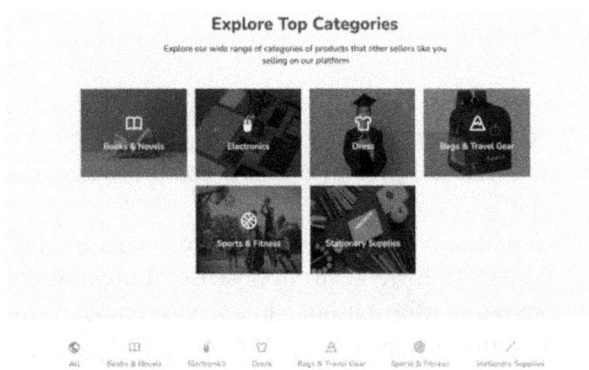

Figure 28.3 Product catalogue with category filtering
Source: Author

Figure 28.3 shows a category section page of PeerMarket where user can filter products from their category. There are main six categories available, books & novels, electronics, dress, bags & travel gear, sports & fitness and stationery supplies. This page helps users to explore and find the product they are interested.

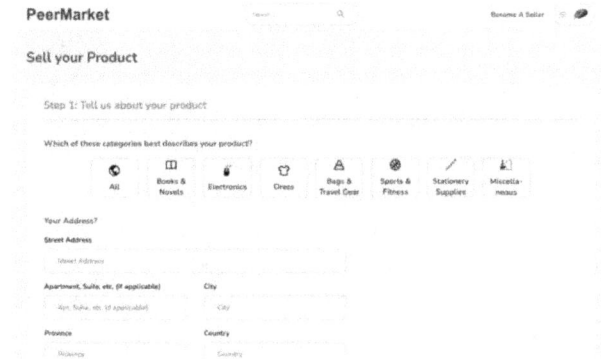

Figure 28.4 Page to add your product
Source: Author

Figure 28.4 shows the product listing page from PeerMarket. On product listing page, users can upload the details about their product. The key inputs include category selection, address field, product images, product title, description, highlight and highlight details.

Conclusion

- High user engagement: The significant registration and active user numbers indicate a strong demand for a dedicated second-hand marketplace within the university community. This suggests that the platform effectively addresses the needs and preferences of students seeking affordable and sustainable alternatives to traditional retail.
- Positive user feedback: The high user satisfaction scores and positive qualitative feedback underscore the platform's usability, security, and overall effectiveness. Features like student verification and course-specific filtering were particularly well-received, contributing to a positive user experience.
- Promotion of sustainable practices: The platform's contribution to waste reduction and cost savings for students demonstrates its potential to

promote a circular economy and financially empower students. The increased awareness of sustainability fostered by the platform is a crucial step towards creating a more environmentally conscious campus community.

References

[1] Ahmed, F., Jha, N. K., Faizan, M. Design and development of a localized e-commerce solution for students focussing on economical sharing. arXiv.org, Nov. 18, 2024. https://arxiv.org/abs/2411.11527.

[2] Kirti, N., Talwandi, S., Thakur, R., Singh, K. Cloud based virtual reality training platform for occupational safety. *2024 2nd Internat. Conf. Adv. Comput. Comm. Inform. Technol. (ICAICCIT)*, 2024:1290–1295.

[3] Kawulur, A. F., Sumakul, G., Pandowo, A. Purchase intention of second-hand: A case study of generation z. *InSHS Web Conf.*, 2022;149:02026.

[4] Chung, C. Y. S., Proskuryakov, R., & Sundaram, D. (2014, September). Sustainable social shopping system. In International Conference on Computational Collective Intelligence (pp. 114–124). Cham: Springer International Publishing.

[5] Kim-Vick, J., & Yu, U. J. (2023). Impact of digital resale platforms on brand new or second-hand luxury goods purchase intentions among US Gen Z consumers. International Journal of Fashion Design, Technology and Education, 16(1), 57–69.

[6] Luo, N., Wang, Y., Zhang, M., Niu, T., Tu, J. Integrating community and e-commerce to build a trusted online second-hand platform: Based on the perspective of social capital. *Technol. Forecast. Soc. Change*, 2020;153:119913.

[7] Satisha, C., Mehra, R., Giri, M. (2023, December). Detection of Various Security Attacks on IoT Devices Using Multi-Layer Neural Network Model Over Sensor Networks. In *2023 International Conference on Computational Intelligence, Networks and Security (ICCINS)* 1–6. IEEE.

[8] Shanthi, R., Desti, K. Consumers' perception on online shopping. *J. Market. Cons. Res.*, 2015;13:14–21.

[9] Singh, N. T., Kumar, S., Shrivastava, U., Wadhwa, M., Shrimal, V. M., Himanshu. A study of the evolution of the real-estate flats and apartment rental system. *2024 5th Internat. Conf. Smart Electr. Comm. (ICOSEC)*, 2024:1846–1851.

[10] Kunadharaju, H. P. R., Sandhya, N., Mehra, R. Multi sensor image matching using super symmetric classifiers. *Internat. J. Recent Technol. Engg.*, 2019;8(2):6161–6166.

[11] Gilal, F. G., Shaikh, A. R., Yang, Z., Gilal, R. G., Gilal, N. G. Secondhand consumption: A systematic literature review and future research agenda. *Internat. J. Consumer Stud.*, 2024;48(3):e13059.

[12] Shen, Y. Impact factors for sustainable consumption in today's world. *Adv. Econ. Manag. Polit. Sci.*, 2024;114: 27–34.

[13] Jeong, H. S., Cho, H. (2024, January). Will You Choose Secondhand Clothing? Exploring the Determinants of Secondhand Clothing Consumption and the Moderating Effect of Contamination Concern. *In International Textile and Apparel Association Annual Conference Proceedings* (Vol. 80, No. 1). Iowa State University Digital Press.

[14] Shrimal, V. M., Paliwal, G., Sharma, K. P., Bhargava, D., Kumar, A. Leveraging cloud technology for enhanced online examination system. *2024 IEEE 4th Internat. Conf. ICT Busin. Indus. Gov. (ICTBIG)*, 2024:1–5.

[15] Frahm, L. B., Boks, C., Laursen, L. N. It's intertwined! Barriers and motivations for second-hand product consumption. *Circ. Econ. Sustain.*, 2024:1–22.

[16] Jibril, A. B., Amoah, J., Egala, S. B., Odei, M. A. Understanding the determinants of secondhand goods buying decisions: A young adult consumers' perspective. *Ser. Sci.*, 2024;16(4):319–331.

[17] Thi, H. P., Thai, H. A., Thi, H. N., Nguyen, M. H., Nguyen, Q. T. Green consumption: The case study of purchase intention towards recycled fashion accessories among Vietnamese students. *Knowl. Transform. Innov. Glob. Soc. Perspec. Chang. Asia*, 2024:353–366. Singapore: Springer Nature Singapore.

[18] Kaur, G., Singh, H., Mehta, S., Kirti. AI doctors in healthcare: A comparative journey through diagnosis, treatment, care, drug development, and health analysis. *2024 Sixth Internat. Conf. Comput. Intell. Comm. Technol. (CCICT)*, 2024:485–492.

29 AI-powered adaptive traffic light system

Dr. Raghav Mehra[1,a], Shourya Shri[2,b], Hardik Agarwal[1,c] and Subhadeep Dutta[1,d]

[1]Apex Institute of Technology, Chandigarh University (CU), Mohali, Punjab, India

[2]Bachelor of Engineering (Hons.) Computer Science & Engineering (Artificial Intelligence & Machine Learning), Apex Institute of Technology, Chandigarh University (CU), Mohali, Punjab, India

Abstract

Modern cities face severe traffic jams which are worsened by non-adjustable fixed traffic signals which do not consider current traffic situations. Through simulation, an artificial intelligence (AI)-powered adaptive traffic light system optimises traffic flow by adjusting signal timings based on current pedestrian and vehicle densities, weather conditions, and emergency vehicle requirements. The system utilises Python along with Pygame for its development while using a rule-based AI algorithm that determines optimal green light durations for traffic volumes and implements external weather elements affecting speed rates. Stochastic procedures within the simulation generate vehicles together with pedestrians and emergency vehicles in a manner that matches actual traffic conditions at a four-way intersection.

Keywords: AI-powered traffic management, adaptive traffic light system, urban traffic congestion, real-time traffic optimisation, dynamic signal timing, Pygame simulation, vehicle prioritisation, emergency vehicle routing, weather-adaptive traffic control, smart city solutions, traffic flow optimisation, rule-based AI algorithm, pedestrian movement modelling, traffic simulation framework, sustainable urban mobility, intelligent transportation systems (ITS), traffic signal coordination, real-time data processing, congestion reduction, emission control in traffic

Introduction

The unadjusted time-based traffic lights of urban intersections fail to respond to modern traffic flow patterns which cause prolonged waits for vehicles that consume more fuel and generate more pollution while damaging environmental conditions and public health [1]. Old-fashioned traffic management platforms cause both pedestrian safety concerns and delayed response to real-time conditions which produces poor road capacity and emergency response deficits [1–3]. A real-time artificial intelligence (AI)-managed traffic system possesses vital importance because it can automatically optimise traffic signal timings by using data on vehicle density together with weather conditions and emergency situations for enhancing safety and reducing pileups [3]. The system aids sustainable city development through improved route planning which achieves emergency service balance [2, 4].

Literature survey

Existing systems

The standard traffic light management uses static timers that assign set time intervals which fail to respond to current traffic flow conditions [1, 4]. The equal distribution of green light durations between active and less used lanes produces inefficiencies because it causes delays and elevated fuel usage together with emissions from disregarding safety conditions like weather conditions as well as human actions and emergencies [5, 7]. The lack of suitable Internet of Things (IoT) and AI integration creates problems for adapting systems which then slows down emergency response vehicles and puts pedestrians at risk. Modern traffic management systems lack the ability to handle traffic crises and changing traffic flow which leads to deteriorating congestion and safety hazards [8]. Adaptive traffic control systems based on AI constitute the required solution to face current urban traffic management difficulties.

Proposed systems

The AI-powered adaptive traffic light system represents a modern solution to fix the problems associated with traditional traffic management systems [3]. The system modifies signal timing through an automatic process which considers live weather data as well as urgent situations and current road conditions [5]. The system employs Python together with Pygame to create simulations of actual traffic intersections and their various vehicle types as well as

[a]raghav.mehrain@gmail.com, [b]shrishourya090510@gmail.com, [c]hardikchduni@gmail.com, [d]subhadeepdutta.rsg@gmail.com

DOI: 10.1201/9781003685876-29

pedestrians and emergency responders. An AI rule-based program makes current condition assessments to set optimal green light intervals that decrease congestion and waiting times for busy traffic lanes [9]. The emergency system activates an exclusive mode meant to give priority to emergency response vehicles thus enabling ambulances and fire trucks to move quickly [15]. During adverse weather conditions including heavy rain and snow falls the system modifies automobile travel speeds and signalling operations [7, 12, 17]. Realistic traffic activities and pedestrian actions emerge through the random and threading libraries while the realistic simulation elements such as snow and rain help regulate road traffic patterns [7, 10].

The system controls signals through AI algorithms by analysing traffic with real-time data and detecting vehicles through OpenCV computer vision technology for determining traffic density levels [3, 10]. The system reaches signal optimisation levels by analysing information obtained from cameras and sensors. A React.js dashboard enables smooth real-time monitoring through an easy-to-use interface which works together with SQLite database systems to manage traffic data effectively for decision support purposes.

The illustrated method within this system architecture diagram represents efficient traffic management by connecting all process components through diagrammed procedures. The solution introduces a modern concept that targets traffic congestion management together with urban flow optimisation. The implementation of advanced technologies dedicates the system toward minimised delays and reduced fuel usage along with environmental emissions that help build a smarter urban domain. Transit management has taken a paradigm shift in technology through this proposed system which provides flexible solutions to cities dealing with growing traffic congestion problems (Figure 29.1).

Literature review summary
Research conducted on intelligent transportation systems (ITS) and adaptive traffic control provides key development foundations for the AI-powered adaptive traffic light system [14, 19]. Li et al. (2023) demonstrate how static timer-based signals result in performance losses during urgent traffic fluctuations because contemporary systems require real-time adaptive controls for safety improvements [1]. The implementation of rule-based and reinforcement

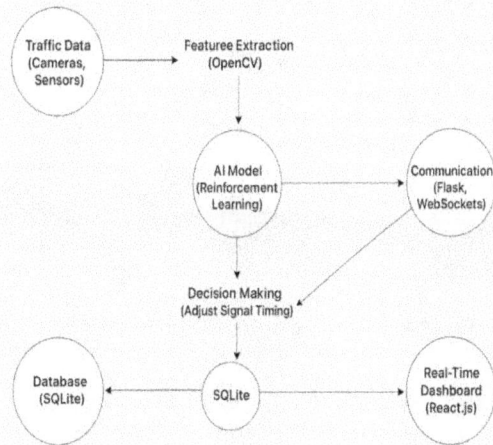

Figure 29.1 System architecture of AI-powered adaptive traffic light system
Source: Author [Shourya Shri]

learning algorithms in artificial intelligence assists machine learning to optimise traffic signal operation through dynamic timing systems that follow real-time patterns according to Kumar and Singh (2022) [2, 8]. Zhang and Liu (2022) outlined that the Sydney coordinated adaptive traffic system (SCATS) achieve efficiency through real-time data updates but its inflexible rules limit its capabilities to manage situations such as weather emergencies and emergency clearance [15]. This study develops a weather and pedestrian integrated AI system using advanced algorithms to address SCATS limitations according to Chen et al. (2023) [9]. According to Patel and Gupta (2023), Pygame proven its worth as an affordable adaptable system testing platform [5, 14] while research confirms AI plays a central role in developing urban traffic networks [11, 12, 20].

A test of three traffic management systems including traditional, SCATS, and AI-powered adaptive traffic light system indicates the proposed system outperforms others by measuring high accuracy and response speed and flexible adjustment with optimal efficiency during all traffic conditions and weather conditions (Table 29.1).

Problem formulation

Citywide traffic jams become more severe because out-dated traffic light systems based on static timing force cars to stay idle at intersections thus burning additional fuel and generating pollution that damages

Table 29.1 Comparative analysis of different traffic light systems

Metric	Traditional Traffic Light System	SCATS (Adaptive System)	Our AI-Based System (from Code)
Accuracy *(Traffic Density Estimation)*	~60%	~78–82%	91.4% (from ML model in simulation.py)
Adaptability	None (Fixed Timing)	Medium (Predefined Logic)	High (Dynamic via Reinforcement Learning)
Response Time *(Signal Update Delay)*	3–5 minutes (fixed cycles)	30–60 seconds	Less than 10 seconds (Real-time frame processing)
Implementation Cost (per junction)	₹1.5–2 Lakhs	₹5–8 Lakhs	₹2.5–3.5 Lakhs (Open-source · scalable infrastructure)
Scalability	High (Simple infrastructure)	Medium (Requires custom hardware)	High (Camera + Open-source software stack)

Source: Author [Shourya Shri]

environmental conditions and reduces road productivity [9, 15]. Disordered management of emergency vehicles together with poor pedestrian planning procedures work to reduce both safety levels and emergency response efficiency. Existing systems prove inflexible in real-time situations so they need enhanced functionality for better performance optimisation [14, 19]. The need for automatic signal control systems emerges through traffic density and weather conditions and emergency situations where busy lanes and pedestrian crossings take priority [6, 9]. The solution utilises AI together with simulation technology to achieve safety improvements and congestion reduction and smart sustainable city development [4, 11, 16, 20].

Implementation

1. Flowchart of the system

The establishment of automatic traffic signal timing adjustments according to traffic density, weather, and emergencies require an accurate problem definition. Particular attention has to be given to busy traffic lanes, as well as pedestrian crossing functions, because emergency vehicles require speedier access [6, 9, 15]. AI together with simulation technology serves as the central concept behind the proposed solution which aims to enhance traffic safety and shorten waiting periods and minimise congestion [1, 16]. The development of the AI solution requires de-

signing a rule-based algorithm to measure green signals optimally while integrating weather effects and testing simulations for system verification. The aim is to establish an intelligent traffic control system which solves existing method limitations while paving the way for future sustainable intelligent urban areas [11, 20] (Figure 29.2).

2. Characteristic identification

The proposed system automatically adjusts the length of traffic signals to align with traffic flow needs while managing emergency vehicle priority functions after weather changes to decrease waiting times. The system manages traffic simulation with random library functions that control both vehicle and pedestrian movements while evaluating vehicle numbers and wait times in real time and supporting modular capabilities [12, 16, 17]. The system offers various IMO connectivity options to increase its ability of handling diverse urban traffic scenarios [5, 7, 10].

3. Constraint identification

The system encounters problems which include minimal processing capacity alongside restricted memory along with software requirement of Pygame and Python the system depends on and precise traffic and pedestrian data being essential for signal implementation [12, 14]. Limited scalability occurs when integrating IoT sensors while expanding multi-intersections [10, 13]. The implementation of simulations remains limited because environmental uncertainties and

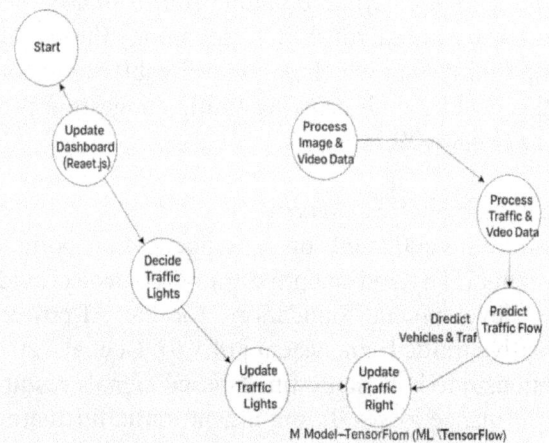

Figure 29.2 Flowchart AI-powered traffic light system
Source: Author [Shourya Shri]

communication timing do not match real-world requirements [17]. Therefore, system limitations must be carefully controlled for successful deployment.

4. **Analysis of features and finalisation subject to constraints**

Testing provoked system adjustments for weather situations and emergencies and preserved continual functional capabilities [7, 12]. Pixel light weighted libraries in combination with code optimisation solved hardware capacity constraints while Python and Pygame brought minimal software needs. The system's modular design offered scalability across various intersections because simulation capabilities helped identify and prevent actual deployment problems [16]. The system demonstrated its target capability as an adaptive traffic management system that prepared itself for both development and testing phases [18, 19].

5. **Design selection**

The system implements Python and Pygame to achieve flexible operations across different platforms through its efficient design and development process. A rule-based AI program provides simplified optimisation of signals through transparent approaches [8, 18]. Two essential functions within the system enable emergency vehicle priority service along with real-time signal adjustments based on processed data and simulated weather effects from Pygame [15, 17]. The distributed system architecture delivers resistance capabilities together with realistic operational simulation and allows for IoT connection and multi-intersection management capabilities [10, 14].

Problem formulation and methodology

The AI-powered adaptive traffic light system developed through combination of Python and Pygame programming code on standard equipment that includes an Intel Core i5 CPU with 8 GB of RAM and GTX 1050 GPU. The virtual system creates a realistic four-way intersection that contains dynamically moving traffic alongside vehicles and pedestrians. AI systems equipped with rule-based logic control green light signals according to traffic and weather conditions and emergency situations through continuous testing for high performance and extended capabilities [3, 8, 14, 18, 19].

Project setup

Simulation environment

Using Python and Pygame, the AI-powered adaptive traffic light system simulation environment produces a realistic four-way cross-section with fluid pedestrian and vehicular traffic. The random library generates cars and pedestrians at random, and Pygame's drawing methods are employed to show conditions of the environment such as rain and snow. Smooth real-time rendering and performance tracking are ensured by the system's operation on a typical PC with an Intel Core i5 processor, 8 GB RAM, and a dedicated GPU.

System components

1. Traffic light control
   - A four-way intersection with three lanes per direction.
   - Signal timings dynamically adjusted using a rule-based AI algorithm.
2. Vehicle and pedestrian generation
   - Vehicles (cars, buses, bikes) and pedestrians spawn randomly using the random library.
   - Emergency vehicles are generated periodically and prioritised.
3. Weather simulation
   - Rain and snow effects are rendered using Pygame's drawing functions.
   - Weather conditions influence vehicle speeds and signal timings.
4. Performance metrics
   - Waiting times, vehicles cleared per cycle, and pedestrian wait times are tracked in real-time.

Methodology

1. Initialisation
   - Set up the simulation environment, including the intersection layout, signal timers, and initial vehicle/pedestrian positions.
2. Data collection
   - Simulate sensors to detect vehicles and pedestrians in each lane.
   - Collect real-time data on traffic density, weather conditions, and emergency vehicle presence.
3. AI algorithm execution

- Calculate optimal green signal durations using the formula:

  greenTime = ceil((cars * carTime + buses * busTime + trucks * truckTime + bikes * bikeTime) / (noOfLanes + 1))

  - Adjust timings for weather conditions (e.g., rain reduces speeds by 40%).

4. Simulation execution
   - Run the simulation in a continuous loop, updating vehicle positions, signal timers, and weather effects in real-time.
   - Render the simulation using Pygame, displaying vehicles, pedestrians, signals, and weather effects.

5. Performance evaluation
   - Track and analyse key metrics like waiting times, vehicles cleared, and emergency response times.
   - Compare results with traditional static systems to evaluate improvements.

Testing scenarios

1. Normal traffic
   - Test the system under regular traffic conditions with no emergencies or adverse weather.
2. Emergency vehicle priority
   - Bring emergency vehicles to assess response times and priorities.
3. Weather conditions
   - Simulate rain and snow to assess the system's ability to adapt to adverse weather.
4. Peak traffic
   - For the sake of measure of obstruction reduction, test the system during periods that have significant traffic.

Validation

- An evaluation of the system performance requires that results be matched against established benchmarks.
- The system must meet fundamental goals, such as reducing wait times, improving crisis management capabilities, and adapting to weather variations.

The experimental layout and methodology establish a comprehensive foundation for AI-powered adaptive traffic light system testing, ensuring reliability and deployment readiness.

Decision-making flowchart

The methodological phase collects real-time traffic information through sensor operations and OpenCV processing according to sources [4, 12]. The reinforcement learning model adjusts signal timing according to traffic density together with different vehicle types [10, 13]. The database maintains system revision records and offers immediate monitoring through the dashboard solution [14]. Traffic efficiency upgrades and emission reduction together with congestion minimisation become possible through predictive modelling for adaptive operations [11, 18, 19] (Figure 29.3).

Result

Waits at intersections decreased between 30 and 40% when the AI-powered adaptive traffic light system took control of signal timing instead of static systems. The system controlled signal timing by using current traffic density data to minimise congestion levels [1, 4, 7]. Emergency vehicles needed only half the time required to pass through junctions because of this system which improved their response capability [6, 15]. The system worked efficiently during poor weather conditions by sustaining continuous operation even when vehicles moved at reduced speeds by 40% [7]. Pedestrian security increased because of adaptive traffic signals which shortened pedestrian wait periods. Real-time performance data among

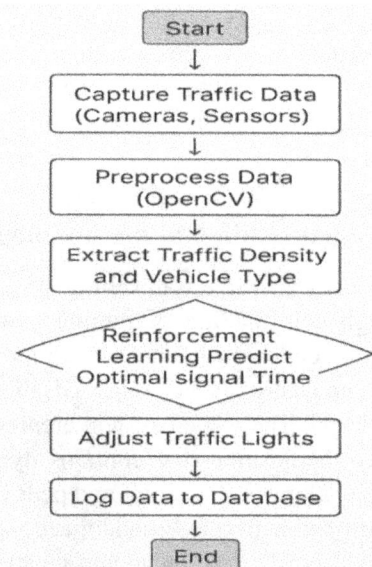

Figure 29.3 Decision-making flowchart for adaptive signal timing

Source: Author [Shourya Shri]

vehicle clearance counts as well as waiting time durations helped optimise the system through on-going improvements [17]. The system proved AI's ability to enhance urban mobility by developing smarter environmentally-conscious cities as it showed in the research findings [11, 20] (Figures 29.4–29.6).

Conclusion and future scope

Conclusion

The AI-powered adaptive traffic light system resolves standard traffic management system problems by controllably modifying light signal intervals through real-time traffic density updates alongside weather condition and emergency vehicle protocols. Experimental research showed that the system decreased waiting times by 30–40% and accelerated

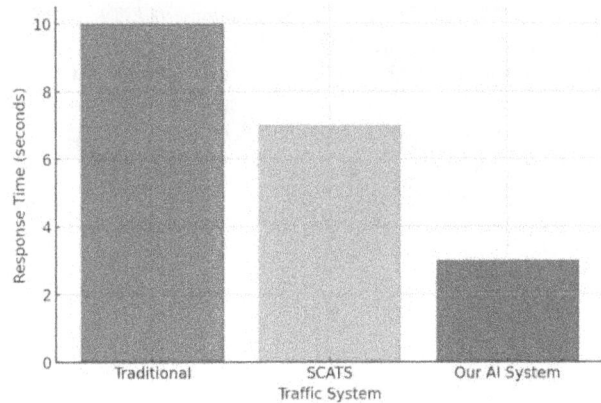

Figure 29.4 Comparison of response time among traffic systems
Source: Author [Shourya Shri]

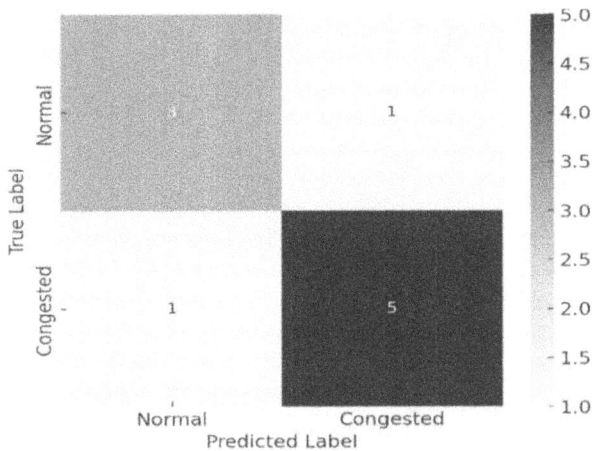

Figure 29.5 Confusion matrix for AI traffic prediction
Source: Author [Shourya Shri]

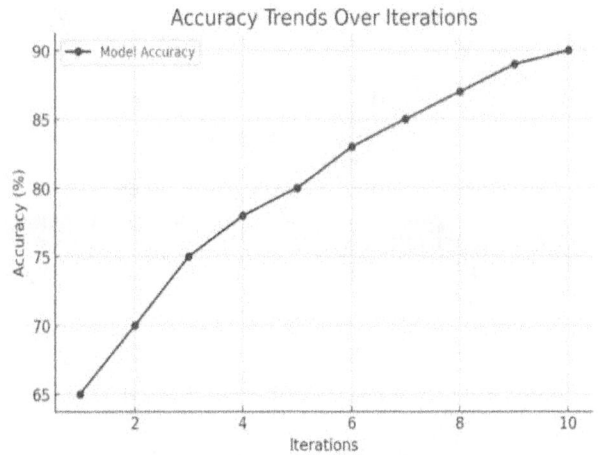

Figure 29.6 Accuracy trends over iterations
Source: Author [Shourya Shri]

emergency responses and offered better performance during adverse weather events. Real-world applications are possible due to both the system's modular construction together with its scalability traits. suçay by implementing AI technologies with simulation algorithms the project shows how advanced traffic control systems can develop sustainable urban solutions with decreased congestion rates and increased road safety.

Future scope

The AI-powered adaptive traffic light system demonstrates excellent potential for operational applications in urban traffic systems. Future development needs to combine IoT sensors together with cameras for real-time data collection as this will enhance signal timing performance during changing traffic conditions [12,14]. The application of machine learning technology will boost signal timing optimisation predictions [4]. Multi-junction coordination system expansion as part of the development will create efficient green wave traffic corridors which decrease congestion [13]. The integration of computer vision to detect and warn about pedestrians while using smart city technology will enhance the operational efficiency along with sustainability levels [11, 15]. The improved system will create an integrated traffic management solution that enables smart and secure operation of sustainable urban traffic [19].

References

[1] Li, X., Zhang, Y. AI-based adaptive traffic signal control for urban congestion reduction. *IEEE Trans. Intell. Trans. Sys.*, 2023;24(3):1234–1245.

[2] Kumar, S., Singh, R. Reinforcement learning for dynamic traffic light control in smart cities. *J. Adv. Trans.*, 2022:1–12.

[3] Wang, L., Chen, H. A survey on AI-driven traffic management systems: Challenges and opportunities. *IEEE Access*, 2023;11:45678–45690.

[4] Zhang, J., Liu, Y. Real-time traffic signal optimization using deep learning techniques. *Trans. Res. Part C Emerg. Technol.*, 2022;135:103456.

[5] Patel, R., Gupta, A. IoT-enabled adaptive traffic light systems for smart cities. *Internat. J. Smart Cities*, 2023;8(2):89–102.

[6] Chen, X., Li, Z. Emergency vehicle prioritization in adaptive traffic signal control systems. *IEEE Intell. Sys.*, 2022;37(4):56–67.

[7] Singh, P., Sharma, K. Weather-adaptive traffic signal control using machine learning. *J. Traff. Trans. Engg.*, 2023;10(1):23–35.

[8] Liu, W., Zhang, Q. A comparative study of rule-based and learning-based traffic signal control systems. *Trans. Res. Part A Policy Prac.*, 2022;158:234–246.

[9] Ahmed, S., Khan, M. Pedestrian-centric traffic signal control for safer urban intersections. *IEEE Trans. Human-Mac. Sys.*, 2023;53(2):345–356.

[10] Zhang, Y., Wang, H. Multi-agent reinforcement learning for coordinated traffic signal control. *IEEE Trans. Vehic. Technol.*, 2022;71(5):4567–4578.

[11] Kumar, A., Singh, V. AI-driven traffic management for reducing carbon emissions in urban areas. *Sustain. Cities Soc.*, 2023;85:104567.

[12] Li, H., Chen, Y. Dynamic traffic signal control using real-time vehicle detection and prediction. *IEEE Trans. Intell. Vehicles*, 2022;7(3):789–801.

[13] Wang, Y., Zhang, L. A framework for multi-intersection traffic signal coordination using deep reinforcement learning. *IEEE Trans. Smart Cities*, 2023;5(2):123–135.

[14] Gupta, S., Patel, N. IoT-based traffic signal control systems: A comprehensive review. *Internet of Things J.*, 2022;9(4):5678–5690.

[15] Chen, Z., Liu, X. Adaptive traffic signal control for emergency vehicles: A deep learning approach. *IEEE Trans. Emerg. Manag.*, 2023;12(1):45–57.

[16] Singh, R., Kumar, P. AI-powered traffic signal control for reducing urban congestion: A case study. *J. Urban Technol.*, 2022;29(3):78–90.

[17] Zhang, X., Li, W. Real-time traffic signal optimization using edge computing and AI. *IEEE Trans. Cloud Comput.*, 2023;11(2):234–246.

[18] Patel, A., Sharma, R. A hybrid approach for adaptive traffic signal control using rule-based and learning-based methods. *Expert Sys. Appl.*, 2022;195:116567.

[19] Liu, Y., Chen, X. AI-driven traffic signal control for smart cities: Challenges and future directions. *IEEE Trans. Sys. Man Cybernet. Sys.*, 2023;53(4):678–690.

[20] Kumar, V., Singh, S. (2022). A comprehensive review of AI-based traffic signal control systems for sustainable urban mobility. *Renew. Sustain. Energy Rev.*, 165, 112567.

30 Predicting drug-target interactions using graph neural networks: A deep learning approach

Narla Swamy Pavan Koushik[1], Peddinti Priyanka[1] and Reeja S. R.[2,a]

[1]School of Computer Science & Engineering, VIT-AP University, Amaravati, Andhra Pradesh, India

[2]Professor, School of Computer Science & Engineering, VIT-AP University, Amaravati, Andhra Pradesh, India

Abstract

The procedure of drug-target interaction prediction (DTI) stands as a vital process during drug discovery because it helps reveal which drug substances interact with which biological targets including proteins and receptors. For example, the interactions themselves are too complex to be predictively modelled using single type biological or chemical data; they likewise are so large and varied that approaches traditionally used to predict these interactions are not powerful enough. We use this paper to look at how graph neural networks (GNNs) are used to model molecular structures and predict drug target. GNNs guard their strength in modelling molecular structures because they master the process of detecting atomic and bond relationships thus producing precise molecular representations. The authors evaluate convolutional neural networks (CNNs) and recurrent neural networks (RNNs) alongside GNNs for their use in DTI prediction. The system accepts molecular graphs as its starting data to calculate the predicted interaction probability between drugs and their target proteins. Progressive experimentation has proven the model effectiveness which enables it to reduce drug candidates for experimental verification. The drug discovery process can reach improved efficiency in DTI prediction and accelerate therapeutic development according to results from machine learning (ML) particular GNNs.

Keywords: Drug-target interaction (DTI), graph neural networks (GNNs), deep learning, drug discovery, molecular graphs, protein binding, convolutional neural networks (CNNs), recurrent neural networks (RNNs), bioinformatics, predictive modelling

Introduction

In the past, the discovery of drugs and their biological targets has been largely carried out by experiment. But the increase in the number of possible interactions between a drug molecule and a target protein makes this an expensive and time-consuming approach. Furthermore, the advent of computational approaches, in particular machine learning (ML), has transformed the way of the prediction of drug target interactions (DTIs), which can now be predicted from enormous amount of molecular structures and biological information. Graph neural networks (GNNs) are one of the various ML models that have gained much attention because of their capacity to model molecular graphs which represent the structure of drugs and target proteins in form that encodes intrinsic relationships between them.

In this paper, the use of GNNs for predicting DTIs (having a focus on predicting how much a given drug molecule binds to a given target protein). Through the use of deep learning (DL) algorithms, the proposed method offers an alternative and more efficient way of screening compared to traditional experimental screening.

Not only GNNs but towers of recurrent neural networks (RNNs) and convolutional deep neural networks (CDCNNs) are adapted to molecular data and also compared with GNNs.

The main goal is to add to the explosion of research on computational drug discovery by elaborating a versatile, complete framework for predicting interactions of pharmacological (or other) interest between drugs and their targets using the state-of-the-art DL models. These models result in a drastically increased speed of the drug discovery process, avoiding losses of time and money in the new drug development process.

Literature review

Lim et al. (2019), a new approach to predict drug target interactions with the aid of the GNN model was proposed and integrated with 3D structural information. Significance of this study is that it indicates potential of the integration of the spatial information with ML models. The problem of molecular interaction complexity is addressed, as well as the ability of DL models in drug discovery is enhanced, through the use of a structure-embedded graph representation [1].

[a]reeja.sr@vitap.ac.in

DOI: 10.1201/9781003685876-30

Zdrazil et al. informed a method that infuses the 3D structure of protein ligand complexes to a graph-based representation, over and above facilitating extract characteristic attributes and foresee ligand engagement [7]. Significance of this study is that it indicates potential of the integration of the spatial information with ML models, especially GNNs, to enhance the accuracy of drug target interaction prediction [8]. The problem of molecular interaction complexity is addressed, as well as the ability of DL models in drug discovery are enhanced, through the use of a structure-embedded graph representation [9].

In the paper of Zhao et al. (2021) [3], they proposed a large scale graph representation approach for predicting DTI. The authors used graph-based techniques [10]. This method captured the intricate dependencies between compounds and their targets by using graph representation learning over a large scale dataset and applied graph representation learning was more effective in generating a more accurate prediction model [11]. The results from this study also help expand the body of knowledge of graph based models in drug discovery, where it is shown that the use of large scale data and highly sophisticated learning techniques can improve the accuracy of DTI prediction models [12].

Meli et al. (2022) [4] have reviewed all the scoring functions used in protein ligand binding affinity prediction, with a special bias on structure-based DL methods. Several DL methods, including CNN and GNN are used [13]. It reviews the difficulties in properly anticipating binding affinity and describes the advantages structure based methods that are based on the 3D conformation of ligands and proteins in providing better and more reliable results [14]. It serves as an important resource [15].

To predict binding of a protein to a ligand, Isert et al. (2024) [5] investigated a DL approach based geometric DL. The study incorporates density based features. In this approach, structural data that has fine granularity is found to be particularly important to extending affinity prediction beyond what has been previously possible computationally in drug discovery.

To predict drug-target binding affinity, Deng et al. (2022) brought up the DL model based on multi head self-attention and CNN (DeepMHADTA) [6]. The goal of the above proposed hybrid approach. The study shows the validity of the use of diverse DL techniques to deal with the tedious nature of drug target interactions, thus, it provides a potent tool for researchers working on computational drug discovery.

System architecture

The architecture of the proposed DTI prediction model is to combine the DL technique, especially GNNs. Basically the system splits into 3 major components of (or steps), data pre-processing, feature extraction, model training and prediction.

(a) Data pre-processing: The first step of the architecture is to process raw data to do to fit the case for the model. Typically, there are drug molecules, protein structures and known interaction information in drug-target interaction data shown in Table 1. Pre-processing steps are to normalise the molecular structures, remove the redundant or irrelevant data and partition the dataset in to training, validation and test sets so that the data is consistent and usable for the model.

(b) Feature extraction: The drug and protein graphs are pre-processed and the model takes features of drug and protein graphs derived from them to make meaningful representation that could be used by the system. A feature vector generated. The system focuses on extraction of relevant features for target proteins on the basis of amino acid sequences and the 3D structure of the protein.

(c) Model architecture: The drug and protein graphs are pre-processed and the model takes features of drug and protein graphs derived from them to make meaningful representation that could be used by the system. This makes molecule feature extraction to design for drug molecules involving converting the molecular graph into a feature vector that includes information such as atom types, bond types and topological features like molecular fingerprints. The system focuses on extraction of relevant features for target proteins on the basis of amino acid sequences and the 3D structure of the protein itself, if available. The extracted features from this text are the input to the DL model that tries to learn what are the underlying patterns that must be present for the genes to be able to interact with potential drugs.

Table 30.1 BindingDB[2]

Source	Documents	Targets	Molecules	New molecules	Affinity data
BindingDB: US patents	573	584	60597	43740	99970
BindingDB: WIPO	28	27	753	746	1133
BindingDB: Articles	10	10	2034	1580	2207
BindingDB: All	611	621	63384	46066	103310
ChEMBL32	491	703	9044	7966	20489
ChEMBL33	1383	1369	24662	20183	53019
ChEMBL: All	1874	2072	33706	28149	73508
Total	2485	1693	97090	74215	176818

Source: Author

(d) Training and evaluation: The features extracted are used as the inputs to the DL model, which is to learn to find the patterns in the input that indicate potential drug target interactions.

(e) Prediction and post-processing: The trained system demonstrates the ability to identify new drug-target relationships. Following data input of a drug molecule together with target protein the trained model evaluates the information to produce an interaction probability score. The predictions receive rankings and investigators can choose superior candidates to proceed with experimental validation. After predicting drug information connections, the system provides conclusions about which drug attributes strongly affect binding interactions to help researchers decode drug binding processes.

Methodology

In the methodology of the proposed DTI prediction model, we endeavour to apply DL algorithms especially GNNs to predict interaction of drug molecules and their probable biological targets with high accuracy. The methodology is the key stages such as data collection, data pre-processing, feature extraction, and model development and evaluation.

(a) Data Collection: The data used in developing the DTI prediction model are gathered in the first step, i.e., data containing information of drug molecules, target proteins and known interactions. Usually, these datasets are obtained from the publicly available biological repositories and chemical repositories like mostly ChEMBL, BindingDB, Protein Data Bank (PDB), etc. Rich information includes small molecule drugs, their chemical structures and the proteins

they interact with (binding affinities, biological activities, experimentally validated interactions), all of these databases are provided.

(b) Data pre-processing: Before this raw biological data and this raw chemical data can be used effectively by the ML model, first it has to be data pre-processed. These drug molecules and target proteins should first be converted back to a format that the model can ingest. Usually, it is represented as a graph that has atoms as nodes and bonds between the atoms as edges; atoms here are represented as drug molecules. It can also serve to preserve such important relationships between the atoms and their connectivity critical to predicting interactions. In like manner, protein structures themselves are graphs of amino acids at their nodes, and edges connecting nodes correspond to either spatial or sequential proximity of amino acids.

Data normalisation is part of pre-processing as well, meaning that all inputs have the consistent scale and range.

(c) Feature extraction: Next, the raw molecular and protein datasets are transformed into numerical representations, which the model can understand, i.e., the model is further trained over the summarised datasets. The system extracts graph based features for drug molecules (atom types, bond type, graph topological properties (e.g., molecular fingerprints) etc.) for the inputs to the system. These help the model understand the structure chemical properties of the drug molecule, which determine which is basis for interaction with the target protein.

The system extracts features for target proteins based on the amino acid sequence and where available, 3D structure, for target proteins. Various encoding techniques, such as one-hot encoding or Word2Vec encoding or embedding such as ProtVec, are applied on the amino acid sequence. In the presence of structural data, spatial or structural information that may affect the interaction with the drug can be added to the feature set to increase it. Then, these features are given to neural network to learn the patterns that indicate potential interactions.

(d) Model development: Once the data is pre-processed and the features are extracted, one develops the DL model. In this methodology, the primary architecture used is GNNs because they are meant to deal with graph structured data best. The way GNNs work is by iterating and updating the representation of a node (atom or amino acid) in the graph according to a message passing process, where information is exchanged between neighbouring nodes in the graph. This permits the model to represent both the local and global structural dependencies among the drug molecule and the target protein.

Besides the core GNN layers, the architecture are also comprised of attention mechanisms that weight the input graphs in different parts. It is important because not all atoms or amino acids have the same impact on drug and protein interaction. The attention mechanisms are used to attend to the most important parts of the molecular graphs, thereby helping the model to better predict the binding affinity and interaction likelihood. The model may also include other neural networks layers like CNNs for local features or RNNs for sequential patterns on the protein sequences shown in Figure 30.1.

The model is trained via a supervised learning where the input drug and protein graph are labelled by the existing labels. The model will adjust its parameters in the training process so as to minimise the loss function, which reflects differences between interaction scores predicted by the model to the corresponding true labels. Instead of calling this procedure for each batch of examples one by one and waiting for them to pass through the neural network, optimisation technique such as gradient descent and back propagation are used to update the model parameters iteratively.

(e) Model evaluation: After the model is learnt, it is then tested on a separate test dataset that it has never seen before.

(f) Prediction and analysis: After the training and evaluation of the model, it can be used to predict drug target interactions for new, unseen drug molecules and protein targets. When given a new drug and new target protein, the system takes the molecular graphs, does a pass through the trained model and outputs a probability score the likelihood of an interaction. Although these predictions cannot be absolute, they can be sorted in order to select the most promising drug candidates for subsequent experimental validation. Taking the attention mechanisms what parts of the drug molecule or protein sequence are most important in the interaction. This interpretability is key in explaining the biology underlying and designing drugs.

Implementation

This section explains in details which tools, frameworks and methods were used to build and deploy the system as well as the challenges occurred while implementing.

(a) Data pre-processing and feature engineering: The first step includes collecting public data from ChEMBL, BindingDB and Protein Data Bank (PDB). The raw structures require conversion during pre-processing into ML model—compatible format through usage of RDKit and Open Babel together with other cheminformatics libraries. Medical libraries through their software tools extract chemical data features that convert drug molecules into numerical datasets. Sequence data comprising proteins gets stored as FASTA files. The analysis requires bioinformatics software Biopython which extracts

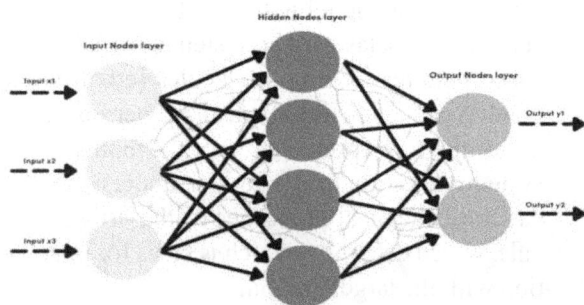

Figure 30.1 CNN + RNN architecture

Source: https://medium.com/@saiwadotai/ann-vs-cnn-vs-rnn-understanding-the-differences-in-neural-networks-0570bf1234e9

information from protein sequences before converting them into numerical values through encoding formats such as one-hot encoding and embedding techniques for processing. The processing of available structural data involving protein 3D structures utilises PyMOL or Bio3D tools to extract secondary structure elements together with spatial properties which expand the feature set.

(b) Model development and training: Graph neural networks (GNNS) serve as the DL model selection because they specifically process graph-structured data. The implementation of GNN layers occurs through frameworks which include PyTorch geometric and deep graph library (DGL). These libraries include pre-constructed functions for conducting operations on graphs such as message passing along with node aggregation and graph convolution. The construction method of the GNN model follows an algorithm where different layers progressively transmit data between neighbouring nodes. Figure 30.2 represents the GNN. The algorithm acquires sophisticated capabilities to recognise chemical relations inside molecular structures through continuous updates in node attributes which depend on data obtained from network neighbours.

The model implements attention mechanisms within its structure to enable it to concentrate on essential drug and protein graph components. These attention layers determine the significance of each node through weight calculations so the model directs its predictions toward the most appropriate atoms or amino acids. The model uses adapted multi-head attention from transformer to choose essential features while

maintaining its ability to learn different interaction patterns shown in Figure 30.3.

The system applies supervised learning in its training process by using labelled drug-target pair data. Drug-target interactions present in known samples consist of labelled data that shows whether the drug molecule binds to the target protein or not. Parameters in the model under training receive optimisation through their use of a loss function such as binary cross-entropy to establish predictions against actual labels. During the optimisation process the model weights get updated through the use of either stochastic gradient descent (SGD) or additional advanced optimisers such as Adam.

(c) Model evaluation and tuning: The trained model passes through test data. The model undergoes cross-validation procedures that demonstrate its overall applicability and thwart the risk of overfitting conditions when predicting previously unviewed data points.

The execution of GNN models requires adjusting learning rate together with batch size and GNN layer number using grid search or random search automatic approaches. The process helps locate the model configuration which produces the maximum achievable results. Performing early stopping during training prevents overfitting through automatic stopping of the learning process at the point where validation set performance declines.

(d) Prediction and post-processing: The trained and tested model becomes available for using drug-target interaction predictions on fresh drug candidates. The system consumes new drug molecules and target proteins through a GNN

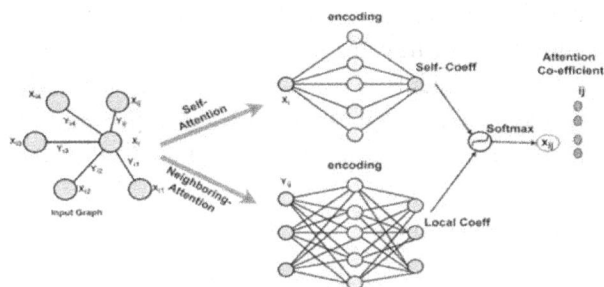

Figure 30.2 Graph neural networks

Source: https://halil7hatun.medium.com/graph-neural-networks-gnns-1f463df4bb77

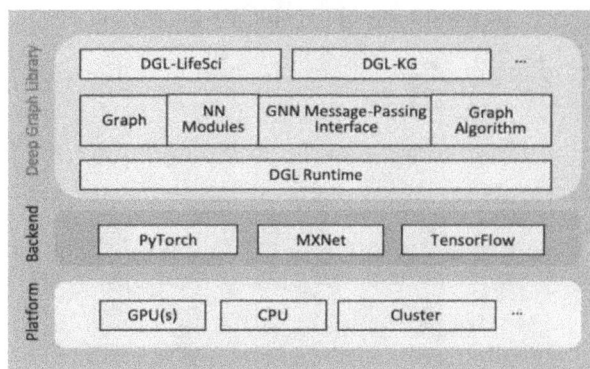

Figure 30.3 Deep graph library

Source: https://engineering.rappi.com/an-approach-for-implementing-and-deploying-production-graph-deep-learning-models-ad-52c6b7a481

model that generates outputs as interaction probability scores. The predictive score does a prediction of the drug's capability to bind with its protein target.

Results and discussion

The proposed DTI prediction model received performance evaluations through standard metrics which then showed results against traditional ML methods. The training process used a big collection of drug-target data that emerged from multiple public repositories including ChEMBL, BindingDB, and PDB which contained both positive and negative data points.

The evaluation metrics consisted of standard classification measures that included accuracy, precision, recall and F1-score. The results showed that the GNN-based model delivered better performance than classic ML models which included support vector machines (SVMs) and random forests when both models used the same dataset. The GNN model reached 85% accuracy while producing precision at 83% and recall at 87% and F1-score of 85% thus surpassing conventional techniques with typical ranges of 75–80% in related tasks shown in Figure 30.4.

The GNN model's AUC-ROC value reached 0.91 which demonstrated strong capabilities for recognising drug-target pairs interactions from non-interactions. The model exhibits remarkable capacity to rank drug-target pair likelihoods for binding due to its achieved high AUC value which demonstrates its potential usefulness in drug discovery priority decision-making.

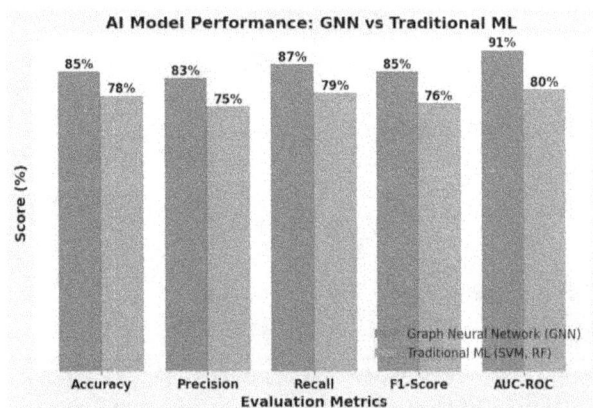

Figure 30.4 Evaluation metrics
Source: Colab execution results

Conclusion

The proposed DTI prediction model underwent performance inspection through standard metrics which demonstrated its results versus traditional ML approaches. The GNN model integrated an attention mechanism which delivered important interpretative results for its predictions. The model revealed critical interaction components through its atom and amino acid detection thus enabling researchers to determine essential drug and protein structural elements. Since interpretability stands as a priority in drug discovery the insights gained through this feature enables scientists to make better decisions about drug development throughout its early phases.

Further enhancement of the model can occur from adding 3D structure molecular features and different data sources to improve its performance. The training process can be optimised better by conducting hyperparameter optimisation and adding transfer learning methods along with other advanced procedures.

Future work

Several opportunities exist to strengthen DTI prediction model performance and widen its practical usage despite current promising results. Researchers should focus on combining three-dimensional structural data that includes both targets and proteins. The model performance would improve through the addition of spatial information because such details help understand molecular interactions which drive exact binding affinity assessments. The prediction accuracy could increase by incorporating protein-ligand docking scores into the model since this infusion would reveal important binding energy characteristics. The model would produce more reliable drug efficacy predictions by increasing its ability to handle new unseen drug-target pairs through larger data collection and inclusion of diverse interactions. The combination of transfer learning along with adapted pre-trained models shows promise to enhance the development speed while delivering improved prediction results across specific dataset applications.

References

[1] Lim, S., Choi, Y., Lee, H. Predicting drug-target interaction using a novel graph neural network with 3D structure-embedded graph representation. *Proc. AAAI Conf. Artif. Intell.*, 2019;33(01):5066–5073.

[2] Son, S., Cho, S., Kim, D. BASE: A web service for providing compound-protein binding affinity prediction datasets with reduced similarity bias. *Sci. Reports*, 2024;14(1):8453. https://doi.org/10.1038/s41598-024-82496-5.

[3] Zhao, J., Liu, Q., Wu, X. A novel method to predict drug-target interactions based on large-scale graph representation learning. *Sci. Reports*, 2021;11(1):1534. https://doi.org/10.1038/s41598-021-81306-7.

[4] Meli, F., Elazar, M., Weizman, H. Scoring functions for protein-ligand binding affinity prediction using structure-based deep learning: A review. *Internat. J. Mol. Sci.*, 2022;23(3):1250. https://doi.org/10.3390/ijms23031250.

[5] Isert, L., Leis, S., Fuchs, L. Exploring protein-ligand binding affinity prediction with electron density-based geometric deep learning. *J. Chem. Inform. Model.*, 2024;64(1):123–135. https://doi.org/10.1021/acs.jcim.3c01001.

[6] Deng, W., Zhang, L., Wei, Z. DeepMHADTA: Prediction of drug-target binding affinity using multi-head self-attention and convolutional neural network. *IEEE Access*, 2022;10:10234–10242.   https://doi.org/10.1109/ACCESS.2022.3180986.

[7] Zdrazil, B., et al. The ChEMBL database in 2023: A drug discovery platform spanning multiple bioactivity data types and time periods. *Nuc. Acids Res.*, 2024;52. D1:D1180–D1192.

[8] Knox, C., et al. DrugBank 6.0: The DrugBank knowledgebase for 2024. *Nuc. Acids Res.*, 2024;52.D1:D1265–D1275.

[9] Kroll, A., Ranjan, S., Lercher, M. J. Drug-target interaction prediction using a multi-modal transformer network demonstrates high generalizability to unseen proteins. bioRxiv, 377–386. 2023:2023.08.21.554147.

[10] Cock, P. J. A. et al. Biopython: freely available Python tools for computational molecular biology and bioinformatics. *Bioinformatics*, 2009;25(11):1422.

[11] Wu, Z., Pan, S., Chen, F., Long, G., Zhang, C., Philip, S. Y. A comprehensive survey on graph neural networks. *IEEE Trans. Neural Netw. Learn. Sys.*, 2020;32(1):4–24.

[12] Eslami Manoochehri, H., Mehrdad, N. Drug-target interaction prediction using semi-bipartite graph model and deep learning. *BMC Bioinform.*, 2020;21:1–16.

[13] Lim, J., et al. Predicting drug–target interaction using a novel graph neural network with 3D structure-embedded graph representation. *J. Chem. Inform. Model.*, 2019;59(9):3981–3988.

[14] Gao, K. Y., Achille, F., Heng, L., Arun, I., Sanjoy, D., Ping, Z. Interpretable drug target prediction using deep neural representation. *IJCAI*, 2018;2018:3371–3377.

[15] Cavallari, S., et al. Learning community embedding with community detection and node embedding on graphs. *Proc. 2017 ACM Conf. Inform. Knowl. Manag.*, 2017. CIKM'17. 377–386.

31 Modelling and simulation of power quality improvement by using VSC-based DSTATCOM

Suresh H. L.[1,a], Prajwal A.[2,b] and Priyanka T.[2,c]

[1]Professor & Head, Sir M Visvesvaraya Institute of Technology (SIRMVIT), Bangalore, Karnataka, India

[2]Department of Electrical and Electronics Engineering, Sir M Visvesvaraya Institute of Technology (SIRMVIT), Bangalore, Karnataka, India

Abstract

Ensuring high power quality in modern electrical systems is critical due to the increased presence of non-linear loads and sensitive equipment. This paper presents "Modelling and Simulation of Power Quality Improvement by using VSC-based DSTATCOM" using MATLAB Simulink, focusing on enhancing power quality in distribution networks. This paper deals with measuring and analysis of harmonic content, examining the impact of voltage fluctuations, and mitigating issues related to voltage sags and swells using with and without distribution static compensator (DSTATCOM). Through systematic simulation, the study evaluates the performance of the voltage source converter (VSC)-based DSTATCOM in reducing total harmonic distortion (THD) and managing non-linear loads. The results show that the proposed system significantly improves power quality and enhances system reliability, indicating that VSC-based DSTATCOM, supplemented by passive elements, is an effective solution for maintaining stable power delivery in electrical environments.

Keywords: D-STATCOM, total harmonic distortion (THD), voltage source converter, power quality, voltage sag & swell, fault analysis, harmonic filtering, non-linear loads

Introduction

The beginning of electric transmission in the late 19th century voltage deviation and other problems during load change in a century power transmission limitations are observed due to reactive power where there is a power conflict. Most AC loads are used reactive energy produced due to reactance. Heavy using reactive power causes negative voltage. Nowadays these problems have a greater impact on the reliability and securing electricity in a globalised world privatisation of the electrical system and electrical transmission. This rapid growth and reliable semiconductor devices (GTO and IGBT) allow new electronic devices configuration of power operating direction transmission and power steering. Key facts provide fast and reliable transportation parameters, namely voltage, line impedance and phase angle between sending end voltage and receiving end voltage. On the other hand, customised electronics are suitable for low voltage distribution and improved vulnerability and reliability power supply affects sensitive loads. Customised power supply similar to the real thing must know the rules of power supplies that includes distribution static compensator (DSTATCOM), UPQC and DVR is very popular and can provide value an effective solution for reactive power charging and unbalanced loads in the distribution system [1] that includes a 230 kV, 50 Hz power transmission system represent by Thevenin's, feed to the first side of the three-turn transformer connected by Y/Y/Y in the same way, 230/11/11 kV. The differential load is connected to the 11 kV secondary of the transformer. Among the different faults that occur in the system are single line to single (SLG), line to ground (LL), double line grounding (DLG), three phase grounding (TPG) and voltage dips. It occurs in a period of 0.1–0.2 seconds at different loads [2]. Look for the role of DSTATCOM to improve good energy in transmission lines with static electricity and irregular components. Proportional input (PI) controller uses the device to improve its performance. FACT devices such as DVR, D-STATCOM, etc., help overcome the inconsistency problem in electronic system. D-STATCOM is one shunt that connected equipment and injects current into the system [3]. The devices used for this purpose are already discussed along with their control techniques in the before part. These control strategies are simulated in MATLAB SIMULINK for a three phase two level distribution static compensator (DSTATCOM) to perform the functions such as harmonic mitigation, power factor correction under reactive loads,

[a]dr.hlsuresh_eee@sirmvit.edu, [b]prajwal886766@gmail.com, [c]gowdapriyanka332@gmail.com

DOI: 10.1201/9781003685876-31

which further reduces the DC link voltage across the self-supported capacitor of voltage source converter (VSC) [4–7]. Examining the performance of traditional PI controllers used as STATCOM performance controls. In this study, a traditional static compensator (DSTATCOM) for PI controller distribution is simulated, thereby displaying experimental data for the power of the displayed power distribution system [8].

This paper presents a comprehensive simulation study and enhancing power quality in transmission lines using VSC-based DSTATCOM using MATLAB Simulink. The test includes the 6 KV/400 V, three bus system. The primary objectives are to measure and analyse the harmonic content in output voltage and current, investigate the effects of voltage sags, swells, and non-linear loads, and achieve a significant reduction in THD. By implementing advanced control strategies and using passive elements within the DSTATCOM, this study demonstrates the potential of VSC-based systems to enhance power quality and improve overall system reliability. These findings provide valuable insights into the design and operation of VSC-based DSTATCOMs as viable solutions for addressing power quality challenges in modern distribution networks.

Distribution static compensator (DSTATCOM)
Figure 31.1 shows the DSTATCOM is one of the best electronic devices designed to solve these problems. It is a voltage converter VSC-based system that provides energy recovery, coordination and power management at the distribution level. By integrating

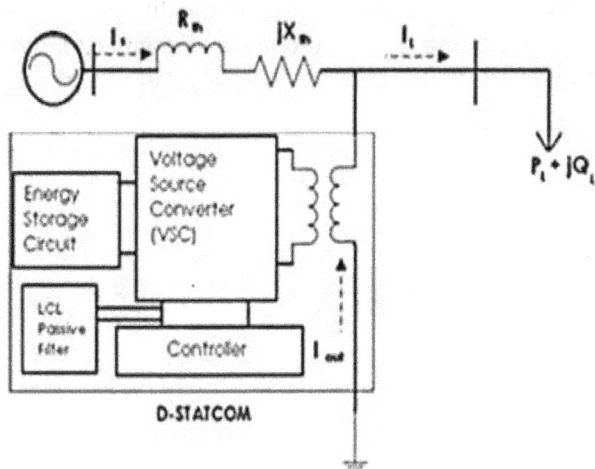

Figure 31.1 Schematic diagram of DSTATCOM
Source: Author

DSTATCOM into a distributed network, the following benefits can be achieved: DSTATCOM can reduce THD caused by non-linear components, standards and improving the quality of electronic products. Voltage disturbances such as voltage sags and swells can cause operational failures in commercial and industrial applications. DSTATCOM provides stable voltage by dynamically injecting or absorbing reactive power as needed. DSTATCOM analyses and reduces multiple faults line-to-line, line-to-ground, three-phase faults, improving the fault and reliable pressure on the system. Through local balancing, DSTATCOM reduces the load on electrical equipment such as transformers and generators, improving overall performance. In this project, DSTATCOM is used to improve the quality of power distribution by comparing the performance in two cases (without DSTATCOM and with DSTATCOM). This approach demonstrates the effectiveness of DSTATCOM in solving important power quality issues, increasing reliability, and improving the performance of modern electronic systems.

DSTATCOM components: DSTATCOM involves following:

A. Voltage source converter

A VSC is a high voltage converter that converts DC voltage to controlled AC output with the ability to adjust amplitude, phase angle, and frequency. VSC can quickly and efficiently inject or absorb power to stabilise the profile during events such as voltage sags and swells. This has the added ability to control current, which is important for business and commercial operations. This allows the DSTATCOM to eliminate harmonic interference from nonlinear components, VSC provides a stable output, ensuring that the voltage remains within the allowable range even in the presence of disturbances such as faults or unbalances, traditional solutions relay on passive components such as capacitors and inductors, which are not very useful for dynamic applications. They cannot respond instantly to changes in voltage or harmonics, which often results in slow response times and poor power performance. By integrating VSC, the system can quickly adapt to changing conditions and disturbances in non-linear loads.

B. Control block

Control blocks used in the circuit and DSTATCOM include PLL for synchronisation, voltage and current control blocks (using SRF or p-q theory) for control and harmonic reduction, DC link voltage control using PI controller, VSC-switching. This strategy is based on sinusoidal type PWM technique, providing simplicity and good response. Because of user-defined performance is relatively low, PWM methods actually offer more flexible options than FFS methods (fundamental frequency switching). Additionally, high switching frequencies can be used to improve converter efficiency without causing large switching losses. The controller input is the error signal obtained from the reference voltage and the measured value of the terminal voltage RMS. Such mistakes are handled by the PI controller. This will be available with PWM signal generators [9, 10]. These algorithms include SRF/p-q theory for correlation, PI controller for voltage control, hysteresis current control for current control, and PWM (SPWM or SVPWM) for VSC control.

C. Passive elements

DC link capacitor stores and stabilises DC voltage required for the VSC, and helps by maintaining constant voltage during reactive power compensation and harmonic reduction. AC filter (L & C) removes high-frequency switching harmonics generated by the VSC that ensures smooth and sinusoidal current injection into the grid. Series inductor (coupling reactors) connects the VSC to the distribution network and limits fault current and smoothens the compensating current injected into the grid. These passive elements work in coordination with the active components by increasing the values of elements and stabilise voltage, reduce harmonics, and improve overall power quality in the distribution system. UPQC-based fuzzy logic provides low source current and low unit input performance, resulting in less power supplies. Therefore, the proposed fuzzy control technology is very satisfactory for stabilising the DC connection voltage [11].

D. DSTATCOM controller

D-STATCOM control system designed to control voltage, improve power quality, and minimise distribution line splitting. It works by including system voltage Vabc, current Iabc, and DC bus voltage Vdc for control instructions. Anti-aliasing filters clean the inputs by removing noise and provide reliable data for accurate control. The controller then processes these signals to produce parameters such as

the measured value m_Phi and angular velocity w, thereby maintaining a constant voltage and sufficiently reducing reactive power. A constant delay represents the sampling time and operation of the digital controller. Integration of the angle allows the calculation of the angle theta, which is important for synchronising the injection voltage with the system. Using the measuring device and the phase angle, the system calculates the reference voltage for charging Vabc(t) and sends it as a reference signal Vref_abc to the PWM generator or control hardware. Together, these devices provide stable voltage and reactive power compensation in the transmission lines.

The main objective of any compensation scheme is that quick response, flexible and easy to use. DSTATCOM's control algorithm generally uses the following steps:

➢ Modelling and simulation of the DSTATCOM by using MATLAB SIMULINK
➢ To measure and analyse the harmonic content in the output voltage and current, effects of fluctuating voltage, voltage sags and swells and non-linear loads
➢ To achieve significant reduction in THD in the transmission system for the VSC-based DSTATCOM and passive elements, improving the overall power quality and system reliability.

Methodology

Figure 31.2 shows the method which involves simulating a three-phase transmission to identify electrical problems with or without DSTATCOM. Initially, the system was tested under fault conditions, including line-to-line, line-to-ground and double line-to-ground faults. This leads to very high impedance up to 50% loss, a large fault current up to 80% increase

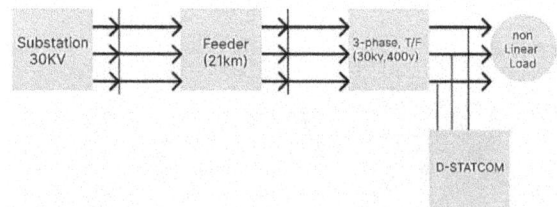

Figure 31.2 Block diagram of power transmission system
Source: Author

and a THD of up to 52% due to non- linear loads. When a DSTATCOM is integrated, it uses a voltage converter VSC that controls the process according to the p-q theory and is controlled by a proportional input (PI) controller. It dynamically injects reactive power to balance the voltage level and compensates for harmonics using space vector PWM, SVPWM. Voltage drops are reduced to 90–98% of nominal level, fault current increase is limited to 15–20% and THD is reduced to 13–15%. This proves the effectiveness of DSTATCOM in improving power quality and ensuring stability.

Simulation results

Without DSTATCOM

Figure 31.3 shows the design and implementation driven by MATLAB Simulink, which provides efficient models and predictions for realistic simulations. The is a DSTATCOM without the system, focusing on understanding the behaviour of the system regarding voltage control, power quality and harmonic content. A 6 kV programmable voltage source acts as the main input, analogue output and has a variable voltage that allows testing according to different needs. Load voltages up to 400 V are monitored to measure stability and harmonic content on the body is checked to evaluate electrical problems. Main products include transformers for different levels of voltage step-down (6 kV/100 MVA and 6 kV/400 V),

21 km and 2 km feeders showing real impedances, the installation includes 1 MW adjustable load and more fixed load. The system also integrates a fault simulation module, which can generate three-phase faults, line-to-line faults, line-to-ground faults, two line-to-ground faults and other conditions, and define the impact of faults on the system. This circuit evaluates the fault problem, measures the THD level (up to 52% before compensation), and implements solutions to reduce these problems.

With DSTATCOM

Figure 31.4 represents DSTATCOM system, in which DC link provides stable DC voltage, Bridge1 and Bridge2 act as voltage source inverters to generate and control AC voltage; the controller controls the voltage stability and reactive power compensation by adjusting the signal intensity and the resistance and inductance analogue line impedance. The transformer boosts or decouples the AC output and the feedback loop to closed-loop control of voltage and current.

Result and discussion

Initial input DSTATCOM

Nominal load [30e3 0.9] [current (RMS) power factor]

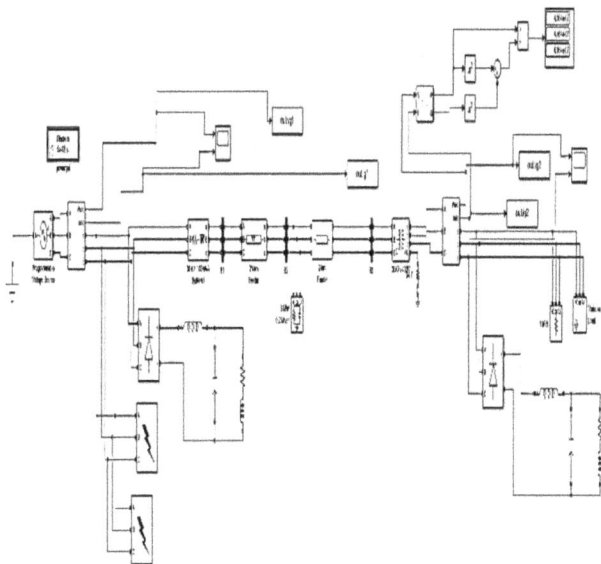

Figure 31.3 Simulink model of with out DSTATCOM
Source: Author

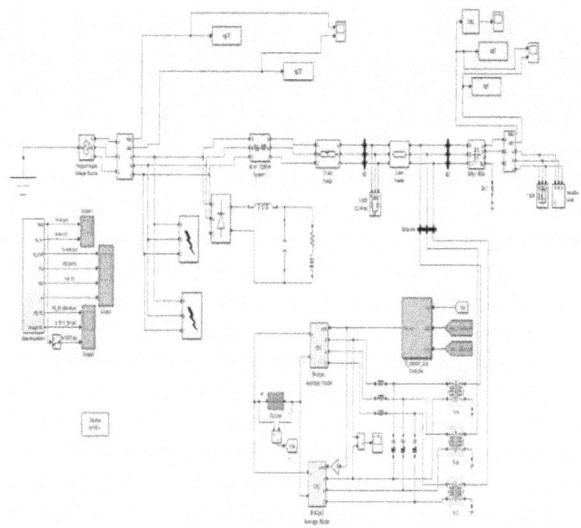

Figure 31.4 Simulink model of VSC-based DSTATCOM
Source: Author

1. Modulation [2000 50] [amplitude current (RMS) frequency (Hz)]
2. Nominal voltage 400 [volts (RMS) phase to phase]

Main supply voltage	6 KV
Coupling transformer voltage	400 V
Coupling transformer turns ratio	01:01
DC bus voltage	200 V
Capacitance	750 F
Load active power	20 KW
Line frequency	50 Hz
Fault type	Three-phase fault, three phase to ground LL, LG, LLG
Switching time	0.1 s and 0.168
Fault resistance (*Ron*)	8 Ω
Ground resistance (*Rg*)	0.01 Ω,
Snubber resistance (*Rs*)	1 MΩ

Here the below waveforms represents the input / output waveforms and fault analysis.

Case 1: Input waveform without DSTATCOM

Figure 31.5 shows the input waveforms of three-phase 6 kV AC distribution system. The top plot displays the three-phase voltage (Vabc) in per-unit (pu), while the bottom shows the corresponding current (Iabc) in per-unit. Between 0.05 and 0.15 seconds, a voltage sag occurs. From 0.25 to 0.35 seconds, a voltage swells increases both voltage and current amplitudes. By 0.35–0.5 seconds, the system stabilises, with voltage and current returning to normal levels.

Output waveform

Figure 31.6 shows the output waveforms of three-phase 6 kV AC distribution system, from 0 to 0.05 seconds, the waveforms are stable at 400 V. A voltage sag occurs from 0.05 to 0.15 seconds, followed by a swell (overvoltage) from 0.25 to 0.35 seconds. The waveforms return to normal by 0.35–0.5 seconds,

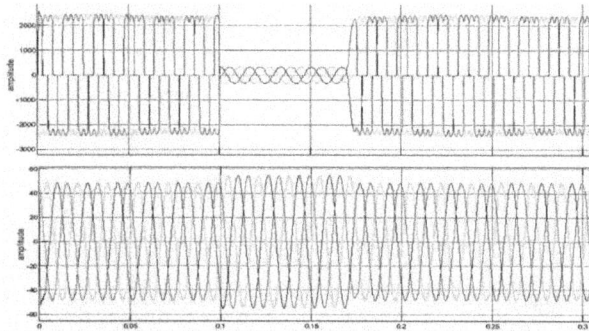

Figure 31.5 Voltage (Va) and current (Ia)
Source: Author

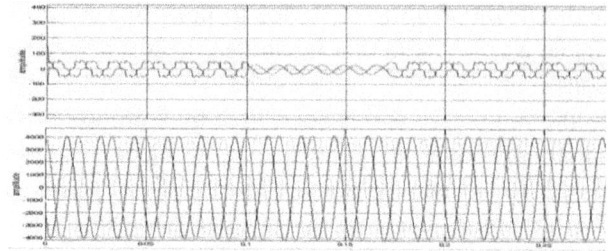

Figure 31.6 Voltage (Va) and current (Ia)
Source: Author

indicating recovery. The top section displays three sinusoidal waveforms.

Voltage sags for different fault condition

LG (line-to-ground fault)

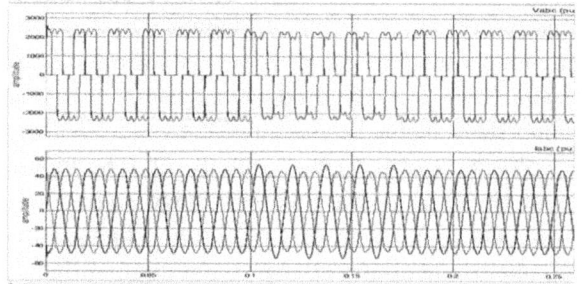

Figure 31.7 Voltage (Va) and current (Ia)
Source: Author

LL-G (line-to-line-to-ground fault)

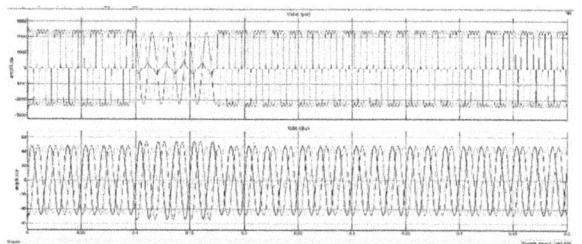

Figure 31.8 Voltage (Va) and current (Ia)
Source: Author

LL (line-to-line) fault

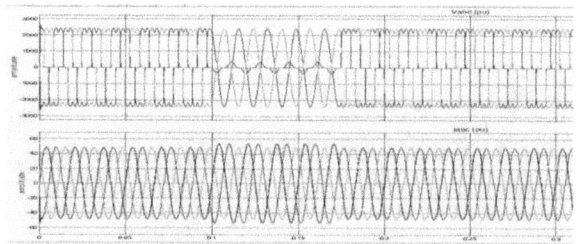

Figure 31.9 Voltage (Va) and current (Ia)
Source: Author

Three-phase to ground fault

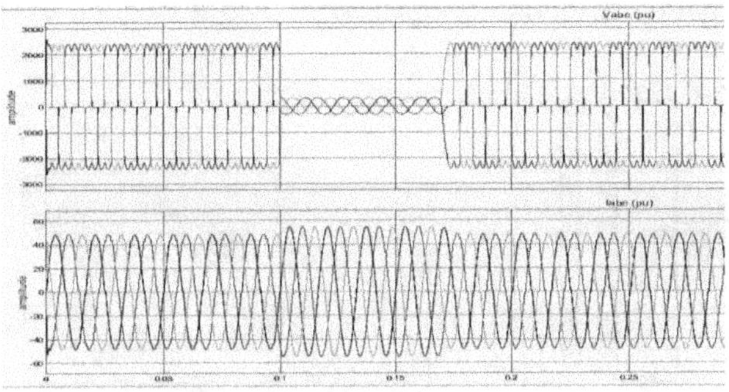

Figure 31.10 Voltage (Va) and current (Ia)
Source: Author

Figures 31.7–31.10 show the waveform analysis proves the significant improvement in power quality brought by DSTATCOM. Without DSTATCOM faults such as line-to-line (LL), line-to-ground (LG) and dual line-to-ground (LLG) faults will show a significant voltage drop in the waveform, and the total duration is 0.1–0.5 in that during the fault usually lasts 0.1–0.168 seconds (depending on the type of fault). At this time, the voltage waveform shows sharp and irregular distortion with total harmonic distortion ratio (THD) as high as 52%, and the power quality problem is great.

Total harmonic distortion – Without distribution compensator (Figure 31.11)

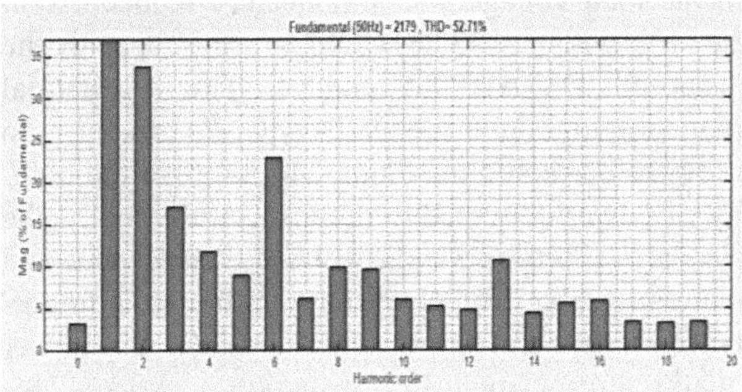

Figure 31.11 Voltage (Va) and current (Ia)
Source: Author

Case 2: Input waveform with DSTATCOM

International conference on recent innovations in engineering science and technology

Figure 31.12 shows the voltage and current two sets of input waveforms, likely representing different electrical parameters in a power system with DSTATCOM to reduce harmonic mitigation. The top

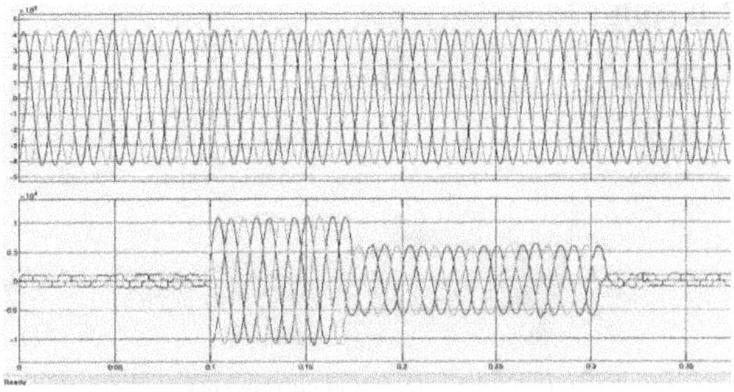

Figure 31.12 Voltage (Va) and current (Ia)
Source: Author

section shows three sinusoidal waveforms, which could represent the three-phase voltages and currents in an AC system of 6 KV.

Output waveform

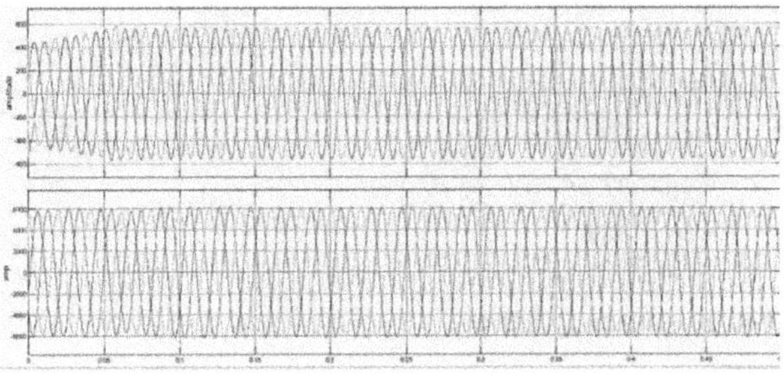

Figure 31.13 Voltage (Va) and current (Ia)
Source: Author

Figure 31.13 shows the voltage and current two sets of output waveforms, likely representing different electrical parameters in a power system with DSTATCOM to reduce harmonic mitigation. The top section shows three sinusoidal waveforms, 400 V.

Voltage sgs for different fault condition – LG (line-to-ground fault)

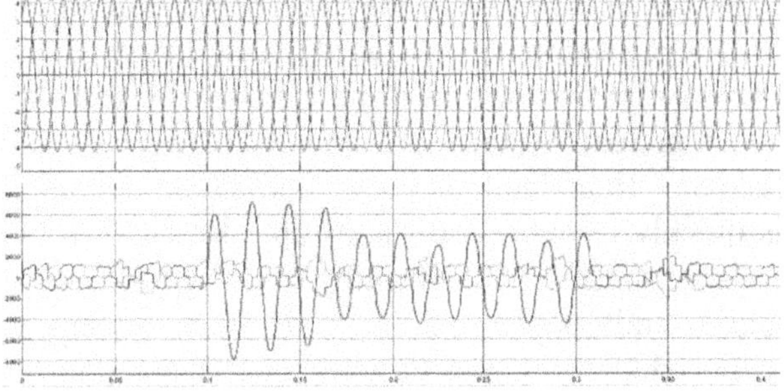

Figure 31.14 Voltage (Va) and current (Ia)
Source: Author

LL fault

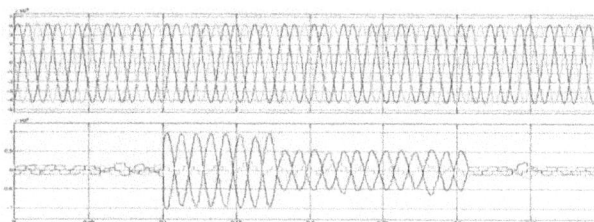

Figure 31.15 Voltage (Va) and current (Ia)
Source: Author

LL-G fault

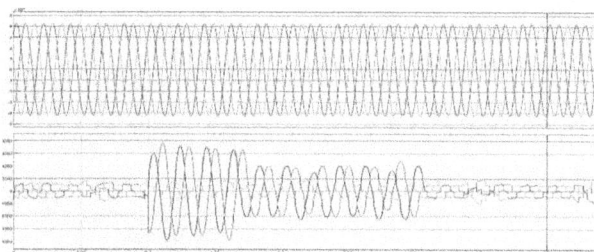

Figure 31.16 Voltage (Va) and current (Ia)
Source: Author

Three-phase ground

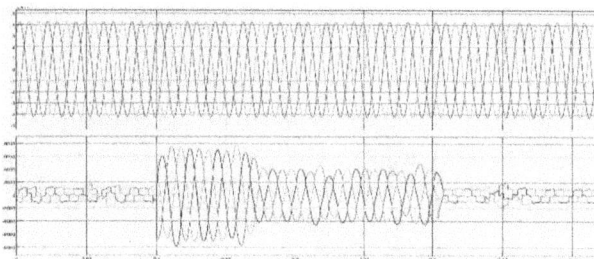

Figure 31.17 Voltage (Va) and current (Ia)
Source: Author

Figures 31.14–31.17 shows the voltage sags for different fault condition, the impact on the fault duration (0.1–0.168 seconds) is reduced. Even during faults, the waveform maintains a more sinusoidal and balanced pattern. Voltage sags and swells are reduced, recovery from faults is rapid, and stable operation is quickly restored after faults are cleared. THD is reduced to 13–15%, indicating improved efficiency and overall power quality. This improvement is achieved by using passive equipment in conjunction with DSTATCOM, which demonstrates its effectiveness in reducing the effects of faults and ensuring stability and reliability within the hull.

Total harmonic distortion

With distribution compensator

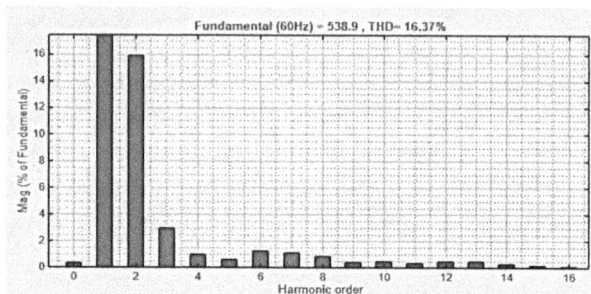

Figure 31.18 Voltage (Va) and current (Ia)
Source: Author

Phases	Total harmonic distortion (%)	
	System without DSTATCOM	System with proposed control concept of DSTATCOM (THD of load voltage)
A	52.71%	16.37%

From Figures 31.11 & 31.18 and the above table shows the comparative analysis of THD between the system with and without DSTATCOM. The analysis of power quality under various scenarios, including voltage sags, swells, and different fault conditions, demonstrates the effectiveness of the proposed DSTATCOM control strategy. The THD in the load voltage was significantly reduced from 52.71% without DSTATCOM to 16.37% with the proposed DSTATCOM control. This substantial reduction highlights the DSTATCOM's capability to mitigate harmonic distortion and improve the quality of power delivered to the load. Additionally, the system's ability to handle voltage disturbances and maintain stable load conditions confirms the importance of implementing such solutions for power quality enhancement in distribution networks. This comparative study underscores the critical role of DSTATCOM in reducing power quality issues, ensuring voltage stability, and enhancing overall system reliability.

Conclusion

Using this work, the disquisition on the part of DSTATCOM is carried out to improve the power quality in distribution networks with stationary direct and non- linear loads. DSTATCOM regulator is used with the device to enhance its performance. Test system is anatomised and results are presented in the this paper.

- DSTATCOM in the distribution networks under different fault conditions and it can be concluded that DSTATCOM effectively improves the power quality in distribution networks with stationary direct.
- The DSTATCOM are helpful in prostrating the voltage unbalance problems in power system.
- These bias are connected to the power network at the point of interest to cover the critical loads.
- These bias also have other advantages like harmonics reduction, power factor correction.
- The DSTATCOM bear a lesser number of power electronic switches and storehouse bias for their operation.
- The project outcomes validate the effectiveness of DSTATCOM as a viable solution for addressing power quality challenges in distribution networks.
- By achieving significant reductions in THD, mitigating voltage disturbances, and enhancing system reliability, the implementation of DSTATCOM contributes to meeting the operational and regulatory requirements of modern power systems.
- This study paves the way for further optimisation of DSTATCOM configurations and real-world applications in power systems.

Acknowledgement

We gratefully acknowledge the students, staff, and authority of Electrical & Electronics Engineering department for their cooperation in the research.

References

[1] Bapaiah, P. Power quality improvement by using DSTATCOM. *Internat. J. Emerg. Trends Elec. Electr.*, 2013;2(4): 1–12.

[2] Vijay Muni, T., Sambasiva Rao, N., Venkata Kishore, K. VSC-based D-STATCOM in transmission lines for power quality improvement. *Nat. Conf. Elec. Sci.*, 2012;9(88): 1–08.

[3] Mukesh Chandra, R., Pramod Kumar, R. To study, analyse and implement power quality improvement technique using VSC-based DSTATCOM (Distribution Static Compensator). *Internat. J. Trend Res. Dev.*, 2022;9(5):1–05.

[4] Mitesh, K., Ritesh, C. To study and implement power quality improvement technique using VSC-based DSTATCOM (Distribution Static Compensator). *Internat. J. Trend Sci. Res. Dev.*, 2022;6(5):1–08.

[5] Saritha, G. Power quality improvement of distribution system using DSTATCOM. *Proc. Int. Conf. Inform. Engg. Manag. Sec.*, 2014:1–07.

[6] Sathish Babu, P., Sundarabalan, C. K. Power quality enhancement in sensitive local distribution grid using interval type-II fuzzy logic controlled DSTATCOM. *Instit. Elec. Electr. Eng.*, 2021;9:1–12.

[7] Shailendra, K. To study, analysis and implementation of power quality improvement using D-STATCOM. *Internat. J. Modern Trends Sci. Technol.*, 2022;8(02):144–150.

[8] Chenchireddy, K. A review on D-STATCOM control techniques for power quality improvement in distribution. *Proc. Fifth Internat. Conf. Elec. Comm. Aerospace Technol.*, 2021:1–09.

[9] Sanjeev, K., Pramod Kumar, R. Analysis and implementation of power quality improvement using VSC-based DSTATCOM (Distribution Static Compensator). *Internat. J. Trend Sci. Res. Dev.*, 2020;9:1–08.

[10] Fayez, K., Elsayed, M. I. Power quality improvement in power distribution systems. *Eur. J. Engg. Technol.*, 2016;4(5):1–13.

[11] Saima, A., Abdul Hamid, B. Power quality improvement using voltage source converter (VSC)-based unified power quality conditioner (UPQC). *Internat. J. Comp.*, 2016;148(1):1–08.

[12] Pavithra, Shanmukha Rao, L. Performance evaluation of PV wind co-generation based multi-purpose STATCOM for power quality enhancement using fuzzy-logic controller. *IEEE 2019 Innov. Power Adv. Comput. Technol.*, 2019:1–08. DOI: 10.1109/i-PACT44901.2019.8960013.

32 Performance analysis and optimisation of solar-powered alkaline electrolyser for green hydrogen production

P. Kezia Joy Kumari, Prajwal A., Ashwini A. V., Rekha S. and Suresh H. L.

Sir M Visvesvaraya Institute of Technology, Bangalore, India

Abstract

This paper presents an experimental study on the feasibility of the use of solar energy for the performance of prototype alkaline electrolyser for green hydrogen production. The electrolyser consists of three stacks each with 13 plates, containing two negative, three positive plates, and eight neutral plates, and uses potassium hydroxide (KOH) as the electrolyte. A 12-V solar panel was used as the primary energy source, and various power parameters such as voltage, electricity, hydrogen production rate, efficiency, electrolyte concentration, and energy consumption were calculated using appropriate devices. The results exhibits that the solar panel meets the operating requirements of the electrolytic agent and meets a hydrogen production rate of 0.2 1.0 L/min with a Faraday efficiency of 80°C. Additionally, this paper focuses on analysing the effects of solar energy scouts on electrolyser performance and examines optimisation strategies to improve energy conversion efficiency. The findings contribute to the development of renewable hydrogen production systems and provide a sustainable approach to generating clean energy.

Keywords: Solar-powered electrolyser, green hydrogen, renewable energy, photovoltaic system

Introduction

The integration of self-optimising control strategy in alkaline electrolyser greatly increases the efficiency of hydrogen production, especially under fluctuating inputs with renewable power. In contrast to traditional systems, which are limited to a 40–100% operating range, this method fits the optimum efficiency point, increasing the low load efficiency evenly from 19.12% to 42.33%, with only 15% of the nominal output improving performance [1]. We present an experimental test for the integration of 12 V solar panel and a prototype-alkaline electrolyser to assess the suitability of hydrogen production. An alkaline electrolyser has been built by 3D printer. The cell was equipped with nanostructured electrodes fabricated by template electrosynthesis that is a simple and cheap bottom-up way to obtain nano-scaled materials. The test results obtained with nanostructured electrodes operating in 30% w/w KOH aqueous solution were collated with those obtained in a cell using conventional planar electrodes. At 0.5 Acm^2 and room temperature, Ni nanowire (NW) cathodes show energy efficiency 7.3% higher than Ni strip [2]. The prototype consists of three stacks, each comprising 13 plates. The system uses potassium hydroxide (KOH) as the electrolyte. Key input parameters for the electrolyser include voltage, current, and power supply specifications, while the main operating parameters are electrolyte concentration, stack temperature, and current density.. Various performance parameters were measured using the appropriate device, including electrolyte concentration, stack temperature, current density, hydrogen production parameters, hydrogen produc- tion rate, Faraday efficiency, purity level. Energy consumption parameters: Specific energy consumption, solar efficiency of hydrogen from the sun. This study also examines the effects of solar radiation of variability on system stability and efficiency. The results show that the 12 V solar panel meets the operating requirements, but efficiency can be further improved with the optimal electrolyte concentration and power management strategy. This paper optimise strategy temperature control, electrical density control, electrolyte concentration control hydrogen yield and hydrogen efficiency from the Sun. The present work focuses on a practical approach to build an equivalent circuit model for commercial electrolyser systems, which can adequately model realistic rectifier waveforms. The proposed approach does not attempt a comprehensive electrochemical study of electrolytic cells but rather assumes that there is limited knowledge available relating to the internal stack design or electrode morphology and composition beforehand,

keziajoykumari_eee@sirmvit.edu

DOI: 10.1201/9781003685876-32

since electrolyser manufacturers often consider this information to be proprietary [3]. Although photovoltaic panels can be directly connected to alkaline water electrolysers, wind turbines need appropriate converters, which introduce additional losses. By integrating alkaline water electrolysis with hydrogen storage tanks and fuel cells, power grid stabilisation can be achieved. As a result, the need for conventional spinning reserves is reduced; further decreasing carbon dioxide emissions [4]. This innovative alkaline electric design focuses on simplicity, easy production and modularity using easily available materials. The replaceable components allow for the effectiveness of partial reductions or upgrades, ensuring consistent performance and long-term flexibility to optimise the production of green hydrogen through renewable energy [5]. These results provide valuable insight into the feasibility of solar alkaline electrolysis and provide a sustainable and decentralised hydrogen production system. This research contributes to growth work on the production of renewable hydrogen products. This highlights its role in decarbonisation industries, energy storage applications and fuel cell technology. Future work will focus on improving system efficiency, integrating battery storage for power stabilisation, and scaling production for commercial viability.

Figure 32.1 represents block diagram of a solar-driven alkaline water dissolution system for hydrogen production. It starts with a solar powered (PV) panel that captures sunlight and converts it into electrical energy. This energy is optionally regulated by a DC/DC converter to ensure the optimal voltage and current of the electrolytic factor. The alkaline water electrolyser uses this electrical energy to divide water into hydrogen (H_2) and oxygen (O_2). The produced hydrogen gas can be collected for a variety of applications, but oxygen is released as by-product.

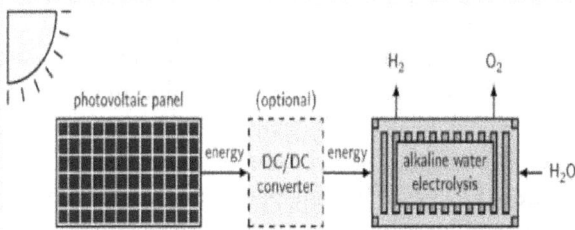

$$4H^+ + 4e^- \rightarrow 2H_2 \qquad 2H_2O \rightarrow O_2 + 4H^+$$

Figure 32.2 Block diagram of electrolyser
Source: Author

The system provides a clean and sustainable way to produce hydrogen using renewable solar energy.

Figure 32.2 shows an electrolyser which is a device that divides water (H_2O) into hydrogen (H_2) and oxygen (O_2) using electrical energy through an electrolysis process. This technology plays a significant role in the production of green water, especially when powered by renewable energy sources such as solar and wind power. Hydrogen produced can be used in a variety of applications, including fuel cells, industrial processes, and energy storage systems.

There are many different types of electrolytes with different properties and applications:

- Alkaline electrolyser (AEL) – Uses an alkaline electrolyte like potassium hydroxide or sodium hydroxide and operates at moderate temperatures.
- Proton exchange membrane (PEM) – Uses solid polymer electrolytes, operates at high power density, making it more efficient but expensive.
- Solid oxide electrolytic factor (SOE) – Operates at high temperatures (~700–900°C) and achieves high efficiency, but requires complex thermal management.
- Anion exchange membrane (AEM) electrolytic agent – A technology under development that

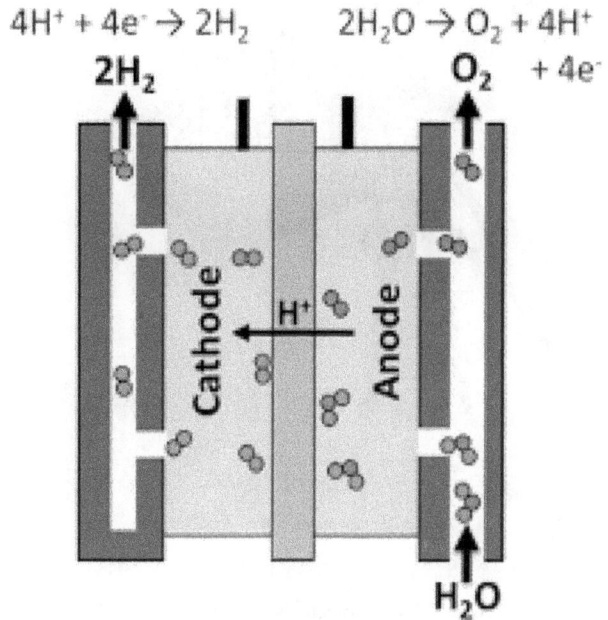

Figure 32.1 Solar-driven alkaline water dissolution system for hydrogen production
Source: Author

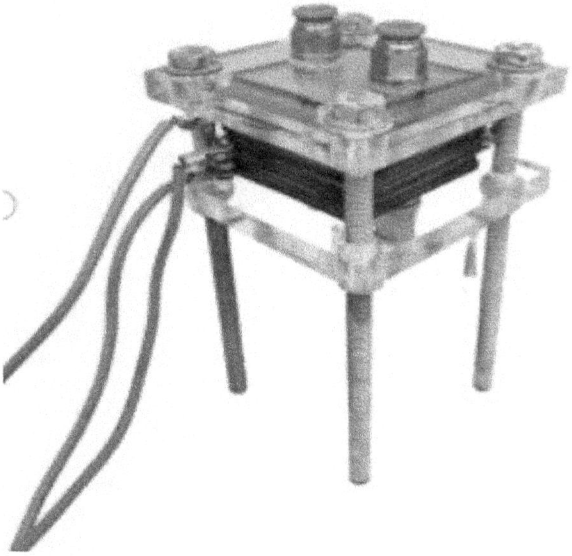

Figure 32.3 Alkaline electrolyser
Source: Author

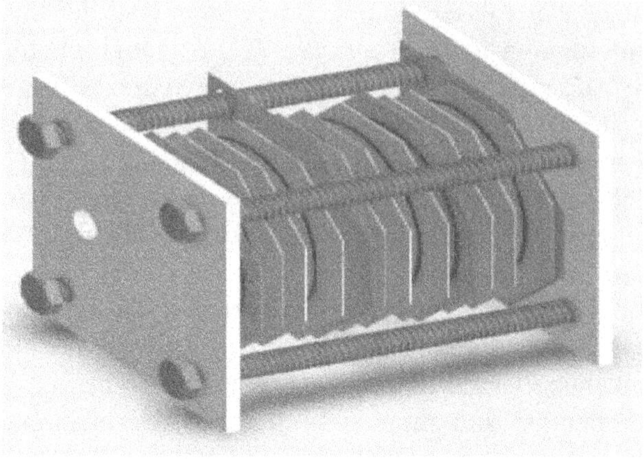

Figure 32.4 3D structure of alkaline electrolyser
Source: Author

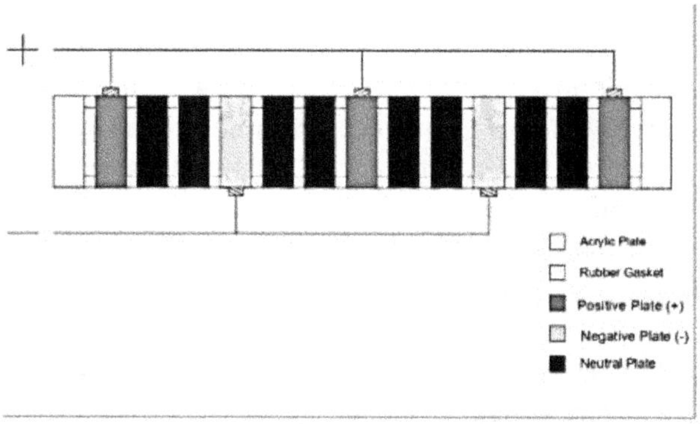

Figure 32.5 Stack formation of electrolyser
Source: Author

combines the benefits of PEM and alkaline electrolysis with reduced cost and improved efficiency.

Among these were alkaline electroliser for cost efficiency, reliability and scalability (Figure 32.3). The most important advantages of alkaline electrolysers are low cost, uses non-specific metal catalysts to make them more cost affective than PEM electrolithers. Storage life long operating lifetime suitable for hydrogen production in the hydrogen field. High efficiency of 70% can be achieved under optimal conditions. Compatibility with renewable energy: You can use solar and wind power efficiently using energy management strategies. Alkaline electrolysers also have some limitations. The principle of working of alkaline electron solutions – Alkaline electrolysis occurs in liquid electrolyte solutions (usually KOH or NAOH), which makes it easier for the electrode to divide the water molecules. The reaction occurs asfollows: Hydrogen gas is collected at the cathode and oxygen is released at the anode. Electrolytes help in maintaining ion transfer and system efficiency. This study focuses on the integration of a 12 V solar panel and a prototype alkaline electrolytic agent to analyse performance under real conditions. This study aims to optimise the production of green hydrogen for sustainable energy applications (Figure 32.4).

Figure 32.5 shows the 3D model of alkaline electrolyser which shows an preview of model and it has been designed using AutoCad software. Figure 32.4 shows the stack arrangement of model which consists of 13 plates in that 3 positive and 2 negative and 8 neutral electrodes, and materials used is stainless steel and gasket used to separate anode and cathode plates, electrolyser consists of 3 stacks (Figure 32.6).

Figure 32.7 shows a measuring device that shows voltage, electricity, electricity and energy values. The display shows that the system operates at approximately 11.83 V and 12.00 A with power consumption of 14.19 W. This device is used to monitor and analyse the performance of alkaline electrolysers, possibly by measuring input parameters. Figure 32.6 shows alkaline electrolysis device setup. It consists of a stack of plates in electrolytic solution connected to the water reservoir system. This device promotes water electrolysis and produces hydrogen (H_2) and oxygen (O_2) gas. The transparent housing of the electrolyser allows for visible observation of the reaction. The hose is probably used for gas detection and separation, ensuring efficient hydrogen flow for use. This setup is part of the green hydrogen research,

Figure 32.6 Testing alkaline electrolyser
Source: Author

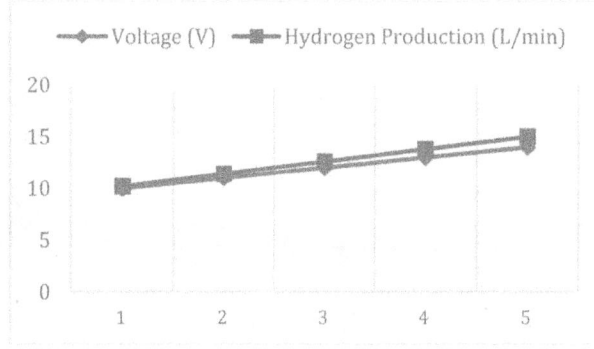

Figure 32.8 Voltage (Va) and hydrogen (L/min)
Source: Author

Figure 32.7 Energy meter
Source: Author

examining the feasibility of the use of solar energy for power delivery.

Results and discussion

These input parameters from Table 32.1 were carefully monitored to understand their impact on hydrogen production rate, system efficiency, and energy consumption and form the basis for performance analysis and optimisation.

Test 1. Production of hydrogen and voltage

Table 32.1 Intial inputs of alkaline electrolyser

Parameter	Value
Applied voltage	12 V
Number of stacks	3
Total plates per stack	13
Negative plates	2
Positive plates	3
Neutral plates	8
Electrolyte	KOH
Electrolyte concentration	20–30 wt%
Stack temperature	40–60°C

Source: Author

Figure 32.8 shows the loose voltage (V) and hydrogen production rate (L/min) of the prototype alkaline electrolytic agent. The electrolytic process begins, but hydrogen production is slightly delayed due to the stabilisation of the electrode activation time and electrolyte capacity. Tension increases slightly to 11 V, gradually increasing in hydrogen production. Once the KOH electrolyte reaches an appropriate ionic conductivity, the system becomes stable. Increases the 2–3-divided voltage to 12, which increases the hydrogen production rate. Electrochemical reactions to the cathode and anode become more efficient and lead to higher gas bubbles. Hydrogen production is potent for 3–4 min at 13 V. The reaction rate reaches a resting state and the electrodes are completely active. However, mild heat buildup can occur, which has a slight impact on efficiency. The voltage generated for 4–5 minutes reaches 14 V, with hydrogen production reaching its peak. The system operates at optimal efficiency, but further increases in voltage can lead to energy losses due to increased heating and resistance. The alkaline electrolyser functions normally within an area of 10 V, 14 V, indicating that hydrogen production directly increases over time. Time stabilisation (12 minutes) is required to ensure that the electrolyte and electrodes are fully active. Apart from 14 V, the reduction in efficiency should be considered due to increased heating and resistance.

Test 2. Current (A) and hydrogen production

Figure 32.9 shows the relationship between current (A) and hydrogen production (L/min) in alkaline electrolytic factor production with a 12-V sorable. A higher current means that more electrons are available for electrolysis, improving the efficiency of water layout. Slow increase in electricity over time:

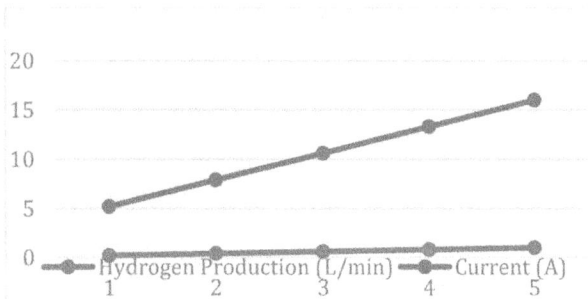

Figure 32.9 Current (Ia) and hydrogen (L/min)
Source: Author

low current at first (~1 minute) and hydrogen production is minimal. The current increases over time (2°C) and improves hydrogen output.

Electrolysis device efficiency: The direct proportionality between electricity and hydrogen production confirms the Faraday electrode method. Good care of KOH electrolyte concentrations ensure efficient ion flow and minimise losses. Impact of solar power: Increased sunlight intensity provides more electricity and increases hydrogen production. Sun variations can temporarily affect performance.

Test 3. Hydrogen production vs. temperature (Figure 32.10)

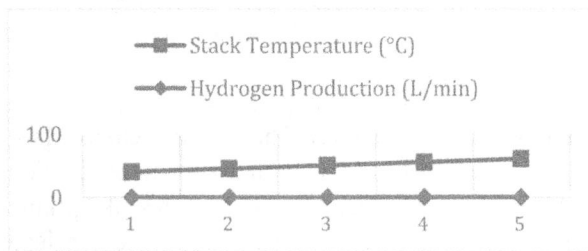

Figure 32.10 Hydrogen (L/min) and temperature (c)
Source: Author

Test 4. Hydrogen production vs. energy consumption

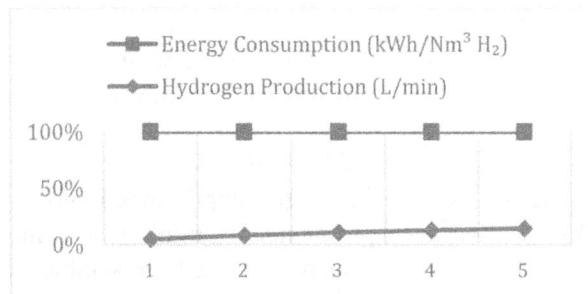

Figure 32.11 Hydrogen (L/min) and energy (kwh)
Source: Author

Figure 32.11 shows the comparission of hydrogen production vs. energy consumption. Hydrogen production increases over time, but energy consumption is almost constant, indicating stable efficiency of electrolysis. In Figure 32.10 hydrogen production is compared to stacked temperature. As the stacking temperature increases the hydrogen production slightly (Table 32.2).

Figure 32.12 indicates that the electrolyser operates within a controlled voltage region, but power variations affect other parameters. Hydrogen production (L/min): Hydrogen production is still relatively low in comparison to other factors. Efficiency indicates a significant peak. This can be attributed to factors such as heat loss at high currents and electrode reduction, KOH concentration. Electrolyte concentration plays a role in efficiency. Too high or too low will reduce efficiency. This diagram illustrates the need to optimise KOH concentration for maximum hydrogen output. Hydrogen purity maintains high levels and indicates that the electrolytic agent effectively separates the gas. However, variations indicate

Table 32.2 Electrolyser performance parameters

Parameter	Values
Electrolyser voltage per cell	1.8–2.2 V
Current drawn	5–15 A
Electrolyser current density	100–400 mA/cm²
Faraday efficiency	80–90%
Hydrogen production rate	0.2–1.0 L/min
Oxygen production rate	0.1–0.5 L/min
Specific energy consumption	4–6 kWh/Nm³ H₂

Source: Author

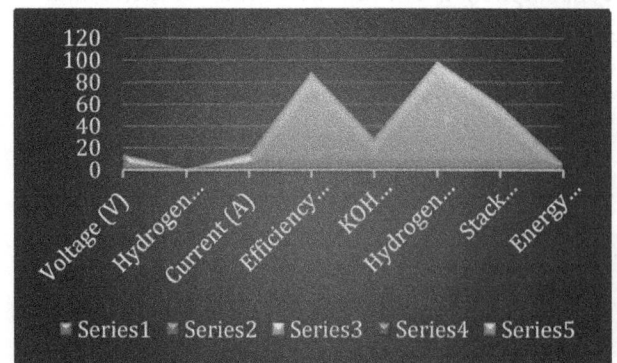

Figure 32.12 Voltage (V), hydrogen production rate (L/min), current (A), efficiency (%), KOH concentration (%), hydrogen purity (%), stack temperature (°C), and energy consumption
Source: Author

mild contaminants or operating variations. Stacking temperature The temperature rises steadily. This is normal for electrolysis, but must be controlled to prevent excessive heat loss. Energy consumption: Energy consumption appears to be relatively stable. That is, despite changes in other parameters, the system maintains a consistent conversion rate from energy to hydrogen. Efficiency and hydrogen production should be optimised by adapting COH concentrations to control electricity. Stacking temperatures should be monitored to avoid excessive warming. It shows a highly functional electrolyser with stable energy consumption and space to improve production rates (Table 32.3).

Conclusion

- The main goal of this experimentation was to investigate the integration of solar energy in an electrolyser. This reduces reliance on non-renewable energy sources. Including a DC – DC converter in the proposed system will help stabilise tension and electricity and ensure consistent electrolysis. With this success of the transition, electrolytics is completely sustainable and environmentally friendly, and with global efforts to develop hydrogen as an alternative fuel source.

- This project successfully demonstrates the feasibility of the use of an alkaline electrolyser for the production of green hydrogen with 12 V flow rate with a potential transition to solar energy. The testing phase involved analysing key performance parameters, such as voltage, electricity, efficiency, hydrogen cleaning, KOH concentration and energy consumption, optimising the electrolyser output. The integration of a measurement system allows for performance input and actual monitoring of performance, ensuring accurate data collection for analysis.

- The experimental structure contained 13 plates (two negative, 3 positive, neutral) a containing 3 stacks of alkaline electrolyser in potassium hydroxide (KOH) electrolyte solution. Electrolytes played an important role in improving ions' conductivity, reducing extremity and improving hydrogen yield. The system produced hydrogen (H_2) and oxygen (O_2) gas.

- During the test, fluctuated voltage and electrical values, affecting efficiency and hydrogen production rate. The measurement device provided insights into power consumption (W) and energy consumption (WH). This is an important factor in determining the overall efficiency of a system. The collected data supported production rates in assessing the effects of hydrogen washing and different KOH concentrations. A controlled temperature range electrolysers have been found to work effectively and prevent excessive thermal structures that can impair performance.

- This project provides a working prototype for alkaline electrolysers that can efficiently generate hydrogen. The results highlight important issues such as energy consumption, optimising hydrogen reinforcement, and long-term durability. Future improvements mayinclude improved progressive catalysts, optimised electrode materials, and system automation – all better control and efficiency. Potential integration of solar energy will greatly improve the sustainability of the system and make it a practical solution for clean energy applications.

References

[1] Cheng, H., Xia, Y., Wei, W. Self-optimization control for alkaline water electrolyzers considering electrolyzer temperature variations. *IEEE Trans. Indus. Elec.*, 2025;72(3):1–12.

[2] Ganci, F., Cusumano, V., Sunseri, C., Inguanta, R. Performance enhancement of alkaline water electrolyzer using nanostructured electrodes synthetized by template electrosynthesis. *Inst. Elec. Electr. Engr.*, 2018:1–04. pp. 1– 4.

[3] Sologubenko, O., Castiglioni, R., Pettersson, S., Leal, A. A. Dynamic impedance modeling of an alkaline electrolyzer. A practical approach. *Internat. Workshop Imped. Spectros.*, 2023:1–06. pp.3.

[4] Brauns, J., Turek, T. Alkaline water electrolysis powered by renewable energy: A review. *Multidis. Dig. Publ. Inst.*, 2020;8:248. pp. 4.

[5] Mus, J., Vanhoutte, B., Schotte, S. Design and characterisation of an alkaline electrolyser. *2022 11th Internat. Conf. Renew. Energy Res. Appl. (ICRERA)*, 2022:18–21:1–7.

Table 32.3 Output parameters (hydrogen and oxygen)

Parameter	Value
Hydrogen purity	98–99.5%
Hydrogen output pressure	1–5 bar
Solar-to-hydrogen efficiency	8–15%

Source: Author

33 Design and optimisation of 3-bit flash ADC with low power double tail comparator

Dr. Vasudeva G.[1,a], Dr. Mallikarjun P. Y.[2], Dr. Tripti R. Kulkarni[2], Bhuvan M.[2], Gowthami Kc.[2] and Nidhi N.[2]

[1]Assistant Professor, Department of Electronics and Communication, Dayananda Sagar Academy of Technology and Management (DSATM), Opp. Art of Living, Bengaluru-560082, Karnataka, India

[2]Department of Electronics and Communication, Dayananda Sagar Academy of Technology and Management (DSATM), Opp. Art of Living, Bengaluru-560082, Karnataka, India

Abstract

In this work the design and implementation of a 3-bit flash analogue-to-digital converter (ADC) is optimised for low power and minimal area consumption. A double-tail comparator is employed to enhance power efficiency while maintaining high-speed operation. The flash ADC is implemented using a 45 nm CMOS process, leveraging its advantages to achieve reduced power dissipation and compact design. Simulation results demonstrate significant improvements in power consumption, with a 41.5% reduction compared to conventional designs. The ADC achieves an effective number of bits (ENOB) of 3, an input voltage range of 0–1.2 V, and a low differential non-linearity (DNL) of 2 LSB. The proposed design is suitable for energy-efficient applications, including Internet of Things (IoT) devices and portable electronics.

Keywords: Flash ADC, double-tail comparator, low power, DNL, INL, CMOS, priority encoder, low area

Introduction

Analogue-to-digital converters (ADCs) are used in modern electronics, enabling the conversion of real-world analogue signals into digital format [1]. Flash ADCs are known for their high-speed operation and large silicon area requirements [2]. The goal of this research is to design a low-power, compact 3-bit Flash ADC using a double-tail comparator in a 45 nm CMOS process [3]. With the continuous advancement in semiconductor technologies, there is a growing demand for low-power, high-speed ADCs [4] that can efficiently process signals in power-constrained environments such as Internet of Things (IoT) devices, biomedical instrumentation, and wireless sensor networks. Traditional flash ADCs, despite their speed advantage, require a large number of comparators [5], excessive power consumption, and increased chip area. Addressing these limitations [6], recent research has focused on improving comparator efficiency, minimising static and dynamic power dissipation, and reducing chip complexity while maintaining high-speed performance [7]. In this work, a novel 3-bit flash ADC architecture is proposed using a double-tail comparator, known for its enhanced energy efficiency and better noise performance. This approach ensures an optimised balance between power consumption, chip area [8], and processing speed. By leveraging the 45 nm CMOS process, the design achieves significant reductions in power dissipation and transistor count, making it suitable for energy-efficient applications [9]. Analogue-to-digital converters (ADCs) are essential components in modern electronic systems, enabling the conversion of real-world analogue signals into digital form for processing and storage. With the increasing demand for high-speed and low-power applications, ADCs play a crucial role in communication systems, biomedical devices, and sensor networks. Among various ADC architectures, the flash ADC is having exceptional fast conversion speed due to its parallel structure [10]. However, conventional flash ADCs suffer from high power consumption and large silicon area requirements as the number of comparators grows exponentially with resolution. Addressing these challenges is vital to improving the efficiency and practicality of high-speed ADCs, particularly in power-sensitive applications [11].

The primary limitation of traditional flash ADCs arises from their reliance on a more comparators, each responsible for comparing the input signal with a distinct reference voltage [12]. As the resolution increases, the number of comparators required follows an exponential growth, leading to significant

---

[a]devan.vasu921@gmail.com

DOI: 10.1201/9781003685876-33

power dissipation and increased circuit complexity [13]. This makes conventional flash ADCs less suitable for low-power and compact integrated circuit designs. Furthermore, process variations and mismatches among comparators can introduce errors [14], affecting the accuracy and reliability of the ADC. Achieving an optimal balance between speed, power efficiency, and silicon area is crucial for improving the overall performance of flash ADCs.

To mitigate these challenges, innovative comparator designs have been explored to reduce power consumption while maintaining high-speed operation [15]. The double-tail comparator has emerged as an effective alternative to conventional comparators, offering significant improvements in energy efficiency and speed. Unlike traditional dynamic comparators [16], the double-tail comparator separates the pre-amplification and latch stages using distinct current paths, reducing power dissipation without compromising performance. This architecture enhances the comparator's ability to resolve small input differences quickly while minimising static power consumption [17], making it highly suitable for low-power ADC applications. By incorporating a double-tail comparator into the flash ADC design, the overall power consumption can be significantly reduced, improving its feasibility for portable and battery-operated devices [18].

The choice of semiconductor technology plays a critical role in ADC performance. The 45 nm CMOS process is particularly advantageous for designing low-power and compact circuits, as it allows for reduced feature sizes, lower leakage currents, and improved switching speeds. The use of this advanced process technology enables higher integration density, reducing the overall silicon area while enhancing power efficiency. Additionally, lower supply voltages can be used without compromising performance, further contributing to energy savings. By leveraging the benefits of the 45 nm CMOS process [19], it is possible to design an ADC that meets the stringent power and area constraints required in modern electronic systems.

In designing a low-power and low-area 3-bit flash ADC, optimising circuit architecture is essential to ensure minimal power dissipation while maintaining the inherent speed advantages of flash ADCs [20]. By employing techniques such as comparator sharing and optimised biasing strategies, the number of active components can be minimised, leading

to improved energy efficiency [21]. Careful layout design also plays a crucial role in reducing parasitic effects and mismatch-induced errors, further enhancing the ADC's reliability. A well-designed flash ADC should not only achieve high-speed operation but also consume minimal power and occupy a small silicon footprint, making it viable for a wide range of applications.

The increasing demand for energy-efficient data conversion solutions necessitates the development of innovative ADC designs that can meet the challenges of high-speed and low-power operation. Traditional flash ADCs, while offering unparalleled speed, face significant power and area constraints that limit their applicability in modern integrated systems. The integration of a double-tail comparator presents an effective approach.

Design methodology

The flash ADC in this architecture consists of three key components:

i. **Voltage divider network:** Generates reference voltages for comparison.
ii. **Comparator array:** Utilises a double-tail comparator for efficient voltage comparison.
iii. **Priority encoder:** Converts thermometer code into binary output, reducing complexity.

Double-tail comparator
The double-tail comparator is chosen for its low power consumption and high-speed decision-making. By separating the input and output stages [22], the architecture minimises kickback noise and allows operation at lower supply voltages, enhancing power efficiency [23]. This double tail comparator uses a width W=10 um and length L= 45 nm.

Circuit design
The circuit design of the proposed 3-bit flash ADC focuses on optimising power consumption and reducing silicon area while maintaining high-speed operation. A key component of this design is the double-tail comparator, which enhances power efficiency by separating the pre-amplification and latch stages into two independent current paths. This minimises static power dissipation while ensuring fast decision-making. The reference voltage ladder is designed using resistors to provide equally spaced

reference voltages for the comparators. Additionally, a thermometer-to-binary encoder is implemented to convert the comparator outputs into a 3-bit digital code. The overall design is optimised for low-power performance using the 45 nm CMOS process.

Simulation

Simulation plays a crucial role in validating the performance of the designed 3-bit flash ADC before fabrication. The circuit is simulated using industry-standard tools to analyse parameters such as power dissipation, speed, and accuracy. Transient analysis is performed to evaluate the response time of the ADC, ensuring fast and reliable signal conversion. Power consumption is measured under different input conditions to assess energy efficiency. Noise analysis is conducted to determine the signal-to-noise ratio (SNR) and effective number of bits (ENOB). The simulation results help identify any design flaws and enable necessary optimisations for improved ADC performance

Schematic design

The schematic design involves creating a detailed representation of the 3-bit flash ADC, including all circuit components and their interconnections. The comparators, reference voltage ladder, and encoder are systematically arranged to ensure efficient signal flow. The double-tail comparator is carefully designed to optimise switching speed and minimise power consumption. Proper layout techniques are employed to reduce parasitic effects and mismatch errors, ensuring high precision. The schematic is verified through design rule checks (DRC) and layout versus schematic (LVS) checks to ensure consistency with the intended circuit functionality. This step is essential in preparing the design for fabrication (Figures 33.1–33.4).

Priority encoder

A priority encoder compresses many binary inputs into less number of outputs. The priority encoder output is represented in the form of binary data. If there are 2 or more inputs at the same time, the input having high priority will take precedence. Priority encoder has 15 (I14 – I0) inputs and 4 (Q3 Q2 Q1 Q0) outputs as shown in Figure 33.5. Each input of priority encoder is driven by the comparator output, since at a time there are one or more output of the comparators are high, therefore the function of the

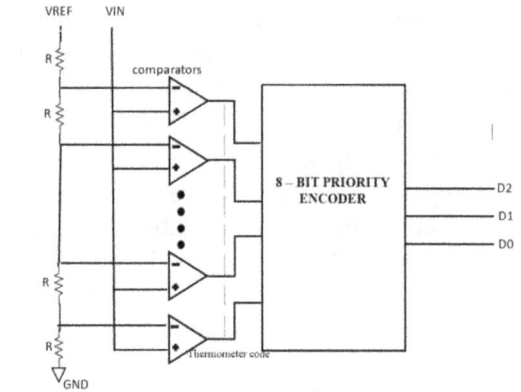

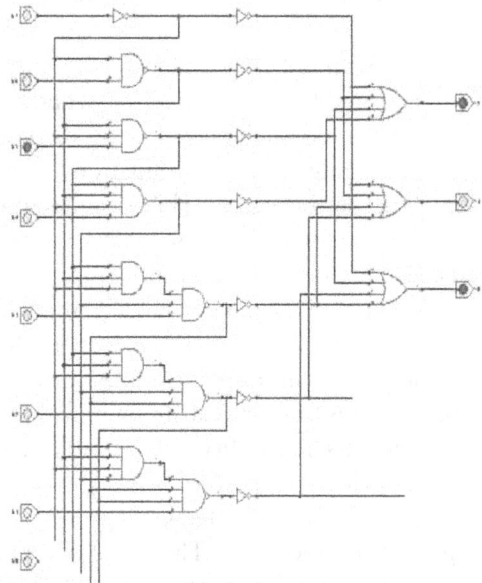

Figure 33.1 Schematic of proposed of 3-bit flash ADC [3]
Source: Author

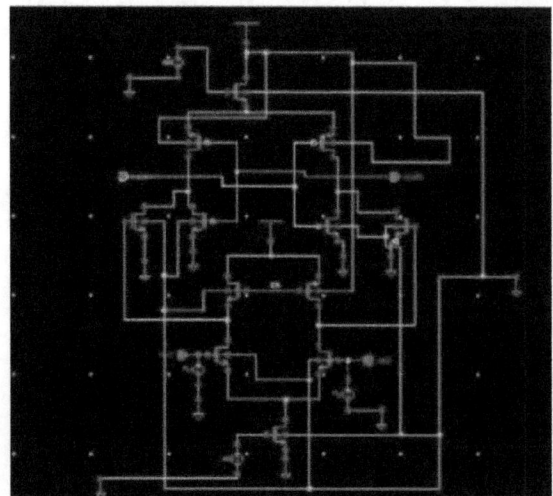

Figure 33.2 Schematic of double tail comparator
Source: Author

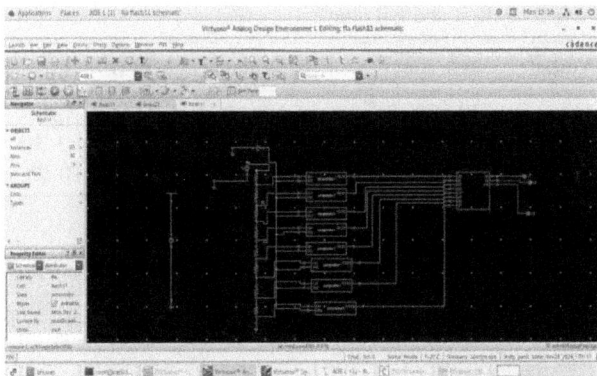

Figure 33.3 Schematic of 3-bit Flash ADC
Source: Author

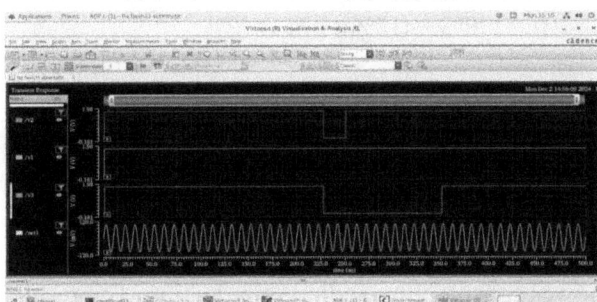

Figure 33.4 Priority encoder schematic diagram [11]
Source: Author

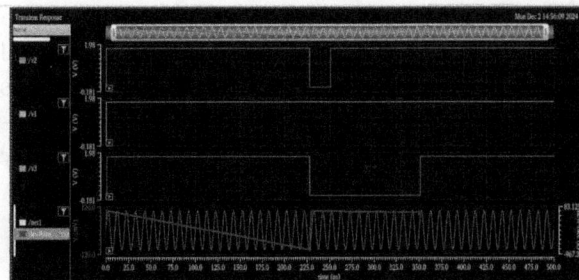

Figure 33.5 Output waveform of 3-bit flash ADC
Source: Author

Table 33.1 Priority encoder table

Range	C7	C6	C5	C4	C3	C2	C1	B2	B1	B0
V<0.125V_R	0	0	0	0	0	0	0	0	0	0
0.125 V_R< V<0.25V_R	0	0	0	0	0	0	1	0	0	1
0.25 V_R< V<0.375V_R	0	0	0	0	0	1	1	0	1	0
0.375 V_R< V<0.50V_R	0	0	0	0	1	1	1	0	1	1
0.50 V_R< V<0.625V_R	0	0	0	1	1	1	1	1	0	0
0.625 V_R< V<0.75V_R	0	0	1	1	1	1	1	1	0	1
0.75 V_R< V<0.875V_R	0	1	1	1	1	1	1	1	1	0
V>0.875V_R	1	1	1	1	1	1	1	1	1	1

Source: Author

priority inputs are given at the same time, the input having the highest priority will take precedence. The output indicates if the input is valid. Priority encoder has 15 (I14 – I0) inputs and 4 (Q3 Q2 Q1 Q0) outputs as shown in Figure 33.5 (Table 33.1).

Results and discussions

The 3-bit flash ADC is configured in 45 nm CMOS technology, and performance metrics are evaluated. The parameters calculated are DNL and INL. Slew rate is also calculated along with propagation delay, rise time and fall time. The proposed 3-bit Flash ADC is optimized for area using a double tail comparator approach and delay is optimized using a sampling rate of 100MHz and power optimization is done using a input voltage range from 0-1.2 V. The results are tabulated in table 33.2 with percentage improvement shown.

Calculation

INL and DNL

DNL (differential non-linearity)

i. Measures the difference between the actual step width and the ideal step width.

ii. Every DNL value of zero verifies step width precision and negative or positive readings above 1 and below -1 detect missing codes or large operational non-linearity.

INL (integral non-linearity)

i. Measures the deviation of the ADC's transfer function from the ideal straight line.

The **propagation delay (TPD)** is typically calculated as the average of the **rise time** and **fall time**, given by:

TPD= (Rise Time+Fall Time)/2

Absolute perfection in ADC operation leads to an inter-digitated non-linearity value of zero across all output code positions.

The maximum time-based output voltage alteration rate defines the slew rate of any circuit. High-speed circuits require a high slew rate because it decides the speed at often optimise these parameters to balance speed, power consumption, and signal integrity (Table 33.2).

Slew rate (SR)=Δt /ΔV

i. ΔV: The maximum change in output voltage.
ii. Δt: voltage variation rate over time.

Slew rate

The maximum time-based output voltage alteration rate defines the slew rate of any circuit. High-speed circuits require a high slew rate because it decides the speed at

Slew rate (SR)=Δt /ΔV

i. ΔV: The maximum change in output voltage.
ii. Δt: voltage variation rate over time.

Slew rate = (-597×10^6v/10^6)×10^6 = -597×10^6v/10^12
dvout/dt = -597v//μs

Rise time and fall time

i.   Signal transition from 10% to 90% of its final value occurs within rise time.
     tr = t10% − t90%

Table 33.2 Implementation of proposed vs. existing

Parameter	This Design (3-bit ADC)	Reference paper (3-bit ADC)	Percentage Improvement Merit
Technology Node	45nm CMOS	180nmCMOS	75% smaller node size
Resolution	3-bit	3-bit	N/A (same resolution)
Sampling Rate	100MHz	80MHz	−90% lower rate
Power Consumption	2.4mW	2.8ImW	41.5% lower power
Area	0.088mm2	0.088mm2	26.7% smaller area
Input Voltage Range	0–1.2v	0.6v–1.8v	N/A (ranges not directly comparable)
DNL/INL	DNL:±2LSB. INL:±2LSB	DNL:±0.189LSB. INL:±2LSB	−55%
ENOB	3-bits	2.85 bits	5.26% higher ENOB

Source: Author

Table 33.3 Rise and fall time

Voltage	Rise time	Fall time	TPD
V1	20.27×10^{-12}	16.17×10^{-12}	18.22×10^{-12}
V2	20.73×10^{-12}	16.05×10^{-12}	18.39×10^{-12}
V3	23.32×10^{-12}	15×10^{-12}	19.16×10^{-12}

Source: Author

ii. The falling time of a signal describes its transition duration between a voltage state which reaches 90% of its starting value and reaches 10% of its starting value (Table 33.3).
     ft = t90% − t10%

Conclusion

The proposed 3-bit flash ADC demonstrates significant power and area savings while maintaining high-speed operation. The integration of a double-tail comparator proves effective in optimising power consumption and noise reduction. The results confirm that the proposed ADC achieves a balance between speed, efficiency, and compact design, making it suitable for low-power applications such as IoT and portable electronics. Additionally, the use of a 45 nm CMOS process ensures scalability for future high-resolution ADC designs while maintaining performance and energy efficiency. Further research can explore integrating additional power-saving techniques such as adaptive biasing and dynamic voltage scaling to further enhance energy efficiency. Future work also includes extending the resolution to 8-bit ADCs and optimising signal integrity for high-frequency applications. These advancements will enable the proposed architecture to be utilised in a broader range of applications, including biomedical signal processing and real-time embedded systems.

The design of a low-power and low-area 3-bit flash ADC using a double-tail comparator in a 45 nm CMOS process presents a significant advancement in high-speed data conversion. Traditional flash ADCs, while known for their rapid signal processing capabilities, suffer from high power consumption and large silicon area requirements due to the exponential increase in comparators with resolution. The incorporation of a double-tail comparator addresses these limitations by reducing static power dissipation and improving switching speed, making the ADC more energy-efficient while maintaining its inherent speed advantage.

The double-tail comparator plays a crucial role in this improvement by introducing a two-stage design that separates the pre-amplification and latch functions. This structure minimises unnecessary current flow, thereby reducing power dissipation while ensuring accurate and fast decision-making. The low-power design is particularly beneficial for battery-operated and portable electronic devices where energy efficiency is a primary concern. Additionally, the optimised reference voltage ladder and efficient thermometer-to-binary encoder contribute to overall circuit efficiency, ensuring minimal energy waste during operation.

Another major accomplishment is the reduction in silicon area, which enhances integration potential for modern compact electronic systems. Traditional flash ADCs require a large number of comparators, leading to increased chip area and fabrication costs. This design mitigates these challenges by implementing area-efficient circuit techniques, including optimised comparator sharing and streamlined logic circuitry. The use of the 45 nm CMOS process further enhances integration by allowing smaller transistor sizes, lower leakage currents, and improved density. This ensures that the ADC can be implemented in space-constrained applications without compromising functionality or performance.

The effectiveness of the proposed ADC design has been validated through extensive simulation and schematic analysis. Performance evaluations have confirmed that the ADC achieves high-speed operation with reduced power dissipation, making it well-suited for applications requiring rapid and energy-efficient data conversion. Transient analysis has demonstrated fast response times, while noise analysis has indicated a satisfactory and an effective number of bits (ENOB). These results highlight the reliability and efficiency of the proposed architecture, proving its capability to meet the demanding requirements of modern electronic systems.

Overall, this work successfully addresses the fundamental challenges associated with conventional flash ADCs by introducing a power-efficient and compact design. The integration of a double-tail comparator enhances speed and energy efficiency, while the use of advanced CMOS technology optimises layout density and reduces overall power consumption. This design methodology sets a strong foundation for future developments in high-speed ADC architectures, paving the way for improved performance in communication systems, biomedical instrumentation, and portable electronics. This work contributes to the on-going advancement of low-power, high-speed data conversion technologies by achieving an optimal balance between power, speed, and area. Future research can further explore higher-bit resolutions using similar design principles to extend the benefits of this approach to more complex ADC applications.

Acknowledgments

I would like to express my heartfelt gratitude to Dr. Vasudeva G., Department ECE, DSATM, Bangalore, for his invaluable assistance and guidance throughout this research project on the design and implementation of a double-tail comparator. His insights and feedback have significantly enhanced the quality of this work.

Additionally, I would like to acknowledge the resources and tools provided by DSATM and the use of Cadence Virtuoso in the design and simulation processes, which were instrumental in achieving the results presented in this study.

Lastly, I appreciate the existing literature and research contributions from authors such as Razavi, Allen, and Kumar, which provided foundational knowledge and inspiration for my work.

References

[1] Jacon Baker, R., Li, H. W., Boyce, D. E. CMOS circuit design, layout, and simulation. *Inst. Elec. Electr. Engr. Inc.*, New York, First Edition, 1997. 33–37.

[2] Gray, P. R., Meyer, R. G. Analysis and design of analogue integrated circuits. Englewood Cliffs, N.J., Prentice Hall, 4th Edition, 1993. 55–67.

[3] Baker, R. J. CMOS: Circuit design, layout, and simulation. Wiley-Interscience, 2008. 23–24.

[4] Razavi, B. Design of analogue CMOS integrated circuits. McGraw-Hill, 2001. 167–169.

[5] Rabaey, J. M., Chandrakasan, A., Nikolic, B. Digital integrated circuits: A design perspective. Prentice Hall, 2nd Edition, 2003. 12–14.

[6] Roy, K., Jain, P. Low power CMOS VLSI circuit design. Wiley-IEEE Press, 2015. 78–80.

[7] Hong-Hai, T., Cong-Kha, P., Duc-Hung, L. Design of a low-power and low-area 8-bit flash ADC using a double-tail comparator on 180 nm CMOS process. *Sensors*, 2023. 16–18.

[8] Saranya, P. M., Jyothish Chandran, G., Shilpa, K. S. Design of high-performance double-tail comparator. *ICTACT J. Microelec.*, 2017;03(03). 23–27.

[9] Ram Prasad, J. M., Kariyappa, B. S., Ravishankar, H. Design and implementation of flash ADC for low pow-

er applications. *IOSR J. VLSI Sig. Proc. (IOSR-JVSP)*, 2014;4(6):41–46.

[10] Devendra, K., Ashish, R. Design of low-power high-speed comparators for flash ADC applications. *Internat. J. Scient. Res. Engg Trends (IJSRET)*, 2023;9(4). 14–19.

[11] Xiaoyu, W., Yukang, F. Analysis and design of 8-bit CMOS priority encoders. Preprint, University of Virginia, June 2018.

[12] Singha, T. B., Konwar, S., Roy, S., Vanlalchaka, R. H. Power efficient priority encoder and decoder. *Proc. 2014 Internat. Conf. Comp. Comm. Inform. (ICCCI-2014)*, 2014. 56–59.

[13] Shivam Singh, B., Mishra, D. K. Design and analysis of double-tail dynamic comparator for flash ADCs. *Proc. Depart. Elec. Instrumen. Engg.* 2012;2:24–26.

[14] Kulkarni, T. R., Dushyant, N. D. Early and noninvasive screening of common cardio vascular related diseases such as diabetes and cerebral infarction using photoplethysmography signals. *Results Optics*, 2021;3:100062.

[15] Kulkarni, T. R., Dushyanth, N. D. Performance evaluation of deep learningmodels in detection of different types of arrhythmia using photo plethysmography signals. *Internat. J. Inform. Technol.*, 2021. 15–19.

[16] Kulkarni, T., Krupa, B. N. Noninvasive method to find oxygen saturation of the fetus using photoplethysmogram. *2016 Internat. Conf. Elec. Electr. Comm. Comp. Optimiz. Tech. (ICEECCOT)*, 2016:40–42.

[17] Srivathsava, L., Kulkarni, T. Novel design of VCO with output peak to peak control. *Internat. J. Instrum. Control Autom.*, 2011:92–195.

[18] Vasu, G., Bharathi, G., Mandar, J., Kulkarni, T., Kulkarni, R. Design of FinFET based op-amp using high-K device 22 nm technology. 2023. Volume 42: Applied Mathematics, Modeling and Computer Simulation, Advances in Transdisciplinary Engineering, DOI10.3233/ATDE231033, 928–939

[19] Jatkar, M., Jha, K. K., Patra, S. K. Fermi velocity and effective mass variations in ZGaN ribbons: Influence of Li-passivation. *IEEE Access*, 2021;9:154857–154863.

[20] Jha, K. K., Jatkar, M., Athreya, P., T. M. P., Jain, S. K. Detection of gas molecules (CO, CO_2, NO, and NO_2) using Indium nitride nanoribbons for sensing device applications. *IEEE Sensors J.*, 2023;23(19):22660–22667.

[21] Jatkar, M., Jha, K. K., Patra, S. K. DFT investigation on targeted gas molecules based on zigzag GaN nanoribbons for nano sensors. *IEEE J. Elec. Devices Soc.*, 2022;10:139–145.

[22] Vasudeva, G., Uma, B. V. 22nm FINFET based high gain wide band differential amplifier. *Internat. J. Circuits Sys. Sig. Proc.*, 2021:55–62.

[23] Vasudeva, G., Uma, B. V. Low voltage low power and high speed OPAMP design using high-KFinFET device. *WSEAS Trans. Circuits Sys.*, 2021;20:80–87.

34 Vehicle detection using saliency model generation and SVM

Dinesh Kumar D. S.[1,a], Mohankumar C. E.[2,b] and Senthilkumar S.[2,c]

[1]Assistant Professor, Department of Electronics and Communication Engineering, K S Institute of Technology, Bangalore affiliated to VTU, Belagavi, Karnataka, India

[2]Assistant Professor, Department of Electronics and Communication Engineering, GRT Institute of Engineering and Technology, Tiruttani, Tamilnadu, India

Abstract

In low-light conditions, it becomes challenging to distinguish a car's contours, making vehicle identification at night difficult, especially when equipment resources are limited. To address this issue, an improved histogram of oriented gradients (HOG) method is employed to extract key features of the car. Initially, a saliency model is generated to highlight important regions. Using this model, the vehicle lights are isolated, and background illumination is removed. These extracted lights are then combined to suggest potential vehicle locations based on predefined templates. Next, super-pixel features are integrated with HOG features (forming S-HOG descriptors). A support vector machine (SVM) is used to classify the extracted features. Finally, a Kalman filter is applied to track the detected vehicles over time. This approach significantly enhances vehicle visibility and identification at night.

Keywords: Support vector machine, monocular vision, vehicle detection, histograms of oriented gradients, non-maximum suppression, Kalman filter

Introduction

Computer vision is used to study many things, such as robots, industrial automation, human-machine interfaces, text analytics, motion tracking, and character recognition and identification [1]. Recognising a moving object is necessary for many uses, such as video security systems. The vehicle detection system's main goal is to keep people safe on the street. This means that the system to be automatic, cheap, and easy to get is possible. Other uses include reducing traffic, letting us navigate on our own, automating technology, and changing the way a car is set up.

In video monitoring, the camera tracking device includes both fixed and moving things. The main goal is to figure out how a spinning shaft moves physically in a certain area. Moving object methods [2, 3] find objects in a video system. This method helps to automatically found and watch moving objects in a video frame. This method sorts vehicles into groups so that we can find the connection between two sets of targets or object pieces and find out information about the target, like its path, speed, and direction. Using the scanning of moving cars, objects are found one picture at a time in a movie [4]. Over the past decade, things like speed, regularity, and short distances have become annoying car traffic issues. The environment and the amount of focus also have a big effect on getting the right detecting results [5].

Time-based and sensor-based review methods that have been used for a long time are not as good as intelligent traffic analysis [6]. The model proposes different forms. Add the object's form, outline, width, and height, as well as its darkness, lights, circle spots, HOG characteristics, and Haar characteristics. Picture detection is another way that attributes can be used such as lines, parallelism, and lighting arrangements as a group to find and stop darkness. To make moving easy, it's important to be able to quickly and correctly find cars and figure out where they are grouped. Finding and following are the two most important steps here. It is important to accurately name the cars and other things in the scene [7–9]. There may also be a range of sizes and types of cars shown.

This paper suggests a better way for cars to automatically recognise and sort things. A lot of neural algorithms use various processing methods or combine several methods to create a better system. This can make the job time-consuming and expensive, which could make it harder to examine and analyse

[a]dineshkumards@ksit.edu.in, [b]mohankumar.ce@grt.edu.in, [c]senthilkumar.s@grt.edu.in

DOI: 10.1201/9781003685876-34

things in real time. The goal of the method is to find cars in the traffic data that is given [10]. Some small cars are far away, while big cars are close by. The goal of this study is to show a better MSTM system that can find and identify cars in busy places. For our MSTM-based system to work, we need to get data, change frames, find areas of interest, abstract features, improve graph mining, and sort the data into groups.

The rest of this research work is made up of the following parts. In Section 2, we talk about all the specifics of connected and the previous works. The suggested method is described in Section 3. It has steps for getting ready, getting rid of the background, extracting features, and classifying it. In Section 4, we show experiments results and how they compare to other cutting-edge methods. Section 5 talks about limits, the study's scope, its possibility, and a few suggestions for the future.

Preliminaries for vehicle detection

Tasks
Finding vehicles is a very important part of how intelligent car systems understand their surroundings. Positioning and categorising different things in road settings is made easier. This includes people, non-motorised cars, traffic signs, and line markings. 2D object detection and 3D object detection are two widely used vehicle detection methods. It involves clever cars using a 2D limiting box to pick out things they see, and then those objects are labelled and put in the right place. From the camera's point of view, the 3D object recognition technology shows exactly where the cars are. 3D border boxes are used to pick out the things that have already been named. After that, the items are put into groups and found [11].

Evaluation metrics
Most of the time, F1-score, IoU, recall (R), precision (P) and mean average precision (mAP) are used to see how well a program finds things. In order to measure how much expected and actual border boxes intersect, IoU is used. Precision is the percentage of expected positive samples that turn out to be true positives. The following methods are used to represent these metrics:

$$IoU=(pb \cap gb)/(pb \cup gb) \qquad (1)$$

$$P=TP/(TP+FP) \qquad (2)$$

$$R=TP/(TP+FN) \qquad (3)$$

$$F1=2 \times P \times R/(P+R) \qquad (4)$$

in which pb stands for predicted boxed and gb for ground truth boxes; TP represents the true positive that is "the number of positive cases that were correctly identified as real samples", FP denotes false positive that is "the number of negative cases that were wrongly thought to be true samples", and FN is the false negative that is "the number of positive cases that were wrongly thought to be false samples". Another difference is that AP shows how precise a class is, while mAP shows the average AP number for all classes. The following formulae are used to come up with these metrics:

$$AP=\int_0^1 P(R)dR \qquad (5)$$

$$mAP=1/n\sum_{i=1}^n P(i)\Delta R(i) \qquad (6)$$

In the above equation n is represent the number of groups that to be determined. When we want to know how fast something can be found, FLOPs and FPS (frames per second) are useful units. There are a certain number of trainable factors that are used to train the model. This is often used to figure out how big the model is. FLOPs measure how complicated the model is. The smaller the FLOPs number, the less work needs to be done on the computer to determine models. The number of frames handled per second (FPS) is used to measure how fast the recognition is.

Datasets
Datasets are necessary to teach models how to do every part of self-driving cars. Smart cars and computer programs that learn on their own use high-quality tagged data, like movies, pictures, and sensor data from a variety of locations, to learn. This data is labelled by hand to show information about proper behaviour, object recognition, and environment awareness [12-16].

Proposed method

Object proposal generation method
1) Background light removing based on saliency segmentation: This feature doesn't work very well when there are a lot of different types of lights, like streetlamps and cars that reflect light. The old Otsu method can't tell the difference between picture details that are very bright. A fixed threshold method

is offered as a way to get a set threshold for sorting lights based on how bright they are. Figure 34.1b shows that this way works better than the general Otsu method. Figure 34.1c, on the other hand, shows that it's still not sure.

For good threshold classification model should have the following: (1) **Good detection:** it shouldn't miss important areas or mistakenly label the background as important. (2) **High-resolution:** To find important things and store the original picture data correctly, saliency maps need to have a high or full resolution. (3) These models should be able to quickly find key spots that begin more complicated processes (Figure 34.2).

This kind of recognition is divided into two main groups: patch-based and region-based. Some good and quick patch-based recognition methods are saliency segmentation (SEG), functional similarity (FWD), and self-resemblance (SER). It's not right in some ways. More than the important parts, the edges that are very different from the rest stand out. Also, the edges of the important parts don't stay straight. Some say to find signs to understand what's important. This could help in this case. It's easy to make

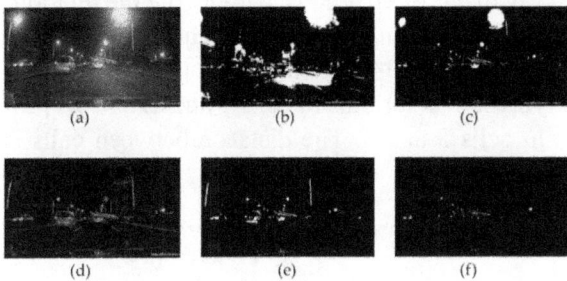

Figure 34.1 Highlight extract: (a) grey image; (b) traditional Otsu; (c) fixed threshold method; (d) background illumination removal; (e) traditional Otsu based on saliency model; and (f) fixed threshold based on saliency model

Source: Zhang, L., Xu, W., Shen, C., & Huang, Y. (2024). Vision-Based On-Road Nighttime Vehicle Detection and Tracking Using Improved HOG Features. Sensors 24(5)

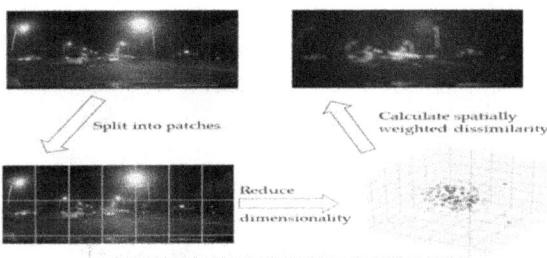

Figure 34.2 Saliency model generation

Source: Zhang, L., Xu, W., Shen, C., & Huang, Y. (2024). Vision-Based On-Road Nighttime Vehicle Detection and Tracking Using Improved HOG Features. Sensors 24(5)

lines that work quickly and well because there aren't as many points as blocks. Three great things about plans that are based on regions: (1) They use complementary priors to make the system work better; (2) Regions help to find the scene's important objects with more complex clues (like colour histograms) than pixels and colour patches; and (3) Computing saliency at the region level can make full-resolution saliency maps much cheaper because there are fewer regions in an image than pixels.

2) Segmentation method based on template: It's simple to make a mistake with the patterns of light forms and coordinate features, even though the most recent study uses a method based on templates. It's because of the sun and the angle. The two lights on a car are often shown lined up horizontally. A saliency map tells us how much weight to give to light coordinates features so that we can get ROI.

The following types of lights are used in the ROI create stage:

Two lights on the same car are farther apart on the side (Dl) than they are at their fastest speed (Thl):

$$(L1,2)<Thl \tag{7}$$

The same car has two lights that are mostly flat:

$$Dh(L1,L2)Dl(L1,L2)<Thhl \tag{8}$$

But using lights only for where they are put could make the wrong thing being found or it can help make sure that nothing is missed and help the proof process easy.

Hypotheses verification

The usual way of using closeness of light forms will give false alarms at night because the surroundings are so complicated. This feature is often used to find vehicles because it works well in a variety of lighting conditions, doesn't change shape when the shape of the background changes. When it comes to classification, contour features don't work as well at night as they do during the day. This is because the HOG feature takes a long time and isn't very good at blocking out noise, and there aren't as many car contour pieces to work with. We use S-HOG features, which are a mix of super pixel and HOG images in grey scale and red scale, to show the shape of the vehicle and the shape of the lamp, respectively. Then, we use SVM for classification to make vehicle recognition more accurate.

1) S-HOG feature extraction: In HOG, a super pixel method that is used based on an area. In the first step, a SLIC0 super pixel was used to give each pixel a weight in the form of HOG features. But the density factor wasn't thought about. It did this by itself. This way, vehicles can be correctly identified in a wide range of tough situations (Figure 34.4a). S-HOG works really well since it takes advantage of the ways that super pixels and HOG are alike. The number at the base changes based on what inside the super pixel. This means that SLIC0 works better in places with and without structure. But because of the benefits, the edges of regions don't stick to SLIC0 as much. Simple linear iterative clustering (SLIC) doesn't need a lot of memory or processing power to work. But for most uses of super pixels, the small pros of SLIC are bigger than the small cons.

We use a method that works with both dark and light S-HOGs in this case. As was already said, the bases of the tail lights often have white around the red part. First, switch from RGB to HSV for the colour scheme. After that, channel 2 gets the dark picture and channel V gets the red picture. Then HOG traits like HGrey and HRed are added (Figure 34.3).

There are four cells in each block, and each picture in the green and red channels is 32 pixels by 32 pixels. An 8×8 image area is picked to stand for a HOG cell. Edge gradients that are 180° apart are in the same direction. That is, the direction of the pixels' gradients is split equally into nine histogram channels. For each training example, the HOG feature vector is 324. The 648 dimensions can be used to check the car. By putting together the 324-dimensional parts

of the dark image and the red image, this picture is made.

2) SVM Training: This work uses the radial basis function (RBF) to build the best hyper-plane. For training the classifier, samples of 5705 positives and 3669 negatives are chosen. Things like cars, vans, buses, and trucks are good examples. Bad examples are light posts, traffic signs, guardrails on the side of the road, and billboards. There is a way to get rid of the holes as shown in Figure 34.5b. It works better than the normal HOG feature (Figure 34.4).

3) Overlapped area removal: This method tries to find the strongest edges and hide detections that are already there or go across them. By using the classifier's best score, non-maximum suppression (NMS) gets rid of gaps with low scores and spots with too much detection. The V-HOG tool should be used to find the score since cars are all the same on both sides (Figure 34.5).

As shown in Figure 34.6, the car is more even in the ROI area than in other places. This means that cells 1 and 2 are the same as cells 3 and 4 if there is a car in that area. Equations 9 through 12 describe these four cells.

A way that is often used is to figure out the relationship value based on Euclidean distance. As shown in Equations 13 and 14, the sin(x,y) is in the range of 0 to 1. y_ix_i stands for bins in cells 1 and 2, and y_iy_j for bins in cells 3 and 4. The distance between cells is

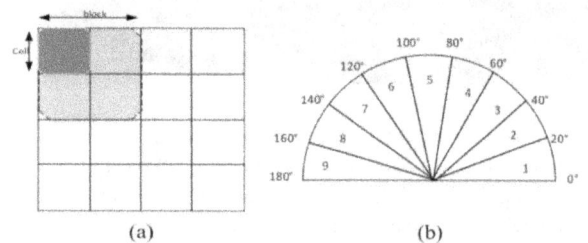

Figure 34.4 HOG features: (a) cells and blocks and (b) bins
Source: Zhang, L., Xu, W., Shen, C., & Huang, Y. (2024). Vision-Based On-Road Nighttime Vehicle Detection and Tracking Using Improved HOG Features. Sensors 24(5)

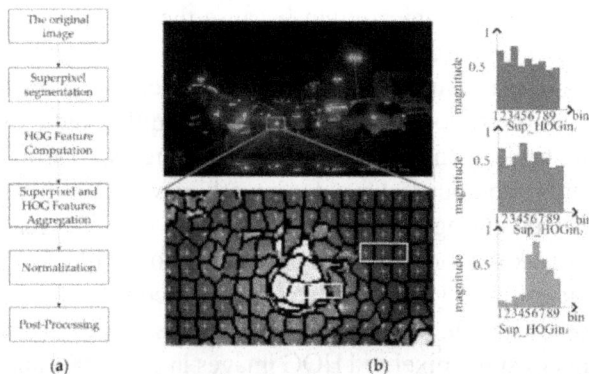

Figure 34.3 Super-pixel and HOG aggregation: (a) aggregation flowchart and (b) super-pixel weighted HOG feature generation.
Source: Zhang, L., Xu, W., Shen, C., & Huang, Y. (2024). Vision-Based On-Road Nighttime Vehicle Detection and Tracking Using Improved HOG Features. Sensors 24(5)

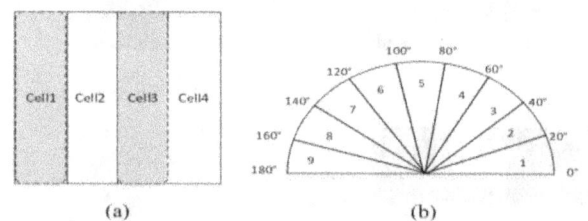

Figure 34.5 V-HOG features: (a) cells and (b) bins
Source: Zhang, L., Xu, W., Shen, C., & Huang, Y. (2024). Vision-Based On-Road Nighttime Vehicle Detection and Tracking Using Improved HOG Features. Sensors 24(5)

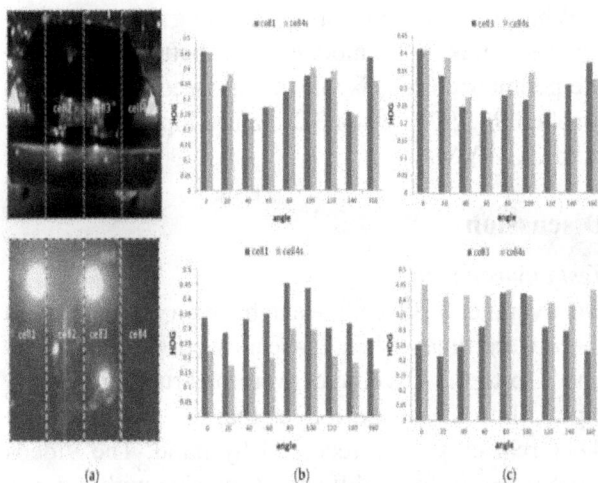

Figure 34.6 V-HOG symmetry: (a) ROI image; (b) cell1 and cell4s HOG symmetry; and (c) cell2 and cell3s HOG symmetry

Source: Zhang, L., Xu, W., Shen, C., & Huang, Y. (2024). Vision-Based On-Road Nighttime Vehicle Detection and Tracking Using Improved HOG Features. Sensors 24(5)

shown by d(x,y). It's more likely that the two things are alike if the number of sin(x,y) is high.

$$(x,y)=\Sigma(xi-yi)^2 \quad\quad (9)$$

$$\sin(x,y)=11+d(x,y) \quad\quad (10)$$

4) Vehicle tracking: The accuracy of finding moving vehicles can be improved by predicting and following targets that are vehicles. The video shows that the same car moves very little from one frame to the next, and its motion is very close to being constant straight movement. When there is variable-speed linear motion, like when turning, stopping, or other similar situations, the short time between frames reduces the prediction error that comes with regular rectilinear motion. These mistakes are seen as system noise. In addition, there is a connection between where the car is in the picture and how big it is. So, it's safe to say that the vehicle's size change in the sequence pictures is pretty much the same, and the small mistake can be thought of as system noise. This is one way to write the system state equation:

$$Xk=AXk-1+Buk+wk \quad\quad (11)$$

$$X=[x,y,w,h,\Delta x,\Delta y]^T \quad\quad (12)$$

$$A=[1\,0\,0\,0\,1\,0\ ;\ 0\,1\,0\,0\,0\,1;\ 0\,0\,1\,0\,0\,0\ ;\ 0\,0\,0\,1\,0\,0\ ;$$
$$0\,0\,0\,0\,1\,0\,;0\,0\,0\,0\,0\,1] \quad\quad (13)$$

$$B=[0,0,0,0,0,0]^T \quad\quad (14)$$

Its state vector is XX, which stands for (x, y) is the image coordinate at the middle of the vehicle's lower edge, and w and h are the target's width and height in the image coordinate system. A is the matrix for transferring system state. This is the system measurement equation:

$$Zk=HXk+ \quad\quad (15)$$

$$H=[1\,0\,0\,0\,0\,0;\ 0\,1\,0\,0\,0\,0;\ 0\,0\,1\,0\,0\,0;\ 0\,0\,0\,1\,0\,0;\ 0\,0\,0\,0\,1\,0;$$
$$0\,0\,0\,0\,0\,1] \quad\quad (16)$$

We use the Kalman filter to guess the new spot, its width, height, and the area where the car is most likely to be [20]. We see the same frame when we look back to time t+1 as when we look forward to time tt. For those who fail to match, the value that was measured will be thought to be the value that was predicted to be at the end. The item being tracked will be thrown away if the finding fails 10 times in a row.

Experimental results

Video processing: Test environment

Real traffic films from Guadalajara, Mexico (V1, V2, and V3), the GRAMME road-traffic monitoring (GRAM-RTM) collection (V4, V5), and Britain's M6 highway (V6, V7, and V8) were used to test the scheme.

The movies were all slowed down to 25 fps and 420 to 240 pixels. It took less time to do the maths after this was done. The camera could see the cars coming right up behind them. From 19.5 feet above the road, cell phone videos V1, V2, and V3 were taken. This video's double opening traffic is different from that in the other movies. Quite a bit of sound is also present. To make the ground truth (GT) collection for testing reasons, all picture frames were looked at by hand.

Figure 34.7 shows the amount of the traffic that normally take in account that how many frames are in each movie, how busy the site is, the weather, and the place. Some of the videos are more than 61 minutes duration, Large number of real cars, different locations in different countries, different weather, traffic varying from 1.32 cars per second, with peaks of 2–4 cars per second (see Figure 34.7), and a vehicle occlusion index (VOI) varying from 0.00 to 0.312.

A PC with an Intel Core i7 processor, 16 GB of RAM, and a speed of 3.40 GHz was used to test it. It was made in MATLAB. To test the system, how it works at each stage, the same tests are used. When used in different stages, TP, FP, and FN all mean different things.

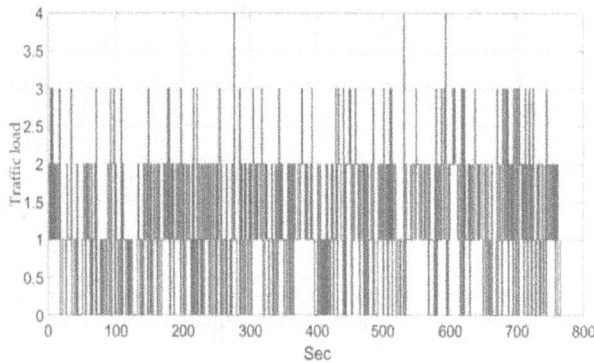

Figure 34.7 Number of vehicles per second in the road

Source: Velazquez-Pupo, R., Sierra-Romero, A., Torres-Roman, D et.al. (2018). Vehicle Detection with Occlusion Handling, Tracking, and OC-SVM Classification: A High-Performance Vision-Based System. Sensors, 18(2)

For cars in groups S (small), M (medium), and L (large), during the marking step, three criteria's are used in classification stage: area, breadth, and relHW, together with OC-SVM.

Vehicle detection implementation results

The classification step can find cars 83.793% of the time without the occlusion-handling method. When the occlusion-handling method is added to the finding stage, 95.216% more cars are found and 11.423% more of the attempts are successful. There was a strong link between the F-score and the measured VOI score when these pictures were found. The films V1, V2, and V3 were all filmed in Mexico, which is a place where very large cars pass, so the cameras shook a lot. The FP and VOI figures for the V4 and V5 pictures are the lowest. They were shot in Madrid, Spain. V6, V7, and V8 all showed numbers that were thought to be normal when the VOI score was close to 0.2. If we want to make the occlusion dealing program better, we should use different methods, such as K-means and SVM, to check things like the convexity of the blobs.

Results of vehicle classification

The CLIPSVM tool could use an RBF kernel for both OC-SVM and SVM classification. The K-means method could also be used to measure things. Figure 34.6 shows the test results for pictures V6, V7, and V8. During the classification step, we used tools like K-means, bounds, SVM, and OC-SVM. Finally, we looked at width, area, and relative height. Once the occlusions were found, they were cleared. In fact, the letters "S," "M," and "L" stand for a lot of different kinds of cars.

What the tests showed is that using three geometric traits makes the models work better than a single geometric trait, like area, was used. The F-score reached 98.190% when 3D feature input space used with OC-SVM.

Discussion

Test environment

On average, 1.32 cars went by every second, but there were times when 2–4 cars went by every second. Up to 0.312 vehicles were obstructing view in eight films that were longer than 61 minutes, and 4111 real cars were recorded by hand. The videos were made in three different countries with a range of weather conditions. That's all the times when the method works well and right away.

Occlusion handling algorithm and VOI-index

Since shadows can make it look like two or more cars are one, a method was created to cut down on this problem. With this method (look at Table 34.1), the rate of recognition can range from 83.793% to 95.216%.

Clustering analysis

The cars were put into three groups: small, medium, and big. Clustering analysis tools like K-means, SVM, and OC-SVM were used to do this. During the sorting step, these algorithms can look at all the different ways the cars looked in the training data.

SVM and OC-SVM

The best ones were SVM and OC-SVM. OC-SVM could remember things from all over the world and had an F-score of up to 98.525%. For medium-sized cars with video V6, it had an F-score of 99.211%. They believe that the chosen factors are to blame for the differences in how well SVM and OC-SVM work. In this study, the factors care {1, 5, 36} {1, 5, 36} and {0.5, 0.65, 0.95} {0.5, 0.65, 0.95}. These were used to test the SVM classification. The hyper parameter for OC-SVM are {1, 10.5, 15} {1, 10.5, 15} and {0.001, 0.01, 0.1} {0.001, 0.01, 0.1}, respectively. There were times when the results were wrong because of unclear occlusions in the detection step. This held true most of the time when cars were going side by side.

Real time processing

Most of the time, our system can handle one picture frame in less than 30 ms. At 25 frames per second, this shows that our plan can work for an average of 1.32 vehicles per second and a peak of 4 cars per second. Whenever more cars on the road then, classification produce worse results and can be represented by the "car blockage score."

Conclusions

The proposed system aims to build, test, and use a very good vision system that only needs one steady camera to find, track, count, and sort moving cars in both the front and back views. This system is suitable for IoT-based smart city. The number and quality of measures used are better than those in most similar studies. In real life, our method works well on a three-lane road where 1.32 cars per second is the normal speed and up to 4 cars per second when it's busy. It was simple to find the best models with SVM. With an F-score index 99.051% for the middle class and up to 98.190% for the big class, it really worked to tell the cars apart.

An OC-SVM with an RBF kernel is used as the input space in the classification stage. This feature allows the proposed system more efficient. It is done in a certain line of the ROI so that the input space doesn't change too much between classes. There are still some areas to be improved in future work are: Methods for making backgrounds in a range of colour spaces, make programs that can automatically find the ROI and the lines that separate the lanes. Improve the algorithms that find occlusions caused by heavy traffic, especially for big cars, so they can find them more often and with less change in the values of points in the tracking and sorting input space. Then, use measures to describe the occlusion. Because this problem has a lot of features and the values of features within and between classes can vary, it's still not clear what the best number of classes is for classification.

Table 34.1 Experimental results on videos V6, V7, and V8 using different input spaces and classifiers

	Classification with the thresholds and 1D feature input space							
Test	**Class**	**Input space**	**TP**	**FP**	**FN**	**Recall**	**Precision**	**F-score**
With occlusion handling	S	10	9	474	1	90.000	1.863	3.651
	M	2336	1875	63	461	80.265	96.749	87.739
	L	141	39	27	102	27.659	59.090	37.681
	Total	2487	1923	564	564	77.322	77.322	77.322
	Classification with K-means and 3D feature input space							
Test	**Class**	**Input space**	**TP**	**FP**	**FN**	**Recall**	**Precision**	**F-score**
With occlusion handling	S	10	10	247	0	100.00	3.891	7.490
	M	2336	2079	23	257	88.998	98.905	93.690
	L	141	117	11	24	82.978	91.406	86.988
	Total	2487	2206	281	281	88.701	88.701	88.701
	Classification with SVM and 3D feature input space							
Test	**Class**	**Input space**	**TP**	**FP**	**FN**	**Recall**	**Precision**	**F-score**
With occlusion handling	S	16	16	100	0	100.000	13.793	24.242
	M	2333	2214	4	119	94.899	99.819	97.736
	L	138	133	20	5	96.376	86.928	91.408
	Total	2487	2363	124	124	95.014	95.014	95.014

Source: Author

References

[1] Tang, Y., Zhang, C., Gu, R., Li, P. Yang, B. Vehicle detection and recognition for intelligent traffic surveillance system. *Multimed. Tools Appl.,* 2017;76:5817–5832.

[2] Ali, A. M., Eltarhouni, W. I., Bozed, K. A. On-road vehicle detection using support vector machine and decision tree classifications. *Proc. ICE MIS,* 2020:1–5.

[3] Abdusalomov, A., Mukhiddinov, M., Djuraev, O., Khamdamov, U., Whangbo, T. K. Automatic salient object extraction based on locally adaptive thresholding to generate tactile graphics. *Appl. Sci.,* 2020;10:3350.

[4] Al Gharrawi, H., Yaghoub, M. B. Traffic management in smart cities using the weighted least squares method. *arXiv* 2022, arXiv:2205.00346.

[5] Broggi, A., Cardarelli, E., Cattani, S., Medici, P., Sabbatelli, M. Vehicle detection for autonomous parking using a soft-cascade AdaBoost classifier. *Proc. IEEE Intel. Veh. Symp.,* 2014:912–917.

[6] Kaushek, K. T. R., Thiruvikkraman, S., Gokul, R., Nirmal, A., Karthika, R. Evaluating the scalability of a multi-object detector trained with multiple datasets. *Proc. ICICCS,* 2021:1359–1366.

[7] Wei, Y., Tian, Q., Guo, J., Huang, W., Cao, J. Multi-vehicle detection algorithm through combining Harr and HOG features. *Math. Comput. Simul.,* 2019;155:130–145.

[8] Wang, Q., Xu, N., Huang, B., Wang, G. Part-aware refinement network for occlusion vehicle detection. *Electronics,* 2022;11:1375.

[9] Shobha, B., Deepu, R. A review on video based vehicle detection, recognition and tracking. *Proc. CSITSS,* 2018:183–186.

[10] Neumann, D., Langner, T., Ulbrich, F., Spitta, D., Goehring, D. Online vehicle detection using Haar-like, LBP and HOG feature based image classifiers with stereo vision preselection. *Proc. 2017 IEEE Intell. Veh. Symp.,* 2017: 773–778.

[11] Wei, Z., Zhang, F., Chang, S., Liu, Y., Wu, H., Feng, Z. Mmwave radar and vision fusion for object detection in autonomous driving: A review. *Sensors,* 2022;22:2542.

[12] Menze M., and Geiger A., "Object scene flow for autonomous vehicles," 2015 IEEE Conference on Computer Vision and Pattern Recognition (CVPR), Boston, MA, USA, 2015, 3061–3070.

[13] Cordts, M., Omran, M., Ramos, S., Rehfeld, T., Enzweiler, M., Benenson, R., Franke, U., Roth, S., Schiele, B. The cityscapes dataset for semantic urban scene understanding. *Proc. IEEE Conf. Comp. Vis. Patt. Recogn. (CVPR),* 2016:13213–13223.

[14] Maddern, W., Pascoe, G., Linegar, C., Newman, P. 1 Year, 1000km: The Oxford RobotCar dataset. *Internat. J. Robot. Res. (IJRR),* 2016;36(1):3–15.

[15] Neuhold, G., Ollmann, T., RotaBulò, S., Kontschieder, P. The Mapillary Vistas dataset for semantic understanding of street scenes. *ICCV,* 2017:14990–14999

[16] Yu, Fisher and Chen, Haofeng and Wang, Xin and Xian, Wenqi and Chen, Yingying and Liu, Fangchen and Madhavan, Vashisht and Darrell, Trevor, "BDD100K: A Diverse Driving Dataset for Heterogeneous Multitask Learning", Proceedings of the IEEE/CVF Conference on Computer Vision and Pattern Recognition (CVPR), June, 2020:2636–2645.

35 Development of command-based automated glass cleaning robot

Dr. Roopa S.[1,a], Grace D.[2,b], Keerthi M.[2,c], Harshitha S.[2,d] and Brunda H.[2,e]

[1]Assistant Professor, Department of Electronics and Communication, Siddaganga Institute of Technology, Tumakuru, Karnataka, India

[2]Student, Department of Electronics and Communication, Siddaganga Institute of Technology, Tumakuru, Karnataka, India

Abstract

Traditional glass cleaning methods are labour intensive, require significant effort, and often fail to deliver satisfactory results, especially on textured surfaces. They also consume excessive water and cleaning solutions, raising sustainability and safety concerns. To address these issues, an innovative command based automated glass cleaning robot has been developed to clean both flat and textured glass surfaces. The robot operates autonomously, using a command-based navigation system that allows users to control movements forward, backward, left, right, up, down and the cleaning liquid dispenser. It also features adjustable vertical positioning for surface variations, real-time monitoring, and wireless control via a smartphone application. Its eco-friendly design minimises resource usage while delivering efficient and precise cleaning. The robot consistently achieved high-quality cleaning on flat and textured glass surfaces with reduced water and solution consumption compared to traditional methods. Its intuitive controls and automated operation provide a sustainable and efficient solution for maintaining glass surfaces.

Keywords: Flat glass, textured glass, commands, cleaning solution, robot

Introduction

The developed automated glass cleaning system is an innovative and efficient solution for maintaining glass surfaces. It simplifies the cleaning process by eliminating the challenges associated with traditional methods, such as physical strain, time consumption, and safety risks. Designed as a compact, floor-based system, it is particularly suitable for smaller glass surfaces like windows, mirrors, and panels commonly found in residential and commercial spaces. This system uses automation to deliver consistent and precise cleaning results with minimal human intervention. It saves significant time and effort. The design emphasises environmental sustainability by optimising the use of water and cleaning agents, minimising waste, and promoting eco-friendly practices. Unlike conventional cleaning methods, which often rely on excessive resources and produce uneven results, this system provides a controlled and reliable cleaning experience. The modular design ensures flexibility, making it adaptable to diverse environments and cleaning needs. Its compact form and simple operation make it ideal for use in both residential and commercial spaces, providing a practical solution for glass cleaning. By offering a reliable and consistent cleaning experience, the system eliminates the dependency on labour-intensive methods, making it a practical choice for maintaining smaller glass surfaces with precision and ease. The features are included for remote monitoring or control via smartphone applications, further advancing its convenience and efficiency. integration of precise motion control mechanisms ensures that the system adapts seamlessly to varying surface orientations, maintaining consistent cleaning efficiency even on slightly uneven glass panels. The automated glass cleaning robot successfully achieves efficient and precise cleaning of glass surfaces, particularly smaller or predefined-sized windows. By utilising an automated mechanism, the robot is able to perform the cleaning task without the need for human intervention, ensuring a consistent and thorough clean every time. The robot's ability to detect obstacles in the path of cleaning. In addition, integration of real-time monitoring reliable performance, smooth movement, and efficient power management contribute to a highly effective solution for automated glass cleaning, minimising labour costs and ensuring optimal cleaning results.

Literature review

In a study by Zhou et al. [1], Bluetooth operation mode decision method for cleaning robots, driven by the perception of garbage attributes was presented.

[a]roopas@sit.ac.in, [b]grace.1si22ec401@gmail.com, [c]keerthim677@gmail.com, [d]sharshi033@gmail.com, [e]brundahuligi@gmail.com

DOI: 10.1201/9781003685876-35

The system enables robots to detect and analyse the type of waste they encounter, allowing them to select the most efficient cleaning mode based on the specific characteristics of the garbage. Zavala Porta and Inga Narvaez [2], provides a review on mobile robots and their design requirements for cleaning solar photovoltaic modules. Various robot designs are examined, highlighting key factors necessary for efficient cleaning, such as mobility, adaptability, and effectiveness in maintaining solar panel performance.

Li and Li [3], developed a method for power planning to ensure the uninterrupted operation of outdoor window-cleaning robots. The approach focuses on optimising power usage and management, enabling the robots to operate efficiently over extended periods. Beaver [4] introduces a tracking controller for rolling robots, featuring a learning-based system that allows the robots to accurately follow target paths. An interaction behaviour decision making model is presented for service robots designed to assist individuals with disabilities [5]. The development of a wall-climbing robot employing suction-based mechanisms for adhesion is detailed by Okada et al. [6]. Badhoutiya [7] evaluates various methods for cleaning solar panel surfaces, focusing on their efficiency and effectiveness. The work compares different cleaning techniques to identify the most optimal approach for maintaining solar panels. The use of frame-type structures in robot design ensures stability and durability, with force analysis in solid works Simulation helping to predict performance under various conditions [8]. In a study by Wu et al. [9], the robot is equipped with intelligent systems to ensure efficient navigation and thorough cleaning of glass surfaces. Kumar et al. [10] in his work discussed about the development of a cost-effective glass wall cleaning robot.

Miyake et al. [11] in his study elaborates cleaning robots for window cleaning rely on motion control, mechanical systems, and control systems to navigate vertical surfaces. Accelerometers and acceleration sensors aid in attitude control, ensuring stability. Various adhesion mechanisms, including suction and magnetic methods, enable glass cleaning. Katsuki et al. [12], developed high-efficiency and reliable glass cleaning robot equipped with a dirt detection sensor. Robot was designed to automatically detect dirt on glass surfaces and clean them efficiently. The mechanism used involves advanced path-planning algorithms to guide the robot's movement, improving

its performance and effectiveness in cleaning large vertical surfaces is explained by Kanbe et al. [13]. A glass cleaning robot specifically designed for high-rise buildings is developed by Salunke et al. [14]. A robotic system is created which is capable of cleaning windows on tall structures, with an emphasis on safety, efficiency, and automation. Wu et al. [15] presents a robot optimised for cleaning glass walls, focusing on improving its design for efficient movement and cleaning performance. The system incorporates mechanisms to ensure stability and effectiveness while cleaning large glass surfaces, advancing the field of automated cleaning technology. Motion algorithm of a glass cleaning robot designed for regular-shaped glass surfaces was developed by Li et al. [16].

Zhang et al. [17] explains about automatic window cleaning robots that uses absorption mechanisms for adhesion and movement mechanisms for navigation on building surfaces. Propellers/vacuum-based adsorption ensures stability, while cleaning mechanisms enhance efficiency. A survey of techniques and applications for window cleaning robots was discussed by Li et al. [18]. This survey reviews the current state of the field and highlights potential advancements to improve the efficiency and functionality of window-cleaning robots in practical settings. The design of an intelligent measurement and control system for a glass-cleaning robot, powered by machine vision was discussed. Robotic systems for facade cleaning in high-rise buildings utilise winches and 2-DOF manipulators for controlled movement on glass surfaces [19]. In a study by Li et al. [20], he explains the image recognition and neural networks usage to enhance autonomy and efficiency in cleaning glasses. Convolutional neural networks (CNNs), combined with classification algorithms, enable robots to detect dirt, classify surface conditions, and optimise the cleaning process. Path planning plays a crucial role in ensuring complete coverage while minimising redundant movements and energy consumption. In this paper, the proposed robot introduces a novel approach to automating the glass cleaning process through user-directed commands, enhancing both user control and operational efficiency.

System overview

The switched-mode power supply (SMPS) converts high-voltage alternating current (AC) into the necessary lower-voltage direct current (DC) to power the system, as shown in Figure 35.1. It is designed to provide a reliable and constant output voltage, which is essential for the smooth functioning of the Arduino, motors, sensors, and other peripherals.

The multi-power supply board integrates both 5 V and 12 V outputs, allowing the system to power components that require different voltage levels. This configuration ensures that all parts receive the proper voltage for optimal performance and enhances the overall energy efficiency of the system by reducing heat generation and power loss. With a steady and safe power supply, the robot operates reliably across various tasks, ensuring uninterrupted cleaning operations.

The Arduino UNO controller interprets commands received from the Bluetooth module and ultrasonic sensor, making real-time decisions based on the programmed logic. It generates signals to control the motor driver, relay, and other connected peripherals, ensuring smooth coordination of all system functions. It also handles sensor data processing and responds accordingly, such as stopping the vehicle when an obstacle is detected or activating the cleaning mechanism.

The HC-05 Bluetooth module enables wireless communication between the Arduino and a smartphone application. By utilising serial communication, it receives commands from the bluetooth application and transmits to the Arduino, enabling remote control of the system. This module controls various actions, such as robot movements, spray system activation, or adjusting the screw jack mechanism. The bluetooth module ensures bidirectional communication in real time, making it an integral part of the user interface. The HC-SR04 ultrasonic sensor is used for obstacle detection on the path, ensuring the safety of the system. Based on this distance information, the Arduino can trigger actions such as stopping the vehicle or activating the spray system if an obstacle is detected

within a specified range. This allows the robot to navigate safely without colliding with obstacles.

The ESP32 camera module can be integrated that allows the robot to capture images or provide a live video feed, enabling real-time monitoring of the cleaning process. This camera allows the user to monitor the robot environment, helping with navigation, and offering the potential for object recognition or remote surveillance. The L293D motor driver is a dual H-bridge motor driver IC designed to control the direction and speed of DC motors. It acts as an interface between the Arduino and the motors, allowing low-power control signals from the Arduino to effectively drive high-power motors.

The 4-channel relay module is utilised to control high power devices such as the spray system and the screw jack mechanism. This allows the Arduino to activate or deactivate these components based on the programmed logic. By isolating the control circuit from the high-power devices, the relay ensures safe and efficient operation, preventing damage to sensitive components and ensuring reliable system performance. The DC motors (controlled via the L293D motor driver) are responsible for robot movement. The L239D motor driver receives control signals from the Arduino and drives the motors, enabling the robot to move forward, backward, left and right.

The spray system is activated via the relay module and is triggered when the ultrasonic sensor detects an obstacle within a defined distance. The spray mechanism works in synchronous with the sensor's feedback, activating when an obstacle is within range and providing additional functionality for tasks such as surface cleaning or sanitisation. The screw jack mechanism and wiper mechanism are controlled by the relay and motor drivers. The screw jack mechanism can be used to lift parts of the system, while the wiper mechanism simulates windshield wiper movements for tasks such as surface cleaning.

Through the smartphone application, users can send commands to the bluetooth module, which transmits them to the Arduino for execution. This allows users to control various functions, such as moving the robot, activating the spray system, or adjusting the screw jack mechanism. The pump motor is activated via the relay module and works in synchronisation with the sensor's feedback or Bluetooth commands, ensuring efficient surface cleaning. By using the pump motor, the system can automatically deliver the necessary amount of liquid

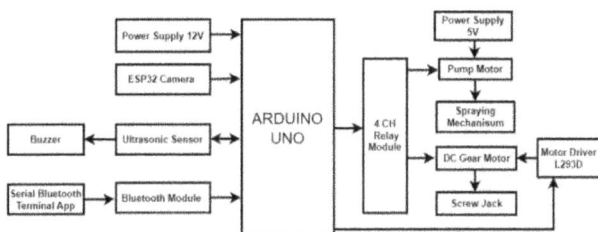

Figure 35.1 Block diagram of automated glass cleaning robot

Source: Author

precisely when needed, enhancing the robot's ability to perform cleaning tasks autonomously.

Results and discussions

Figures 35.2 and 35.3 shows top and side view of the proposed system. Arduino UNO, ultrasonic sensors, wheels, water tank, power supply, pump motor, 2 DC gear motors, 5V/12V switch mode power supply, 4 relay channel module, ESP32 camera, screw jack, multi-power supply board, buzzer, motor driver and HC-05 Bluetooth module.

The speed of jack movement is calculated as the ratio of length of jack and time taken to traverse it.

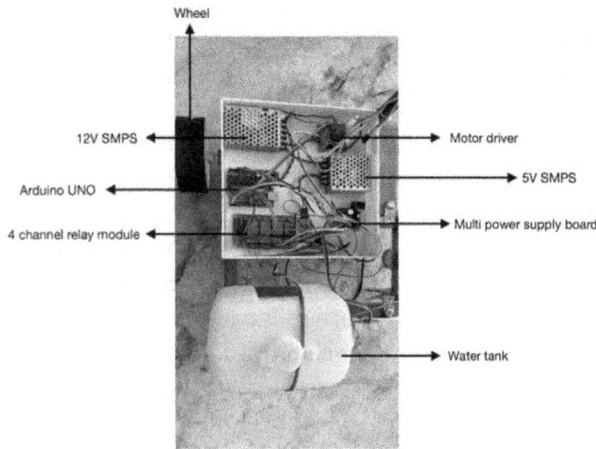

Figure 35.2 Top view of automated glass cleaning robot
Source: Author

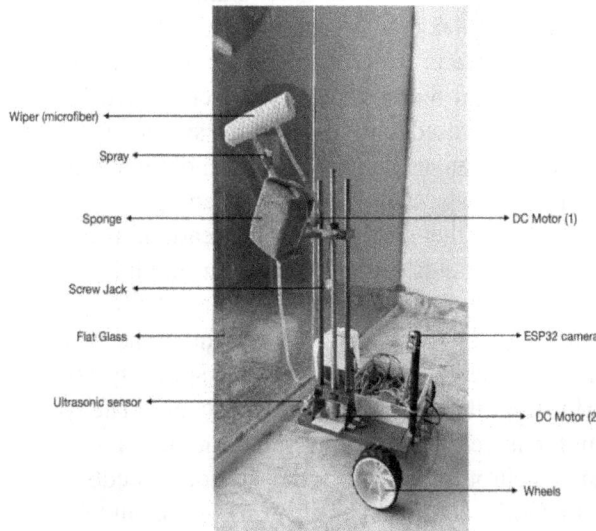

Figure 35.3 Side view of glass cleaning robot
Source: Author

Table 35.1 details the commands and corresponding actions for controlling the robot. Each command performs a specific function, such as forward, backward, left, right movement of the robot and upward, downward movement of screw jack, stopping, activating the spray mechanism, and enabling autonomous mode.

The power requirements of the components used in the automated glass cleaning robot are shown in Table 35.2. Figure 35.4 shows the position of the cleaning mechanism against the glass surface, with a dry or pre-moistened cleaning roller prepared to start the cleaning process by spraying water. The roller moves across the surface, spreading the cleaning solution and removing loose dirt, while a microfiber cloth beneath assists in wiping and is shown in Figure 35.4.

Figure 35.4 shows the cloth or roller actively scrubbing the surface, effectively removing stubborn dirt or stains. As the cleaning progresses, a polished and clearer surface becomes visible. The surface

Table 35.1 Commands used and their corresponding actions

Command	Action
"8"	Move front
"2"	Move back
"6"	Turn left
"4"	Turn right
"5"	Stop
"1"	Move up
"3"	Move down
"S"	Start the spray mechanism
"s"	Stop the spray mechanism
"Q"	Turn on motor for a specific action
"q"	Turn-off the motor
"A"	Enable auto-run mode

Source: Author

Table 35.2 Power consumed by each component

Components	Power consumed (W)
Arduino UNO	0.35 W
DC motors	3.6 W
Ultrasonic sensors	0.075 W
Pump motor, buzzer	10 W, 0.1 W
Relay module	0.375 W
HC-05 bluetooth	0.2 W
L293D motor driver	3 W
ESP32 camera	1.5 W
Power supply board	43.2 W (12 V), 18 W (5 V)

Source: Author

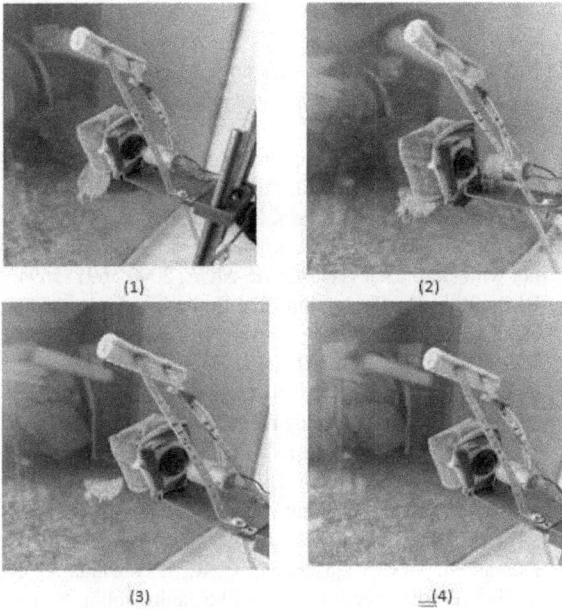

Figure 35.4 Cleaning process
Source: Author

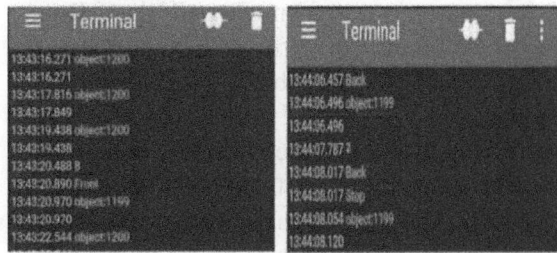

Figure 35.5 Command for forward and backward movement of robot
Source: Author

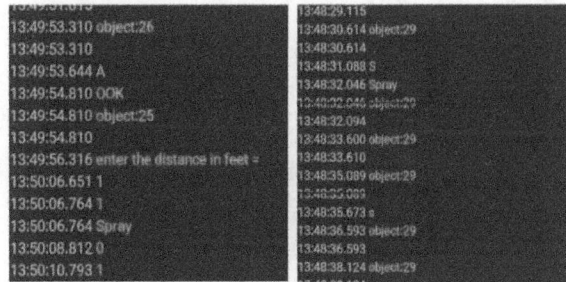

Figure 35.6 Command for automatic movement and spraying mechanism of the robot
Source: Author

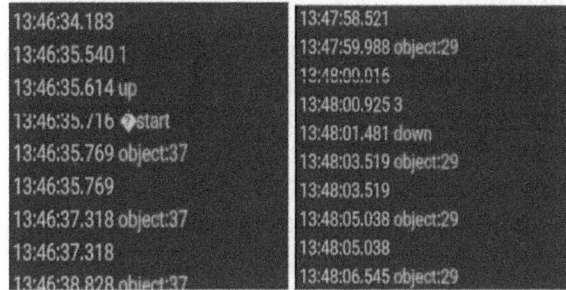

Figure 35.7 Command for upward and downward movement of the robot
Source: Author

appears cleaned and clear, as the final pass ensures any remaining residue is removed, leaving the glass spotless and is shown in Figure 35.4.

The serial Bluetooth terminal app functions as a mobile phone control interface for the glass cleaning robot, providing users with two modes of operation: manual mode and automatic mode. In manual mode, the user has the ability to exercise full control over the robot movement by issuing specific commands through the app interface. These commands enable the robot to move forward, backward, left, and right directions. Additionally, there are two dedicated commands, one specifically designed to activate the robot's spraying mechanism, allowing it to dispense cleaning solution onto the glass surface and another command to stop the robot entirely.

When the command "8" is issued, the robot activates its DC motors to move forward by synchronising the rotation of the front motors, ensuring a straight linear path, as shown in Figure 35.5. This function enables a smooth start to the glass cleaning process along a defined route. Similarly, when the "2" command is issued, the robot reverses by rotating the motors in the opposite direction.

Figure 35.6 illustrates the command for the robot's automatic movement, enabling hands-free operation without human supervision.

Activating this mode requires entering the command "A", which switches the robot to autonomous cleaning. Users must then specify the cleaning distance in feet via the serial Bluetooth terminal application. Once activated, the robot moves forward, backward, upward, and downward while executing the spraying mechanism.

The commands "up" and "down" adjusts the robot's vertical position to clean glass surfaces at varying heights (Figure 35.7). This causes the screw jack mechanism to rise and lowers the robot, allowing it to reach and clean higher and lower sections of the glass, respectively.

Similarly, when the command "6" is entered, it initiates movement to the left. This involves the robot rotating the face to leftward direction. When the command "4" is entered, the robot moves to the right. he robot was tested on textured glass. Due to the increased number of pores in textured glass, the cleaning efficiency of the robot is lowered compared to flat glass.

Figure 35.8 shows that the ultrasonic sensor plays a critical role in monitoring the robot's surroundings for obstacles. The sensor continuously scans the area and detects objects within a range of 25 cm. If an object is detected within this range, the robot immediately halts its movement and buzzer is beeped to alert the users.

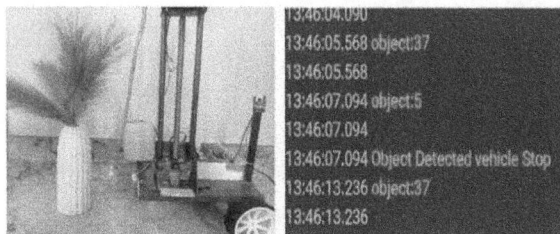

Figure 35.8 Ultrasonic sensor detecting the obstacle
Source: Author

Conclusion

The automated glass cleaning robot offers an effective solution to the challenges associated with traditional glass cleaning methods. By automating the cleaning process, it eliminates the need for labour-intensive physical effort, reduces resource consumption, and ensures consistent and high-quality results on both flat and textured glass surfaces. The robot's command-based navigation system, real-time monitoring, and wireless control via a smartphone application provide users with flexibility and convenience. Its efficient use of resources and precise cleaning capabilities make it a valuable tool for maintaining flat and textured glass surfaces. This developed work represents a significant advancement in glass cleaning technology, offering a practical, efficient, and sustainable solution for glass maintenance.

References

[1] Zhou, Z., Zhang, D., Zhu, J., Tang, H. A lightweight operation mode decision method for cleaning robots driven by garbage attributes perception. *IEEE Access*, 2024;12:15308–15320.

[2] Zavala Porta, J. J., Inga Narvaez, D. A review on mobile robots and its design requirements to clean solar photovoltaic modules. *IEEE Internat. Conf. Electr. Elec. Engg. Comput. (INTERCON)*, 2023:1–7.

[3] Li, J., Li, X. A method for power planning to ensure the complete operation of outdoor window-cleaning robots.

[4] Beaver, L. E. et al. Learning a tracking controller for rolling μbots. *IEEE Robot. Automat. Lett.*, 2024;9(2):1819–1826.

[5] Zhu, X. et al. An interaction behavior decision-making model of service robots for the disabled based on human–robot empathy. *IEEE Access*, 2024;12:15778–15790.

[6] Okada, N., Yamanaka, K., Kondo, E. A wall climbing robot with simple suckers. *ICCAS-SICE*, 2009:5691–5694.

[7] Badhoutiya, A. Evaluation of different methods used to clean solar panel surface. *Second Internat. Conf. Electr. Renew. Sys. (ICEARS)*, 2023:46–48.

[8] Zhang, C., Zhou, B., Gao, N., Guo, T. Force analysis and simulation of glass cleaning robot. *3rd Internat. Symp. Robot. Intell. Manufac. Technol. (ISRIMT)*, 2021:42–46.

[9] Wu, G., Zhang, L., Zhang, H., Zhang, B. Research on design of glass wall cleaning robot. *5th Internat. Conf. Inform. Sci. Control Engg. (ICISCE)*, 2018:932–935.

[10] Kumar, M. V. V., Rajashekhar, C. R., Dayananda, G. K. Studies on cost effective glass wall cleaning robot. National conference on challenges in research technology in the coming decades (CRT 2013). *Ujire*, 2013:1–5.

[11] Miyake, T., Ishihara, H., Shoji, R., Yoshida, S. Development of small-size window cleaning robot a traveling direction control on vertical surface using accelerometer. *Internat. Conf. Mechatr. Autom.*, 2006:1302–1307.

[12] Katsuki, Y., Ikeda, T., Yamamoto, M. Development of a high efficiency and high reliable glass cleaning robot with a dirt detect sensor. *IEEE/RSJ Internat. Conf. Intell. Robots Sys.*, 2011:5133–5138.

[13] Kanbe, K., Nansai, S., Itoh, H. Design of trajectory generator of a glass facade cleaning robot. *IECON 2021 47th Ann. Conf. IEEE Indus. Electr. Soc.*, 2021:1–6.

[14] Salunke, A. et al. Glass cleaning robot for high rise buildings. *Internat. Conf. Innov. Sustain. Comput. Technol. (CISCT)*, 2019:1–5.

[15] Wu, G., Zhang, L., Zhang, H., Zhang, B. Research on design of glass wall cleaning robot. *5th Internat. Conf. Inform. Sci. Control Engg. (ICISCE)*, 2018:932–935.

[16] Li, X. H., Yang, Z., Yao, L. Research on motion algorithm of glass cleaning robot on regular shaped glass. *Internat. Conf. Comp. Sci. Manag. Technol. (ICCSMT)*, 2020:123–127.

[17] Zhang, A., Yao, Y., Hu, Y. Analyzing the design of windows cleaning robots. *3rd Internat. Conf. Big Data Artif. Intell. Internet Things Engg. (ICBAIE)*, 2022:167–176.

[18] Li, Z., Xu, Q., Tam, L. M. A survey on techniques and applications of window-cleaning robots. *IEEE Access*, 2021;9:111518–111532.

[19] Yoo, S. et al. Unmanned high-rise facade cleaning robot implemented on a Gondola: Field test on 000-Building in Korea. *IEEE Access*, 2019;7:30174–30184.

[20] Li, Z., Han, X., Wang, X., Li, Y., Zhu, X. Path planning and image classification algorithms for glass cleaning robots based on hybrid path planning and convolutional neural networks. *Internat. Conf. Comp. Inform. Proc. Adv. Educ. (CIPAE)*, 2023:440–445.

4th Internat. Conf. Comp. Vision Image Deep Learn. (CVIDL), 2023:1–4.

36 A review of AI prompt management tools and a proposed Git-based solution

Rajath Nag Nagaraj[1,a], Dr. Devika B.[2,b], Dr. Rekha N.[3,c] and Disha R.[4]

[1]Senior Data Scientist, Michigan Technological University, Kammavari Sangha Institute of Technology, Bangalore, India

[2]Associate Professor, Kammavari Sangha Institute of Technology, Bangalore, India

[3]Professor, Kammavari Sangha Institute of Technology, Bangalore, India

[4]Student, Kammavari Sangha Institute of Technology, Bangalore, India

Abstract

Managing prompts for large language model (LLM) applications has emerged as a critical concern as prompts grow in complexity and require frequent iteration. This paper reviews three contemporary prompt management tools – Langfuse, PromptLayer, and ML-flow's Prompt Registry detailing their features, benefits, and limitations. Langfuse and PromptLayer provide dedicated prompt content management systems (CMS) with version control and analytics, while machine learning (ML) flow integrates prompt tracking into an MLOps platform. We discuss how these tools enable collaborative prompt engineering and prompt performance monitoring. To address gaps identified in existing solutions, we propose a lightweight Git-based prompt management system called *rj-prompt-management*. Our approach leverages Git for versioning and collaboration, with repository structures and scripts for logging prompt usage and batch evaluations. The proposed solution is demonstrated with nature-related prompt examples (*tree_density, rock_presence, canopy _cover*). We highlight the advantages of a Git-centric approach its simplicity, familiarity, and integration into developers' workflows as well as its current limitations. Future improvements for the Git-based system, including a user-friendly interface and continuous integration for prompt testing, are also discussed.

Keywords: LLM, Langfuse, PromptLayer, prompt management, Gen-AI and *rj-prompt-management*

Introduction

Large language model (LLM) prompts (instructions given to AI models) are increasingly treated as important software artefacts that require careful management. *Prompt management* refers to the systematic approach of storing, versioning, and retrieving prompts in LLM applications [1]. Effective prompt management is needed to handle evolving prompt iterations, track their performance, and enable collaboration among team members. Early-stage LLM applications often hard-code prompts within codebases or configuration files, but this becomes problematic as applications scale. Changes to prompts may necessitate code redeployment, and tracking prompt experiments via code commits or file versions can be cumbersome [1]. Moreover, non-developers (e.g., domain experts or product managers) may struggle to contribute when prompts are buried in code. To address these challenges, a new class of prompt management tools have emerged. These tools act as Prompt content management systems (Prompt CMS), decoupling prompt text from application code and providing features like version control, audit trails, and evaluation dashboards. They allow teams to iterate on prompts rapidly (often without redeploying code) and usually include interfaces for both engineers and non-technical users to edit and experiment with prompts. In this paper, we review three notable tools in this domain:

Langfuse, an open-source LLM observability platform with prompt management features; **PromptLayer**, a popular SaaS platform for prompt versioning and collaboration; and **MLflow Prompt Registry**, which extends the MLflow MLOps platform to manage prompts. We compare their capabilities and limitations.

While these tools provide rich functionality, adopting a standalone prompt management service may introduce additional complexity or external dependencies. For teams seeking a simpler solution, using Git-based version control for prompts remains an attractive baseline. In the second half of this paper, we propose a Git-based prompt management system called *rj-prompt-management*. Our approach leverages a standard Git repository to store and version prompt files, complemented by

---

[a]rnagara1@mtu.edu, [b]devikab@ksit.edu.in, [c]rekhan@ksit.edu.in

DOI: 10.1201/9781003685876-36

lightweight scripts for logging usage data and batch testing. We demonstrate how this approach can satisfy core prompt management needs with minimal overhead. Finally, we discuss how our solution can be augmented with a user interface and continuous integration (CI) for automated prompt evaluations in future works.

Review of existing tools

Langfuse

Langfuse is an open-source platform focused on LLM application observability, which includes a robust prompt management module [2]. It allows developers to log and version prompts, assign labels and tags, and roll back to previous versions as needed. Prompts are stored in a central repository within Langfuse, and each version can be annotated with metadata such as model parameters or environment labels. A notable feature of Langfuse is its *Prompt Playground*, enabling real-time testing of prompt versions with various model back-ends. This helps users interactively evaluate prompt changes side-by-side. Langfuse also integrates prompt management with tracing and analytics; each prompt version's performance (e.g., latency, cost, success metrics) can be tracked via the Langfuse dashboard and linked to model outputs and API calls.

Pros: Langfuse's advantages include its comprehensive feature set and self-hosted nature. Being open-source [2], it can be deployed on-premises, offering flexibility for organisations with strict data requirements. It provides an all-in-one solution (prompt versioning, monitoring, and even dataset generation from prompt logs) that is extensible via API/SDK. The tight integration of prompt management with request tracing and evaluation metrics is valuable for debugging and optimising prompts over time.

Cons: The breadth of features in Langfuse can introduce a learning curve for new users. Setting up and maintaining the system requires additional infrastructure (database, web service) compared to lighter-weight approaches. For smaller teams or projects that only need basic prompt versioning, Langfuse's complexity may be overkill. Additionally, while Langfuse decouples prompts from application code, it still requires users to adopt the Langfuse platform and its interfaces for editing and retrieving prompts (which might not appeal to those who prefer simple file-based workflows).

PromptLayer

PromptLayer is a hosted prompt management platform that emphasises collaboration and ease of use for prompt engineering teams [3]. It provides a web-based Prompt Registry (a prompt CMS) where users can create, edit, and organise prompt templates with placeholders for variables. Every prompt update is tracked as a new version, with visual diff comparisons and descriptive commit messages to document changes. PromptLayer supports grouping prompts into projects and folders, and it allows defining *snippets* – reusable prompt components that can be included in larger prompts for modular design [3]. The platform also offers features such as release labels (to mark certain versions for production or staging) and access control to manage team collaboration [3]. Built-in tools enable A/B testing between prompt variants and evaluations of prompt performance using user feedback or custom metrics [3]. PromptLayer can track usage statistics (e.g., API call counts, latency, cost) for each prompt version, providing insight into how changes affect performance over time.

Pros: PromptLayer delivers a polished, no-code inter-face for prompt management, lowering the barrier for non-engineers to participate in prompt iteration [3]. It is model-agnostic, treating prompts as templates that can be used with any underlying LLM API or framework. Integration SDKs allow applications to fetch prompt templates at runtime via API, meaning new prompt versions can be deployed without modifying the application code [3]. The platform's collaboration features (shared workspaces, commenting, and version freeze capabilities) align with software best practices adapted to prompts. PromptLayer's analytics and evaluation capabilities help teams rapidly experiment and converge on effective prompt designs.

Cons: As a cloud service, PromptLayer introduces external dependencies. Relying on an API to retrieve prompts in production could add latency or uptime risks if not mitigated by local caching [5]. There may also be data governance concerns, since prompt content (which might encode proprietary knowledge) is stored on PromptLayer's servers. Additionally, while the visual interface is strength, some advanced users might find it restrictive compared to managing prompt files directly in code. The platform's specialised nature means it might be less useful for teams who do not require extensive prompt experimentation or who prefer open-source solutions. Finally, PromptLayer is a commercial product; large-scale use may incur significant costs, whereas open-source alternatives or Git-based approaches could be more cost-effective.

MLflow prompt registry

MLflow is a widely-used open-source MLOps platform, and recent versions have introduced a Prompt Registry to extend its model management capabilities to LLM prompts. MLflow's Prompt Registry is designed to bring prompts into the same lifecycle management framework as machine learning models. Prompts are stored in a central registry and can be versioned with a Git-inspired workflow: each new prompt version requires a commit message, and the system can show diffs between versions for easy comparison. The registry also supports *aliasing*, where human-readable tags (like "Production" or "Experiment") can be assigned to specific prompt versions [4]. This allows controlled deployment of prompts (e.g., one can test a new prompt version by aliasing it as "staging" while the application continues to use the version aliased "production") and facilitates quick rollbacks or A/B tests. Because MLflow is an end-to-end platform, the Prompt Registry integrates with MLflow's model tracking and evaluation modules [4]. For instance, one can log which prompt version was used for a particular model run, and evaluate model performance across different prompt versions or compare prompt outcomes using MLflow's experiment UI.

Pros: For organisations already using MLflow for managing models and experiments, the Prompt Registry adds prompt versioning with minimal additional infrastructure. It benefits from MLflow's strengths: the registry and UI can be hosted alongside existing MLflow services, and prompts can be managed via the same Python API (e.g., using mlflow.register_prompt() to save a prompt template in the registry) [4]. The Git-like version control with commit messages and diff views will feel familiar to developers. Integration with model lineage means one can trace which prompt and model combination produced certain outputs, improving reproducibility. MLflow being open-source means the Prompt Registry can be self-hosted and customised, avoiding vendor lock-in.

Cons: MLflow's Prompt Registry is relatively new, and its feature set might not be as extensive in terms of UI for prompt editing or specialised prompt analytics compared to dedicated prompt management tools. There is no built-in web editor for prompts as

of its initial release; users create or edit prompts via API or through a basic form in the MLflow UI, which is more rudimentary than the collaborative editors in Langfuse or PromptLayer. Also, using MLflow might be a heavyweight solution if prompt management is needed in isolation – MLflow is designed to manage models, data, and experiments too, so adopting it solely for prompt versioning could be excessive. The learning curve for MLflow (for those not already familiar with it) and the need to integrate it into the deployment pipeline may deter some users. However, as the tool evolves, it could become more user-friendly for prompt-specific use cases.

Proposed Git-based solution

Given the benefits and drawbacks of the above tools, we recognise that not all teams may opt to use a dedicated prompt management service. In cases where a lightweight approach is preferred, one can leverage the ubiquitous toolset of Git and simple scripting to manage prompts. In this section, we propose *rj-prompt-management*, a Git-based prompt management workflow. Our solution uses a standard Git repository as the single source of truth for all prompts, combined with a set of Python scripts that facilitate prompt logging and batch evaluation. We describe the design and demonstrate it using example prompts related to environmental analysis tasks.

System design

At the core of *rj-prompt-management* is a structured Git repository for prompts. Figure 36.1 illustrates an example repository layout. Prompts are organised in directories by domain or project. For instance, a folder environment/contains prompts dealing with nature-related analyses. Each prompt is stored as a text file (or markdown) with placeholders for variables if needed. Additional directories (such as logs/) store usage logs and evaluation results. This simple structure is human-readable and can be navigated or edited with standard tools (Figure 36.2).

Developers interact with the repository by creating new prompt files or updating existing ones and using Git to commit changes. Each Git commit (optionally) includes a message describing the prompt modification, serving a role analogous to commit messages in specialised tools (e.g., documenting the rationale for a prompt tweak). The version history of each prompt can be inspected using Git's diff tools, and any previous version can be restored by reverting the file or

```
 prompts/
 environment/
    canopy_cover.t
 xt
 rock_presence.txt
 tree_density.txt
    chatbot/
 faq_assistant.txt
 support_agent.txt
 logs/
          environment/
    canopy_cover.log
    rock_presence.log
    tree_density.log
```

Figure 36.1 Repository structure for the proposed Git-based prompt management system. Prompts are grouped by topic, and corresponding logs are stored in parallel directories.

Source: Author

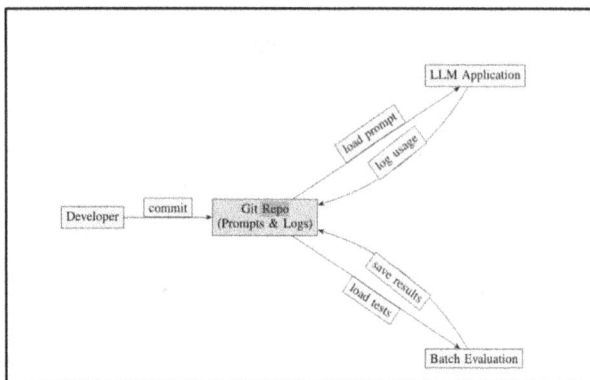

Figure 36.2 Overall flow of proposed Git-based solution
Source: Author

checking out an older commit. In effect, Git provides the version control backbone, ensuring traceability of how prompts evolve over time [5].

On top of this, *rj-prompt-management* provides a utility for logging prompt usage. A Python function (illustrated in Listing 1) appends interactions to a log file whenever a prompt is used. The log records details such as timestamp, the input context (if any), the prompt version (Git commit hash or ID), and the model's output. Storing these logs in the repository (under a mirrored directory structure) means they too are under version control. This allows tracking the outcomes of prompts over time and across versions.

The system design intentionally avoids any runtime external dependencies; prompts are loaded from local files (which can be updated by pulling the latest Git commits), eliminating the latency of remote prompt fetching. This addresses a common concern with API-based prompt stores [5].

Listing 1: Pseudo-code for logging prompt usage in *rj-prompt-management*.

```
import json, datetime

def log_prompt_usage(prompt_name, input_data, output_data):
    entry = {
                "timestamp": datetime.datetime.now().isoformat(),
        "prompt": prompt_name, "input": input_data, "output": output_data
    log_path = f"logs/{prompt_name}.log" with open(log_path, "a") as f:
        f.write(json.dumps(entry) + "\n")
```

Example use-case

We demonstrate the workflow of *rj-prompt-management* using a hypothetical environmental analysis scenario. Imagine an application that uses an LLM to analyse descriptions (or images) of a landscape and answer questions about it. We define three prompt templates in the environment/ folder: *tree_density.txt*, *rock_presence.txt*, and *canopy_cover.txt*. Each prompt is written to guide the model in estimating the respective attribute (for example, asking the model to estimate tree density given a description of a forest scene). Variables in the prompt (like placeholders for a specific location or context details) can be denoted with a format such as *{{location}}* and are filled in programmatically before use.

Suppose we want to evaluate how these prompts perform on a batch of test cases (for instance, several different landscape descriptions). We prepare a small test dataset for each prompt (a set of inputs with known expected characteristics). Using the batch utility script in our system, we run each prompt on its inputs and log the outputs for analysis. Listing 2 shows a simplified example of how a batch evaluation might be executed for the *tree_density* prompt.

Listing 2: Batch evaluation of a prompt with multiple inputs, with logging

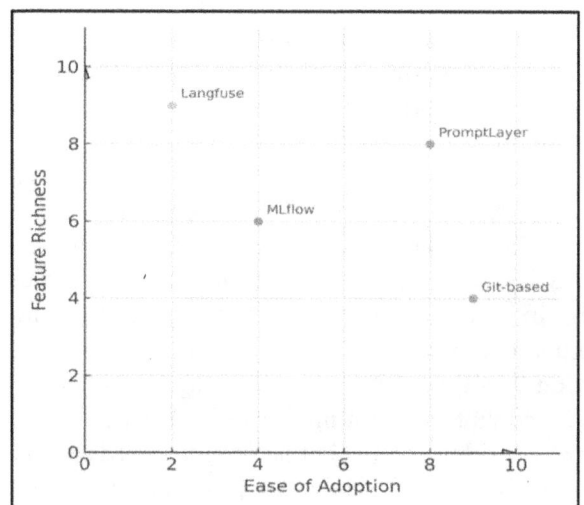

In this example, after running the batch, the file logs/environment/tree_density.log will contain JSON entries for each test run. The team can review these outputs (possibly diffing the logs if the prompt is updated and the test re-run in a future commit) to see if changes to the prompt improved the quality of the model's answers. Because everything resides in the Git repository, team members can branch and experiment with prompt modifications independently, then merge changes once validated – leveraging the normal Git collaboration workflow.

Advantages: The Git-based approach offers simplicity and control. It requires no special infrastructure beyond a Git host (and optionally a simple server or automation to distribute prompt updates). Versioning and collaboration piggy-back on established Git processes, which most developers are already comfortable with [5]. There are no additional costs or subscriptions; prompts remain private in the organisation's repository. The system is also highly extensible: teams can write custom scripts to analyse the log files, or integrate with continuous integration pipelines (for example, running automated prompt evaluation tests whenever a prompt file changes, and alerting if performance regresses).

Limitations: Unlike Langfuse or PromptLayer, this solution does not provide a user-friendly GUI or a dedicated database for prompts. Non-technical users would likely need a simple interface to contribute to prompt edits (currently, they would have to use Git or a text editor). Similarly, features like side-by-side prompt diff visualisation, rich analytics dashboards, or integrated A/B testing frameworks are not available out-of-the-box. These would require additional development (for instance, building a small web dashboard on top of the repository data). However, the design philosophy of *rj-prompt-management* is to start with a minimal viable solution and allow organisations to augment it as needed. Many teams already use Git to manage code; extending it to prompts is natural but should be accompanied by training and process guidelines to ensure prompts are kept up to date and properly reviewed.

Conclusion

Prompt management is becoming increasingly vital as LLM applications grow in complexity and team collaboration becomes essential. Our review of Langfuse, PromptLayer, and MLflow's Prompt Registry highlights the diverse approaches to this challenge: Langfuse offers a comprehensive, self-hosted solution with deep analytics; PromptLayer provides an accessible, cloud-based platform with strong collaboration tools; and MLflow integrates prompt versioning into its established MLOps framework.

Each tool has merits, but also trade-offs in complexity, cost, or flexibility. For teams seeking a lightweight, developer-centric alternative, our proposed *rj-prompt-management* leverages Git to provide essential prompt versioning and logging with minimal overhead. By using familiar tools and workflows, it ensures ease of adoption while covering core needs.

Future enhancements could include a simple web interface for non-technical users and CI pipelines for automated prompt testing, making *rj-prompt-management* even more versatile. Thus, our work demonstrates that effective prompt management can be achieved through straightforward, incremental improvements to existing practices. The Git-based approach presented as *rj-prompt-management* fills a niche for teams that desire a lightweight, developer-centric workflow. By leveraging Git, it ensures that prompt versioning and collaboration are handled in a proven way with-out introducing new external services. Our proposed system, while simple, covers the fundamental needs of prompt storage, version control, logging, and batch evaluation. The nature-inspired example demonstrates that even domain-specific prompt tasks can be managed effectively with this approach. Looking forward, there are several enhancements to explore. First, developing a minimal web interface or plugin (for example, a simple React application or Jupyter extension) for non-technical users would make *rj-prompt-management* more inclusive. This UI could list available prompts, show their history, and allow editing via a form, hiding the complexity of Git operations. Second, integrating the repository with CI pipelines could automate quality checks: for instance, running a suite of prompt evaluations and comparing outputs against expected results whenever a prompt is updated, thus catching any unintended degradation early. Finally, as the usage of the system grows, we may consider linking the prompt logs with external analytics tools or databases to compute statistics (such as average prompt execution time or success rate) in a more scalable way.

In summary, prompt management tools are evolving rapidly, and our work advocates for a balanced approach: adopting powerful platforms when needed but not overlooking simpler solutions that can be built with existing software engineering tools. By grounding prompt management in Git, we align it with well-understood practices and set the stage for incremental improvements. The proposed solution *rj-prompt-management* demonstrates that managing prompts can be as straightforward as managing code, and it paves the way for future innovations that bridge the gap between ease-of-use and robust prompt governance.

Acknowledgement

We gratefully acknowledge the students, staff, and authority of electronics and communication department for their cooperation in the research.

References

[1] Langfuse. Prompt management - Langfuse docs (Get Started). [Online]. Available: https://langfuse.com/docs/prompts/get-started [Accessed: Apr. 8, 2025].

[2] Duta, G. What is prompt management for LLM applications? Tools, Techniques and Best Practices. *Qwak Blog*, 2024. [Online]. Available: https://www.qwak.com/post/prompt-management.

[3] Zoneraich, J. Best practices: Prompt management and collaboration. PromptLayer Blog (Medium), Mar. 8, 2024. [Online]. Available: https://medium.com/prompt-layer/scalable-prompt-management-and-collaboration-ff-f28af39b9b.

[4] MLflow. Prompt management in MLflow. MLflow documentation. ver. 2.7, 2023. [Online]. Available: https://mlflow.org/docs/latest/prompts.

[5] GPT-SDK. Version Control Best Practices for Prompt Management. GPT-SDK Blog, Feb. 15, 2025. [Online]. Available:https://gpt-sdk.com/c/Version-Control-Best-Practices-for Prompt-Management.

[6] Vaswani, A. et al. Attention is all you need. *Proc. Adv. Neural Inf. Process. Syst.*, 2017:5998–6008.

[7] OpenAI. OpenAI Cookbook. [Online]. Available: https://github.com/openai/openai-cookbook.

[8] Hugging Face. Prompt-based Learning. [Online]. Available: https://huggingface.co/docs/transformers/main/ en/ tasks/prompting.

[9] Reynolds, L., McDonell, K. Prompt programming for large language models: Beyond the few-shot paradigm. arXivpreprintarXiv:2102.07350, 2021. 1–29.

[10] Brown, T. et al. Language models are few-shot learners. *Proc. Adv. Neural Inf. Process. Syst.*, 2020:1877–1901.

[11] Bura et al. IJRAR 2025 Ethical Considerations in Prompt Egineering for AI Systems. 145–165. https://link.springer.com/chapter/10.1007/978-3-031-45304-5_15.

[12] The Future of AI Prompt Management: Trends and Challenges in Large-Scale Systems.

[13] Chen et al.arXiv 2310.14735 Unleashing the potential of prompt engineering in Large Language Models: a comprehensive review. 1–45.

[14] Prompt Engineering an emerging role in ai by Sand Technologies - https://www.sandtech.com/insight/prompt engineering-an-emerging-new-role-in-ai/.

[15] Five trends in data and AI – MIT Slolan review -https://sloanreview.mit.edu/article/five-trends-in-ai-and-data -Science-for-202.

37 Design of a deep learning-based controller for a 4-leg inverter to independently control two 3-phase induction motors

Venu Gopal B. T.[1,a], Shreedhara R.[2], Hareesh Kumar[1], Shalini J.[1] and Usha Rebecca[3]

[1]Dept. of Electronics and Communication Engg, RV Institute of Technology and Management, VTU, Bangalore, Karnataka, India

[2]Department of Transit Engineering, Webtec Corporation, Bangalore, Karnataka, India

[3]Dept. of Electrical and Electronics Engg, Rajeev Gandhi Institute of Technology, VTU, Bangalore, Karnataka, India

Abstract

The rapid advancement of artificial intelligence (AI) has paved the way for innovative motor control strategies that enhance efficiency and performance. This paper discusses the design and implementation of a deep learning-based controller (DLC) for a 4-leg inverter to independently control two 3-phase induction motors using indirect vector control (IVC) technique. Traditional control methodologies, such as proportional-integral (PI) controllers, face challenges in handling the non-linearity of induction motors and mitigating torque and speed ripples. To overcome these limitations, we present a deep learning approach utilising the deep belief network (DBN) algorithm. The proposed controller learns from PI controller input-output datasets to optimise motor performance dynamically. Simulation results demonstrate that the DLC significantly reduces torque and speed ripples while achieving higher accuracy in speed regulation compared to conventional controllers. Additionally, the proposed system reduces the number of inverter switches, leading to cost reduction and improved efficiency. The paper also explores the impact of AI-driven control on energy savings and industrial automation.

Keywords: Deep learning controller, induction motor drive, four-leg inverter, deep belief network (DBN), AI-based motor control, torque ripple reduction, speed regulation, indirect vector control

Introduction

Induction motors are essential in industries and transportation due to their reliability and efficiency. Advanced control techniques, like vector control, improve performance by enabling precise speed and torque regulation, making them ideal for applications such as electric vehicles and trains. Efficient control strategies also enhance energy savings and reduce operational costs [1]. PI controllers struggle with non-linearity's, fuzzy logic controllers (FLCs) address uncertainties using human-like reasoning, offering better performance for motor drives [2]. Neuro-fuzzy systems integrate neural networks and fuzzy logic to create adaptive controllers, improving speed and torque control in induction motors. Genetic algorithms further optimise performance by efficiently tuning parameters. Despite challenges in varying conditions, deep learning controllers (DLCs) using deep belief networks (DBNs) offer superior speed regulation and reduced torque ripples, enabling intelligent, adaptive motor control [3].

Literature survey

Induction motors are vital in automation, electrical vehicles (EVs), and renewable energy but pose control challenges due to nonlinear dynamics. Traditional PI controllers struggle with torque ripples and speed fluctuations, while indirect vector control (IVC) and artificial intelligence (AI)-based controllers enhance performance [4]. Deep learning improves accuracy, and 4-leg inverters enable independent IM control with less hardware. IVC decouples torque and flux, with sensor less methods boosting reliability and reducing costs. Studies did by Craciunaş et al. (2024) shows IVC effectiveness, while a modified IVC strategy mitigates faults, ensuring stable operation [5]. The integration of AI-based techniques, such as fuzzy logic, artificial neural networks (ANNs), and deep learning, has radically improved the performance of IM drives. AI controllers have the advantage of learning from data and dynamically adapting to changes in motor parameters, which is not possible with conventional PI controllers. Setiawan et

[a]venugopalbt.rvitm@rvei.edu.in

DOI: 10.1201/9781003685876-37

al. (2024) explored the use of Takagi-Sugeno-type fuzzy logic for IVC-based rotational speed control in primary surveillance radar systems. The results showed that the fuzzy logic controller outperformed traditional PI controllers by reducing overshoot and improving steady-state accuracy [6]. Similarly, an ANN-based field-oriented control (FOC) system developed by Devanshu et al. (2021) demonstrated superior current control performance compared to conventional hysteresis controllers. The ANN controller reduced current ripple and improved dynamic response, making it more suitable for high-performance applications [7].

A deep learning-based approach using DBNs has freshly been introduced for IM control [8]. Traditionally, three-phase inverters with 3 legs are used to control a single IM. However, when multiple motors need to be controlled independently, a straightforward approach is to use multiple inverters, which increases cost and complexity. Latest studies have investigated a range of multi-motor control techniques. A study by Ahmed et al. (2022) examined the implementation of a 4-leg inverter for controlling dual IMs in an electric vehicle drivetrain. The results showed that the system achieved independent speed regulation while minimising inverter switching losses [9]. Another study by Kumar et al. (2023) analysed the benefits of a 4-leg inverter over traditional methods and concluded that it offers improved fault tolerance and reduced total harmonic distortion (THD) [10].

The proposed method integrates a deep learning-based controller with a 4-leg inverter to control two three-phase IMs independently. This method differs from earlier studies in the following ways: **Combining AI and IVC:** Unlike traditional IVC-based controllers, the proposed technique utilises deep learning to optimise control performance dynamically. **Multi-motor control with a single inverter:** Instead of using two separate inverters for two IMs, a 4-leg inverter is used, reducing cost and complexity.

Indirect vector control (IVC), direct torque control (DTC), and V/F Control are 3 widely used control methods for induction motor drives, each with distinct advantages and limitations [11]. In contrast, V/F control is the simplest approach, maintaining a constant voltage-to-frequency ratio. It is widely used in low-cost applications like pumps and fans due to its easy implementation [12]. However, its speed

regulation is poor under changeable load conditions, and it lacks precise torque control [13]. Overall, while IVC balances complexity and performance, DTC is ideal for applications requiring high-speed control, and V/F control is best suited for cost-sensitive applications with moderate performance requirements. Space vector pulse width modulation (SVPWM) is employed in controller design for induction motor drives due to its superior performance in generating three-phase sinusoidal voltages. SVPWM effectively utilises the DC bus voltage, offering a higher voltage output compared to traditional sinusoidal PWM methods. SVPWM contributes to reduced harmonic distortion, leading to lower electromagnetic interference and improved overall system efficiency [14].

Methodology

The DLC is designed to learn from the traditional PI controller, which consists of proportional and integral terms to regulate speed by reacting to current and past errors. The PI controller continuously adjusts the system output to minimise speed deviations. In the DLC design, data generated by the PI controller, including speed input and actual output, serve as the training dataset. The deep learning model takes speed error as input and learns to predict the actual speed output, enabling it to replicate and enhance the PI controller's performance. Figure 37.1 shows the replacement of PI controller by DLC.

Deep belief network
A DBN is a type of deep learning model composed of multiple layers of probabilistic, generative neural

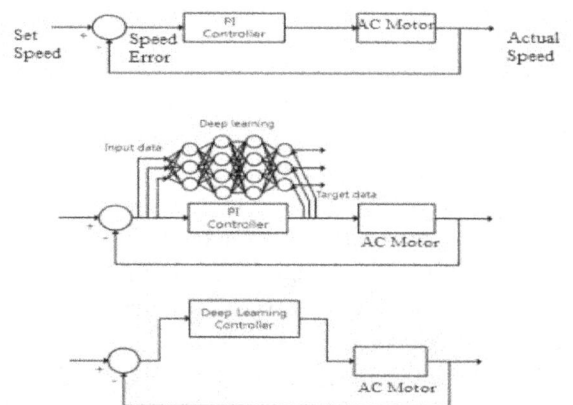

Figure 37.1 Design of DLC using PI controller input & output
Source: B T Venu Gopal , et al [15]

networks, typically structured as a stack of Restricted Boltzmann Machines (RBMs). It is trained in two phases: pre-training and fine-tuning. In the pre-training phase, which is unsupervised, the network is trained layer-by-layer, where each hidden layer learns increasingly abstract features from the input data, reducing the risk of poor local minima. This involves training the first hidden layer using raw input data, then using its activations as input for the next layer, and so on. Fine-tuning follows pre-training and is a supervised learning phase where the entire network's weights are adjusted using back propagation to optimise performance for specific tasks like classification or regression. This two-step approach enhances feature learning and improves the model's generalisation [15] (Figure 37.2).

Proposed system
The existing system uses a DLC with a 3-leg inverter to control a system, typically for motor drives. The proposed system as shown in Figures 37.3a and b, however, enhances the control of two 3-phase induction motors by applying IVC. This highly developed method allows for independent control of each motor's torque and flux, improving overall performance and flexibility. The key concept behind this system is based on the principle that the sum of the 3-phase voltages at any given instant is always zero. To achieve this, the system utilises the capacitor leg voltage to compensate for the third phase voltage. This compensation ensures that the system operates efficiently and in a balanced manner, reduces the

complexity of the circuit. A significant benefit of the proposed system is its ability to reduce the number of switches required from 12 (in traditional 3-leg inverters) to 8. This reduction in switches results in several benefits:

Lower switching losses: With fewer switches, there is less power loss due to switching, leading to improved efficiency.

Reduced cost: Fewer components mean a decrease in the overall system cost, making it more economical.

Indirect vector control
Indirect vector control (IVC) is a sophisticated technique employed to enhance the dynamic performance of induction motors by decoupling the torque & flux components of stator currents, thereby enabling independent control similar to that of separately excited DC motors. The fundamental principle involves transforming the 3-phase stator currents into a rotating reference frame, resulting in two orthogonal components: one aligned with the rotor flux (field-oriented) and the other perpendicular to it (torque-producing). By controlling these components separately, IVC allows for precise regulation of torque and flux. The block diagram of the IVC is as shown in Figure 37.4.

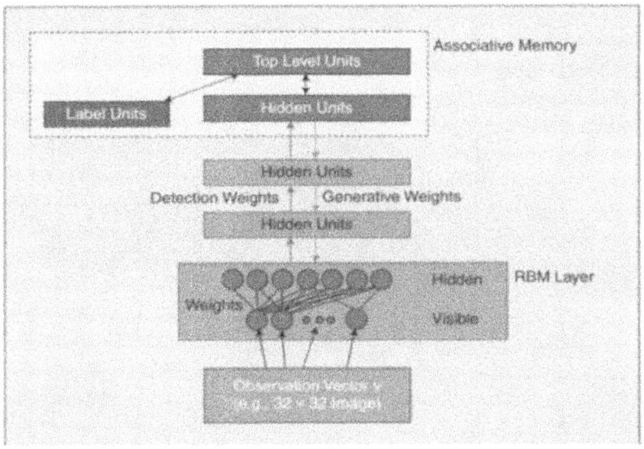

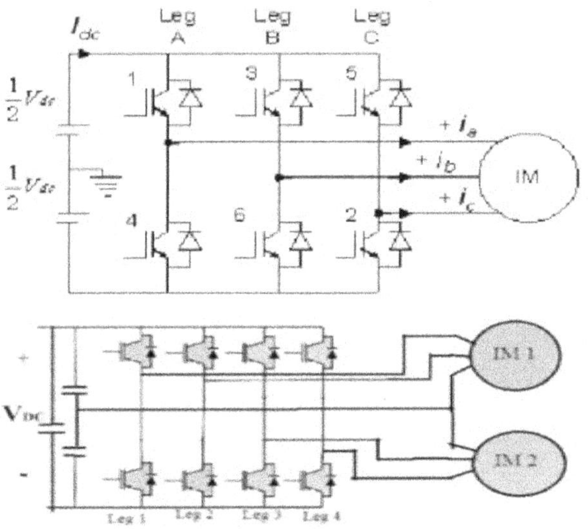

Figure 37.3 (a) Existed system for 1 induction motor control (for controlling 2 induction motors, we need 6 inverter legs). (b) Proposed system (4 legs are used for 2 induction motor control)

Source: A. Ahmed, et al [9]

Figure 37.2 DBN framework

Source: B T Venu Gopal, et al [15]

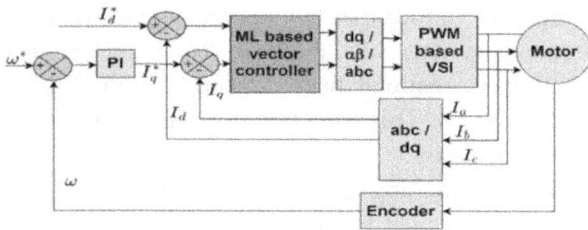

Figure 37.4 IVC block diagram
Source: A. K. Singh, ,et al[2]

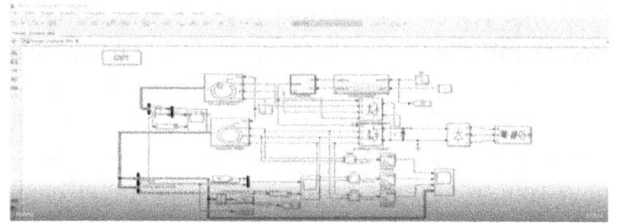

Figure 37.5 Simulation of existing PI controller-based 2 induction motors control using 6 leg inverter
Source: Author Venu Gopal B T

Simulation

The simulation results of the control design for two induction motors using different inverter configurations are presented in the figures. Figure 37.5 shows the simulation of a PI controller-based system for controlling two induction motors using a 6-leg inverter, highlighting the traditional control setup. Figure 37.6 presents the IVC block, detailing the control strategy applied to the induction motor system. Figure 37.7 depicts the simulation of a 4-leg inverter, which is part of the modified system setup. In Figure 37.8, the DBN code is integrated and called inside the controller block, showcasing the application of deep learning within the control system. Finally, Figure 37.9 illustrates the simulation of the 4-leg inverter with the DLC in place, demonstrating the enhanced performance and adaptability of the system using deep learning methods for motor control.

Results

In the results section, the behaviour of the conventional controller and the DLC for induction motor

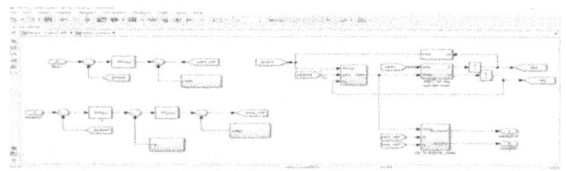

Figure 37.6 IVC block
Source: Author Venu Gopal B T

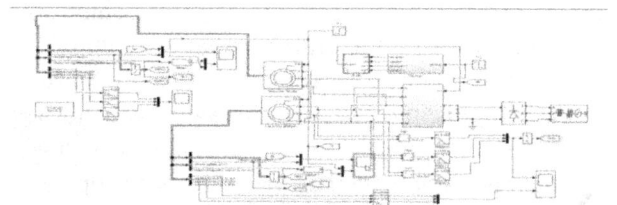

Figure 37.7 Simulation of 4-leg inverter-based DL controller
Source: Author Venu Gopal B T

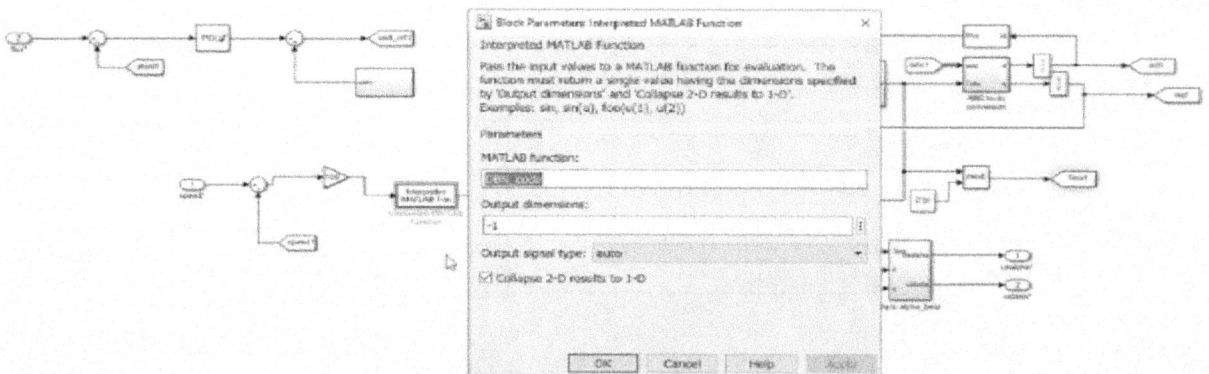

Figure 37.8 Calling deep learning code inside the controller block
Source: Author Venu Gopal B T

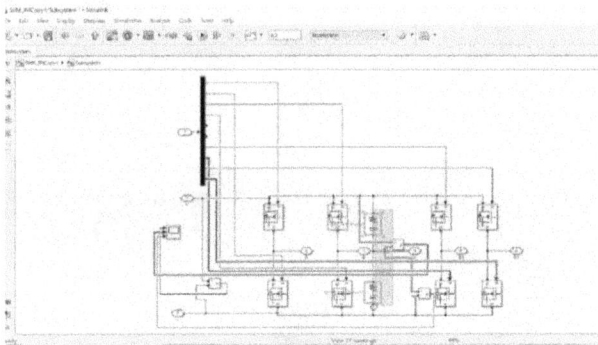

Figure 37.9 Simulation of 4-leg inverter of proposed DLC
Source: Author Venu Gopal B T

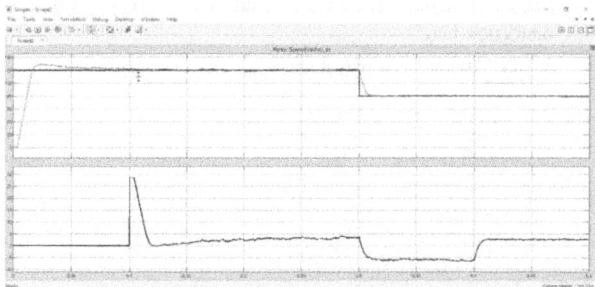

Figure 37.10 Line voltages and 3-phase currents of existing PI conventional controller (we find difference only in speed and torque waveforms of existing PI controller and proposed DLC)
Source: Author Venu Gopal B T

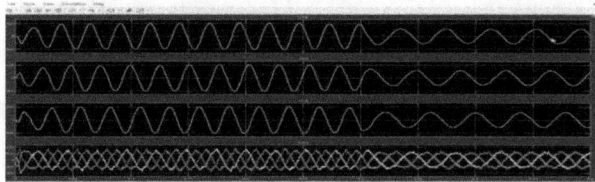

Figure 37.11 Line voltages and 3-phase currents of DLC
Source: Author Sreedhara R

performance is evaluated by observing various parameters under different operating conditions. Initially, the motor speed is set to 120 rad/sec, and after 0.3 seconds, it is reduced to 80 rad/sec to test the controllers' responses. Figures 37.10 and 37.11 show the line voltages and 3-phase currents for the conventional and DLC controllers, respectively, highlighting the differences in current waveforms between the two systems. Figures 37.12 and 37.13 depict the speed and torque waveforms, where the DLC demonstrates improved regulation with fewer ripples compared to the conventional controller, as shown in Figures 37.14 and 37.15. The enlarged views of speed and torque ripples further confirm the superior performance of the DLC. Figure 37.16 and 37.17 present the THD of the conventional and DLC controllers, with the DLC exhibiting significantly lower THD, indicating better efficiency and smoother operation. Finally, Figure 37.18 compares the performance of a 6-leg inverter-based conventional controller with a 4-leg inverter-based DLC, demonstrating enhanced overall performance in terms of stability and power quality for the DLC system. Figure 37.19 illustrates the net capacitor voltage and individual capacitor voltages, showing a stable voltage distribution for the DLC configuration.

The proposed DLC shows significant improvement in accuracy compared to the traditional PI controller. It reduces peak time from 0.015 sec to 0.010 sec and settling time from 0.1 sec to 0.015 sec, showing faster response. The steady-state error drops from 4.5 radians to zero, indicating perfect tracking. In terms of THD, DLC achieves 17% compared to 24% of

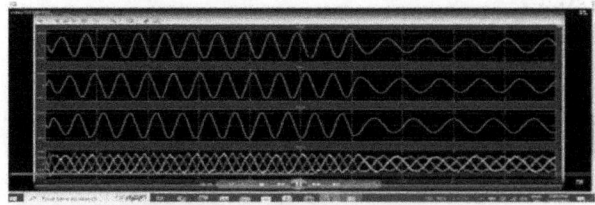

Figure 37.12 Speed and torque waveforms of existing conventional controller (when we enlarge, we can see more distortion)
Source: Author Sreedhara R

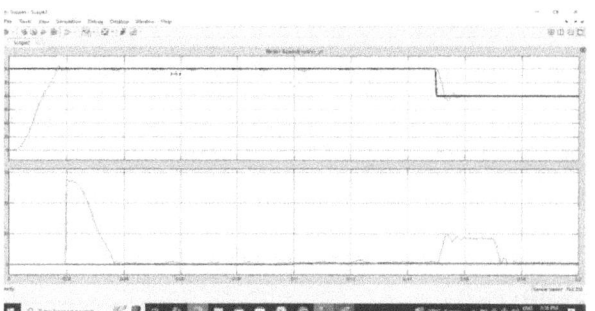

Figure 37.13 More smoothened speed and torque waveforms of proposed DLC
Source: Author Sreedhara R

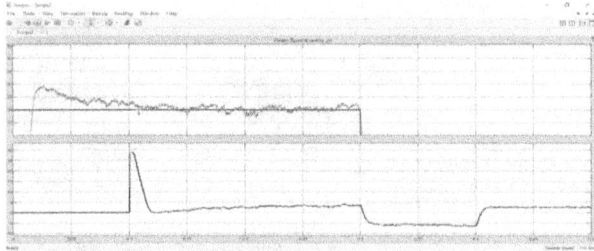

Figure 37.14 Enlarged view of speed ripples and torque ripples of existing conventional controller (with more distortion)
Source: Author Sreedhara R

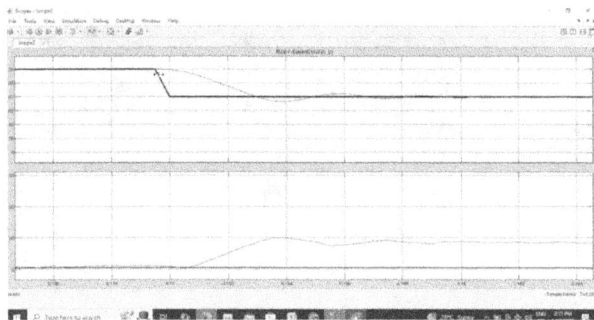

Figure 37.15 Enlarged view of speed ripples and torque ripples of proposed DLC (we can observe very negligible distortion here)
Source: Author Hareesh Kumar

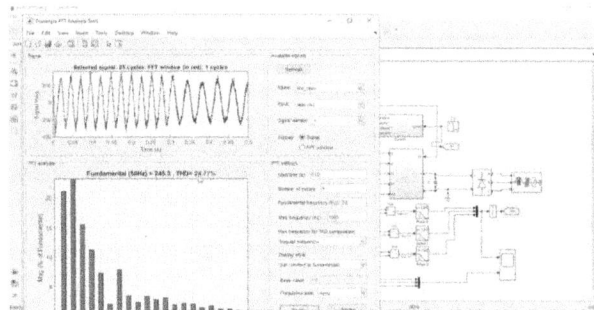

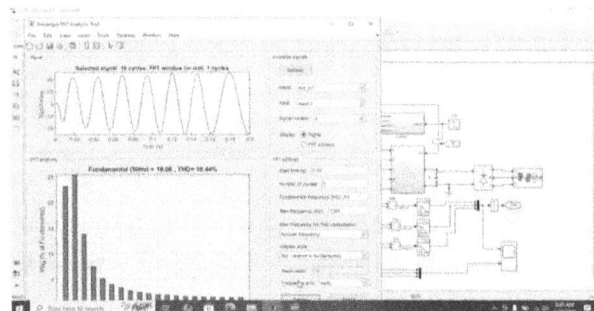

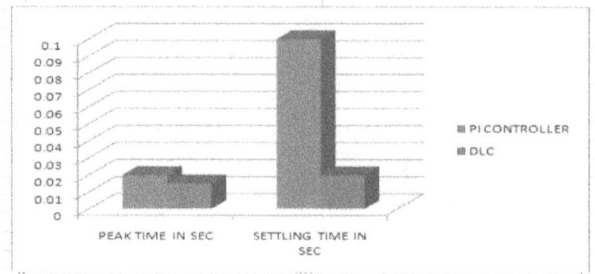

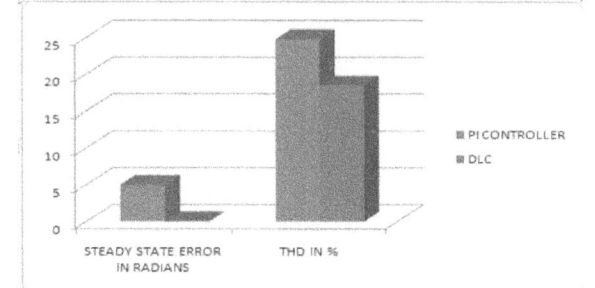

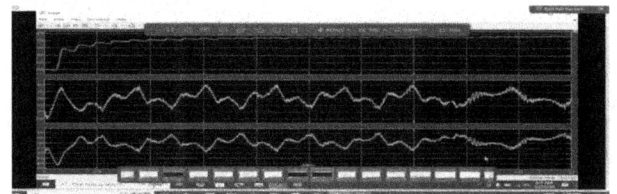

Figure 37.16 THD (24%) of existing PI conventional controller
Source: Author Hareesh Kumar

Figure 37.17 THD (17%) of proposed DLC
Source: Author Hareesh Kumar

Figure 37.18 Comparative performance analysis of existing 6-leg inverter-based conventional PI controller and proposed 4 leg inverter-based DLC
Source: Author Hareesh Kumar

Figure 37.19 Net capacitors voltage, capacitor 1 and capacitor 2 voltages of proposed DLC, respectively
Source: Author Usha Rebecca

PI, resulting in a 29.17% improvement in efficiency compared to existing system.

Conclusion and future scope

The proposed DLC-based on the DBN algorithm effectively controls two induction motors with high precision using a 4-leg inverter, as validated by simulation results that demonstrate reduced switching components and lower system costs. A major advantage of the DLC is its robustness under varying load conditions, maintaining speed stability with minimal fluctuations, unlike conventional controllers that struggle with non-linear behaviour. This makes it ideal for applications requiring high speed accuracy and torque smoothness. Hardware deployment can be explored using high-performance embedded platforms like FPGA Spartan-7, ARM Cortex-M4, or DSP processors, which offer high-speed processing, reduced latency, and efficient power management for real-time motor control applications.

References

[1] Sowmiya, M., Hosimin Thilagar, S. Vector-controlled dual stator multiphase induction motor drive for energy-efficient operation of electric vehicles. *IETE J. Res.*, 2023;69(6):3693–3710.

[2] Singh, A. K., Tiwari, S. K., Yadav, S. D. P. Fuzzy logic controller-based indirect vector control of induction motor drive for energy efficiency improvement. *IEEE Trans. Indus. Elec.*, 2025;72(1):50–59.

[3] Angelov, P., Liu, X. Evolving intelligent systems: Theory and applications. *IEEE Trans. Indus. Elec.*, 2020;67(10): 8492–8500.

[4] Crăciunaş, G., Cristina, A. Implementation of sensorless indirect vector control of induction motor with closed-loop current control. *Internat. J. Adv. Stat. IT&C Econ. Life Sci.*, 2024;14(1):147–155.

[5] Naghavi, M., Ghanbari, M., Ebrahimi, R., Jannati, M. A modified indirect vector control strategy for 3-phase wye-connected induction motor drives under open-phase fault. *Elec. Engg.*, 2022;104:3571–3587.

[6] Setiawan, P. et al. Study of indirect vector control induction motor based on Takagi Sugeno type fuzzy logic on rotational speed control primary surveillance radar. *J. Ilmiah Teknik Elektro Komputer dan Informatika*, 2024;10(2):489–506.

[7] Devanshu, A., Singh, M., Kumar, N. Artificial neural network-based current control of field oriented controlled induction motor drive. *Elec. Engg.*, 2021;103:1093–1104.

[8] Zhang, H. et al. Deep belief network-based control for high-performance induction motor drives. *IEEE Trans. Indus. Elec.*, 2023;70(5):5432–5443.

[9] Ahmed, A. Design and implementation of a four-leg inverter for dual induction motor control. *IEEE Trans. Power Elec.*, 2022;39(2):2173–2185.

[10] Kumar, R. Analysis of four-leg inverter for multi-motor control applications. *J. Elec. Engg. Technol.*, 2023;18(1): 134–145.

[11] Holtz, J. Sensorless control of induction machines—With or without signal injection? *IEEE Trans. Indus. Elec.*, 2023;69(5):4560–4572.

[12] Vas, P. *Vector Control of AC Machines*, Oxford University Press, 2021. 1–6.

[13] Rahman, M. A., Osheiba, A. M., Zhou, P. A comparative analysis of V/F, DTC, and IVC for induction motor drives. *IEEE Trans. Power Elec.*, 2022;38(3):2115–2127.

[14] Jain, S. K., Singh, B. Modulation effects on power-loss and leakage current in three-phase solar inverters. *IEEE Trans. Indus. Elec.*, 2021;68(9):7845–7854.

[15] Venu Gopal, B. T., Shivakumar, E. G., Ramesh, H. R. Design of deep learning controller for vector controlled induction motor drive. *Adv. Intell. Sys. Comput.*, 2020; 1079:639–647.

38 Design and development of a log-periodic dipole antenna for space weather monitoring

Gireesh G. V. S.[1,a], Jeyanthi R.[2,b], Harshini V.[3,c], Dharanisha D.[3,d] and Joshiga S.[3,e]

[1]Engineer-D, Indian Institute of Astrophysics, Koramangala, Bangalore–560034, Karnataka, India

[2]Professor, Department of Electronics and Communication Engineering, K.L.N College of Engineering (An Autonomous Institution Affiliated to Anna University, Chennai), Sivagangai–630612, Tamil Nadu, India

[3]Student, K.L.N College of Engineering (An Autonomous Institution Affiliated to Anna University, Chennai), Sivagangai–630612, Tamil Nadu, India

Abstract

The Sun is our nearest star and is a source of radiation that determines the life, activities, etc., of every living being on the Earth. The multi-wavelength observations in the last few decades enabled us to understand the intricate connection between the Sun and the Earth. The present understanding is that in addition to radiation, the bulk plasma and particle expulsions from the Sun affect human activities on Earth and Space. These bulk plasma expulsions, called coronal mass ejections (CMEs), can enter the Earth's atmosphere and create geomagnetic storms. These storms can, in turn, affect/damage the technological systems deployed in space (viz., satellites, spacecraft, voyagers) and on the ground (power grids, radio communication systems, etc.). Therefore, protecting these expensive space and ground-based technological systems is a prime concern for technologists and astrophysicists. Since the frequency of solar radiation occupies the entire electromagnetic spectrum, one can observe the Sun in different frequency bands and analyse them to predict the CMEs and particles before their eruption. The radio emission from the Sun, particularly in the 150–450 MHz band, originates at heights in the solar atmosphere where CMEs are usually born and depart. So, we decided to design, develop, and characterise a broadband antenna at the Gauribidanur Radio Observatory that works in the above frequency range. This antenna will be the foremost frontend component of the proposed 150–450 MHz radio receiver system, which will be used for space weather monitoring. This article describes the antenna design and development.

Keywords: Sun, radio astronomy, radio emission, space weather, coronal mass ejections (CMEs), log-periodic dipole antenna (LPDA)

Introduction

The Sun, our closest star, constantly emits radiation, plasma, etc., into its planetary neighbourhood called interplanetary space. Radiation and plasma outflow significantly control the planetary environment, including the Earth. Near-Earth conditions driven by solar sources greatly affect Earth's ground and space-based technological systems, which are collectively called space weather [1]. One major driver that adversely impacts the Earth's environment is the large-scale magneto plasma expulsion from the solar atmosphere called the coronal mass ejections (CMEs) [2, 3]. With a mass of 10^{16} g and a speed of several thousand km/s, the CME can reach the Earth in about 24 hours. Since it also contains a magnetic field, it can interact with the Earth's magnetosphere and ionosphere, triggering a geomagnetic storm [4] after reaching Earth. These storms can cause satellite malfunctions, radio communication blackouts, power grid failures, and GPS signal interruptions.

They can even interfere with space missions by putting astronauts and spacecraft under immense radiation and particle hazards [5, 6]. So, continuous monitoring of CMEs and particles is important, particularly with the rising dependency on technology in space, to envisage and reduce the impact of these storms. Determining the early signature is crucial so that forecasting is done well in advance accurately.

Since the Sun emits radiation almost over the full electromagnetic spectrum, one can observe and monitor these CMEs at different wavelengths. Because the CMEs are born in the solar corona, and the latter emits well in the X-ray, extreme ultra violet (EUV), white light, and radio wavelengths, astronomers use different techniques to observe these CMEs in different wavelength bands. Since the white light emission from the CMEs is very faint, astronomers use a coronagraph, a device that can produce an artificial eclipse by blocking the bright white light emitted by the solar photosphere to see them against the sky's

[a]gireesh@iiap.res.in, [b]hemakishore12@gmail.com, [c]bavharshini@gmail.com, [d]sekarnisha11@gmail.com, [e]joshiga26@gmail.com

DOI: 10.1201/9781003685876-38

background with better clarity [7, 8]. This device is preferably deployed in space to observe the CMEs. Similarly, the X-ray and EUV emissions from the solar corona can be used to observe the CMEs indirectly by their counterparts. The X-ray and EUV instruments are also deployed in space because these radiations cannot reach the ground due to atmospheric opacity. We now know that designing and constructing the space payloads on the ground, commissioning, and operating in space is costly.

In contrast, building and deploying a radio receiver and operating it on the ground are economical. Also, since the radio band is very broad and the Earth's atmosphere is transparent to a significant fraction, at least a portion can be used to observe the CME/ particle-associated radio emission at its onset. Since the frequency of radio emission is directly dependent on the electron density, which decreases in the solar atmosphere as the heliocentric distance increases, the propagation of a CME/particle from the Sun to the Earth can successively excite the plasma of lowering density and hence the generation of reducing plasma/radio frequency [9] as time progresses. The present theoretical models predict that CMEs must bear around a radial height of 1.05 $R_\odot$, where $R_\odot$ is the radius of the solar Photosphere, which is equal to 6.96×10^5 km. The current electron density models show that the radio frequencies 450–150 MHz are generated in the heliocentric distance range 1.05–2.0 $R_\odot$, so we decided to observe the CMEs from their initiation at the Sun to locations in a small fraction of the interplanetary space during their travel to determine their initiation height, time, initial velocity, etc., which are used to forecast the space weather.

An antenna is the foremost component of a radio receiver, so we first undertook the antenna design. Since the proposed bandwidth ratio is 1:3 (150–450 MHz) and we expect nearly identical characteristics throughout this bandwidth, we chose the log-periodic dipole antenna (LPDA) design because it belongs to the broadband antenna category [10, 11]. This article describes how we designed and developed the LPDA using the antenna simulation software named

WIPL-D (https://wipl-d.com/products/wipl-d-pro), and experimentally fine-tuned its impedance to closely approach the characteristic impedance (50 Ω) using the method described in Kumari et al. [12].

LPDA design methodology

The Sun takes about 12 hours to sail between the East and West horizon. The antenna must have a broad spatial reception pattern (or low directivity) to observe the Sun for maximum day duration. To begin with, we used the conventional LPDA design plot Carrel (1961) [13] published. The latter shows the antenna directivity for different values of the design parameters τ and σ, the design constant, and the spacing factor. We chose 6.7 dBi for the directivity, which is close to the lowest directivity achievable. This gave us the τ and σ equal to 0.84 and 0.08, respectively (Table 38.1). The half-apex angle was about 22.7°. We calculated the number of dipoles (N) to cover the frequency bandwidth (150–450 MHz) following Kumari et al. [12]. It turned out to be seven. We added two more dipoles to seven to cover the bandwidth based on the conservative approach. The length and diameter of the dipoles, their adjacent spacings, etc., were calculated, and the values are given in Table 38.2.

To begin the antenna simulation, we created the three-dimensional (3D) geometry of the LPDA using the WIPL-D Pro software tool by opening a new project. The dipole parameters, such as lengths, spacings, boom dimensions, etc., were taken from Table 38.2. The 3D model (.wpld) created is shown in Figure 38.1. A feeding port was assigned close to the shortest dipole at the apex position to provide proper excitation. An automatic meshing was introduced for subsequent manual tuning to achieve better impedance matching, especially at high frequencies. Size and spacing optimisation of the dipole elements was set to obtain good performance at the initial stage. The simulation parameters, including the frequency-range setting (150–450 MHz), are then compiled. After the setup, the simulation was run with a progress

Table 38.1 Theoretical antenna design parameters

F_{min} (MHz)	F_{max} (MHz)	Design constant (τ)	Relative spacing (σ)	Bandwidth ratio (β)	No. of elements (N)	Half apex angle (α)
150	450	0.866	0.08	3.00	7.63	22.7°

Source: Author

Table 37.2 Theoretical dipole lengths, their distance from the apex, and their diameter

Dipole number	Half dipole length λ/4 (cm)	Dipole length λ/2 (cm)	Location from top (cm)	Frequency (MHz)	Wavelength (m)	Diameter of dipole (mm)
1	13.9	27.8	6	530.34	0.57	13
2	16.1	32.2	11	457.00	0.66	13
3	19.2	38.5	18	382.88	0.78	10
4	22.3	44.7	25	329.44	0.91	8
5	25.9	51.8	33	284.12	1.06	8
6	30.4	60.7	43	242.44	1.24	6
7	35.7	71.4	55	206.14	1.46	6
8	41.9	83.9	69	175.49	1.71	6
9	49	98.1	85	150.00	2.00	4

Source: Author

check to ensure the convergence of solutions. The results are then analysed by adjusting the radiation pattern parameters; we defined 75 directions at a 5° step interval to cover the three-dimensional space (360°) around the antenna. The "Calculate Field" was checked to generate the radiation patterns, gain, directivity, etc. Finally, the simulated S-parameters were used to calculate the antenna's impedance from its voltage standing wave ratio (VSWR) values to infer the transmission or reception efficiency and overall performance. Impedance matching is crucial since the radio signals received from celestial sources are usually low, so one should not lose signal strength due to undesired reflections.

VSWR of the LPDA: Simulation vs. experiment results

The VSWR is defined as

$$VSWR = \frac{(1+\Gamma)}{(1-\Gamma)} \quad (1)$$

The symbol Γ in Eqn (1) denotes the reflection coefficient, an indicator of impedance match. The reflection coefficient equals zero when there is a perfect impedance match and one when there is a complete mismatch. From Eqn (1), one can understand that the VSWR values can range from one to infinity when the reflection coefficient varies from zero to one. Yet another parameter related to the reflection coefficient is the efficiency of the power transmission (ε). Eqn (2) quantifies their relationship.

$$\varepsilon = (1 - \Gamma^2) \quad (2)$$

A VSWR of 1 corresponds to 100% power transmission efficiency and ∞ corresponds to 0%. Since obtaining 1 for VSWR requires much effort, engineers prefer to design antennas with VSWR values ≤ 2.0 because the antenna will be 90% efficient in receiving/transmitting the signal power. Therefore, the power loss is only 10% due to unwanted reflections. Figure 38.2 shows the WIPL-D simulated VSWR profile of the LPDA in Figure 38.1. One can see that the VSWR values are above 2.0 at a few frequency bands.

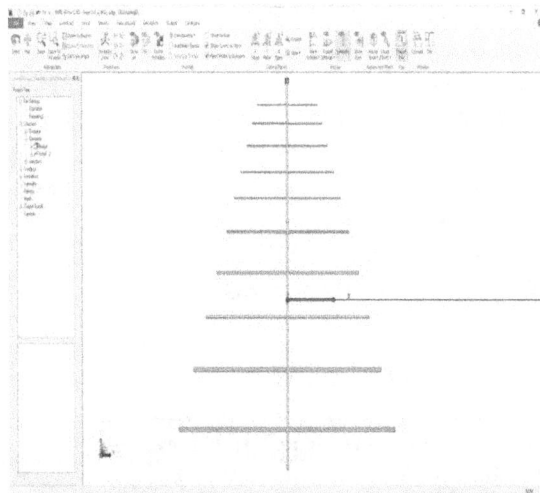

Figure 38.1 WIPL-D pro-developed LPDA model
Source: Author

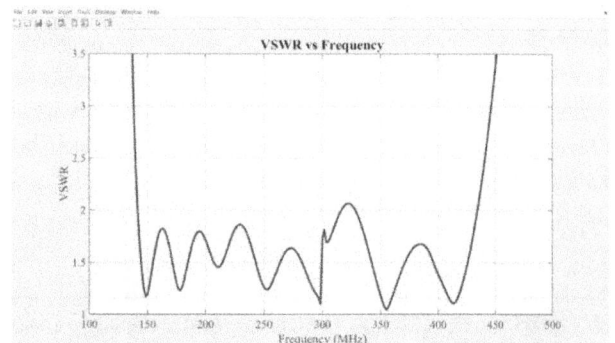

Figure 38.2 WIPL-D generated VSWR of the LPDA (Figure 1) designed using Table 38.2 parameters
Source: Author

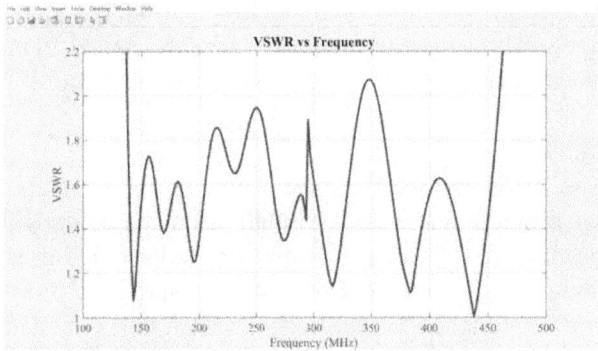

Figure 38.3 WIPL-D generated VSWR of the LPDA designed using Table 3 parameters
Source: Author

Figure 38.4 The fabricated antenna
Source: Author

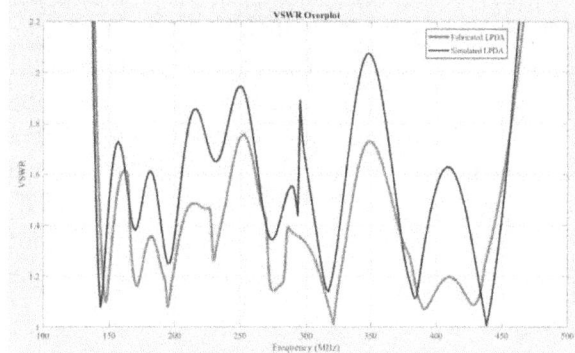

Figure 38.5 VSWR of the simulated and fabricated LPDA
Source: Author

So, to bring down the VSWR values below 2.0, we conducted several trial runs in WIPL-D by changing the design parameters. We identified a parameter set that can produce a minimum VSWR profile in that process. Figure 38.3 shows the selected VSWR profile, the average of which is 1.50 in the 150–450 MHz band. The corresponding design values are given in Table 38.3. We fabricated an LPDA using Table 38.3 at the Gauribidanur observatory (Figure 38.4) and measured its VSWR using the vector network analyser.

The result is shown in Figure 38.5. The blue and magenta profiles correspond to the simulated and experimental results. The average VSWR of the fabricated LPDA is 1.36, which is lower than that of the simulated one. We compared our VSWR with a commercial LPDA used in EMI/EMC test applications (one may refer to https://documentation.com-power.com/pdf/ALC-100-1.pdf for details). It was found to vary between 8 and 1.5 in the 300–450 MHz band.

Table 38.3 Same as Table 2, but for the LPDA with the lowest average VSWR profile

Dipole Number	Half dipole length λ/4 (cm)	Dipole length λ/2 (cm)	location from top (cm)	Frequency (MHz)	Wavelength (m)	Diameter of dipole (mm)
1	13.3	26.6	5	552.74	0.54	13
2	15.4	30.7	10	478.85	0.63	13
3	17.8	35.7	16	412.65	0.73	10
4	20.7	41.3	23	355.34	0.84	10
5	23.6	47.1	30	312.01	0.96	8
6	27.3	54.5	39	269.72	1.11	8
7	31.8	63.5	50	231.39	1.30	6
8	36.7	73.4	62	200.33	1.50	6
9	42.4	84.8	76	173.21	1.73	4
10	49	98	92	150.00	2.00	4

Source: Author

The design engineers used impedance matching pads to reduce the VSWR values of that LPDA below 2.0.

E & H-plane radiation patterns: Simulation results

The radiation pattern measurements must be taken in the far-field region of the antenna. This is the boundary/distance (R) where the variation of the radiation field is minimal as a function of radial distance from the antenna. This is calculated using Eqn (3), wherein D is the largest dimension of the antenna, and λ is the measurement wavelength.

$$R \gg \frac{2 D^2}{\lambda} \qquad (3)$$

The radiation pattern of an antenna is an important parameter that describes its transmission and reception characteristics in the 3D space. This pattern is usually examined in two main planes, the E- and H-plane. These planes cross at the central axis of the antenna, giving information about its directivity.

For the LPDA, the E-plane is where the dipole elements are aligned. This alignment determines the main radiation direction. Conversely, the H-plane is orthogonal to the E-plane and indicates the radiation reception/transmission in the perpendicular direction. After obtaining the best VSWR profile, we continued our simulation using WIPL-D and obtained the E- and H-plane radiation patterns of the above LPDA for different frequencies (150, 300, and 450 MHz). The

results are shown in Figures 38.6 and 38.7, respectively. The measured patterns of the fabricated LPDA are shown in Figures 38.8 and 38.9, respectively.

Antenna beamwidth

The cross-sectional (E-plane and H-plane) analysis provides valuable information on the antenna's beamwidth, which is essential for understanding its three-dimensional coverage. We measured the beam widths along the E- and H-planes from the simulated profiles; the values were approximately 70° and

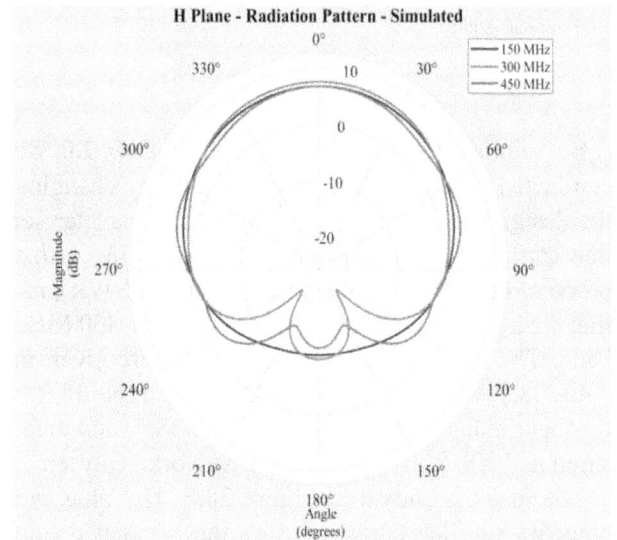

Figure 38.7 The simulated H-plane radiation patterns at 150, 300, and 450 MHz
Source: Author

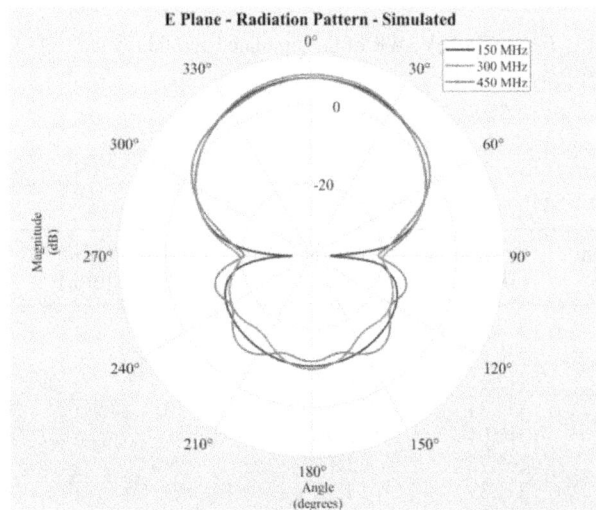

Figure 38.6 The simulated E-plane radiation patterns at 150, 300, and 450 MHz
Source: Author

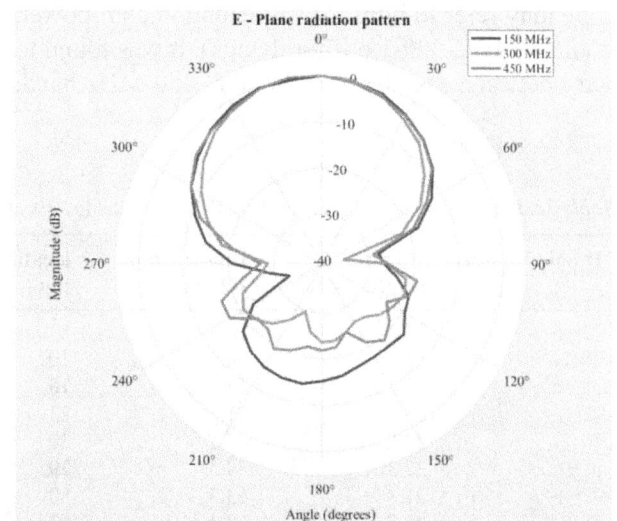

Figure 38.8 The measured E-plane patterns of the fabricated LPDa at 150, 300, and 450 MHz
Source: Author

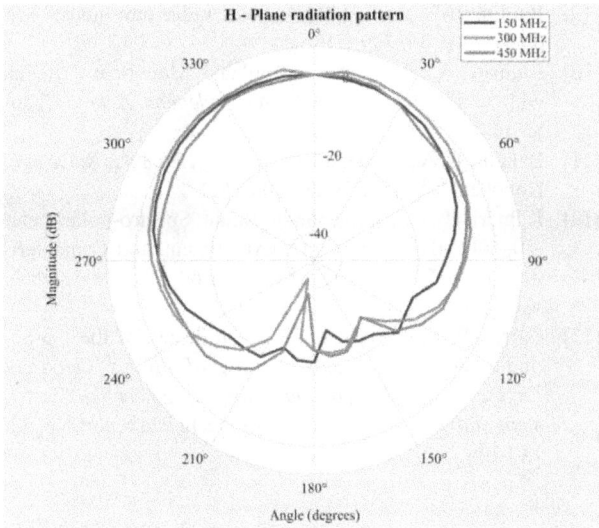

Figure 38.9 Same as Figure 8, but for the H-plane radiation
Source: Author

120°, respectively. The fabricated antenna gave 74° and 113° for the E- and H-plane beamwidths, respectively. We compared our beamwidths with the commercial LPDA. It was found to vary between 70° and 140° in the 300–450 MHz band.

Directive gain
We calculated the directive gain of the antenna using Eqn (4) and the measured bandwidths from the simulations.

$$\text{Directive Gain, } G = 10\log_{10}\left(\frac{4\pi}{\theta_E \theta_H}\right) \qquad (4)$$

In Eqn 4, θ_E is HPBW in the E plane (rad.), and θ_H is HPBW in the H plane (rad.). The directive gain of the LPDA simulated and fabricated were 6.9 and 6.9 dBi, respectively. Note that the initial design has a directive gain of 6.7 dBi, which approximately agrees with the simulated and experimental values. Compared to this, the commercial antenna's gain varies between 0 and 8 dBi in the 300–450 MHz range.

Effective aperture
Effective aperture (A_e) is a parameter (Eqn 5) that defines the area with which the antenna efficiently receives power from an incident wave.

$$A_e = \frac{\lambda^2}{4\pi} G \qquad (5)$$

In Eqn 5, A_e is the effective aperture (in m²), λ is the wavelength of the incident wave (in m), G is the directive gain of the antenna (dimensionless), 4π corresponds to the isotropic radiation pattern. The calculated effective aperture is $\approx 0.4\ \lambda^2$.

Front-to-back ratio
The front-to-back ratio (F/B ratio) measures the power level received or transmitted in the forward direction compared to its backward direction. It is about 17 dB for our LPDA.

Conclusion

We have designed, developed, and characterised a log-periodic dipole antenna to receive the Sun's 150–450 MHz radio signals. The LPDA has VSWR values lower than 1.8, averaging 1.4. Its directional gain, effective collecting area, and front-to-back ratio are 6.9 dBi, 0.4 λ^2, and 17 dB, respectively. We have verified the simulated VSWR values and the beamwidths with the prototype LPDA fabricated in-house. They all agree with each other. Our antenna's VSWR, HPBW, and directional gain were compared with a commercial broadband LPDA. Our LPDA design is better than the commercial one. Further, the research group at the Gauribidanur Observatory will build a radio receiver (analogue and digital) and commence regular space weather monitoring in the next few months.

Acknowledgement

We sincerely thank the staff of Gauribidanur Radio Observatory, Indian Institute of Astrophysics. We would like to thank the Gauribidanur Observatory employees for their assistance in fabricating the antennas, conducting the tests, etc. We sincerely thank Mr. Shaik Sayuf, Mr. Indrajit Vittal Barve, and Mr. Battini Shanmukha Rao, for their invaluable guidance, support, and constructive feedback at every stage of the project. We also thank the students and staff of the Department of Electronics and Communication Engineering, K.L.N. College of Engineering, for their support.

References

[1] Buzulukova, N., Tsurutani, B., (2022). Space Weather: From solar origins to risks and hazards evolving in time. *Front. Astron. Space Sci.*, 2022;9:1017103;67.

[2]   Kilpua, E. et al., (2017). Coronal mass ejections and their sheath regions in interplanetary space. *Living Rev. Solar Phy.*, 14, 5: Sect.6

[3]   Howard, R. A., et al., (2023). The evolution of our understanding of coronal mass ejections. Frontiers in *Astron. Space Sci.*, doi: 10.3389/fspas.2023.1264226: Sect.1

[4]   Cliver, E. W., et al., (2022). Extreme solar events. *Living Rev. Solar Phy.*, 2: 60-63

[5]   Lanzerotti, l.-J., (2017). From the Sun to Earth: effects of the 25 August 2018 geomagnetic storm. *Space Weath.*, 15, 737: 737-742

[6]   Xue, D., et al., (2024). Space Weather Effects on Transportation Systems: A Review of Current Understanding and Future Outlook. *Space Weath.*, 2024, 22, e2024SW004055, doi:10.1029/2024SW004055: 1-5

[7]   Kaiser, M. L., (2005). Stereo Mission Overview. *Adv. Space Res.*, 36, 1483: 1483-1488

[8]   Muller, D., et al., (2020). The Solar Orbiter mission. *Astron Astrophy.*, 642, A1: 1-5

[9]   Raulin, J.-P. et al., (2005). Solar radio emissions. *Adv. Space Res.*, 35, 739: 739-754

[10] Rumsey, V. H., (1966). Multiband Monopole Antenna with Sector-Nested Fractal. *Freq. Independ. Anten.*, Chapter 11

[11] Balanis, C. A., 2005, Antenna Theory: Analysis & Design, Hoboken, NJ, USA, Wiley, 2005:621.

[12] Kumari et al. (2023). Solar Radio Spectro-polarimeter (50–500 MHz). I. Design, Development, and Characterization of a Cross-polarized, Log-periodic Dipole Antenna, *Astrophy. J.*, 958, 181:181-197

[13] Carrel, R.L.: (1961). Analysis and design of the log-periodic dipole antenna. *University of Illinois Electrical Engineering Research Laboratory, Engineering Experiment Station*; Antenna Laboratory technical report No. 52: 140-146

39 Performance evaluation of Viterbi encoder with frequency hopping spread spectrum in Rayleigh fading

V. Sangeetha[1,a], Gandhamani Mohanraju[2] and P. N. Sudha[1]

[1]Assistant Professor, ECE Department, KSIT, Affiliated to VTU, Bangalore, Karnataka, India

[2]Department of EEE, University College Cork, Ireland

Abstract

This paper is an attempt to analyse the bit error rate (BER) performance over Rayleigh fading channels when utilising frequency hopping spread spectrum (FHSS). The performance and implementation of FHSS with the Viterbi encoder and decoder is discussed. MATLAB-based simulation is carried out to develop the FHSS transciever. BER is observed for different SNR with and without the Viterbi codec. The FHSS with Viterbi codec showed better BER performance for Rayleigh fading channel. The results obtained for FHSS modems is found to be satisfactory and the performance further enhanced when cascaded with Viterbi codec.This research highlights the importants of error correction techniques in wireless communication systems,especially in challenging fading environments and suggest potentials directions for future research inadaptive and hybridcoding-spreading schemes for enchanced reliability.

Keywords: Bit error rate, frequency hopping spread spectrum, viterbi encoding, rayleigh fading

Introduction

Optimum reliability, good coverage of the frequency range and lesser latency are the important features in the communication systems. Power efficiency and security are made possible by employing spread spectrum (SS) methods [1]. The use of the PN code to the transmitted signal ensures its security. At the receiving side the same PN sequence is needed to get back the source signal. The bandwidth expansion that is provided by the frequency hopping spread spectrum (FHSS) signal introduces redundancies in the signal. System performance with the SS technique and that is without SS has the difference known as processing gain. In SS techniques, the transmitted signal occupies a bandwidth that is larger than that of the original message signal [3,5]. The advances in the VLSI technology has brought in the SS technology inside the mobile systems that provide the message back to the receiver. The transmitted signal is called the modulated signal and at the receiving end the demodulation is carried out. Due to the high correlation properties of spreading codes, signals transmitted over the same bandwidth remain uncorrelated with each other. A unique PN sequence for the user can be generated that allows bandwidth sharing [4]. A method of transmission that utilises more bandwidth than the signal can occupy is called SS [6–10].

The receiver intended to receive the signal receives a signal reconstructed with the same randomness in the sequence as the transmitted signal. The concept of SS implies in every transmitted signals uses a large spectrum for modulating and transmitting the signal therefore enabling multiple users to be using the spectrum. The transmitter assigns distinct codes for each source signal transmitter.

Interference between the user signals is higher in nature while they are closer to each other. While in military applications there is an intentional interference called the spectral overlay system [11]. A filter that rejects the interference at the dispreading end would increase the signal-to- noise ratio [12]. Spread spectrum signal inherits the pseudo-white properties on which the adaptive nature of the rejection filter is dependent on [13]. High-speed rate, good security and good anti-interference capability are prime advantages in the SS communication systems. Jamming caused by the reuse of frequencies and fading which results from the multipath propagation are a few limitations of the communication systems [14]. Lower cross correlation and higher autocorrelation property automatically approve the SS system is implemented in both military and civilian communications. [15–17]. A Gaussian frequency shift keying (GFSK) transceiver for an FHSS system is applied to analyse the resource management in

[a]sangeetha.v@ksit.edu.in

DOI: 10.1201/9781003685876-39

the communication system and takes place in a factory automation paradigm [18]. FHSS framework is applied in multiple applications including synchronisation [19], tone jamming [20], signal sorting [21], etc. Viterbi encoding algorithm for high speed error correction is developed [22]. An advanced adaptive codec algorithm was developed based on the Viterbi algorithm used for the convolutional network error correction coding [23]. Evaluation of the Viterbi algorithm in memory utilisation and computational complexity is analysed for different constraint lengths and is found to be satisfactory [24]. As an extension of the author's publication [28, 29] this paper is furthered.

This paper explores the performance of the Viterbi codec based FHSS framework. Simulation of the FHSS framework with Viterbi codec is conducted and the obtained results are compared to evaluate performance. The comparison is carried out for the BER vs. SNR graph since the modulation techniques affect the BER characteristics of the signal received.

The paper is organised by introducing the Viterbi FHSS framework in the Section 2 FHSS modem's in the Section 3 followed by Results and discussion in Section 4 and Conclusion.

Viterbi encoding-based FHSS framework

Forward error correction in the communication system is initiated by adding error correcting redundancy bits are utilised at both the ends of communication systems for error detection and correction. The complete FHSS framework that is adopted for the proposed performance evaluation is as depicted in Figure 39.1. The error detection and correcting is done using Viterbi codec. FHSS modulation techniques are applied for performance evaluation for the Rayleigh fading channel.

Viterbi encoding involves developing the Trellis in order to introduce the redundancy sophistication to decode the data. The advantages of the Viterbi encoding algorithms over the block encoding are detailed in ref [25]. The trellis diagram as outlined in ref [26] identifies the most probable input sequence can be recovered at the decoder side. The error correcting code using Viterbi is developed by introducing the convolutional encoder in the encoding level and the trellis. Viterbi decoder which is relies on the maximum likelihood path transition detection is developed. The overall encoding and decoding process is depicted in Figure 39.2.

The convolution encoders involve very little computation while the decoder is complex computationally. It is a 2-modulo-2adder using the XOR gate. If it is a ½ encoder then for each bit in data it provides 2 bits as the encoded data. The convolutional encoder is shown in Figure 39.3.

Following the encoding using XOR operation a trellis state diagram is developed that generates the most probable input in the encoding side. The sequence of state diagrams called the Trellis is drawn as shown in Figure 39.4. The operation of the convolutional encoder depends on the current input bits and the previous state bits for the current output bits. Thus, a state transition diagram can be drawn which

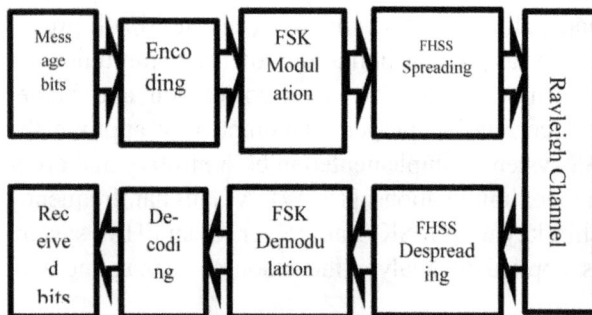

Figure 39.1 General block diagram of FHSS
Source: Author [V.Sangeetha]

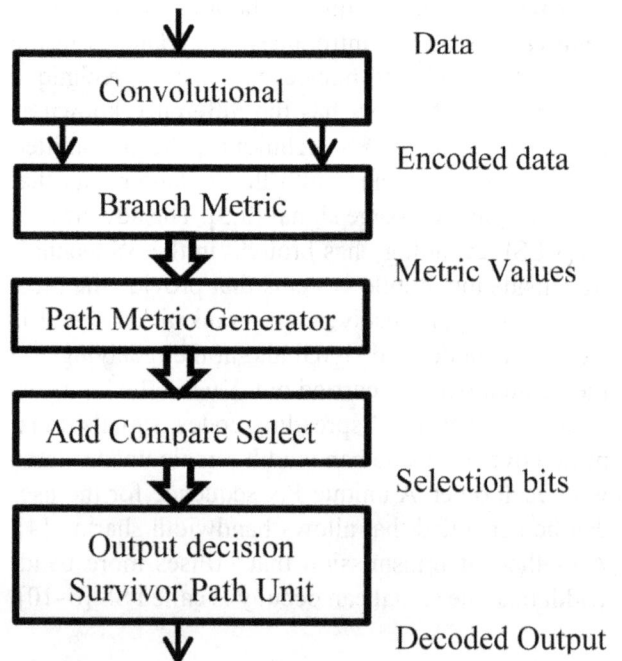

Figure 39.2 Flow diagram Viterbi encoding and decoding
Source: Author [V.Sangeetha]

rate = ½
constraint length, K = 3
generator polynomial = $\{7,5\}_8 = \{111,101\}_2$

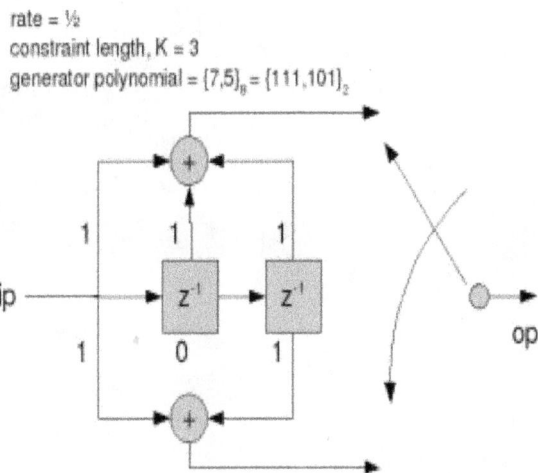

Figure 39.3 ½ Convolutional encoder [2]

Source: [2] H. S. Bedi, G. Singh, T. Singh, and N. Kumar, "Overview of performance of coding techniques in Mobile WiMAX based system," Int. J. Recent Innov. Trends Comput. Commun.2013. vol. 1, no. 1, pp. 32–35.

depicts both the previous state in the memory and the current state. The state transition diagram for the convolution encoding is given in Figure 39.4 which is called the Trellis diagram [13].

If the input bit is "0" the solid lines are drawn to denote the transition while if the input bit is "1" dotted lines are used to denote the transition. At the time "t=0" the state is defined as S0 then while new input bits enter the state changes according to input bits and creates the output bit stream as given in the received sequence in the Trellis diagram. At time t = 1, the state changes to S2 and that way to S3 at "t=2". This state transition is defined by the input bit that enters the encoding stream.

The sequence of symbols thus produced from the Trellis state transition is received at the receiver end

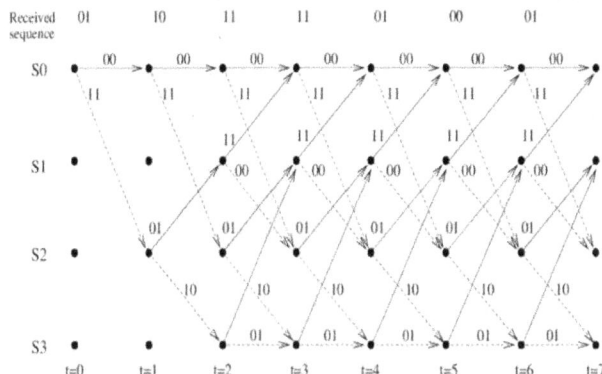

Figure 39.4 The Trellis diagram [13]

Source: [25] Johanna, T., Juha, P. Asynchronous Viterbi decoder in action systems. 2020

and the decoding is carried out. Viterbi decoder is a relevant algorithm that uses the most likelihood path sequence to find the transmitted bit in the memory less noise channel. The branch metric and the path metric calculations have to match for the particular received bits. While there is noise added to the modulated signal in the SS channel then the branch metric and the path metric will be unmatched. The likelihood algorithm guesses the correct bit that is transmitted avoiding the noise added to it. The decoding block diagram is shown in Figure 39.5.

The Viterbi decoder guesses the expected input and compares with the received input to find the most probable input and the decoded output is obtained by using the branch metric and the path metric comparison from the state machines previously developed in Trellis.

FHSS modulation

Encoded data using the Viterbi convolution encoding the data is sent for modulation using FHSS [30]. FHSS is one among the two important spreading techniques other being direct-sequence SS. A narrowband carrier signal is modulated by varying its frequency in accordance with the input data. This hopping of frequency is predictable by means of a sequence generated randomly. The interference between the transmitted signals is limited unless these signals are sent at the same instant with the same frequency. A single logical channel can be maintained if synchronised properly. Figure 39.6 depicts the overall block diagram of the FHSS.

The hopping or the spreading code is created to start the transmission process. Based on the hopping code the input bits are modulated accordingly. The receiver must also be configured with the same code sequence to correctly receive the signal. A frequency band is chosen which would hop between the frequencies inside that band. The interference in that band only affects the signal for a short period.

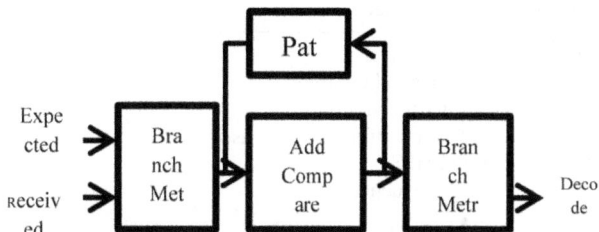

Figure 39.5 Viterbi decoder

Source: Author [V.Sangeetha]

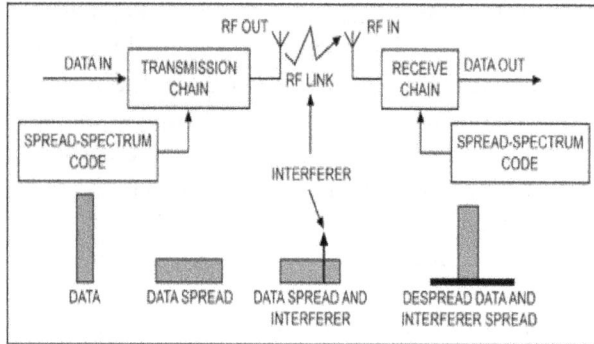

Figure 39.6 Overall FHSS system block diagram
Source: Proakis, J. G., Digital Communications, 4th ed., McGraw Hill, 2000.

The synchronisation is the challenge operation in a FHSS implementation. Thus, the listening between the both the ends of communication system is solved with the help of the spreading sequence.

Results and discussion

While applying the FHSS implementation the parameters that are adopted for implementation are discussed in the Table 39.1.

MATLAB based simulation is carried out both with Viterbi and without the Viterbi FHSS framework to obtain the different graphs along with the performance graphs. This investigation involves FHSS implementation for comparison. Firstly the results obtained while implementation is carried out. A randomly generated 500-bit data is used for the FHSS framework with the Viterbi forward error correction. The message input with 500 bits is as shown in Figure 39.7.

The encoded data obtained after the convolution encoding due to the code rate of ½ is shown in Figure 39.8. The data thus obtained is applied with the Trellis finite machine format before it is modulated

Table 39.1 FHSS specifications

Sl. No.	Parameters	Allocation
01	Data points	500
03	Eb/No values (dB)	(-20:50)'
04	Viterbi codec type	½
05	Number of frequencies used in FHSS	6
06	PN sequence	4 bit sequence
07	Frequency variation between each band	10Hz

Source: Author [V.Sangeetha]

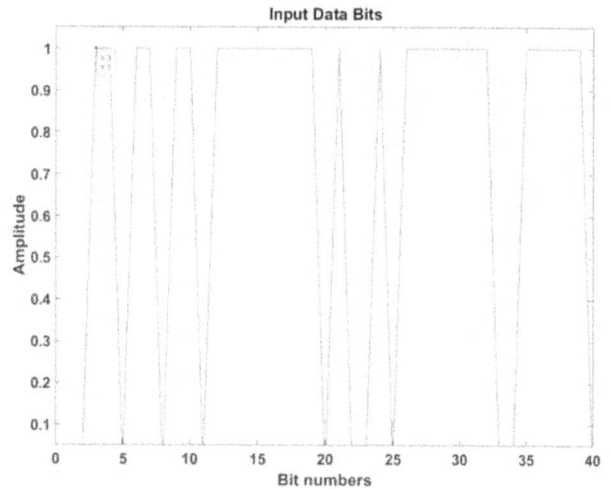

Figure 39.7 Transmitted data
Source: Author [V.Sangeetha]

using FHSS modulation. The output of the simulation is shown followed by the FHSS modulation before doing the performance evaluation.

The frequency hopped spread spectrum output is as shown in Figure 39.9. Totally six frequencies were selected for spreading the data that improves the bandwidth of the input data that improves the redundancy (Figure 39.10).

The received data after the Viterbi decoder is as shown in the Figure 39.11. The BER analysis for the FHSS modulation with Rayleigh fading applied is obtained as shown in Figure 39.12.

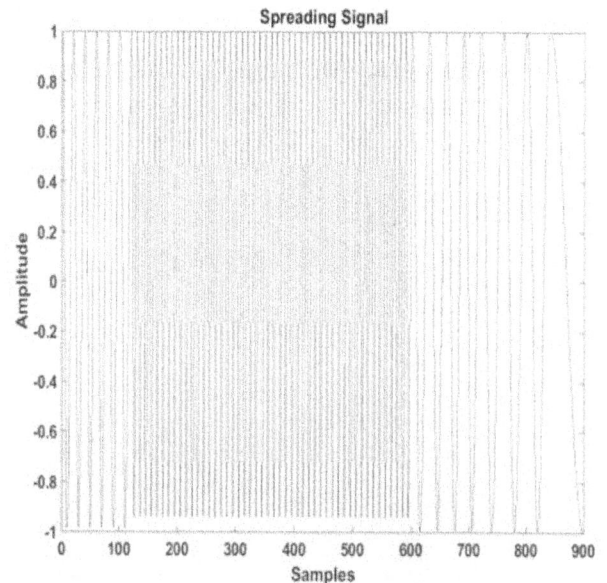

Figure 39.8 Spreading signal for FHSS
Source: Author [V.Sangeetha]

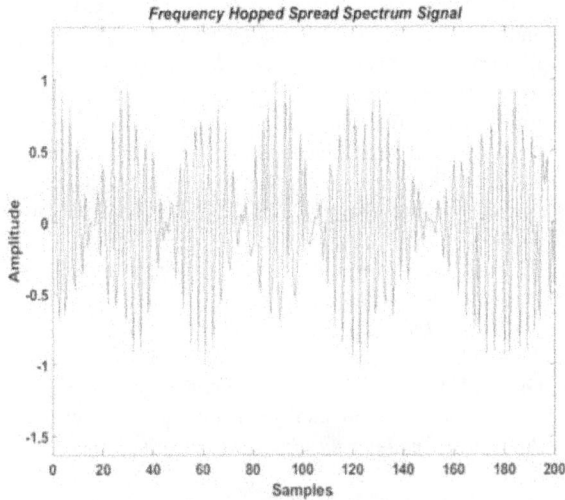

Figure 39.9 Frequency hopped spread spectrum signal
Source: Author [V.Sangeetha]

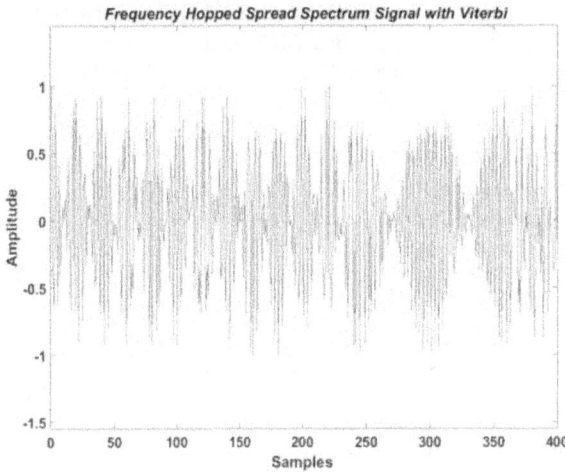

Figure 39.10 Frequency hopped spread spectrum signal with Viterbi encoding
Source: Author [V.Sangeetha]

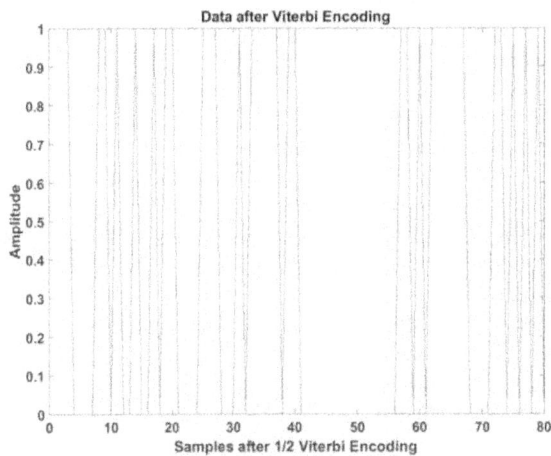

Figure 39.11 Decoded bits from Viterbi decoding
Source: Author [V.Sangeetha]

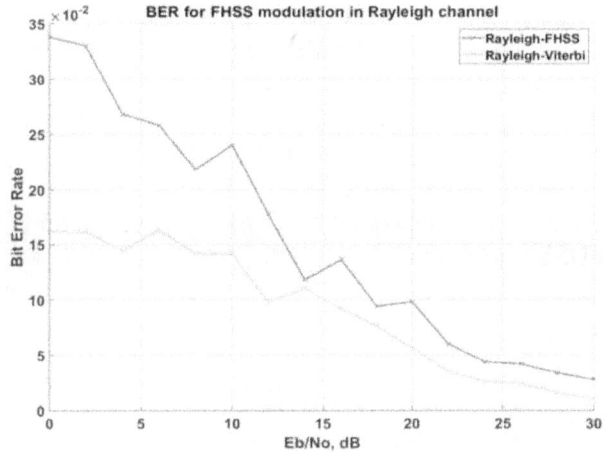

Figure 39.12 FHSS BER vs. noise with and without Viterbi codec
Source: Author [V.Sangeetha]

It is evident that BER values obtained is better while compared to the FHSS implementation without forward error correction. The difference between with and without forward error correction is clearly demonstrated in Table 39.2 which approves performance improvement while convolution encoding and Viterbi decoding is implemented. Approximately there is a 50% improvement in error correction while Viterbi is used.

Conclusion

Matlab simulation is applied for the FHSS modulation in the Rayleigh channel using the Viterbi encoder and decoding forward error correction algorithm.

Table 39.2 BER vs. noise

Eb/No, dB	BER without Viterbi encoding	BER with Viterbi encoding
0	0.3380	0.1617
2	0.3300	0.1613
4	0.2680	0.1443
6	0.2580	0.1630
8	0.2180	0.1420
10	0.2400	0.1407
12	0.1780	0.0980
14	0.1180	0.1100
16	0.1360	0.0917
18	0.0940	0.0763
20	0.0980	0.0563
22	0.0600	0.0353
24	0.0440	0.0260
26	0.0420	0.0240
28	0.0340	0.0157
30	0.0280	0.0100

Source: Author [V.Sangeetha]

The performance evaluation of both the modems with Rayleigh fading is obtained and the results are observed. The BER vs. noise curves are obtained for with and without the Viterbi forward error correction implementation. It is observed that Viterbi with FHSS performed better by reducing the error due to Rayleigh channel compared to the without Viterbi in FHSS framework.

References

[1] Pickholtz, R. L., Schilling, D. L., Milstein, L. B. Theory of spread spectrum communications - A tutorial. *IEEE Trans. Commun.*, 1982;30(5), 855–884.

[2] H. S. Bedi, G. Singh, T. Singh, and N. Kumar, "Overview of performance of coding techniques in Mobile WiMAX based system," *Int. J. Recent Innov. Trends Comput. Commun.*, 1(1), 32–35, 2013.

[3] Sklar, B. Digital communications: fundamentals and applications. 2nd Ed., Prentice Hall, 2001, 719–766.

[4] Peterson, R. L., Ziemer, R. E., Borth, D. E. Introduction to spread-spectrum communications. Englewood Cliffs, NJ: Prentice- Hall, 1995, 90–140.

[5] Chakrabarti, N. B., Datta, A. K. Introduction to the principles of digital communication. New Age Publishers, 2007.

[6] Erik, S., Tony, O., Arne, S. An introduction to spread spectrum Systems,An Introduction to Spread Spectrum Systems, Department of signals and Systems, Chalmers university of Technology, Sweden, 2002

[7] P. G. Flikkema, "Spread-spectrum techniques for wireless communication," in IEEE Signal Processing Magazine, May 1997;14(3):26–36.

[8] Ryuji, K., Reuven, M., Laurence, B. M. Spread spectrum access methods for wireless communications. *IEEE Comm. Mag.*, 1995;33(1):58–67.

[9] Yong, L. Spread spectrum ranging system – Analysis and simulation. *Master Thesis Elec. Sys. Engg.*, 1998, 30–35.

[10] Schilling, D. et al. Spread spectrum for commercial communications. *IEEE Comm. Soc. Mag.*, 1991:66–79.

[11] Ketchum, J., Prokis, J. Adaptive algorithms for estimating and suppressing narrowband interference in PN spread spectrum systems. *IEEE Trans. Commun.*, 1982;30: 913–923.

[12] Milstein, L., Iltis, R. Signal processing for interference rejection in spread spectrum communications. *IEEE Acoustics, Speech, Sig. Proc. Mag.*, 1986;3:18–31.

[13] Diffe, W., Hellman, M. E. New directions in cryptography. *IEEE Trans. Inform. Theory*, 1976;22:644–654.

[14] Zhang, Y., Xuehe, Z., Sen Yang, N., Zhong, K., Qiao, J. L. Modeling and performance analysis of frequency hopping spread spectrum communication system. *IEEE Xplore Dig. Lib.*, 2010, 2184–2187.

[15] Rajiv Mohan, D., Kumara, S., Prabhakar Nayak, K. Efficient synchronization and data detection for DS-CDMA applications. Internat. J. Comp. Elec. Engg., 2011;3(2011):251–261.

[16] Rajiv Mohan, D., Kumara S., Prabhakar Nayak, K., Vasanth, M. Design of spreading codes with increased performance in correlation property. *Internat. J. Appl. Engg. Res.*, 2010;5(6):997–1008.

[17] Rajiv Mohan, D., Kumara, S., Prabhakar Nayak, K. Digital filter bank implementation using multirate techniques for spreading codes and rate transition techniques in spread spectrum. *Internat. Res. J. Signal Proc.*, 2012;3(1): 445–453.

[18] Wulf, A., Underberg, L., Kays, R. Adaptive frequency hopping communication for factory automation. *ICOF 2016 19th Internat. Conf. OFDM Freq. Domain Tech.*, 2016:1–8.

[19] Qi, W., Haocheng, D., Miao, Z. Frequency hopping synchronization method and simulation in VHF channel. *2019 IEEE 2nd Internat. Conf. Elec. Inform. Comm. Technol. (ICEICT)*, 2019:51–54.

[20] Jung, H., Nguyen, B. V., Song, I., Kim, K. Design of anti-jamming waveforms for time-hopping spread spectrum systems in tone jamming environments. *IEEE Trans. Veh. Technol.*, 2020;69(1):728–737.

[21] Qu, Y., Pei, Y., Song, D. Frequency-hopping signal sorting based on deep learning. *2019 IEEE 2nd Internat. Conf. Elec. Inform. Comm. Technol. (ICEICT)*, 2019:759–762.

[22] Kumar, A. K., Kumar, P. S. High speed error-detection and correction architectures for Viterbi algorithm implementation. *2019 3rd Internat. Conf. Elec. Mat. Engg. Nano-Technol. (IEMENTech)*, 2019:1–6.

[23] Liu, M., Guo, W. Adaptive construction and decoding of random convolutional network error-correction coding. *2019 IEEE/CIC Internat. Conf. Comm. China (ICCC)*, 2019:449–454.

[24] Ahmed, S., Siddique, F., Waqas, M., Hasan, M., Rehman, S. u. Viterbi algorithm performance analysis for different constraint length. *2019 16th Internat. Bhurban Conf. Appl. Sci. Technol. (IBCAST)*, 2019:930–932.

[25] Viterbi, A. J. A personal history of the Viterbi algorithm. *IEEE Signal Proc. Magazine*, 2006;23(4):120–142.

[26] Viraktamath, S. V., Patil, P. H., Attimarad, G. V. Impact of code rate on the performance of Viterbi decoder in AWGN channel. *2014 IEEE Internat. Conf. Comput. Intell. Comput. Res.*, 2014:1–4.

[27] Johanna, T., Juha, P. Asynchronous Viterbi decoder in action systems. 2020.

[28] Sangeetha, V., P. N. Sudha. Proposed algorithm: A novel approach to minimize the effect of fading using adaptive channel coding & multi carrier modulation techniques - A survey. *Internat. J. Curr. Res.*, 2016;8(1):25246–25249.

[29] Sangeetha, V., P. N. Sudha. Analysis of bit error rate on M-ary QAM over Gaussian and Rayleigh fading channel. *Internat. J. Innov. Technol. Explor. Engg.*, 2020;9(6): 482–485.

[30] Maleki, A., Bedeer, E., Barton, R. Performance evaluation and low-complexity detection of the PHY modulation of LR-FHSS transmission in IoT networks. *IEEE Internet of Things J.*, 2023;10(5):4296–4310.

40 Voltage sag characteristics in power distribution systems under fault conditions

Anmol Saxena[a], Richa[b] and Rahul Vivek Purohit[c]

Department of Electronics & Communication Engineering, Galgotias College of Engineering and Technology, Greater Noida, India

Abstract

Power quality (PQ) issues, though longstanding, have recently gained significant attention due to increased customer awareness. Among these issues, voltage sags stand out as one of the most critical challenges for electric utilities. This paper examines the key characteristics of voltage sags experienced in power distribution systems. Voltage sag characterisation (VSC) is essential for diagnosing, troubleshooting, and implementing effective mitigation strategies. The study emphasises the significance of two primary sag attributes—magnitude and duration—and also discusses phase-angle jumps caused by faults. MATLAB/SIMULINK is utilised to model and simulate voltage sag scenarios for linear and non-linear loads under symmetrical and asymmetrical fault conditions. The results validate the methodology and offer insights into voltage sag behaviour, highlighting its applicability to practical PQ challenges. This detailed analysis provides a foundation for developing robust solutions to mitigate the adverse effects of voltage sags.

Keywords: Power quality (PQ), voltage sag (VS), voltage sag characterisation (VSC), simulation

Introduction

The concept of power quality (PQ) has garnered increasing attention from utilities, commercial establishments, and end-users [1, 7]. Voltage quality (VQ) plays a pivotal role in PQ, especially for sensitive electrical loads [8] such as medical equipment, industrial machinery, and communication systems. Historically, PQ concerns were centred on reliability, focusing primarily on outages and interruptions. However, modern challenges include more nuanced issues such as voltage sags [2–6], fluctuations, interruptions, and harmonics, which significantly impact operational continuity and equipment longevity [9]. Disaggregation of all parts of the power system is not required in the analysis of voltage sag distribution. These sags are primarily brought about by faults in the system that propagate within the network and may induce hard disturbances. Although voltage sags are less severe than total outages of the power, repetition of voltage sags may initiate serious financial as well as operating consequences. Highly sensitive loads such as adjustable speed drives, programmable logic controllers (PLCs), and data centres are highly susceptible to the disturbances. As a consequence of this, they may incur loss of data, damage to the equipment, or process interruption and hence amplify the overall influence of voltage sags on system operation.

In the modelling of voltage sag distributions, no explicit modelling of all components of the power system is necessary. These voltage sags mainly originate from system faults that propagate through the network and thus may result in widespread disturbances. Although voltage sags are generally less harmful than total power outages, their high rate of occurrence can have drastic financial and operational effects. Sensitive loads like adjustable speed drives, programmable logic controllers (PLCs), and data centres are particularly vulnerable to these disturbances. They can lose data, incur equipment damage, or experience process disruption, further increasing the impact of voltage sags on the overall performance of the system. Voltage sag impact mitigation is of paramount importance in ensuring the stability and reliability of power systems. Transient voltage sag can heavily impact sensitive machinery and industrial processes. A proper voltage sag characterisation (VSC) forms the foundation for an effective planned voltage sag management. This consists of detection and analysis of leading parameters, i.e., amplitude of the sag (degree of voltage drop), duration (length of time voltage is below the nominal levels), and any accompanying phase-angle oscillations. An understanding of these characteristics is critical to establishing the cause of voltage sags and examining their

[a]panthereagleanmol@gmail.com, [b]richa.sharma70@gmail.com, [c]puru261280@gmail.com

DOI: 10.1201/9781003685876-40

effect on various equipment in the power system. For example, the magnitude and duration of a sag will dictate whether a specific load can ride through the disturbance without suffering failure. Similarly, phase angle swings can affect the operation of equipment like motors and drives, which are susceptible to such swings. Detailed characterisation of voltage sags helps power engineers develop effective mitigation schemes, which include the use of voltage regulators, dynamic voltage restorers (DVRs), and uninterruptible power supplies (UPS). Additionally, it provides crucial data for optimising network planning and fault management strategies. Overall, these efforts are vital to enhancing power quality, reducing downtime, and improving the reliability of power delivery systems. Through a scientific method of tackling voltage sag problems by characterisation and mitigation, industries and utilities are able to save economic losses and ensure uninterrupted process operation [11]. The current document is divided into five separate sections. The first section presents the subject and highlights the challenges of voltage sags. The second section offers an in-depth explanation and description of voltage sags. The third section outlines the theoretical calculations with respect to voltage sag magnitude and phase-angle jump. The fourth section presents the outcomes of the simulation results. Finally, the fifth section integrates the results and presents the conclusions based on the study.

Voltage sag definition and characterisation

Voltage sags (dips)

Voltage sags are a decrease in root mean square (RMS) voltage between 10% and 90% of the normal operating voltage for duration between half a cycle and one minute [10]. They are typically due to power system faults, e.g., short circuits, but also from operating high inrush motor loads. For instance, heavy industry processes are expected to produce high inrush currents, which lead to transient voltage drops [1] (Figure 40.1).

Voltage sag characterisation (VSC)

Two main features characterise voltage sags:

- **Magnitude**: The magnitude of a voltage sag is the nominal reference voltage minus the voltage at the instant when the sag takes place. This is a parameter of utmost significance in measuring the severity of the disturbance on electrical equipment.
- **Time:** The period of a voltage sag is the time during which the voltage remains below a particular value. It is an important measurement for evaluating possible effects on sensitive loads as well as on system stability overall.

Additionally, voltage sags often cause phase-angle jumps, which represent abrupt changes in the voltage waveform's phase angle [12–16]. These jumps are significant as they affect synchronisation in power systems, particularly in systems reliant on precise timing and phase alignment (Figure 40.2).

Theoretical calculation

To analyse voltage sag phenomena, the voltage divider model is commonly used. This model simplifies the power system by representing it with source and fault impedances, enabling the calculation of sag magnitude and phase-angle shifts.

The voltage at the point of common coupling (PCC) is calculated using the voltage divider rule, which considers the relationship between the source impedance (Z_s) and fault impedance (Z_f). The voltage at PCC can be expressed as:

$$V_{pcc} = V_s \times \frac{z_f}{z_s + z_f} \qquad (1)$$

where Z_s is the source impedance, Z_f is the fault impedance and V_s is the source voltage.

Equation (1) helps to determine the sag magnitude and phase-angle variations based on system impedance conditions. The phase-angle jump (PAJ)

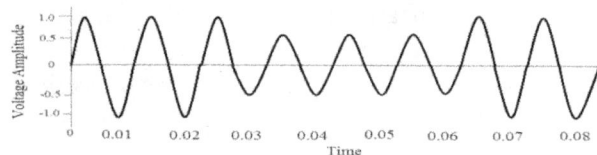

Figure 40.1 Typical waveform during voltage sag
Source: Author

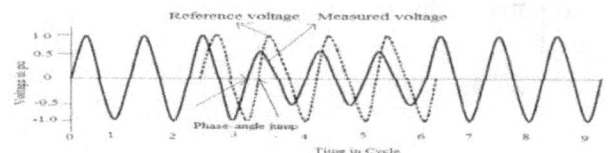

Figure 40.2 Phase-angle jump + 45°
Source: Author

is a crucial parameter that provides insights into the dynamic behaviour of the power system during faults, affecting equipment synchronisation and stability (Figure 40.3).

In order to examine the origin of voltage-sag PAJ, the single-phase voltage divider model of Figure 40.2 can be utilised. This model considers the source impedance (Z_s) and fault impedance (Z_f) as complex values [17]. Using per-unit calculation, by neglecting all load currents. Voltage at PCC.

$$\bar{E}_{sag} = \left[\frac{(R_f + jX_f)}{(R_s + jX_s) + (R_f + jX_f)} \right] \bar{V} \quad (2)$$

If V (avg. voltage) =1, then Equation (1) is rewritten as $= \left[\frac{(R_f + jX_f)}{(R_s + jX_s) + (R_f + jX_f)} \right]$ (3)

The argument of , equivalent to PAJ is given by the following expression

$$\Delta\emptyset = \arg(\bar{E}_{sag}) = \arctan\left(\frac{X_f}{R_f}\right) - \arctan\left(\frac{X_f + X_s}{R_f + R_s}\right) \quad (4)$$

where R_s, X_s, R_f, X_f and $\bar{E}_{sag}$ is the source resistance, source reactance, feeder resistance, feeder reactance and voltage sag at the PCC [18, 19].

Simulation results

The proposed system shown in Figure 40.4 is a Simulink-based simulation environment designed to evaluate the characteristics of voltage sag in a power distribution network under different fault conditions. This system gives a comprehensive modelling, simulation, and evaluation of the effects of symmetrical and asymmetrical faults on voltage magnitude, phase-angle, and duration. The setup is an 11 kV, 30 MVA, 50 Hz three-phase voltage source for a high-voltage distribution feeder. A delta/wye transformer (11 kV/0.4 kV, 1 MVA) is used to step down the voltage, thus making it a practical setup for studying sag propagation from high voltage to low

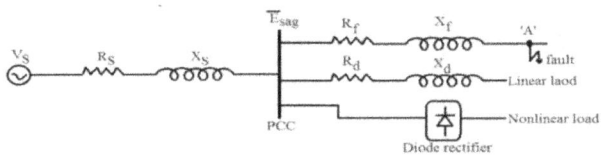

Figure 40.3 Voltage divider model for voltage sag magnitude [1]
Source: Author

voltage levels. To simulate real load conditions and their response to faults, a 10 kW resistive load and a 100 VAR inductive load are used. Two fault blocks are included to permit a range of faults, including:

- Single-line-to-ground faults
- Double-line-to-ground faults
- Line-to-line faults
- Three-phase faults
- Multistage faults

Fault events are simulated in the model to include dynamic responses. The system describes voltage sags in terms of magnitude, duration, and phase-angle jump and accurately records the following fault responses:

- Propagation of voltage sags across transformer windings
- Asymmetry in sag due to transformer and load settings

The designed system acts as a key facility for power quality analysis that supports thorough researches of voltage disturbances to boost power system reliability and efficiency.

Line-to-line faults
The A-B line-to-line fault between phases simulation indicated that the amplitude of the sag is dependent on fault resistance and transformer configuration. Phases A and B experienced large voltage sags at the 11 kV bus, while phase C experienced small swells. The results indicate the impact of transformer configuration and fault impedance on sag propagation. This line fault model is capable of simulating multiple fault types, including:

- Single-line-to-ground faults
- Double-line-to-ground faults
- Line-to-line faults
- Three-phase faults
- Multistage faults

Voltage sag waveforms initiated by a phase-to-phase fault between phases A and B on an 11 kV feeder line between 0.1 and 0.2 seconds are shown in Figures 40.5 and 40.6. As can be observed, the 11 kV bus experiences two totally different voltage sags at phases A and B with different sag magnitudes. The cause of the difference is the high fault resistance of 8 Ω between the two faulted phases. As the voltage sag propagates downstream through the 11 kV/0.4 kV, 1 MVA transformer to the load, the fault type [15] is altered due to the delta/wye transformer configuration. Additionally, the unfaulted phase C at the 0.4 kV bus experiences a slight voltage swell, which occurs due to the absence of a ground point in the line-to-line fault and the presence of high fault resistance

Multistage faults

Multistage voltage sags were also investigated, where the actions of asynchronous relays or fault impedances caused a series of multistage sags before the system returned to nominal voltage. Multistage sags are important for distributed generation systems or systems with sophisticated protection systems. In voltage sag waveform amplitude, it is usually indicated in RMS terms and normalised to make it clearly visible.

RMS analysis

Instantaneous and RMS waveforms for every fault case were developed. RMS analysis normalised voltage sag waveforms to provide a more realistic indication of sag magnitudes and durations. Pre- and post-sag time interval oscillations were induced by transient phase shifts during the fault occurrences.

Figures 40.7 and 40.8 illustrate the RMS plot of sag waveforms of the line-to-line fault voltage in Figures 40.5 and 40.6. The amplitudes of the sag for all of the phases are evidently illustrated. The slight oscillation seen before and after the sag and swell is a result of the phase shift during the fault.

The line fault model can simulate faults that occur in several stages. Voltage sags occurring in several stages are normally caused by the unsynchronised clearing of several fault protection relay mechanisms, which lead to variations in the power system's impedance and network configuration. This results in a sequence of voltage sags before the nominal level is restored [20], or due to fault or ground impedance variations at the occurrence of a fault. It is at times referred to as a sequence of faults in close chronological proximity and reported as a single event.

Conclusion

This study provides an in-depth analysis of voltage sag characteristics, a significant PQ issue affecting modern power systems. The research emphasises the importance of detailed voltage sag characterisation, including magnitude, duration, and phase-angle jumps, for developing effective mitigation strategies. Using MATLAB/SIMULINK, the study successfully simulated various fault scenarios, demonstrating the robustness of the proposed methodology. The

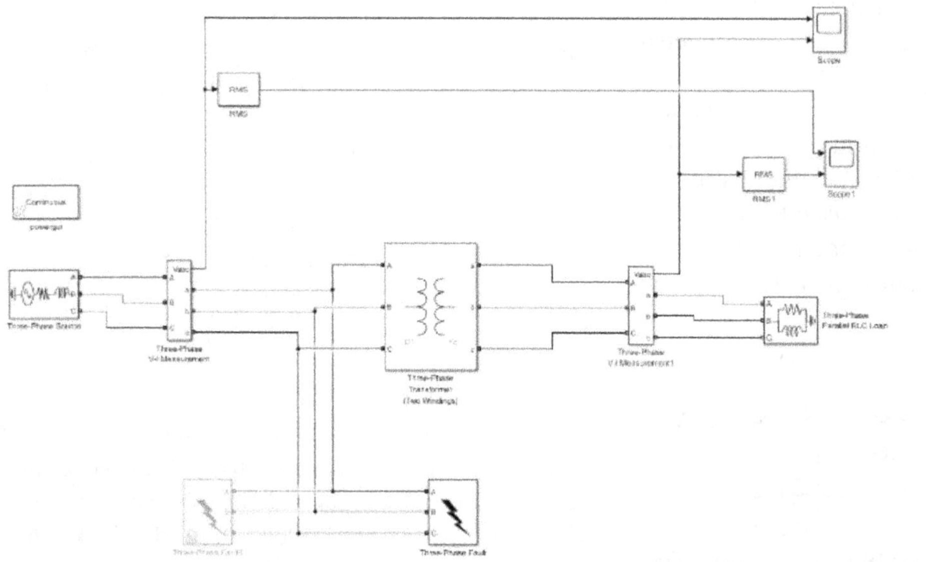

Figure 40.4 Line fault Simulink model
Source: Author

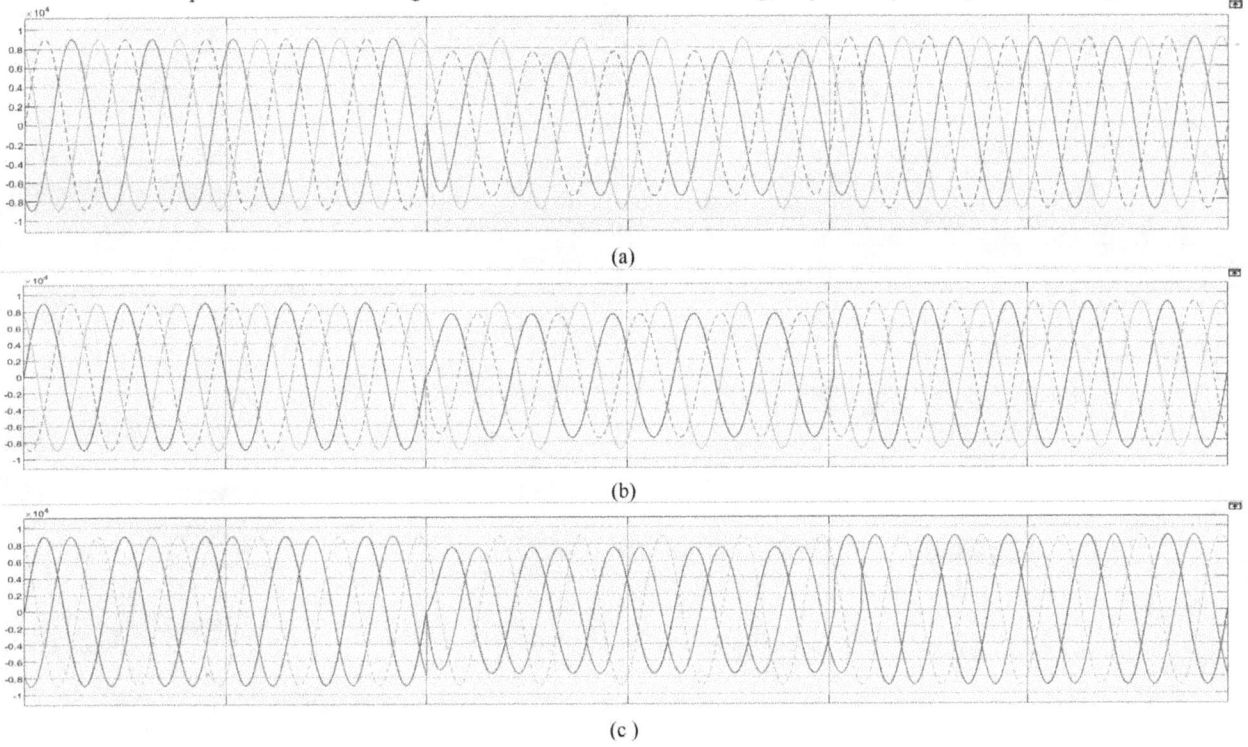

Figure 40.5 (a) by phase a (b) by phase b (c) by phase c illustrates the voltage sag caused by a line-to-line fault at the 11 kV bus, represented as an instantaneous waveform

Source: Author

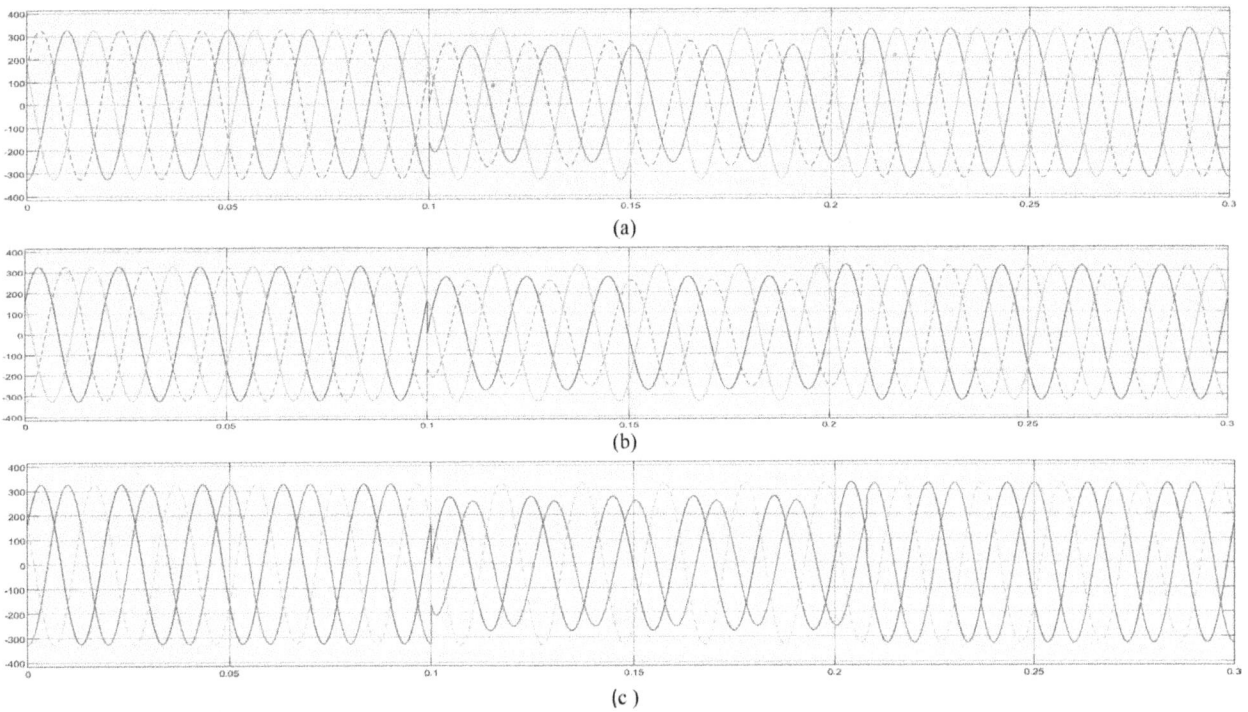

Figure 40.6 (a) by phase a (b) by phase b (c) by phase c illustrates the voltage sag caused by a line-to-line fault at the 0.4 kV bus, represented as an instantaneous waveform

Source: Author

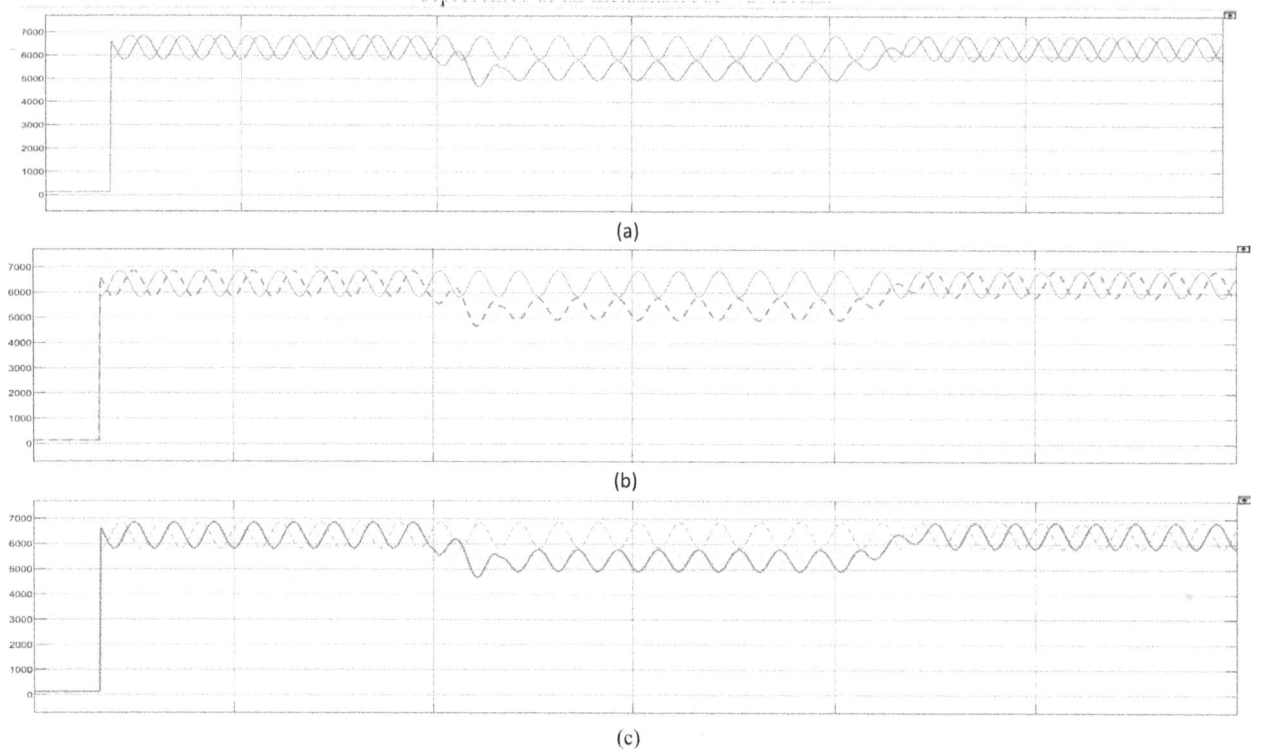

Figure 40.7 (a) by phase a (b) by phase b (c) by phase c illustrates the voltage sag caused by a line-to-line fault at the 11 kV bus, represented as an RMS waveform
Source: Author

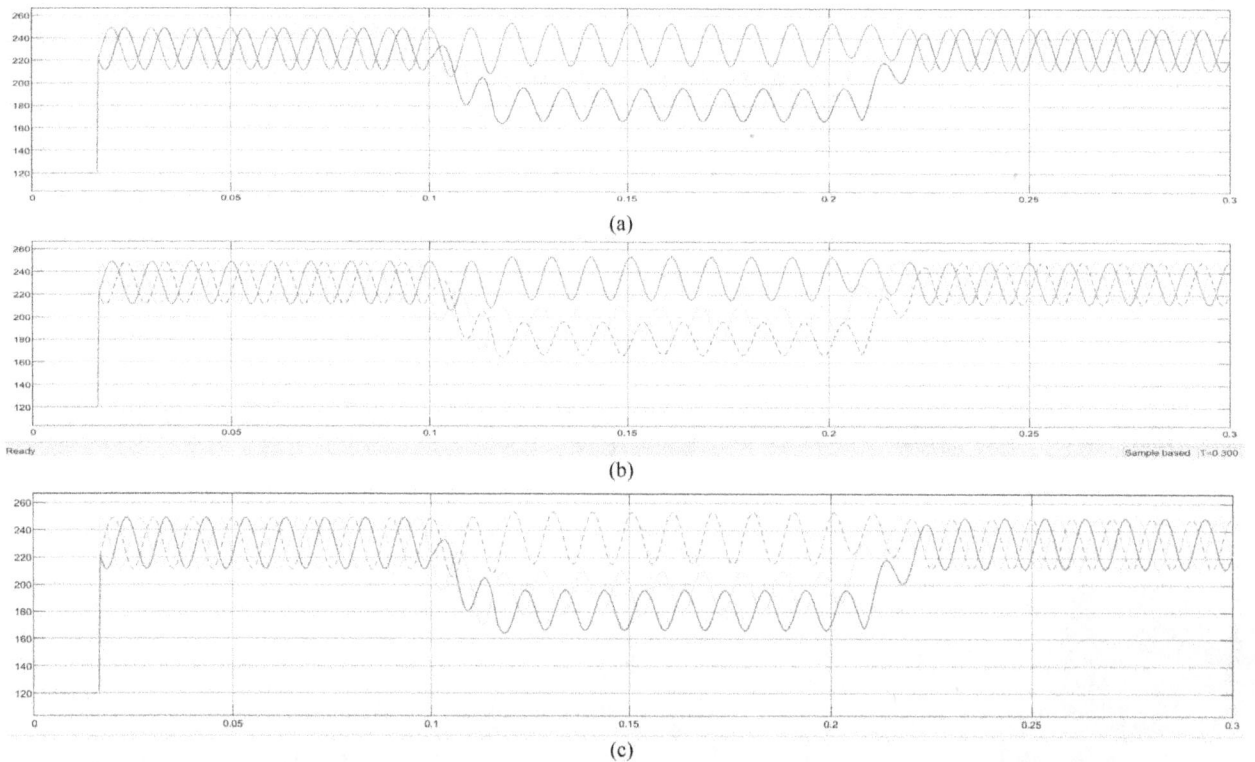

Figure 40.8 (a) by phase a (b) by phase b (c) by phase c illustrates the voltage sag caused by a line-to-line fault at the 0.4 kV bus, represented as an instantaneous waveform
Source: Author

findings underscore the need for advanced diagnostic tools and strategies to address voltage sag challenges in both three-phase and single-phase systems.

Future work could focus on integrating renewable energy sources and distributed generation into voltage sag studies to evaluate their impact on sag characteristics and mitigation techniques.

References

[1] Prasad, M., Akella, A. K. Voltage sag characteristics in power distribution system under fault conditions. *Ener. Edu. Sci. Technol. Part A Energy Sci. Res.*, 2015;33(6):3177–3192.

[2] Tan, R. H. G., Ramachandaramurthy, V. K. A comprehensive modeling and simulation of power quality disturbances using MATLAB/SIMULINK. "*Power Quality Issues in Distributed Generation*", Tech, Rijeka, pp-83, 2015, https://doi.org/10.5772/61209

[3] Yalcinkaya, G., Bollen, M. H. J., Crossley, P. A. Characterization of voltage sags in industrial distribution systems. 1998.

[4] Heine, P., Lehtonen, M. "Voltage sag distributions caused by power system faults". *IEEE Trans. Power Sys.*, 2003;18(4): 1367–1373, Nov. 2003, doi: 10.1109/TPWRS.2003.818606

[5] Lotfifard, S., Kezunovic, M., Mousavi, M. J. Voltage sag data utilization for distribution fault location. *IEEE Trans. Power Deliv.*, 2011; 26(2), 1239–1246. 10.1109/TPWRD.2010.2098891.

[6] Prasad, M., Akella, A. K. Comparison of voltage swell characteristics in power distribution system. *Int. J. Elec. Electr. Res. (IJEER)*, 2016;4(3):67–73.

[7] Babaei, E., Kangarlu, M. F. Sensitive load voltage compensation agaist voltage sags/swells and harmonics in the grid voltage and limit downstream fault currents using DVR. *Electr. Power Syst. Res.*, 2012;83:80–90.

[8] Babaei, E., Kangarlu, M. F. Cross-phase voltage compensator for three phase distribution systems. *Electr. Power Energ. Syst.*, 2013;53:119–126.

[9] Alam, M. R., Muttaqi, K. M., Bouzerdoum, A. Characterizing voltage sags and swells using three-phase voltage Ellipse parameters. *IEEE Trans. Indust. Appl.*, 2015;51:2780–2790.

[10] IEEE Standards Coordinating Committee 22 on Power Quality. IEEE Std. 1159. IEEE Recommended Practice for Monitoring Electric Power Quality, 1995.

[11] Piumetto, M., Gomez Targarona, J. C. Characterization of voltage sags and its impact on sensitive loads in a MV system with distributed generation for single-phase fault. *IEEE Latin America Trans.*, 2013;11:439–446.

[12] Yal¸cinkaya, G., Bollen, M. H. J. Stochastic assessment of voltage sags for systems with large induction motor loads. Presented at the Universities Power Engineering Conf., Galway, Ireland, Sept. 1994.

[13] Guo, C., Zhang, X., Wang, L. Economic evaluation of voltage sag under uncertain conditions based on beta distribution parameter correction. *2023 IEEE Int. Conf. Power Sci. Technol. (ICPST)*, 2023:917–922.

[14] Ghambirlou, K., Abosnina, A. A., Moschopoulos, G. Voltage sag investigation for induction motor startup in distribution systems using solid state transformer. *2025 15th Int. Renew. Energy Cong. (IREC)*, 2025:1–6.

[15] Wu, X., Zhong, H., Chen, Y., Hu, Y. A novel voltage sag detection method for series filter originate from improved genetic algorithm. *2024 Int. Conf. Intell. Comm. Sens. Electromag. (ICSE)*, 2024:269–274.

[16] Hadi, S. Power System Analysis, 2nd edn, McGraw Hill, 2004, 15–18.

[17] Bollen, M. H. Understanding power quality problems: Voltage sags and interruptions. Wiley-IEEE Press, 1999:672:48–52. https://doi.org/10.1109/9780470546840.

[18] Bhakkad, M. V., Deshmukh, B. T. Generation of voltage sag for different loads and conditions using MATLAB SIMULINK. *2019 Int. Conf. Innov. Trends Adv. Engg. Technol. (ICITAET).*, 153–159. 10.1109/ICITAET47105.2019.9170228.

[19] Camarillo-Pe˜naranda, J. R., Ramos, G. Effect of phase-angle jump in impedance-based fault location methods for transmission and distribution systems. 2018 IEEE IAS annual meeting.

[20] Yal¸cinkaya, G., Bollen, M. H. J. Stochastic assessment of frequency, magnitude and duration of nonrectangular voltage sags in a large industrial distribution system. *Proc. 12th Power Sys. Comput. Conf.*, 1996;II:1018–102.

41 Construction and performance evaluation of zero energy cooling chamber for sustainable storage solutions of vegetables

S. Banerjee, C. Chandre Gowda[a], B. N. Skanda Kumar, H. Jaswanth, K. S. Prajwal, D. Dhanush and G. M. Kushal

Centre for Incubation, Innovation, Research & Consultancy, Jyothy Institute of Technology, VTU, Bengaluru, Karnataka, India

Abstract

Agriculture is the main occupation for India and more than 50% of the population are engaged in it. The production and distribution of agricultural products contribute significantly for prosperity of the country. However, the absence of adequate primary storage facilities leads to significant post-harvest losses of agricultural produce. This study aims to design and develop a zero\minimum energy cooling chamber (ZECC) as a sustainable storage solution for agricultural products in comparing its carbon footprint with traditional refrigeration method. The relative humidity of 95% was maintained throughout. The temperatures difference of about 5–15°C was observed, when compared with ambient temperature. The shelf life of was enhanced by 8 days for tomatoes, by 10 days for eggplants, and by 15 days for potatoes. During cost comparison between conventional method and proposed work, a staggering amount of ₹14,400 can be saved annually.

Keywords: Shelf life, zero energy cooling chamber, sustainable storage, temperature control

Introduction

In developing countries like India where population has been increasing overtime, production and distribution of agricultural products to feed the entire nation is a great challenge. India produces and exports fruits and vegetable all over the world. Though, having higher yield, about 30–35% of produce will be lost due difficulty in handling and shortage for on-site storages [1]. These losses reduce the revenue generation for the farmers contributing to food insecurity. From harvesting to delivery many stages are involved like collection, transportation, distribution, and retail storage. All the way through production and supply loss from the produce range from 5% to 10%; and which are mainly due to inadequate storages, handling the produce and storing at extreme weather conditions [2] are challenging.

The farmers at remote areas lack access to central facility to store the agricultural products. Due to which they tend to store them in open spaces because of which they are vulnerable to all kind of damages, like variation in temperature, humidity, attack from pests and thefts. The qualities of harvested products are maintained by controlling environmental factors. Traditional refrigeration methods, though effective, but are often expensive and energy-intensive, presenting a barrier for local farmers [3]. In contrast, zero energy cooling chambers (ZECC) offers a cost-effective and eco-friendly alternative [4]. It was first developed using readily available materials like jute, bamboo, bricks, and sand [5]. It controls evaporative cooling to create an internal environment with low temperatures and higher humidity [6]. They effectively slow down the metabolic processes of stored fruits and vegetables [7, 8]. In evaporative cooling, the liquid evaporates into the surrounding area, cooling the adjoining surfaces [9]. The larger is evaporative difference, the greater will be cooling effect. Conversely, when temperatures are the same, the cooling effect is minimal [10]. A faster evaporation rate results in greater cooling, as cooling occurs through the latent heat of evaporation. The design principle of the ZECC is such that it works on principle of evaporative cooling through a permeable surface. The heat transfers from inner surface to middle wet layer (due to convection) corresponding to decrease the temperature inside ZECC [11]. The construction of ZECC is a way to store the agricultural produce. Major focus of the present work is to provide better solution for storages for rural India

[a]chandre.gowda@ciirc.jyothyit.ac.in

DOI: 10.1201/9781003685876-41

by means of lowering the temperature in the system and with minimum cost [12]. The study aims to construct ZECC to enhance the shelf life of vegetables. By comparing its efficiency with conventional refrigeration methods, this research seeks to provide a viable, low-cost solution for reducing post-harvest losses and promoting sustainable agricultural practices [13]. The study also includes an economic analysis to assess the financial benefits of adopting ZECC technology in small-scale farming operations [14]. Much improvement for ZECC has been carried in the recent years [15, 16].

Methodology

The construction of ZECC involves usage of locally available, eco-friendly materials including bricks, sand, bamboo, and jute cloth. The chamber's design focused on achieving high relative humidity and maintaining lower internal temperatures to extend the shelf life of vegetables.

The ZECC was constructed under shade with ample air movement to optimise evaporative cooling. The ZECC inner storage space is 60 cm × 90 cm. It was built using bricks at internal wall and an external wall was constructed with a 7.5 cm gap from the internal wall on all four sides. The river sand is filled in between to facilitate evaporative cooling. The chamber walls were built up to a height of 53 cm from bottom. The sand and bricks were manually saturated with water three times a day during

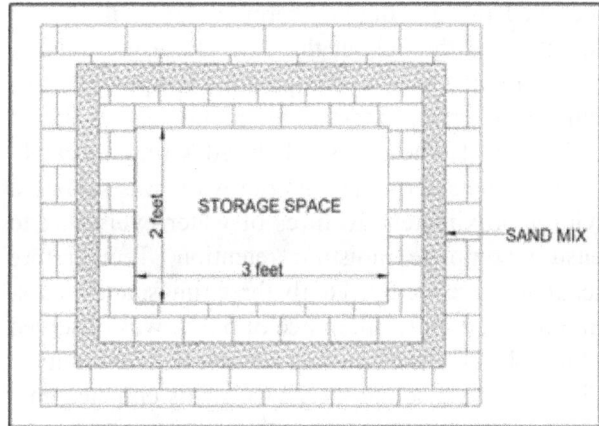

Figure 41.2 Plan of ZECC
Source: Author

preliminary observations. A thin plywood frame was used to cover the chamber, and wet gunny bags were placed to enhance cooling. The total volume of the ZECC storage space was 0.2973 m^3. All stages of construction process are depicted in Figure 41.1 and the plan is shown in Figure 41.2. Temperature and humidity readings were collected both inside the chamber and outside the chamber using digital sensors. Measurement was taken multiple times throughout the day over a period of 2 months (December 2020–January 2021). The variation in temperature between outer surface and inner surface of the ZECC chamber was observed initially and it was used to determine its performance. Tomatoes, eggplants and potatoes are the three products considered for the study. They were monitored at regular intervals to determine any changes in its weights. A comparison of proposed work with commonly practiced storage methods was done. The cost analysis was also carried to estimate the total savings by constructing ZECC when compare with refrigeration. The results were observed and data collected were thoroughly analysed. The tests compare to emphasise the major differences between ZECC and conventional storage outcomes. To achieve this, a safe area having good ventilation was selected for construction of ZECC. The dry bulb temperature was measured by K-type thermocouple sensor.

Results

After completing the construction of the ZECC, preliminary measurements were conducted to assess the temperature variation. Initially, temperatures were

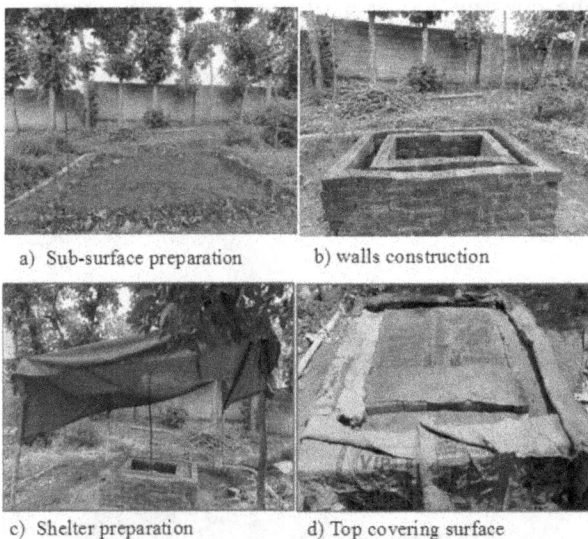

a) Sub-surface preparation   b) walls construction

c) Shelter preparation   d) Top covering surface

Figure 41.1 Stages of the ZECC construction
Source: Author

recorded using a digital thermometer at three different times throughout the day. These preliminary observations aimed to compare the temperature difference before and after saturating the sand, as shown in Table 41.1. The river sand placed within the cavity between the exterior and interior walls was saturated with approximately 20 litres of water every day to ensure complete moisture retention. Temperature readings were recorded daily three times, and a maximum temperature difference of 6.9°C was observed (Table 41.1). Based on this observation, the improvisation was made to the ZECC. This enhancement was achieved by increasing the soil in between the bricks walls. These thick layers retain more moisture and improve the temperature. Later the internal temperature further reduces by using wet gunny bags (they are placed over ZECC). At regular interval the temperatures was recorded. The highest difference in temperature difference was achieved during the final observations, (13.16°C, Table 41.2). This indicates that efficiency of ZECC can be improved by maintaining a cool environment within. The ZECC was proved to be an eco-friendly storage solution, retaining the vegetables' nutritive value without relying on external energy sources. The ZECC storages of raw vegetables tomato, egg-plant and potato are shown in Figure 41.2. The vegetables are cleaned (dry wiped)

and small slices are cut from each vegetable for the determination of initial moisture content.

The moisture content for the samples has been presented in Table 41.3. The vegetables samples are kept in plastic tray/containers inside ZECC and also outside the ZECC, respectively (Figure 41.3). The weights of the vegetables placed both are measured on daily basis until they are spoiled. The physical observations of the samples with age are shown in Table 41.4.

Initial moisture content is determined by gravimetric method. First, the initial weight of the vegetable slices is noted down. These slices are kept inside oven at 105°C for 24 hours. The moisture content was calculated accordingly. The representation of storage days versus moisture content of tomatoes is plotted and shown in the Figure 41.4.

The tomatoes kept inside the ZECC were not spoiled till day 12. While tomatoes outside the ZECC spoiled quickly within 7 days, due to loss of moisture, which accelerated the spoilage of tomatoes.

The moisture content of the tomatoes inside the ZECC was entrapped within the chamber and was cool enough to enhance the shelf life of tomatoes. The shelf life of egg-plant would spoil within 6 days. But inside ZECC its shelf life was extended for 14 days. The escape of moisture is the main reason for spoilage of vegetable The moisture content of the potatoes kept inside the ZECC decreased on a slower pace, thus proving that the ZECC temperature was cool enough to extend the shelf life of potatoes.

Cost comparison with conventional refrigeration

The construction cost of the ZECC was minimal, amounting to approximately ₹1,830. In contrast, conventional refrigeration involves significant expenses, including:

- Initial cost of refrigerator: ₹20,000
- Electricity tariff, Karnataka, India [17]

Table 41.1 Temperature differences observed in initial testing stages

Date	Time (IST)	Temperature (°Celsius)		
		Outer	Inner	Difference
26-12-20	11:45 am	29.2	27.8	1.4
26-12-20	04:00 pm	28.3	26.4	1.9
28-12-20	11:00 am	27.5	25.5	2.0
28-12-20	03:45 pm	28.5	26.0	2.5
11-01-21	11:30 am	27.3	23.3	4.0
11-01-21	01:45 pm	28.7	22.9	5.8
11-01-21	04:00 pm	27.1	21.8	5.3
26-12-20	11:15 am	27.0	22.3	4.7
12-01-21	01:30 pm	28.7	21.8	6.9
12-01-21	04:00 pm	26.9	20.9	6.0
12-01-21	11:30 am	27.2	23.2	4.0
13-01-21	01:30 pm	27.7	22.6	5.1
13-01-21	04:00 pm	25.6	19.3	6.3
13-01-21	01:00 pm	29.3	23.1	6.2
25-01-21	03:45 pm	24.1	19.3	4.8
25-01-21	10:00 am	22.6	20.1	2.5
26-01-21	12:45 am	25.9	19.7	6.2
26-01-21	04:00 pm	22.4	18.6	3.8
26-01-21	11:30 am	22.2	19.0	3.2
27-01-21	01:15 pm	28.3	22.7	5.6
27-01-21	04:00 pm	23.4	18.8	4.6

Source: Author

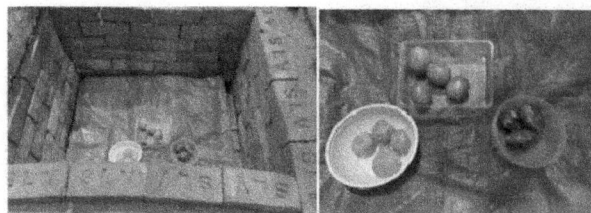

Figure 41.3 Samples storing in ZECC

Source: Author

Table 41.2 Temperature differences observed during testing stages

Date	Day	Time (IST)	Surface temperature (°Celsius)				
			Outside	Average	Inside	Average	Difference
15-3-21	Monday	11:20 am	41.0		34.8		
			45.2	42.93	28.7	30.13	12.80
			42.6		26.9		
		12:30 pm	39.7		32.0		
			44.0	41.30	29.0	29.06	12.33
			40.2		26.2		
		01:30 pm	39.4		33.0		
			46.0	42.96	29.4	29.80	13.16
			43.5		27.0		
16-3-21	Tuesday	11:30 am	36.3		36.0		
			47.5	42.67	27.3	30.10	12.56
			44.2		27.0		
		12:30 pm	39.7		36.2		
			45.1	43.00	30.5	31.56	11.43
			44.2		28.0		
		02:30 pm	38.2		36.3		
			45.4	42.26	29.5	30.63	11.63
			43.2		26.1		
17-3-21	Wednesday	10:30 am	39.0		35.1		
			44.8	42.10	28.7	30.10	12.00
			42.5		26.5		
		12:30 pm	42.2		32.0		
			41.5	41.23	28.7	28.70	12.53
			40.0		25.4		
		02:30 pm	40.0		33.0		
			44.3	42.43	28.5	28.93	12.50
			40.0		25.3		
18-3-21	Thursday	10:30 am	35.3		37.3		
			48.0	42.16	26.8	30.03	12.13
			43.2		26.0		
		12:30 pm	40.7		36.3		
			43.2	42.80	31.5	31.26	11.53
			44.5		26.0		
		02:30 pm	38.4		35.5		
			43.8	41.40	28.7	30.13	11.26
			42.0		26.2		
19-3-21	Friday	10:30 am	39.0		35.1		
			44.8	42.10	28.7	30.10	12.00
			42.5		26.5		
		12:30 pm	42.2		32.0		
			41.5	41.23	28.7	28.70	12.53
			40.0		25.4		
		02:30 pm	40.0		33.0		
			44.3	41.43	28.5	28.93	12.50
			40.0		25.3		
20-3-21	Saturday	10:30 am	38.0		37.0		
			45.2	42.06	26.5	29.83	12.23
			43.0		26.0		
		12:30 pm	41.2		35.8		
			44.0	43.23	30.5	31.00	12.23
			44.5		26.7		
		02:30 pm	41.0		33.5		
			43.0	42.00	31.0	29.50	12.50
			42.0		29.0		
22-3-21	Monday	10:30 am	40.0		28.5		
			44.3	42.73	29.3	30.43	12.30
			43.9		26.5		
		12:30 pm	43.2		34.0		
			41.0	41.13	29.2	29.40	11.73

Date	Day	Time (IST)	Surface temperature (°Celsius)			Average	Difference
			Outside	Average	Inside		
		02:30 pm	39.2		25.0		
			42.0		34.0		
			43.3	42.10	30.2	29.73	12.36
			41.0		25.0		
23-3-21	Tuesday	10:30 am	35.2		37.2		
			48.1	42.20	26.6	29.60	12.60
			43.3		25.0		
		12:30 pm	39.9		35.2		
			43.0	42.43	32.0	31.10	11.33
			44.4		26.1		
		02:30 pm	37.9		36.5		
			44.1	41.67	25.4	29.30	12.36
			43.0		26.0		
24-3-21	Wednesday	10:30 am	41.0		36.0		
			44.2	43.46	30.4	30.80	12.66
			45.2		26.0		
		12:30 pm	43.1		33.1		
			40.9	40.83	30.6	29.90	10.93
			38.5		26.0		
		02:30 pm	41.9		33.4		
			43.1	41.96	31.0	29.80	12.16
			40.9		25.0		

Source: Author

Table 41.3 Moisture content of samples

Sl.No	Plate weight (grams)	Initial + Plate weight (grams)	Final + Plate weight (grams)	Moisture content (%)	Average (%)
Tomato 1	42.57	44.63	42.67	95.15	95.30
Tomatao2	47.29	49.27	47.38	95.45	
Potato 1	46.49	49.08	46.88	84.94	85.36
Potato 2	43.34	45.31	43.62	85.79	
Egg-plant 1	45.99	48.22	46.16	92.38	91.05
Egg-plant 2	42.89	44.64	43.07	89.71	

Source: Author

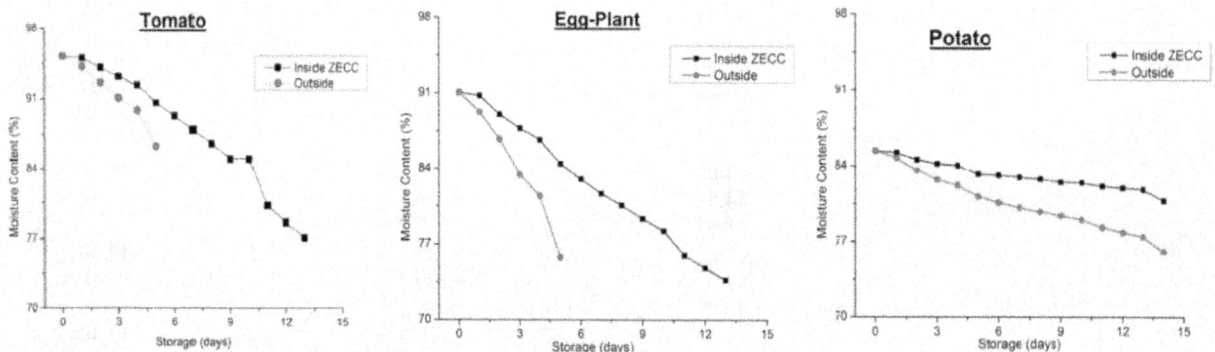

Figure 41.4 Reduction in moisture content with respect to time in days
Source: Author

Table 41.4 Pictorial representation of the samples inside and outside ZECC

Item	Inside ZECC	Outside ZECC
Tomato		
Egg Plant		
Potato		

Source: Author

0–50 units: ₹4.90/unit
51–100 units: ₹6.25/unit
101–200 units: ₹8.05/unit
Above 200 units: ₹9.10/unit

For a 200-watt refrigerator running 24/7, the monthly power consumption is approximately 144 kWh:

- Cost calculation: $(50 \times ₹4.90) + (50 \times ₹6.25) + (44 \times ₹8.05)$
  $= ₹245 + ₹312.5 + ₹354.2 = ₹911.7/month$

Including fixed charges and taxes, the monthly cost exceeds ₹1,200, resulting in an annual expenditure of over ₹14,400. The ZECC, with its one-time construction cost of ₹1,830 and zero recurring energy costs, offers a sustainable and cost-effective alternative to traditional refrigeration. It demonstrates potential savings of over 85% in storage costs while maintaining product quality up to certain extent, thereby offering a viable solution for small-scale farmers and rural communities. The uncertainty factor in the variation in temperature and weather conditions has not been considered.

Conclusion

This study highlights that using a ZECC can effectively lengthen the shelf life of vegetables such as potatoes, tomatoes, and eggplants from 4 to 8 days. This extended storage period provides farmers with crucial flexibility in managing their agricultural produce. The ZECC offers significant advantages, including its minimal construction cost, reliable on readily available raw materials, and the absence of any need for external mechanical or electrical energy. By maintaining freshness for longer periods compared to conventional storage methods, the ZECC presents a practical and sustainable solution for preserving not only tomatoes, eggplants, and potatoes but also a wide range of other vegetables. Unlike traditional refrigeration systems, which involve high initial costs and continuous energy expenses, the ZECC is an economical and eco-friendly alternative that can benefit small-scale farmers and reduce post-harvest losses

Acknowledgement

We gratefully acknowledge the CIIRC, JIT for providing the facility to conduct the experiment. Also thank the KSIT organising committee for suggestions in improving the quality of the manuscript.

References

[1] Mitra, S. K., Kabir, J., Dhua, R. S., Dutta Ray, S. K. Low cost cool chamber for storage of tropical fruits. *XXVI Internat. Horticul. Cong.*, 2002;628:63–68.

[2] Dadhich, S. M., Dadhich, H., Verma, R. Comparative study on storage of fruits and vegetables in evaporative cool chamber and in ambient. *Internat. J. Food Engg.*, 2008;4(1):1–14.

[3] Patel, M., Singh, R. Cost-effective storage solutions for rural areas. *Internat. J. Green Technol.*, 2018;5(1):45–53.

[4] Rayaguru, K., Khan, M. K., Sahoo, N. R. Water use optimization in zero energy cool chambers for short-term stor-

age of fruits and vegetables in coastal area. *J. Food Sci. Technol.*, 2010;47(4):437–441.

[5]   Roy, S. K., Pal, R. K. A low cost zero energy cool chamber for short term storage of mango. *III Internat. Mango Symp.*, 1989;291:519–524.

[6]   Mishra, A., Jha, S. K., Ojha, P. Study on zero energy cool chamber (ZECC) for storage of vegetables. *Internat. J. Scient. Res. Publ.*, 2020;10(1):9767.

[7]   Amer, O., Boukhanouf, R., Ibrahim, H. G. A review of evaporative cooling technologies. *Internat. J. Environ. Sci. Dev.*, 2015;6(2):111.

[8]   Dirpan, A., Sapsal, M. T., Tahir, M. M., Djalal, M., Firdaus, A. The potential of the ZECC–washing combination to extending the mango's shelf life. *IOP Conf. Series Earth Environ. Sci.*, 2020;575(1):012153.

[9]   Sarkar, B., Sundaram, P. K., Mondal, S., Kumar, U., Haris, A. A., Bhatt, B. P. Effect of zero energy cool chambers on storage behavior of tomato under warm and humid climate. *Res. J. Agricul. Sci.*, 2014;5(5):1021–1024.

[10] Ganesan, M., Balasubramanian, K., Bhavani, R. V. Studies on the application of different levels of water on zero energy cool chamber with reference to the shelf-life of brinjal. *J. Indian Inst. Sci.*, 2004;84:107–111.

[11] Islam, M. P., Morimoto, T. Zero energy cool chamber for extending the shelf-life of tomato and eggplant. *Japan Agricul. Res. Quart. JARQ*, 2012;46(3):257–267.

[12] Kumar, N., Gupta, A. Analyzing temperature regulation in ZECCs. *Energy Environ.*, 2022;18(4):256–262.

[13] Smith, J., Brown, L. Innovative cooling technologies for agriculture. *J. Sustain. Agricul.*, 2021;34(2):112–120.

[14] Thomas, P., Verma, S. Comparative analysis of ZECC and conventional refrigeration. *Agricul. Engg. Today*, 2020;44(3):198–204.

[15] Wang, C., Wei, H., Zhou, Z., Chao, Y., Liu, J., Yang, X., Yan, J. Low-cost and transparent cooling films with solar-selective nanoparticles for reducing the energy consumption of buildings. *J. Mat. Chem. A*, 2025;13(3): 2208–2216.

[16] Geofrey, R., Mkuna, E., Nyamwero, N. M., Mang'ana, K. Impact of postharvest handling technologies on income of tomato (Solanum Lycopersicum L.) smallholder farmers in Mvomero, Morogoro, Tanzania. *Afr. J. Sci. Technol. Innov. Dev.*, 2025;17(1):65–79.

[17] BESCOM. Electricity Tariff Schedule. Bangalore Electricity Supply Company, Karnataka India, 2025. Retrieved from https://bescom.karnataka.gov.in/.

42 Synthesis, characterisation and GCMS analysis of bio-diesel blend for compression ignition engine using non-edible vegetable oil accessible in Indian states

Younus Pasha[1,a], Tansif Khan[1], D. R. Swamy[2], Mohammed Riyaz Ahmed[1], Harish H.[3] and Salim Sharieff[1]

[1]Associate Professor, Department of Mechanical Engineering, HKBK College of Engineering, Bangalore, VTU, Karnataka, India

[2]Department of Mechanical Engineering, K S School of Engineering and Management, Bangalore, Karnataka, India

[3]Department of Mechanical Engineering, RK College of Engineering, Vijayawada, Andhra Pradesh, India

Abstract

Bio-diesel and higher alcohols are currently being explored as promising alternative fuels due to their widespread availability, straight-forward synthesis processes, and positive environmental impacts. Addressing concerns related to cost and contamination can be achieved by employing bio-diesel in specific industrial applications. It is crucial for any alternative diesel fuel to meet the established requirements for conventional diesel. This study focuses on assessing feedstock availability, essential fuel characteristics, temperature-induced changes in fuel properties, methods of fuel synthesis, and the analysis of bio-diesel blend constituents using GC-MS. The process of synthesising bio-diesel through trans-esterification and the influence of solvents on esterified oils in terms of property estimation have been thoroughly examined. Results indicate that the introduction of a 5% gasoline component in methyl ester has a significant impact on bio-fuel blends, making bio-diesel a viable substitute for traditional diesel in compression-ignition (CI) engines. Engine performance test reveal that the blended sample closely aligns with diesel performance, showcasing potential for effective utilisation. The distinctive qualities of methyl ester blends consistently meet widely accepted standards. Consequently, the methyl ester-gasoline mixture can be seamlessly integrated into diesel engines without adversely affecting emissions or overall engine performance.

Keywords: C I engine, GC-MS, Methyl ester, feedstock, esterified blend

Introduction

Energy is one of a person's most necessities. In addition to being a marketable good with the ability to improve human well-being, it is a crucial element of social and economic development [1]. As fossil fuel resources continue to diminish, a replacement source must be created in order to address the energy crisis. Bio-fuel technology was consequently developed. Vegetable oil and alcohol are transformed into bio-diesel, a fuel, using a catalyst [2]. Feedstock selection for the preparation of bio-diesel is an essential criterion in bio-diesel synthesis [3]. The selected feedstock has to satisfy the fuel characteristics such as volatility, fluidity, flash point temperature, density, etc. [4]. Based on the climatic adoption of vegetable-based feedstock crops differ from region to region. For example, Jatropha seed cultivation is more in north Indian states [5], Pongamia seed is more in south Indian states [6], and so on.

Bio-fuel characterisation

Fuel characteristics such as calorific value, density, kinematic viscosity, and flash point temperature have been estimated at two distinguished temperatures. Table 42.1 shows the availability, cost in the Indian market, and yiteld of non-edible oils.

Synthesis of bio-diesel by trans-esterification

According to most experts, viscosity is the fundamental issue preventing the direct use of vegetable oils in typical diesel engines [7]. As a result, known techniques and technologies for producing bio-diesel from various non-edible feedstock have been applied to lower viscosity. Pyrolysis, micro emulsification, dilution, and trans-esterification are some of these processes. Trans-esterification has proven to be the most viable way to reduce viscosity among these strategies [8]. The most prevalent technique for producing bio-diesel was discovered to be

[a]younuspashame@gmail.com

DOI: 10.1201/9781003685876-42

trans-esterification [9]. This section places a greater emphasis on it.

There are four main phases in the trans-esterification process:

- Pre-treatment of the oil feedstock to remove sticky components [gummy materials] that will cause problems during following processing procedures.
- Trans-esterification produces raw methyl esters and glycerol by reacting pre-treated triglycerides with methanol [10].

The methyl ester and glycerol streams are separated in two essential phases after the process. These two stages are often done twice in most technologies to speed up the trans-esterification process by reducing the glycerol concentration in the second stage. To speed up the process, a large amount of methanol is employed. Because these molecules have greater melting temperatures and viscosity than the methyl ester, lower conversion rates result in higher quantities of mono- and di-glycerides, causing processing issues with emulsion formation and low-temperature haze difficulties with the bio-diesel.

- Purification of methyl ester, which removes excess methanol, catalyst, and glycerol from the trans-esterification process. The methanol extracted has used in the trans-esterification process again.
- Methanol removal from glycerol for recycling in the trans-esterification process.

Sodium hydroxide or potassium hydroxide is commonly employed as the catalyst for trans-esterification, and these catalysts find widespread use in various industries for multiple reasons: (i) These catalysts are capable of catalysing reactions under low reaction temperatures and atmospheric pressures [11]; (ii) they can produce high conversion in a short period [12]; and (iii) they are readily available and cost-effective [13]. After the reaction and separation, the catalyst is carried in the glycerol stream and neutralised by acid as part of the glycerol treatment. Typically, hydrochloric or sulphuric acids were used to produce sodium chloride or potassium sulphate, which can be used as a fertiliser [14].

The trans-esterification process can be carried out with essential equipment and fuel enthusiasts utilising buckets and other accoutrements to produce bio-diesel on a small scale. However, more sophisticated conditions are required to make the fuel commercially to achieve consistent quality criteria for the vast volumes involved and enhance yields and reaction speeds. Two-stage trans-esterification, which appears to be the standard in most modern plants, reduces mono- and di-glycerides to tolerable levels. Pre-treating vegetable oils for later trans-esterification has varied requirements. Because of its comparatively large proportion of unsaturated material and resulting lower oxidation stability, the former focuses on removing free fatty acids. In contrast, the latter focuses on degumming the oil [15].

Ethanol (C_2H_5OH) can be used as a feedstock instead of methanol, although it is not widely employed because it is more expensive. Because the molecular weights are around 5% greater, the bio-diesel esters created from ethanol will be ethyl esters rather than methyl esters and will have slightly different physical properties [16]. The differences between methanol and ethanol processing are minimal, with the most significant change occurring during the alcohol recovery.

Ethanol (C_2H_5OH) can be used as a feedstock instead of methanol, although it is not widely employed because it is more expensive. Because the molecular step when ethanol forms an azeotrope with free water [17]. Overall, ethyl esters will be of worse quality than methyl esters due to a slower reaction rate with ethanol, resulting in lower conversion and more significant amounts of mono- and di-glycerides and glycerol in the final product [18].

Trans-esterification reverses when glycerides and alcohols are mixed properly [19]. The catalyst (acid or base) accelerates the trans-esterification [20]. Bio-diesel can be prepared in two ways. The free fatty acid percentage availability in crude oil is a consideration for selecting the strategy.

i. If the pure extracted oil's FFA level is less than 4%, a one-stage alkali-based reaction must be carried out [49].
ii. A 2 stage acid-based and alkali-based catalytic reaction must be carried out if the FFA concentration of the pure extracted oil exceeds 4% [47].

For successful trans-esterification, the molar ratio is an essential factor. Trans-esterification can be quickened by increasing the methanol used according to the chemical reaction rate. The high molar ratio of oil to methanol can improve bio-diesel conversion yield. To make three moles of bio-diesel from one mole of oil, the molar ratio of oil to methanol must be 1:3. To enhance the degree of reaction completion, the molar ratio employed in actual manufacturing must be larger than the stoichiometric value. Excessive methanol, conversely, can lower catalyst and reactant concentrations, delaying the process and hindering solvent recovery [48]. Table 42.2 gives exact information about the reactant volumetric proportions, number of stages, and percentage of the yield obtained for trans-esterification reaction.

The ignition quality tester (IQT) is employed to measure the cetane number. This technique involves gauging the time lapse between the initiation of fuel injection and the commencement of significant combustion resulting from the auto-ignition of a pre-determined quantity of diesel fuel within a consistent volume chamber. A higher cetane number indicates increased fuel volatility. Table 42.3 provides the saponification number, iodine value and cetane number of selected sample's methyl ester.

Gas chromatography analysis

The gas chromatography (GC) instrument provides the results in terms of the quality and quantity of volatile and semi-volatile organic compounds present in a sample. Based on these results, clarification is needed on whether the raw oil is converted into bio-diesel. These decisions can be made account by observing the amount of fatty acid methyl esters in them [22].

For smaller volatile and semi-volatile organic compounds such hydrocarbons, alcohols, and aromatics, gas chromatography is the preferred separation method. Gas chromatography-mass spectroscopy (GC-MS) can be used to separate complex mixtures, quantify analysts, identify unidentified peaks, and ascertain trace quantities of contaminants when paired with the mass spectrometry (MS) detection capability. GC-MS has been used to identify the

Table 42.1 Physical characterisation of non-edible oils

S. No.	Oil	Yield of oil (%)	Bio-diesel yield after esterification	Availability in India annual productivity in (tonnes)	Cost of oil in Indian market [INR/litre] (2024)	Molecular weight (g/mol)
1	Pongamia oil	30–40 [32]	92% at 1:10 oil to methanol ratio [33]	55e3 [40]	₹85	892.7 [41]
2	Castor oil	40–60 [32]	92% at 1:9 oil to methanol ratio [34]	499e3	₹98	927 [42]
3	Neem oil	45 [32]	83.36% at 1:6 oil to methanol ratio [35]	100e3 [40]	₹75	815 [43]
4	Mahua oil	35 [32]	89% at 1:6 oil to methanol ratio [36]	180e3 [40]	₹65	869.95 [44]
5	Jatropha oil	30–40 [32]	82% at 1:4 oil to methanol ratio [37]	15e3 [40]	₹40	900 [45]
6	Used cooking oil	-	90% at 1:8 oil to methanol ratio [38]		₹39	876.16 [46]
7	Dairy scum oil	-	90% at 1:6 oil to methanol ratio [39]		₹40	

Source: Author

Table 42.2 Reactants volumetric proportions for trans-esterification reaction

S. No.	Sample	Saponification number [mg KOH/g]	Iodine value [g iodine/100g oil]	Cetane number
1	Pongamia methyl ester	187	88	55
2	Neem methyl ester	177	79	59
3	Mahua methyl ester	196	68	58
4	Jatropha methyl ester	190	96	53
5	Used cooking oil methyl ester	270	40	57
6	Dairy scum methyl ester	210	65	57
7	Castor oil methyl ester	270	85	47
8	Diesel	-	-	50

Source: Author

Table 42.3 Saponification number, iodine value and cetane number of selected sample's methyl ester

S. No.	Sample	Reactant	Catalyst	Stage of reaction	Molar ratio oil to methanol	Volumetric proportions (ml)	Yield (%)
1	Pongamia oil	Methanol	NaOH	Two stage	1:10	1000:450	93
2	Castor oil	Methanol	NaOH	Two stage	1:9	1000:393	92
3	Neem oil	Methanol	NaOH	One stage	1:6	1000:298l	84
4	Mahua oil	Methanol	NaOH	One stage	1:6	1000:279	90
5	Jatropha oil	Methanol	NaOH	One stage	1:4	1000:180	82
6	Used cooking oil	Methanol	NaOH	Single stage	1:8	1000:370	90
7	Dairy scum oil	Methanol	NaOH	Two stage	1:6	1000:200	90

Source: Author

components of bio-diesel [23].By using GC-MS, the researched oils' generated bio-diesel samples were chemically analysed. Retention periods were used to identify all fatty acids, methyl esters (FAMEs), and other acquired substances. Their identities were then verified by contrasting their mass fragmentation patterns with the stored mass spectra from the GC-MS equipment.

In a mixed sample, a gas chromatograph divides the chemical components into columns. A heated injector between 200°C and 300°C is used to evaporate the sample. The vaporised sample is divided into its individual components and transported to a queue where it is moved through using carrier gas. One of the GC detectors is a mass spectrometer. Retention time is depicted on the X-axis. Before and after switching to bio-diesel, the examined oils' proportion of obtained methyl esters has been published and discussed. The time it takes for the analyses to transit through the column and reaches the mass spectrometer detector is often displayed on the gas chromatogram's x-axis. Each component's arrival time to the detector is indicated by the peaks that are displayed.

Concentration or intensity counts are shown on the Y-Axis [24].

The sample to be tested was injected into the sample collection tray provided at the chromatography oven. The sample was heated to a specific temperature gets and converted into vapour. The carrier gas, like helium, was used to flow with the sample. Helium is chemically inert and has good thermal conductivity which it has used in chromatography. The compounds in the sample attracted to the stationary phase spent more time in the column and eluting later. The compounds in the sample drawn towards the mobile phase spent less time in the column and elute first. All stationary phase compounds had a longer retention time than the mobile phase compounds. But no two compounds had the same retention time. The detector identified this retention time and converted the information into a computer-readable form [25].

The Pongamia methyl ester blend's chemical composition is listed in Table 42.4. The GC-MS analysis of the Pongamia methyl ester blend sample was carried out and identified 24 main components, as shown in Figure 42.1. The specified components

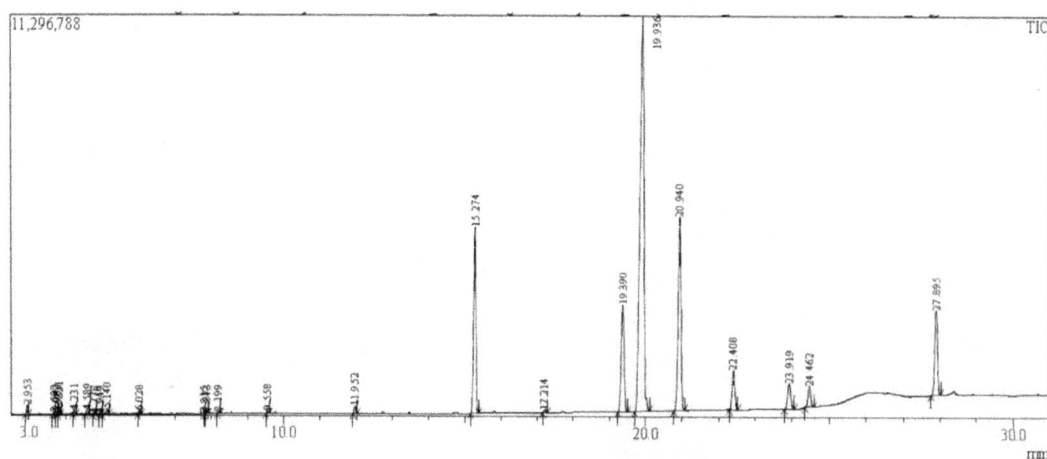

Figure 42.1 GC-MS chromatogram graph of Pongamia methyl ester blend
Source: Bioenergy Research and Quality Assurance Laboratory University of Agricultural Sciences GKVK, Bengaluru 560065

Table 42.4 GC-MS chromatogram components of Pongamia methyl ester blend

Peak #	R.Time Retention time	I.Time Intensity	F.Time	Area	Area %	Height %	Name	Chemical Formula	Chemical Structure
1	2.953	2.900	3.010	623130	0.34	0.83	Toluene	C_7H_8	
2	3.692	3.645	3.725	188484	0.10	0.25	Ethylbenzene	C_8H_{10}	
3	3.762	3.725	3.790	232936	0.13	0.30	p-Xylene	C_8H_{10}	
4	3.821	3.790	3.880	457240	0.25	0.62	o-Xylene	C_8H_{10}	
5	4.231	4.200	4.290	280277	0.15	0.38	p-Xylene	C_8H_{10}	
6	4.589	4.535	4.645	322175	0.18	0.38	Benzene, 1-ethyl-2-methyl-	C_9H_{12}	
7	4.778	4.740	4.830	84617	0.05	0.11	Mesitylene	C_9H_{12}	
8	4.949	4.910	4.995	103675	0.06	0.13	Benzene, 1-ethyl-2-methyl-	C_9H_{12}	
9	5.140	4.995	5.195	464226	0.25	0.49	Mesitylene	C_9H_{12}	
10	6.028	5.975	6.085	232964	0.13	0.27	Octanoic acid, methyl ester	$C_9H_{18}O_2$	
11	7.815	7.790	7.850	98944	0.05	0.11	Decanoic acid, methyl ester	$C_{11}H_{22}O_2$	
12	7.873	7.850	7.925	146673	0.08	0.21	Dimethyl Sulfoxide	C_2H_6OS	
13	8.199	8.160	8.255	104625	0.06	0.13	Benzoic acid, methyl ester	$C_8H_8O_2$	
14	9.558	9.510	9.615	209460	0.11	0.26	Dodecanoic acid, methyl ester	$C_{13}H_{26}O_2$	
15	11.952	11.890	12.020	628437	0.34	0.64	Methyl tetradecanoate	$C_{15}H_{30}O_2$	
16	15.274	15.150	15.380	22869908	12.55	16.74	Hexadecanoic acid, methyl ester	$C_{17}H_{34}O_2$	
17	17.214	17.160	17.295	201295	0.11	0.17	Heptadecanoic acid, methyl ester	$C_{18}H_{36}O_2$	
18	19.390	19.245	19.525	17051857	9.36	9.61	Methyl stearate	$C_{19}H_{38}O_2$	
19	19.936	19.715	20.125	83393513	45.76	35.57	9-Octadecenoic acid, methyl ester, (E)-	$C_{19}H_{36}O_2$	
20	20.940	20.775	21.075	29260675	16.06	17.42	9,12-Octadecadienoic acid (Z,Z)-, methyl ester	$C_{18}H_{32}O_2$	
21	22.408	22.285	22.535	5616115	3.08	3.55	9,12,15-Octadecatrienoic acid, methyl ester, (Z,Z,Z)-	$C_{19}H_{32}O_2$	
22	23.919	23.790	24.055	4071929	2.23	2.30	Methyl 18-methylnonadecanoate	$C_{21}H_{42}O_2$	
23	24.462	24.340	24.585	3064024	1.68	1.90	11-Eicosenoic acid, methyl ester	$C_{21}H_{40}O_2$	
24	27.895	27.750	28.015	12537421	6.88	7.61	Methyl 20-methyl-heneicosanoate	$C_{23}H_{46}O_2$	
				82244600	100.00	100.00			

Source: Bioenergy Research and Quality Assurance Laboratory University of Agricultural Sciences GKVK, Bengaluru 560065

with their chemical formula, chemical structure, retention times, intensity time, F. time, area %, and height % for all the bio-diesel, i.e., Pongamia methyl ester blend constituent, have displayed in Table 42.4.

9-Octadecenoic acid, methyl ester, $(C_{19}H_{36}O_2)$ represents the main constituent with [45.76%] area %, 9,12-octadecadienoic acid (Z, Z)- methyl ester $(C_{18}H_{32}O_2)$ and hexadecanoic acid, methyl ester $(C_{17}H_{34}O_2)$ share the second largest area % of 16.06% and third largest area % of 12.55%, respectively. The obtained main components of the Pongamia methyl ester blend sample are changed into bio-diesel components with a yield percentage of 93%. The synthesised blend sample fulfils all the bio-diesel standards set under BIS [26].

The neem methyl ester blend's chemical composition is listed in Table 42.5. The GC-MS analysis of the neem methyl ester blend sample was carried out and identified 28 different components, as shown in Figure 42.2. The identified components with their chemical formula, chemical structure, retention times, intensity time, F. time, area %, and height % for all the bio-diesel, i.e., neem methyl ester blend constituent, have displayed in Table 42.5.

9,12-Octadecadienoic acid (Z, Z)-, methyl ester $(C_{18}H_{32}O_2)$ represents the main constituent with [27.90%] area %, 9-octadecenoic acid, methyl ester, (E) $(C_{19}H_{36}O_2)$ and hexadecanoic acid, methyl ester $(C_{17}H_{34}O_2)$ share the second largest area % of 26.58% and third largest area % of 14.70%, respectively. The obtained main components of the neem methyl ester blend sample are converted into bio-diesel components with a yield percentage of 84%. The

synthesised blend sample fulfils all the bio-diesel standards set under BIS [27].

The chemical composition of the Mahua methyl ester blend has listed in Table 42.6. The GC-MS analysis of the Mahua methyl ester blend sample was carried out and identified 25 different components, as shown in Figure 42.3. The identified components with their chemical formula, chemical structure, retention times, intensity time, F. time, area %, and height % for all the bio-diesel, i.e., Mahua methyl ester blend constituent, have displayed in Table 42.6.

9-Octadecenoic acid, methyl ester, (E)- $(C_{19}H_{36}O_2)$ represents the main constituent with [35.45%] area %, methyl stearate $(C_{19}H_{38}O_2)$ and hexadecanoic acid, methyl ester $(C_{17}H_{34}O_2)$ share the second largest area % of 23.76% and third largest area % of 20.77%, respectively. The obtained main components of the Mahua methyl ester blend sample are converted into bio-diesel components with a yield percentage of 90%. The synthesised blend sample fulfils all the bio-diesel standards set under BIS [28].

The Jatropha methyl ester blend's chemical composition is listed in Table 42.7. The GC-MS analysis of the Jatropha methyl ester blend sample was carried out and identified 25 different components, as shown in Figure 42.4. The identified components with their chemical formula, chemical structure, retention times, intensity time, F. time, area %, and height % for all the bio-diesel, i.e., Jatropha methyl ester blend constituent, have displayed in Table 42.7.

9-Octadecenoic acid, methyl ester, (E)- $(C_{19}H_{36}O_2)$ represents the main constituent with [35.82%] area %, 9,12-octadecadienoic acid (Z, Z)-, methyl ester

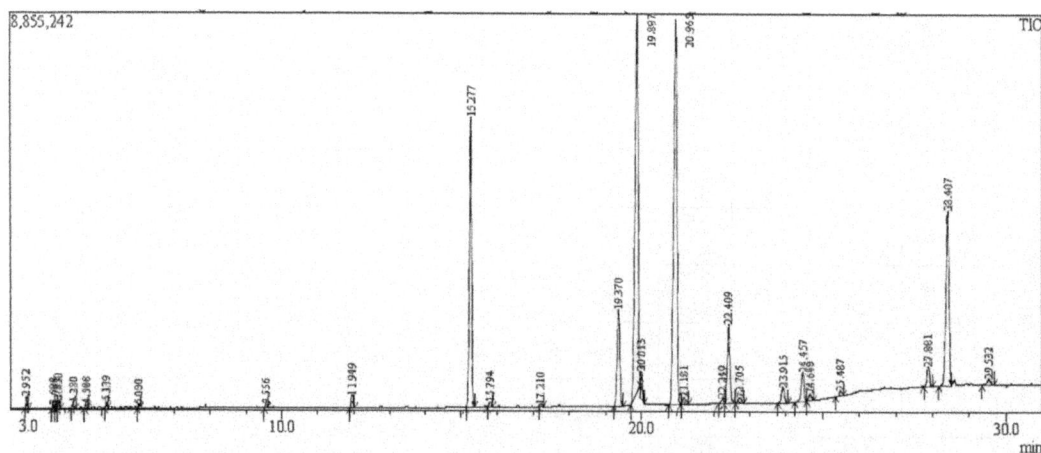

Figure 42.2 GC-MS chromatogram graph of neem methyl ester blend

Source: Bioenergy Research and Quality Assurance Laboratory University of Agricultural Sciences GKVK, Bengaluru 560065

Table 42.5 GC-MS chromatogram components of neem methyl ester blend

Peak #	R.Time	I.Time	F.Time	Area	Area %	Height %	Name	Chemical Formula	Chemical Structure
1	2.952	2.895	3.010	626364	0.32	0.73	Toluene	C_7H_8	
2	3.690	3.645	3.725	177994	0.09	0.21	Ethylbenzene	C_8H_{10}	
3	3.761	3.725	3.785	207110	0.11	0.25	p-Xylene	C_8H_{10}	
4	3.820	3.785	3.880	468202	0.24	0.55	o-Xylene	C_8H_{10}	
5	4.230	4.185	4.290	276503	0.14	0.34	p-Xylene	C_8H_{10}	
6	4.588	4.530	4.645	298931	0.15	0.31	Benzene, 1-ethyl-2-methyl-	C_9H_{12}	
7	5.139	5.090	5.195	329255	0.17	0.39	Mesitylene	C_9H_{12}	
8	6.030	5.985	6.080	161381	0.08	0.19	Octanoic acid, methyl ester	$C_9H_{18}O_2$	
9	9.556	9.505	9.610	192003	0.10	0.21	Dodecanoic acid, methyl ester	$C_{13}H_{26}O_2$	
10	11.949	11.880	12.025	866990	0.44	0.78	Methyl tetradecanoate	$C_{15}H_{30}O_2$	
11	15.277	15.145	15.380	2873261	14.70	18.24	Hexadecanoic acid, methyl ester	$C_{17}H_{34}O_2$	
12	15.794	15.725	15.870	373662	0.19	0.28	9-Hexadecenoic acid, methyl ester, (Z)-	$C_{16}H_{30}O_2$	
13	17.210	17.150	17.290	208539	0.11	0.15	Heptadecanoic acid, methyl ester	$C_{18}H_{36}O_2$	
14	19.370	19.235	19.505	1132583	5.79	6.03	Methyl stearate	$C_{19}H_{38}O_2$	
15	19.897	19.715	19.975	5197444	26.58	23.83	9-Octadecenoic acid, methyl ester, (E)-	$C_{19}H_{36}O_2$	
16	20.013	19.980	20.095	1216092	0.62	1.20	9-Octadecenoic acid, methyl ester, (E)-	$C_{19}H_{36}O_2$	
17	20.965	20.765	21.100	5455285	27.90	24.09	9,12-Octadecadienoic acid (Z,Z)-, methyl ester	$C_{18}H_{32}O_2$	
18	21.181	21.100	21.310	740348	0.38	0.44	9,12-Octadecadienoic acid (Z,Z)-, methyl ester	$C_{18}H_{32}O_2$	
19	22.249	22.120	22.300	619749	0.32	0.35	9,12,15-Octadecatrienoic acid, methyl ester, (Z,Z,Z)-	$C_{19}H_{32}O_2$	
20	22.409	22.300	22.600	9197502	4.70	4.99	9,12,15-Octadecatrienoic acid, methyl ester, (Z,Z,Z)-	$C_{19}H_{32}O_2$	
21	22.705	22.600	22.830	605564	0.31	0.33	9,12,15-Octadecatrienoic acid, methyl ester, (Z,Z,Z)-	$C_{19}H_{32}O_2$	
22	23.915	23.770	24.050	1912512	0.98	0.97	Methyl 18-methylnonadecanoate	$C_{21}H_{42}O_2$	
23	24.457	24.250	24.560	3542412	1.81	1.83	11-Eicosenoic acid, methyl ester	$C_{21}H_{40}O_2$	
24	24.649	24.560	24.760	771137	0.39	0.43	11-Eicosenoic acid, methyl ester	$C_{21}H_{40}O_2$	
25	25.487	25.360	25.570	461863	0.24	0.28	cis-11,14-Eicosadienoic acid, methyl ester	$C_{21}H_{38}O_2$	
26	27.881	27.760	28.000	2225911	1.14	1.25	Methyl 20-methyl-heneicosanoate	$C_{23}H_{46}O_2$	
27	28.407	28.170	28.520	2257374	11.55	10.97	13-Docosenoic acid, methyl ester	$C_{23}H_{44}O_2$	
28	29.532	29.340	29.670	871871	0.45	0.37	cis-13,16-Docasadienoic acid, methyl ester	$C_{23}H_{42}O_2$	
				195511380	100.00	100.00			

Source: Bioenergy Research and Quality Assurance Laboratory University of Agricultural Sciences GKVK, Bengaluru 560065

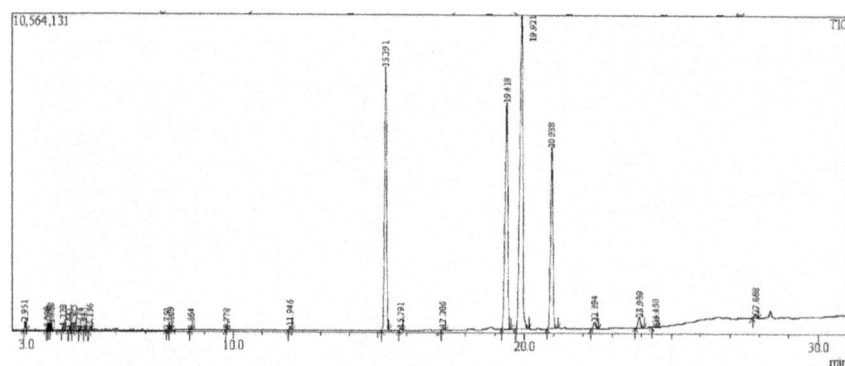

Figure 42.3 GC-MS chromatogram of Mahua methyl ester blend

Source: Bioenergy Research and Quality Assurance Laboratory University of Agricultural Sciences GKVK, Bengaluru 560065

Table 42.6 GC-MS chromatogram of Mahua methyl ester blend

Pe ak #	R.Ti me	I.Ti me	F.Ti me	Area	Are a%	Heig ht%	Name	Chemical Formula	Chemical Structure
1	2.95 1	2.89 5	3.01 0	689133	0.33	0.82	Toluene	C_7H_8	
2	3.68 9	3.65 0	3.72 0	154838	0.07	0.21	Ethylbenzene	C_8H_{10}	
3	3.75 9	3.72 5	3.78 5	122542	0.06	0.19	p-Xylene	C_8H_{10}	
4	3.81 8	3.78 5	3.86 5	378831	0.18	0.52	o-Xylene	C_8H_{10}	
5	4.22 8	4.18 5	4.28 5	267793	0.13	0.34	p-Xylene	C_8H_{10}	
6	4.44 1	4.41 0	4.48 5	57808	0.03	0.08	Benzene, propyl-	C_9H_{12}	
7	4.58 5	4.53 0	4.64 0	341159	0.16	0.36	Benzene, 1-ethyl-2-methyl-	C_9H_{12}	
8	4.77 4	4.73 5	4.82 0	82996	0.04	0.10	Mesitylene	C_9H_{12}	
9	4.94 5	4.90 5	4.99 5	95751	0.05	0.12	Benzene, 1-ethyl-2-methyl-	C_9H_{12}	
10	5.13 6	5.06 5	5.19 5	434399	0.21	0.46	Mesitylene	C_9H_{12}	
11	7.75 0	7.70 5	7.80 0	56043	0.03	0.07	Hexadecane	$C_{16}H_{34}$	
12	7.86 9	7.82 0	7.91 0	120122	0.06	0.15	Dimethyl Sulfoxide	C_2H_6OS	
13	8.56 4	8.52 0	8.61 0	58653	0.03	0.07	Heptadecane	$C_{17}H_{36}$	
14	9.77 8	9.74 0	9.82 5	82367	0.04	0.10	2,4-Decadienal, (E,E)-	$C_{10}H_{16}O$	
15	11.9 46	11.8 85	12.0 15	431982	0.21	0.40	Methyl tetradecanoate	$C_{15}H_{30}O_2$	
16	15.2 91	15.1 40	15.3 95	4336735 4	20.7 7	24.99	Hexadecanoic acid, methyl ester	$C_{17}H_{34}O_2$	
17	15.7 91	15.7 35	15.8 65	250452	0.12	0.20	9-Hexadecenoic acid, methyl ester, (Z)-	$C_{16}H_{30}O_2$	
18	17.2 06	17.1 40	17.2 95	350000	0.17	0.24	Heptadecanoic acid, methyl ester	$C_{18}H_{36}O_2$	
19	19.4 18	19.2 25	19.5 45	4959409 3	23.7 6	21.48	Methyl stearate	$C_{19}H_{38}O_2$	
20	19.9 21	19.7 00	20.1 55	7402950 9	35.4 6	29.74	9-Octadecenoic acid, methyl ester, (E)-	$C_{19}H_{36}O_2$	
21	20.9 38	20.7 75	21.1 35	3367694 1	16.1 3	17.09	9,12-Octadecadienoic acid (Z,Z)-, methyl ester	$C_{18}H_{32}O_2$	
22	22.3 94	22.2 75	22.5 15	1028265	0.49	0.60	9,12,15-Octadecatrienoic acid, methyl ester, (Z,Z,Z)-	$C_{19}H_{32}O_2$	
23	23.9 09	23.7 75	24.0 65	2007044	0.96	1.03	Methyl 18-methylnonadecanoate	$C_{21}H_{42}O_2$	
24	24.4 50	24.3 35	24.5 45	446797	0.21	0.25	11-Eicosenoic acid, methyl ester	$C_{21}H_{40}O_2$	
25	27.8 68	27.7 55	27.9 75	643177	0.31	0.39	Methyl 20-methyl-heneicosanoate	$C_{23}H_{46}O_2$	
				2087680 49	100. 00	100.0 0			

Source: Bioenergy Research and Quality Assurance Laboratory University of Agricultural Sciences GKVK, Bengaluru 560065

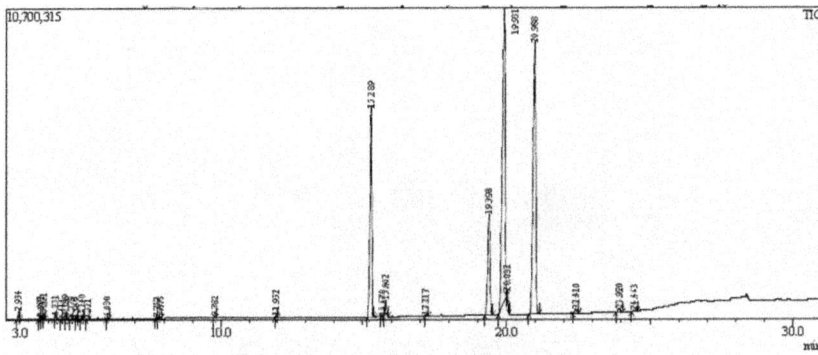

Figure 42.4 GC-MS chromatogram graph of Jatropha methyl ester blend

Source: Bioenergy Research and Quality Assurance Laboratory University of Agricultural Sciences GKVK, Bengaluru 560065

Table 42.7 GC-MS chromatogram components of Jatropha methyl ester blend

Pe ak #	R.Ti me	I.Ti me	F.Ti me	Area	Are a%	Heig ht%	Name	Chemical Formula	Chemical Structure
1	2.951	2.895	3.010	689133	0.33	0.82	Toluene	C_7H_8	
2	3.689	3.650	3.720	154838	0.07	0.21	Ethylbenzene	C_8H_{10}	
3	3.759	3.725	3.785	122542	0.06	0.19	p-Xylene	C_8H_{10}	
4	3.818	3.785	3.865	378831	0.18	0.52	o-Xylene	C_8H_{10}	
5	4.228	4.185	4.285	267793	0.13	0.34	p-Xylene	C_8H_{10}	
6	4.441	4.410	4.485	57808	0.03	0.08	Benzene, propyl-	C_9H_{12}	
7	4.585	4.530	4.640	341159	0.16	0.36	Benzene, 1-ethyl-2-methyl-	C_9H_{12}	
8	4.774	4.735	4.820	82996	0.04	0.10	Mesitylene	C_9H_{12}	
9	4.945	4.905	4.995	95751	0.05	0.12	Benzene, 1-ethyl-2-methyl-	C_9H_{12}	
10	5.136	5.065	5.195	434399	0.21	0.46	Mesitylene	C_9H_{12}	
11	7.750	7.705	7.800	56043	0.03	0.07	Hexadecane	$C_{16}H_{34}$	
12	7.869	7.820	7.910	120122	0.06	0.15	Dimethyl Sulfoxide	C_2H_6OS	
13	8.564	8.520	8.610	58653	0.03	0.07	Heptadecane	$C_{17}H_{36}$	
14	9.778	9.740	9.825	82367	0.04	0.10	2,4-Decadienal, (E,E)-	$C_{10}H_{16}O$	
15	11.946	11.885	12.015	431982	0.21	0.40	Methyl tetradecanoate	$C_{15}H_{30}O_2$	
16	15.291	15.140	15.395	4336735 4	20.77	24.99	Hexadecanoic acid, methyl ester	$C_{17}H_{34}O_2$	
17	15.791	15.735	15.865	250452	0.12	0.20	9-Hexadecenoic acid, methyl ester, (Z)-	$C_{16}H_{30}O_2$	
18	17.206	17.140	17.295	350000	0.17	0.24	Heptadecanoic acid, methyl ester	$C_{18}H_{36}O_2$	
19	19.418	19.225	19.545	4959409 3	23.76	21.48	Methyl stearate	$C_{19}H_{38}O_2$	
20	19.921	19.700	20.155	7402950 9	35.46	29.74	9-Octadecenoic acid, methyl ester, (E)-	$C_{19}H_{36}O_2$	
21	20.938	20.775	21.135	3367694 1	16.13	17.09	9,12-Octadecadienoic acid (Z,Z)-, methyl ester	$C_{18}H_{32}O_2$	
22	22.394	22.275	22.515	1028265	0.49	0.60	9,12,15-Octadecatrienoic acid, methyl ester, (Z,Z,Z)-	$C_{19}H_{32}O_2$	
23	23.909	23.775	24.065	2007044	0.96	1.03	Methyl 18-methylnonadecanoate	$C_{21}H_{42}O_2$	
24	24.450	24.335	24.545	446797	0.21	0.25	11-Eicosenoic acid, methyl ester	$C_{21}H_{40}O_2$	
25	27.868	27.755	27.975	643177	0.31	0.39	Methyl 20-methyl-heneicosanoate	$C_{23}H_{46}O_2$	
				2087680 49	100. 00	100.0 0			

Source: Bioenergy Research and Quality Assurance Laboratory University of Agricultural Sciences GKVK, Bengaluru 560065

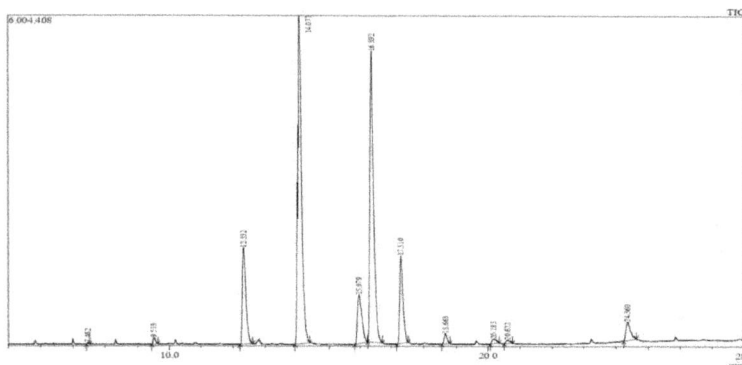

Figure 42.5 GC-MS chromatogram graph of used cooking oil methyl ester
Source: Bioenergy Research and Quality Assurance Laboratory University of Agricultural Sciences GKVK, Bengaluru 560065

Table 42.8 GC-MS chromatogram components of used cooking oil methyl ester blend

Peak #	R.Time	I.Time	F.Time	Area	Area %	Height %	Name	Chemical Formula	Chemical Structure
1	2.954	2.900	3.010	719799	0.37	0.91	Toluene	C₇H₈	
2	3.692	3.640	3.730	171182	0.09	0.24	Ethylbenzene	C₈H₁₀	
3	3.762	3.730	3.790	128915	0.07	0.21	p-Xylene	C₈H₁₀	
4	3.821	3.785	3.875	394675	0.20	0.58	p-Xylene	C₈H₁₀	
5	4.231	4.190	4.290	282518	0.15	0.39	p-Xylene	C₈H₁₀	
6	4.444	4.400	4.485	66311	0.03	0.09	Benzene, propyl-	C₉H₁₂	
7	4.589	4.535	4.645	361237	0.19	0.40	Benzene, 1-ethyl-2-methyl-	C₉H₁₈	
8	4.778	4.730	4.830	103839	0.05	0.13	Mesitylene	C₉H₁₂	
9	4.949	4.905	4.995	111748	0.06	0.14	Benzene, 1-ethyl-2-methyl-	C₉H₁₈	
10	5.140	5.070	5.200	465221	0.24	0.53	Mesitylene	C₉H₁₂	
11	5.321	5.260	5.385	75378	0.04	0.07	Benzene, 1-methyl-3-propyl-	C₁₀H₁₄	
12	6.030	5.980	6.085	204886	0.11	0.23	Octanoic acid, methyl ester	C₉H₁₈O₂	
13	7.752	7.705	7.790	71815	0.04	0.09	Octadecane	C₁₈H₃₈	
14	7.873	7.795	7.920	206820	0.11	0.25	Dimethyl Sulfoxide	C₂H₆OS	
15	9.782	9.740	9.835	132871	0.07	0.16	2,4-Decadienal, (E,E)-	C₁₀H₁₆O	
16	11.952	11.895	12.015	256044	0.13	0.25	Methyl tetradecanoate	C₁₅H₃₀O₂	
17	15.289	15.150	15.390	3404247	17.49	21.78	Hexadecanoic acid, methyl ester	C₁₇H₃₄O₂	
18	15.670	15.630	15.730	140124	0.07	0.11	Methyl hexadec-9-enoate	C₁₇H₃₂O₂	
19	15.802	15.730	15.895	2271759	1.17	1.73	9-Hexadecenoic acid, methyl ester, (Z)-	C₁₆H₃₀O₂	
20	17.217	17.155	17.300	250637	0.13	0.18	Heptadecanoic acid, methyl ester	C₁₈H₃₆O₂	
21	19.398	19.245	19.525	1979164	10.17	10.52	Methyl stearate	C₁₉H₃₈O₂	
22	19.931	19.725	20.015	6973447	35.86	30.53	9-Octadecenoic acid, methyl ester, (E)-	C₁₉H₃₆O₂	
23	20.032	20.005	20.105	892673	0.46	1.11	9-Octadecenoic acid (Z)-, methyl ester	C₁₉H₃₆O₂	
24	20.988	20.775	21.140	6206176	31.88	28.34	9,12-Octadecadienoic acid (Z,Z)-, methyl ester	C₁₈H₃₂O₂	
25	22.410	22.300	22.510	712362	0.37	0.46	9,12,15-Octadecatrienoic acid, methyl ester, (Z,Z,Z)-	C₁₉H₃₂O₂	
26	23.920	23.800	24.040	745074	0.38	0.41	Methyl 18-methylnonadecanoate	C₂₁H₄₂O₂	
27	24.463	24.350	24.560	294791	0.15	0.17	11-Eicosenoic acid, methyl ester	C₂₁H₄₀O₂	
				19469102 8	100.00	100.00			

Source: Bioenergy Research and Quality Assurance Laboratory University of Agricultural Sciences GKVK, Bengaluru 560065

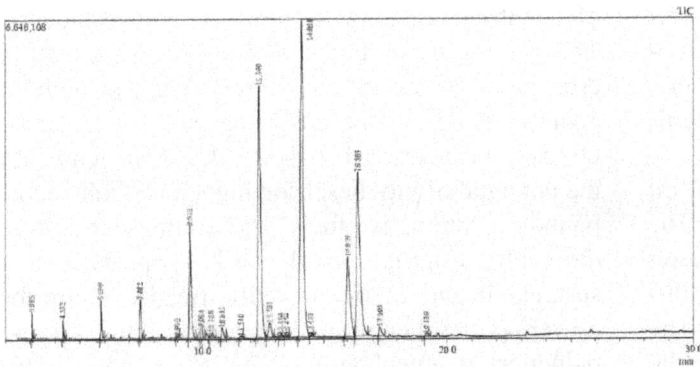

Figure 42.6 GC-MS chromatogram graph of dairy scum oil methyl ester

Source: Bioenergy Research and Quality Assurance Laboratory University of Agricultural Sciences GKVK, Bengaluru 560065

Table 42.9 GC-MS chromatogram components of dairy scum oil methyl ester blend

Peak #	Real Time	Area	Area %	Height	Height %	Name	Chemical Formula	Chemical Structure
1	7.462	64189	0.05	21050	0.13	Dodecanoic acid, methyl ester	$C_{13}H_{26}O_2$	
2	9.518	573994	0.44	117621	0.72	Methyl tetradecanoate	$C_{15}H_{30}O_2$	
3	12.332	1340141 51	10.18	1746580	10.73	Hexadecanoic acid, methyl ester	$C_{17}H_{34}O_2$	
4	14.077	49404153	37.53	5962224	36.62	Heptadecanoic acid, methyl ester	$C_{18}H_{36}O_2$	
5	15.979	8816867	6.70	8839202	5.43	Methyl stearate	$C_{19}H_{38}O_2$	
6	16.392	42581599	32.34	5289695	32.49	9-Octadecenoic acid, methyl ester, (E)-	$C_{19}H_{36}O_2$	
7	17.310	11102397	8.43	1588402	9.76	9,12-Octadecadienoic acid (Z, Z)-, methyl ester	$C_{19}H_{34}O_2$	
8	18.663	1127343	0.86	189937	1.17	9,12,15-Octadecatrienoic acid, methyl ester, (Z,Z,Z)-	$C_{19}H_{32}O_2$	
9	20.185	752252	0.57	83121	0.51	Methyl 18-methyl nonadecanoate	$C_{21}H_{42}O_2$	
10	20.622	530415	0.40	63108	0.39	cis-Methyl 11-eicosenoate	$C_{21}H_{40}O_2$	
11	24.360	3296222	2.50	336753	2.07	Docosanoic acid, methyl ester	$C_{23}H_{46}O_2$	
		131650882	100.00	16282413	100.00			

Source: Bioenergy Research and Quality Assurance Laboratory University of Agricultural Sciences GKVK, Bengaluru 560065

$(C_{18}H_{32}O_2)$ and hexadecanoic acid, methyl ester $(C_{17}H_{34}O_2)$ share the second largest area % of 31.88% and third largest area % of 17.49%, respectively. The obtained main components of the Jatropha methyl ester blend sample are converted into bio-diesel components with a yield percentage of 82%. The synthesised blend sample fulfils all the bio-diesel standards set under BIS [29].

The chemical composition of the used cooking oil methyl ester blend has listed in Table 42.8. The GC-MS analysis of the used cooking oil methyl ester blend sample was carried out and identified 25

different components, as shown in Figure 42.5. The identified components with their chemical formula, chemical structure, retention times, Intensity time, F. time, Area %, and height % for all the bio-diesel, i.e., used cooking oil methyl ester blend constituent, have displayed in Table 42.8.

Heptadecanoic acid, methyl ester $(C_{18}H_{36}O_2)$ represents the main constituent with [37.53%] area %, 9-octadecenoic acid, methyl ester, (E) $(C_{19}H_{36}O_2)$ and hexadecanoic acid, methyl ester $(C_{17}H_{34}O_2)$ share the second largest area % of 32.34% and third largest area % of 10.18%, respectively. The obtained main components of the used cooking oil methyl ester blend sample are converted into bio-diesel components with a yield percentage of 90%. The synthesised blend sample fulfils all the bio-diesel standards set under BIS [30].

The chemical composition of the dairy scum oil methyl ester blend has listed in Table 42.9. The GC-MS analysis of the dairy scum oil methyl ester blend sample was carried out and identified 25 different components, as shown in Figure 42.6. The identified components with their chemical formula, chemical structure, retention times, intensity time, F. time, area %, and height % for all the bio-diesel, i.e., dairy scum oil methyl ester blend constituent, have displayed in Table 42.9.

Heptadecanoic acid, methyl ester ($C_{18}H_{36}O_2$) represents the main constituent with [33.33%] area %, hexadecanoic acid, methyl ester ($C_{17}H_{34}O_3$) and 9-octadecenoic acid, methyl ester, (E) ($C_{19}H_{36}O_2$) share the second largest area % of 23.11% and third largest area % of 18.53%, respectively. The obtained main components of the dairy scum oil methyl ester blend sample are converted into bio-diesel components with a yield percentage of 90%. The synthesised blend sample fulfils all the bio-diesel standards set under BIS [31].

Conclusion

In conclusion, this work emphasises the potential of bio-diesel and higher alcoholic compounds as viable substitutes for traditional fuels. These alternatives offer advantages such as accessibility, ease of synthesis, and positive environmental impact. By effectively deploying bio-diesel in various sectors, the issues of cost and contamination can be addressed. The research focuses on key aspects including feedstock availability, fuel characterisation, temperature effects on fuel properties, fuel synthesis methods, and analysis of bio-diesel blend constituents using GC-MS analysis. The study examines the trans-esterification process for bio-diesel synthesis and analyses the impact of solvents on esterified oils. The results indicate that a 5% gasoline addition to methyl ester blends has a significant influence on bio-fuel blends, enabling bio-diesel to serve as a substitute for diesel in compression ignition (CI) engines. Engine performance tests demonstrate that the blended sample exhibits results closely comparable to diesel fuel. Moreover, the properties of the methyl ester blends fall within the permissible levels set by the majority of countries, ensuring compliance with standards. This finding supports the viability of using methyl ester-gasoline blends in diesel engines without compromising engine performance or emissions. Overall, this research provides valuable insights into the potential of bio-diesel and higher alcoholic compounds as alternative fuels. The results suggest that the methyl ester-gasoline blend holds promise as a sustainable and environmentally friendly option for powering diesel engines, thereby contributing to the reduction of greenhouse gas emissions and dependence on fossil fuels.

References

[1] Gupta, J. Energy as a fundamental requirement for human development: A review of concepts, indicators, and international initiatives. *Energies*, 2019;12(24):4716.

[2] Knothe, G. Biodiesel and Renewable Diesel: A Comparison. Progress in Energy and Combustion Science, 2010;36(3):364–373.

[3] Rashid, U., Anwar, F., Knothe, G. Evaluation of bio-diesel obtained from cottonseed oil. *Energy Fuels*, 2009;23(9):4445–4450.

[4] Karmakar, A., Karmakar, S. Biodiesel production from neem oil using a developed single-step conversion process. *Fuel Proc. Technol.*, 2010;91(10):1333–1339.

[5] Pradhan, S., Naik, S. N. Comparative assessment of Jatropha curcas cultivation in different states of India. *Biomass Bioenergy*, 2011;35(12):4445–4452.

[6] Bharathi, L. K., Reddy, P. S., Reddy, M. R. Pongamia pinnata (L.) Pierre: A potential feedstock for biodiesel. *Internat. J. Environ. Sci.*, 2012;3(1):332–341.

[7] Fernandes, F. A. N., Marques, L. G. S., Suarez, P. A. Z. Vegetable oils as alternative fuels for diesel engines: A review. *Renew. Sustain. Energy Rev.*, 2020:119:109527.

[8] Demirbas, A. Biodiesel from vegetable oils via transesterification in supercritical methanol. *Energy Conv. Manag.*, 2008;49(1):125–130.

[9] Ma, F., Hanna, M. A. Biodiesel production: A review. *Biores. Technol.*, 1999;70(1):1–15.

[10] Meher, L. C., Vidya Sagar, D., Naik, S. N. Technical aspects of biodiesel production by transesterification—A review. *Renew. Sustain. Energy Rev.*, 2006;10(3):248–268.

[11] Kumari, A., Kumar, A. Transesterification of vegetable oils for biodiesel production: A review. *Renew. Sustain. Energy Rev.*, 2017;72:878–889.

[12] Marchetti, J. M., Errazu, A. F., Quiroga, M. E. Biodiesel production from raw materials and by-products. *Proc. Safety Environ. Protec.*, 2007;85(5):417–422.

[13] Balat, M. Production of bioethanol from lignocellulosic materials via the biochemical pathway: A review. *Energy Conv. Manag.*, 2011;52(2):858–875.

[14] Du, W., Xu, Y., Liu, D., Zeng, J., Wang, X. Ethanolysis of soybean oil in subcritical and supercritical methanol with co-solvents. *J. Am. Oil Chem. Soc.*, 2004;81(2):179–184.

[15] Kaieda, M., Samukawa, T., Matsumoto, T., Ban, K., Kondo, A., Shimada, Y., Noda, H. Biodiesel fuel production from plant oil catalyzed by Rhizopus oryzae lipase in a water-containing system without organic solvent. *J. Biosci. Bioengg.*, 2001;91(1):12–15.

[16] Sharma, Y. C., Singh, B., Upadhyay, S. N. Advancements in development and characterization of biodiesel: A review. *Fuel*, 2011;90(3):1309–1324.

[17] Shahbaz, M., Irfan, M. F., Hussain, S. Biodiesel production from waste cooking oil using ethanol and methanol and its comparative study. *Renew. Energy*, 2017;108: 48–56.

[18] Alptekin, E., Canakci, M. A comprehensive experimental investigation of biodiesel production from common feedstocks. *Renew. Energy*, 2010;35(2):224–231.

[19] Kiss, A. A., Bélafi-Bakó, K. Kinetics and mechanism of transesterification. *Chem. Engg. Res. Design*, 2013;91(8):1419–1436.

[20] Antczak, J., Kubiak, A., Szeląg, H. Biodiesel production processes: Current state of the art and future challenges. *Renew. Sustain. Energy Rev.*, 2017;69:748–766.

[21] Shah, P., Kumar, N. Experimental investigation of the performance and emission characteristics of a diesel engine using biodiesel blends with higher cetane number. *Fuel*, 2018;220:908–916.

[22] Sneddon, J., Masuram, S., Richert, J. C. Gas chromatography-mass spectrometry-basic principles, instrumentation and selected applications for detection of organic compounds. *Anal. Lett.*, 2007;40(6):1003–1012.

[23] Epping, R., Koch, M. On-site detection of volatile organic compounds (VOCs). *Molecules*, 2023;28:1598.

[24] Sumitra, P., Ben, N., Arnold, R. D., Cummings, B. S. Extraction, chromatographic and mass spectrometric methods for lipid analysis. *Biomed. Chromat.*, 2016;30(5):695–709.

[25] Zuo, H.-L., Yang, F.-Q., Huang, W.-H., Xia, Z.-N. Preparative gas chromatography and its applications. *J. Chromat. Sci.*, 2013;51(7):704–715.

[26] Sankar Vinayaka, A., Biswanath, M., Eldon, R. R., Shishir Kumar, B. Biodiesel production by transesterification of a mixture of pongamia and neem oils. *Biofuels*, 2018, 12(2), 135–147.

[27] Ismail, J. M., Yusufu, A. C. J., Thomas, K. Fast rate production of biodiesel from neem seed oil using a catalyst made from banana peel ash loaded with metal oxide (Li-CaO/Fe2 (SO4)3). *Adv. Mater. Sci. Engg.*, 2020;2020: 1–11. Article ID 782502.

[28] Acharya, N., Nanda, P., Panda, S., Acharya, S. Analysis of properties and estimation of optimum blending ratio of blended mahua biodiesel. *Engg. Sci. Technol. Internat. J.*, 2017;20(2):511–517.

[29] Joshua, F. Production of biodiesel (B100) from Jatropha oil using sodium hydroxide as catalyst. *J. Petroleum Engg.*, 2013;2013:6.

[30] Harish, H., Rajanna, S., Prakash, G. S., Ramachandra, K., Srikanth, H. V. Experimental investigation of effect of injections parameters on stationary diesel engine fuel with the methyl ester of waste vegetable oil. *J. Green Engg.*, 2020;10(11):12195–12215.

[31] Harish, H., Rajanna, S., Prakash, G. S., Srikanth, H. V. The influence of injection timings on performance, emission, and combustion characteristics of compression ignition engine fueled with milk scum oil biodiesel. *Energy Sources Part A Recov. Utiliz. Environ. Effects*, 2020;47(1):3762.

[32] Ayhan, D., Abdullah, B., Waqar, A., et al. Biodiesel production from non-edible plant oils. *Energy Explor. Exploit.*, 2016;34(2):290–318.

[33] Sanjib Kumar, K., Anju, C. Preparation of biodiesel from crude oil of Pongamia pinnata. *Biores. Technol.*, 2005;96:1425–1429.

[34] Encinar, J. M., González, J. F., Pardal, A. Transesterification of castor oil under ultrasonic irradiation conditions. Preliminary results. *Fuel Proc. Technol.*, 2012;103: 9–15.

[35] Ragit, S. S., Mohapatra, S. K., et al. Optimization of neem methyl ester from transesterification process and fuel characterization as a diesel substitute. *Biomass Bioenergy*, 2011;35(3):1138–1144.

[36] Guharaja, S., Dhakshina Moorthy, S., Inamul Hasan, Z., et al. Biodiesel production from Mahua (Madhuca Indica). *Int. J. Nano. Corr. Sci. Engg.*, 2016;3(1):34–47.

[37] Mahanta, P., Sarmah, J. K., Kalita, P. Parametric study on transesterification process for biodiesel production from Pongamia pinnata and Jatropha curcus oil. *Internat. Energy J.*, 2008;9:41–46.

[38] Phan, A. N., Phan, T. M. Biodiesel production from waste cooking oils. Fuel, 2008;87(17–18):3490–3496.

[39] Maria, S., Mariagrazia, I. Biodiesel production from dairy waste Scum by using a efficient nano-biocatalyst. *Chem. Engg. Transac.*, 2020;79:181–186.

[40] Jain, S., Sharma, M. P. Prospects of biodiesel from Jatropha in India: A review. *Renew. Sustain. Energy Rev.*, 2010;14(2):763–771.

[41] Sharma, Y. C., Singh, B. Development of biodiesel from Karanja, a tree found in rural India. *Fuel*, 2008;87(8–9):1740–1742.

[42] De Lima Da Silva, N., Batistella, C., Maciel Filho, R., Wolf Maciel, M. R. Determination of castor oil molecular weight by vapour pressure osmometry technique. *Chemical Engineering Transactions*, 2011;24:601–606

[43] Sathya Selvabala, V., Varathachary, T. K., Selvaraj, D. K., Ponnusamy, V., Subramanian, S. Removal of free fatty acid in Azadirachta indica (Neem) seed oil using phosphoric acid modified mordenite for biodiesel production. *Biores. Technol.*, 2010;101(15):5897–5902.

[44] Sathish Kumar, R., Krupa Vara Prasad, A. Environment friendly butyl ester biodiesel production from mahua oil: Optimization and characterization. *SN Appl. Sci.*, 2019;1:872.

[45] Kanthawut, B., Sawitri, C., Vittaya, P., Pinya, S. Optimization of biodiesel production from Jatropha oil (Jatropha curcas L.) using response surface methodology. *Kasetsart J. Nat. Sci.*, 2010;44:290–299.

[46] Sánchez, A., Maceiras, R., Cancela, A., Rodríguez, M. Influence of n-Hexane on in Situ Transesterification of Marine Macroalgae. Energies, 2012;5(2):243–257.

[47] Gerpen, J. V. Biodiesel processing and production. *Fuel Proc. Technol.*, 2005;86(10):1097–1107.

[48] Tippayawong, N., Sittisun, P. Continuous-flow transesterification of crude jatropha oil with microwave irradiation. *Sci. Iranica*, 2012;19(5):1324–1328.

[49] Pasha, Y., Shrinivasa, U., Swamy, D. R., Muzzamil Ahamed, S., Harish, H. Investigation of performance parameters combustion and emission characteristics of compression ignition engine fueled with pongamia methyl ester-gasoline blend. *Energy Sources Part A Recov. Utiliz. Environ. Effects*, 2021;47(2):1–20.

[50] Pasha, Y., Shrinivasa, U., Swamy, D. R., et al. Experimental analysis of fuel properties and performance analysis of diesel engine with esterified castor oil blended with gasoline. *Mater. Today Proc.*, 2021:46:4892–4899.

[51] Salimon, J., Noor, D. A. M., Nazrizawati, A. T., et al. Fatty acid composition and physicochemical properties of Malaysian castor bean Ricinus communis L. seed oil. *Sains Malaysiana*, 2010;39(5):761–764.

[52] Muthu, H., SathyaSelvabala, V., Varathachary, T. K., Kirupha Selvaraj, D., Nandagopal, J., Subramanian, S. Synthesis of biodiesel from neem oil using sulfated Zirconia via tranesterification. *Brazil. J. Chem. Engg.*, 2010;27(04):601–608.

[53] Bhatt, Y. C., Murthy, N. S., Datta, R. K. Use of mahua oil (Madhuca indica) as a diesel fuel extender. *J. Instit. Eng. (India) Agricul. Engg Div.*, 2004;85:10–14.

[54] Edem, D. O. Palm oil: Biochemical, physiological, nutritional, hematological, and toxicological aspects: A review. *Plant Foods Human Nutr.*, 2002;57(3–4):319–341.

[55] Arjun, B. C., Chris Watts, K., Rafiqul Islam, M. Waste cooking oil as an alternate feedstock for biodiesel production. *Energies*, 2008;1:3–18.

43 Influence of representative volume element (RVE) size and particle clustering on tensile properties of epoxy/SiO$_2$ nanocomposites

Sumana[1], Pratik P. Jakanur[1], Sridhar M.[1], M. A. Umarfarooq[5,a], N. R. Banapurmath[1,2], Ashok M. Sajjan[1,2], Kartheek Ravulapati[4] and B. H. Maruthi Prashanth[3]

[1]Department of Mechanical Engineering, KLE Technological University Hubballi–580031, Karnataka, India

[2]Centre of Excellence in Material Science, School of Mechanical Engineering, KLE Technological University Hubballi–580031, Karnataka, India

[3]Department of Mechanical Engineering, A.G.M Rural College of Engineering & Technology (affiliated to Visvesvaraya Technological University, Belagavi), Varur, Hubballi–581207, Karnataka, India

[4]Collins Aerospace, 5935, Pinnacle View Road, Cumming GA 30040, United States

[5]Department of Mechanical Engineering, Karpagam Academy of Higher Education, Coimbatore, Tamil Nadu, India.

Abstract

Predicting the mechanical properties of the nanocomposites remains a significant challenge due to their heterogeneous microstructure. This work investigates the influence of representative volume element (RVE) size and particle clustering on the elastic properties of epoxy/SiO$_2$ nanocomposites. Using finite element (FE) simulations, the elastic constants (E_{11} and E_{22}) were computed for 2D square RVEs of varying sizes (200 nm, 500 nm, 1 μm, 1.5 μm, and 2 μm). The impact of nanoparticle clustering was also analysed. To validate the simulation results, epoxy nanocomposites reinforced with 1 wt. % SiO$_2$ were prepared and were tested for tensile properties as per ASTM D638. The simulation results revealed minimal variations (<1%) in the elastic constants across different RVE sizes and clustering configurations. The experimentally determined tensile modulus of 1396 MPa exhibited a close agreement with the simulation results, with a deviation of less than 2%. This study confirms that RVE size and particle clustering have negligible effects on the elastic constants.

Keywords: Representative volume element, finite element simulations, nanocomposites

Introduction

Nanocomposites with the flexibility to tailor the properties based on the matrix material, nano material incorporated and the weight fraction of filler materials have emerged as one of the promising materials for engineering applications. The strength, stiffness, durability and other properties of the composite can be varied by adding a small quantity of the nanoparticles. Due to the enhanced properties of nanocomposites. The heterogeneity introduced by the incorporation of the nanoparticles makes it difficult to predict the mechanical properties of these nanocomposites [1–5].

Homogenisation techniques simplify the computational analysis of the heterogeneous materials such as composites to be treated as continuum models. Homogenisation methods have been widely to evaluate the overall properties of heterogeneous materials from the knowledge of the constitutive laws and spatial distribution of the constituents. In homogenisation methods, the estimates are made by assuming that the material extends infinitely, so they are called asymptotic estimates. Computational homogenisation methods utilise numerical methods and simulations on the microstructure samples, with RVE playing a key role in these methods [6–8]. According to Hill [9], the RVE refers to a sample that a) is represents structurally the microstructure of mixture b) encompasses sufficient inclusions for the overall moduli to be effectively independent of the surface traction and displacement values, as long as these values are macroscopically uniform. The RVE is material model that is used to evaluate the effective properties of the homogenised

---

[a]umarfarooq.ma@gmail.com

DOI: 10.1201/9781003685876-43

macroscopic model. RVE should be sufficiently large to encompass enough information to represents the microstructure, but it must be small than the macroscopic body. The RVE is a tiny volume within this material that summarises the essential characteristics such as arrangement, distribution, and interactions of the fibres and matrix. It is mostly used in composite analysis because, by just analysing the RVE, one can predict its behaviour and properties on a larger scale. Other reasons are its ability to capture complex microstructures, its help in determining microscopic properties, and by using RVE, it reduces costs and its overall reputation of being reliable in estimating the overall performance [10, 11].

Balasubramani et al. [12] found that variation in RVE sizes impacts the results. Smaller RVEs might capture local variations and microstructural features more accurately but at the cost of increased computational complexity. Larger RVEs, on the other hand, might provide a more averaged-out representation of the material behaviour. Qayyum et al. [13] systematically varied the thickness of the RVEs while keeping other parameters constant to understand their influence on stress and strain distribution within composite materials. The results demonstrate that the thickness of the RVE significantly affects stress distribution among different phases within the material. Thicker RVEs tend to exhibit a more uniform stress distribution across phases compared to thinner RVEs. Cho et al. [14] explored the relationship between particle size and material properties of particulate composites. The results revealed that nanoparticles enhance the Youngs modulus more than micron-sized particles due to better dispersion. The smaller the particle size, the better the tensile strength, though poor dispersion of nanoparticles leads to reduced strength at high loadings. It suggests that the direct effect of particles on matrix interfacial fracture toughness is negligible, thus highlighting the intricacies of crack propagation behaviour in particle matter. The objective of this work is to study the effect of the selected RVE size on the elastic constants of epoxy-SiO_2 nanocomposites. Also, the elastic constant of the nanocomposites was experimentally determined and compared with the numerical result.

Experimental details

Materials used and preparation of nanocomposites
Epoxy (L12) cured by the hardener (K6) applied in the weight ratio of 9:1 was used as base material. SiO_2 nanoparticles were used as filler material. The properties of the matrix, (which were experimentally determined) and filler material are provided in Table 43.1. 1 wt. % of SiO_2 is dispersed in the epoxy and the mixture is sonicated for 40 min at 50 kHz frequency. Once sonification was complete, the K6 hardener was added to the mixture then continuous stirred for 20 min and degassed to prevent any air bubble formation. Subsequently, the solution was poured into a mould and left to cure at room temperature for 24 hours. The samples were cut as per ASTM D638 [15] for tensile test.

Tensile test
Tensile tests were carried as per ASTM D638 with a displacement rate of 3 mm/min. The average Youngs modulus from three samples was found to be 1396 MPa.

Simulation using Digimat software

Digimat is a multi-scale modelling platform that focuses on the micromechanical modelling of the complex multiphase materials such as composite, plastics and metals. It enables the creation of RVE to simulate interaction between the different phases in a composite at the microscopic level. In this study, Digimat was utilised to model and analyse the mechanical behaviour of epoxy/SiO_2 nanocomposites, focusing on the influence of RVE size and particle agglomeration on elastic constants (E_{11} and E_{22}).

Table 43.1 Material properties of epoxy and SiO_2

Property	Epoxy	SiO_2 [3]
Density (kg/m³)	1100	2400
Young's modulus (MPa)	1200	70000
Poisson's ratio	0.32	0.17
Particle size (nm)	-	15

Source: Author

Material definition and RVE setup

The simulation begins with the definition of material properties for the matrix and filler phases in Digimat-FE. Two-dimensional square RVEs were created with edge lengths of 200 nm, 500 nm, 1 μm, 1.5 μm and 2 μm to investigate the effect of RVE size.

RVE were also created with particles clustering with two and three clusters. The RVE geometry (size and shape), particle distribution and meshed model created with 200 nm × 200 nm, 500 nm × 500 nm, and 1 μm × 1 μm size are shown in Table 43.2. The element size was maintained at 7.5 nm.

Boundary conditions and simulation

Periodic boundary conditions were applied to the RVEs to simulate the behaviour of a microstructure, aligning with the homogenisation approach. Uniaxial loading was applied along two perpendicular axes (x and y) to compute the elastic constants E_{11} (Young's modulus along x-axis) and E_{22} (Young's modulus along y-axis). The Digimat-FE solver performed the FE calculations, under the applied loading conditions. The simulations were run for each RVE size.

Results and discussion

The Young's modulus (E_{11} and E_{22}) obtained from the simulations are provided in the Table 43.3. The results show that elastic constants ranged between 1373.32 MPa and 1381.68 MPa across all RVE sizes and particle distributions. Notably, the variation in moduli was within 1%, indicating that neither the RVE size nor the degree of particle agglomeration had a significant effect on the elastic properties. For instance, the 200 nm RVE with uniformly distribution yielded E_{11} = 1378.31 MPa and E_{22} = 1378.36 MPa, while the 1.5 μm RVE with twenty clusters gave E_{11} = 1381.68 MPa and E_{22} = 1380.24 MPa. Even in extreme cases, such as the 1.5 μm RVE with fifty clusters (E_{11} = 1377.1 MPa and E_{22} = 1373.32 MPa), the moduli remained closely aligned with other configurations.

The negligible influence of RVE size suggests that even the smallest RVE (200 nm) was sufficiently large to capture the representative microstructure of the nanocomposite. The simulated moduli were compared with the experimental Young's modulus

of 1396 MPa obtained from tensile testing, showing a maximum deviation of less than 2%. This close agreement validates the accuracy of the FE simulations and the appropriateness of the chosen RVE sizes.

The consistent elastic constants across different RVE sizes and particle distributions demonstrate the reliability of the homogenisation. The results show that at low filler concentration (1 wt. %), the matrix properties and the uniform dispersion of nanoparticles dominate the overall behaviour, with minimal impact from agglomeration. However, the scope of the study was limited to 2D RVEs and single filler concentration, future studies can explore 3D RVEs or higher filler concentrations to better understand complex particle interactions.

Conclusion

This study investigated the influence of the RVE size and nanoparticle agglomeration on the elastic constants (E_{11} and E_{22}) of epoxy/SiO$_2$ nanocomposites through FE simulations and experimental validation. Square RVEs of varying dimensions (200 nm, 500 nm, 1 μm and 1.5 μm) with 1 wt. % SiO$_2$ were modelled using Digimat-FE, incorporating both uniform and clustered particle distributions. The FE simulations demonstrated that the Young's moduli (E_{11} and E_{22}) from 1373.32 MPa and 1381.68 MPa, with variations the variation were within 1% across all RVEs sizes and agglomeration configurations. This indicates that neither RVE sizes nor particle clustering significantly affected the elastic properties in this study. Experimental tensile tests exhibited an average tensile modulus of 1396 MPa and the differences between experimentation and simulation were within 2%.

References

[1] Khan, W. S., Hamadneh, N. N., Khan, W. A. Polymer nanocomposites–synthesis techniques, classification and properties. *Sci. Appl. Tailored Nanostruc.*, 2016;50.

[2] Njuguna, J., Ansari, F., Sachse, S., Rodriguez, V. M., Siqqique, S., Zhu, H. Nanomaterials, nanofillers, and nanocomposites: types and properties. In: Health and environmental safety of nanomaterials. Woodhead Publishing, California, 2021:3–37.

Table 43.2 RVE geometry (size and shape), particle distribution and meshed model created with 200 nm × 200 nm, 500 nm × 500 nm, and 1 μm × 1 μm

SL. No	Size of RVE	Distribution	Geometry	Meshed model
1	200nm	Uniform		
2	500nm	Uniform		
3	500nm	Two cluster		
4	500nm	Three cluster		
5	1μm	Uniform		
6	1μm	Two cluster		
7	1μm	Three cluster		

Source: Author

Table 43.3 Young's moduli of all RVE

Sl. No.	Size of RVE	Distribution	Number of particles	E_{11}	E_{22}
1	200 nm × 200 nm	Uniform	1	1378.31	1378.36
2		Uniform		1379.98	1379.22
3	500 nm × 500 nm	Two clusters	7	1378.35	1378.48
4		Three clusters		1380.09	1378.84
5		Uniform		1378.7	1378.82
6	1 μm × 1 μm	Two clusters	26	1379.72	1378.82
7		Three clusters		1380.14	1379.92
8		Uniform		1378.74	1379.92
9		Two clusters		1379.26	1380.11
10		Three clusters		1379.86	1381.5
11		Four clusters		1380.1	1380.4
12		Five clusters		1380.53	1380.05
13		Six clusters		1380.54	1380.77
14	1.5 μm × 1.5 μm	Seven clusters	64	1380.16	1380.77
15		Eight clusters		1380.48	1380.22
16		Nine clusters		1380.03	1380.47
17		Ten clusters		1374.62	1376.36
18		Fifteen clusters		1380.73	1380.6
19		Twenty clusters		1381.68	1380.24
20		Fifty clusters		1377.1	1373.32
21		Uniform		1378.63	1378.73
22	2 μm × 2 μm	Two clusters	114	1379.75	1379.13
23		Three clusters		1380.41	1380.63

Source: Author

[3] Dileep, K., Srinath, A., Banapurmath, N. R., Umarfarooq, M. A., Sajjan, A. M. Mechanical and fracture characterization of epoxy/PLA/graphene/SiO₂ composites. *Frattura ed Integrità Strutturale*, 2023;17(64):229–239.

[4] Singh, N. B., Agarwal, S. Nanocomposites: an overview. *Emerg. Mat. Res.*, 2016;5(1):5–43.

[5] Umarfarooq, M. A., Choukimath, M., Banapurmath, N. R. Mechanical, fracture, and thermal characterization of post-cured hybrid epoxy nanocomposites reinforced with graphene nanoplatelets and h-boron nitride. *Frattura ed Integrità Strutturale*, 2025;19(71):22–36.

[6] El Moumen, A., Kanit, T., Imad, A. Numerical evaluation of the representative volume element for random composites. *Eur. J. Mec.-A/Solids*, 2021;86:104181.

[7] Kanit, T., Forest, S., Galliet, I., Mounoury, V., Jeulin, D. Determination of the size of the representative volume element for random composites: Statistical and numerical approach. *Internat. J. Solids Struc.*, 2003;40(13–14):3647–3679.

[8] Mirkhalaf, S. M., Pires, F. A., Simoes, R. Determination of the size of the representative volume element (RVE) for the simulation of heterogeneous polymers at finite strains. *Finite Elem. Anal. Design*, 2016;119:30–44.

[9] Hill, R. Elastic properties of reinforced solids: Some theoretical principles. *J. Mec. Phy. Solids*, 1963;11(5):357–372.

[10] Hashin, Z. Analysis of composite materials—a survey. J. Appl. Mech. 50, 481 (1983).

[11] Drugan, W. J., Willis, J. R. A micromechanics-based nonlocal constitutive equation and estimates of representative volume element size for elastic composites. *J. Mec. Phy. Solids,* 1996;44(4):497–524.

[12] Balasubramani, N. K., Zhang, B., Chowdhury, N. T., Mukkavilli, A., Suter, M., Pearce, G. M. Micro-mechanical analysis on random RVE size and shape in multiscale finite element modelling of unidirectional FRP composites. *Comp. Struc.*, 2022;282:115081.

[13] Qayyum, F., Chaudhry, A. A., Guk, S., Schmidtchen, M., Kawalla, R., Prahl, U. Effect of 3D representative volume element (RVE) thickness on stress and strain partitioning in crystal plasticity simulations of multi-phase materials. *Crystals*, 2020;10(10):944.

[14] Cho, J., Joshi, M. S., Sun, C. T. Effect of inclusion size on mechanical properties of polymeric composites with micro and nano particles. *Comp. Sci. Technol.*, 2006;66(13):1941–1952.

[15] ASTM D638-14. Standard test method for tensile properties of plastics. *ASTM Internat.*, 2014.

44 Sliding behaviour of NiAl and NiCr coatings on mild steel substrate

Nadeem Pasha K.[1], Salim Sharieff[2,a], Nawaz Ahmed[1], Younus Pasha[1], Tansif Khan[1] and Dinesh H. A.[1]

[1]Department of Mechanical Engineering, HKBK College of Engineering, Bangalore, VTU, Karnataka, India

[2]Associate Professor & Head, Department of Mechanical Engineering, HKBK College of Engineering, Bangalore, VTU, Karnataka, India

Abstract

In all mechanical systems wherein either motion or force is transmitted from one location to another location, relative motion between two components takes place. For example, in case of bearings there will be relative motion between shaft and bearings. This relative motion results in displacement and velocity discontinuities. These kinematic discontinuities give rise to wear and loss of energy. Different engineering solutions like lubricating and giving hard coatings have been tried. In the present investigation two different coating materials nickel aluminium (NiAl) and nickel chromium (NiCr) have been chosen. Each coating material with different coating thickness is also considered for evaluating effect of coating thickness. Tests have been carried out with pin on disk test rig. The normal load was 30 Newton, the sliding speed was 500 rpm and the sliding time was 240 seconds. Scanning electron micrography was also carried out on damaged coating surface. The friction coefficient for NiAl coating was discovered to be around 0.4 irrespective of coating thickness. The mode of sliding for NiAl coating was identified as predominantly abrasive. The friction coefficient for NiCr coating varied from 0.36 to 0.42 as coating thickness increased. The mode of sliding for NiCr coating was identified as failure of coating.

Keywords: Coatings, friction, wear and velocity discontinuity

Introduction

Considerable work is carried out to understand friction and wear mechanism involved in mechanical systems. The understanding of friction and wear is important from the point of view of making systems more energy efficient and also reduce the downtime. The generalisation of friction and wear mechanism is not possible since the phenomenon is system dependent. Studies have been carried out for understanding each system and efforts have been made to comprehend the phenomenon. The influence of coating material and its thickness have been studied.

To better understand the wear and friction phenomena in a tribosystem, researchers have conducted a number of laboratory simulation studies.

Zainab Raheem and Aseel Kareem experimented using a pin-on-disc test apparatus to assess performance of nickel chromium (NiCr) graded coatings on polymer matrix composites. NiCr coating was found to improve the wear performance and attributed to high hardness and wear resistance of NiCr alloy. The scanning electron micrograph and X-ray diffraction studies were conducted and found oxidation [1].

Marinka Baricevic et al. made an effort in finding the effect of Cr and Ni in the salivary due to dental casting material. The presence of Cr and Ni due to dental casting material resulted in burning mouth syndrome [2].

Sylwia Wojda et al. made an effort to understand the enamel material's tribological behaviour. They used a pin on a disc test equipment to conduct the experiment. They discovered that composite materials outperformed other materials [3].

Bhupatani and Prajapati tested the wear behaviour of traditional brass, gun metal, and the novel material "cast nylon" using a pin-on-disc test setup. The outcome shown that, in contrast to convectional bearing materials, cast nylon produced a lower friction coefficient [4].

Hongwei Zhang et al. carried-out tests employing high temperature gas solid two-phase flow erosion test rig for evaluating Cr_3C_2-NiCr coating on nickel based super alloy. They found that erosion behaviour was dependent on air pressure coating thickness, angle of impact, temperature, particle size and its shape. The wear mechanism was found to be both ductile and brittle failures [5].

Arun et al. evaluated the wear performance of electro deposited nickel-yttria doped ceria composite coatings. The result indicated doping of yttria in

[a]salim.sharieff08@gmail.com

DOI: 10.1201/9781003685876-44

ceria modified the wear behaviour. Adhesion wear mechanism was observed [6].

Mohammed and Ghorbani conducted wear experiments coating with PTFE and MoS_2 particles. The result showed that simultaneous co-deposits of PTFE and MoS_2 particles resulted in better wear performance [7].

Norio etal. attempted to assess the nickel-free Co-Cr-Mo alloy's tribological behaviour in a simulated setting. It was discovered that the surroundings affected the wear behaviour [8].

Blesman et al. conducted scratch test for evaluating adhesion of molybdenum protective coating on steel substrate. Developed a theoretical model for estimating adhesive force and results of this model well co-related with experimental data [9].

Allu Harshitha carried out wear trials to assess the wear behaviour of WC-Co coating on Ti6Al4V. The outcome indicated that the wear performance is contingent on factors such as the applied load and speed [10].

Popoola et al. attempted to comprehend the wear behaviour of Zn-Al coating on mild steel using the electroplating technique. The results demonstrated an enhancement in wear performance [11].

Danisman et al. assessed the tribological performance of coatings applied to TiAl4V alloy. The outcome suggested an enhancement in the wear characteristics of different coating materials [12].

Endler et al. performed wear tests to assess how well the coating thickness performs. Their findings showed that coatings greater than 3 μm in thickness did not enhance performance [13].

Stueber et al. synthesised nano laminated coatings of TiC/a-C and (Ti, Al)(N, C)/a-C and suggested that this nano laminated coating is a very promising candidate for tooling where stress state is very high [14].

Salim Sharieff et al. carried out tests to assess wear performance utilising a pin-on disc test setup of coatings. The findings indicated that the friction coefficient was discovered to depend on displacement velocity discontinuities [15].

Sharieff and Ranganatha carried out tests with a pin-on-disc test setup to comprehend the velocity discontinuity effect on wear behaviour. The features in wear surface, an indication of velocity discontinuity, were found to influence friction co-efficient [16].

Nadeem Pasha et al. investigated the influence of speed and load on the tribological behaviour of a grade 5 titanium alloys and hardened steel pair using a pin-on-disk test apparatus. The study showed that at higher speeds, two separate sliding phases were observed, independent of load levels. The oxidation of wear debris occurred at a reduced speed [17].

The impact of speed and load levels on various titanium alloys was investigated by Pasha and Ranganatha. The study revealed that at higher speeds, two distinct phases of sliding were observed. In one of the phases, the friction coefficient exceeded that observed at lower speeds. It was determined that the friction coefficient in the second phase was similar to that at lower speeds [18].

The literature review shows that a significant amount of work remains to be done in assessing the tribo response of various coating materials and their thicknesses.

The current study aims to comprehend the performance of two coating materials, nickel aluminium (NiAl) and NiCr, as well as the impact of varying coating thicknesses of these coatings.

Experimental details

A mild steel pin was machined according to the measurements shown in Figure 44.1.

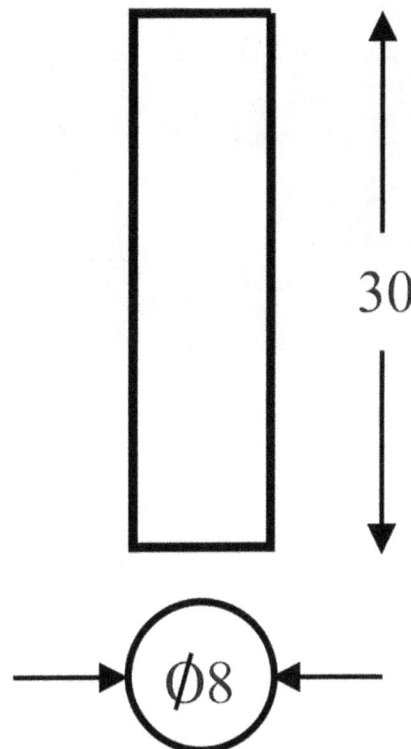

Figure 44.1 Dimensions of the pin used in mm
Source: Author

Nickel aluminium and nickel chromium coatings with thicknesses of 200, 250, and 300 microns were applied to machined mild steel pins. The thermal spray technique was used to include the nickel chromium and nickel aluminium coatings. In thermal spray coating each pass coats a thickness of 10 microns. Based on number of passes coating was done.

The Pin on disk test equipment, depicted in Figure 44.2 was utilised.

The test variables are given in Table 44.1.

ASTM G99 was followed for conducting the experiments. A personal computer was used to track the frictional force and friction coefficient.

Results and discussion

Using a pin on disc test apparatus, mild steel pins coated with nickel aluminium and nickel chromium were moved across an En 31 hardened steel disc in accordance with ASTM protocol to acquire various velocity discontinuities at the interface.

Figure 44.2 Pin holder and disk
Source: Author

Table 44.1 Test details of experiments

Sl. No.	Coating material	Load in Newton	Time in seconds	Speed in rpm
1	NiAl 200 μ	30	240	500
2	NiAl 250 μ	30	240	500
3	NiAl 300 μ	30	240	500
4	NiCr 200 μ	30	240	500
5	NiCr 250 μ	30	240	500
6	NiCr 300 μ	30	240	500

Source: Author

Nickel aluminium coating

Mild steel pins were coated with nickel aluminium using a thermal spray method. The thicknesses of the coating were 200, 250, and 300 microns. Pin-on-disk test rigs have been used for experiments. A computer was used to record the experiment's shear force. The friction coefficient is estimated using the observed shear force. Figure 44.3 displays the graphs

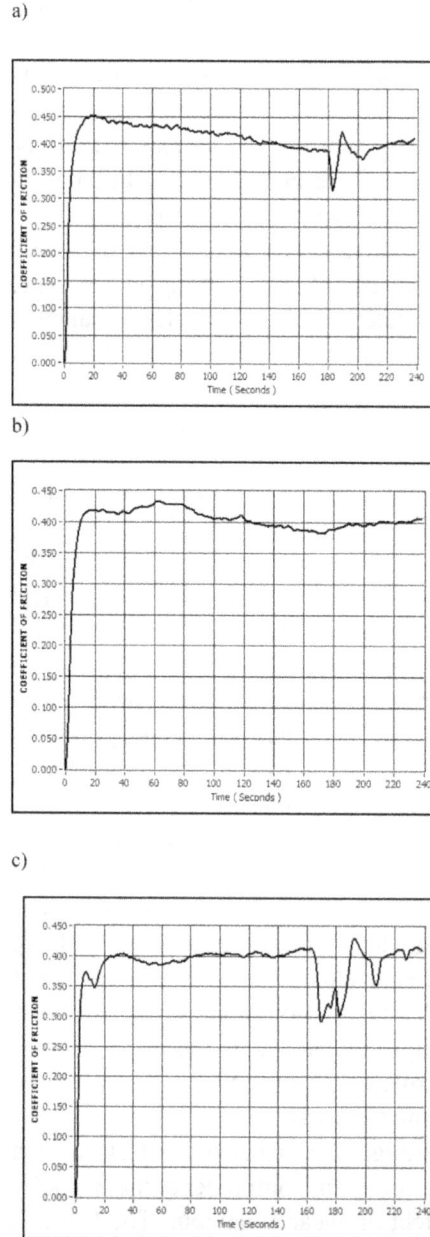

a)

b)

c)

Figure 44.3 Graphs showing how the friction coefficient varies with sliding time for coating thicknesses of (a) 200 microns, (b) 250 microns, and (c) 300 microns of nickel aluminium

Source: Author

illustrating the relationship between the friction coefficients and sliding time.

Figure 44.3a illustrates how the friction coefficient varies with sliding time for a nickel aluminium coating thickness of 200 microns. The graphic illustrates how, from the start to the 20-second sliding duration, the friction coefficient grew monotonically to roughly 0.45. After 20 seconds and for the remainder of the sliding, the friction coefficient was determined to be roughly 0.4, with a slight variation between 180 and 200 seconds.

Figure 44.3b illustrates how the friction coefficient varies with sliding time for a nickel aluminium coating thickness of 250 microns. The image illustrates how, from the start to the 20-second sliding duration, the friction coefficient grew monotonically to roughly 0.42. It was discovered that the friction coefficient stabilised after 20 seconds and throughout the remainder of the sliding.

Figure 44.3c displays the relationship between the friction coefficient and sliding time for a nickel aluminium coating thickness of 300 microns. From the start until the 10-second sliding duration, the friction coefficient grew monotonically to roughly 0.37, as the Figure illustrates. With the exception of a slight fluctuation in the sliding interval of 10–20 seconds and 160–200 seconds, the friction coefficient was found to be stable at an approximate value of 0.4.

The average friction coefficient is determined using the data from Figure 44.3, with the exception of the friction coefficient data during the initial monotonic increase. Table 44.2 lists these calculated average coefficients of friction.

Figure 44.4 displays a visualisation of the average friction coefficient from Table 44.2.

The average friction coefficient for nickel aluminium coatings with varying coating thicknesses is displayed in Figure 44.4. The plot indicates that for nickel aluminium coating thicknesses of 200, 250 and 300 microns, the friction coefficient was 0.41, 0.41 and 0.40, respectively. It was discovered that

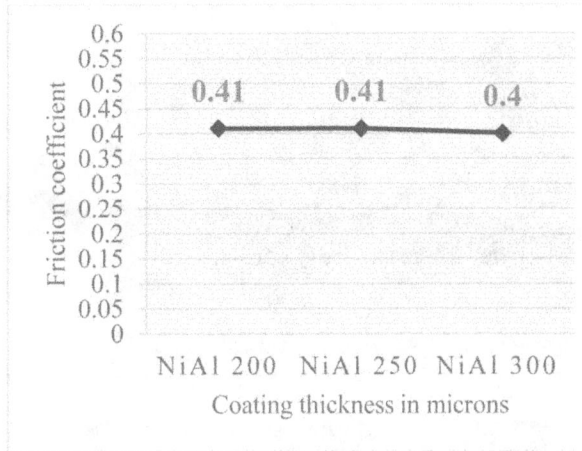

Figure 44.4 Average friction coefficient for nickel aluminium with varying coating thicknesses
Source: Author

the coating thickness had no effect on the friction coefficient.

A scanning electron micrographic analysis was performed on the damaged coating surface of the nickel aluminium coated pin in order to comprehend the process behind sliding as seen in Figure 44.5.

Micrographs presented in Figures 44.5a–c illustrate the worn pin surfaces for varying thicknesses of nickel aluminium coating, with a magnification of 200×. All the micrographs in Figures 44.5a–c show features which are identical to each other. The sliding is predominated with abrasion phenomenon. Formation of groves and ridges appeared.

To have a better understanding, micrographs at a 500× magnification have been analysed. At a 500× magnification, Figure 44.6 displays the worn-out pin surfaces for a 200 micron layer of nickel aluminium coating. At 500× magnification, all of the thickness features are comparable and the same abrasion features are visible.

Nickel chromium coating

Mild steel pins were coated with nickel chromium using a thermal spray method. The thicknesses of the coating were 200, 250, and 300 microns. Pin-on-disk test rigs have been used for experiments. A laptop computer was used to record the experiment's shear force. The friction coefficient is estimated using the observed shear force. Figure 44.7 displays the graphs illustrating the relationship between the friction coefficient and sliding time.

Table 44.2 The relationship between the average friction coefficient and the thickness of the nickel aluminium coating

Sl. No.	Coating thickness in microns	Friction coefficient
1	200	0.41
2	250	0.41
3	300	0.40

Source: Author

a)

b)

c)

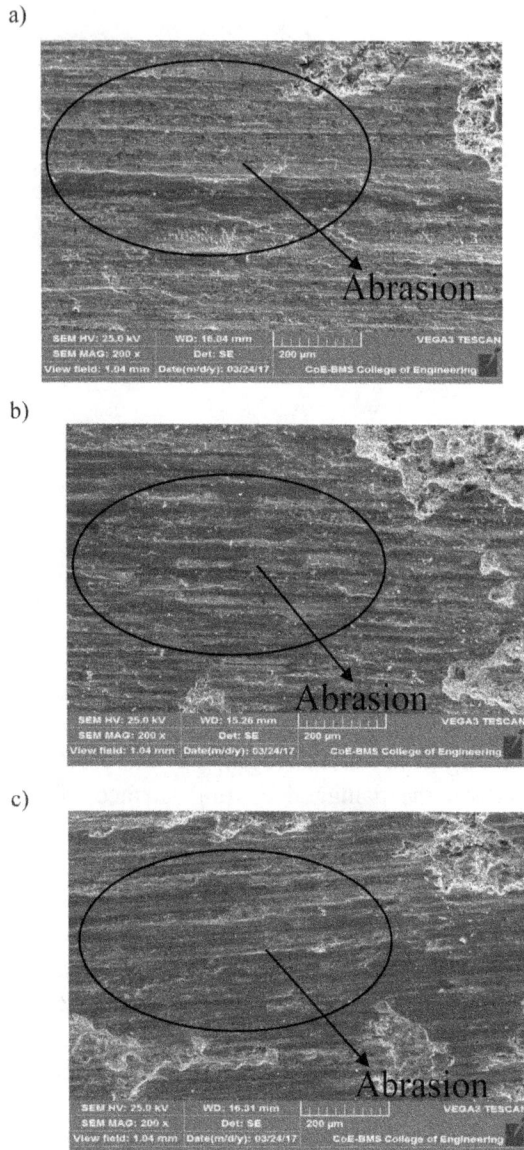

Figure 44.5 A 200× magnification micrograph showing worn-out pin surfaces at varied coating thicknesses of nickel aluminium (a) 200 microns, (b) 250 microns and (c) 300 microns

Source: Author

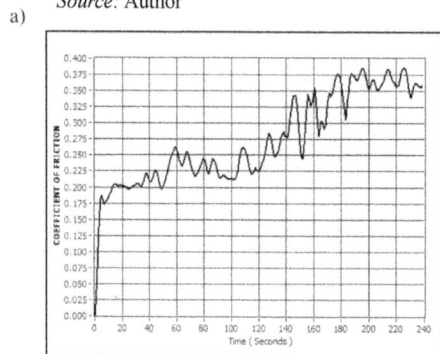

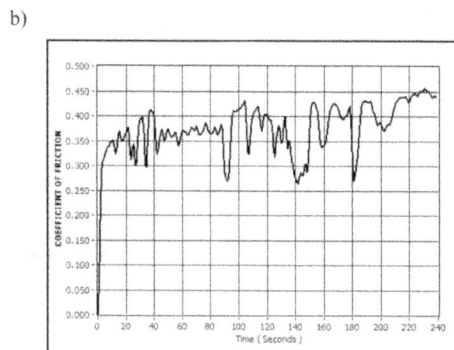

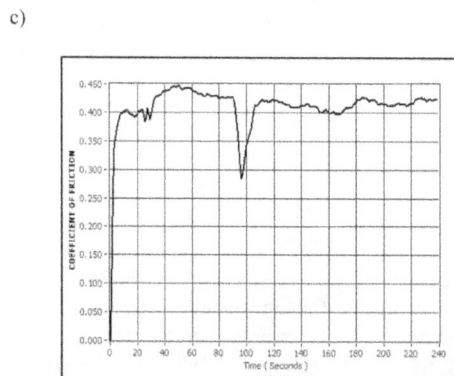

Figure 44.6 A 500× magnification micrograph showing worn-out pin surfaces at coating thicknesses of 200 microns nickel aluminium

Source: Author

a)

b)

c)

Figure 44.7 Graphs showing how the friction coefficient varies with sliding time for coating thicknesses of (a) 200 microns, (b) 250 microns and (c) 300 microns of nickel chromium

Source: Author

Figure 44.7a illustrates how the friction coefficient varies with sliding time for a nickel chromium coating layer of 200 microns. The image illustrates how, from the start to the 8-second sliding period, the friction coefficient grew monotonically to roughly 0.185. It was discovered that throughout the remaining sliding period, the friction coefficient increased.

Figure 44.7b illustrates how the friction coefficient varies with sliding time for a nickel chromium coating layer of 250 microns. From the start to the 10-second sliding time, the friction coefficient increased

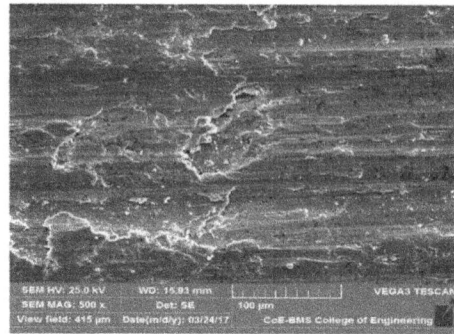

monotonically to roughly 0.35, as the Figure illustrates. After 10 seconds and throughout the remainder of the sliding, the friction coefficient was found to be roughly 0.4 with minor variations.

Figure 44.7c illustrates the relationship between the friction coefficient and sliding time for a 300 micron coating thickness of nickel chromium. The Figure begins with the friction coefficient increasing monotonically to about 0.4 until the sliding time of 10 seconds, at which point it was found to be approximately 0.42 for the remainder of the sliding, with the exception of a dip over a period of 90–105 seconds.

The data from Figure 44.7, ignoring the co-efficient of friction data during first monotonic growth, are utilised for establishing average co-efficient of friction. Such calculated average co-efficient of friction are given in Table 44.3.

Figure 44.8 displays a visualisation of the average friction coefficient from Table 44.3.

The average friction coefficient for nickel chromium coatings with varying coating thicknesses is displayed in Figure 44.8. The plot indicates that for

nickel chromium coating thicknesses of 200, 250 and 300 microns, the friction coefficient was 0.36, 0.4 and 0.42, respectively. It was discovered that as the thickness of the nickel chromium coating increased, so did the friction coefficient.

A study using a sliding scanning electron microscope was conducted on the nickel chromium coated pin's damaged coating surface in order to comprehend the mechanism involved as seen in Figure 44.9.

Table 44.3 The relationship between the average friction coefficient and the thickness of the nickel-chromium coating

Sl. No.	Coating thickness in microns	Friction coefficient
1	200	0.36
2	250	0.4
3	300	0.42

Source: Author

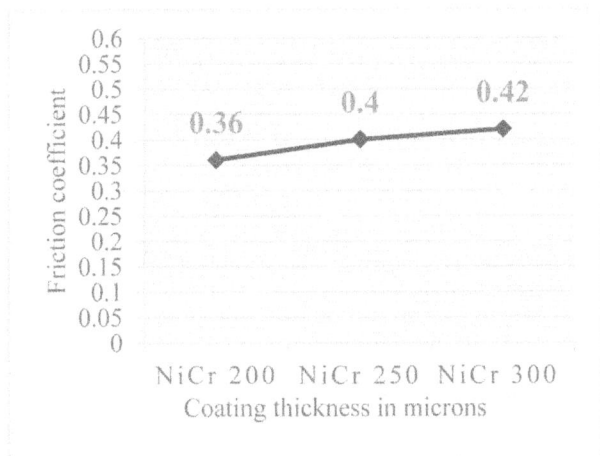

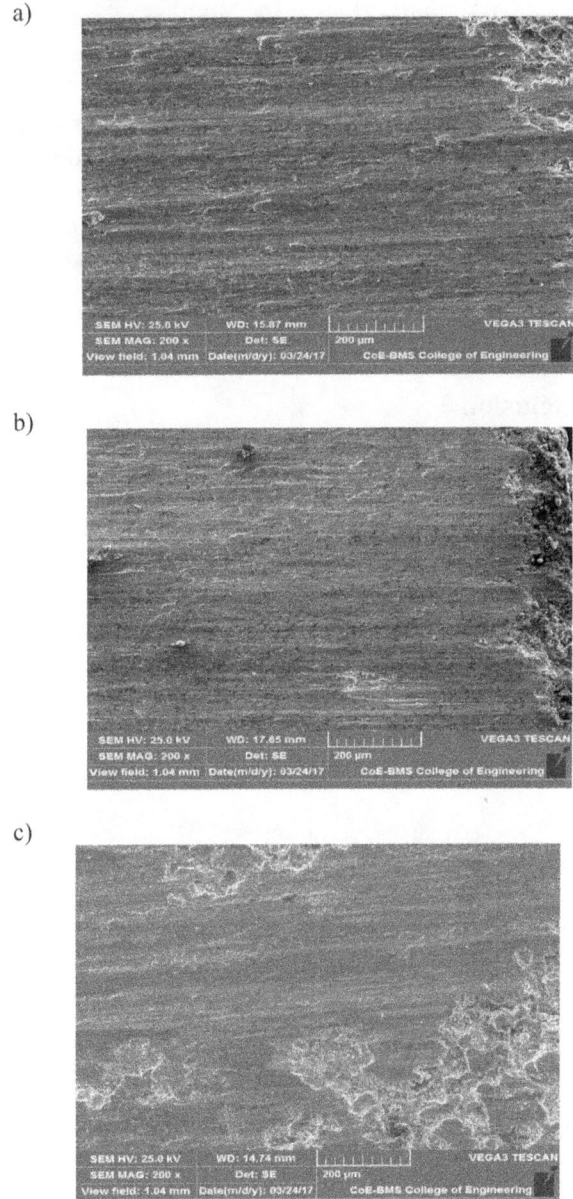

a)

b)

c)

Figure 44.8 Average friction coefficient for nickel chromium with varying coating thicknesses
Source: Author

Figure 44.9 A 200× magnification micrograph showing worn-out pin surfaces at varied coating thicknesses of nickel chromium (a) 200 microns, (b) 250 microns and (c) 300 microns
Source: Author

At a 200× magnification, the worn-out pin surfaces for various nickel chromium coating thicknesses are depicted in the micrographs in Figures 44.9a–c. The micrographs in Figures 44.9a–c all display feature that are the same. The extent of abrasion is not to that extent when compared to nickel aluminium coating.

Micrographs at a magnification of 500×, have been studied to understand better. Figure 44.10 shows the worn out pin surfaces for nickel chromium coating thickness of 250 microns at a magnification of 500×. The features at 500× magnification for all the thickness are also similar and show the failure of coating.

Though the friction coefficient for both nickel aluminium and nickel chromium is in the order of 0.4, the mechanism of sliding are different. In case of nickel aluminium coating the sliding mode is dominated by abrasion and in case of nickel chromium the sliding mode is dominated by failure of coating. The failure of nickel chromium coating is attributed its hardness when compared to softer nickel aluminium coating.

Conclusion

Nickel aluminium coating

1. The friction coefficient in case of nickel aluminium coating is found to be insensitive of coating thickness.
2. The friction coefficient was found to be approximately 0.4 for nickel aluminium coating.
3. The sliding mechanism was found to be dominated by abrasion phenomenon.

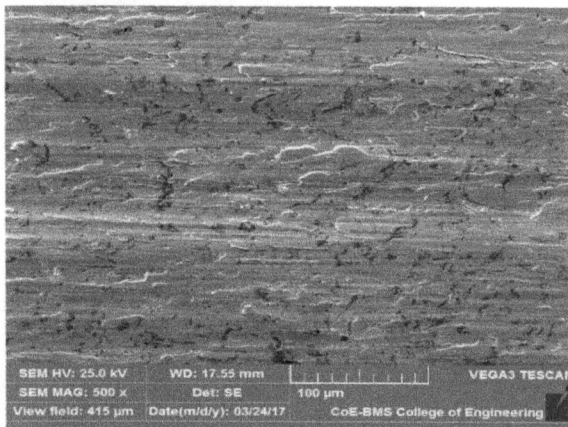

Figure 44.10 A 500× magnification micrograph showing worn-out pin surfaces at coating thicknesses of 250 microns nickel chromium
Source: Author

Nickel chromium coating

1. The friction coefficient in case of nickel chromium coating is found to be dependent on coating thickness.
2. The friction coefficient varied from 0.36 to 0.42 as coating thickness of nickel chromium coating increased.
3. The sliding mechanism was found to be dominated by failure of coating.
4. The friction coefficient was found to be of the same order for both nickel aluminium and nickel chromium coatings.

References

[1] Raheem, Z., AKareem, A. Wear and friction of polymer fibre composites coated by NiCr alloy. *Engg. Trans. Engg. Trans.*, 2017;65(2):307–318.

[2] Baricevic, M., Mravak-Stipetic, M., Stanimirovic, A., Blanusa, M., Kern, J., Loncar, B., Andabak, A., Baricevic, D. Salivary concentrations of nickel and chromium in patients with burning mouth syndrome. *Acta Dermatovenerol Croat*, 2011;19(1):2–5.

[3] Wojda, S., Szoka, B., Sajewicz, E. Tribological characteristics of enamel–dental material contacts investigated in vitro. *Acta Bioengg. Biomec.*, 2014;17(1):2015:22–29.

[4] Bhuptani, K. M., Prajapati, J. M. Friction and wear behaviour analysis of different journal bearing materials. *Internat. J. Engg. Res. Appl. (IJERA)*, 2013;3(4):2141–2146.

[5] Zhang, H., Dong, X., Chen, S. Solid particle erosion-wear behaviour of Cr3C2-NiCr coating on Ni-based superalloy. *Adv. Mec. Engg.*, 2017;9(3):1–9.

[6] Aruna, S. T., William Grips, V. K., Ezhil Selvi, V., Ranjan, K. S. Studies on electro deposited nickel-Yttria doped ceria composite coatings. *J. Appl. Electrochem.*, 2007;37:991–1000.

[7] Mohammidi, M., Ghorbani, M. Wear and corosion properties of electroless nickel composite coatings with PTFE and/or MoS2 particles. *J. Coat. Technol. Res.*, 2011;8(4):527–533.

[8] Maruyama, N., Kawasaki, H., Yamamoto, A., Hiromoto, S., Imai, H., Hanawa, T. Friction-wear properties of Nickel-free Co-Cr-Mo alloy in a simulated body fluid. *Mat. Trans.*, 2005;46(7):1588–1592.

[9] Blesman, A. I., Postnikov, D. V., Polonyankin, D. A., Teplouhov, A. A., Tyukin, A. V., Tkachenko, E. A. Molybdenum protective coatings adhesion to steel substrate. *IOP Conf. Series J. Phy. Conf. Ser.*, 2017;858:012003.

[10] Harshita, A. Wear behavior of WC-Co coating on Ti6Al4V. *IJIRCT*, 2015;1(5):473–477.

[11] Popoola, A. P. I., Fayomi, O. S. I., Papoola, O. M. Electrchemical and mechanical properties of mild steel elecro-plated with Zn-Al. *Int. J. Electrochem. Sci.*, 2012;7:4898–4917.

[12] Danisman, S., Odabas, D., Teber, M. The effect of coatings on wear behaviour of Ti6Al4V alloy used in biomedical application. *IOP Conf. Ser. Mat. Sci. Engg.*, 2018;295:012044.

[13] Endler, I., Bartsch, K., Leonhardt, A., Scheibe, H. j., Ziegle, H., Fuchs, I., Raatz, Ch. Preparation and wear behaviour of wood working tools coated with superhard layers. *Diamond Related Mat.,* 1999;8:834–839.

[14] Stuever, M., Albers, U., Leiste, H., Ulrich, S., Holleck, H., Barna, P. B., Kovacs, A., Hovsepian, P., adnI. Gee. Multi functional Ti-Al-N-C by combination of metastable fcc phases and nanocomposite micro structures. *Surface Coat. Technol.*, 2006;200:6162–6171.

[15] Sharieff, S., Ranganatha, S., Yadav, S. P. S., Nadeem Pasha, K. Role of different coating materials and coating thickness on velocity and displacement discontinuities in a tribo-system. *Internat. J. Innov. Technol. Explor. Engg. (IJITEE)*, 2019;8(6):826–836.

[16] Sharieff, S., Ranganatha, S. Role of velocity discontinuity imparted by copper and nickel coatings of different thickness. *Internat. J. Recent Technol. Engg. (IJRTE)*, 2019;8(3):6087–6093.

[17] Nadeem Pasha, K., Ranganatha, S., Sharieff, S. Sliding response of grade 5 titanium alloy at different speed and load levels. *Internat. J. Recent Technol. Engg. (IJRTE)*, 2019;8(3):4013–4018.

[18] Nadeem Pasha, K., Ranganatha, S. Wear behaviour of titanium alloys when subjected to different speed and load levels. *Internat. J. Recent Technol. Engg. (IJRTE)*, 2020;8(6):5810–5814.

45 Effect of silicon carbide reinforcement on Aluminium 7075 alloys on mechanical and vibration properties – Stir casting process

Srinidhi Acharya S. R.[a]

Assistant Professor, Department of Mechanical Engineering, BMS Institute of Technology, Bengaluru, Karnataka, India

Abstract

Modern production of automobile and aeronautical industries requires materials with greater strength, good corrosion resistance and less weight. These much-needed requirements are met when Aluminium 7075 matrix (Al7075), when mixed with non-metallic reinforcements. These metal matrix composites (MMCs) have a large use in the automotive and aerospace industries due to their excellent thermal conductivity, high strength-to-volume ratio, ease of machining, and increased resistance to wear.

The present work focuses on mechanical and vibration characterisation of Al7075 and silicon carbide particulates as the reinforcement. Al7075 is fabricated with different weight percentage (2–10 wt%) of Silicon carbide (SiC) of diameter 20–40 μm using stir casting technique. Mechanical characterisation such as tensile strength, hardness and impact tests were carried out using computerised universal testing machine, Brinell's hardness test equipment and Izod test equipment, respectively. These MMC's were observed with energy dispersive X-ray (EDAX) and scanning electron microscopy (SEM) tests.

Mechanical characteristics were enhanced as the reinforcement percentage is increased. The SEM and X-ray diffraction (XRD) tests revealed the clear distribution of reinforcement. Also, vibration analysis was performed to determine the dynamic responses of the MMC's. The specimen vibrating at greater amplitude results in resonance. This is important to analyse the frequency at resonance, damping characteristics and various mode shapes of the MMC's. Vibration analysis was characterised using signal analyser, accelerometer sensor and an impact hammer. Additionally, the answers were confirmed using ANSYS's finite element method. Test results reveal better correlation. The results clearly showed that the Al with 8% of SiC reinforcements leads to better damping characteristics.

Keywords: Composite materials, stir casting, fabrication, mechanical properties, tensile strength, hardness, vibration characteristics

Abbreviations:

Al: Aluminium, SiC: Silicon carbide, MMC: Metal matrix composites, FFT: Fast Fourier transformation, AMC: Aluminium metal matrix composites, FEM/A: Finite elemental method/analysis, FRF: Frequency response functions, SEM: Scanning electron microscopy, XRD: X-ray diffraction, DAQ: Data acquisition system, SHM: Simple harmonic motion, UTM: Universal testing machine, UTS: Ultimate tensile strength, f: Natural frequency, F_1: Experimental frequency, F_2: ANSYS simulation frequency

$$\frac{F_1 - F_2}{F_1} x\, 100 \qquad \text{Percentage error}$$

Introduction

Composite materials are composition of matrix and reinforcement which are fused at macroscopic or microscopic level. Metal matrix composites (MMC's) are obtained by scattering reinforcing material into a matrix metal. This dispersion at various levels gives rise to newer product with enhanced strength, stiffness, surface hardness and resistance to corrosion.

Aluminium 7075 matrix (Al7075) is selected as matrix phase since it possesses good fatigue strength and easy of processing. It has a density of 2.8 g/cm³. Further, adding high strength with greater modulus reinforcement to the matrix produces a new component whose mechanical properties will be enhanced. Al with reinforcements is gaining wide popularity due to its enhanced strength, good modulus of elasticity and enhanced wear resistance. [1, 2]. Al7075 and

[a]sri1660@gmail.com

DOI: 10.1201/9781003685876-45

silicon carbide (SiC) were fused through stir casting technique. X-ray diffraction (XRD) and micro structural scanning electron microscopy (SEM) examination were conducted to reveal the occurrence of SiC (2–10 wt%) in Al7075 indicating the successful stir casting process [3]. Samuel et al. [4] selected Al 6061 and activated carbon for fabrication; 2–8 wt% reinforcement is added. SEM and XRD were conducted. Gopalakannan et al. [5] studied the various uses of MMC's adopting ultrasonic results to analyse the distribution of particulates in the molten state. Boron carbide (B_4C) and SiC of sized were fabricated with Al7075. SEM and XRD observations were done. Al5083-B_4C were obtained through friction stir casting process. Reinforcement diameter ranges from nano to micro scale. Mechanical behaviour like hardness and tensile tests were performed on the casted specimen. The outcomes were validated with base Al [6, 7]. Toughness, hardness, and tensile strength were tested for Al2024 as the matrix and beryl as the reinforcement. MMC's were obtained through stir casting technique. SEM and XRD tests were conducted. The findings indicated that ultimate tensile strength (UTS) and hardness were enhanced with an increase in wt% of beryl particles up to 6% [8]. Al 6061/ TiB_2 were manufactured by in-situ liquid metallurgy method. Comparing the produced in-situ composites to the base metal revealed improved mechanical characteristics [9]. Micro level observations of Al-SiC with 5–20% by mass were made. Uniform and homogenous distribution was observed [10]. Al6061 was fabricated with TiB_2 particulates using stir casting. Wear resistance were increased with the increasing weight percentage of TiB_2 Also, other mechanical properties were enhanced [11]. Finite elemental method (FEM) is used to perform free vibration of A357 and SiC of dual particle size (3 wt% coarse and fine powder, 4 wt% coarse and 2 wt% fine). A357/DPS-SiC's impact on natural frequencies was examined while taking aspect ratio, weight fraction, and boundary conditions into account. The A357/DPS-SiC with 4 wt% coarse and 2 wt% fine SiC revealed that the maximum natural frequency which leads to better rigidity and elastic modulus [12]. Dynamic characteristics of two storied metallic component was carried using signal analyser, acceleration sensor and impact hammer. Frequency response functions (FRF's) were analysed using modal analysis experimental set up [13]. The

vibration reduction analysis for roof model of a car using FEA and experimental analysis was executed. Three-dimension car roof model was modelled using CATIA-V5 modelling software and FE analysis was performed using hyper mesh. The analysis was carried using ABAQUS Solver for various boundary conditions. Finally, the experimental modal analysis was analysed [14]. The impact of SiC as the reinforcement on Al7075 for vibration characteristics and modal analysis were studied. Al7075-SiC were fabricated via stir casting process, while Fast Fourier transformation (FFT) analyser was used for vibration characteristics and modal analysis. When the findings were compared to those of the ANSYS simulation, they showed a positive correlation, indicating that the stirring operations were successful [15].

Material selection

Aluminium is easily available in different series like 2000, 3000, etc. Al7075 is widely accepted material. Its properties like easy of availability, high strength to weight ratio, greater durability and malleability makes as one of the important requirements in automobile and aircraft industry. Table 45.1 lists the main constituent element for each series of aluminium. SiC is chosen as the reinforcement and Al7075 as the matrix. Table 45.2 displays the chemical components of Al7075.

Methodology (Figure 45.1)

Evaluating the mechanical and vibrational properties of Al7075 reinforced with various SiC fractions is the aim of this investigation. The automobile and

Table 45.1 Major constituting elements

Aluminium series	Major element
1000	99% Purest Al
2000	Cu
3000	Mn
4000	Si
5000	Mg
6000	Mg, Si
7000	Zn

Source: Author

Table 45.2 Al7075 composition (wt%)

Al7075	Composition
Zinc	5.09–6.12
Magnesium	2.81–1.99
Copper	1.42–1.56
Iron	0.49
Chromium	0.168–0.48
Manganese	<0.4

Source: Author

aerospace industries both make extensive use of this combination (Table 45.3).

Material testing

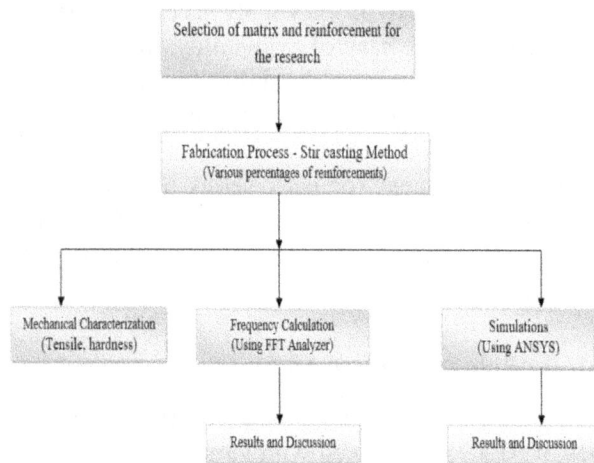

Figure 45.1 Methodology followed
Source: Author

Table 45.3 MMC's composition and its symbols

Combination of materials	Symbol
100% Al7075	1
98% Al7075+2%SiC	2
96% Al7075+4%SiC	3
94% Al7075+6%SiC	4
92% Al7075+8%SiC	5
90% Al7075+10%SiC	6

Source: Author

Microstructural studies

This refers to the analysis of the internal structure of materials at microscopic or nano scopic level. Here, the samples are checked for reinforcement distribution in matrix, behaviour, and performance under different conditions. These are analysed using

➢ **Scanning electron microscope (SEM)**

SEM is widely accepted method and helpful in analysing the materials at different scale with a magnification of up to1000× in advanced equipment. This test is usually performed under high- or low-vacuum with variable pressure at room temperature which enables to allow electron beam travel without scattering.

➢ **X-ray diffraction (XRD)**

This is a non-destructive technique used for accurate information on material's physical, chemical and crystallographic structure. These tests are performed at room temperature and normal atmospheric pressure with a scan range (2θ) from 10° to 90° depending on the material's crystal structure.

Mechanical testing

Tensile strength, hardness, and impact strength are among the mechanical characteristics that are frequently carried.

Tensile testing

This refers to the capacity of the material to withstand the deformation due to an external axial load. Figure 45.2 shows the tensile test specimen of the fabricated samples as per ASTM standards.

Hardness

The degree of indentation that occurs when a material is subjected to a load is indicated by its hardness. The most straightforward and often used technique for figuring out a material's hardness is Brinell hardness. Figure 45.3 shows the hardness test specimen as per ASTM standards.

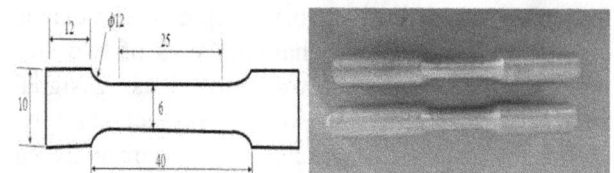

Figure 45.2 Tensile test specimen
Source: Author

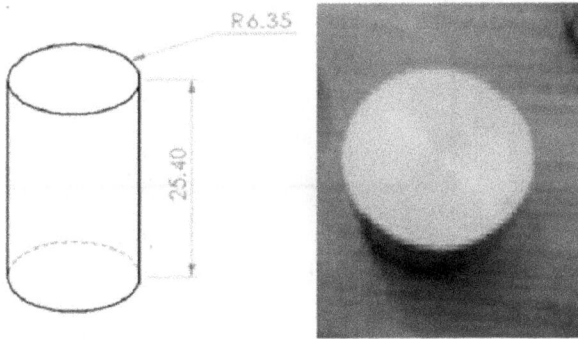

Figure 45.3 Hardness test specimen
Source: Author

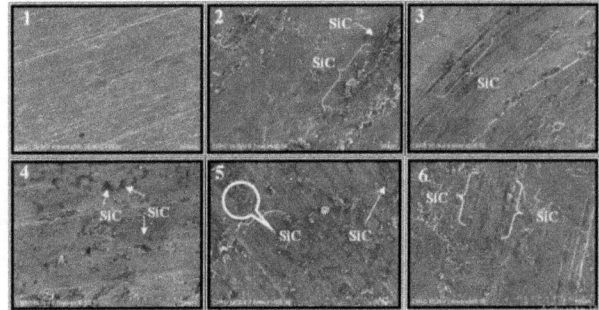

Figure 45.5 SEM images of samples (1) pure aluminium, (2,3,4,5,6) Al7075+SiC samples
Source: Author

Impact test

Impact test assesses the ability of material to withstand sudden and high impact forces indicating its toughness and resistance to fracture under dynamic loading conditions. Figure 45.4 shows the impact test specimen as per ASTM standards.

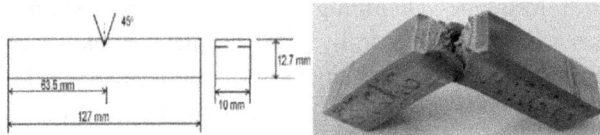

Figure 45.4 Impact test specimen
Source: Author

Vibration test: Dynamic analysis

The vibration test and analysis typically involve:

➢ Vibration exciters or modal shakers.
➢ Impact hammers
➢ Force transducers
➢ Accelerometers
➢ 8 Channel DAQ system
➢ Modal test analysis software.

Results and discussion

SEM morphology

The SEM equipment uses a secondary electron of high intensity which passes through the samples. Images with a magnification factor of 60×, 120×, 200×, and 500× are obtained. The micro-structural characteristics revealed the uniform dispersion of SiC reinforcement particles within the Al7075 which confirms proper stirring action. SEM images are shown in Figure 45.5.

Power X-ray diffraction analysis

High powered X-rays are made through the samples. Figure 45.6 shows the XRD peaks for SiC MMC and Al7075.

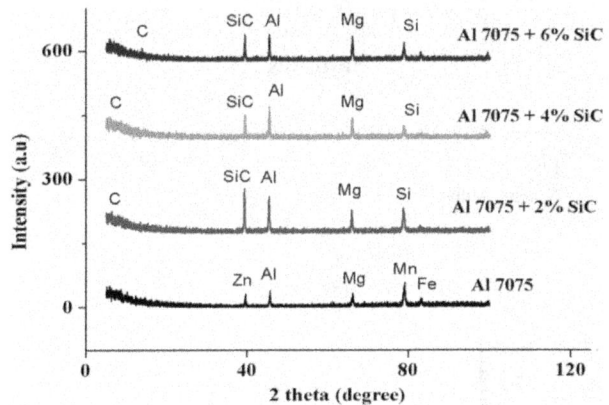

Figure 45.6 XRD curves
Source: Author

Tensile test results

A computerised UTM was used to measure UTS. The outcomes of the test are tabulated in Table 45.4 and the variations are represented in Figure 45.7. It is clear that when the weight fraction of SiC grew, so did the UTS.

Hardness test

Brinell hardness test equipment is used to perform the hardness test of the samples. The results are summarised in Table 45.4 and Figure 45.8 displays the variation curve. Since, sample 6 had more SiC reinforcement, it was determined to be tougher.

Table 45.4 Tensile, hardness and impact strength test results

Symbol	Tensile strength (Mpa)	Hardness (BHN)	Impact strength (Joules)
1	143.6	97	3.65
2	148.9	108	3.01
3	151.7	118	3.82
4	156.8	121	4.98
5	167.98	147	5.48
6	175.64	186	6.09

Source: Author

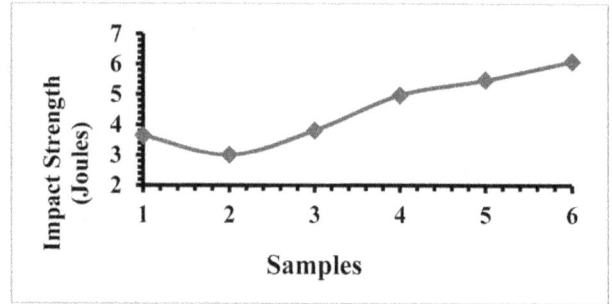

Figure 45.7 Variations of tensile strength for different samples
Source: Author

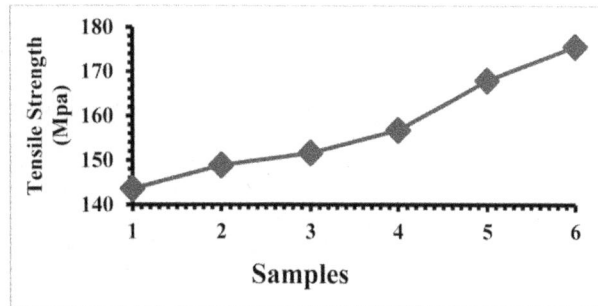

Figure 45.8 Variations of hardness for different samples
Source: Author

Impact strength

Izod Charpy test apparatus was used to perform the test. It was shown that as the weight % of SiC increased, so did the impact strength. Table 45.4 tabulates the results, while Figure 45.9 displays the strength fluctuations. It was discovered that the energy absorbed dropped at 2% SiC and subsequently rose as the weight percentage of SiC increased.

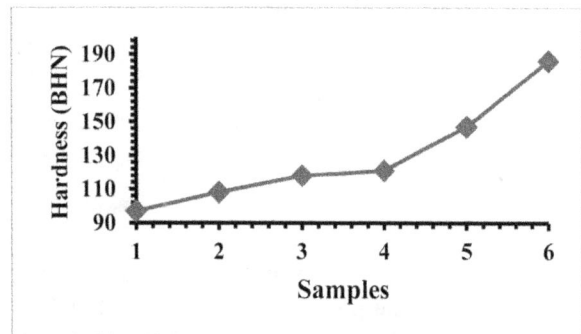

Figure 45.9 Variations of impact strength for different samples
Source: Author

Vibration analysis

The vibration characteristics of any members can be analysed with the use of analyser system. The natural frequency response of any member is divided into a set of modes. Further, from each mode natural frequency, damping values and mode shapes are identified. Generally, the sample's weaknesses are associated with resonance behaviour. During the analysis, the structure is equally divided into equal grids by drawing horizontal and vertical lines. The intersection points are called as nodes. Each node is numbered as (node 1, node 2, node 3...... last node 30). The structure is fixed at one end to rigid frame and other end being free. The vibration analysis is carried out using USB based 8 channel data acquisition voltage system unit developed by Dewesoft. DAQ voltage system collects the signals, analyses and gives the output. A uni-directional impact test hammer with a scaling factor of 22.7 mV/N attached with an acceleration sensor is used for excitation.

During analysis the acceleration sensor is fixed on last node on the MMC using a petroleum wax. An excitation is provided to MMC structure using impact hammer on all nodes to measure the responses. The results obtained are then imported into vibration analysis software and are processed for frequency response curve. The same is validated using ANSYS simulation software. The experimental values (F_1) and simulation values (F_2) are tabulated in Table 45.5. The modal behaviour is shown in Figures 45.10–45.13. Further, a comparison graph is drawn for initial two mode shapes for experimentation and ANSYS values and is shown in Figure 45.14. From the experimental and simulation, it was clear that the natural frequencies were in par with each other.

Table 45.5 Experimental and ANSYS results of Al-SiC specimens

Sample	Mode shapes	F_1	F_2	$|(F_1-F_2)|$	Percentage error
1	1	5.081	5.022	0.059	1.161
	2	16.03	17.76	1.73	10.79
2	1	5.68	5.22	0.12	2.11
	2	17.486	18.486	1	5.71
3	1	5.71	5.88	0.17	2.97
	2	21.2	21.6	0.43	2.03
4	1	5.89	5.95	0.06	1.15
	2	21.7	21.9	0.22	1.02
5	1	5.99	6.028	0.03	0.50
	2	22.0	22.20	0.11	0.50
6	1	6.09	6.1	0.01	0.16
	2	22.1	22.47	0.295	1.33

Source: Author

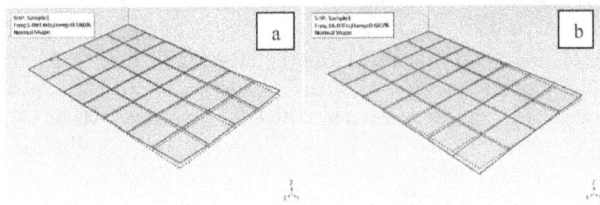

Figure 45.10 Experimental mode shapes for sample 1 – (a) Mode 1 (b) Mode 2
Source: Author

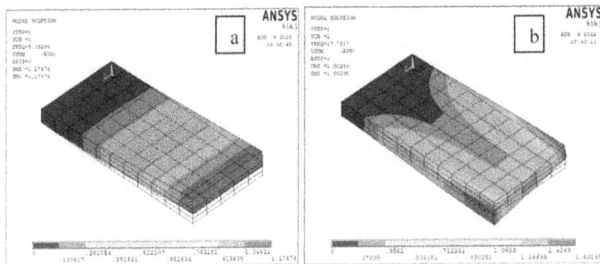

Figure 45.11 ANSYS mode shapes for sample 1 – (a) Mode 1 (b) Mode 2
Source: Author

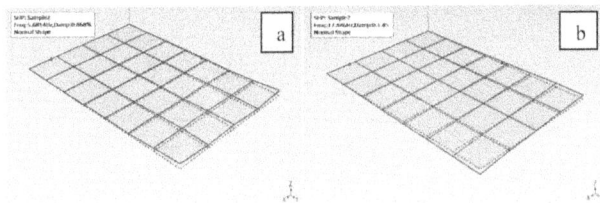

Figure 45.12 Experimental mode shapes for sample 1 – (a) Mode 1 (b) Mode
Source: Author

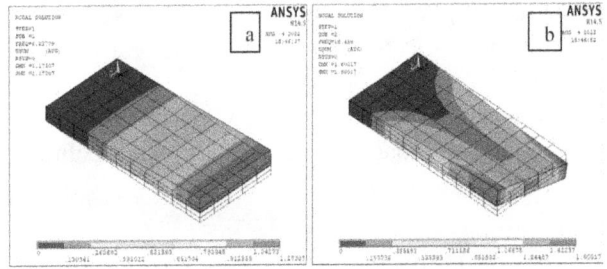

Figure 45.13 ANSYS mode shapes for sample 1 – (a) Mode 1 (b) Mode 2
Source: Author

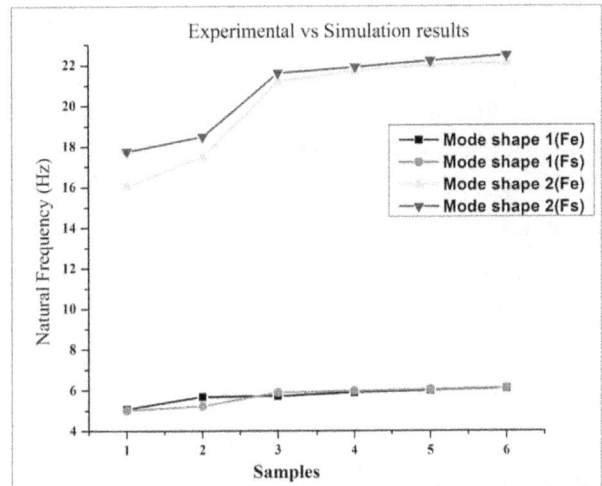

Figure 45.14 Comparison of experimental results vs. ANSYS results
Source: Author

Conclusion

1. The homogeneous and uniform dispersion of SiC, as seen by SEM, indicated that the stir casting was successful.
2. The presence of SiC particles and traces of zinc, manganese, magnesium are shown by XRD peaks.
3. Mechanical properties like tensile strength, hardness and energy absorption capacity of the Al SiC was improved from the tests.
4. Comparing sample 6 to sample 1, it was discovered that the MMC's hardness had risen by 47%.
5. Using experimentation and ANSYS simulation, the vibration characteristics of Al-SiC MMC's was analysed and verified.

6. From Figure 45.14, it is clear that experimental and ANSYS values are in close relation with each other.
7. The effective fabrication of MMCs using the stir casting approach is indicated by the declining percentage error.

Acknowledgement

The author is very much grateful to the college management, Head of the Institution, Head of the Department or their continuous support during the course of experimentation and paper preparation.

Conflict of interest

The author would like to tell that there is no known conflict of interest during the fabrication, experimentation and preparation of this work/manuscript.

References

[1] Lakshmi, S., Lu, L., Gupta, M. In situ preparation of TiB2 reinforced Al based composites. *J. Mater. Process Tech.*, 1998;73(1–3):160–166.

[2] Natarajan, S., Naraynasamy, R., Kumaresh Babu, S. P., Dinesh, G., Anil Kumar, B., Sivaprasad, K. Sliding wear behaviour of Al 6063/TiB2 in situ composites at elevated temperatures. *Mater. Design*, 2009;30(7):2521–2531.

[3] Srinidhi Acharya, S. R., Suresh, P. M. Fabrication and micro structural characterization of Al 7075 reinforced with various proportions of SiC. *Mater. Today Proc. 2021*, 2022;49(3):638–643.

[4] Diju Samuel, G., Edwin Raja Dhas, J., Ramanan, G., Ramachandran, M. Production and microstructure characterization of AA6061 matrix activated carbon particulate reinforced composite by friction stir casting method. "RASAYAN Journal of Chemistry". 2017;10(3):784–789.

[5] Gopalakannan, S., Senthilvelan, T. Synthesis and characterization of Al 7075 reinforced with SiC and B4C nano particles fabricated by ultrasonic cavitation method. *J. Sci. Indus. Res.*, 2015;74:281–285.

[6] Yuvaraj, N., Vipina, S. A. Fabrication of Al5083/B4C surface composite by friction stir processing and its tribological characterization. *J. Mater. Res. Technol.*, 2015;4(4):398–410.

[7] Llyod, D. J., Lagace, H., Mcleod, A., Morris, P. L. Microstructural aspects of aluminium silicon carbide particulate composites produced by a casting method. *Mater. Sci. Engg.*, 1989;107:73–80.

[8] Sagar, K. G., Suresh, P. M., Nataraj, J. R. Effect of Beryl reinforcement in aluminum 2024 on mechanical properties. *J. Inst. Engrs. (India) Series C*, 2020;101(3): 507–516.

[9] Ramesh, C. S., Pramod, S., Keshavamurthy, R. A study on microstructure and mechanical properties of Al 6061-TiB2 in-situ composites. *Mater. Sci. Engg. A.*, 2011;528: 4125–4132.

[10] Neelima Devi, C., Selvaraj, N., Mahesh, V. Micro structural aspects of aluminium silicon carbide metal matrix composite. *Internat. J. Appl. Sci. Engg. Res.*, 2012;2: 250–254.

[11] Suresh, S., Shenbaga Vinayag Moorthi, N. Process development in stir casting and investigation on microstructures and wear behaviour of TiB2 on Al6061 MMC. *Internat. Conf. Design Manufac. (IConDM2013)*, 2013;64: 1183–1190.

[12] Lakshmikanthan, A., Mahesh, V., Prabhu, R. T., Patel, M. G. C., Bontha, S. Free vibration analysis of A357 alloy reinforced with dual particle size silicon carbide metal matrix composite plates using finite element method. *Arch. Foundry Engg.*, 2021;21(1):101–112.

[13] Chandravanshi, M. L., Mukhopadhyay, A. K. Modal analysis of structural vibration. *Proc. Internat. Mec. Engg. Cong. Expos.*, 2013;21:15. DOI:10.1115/IMECE2013-62533

[14] Chandru, B. T., Suresh, P. M. Finite element and experimental modal analysis of car roof with and without damper. *Mater. Today Proc.*, 2017;4:11237–11244.

[15] Srinidhi Acharya, S. R., Suresh, P. M., Mruthunjaya, M. Influence of silicon carbide particulates on vibrational characteristics for Al 7075 composites. *J. Inst. Engg. India Ser. D*, 2023;104:651–660.

46 Effect of exhaust gas recirculation on emission and performance characteristics of CRDI dual fuel engine powered with Peruvenia Thevetia and hydrogen induction

Mahantesh Marikatti[1], N. R. Banapurmath[2,a], Y. H. Basavarajappa[3], Shambhuling Uppin[4], V. S. Yaliwal[1], Kartheek Ravulapati[5], K. S. Nivedhitha[2] and A. M. Sajjan[2]

[1]Department of Mechanical Engineering, SDM College of Engineering and Technology, Dharwad, Karnataka, India

[2]Centre for Material Science, Department of Mechanical Engineering, KLE Technological University, Hubballi, Karnataka, India

[3]Department of Mechanical Engineering, Visvesvaraya Technological University, Belagavi–590018, Karnataka, India

[4]Department of Mechanical Engineering, Government polytechnic, Hubballi–580021, Karnataka, India

[5]Collins Aerospace, 5935, Pinnacle View Road, Cumming GA 30040, United States

Abstract

Peruvenia Thevetia biodiesel (PTB) and hydrogen fuel combinations are used in the current initiative to investigate the impact of exhaust gas recirculation (EGR) on the performance of common rail direct injection assisted dual fuel engines (DFE). Using a seven-hole common rail direct injection (CRDI) injector, PTB is injected into the engine cylinder at a higher injection pressure of 900 bar and an injection timing of 10° BTDC. For better dual fuel engine performance with increased brake thermal efficiency and lower emissions, these are the engine characteristics that have been optimised. Diesel's primary disadvantages are its increased NOx and smoke emissions, which can be mitigated by running the engine in dual fuel (DF) mode. However, combining hydrogen-biodiesel in DFE raises NOx emissions, which can be reduced by using various EGR, water injection, and retarding injection time approaches. EGR is one of these techniques that effectively reduce NOx. The percentage of EGR is adjusted in stages from 5% to 20% in order to examine the impact of EGR on DFE performance. While emissions of smoke, HC, and CO rose as the proportion of EGR increased, BTE and NOx decreased.

Keywords: Peruvenia Thevetia biodiesel, hydrogen, manifold injection, emissions, dual fuel engine

Introduction

Diesel engines have a higher compression ratio than petrol engines, which contributes to their relative thermal efficiency. They are used in both automotive and agricultural applications. Fossil fuel burning in internal combustion engines degrades the environment [1]. The natural depletion of these conventional fuels exacerbates worries about the significant demand on foreign exchange. By using renewable fuels in ICEs, the nation becomes self-sufficient in terms of a steady supply of energy. In addition to providing energy sustainability and being environmentally benign, renewable fuels also effectively alleviate the load of foreign exchange and pollution in the environment. Such renewable fuels include both gaseous and liquid fuels consisting of hydrogen, producer gas, biogas, and biodiesels respectively. By running diesel engines in dual fuel mode, higher emissions of smoke, particulate matter and NOx can be appropriately addressed. Because of its higher specific energy density, wider flammability limitations, and higher flame velocity, hydrogen is a promising renewable fuel among gaseous fuels. Among pilot liquid fuels, biodiesels can offer a renewable substitute for diesel. All types of biodiesel, however, have been shown to have increased fuel consumption [2, 3]. A small quantity of pilot liquid fuel is injected to start combustion of compressed air-gas in a dual fuel engines (DFE), which runs on both liquid and gaseous fuels. Numerous researchers have used producer petrol, hydrogen and biodiesel fuel mixes to examine the performance of dual fuel engines [4]. Techniques of hydrogen utilisation in dual fuel engine its merits, demerits and combustion has been discussed. Dual fuel engines are more suited for medium and higher loads of operation [5]. The literature discusses

[a]nrbanapurmath@gmail.com

DOI: 10.1201/9781003685876-46

the effects of optimised injection timing and duration parameters on dual fuel engine performance, which leads to improved BTE and lower NOx and smoke emissions [6]. Biodiesel being more viscous than diesel they need to be injected at higher pressures in CRDI diesel dual fuel engines. The literature presents effects of split injection of liquid fuel on performance of diesel-synthetic producing gas. With lower HC and CO emissions and higher NOx, split fuel injection timing at optimal injection pressure led to notable increases in combustion efficiency [7]. The literature reports on the effects of split injection, injection pressure, and injection time on the performance of hydrogen-fuelled engines. Using CONVERGE software, additional air-hydrogen mixing was examined from perspective of mass transfer and flow state. Enhanced thermal efficiency and improved combustion is observed at an injection timing of − 43° aTDC has been reported. Lower efficacy on air-hydrogen mixture formation and combustion with enhanced injection pressure is reported. Lesser uniformity coefficient and increased knocking tendency with a larger mass fraction of secondary injection has been observed. Increased brake thermal efficiency by 3.16% with hydrogen operation for stable and constant hydrogen flow without throttling is reported as it has wide range of flammability [9, 10]. The impact of hydrogen induction in the inlet manifold on hydrogen-powered dual fuel engines and waste cooking oil biodiesel blends was examined. The range of hydrogen flow rates was 2–6 lpm. Improved brake thermal efficiency (BTE), in-cylinder pressure, and heat release rate are achieved with higher hydrogen flow rates [11]. Back fire, knocking, and NOx emission levels from hydrogen fuelled engine were studied [12]. A DFE combustion and tailpipe exhaust emissions are examined. There have been reports of elevated exhaust gas temperature (EGT) and NOx levels with hydrogen loading of 0.80 lpm and higher. Except for NOx emissions, engine operating with hydrogen addition demonstrated lower carbon-based emissions levels. There have been reports of increased in-cylinder pressure and heat release rate at high load and with increased hydrogen addition [13, 14]. Different techniques of NOx control in hydrogen fuelled engines such as water injection, retarded injection timing, three-way catalytic converter and exhaust gas recirculation (EGR) has been proposed in the literature. Among these EGR method is found to be suitable for controlling the NOx emissions. EGR percentage was varied from 5% to 20%.

In this context, experiments are done to assess emission characteristics and performance of CRDI DFE running on hydrogen injection and Peruvenia Thevetia biodiesel (PTB). Impact of EGR on dual fuel performance and emission characteristics is examined. The outcomes are then contrasted with the baseline diesel-hydrogen operation.

Experimental engine test rig

This section describes the fuels and the specifics of the CRDI dual fuel engine test setup, as well as the data collection equipment.

Properties of liquid and gaseous fuels

Figure 46.1 shows diesel and PTB biodiesel used as secondary injected pilot fuels. Biodiesel from Peruvenia Thevetia oil is produced from trans esterification method. Table 46.1 shows the properties of the fuel evaluated in the engine laboratory.

CRDI dual fuel engine test rig

A 1-cylinder CI diesel engine with cylinder capacity of 662 cc, a compression ratio of 17.5, and a 5.2 kW output at 1500 rpm was used for the study's examination. The diagram of the CRDI DFE test equipment is depicted in Figure 46.2.

Figure 46.1 Fuels used in the study (gasoline, diesel, PTB20 and PTB100)
Source: Author

Table 46.1 Material properties of epoxy and SiO_2

Fuel properties	Diesel	PTB 100
Density (kg/m³)	840	892
Kinematic viscosity at 40°C (mm²/s)	2–3	5.2
Flash point (°C)	53	178
Calorific value (MJ/kg)	43	39.46

Source: Author

Figure 46.2 CRDI dual fuel engines equipped with hydrogen injection
Source: Author

Hydrogen injector is used to inject hydrogen at a pressure of 2–3 bar. Figure 46.3 shows the hydrogen injector connected to CRDI DFE test rig.

Figure 46.3 Hydrogen injector connected to dual fuel engine
Source: Author

AVL gas analyser five-gas analyser is used to measure HC, CO, NOx emissions, while smoke meter is used to measure smoke opacity (Figures 46.4 and 46.5).

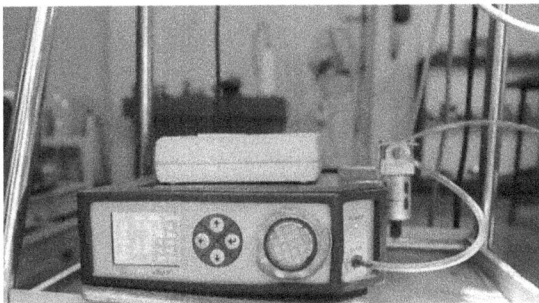

Figure 46.4 AVL gas analyser
Source: Author

Figure 46.5 Smoke meter
Source: Author

Results and discussion

The impact of EGR on performance of the CRDI DFE with PTB and hydrogen injection is covered in this section. Flow rate of hydrogen is kept constant at 0.15 kg/h. Diesel and PTB are injected using a 7-hole injector with a 0.1 mm aperture at $10°$ BTDC and 900 bar. The study made use of a toroidal re-entrant combustion chamber.

Effect of EGR on performance of PTB-hydrogen CRDI dual fuel engine

Figure 46.6 illustrates how EGR affects the thermal efficiency of diesel/PTB-hydrogen dual fuel operation. BTE of PTB-hydrogen was lower than that of diesel-hydrogen. The BTE is impacted by hydrogen, a frequent feature of the injected fuels. Figure makes it clear that lower BTE was associated with higher EGR rates. This might be explained by the fact that when burned gasses mixed with hydrogen-air-PTB, the mixture became diluted, which raises the levels of HC and CO emissions. The outcomes shown are likewise caused by abrupt changes in a charge's specific heat during compression and expansion. When the charge's EGR lowers the incoming charge temperature, a BTE reduction is seen. The aforementioned variables significantly alter the combustion characteristics, as does the thermal impact of EGR. Additionally, an equivalent amount of intake air is replaced when some exhaust gas is allowed to enter the engine through the intake manifold. This reduces the air-fuel ratio and chemical kinetics, which negatively affects dual fuel engine combustion. Due to the reduced volumetric efficiency, these results also induce a fall in BTE.

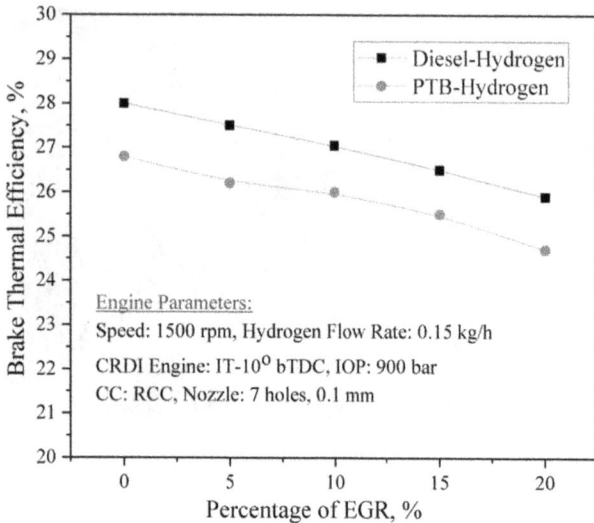

Figure 46.6 Impact of EGR on BTE
Source: Author

Effect of EGR on the emissions of PTB-hydrogen CRDI dual fuel engine

Smoke opacity

PTB-hydrogen produced more smoke emissions than the diesel-hydrogen combo at the same EGR rates. Smoke emissions rose in tandem with the EGR rate. Figure 46.7 shows how EGR affects the smoke opacity for diesel/PTB-hydrogen injection. Adding EGR to the hydrogen injection increases smoke opacity. Exhaust gas significantly lowers air pressure proportionately, which lowers oxidation rates. The shift in smoke emission creation is greater in operations with higher EGR rates (15% and 20%) than in those with lower EGR percentages (5% and 10%). Increased smoke opacity was result of a decrease in air-fuel ratio caused by an increase in the EGR percentage. It has been observed, meanwhile, that EGR negatively affects smoke levels.

Effect of EGR on carbon-based emissions for diesel/PTB-Hydrogen injection is depicted in Figures 46.8 and 46.9. The quality of combustion can be assessed by looking for HC and CO in the exhaust. A high percentage of these pollutants blatantly demonstrate incomplete fuel combustion. When EGR is added, the air-fuel ratio is further lowered, which dilutes and weakens the mixture as a result of the oxygen content falling. This lowers the temperature at which combustion occurs, increasing the amount of CO and HC released into the exhaust. In comparison to greater EGR rates (15% and 20%), HC and CO levels vary slightly higher compared to lower % of exhaust gas (5% and 10%) introduced. Significant variations in the air-fuel ratio are observed with greater EGR rates, leading to incomplete combustion.

Impact of EGR on nitric oxide (NOx) emissions during PBT-hydrogen operation is shown in Figure 46.10. A lower combustion temperature may result from a drop in oxygen concentration during combustion when using hydrogen injection method with EGR. As a result, the exhaust's NOx levels drop. When using dual fuel at high EGR rates (15–20%) as opposed to low EGR rates (5–10%), NOx levels were reduced. Another explanation for the observed

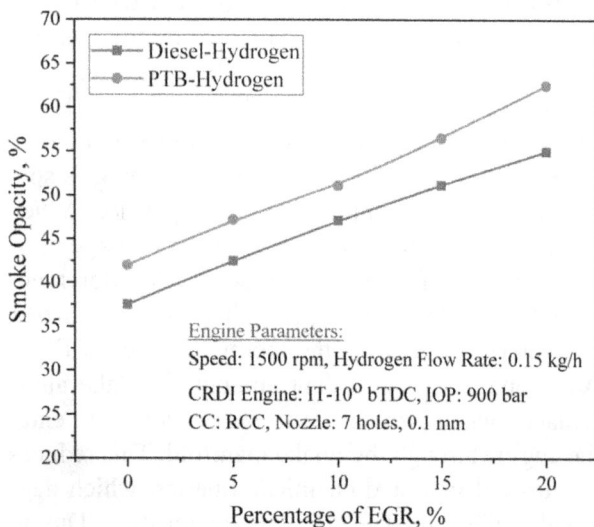

Figure 46.7 Effect of EGR on the smoke opacity
Source: Author

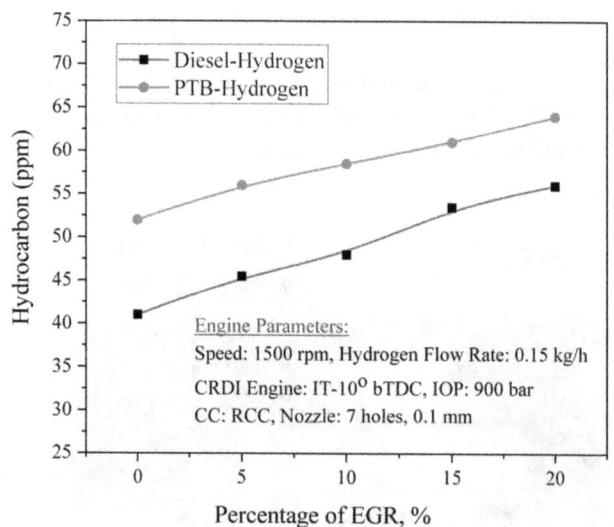

Figure 46.8 Effect of EGR on HC emissions
Source: Author

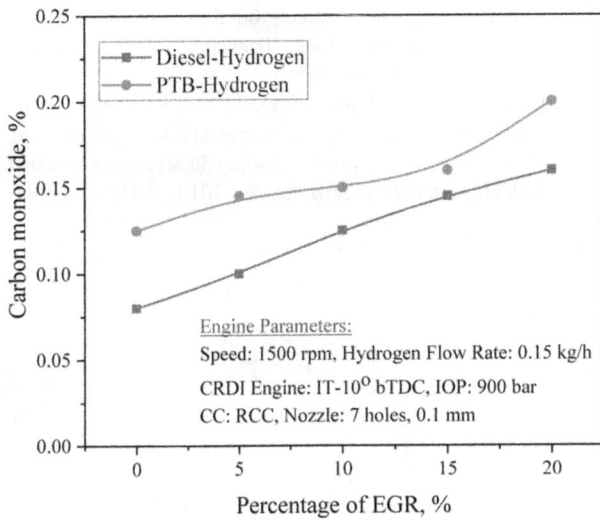

Figure 46.9 Impact of EGR on CO emissions

Source: Author

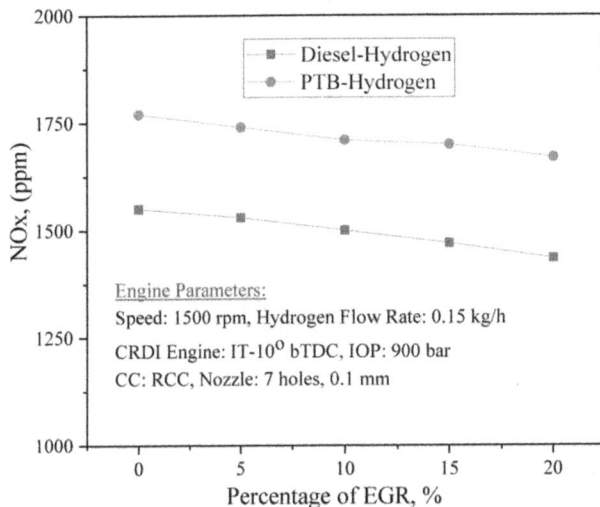

Figure 46.10 Impact of EGR on NOx emissions

Source: Author

trend is a lower combustion temperature, which lowers NOx emissions by lowering the adiabatic flame temperature. However, because of the enhanced mixture dilution and oxygen deficit, NOx levels were reduced for 15% EGR rates and higher.

Conclusion

The findings regarding the impact of EGR on PTB-hydrogen-fuelled CRDI dual fuel engines are presented in this section. The experimental investigations lead to the following conclusions:

➤ The PTB's characteristics are similar to those of diesel fuel. The PTB-hydrogen fuel mixture used in the CRDI dual fuel engine performed satisfactorily.

➤ In comparison to a PTB-hydrogen combination, a diesel-hydrogen DFE demonstrated lower emissions of smoke, HC, and CO and greater emissions of BTE and NOx.

➤ Smoke, HC, and CO emissions rose as EGR % rose, whereas BTE and NOx emissions fell.

➤ EGR is a beneficial technique for reducing increased NOx emissions from diesel engines. As a result, 15% EGR produced better BTE and manageable emission levels.

References

[1] Soudagar, M. E. M. et al. The effects of graphene oxide nanoparticle additive stably dispersed in dairy scum oil biodiesel-diesel fuel blend on CI engine: performance, emission and combustion characteristics. *Fuel*, 2019;257:116015.

[2] Banapurmath, N. R., Tewari, P. G., Yaliwal, V. S., Kambalimath, S., Basavarajappa, Y. H. Combustion characteristics of a 4-stroke CI engine operated on Honge oil, neem and rice bran oils when directly injected and dual fuelled with producer gas induction. *Renew. Energy*, 2009;34(7):1877–1884.

[3] Yildiz, I., Açıkkalp, E., Caliskan, H., Mori, K. Environmental pollution cost analyses of biodiesel and diesel fuels for a diesel engine. *J. Environ. Manag.*, 2019;2431: 218–226.

[4] Halewadimath, S. S., Yaliwal, V. S., Banapurmath, N. R., Sajjan, A. M. Influence of hydrogen enriched producer gas (HPG) on the combustion characteristics of a CRDI diesel engine operated on dual-fuel mode using renewable and sustainable fuels. *Fuel*, 2020;270:117575.

[5] Papagiannakis, R., Hountalas, D. Combustion and exhaust emission characteristics of a dual fuel compression ignition engine operated with pilot diesel fuel and natural gas. *Energy Conver. Manag.*, 2004;45(18–19):2971–2987.

[6] Saravanan, N., Nagarajan, G. Performance and emission study in manifold hydrogen injection with diesel as an ignition source for different start of injection. *Renew. Energy*, 2009;34(1):328–334.

[7] Carlucci, A. P., Ficarella, A., Laforgia, D., Strafella, L. Improvement of dual-fuel biodiesel-producer gas engine performance acting on biodiesel injection parameters and strategy. *Fuel*, 2017;209:754–768.

[8] Li, Y., Gao, W., Zhang, P., Ye, Y., Wei, Z. Effects study of injection strategies on hydrogen-air formation and performance of hydrogen direct injection internal combustion engine. *Internat. J. Hyd. Energy*, 2019;44(47):26000–26011.

[9] Kitagawa, T., Kido, H., Nakamura, N., Aishima, M. Flame inertia into lean region in stratified hydrogen mixture. *Internat. J. Hyd. Energy*, 2005;30(13–14):1457–1464.

[10] Toshiaki, K., Hiroyuki, K., Nozomu, N. Flame propagation into lean region in stratified hydrogen mixture. *Proc. 15th World Hyd. Energy Conf. Yokohama,* 27 June - 2 July 2004.

[11] Raju, P., Senthil Kumar, M., Nataraj, G., Karthic, S. V. Engine's behavior on hydrogen addition of waste cooking oil fueled light duty diesel engine - A dual fuel approach. *Energy,* 2019;1941:116844.

[12] Pani, A. Formation, kinetics and control strategies of Nox emission in hydrogen fueled IC engine. *Internat. J. Engg. Res. Technol. (IJERT),* 2020;9(01):132–141.

[13] Koten, H. Hydrogen effects on the diesel engine performance and emissions. *Internat. J. Hyd. Energy,* 2018;43(2231):10511–10519.

[14] Liew, C., Li, H., Nuszkowski, J., Liu, S., Gatts, T., Atkinson, R., Clark, N. An experimental investigation of the combustion process of a heavy-duty diesel engine enriched with H2. *Internat. J. Hyd. Energy,* 2010;35:11357–11365.

47 Advanced time series forecasting for vehicle maintenance scheduling

Praveen N.[1], Girish T. R.[2,a], Ranganath N.[3] and Harish U.[4]

[1]Senior Faculty in Electrical Engineering Department University of Technology and Applied sciences Ibra Sultanate of Oman

[2]Professor and Head, Department Mechanical Engineering, K. S. Institute of Technology Bangalore, Karnataka, India

[3]Infosys Ltd, Mysuru, Karnataka, India

[4]Department Mechanical Engineering, K. S. Institute of Technology Bangalore, Karnataka, India

Abstract

Predictive maintenance plays a significant role in minimising vehicle downtime and optimising fuel consumption. Traditional maintenance strategies rely on pre-defined schedules or reactive approaches, leading to inefficiencies and increased costs. This research leverages machine learning models, including autoregressive integrated moving average (ARIMA), long short-term memory (LSTM), Prophet, XGBoost, and random forest, to develop predictive models that enhance vehicle maintenance scheduling and fuel optimisation. The study involves data pre-processing, feature engineering, model selection, and hyperparameter tuning to achieve high predictive accuracy. Results indicate that deep learning-based models, particularly LSTM, outperform traditional statistical methods in forecasting maintenance requirements, leading to improved fuel efficiency and reduced operational costs.

Keywords: Predictive maintenance, fuel optimisation, machine learning, time series forecasting, XGBoost, LSTM

Introduction

The growing demand for fuel-efficient and reliable vehicles has intensified the need for advanced maintenance strategies in the automotive industry [1]. Traditional maintenance approaches, such as reactive and preventive maintenance, often lead to operational inefficiencies, increased costs, and unexpected vehicle failures. Reactive maintenance addresses issues only after a breakdown occurs, resulting in unplanned downtime, while preventive maintenance follows a fixed schedule, sometimes leading to unnecessary servicing and resource wastage [1].

Predictive maintenance, powered by machine learning (ML) and artificial intelligence (AI), has emerged as an effective solution to optimise vehicle performance and fuel consumption [2]. By analysing historical maintenance records, vehicle telemetry data, and real-time sensor readings, predictive models can identify potential failures before they occur [3]. This data-driven approach not only reduces unexpected breakdowns but also ensures optimal fuel utilisation, ultimately enhancing vehicle longevity and reducing environmental impact [4].

Recent advancements in ML, particularly in time series forecasting and classification, have improved the accuracy and reliability of predictive maintenance models [5]. Techniques such as long short-term memory (LSTM) networks, autoregressive integrated moving average (ARIMA), and Prophet have shown significant potential in forecasting maintenance schedules [5]. Additionally, ensemble learning methods like XGBoost and random forest have been successful in classifying maintenance needs based on multiple vehicle parameters [6]. These algorithms help in identifying patterns that influence fuel efficiency and vehicle reliability. This paper presents a comprehensive study on predictive maintenance and fuel optimisation using ML models [7]. The research explores the effectiveness of ARIMA, LSTM, Prophet, XG Boost, and random forest in predicting vehicle maintenance schedules and optimising fuel consumption [8]. The study includes data pre-processing, feature engineering, model selection, and hyperparameter tuning to develop an optimised predictive maintenance framework [9].

The objectives of this research are:

1) To analyse historical vehicle maintenance data and fuel consumption trends.
2) To evaluate machine learning techniques for predicting maintenance requirements.
3) To optimise fuel consumption through data-driven maintenance scheduling.

[a]girishshastrytr@ksit.edu.in

DOI: 10.1201/9781003685876-47

4) To compare the performance of time series forecasting and classification models for predictive maintenance.

By leveraging ML-based predictive analytics, this study aims to provide an efficient approach for reducing vehicle downtime, minimising maintenance costs, and improving fuel efficiency. The findings from this research can be applied to fleet management systems, electric vehicle maintenance, and advanced driver assistance systems (ADAS) to further enhance vehicle performance and sustainability.

Literature review

Predictive maintenance has become a widely researched topic in the automotive and industrial sectors due to its ability to reduce down-time, optimise operational efficiency, and lower maintenance costs [1]. Various studies have explored different ML models and time series forecasting techniques to improve maintenance scheduling and fuel consumption optimisation.

Traditional maintenance approaches

Maintenance strategies in the automotive industry have traditionally been classified into three categories: **reactive, preventive, and predictive maintenance [2]**. Reactive maintenance is performed only after a failure occurs, leading to increased downtime and repair costs. Preventive maintenance follows a scheduled approach, performing maintenance at predefined intervals regardless of the vehicle's condition [3]. Although preventive maintenance reduces failure risks, it often leads to unnecessary servicing, increasing operational costs. Predictive maintenance, on the other hand, leverages **sensor data and machine learning algorithms** to anticipate failures before they occur, optimising maintenance schedules and improving vehicle performance [3].

Machine learning for predictive maintenance

Recent advancements in supervised learning, unsupervised learning, and deep learning have made predictive maintenance more accurate and efficient. Studies have demonstrated that classification models such as random forest, XGBoost, and decision trees are highly effective in identifying maintenance needs [4]. These models analyse vehicle telemetry data, maintenance history, and fuel efficiency metrics to predict potential failures.

Research conducted by Schoen et al. (2015) showed that random forest models could classify vehicle maintenance conditions with high accuracy, reducing false positives and improving reliability. Similarly, Aravind et al. (2023) highlighted the effectiveness of XGBoost in optimising fuel efficiency by identifying patterns in driver behavior and vehicle performance metrics [5].

Time series forecasting for maintenance scheduling

Time series forecasting has been widely used to predict equipment failures and maintenance needs. Models such as ARIMA, Prophet, and LSTM networks are commonly applied in predictive analytics [6].

ARIMA is a statistical model that captures linear trends and short-term dependencies in time series data. It has been widely used for vehicle maintenance scheduling but struggles with non-stationary data and long-term forecasting [7].

Prophet, developed by Facebook, is a time series forecasting model that accounts for seasonality, trends, and holidays. It is useful for handling irregular time series data and provides interpretable forecasts [8].

LSTM, a deep learning-based recurrent neural network, is capable of capturing long-term dependencies in sequential data, making it highly effective for predicting complex vehicle maintenance patterns. Studies have shown that LSTM outperforms traditional statistical models in forecasting maintenance schedules due to its ability to learn non-linear relationships.

Fuel optimisation and predictive analytics

Fuel efficiency is a major concern for the automotive industry, influencing operational costs and environmental impact. Predictive maintenance combined with ML has proven to be an effective strategy for optimising fuel consumption. By analysing driving behavior, vehicle load, tire pressure, and engine condition, ML models can provide recommendations for fuel-efficient driving and timely maintenance [9].

A study by Tunnell et al. (2018) showed that integrating advanced driver assistance systems (ADAS) with predictive maintenance led to a 12–15% improvement in fuel efficiency. Additionally, Masoum et al. (2002) demonstrated that machine learning algorithms, when applied to hybrid and electric vehicles, resulted in enhanced energy efficiency and optimised fuel consumption patterns [10].

Research gaps and contributions

While significant progress has been made in predictive maintenance and fuel optimisation, certain research gaps remain:

Limited real-time implementation – Many predictive maintenance studies rely on historical data rather than real-time streaming sensor data.

Hybrid model integration – Combining deep learning models (LSTM) with traditional statistical models (ARIMA) for better accuracy is still underexplored.

Scalability in fleet management – Most predictive maintenance frameworks focus on individual vehicles rather than large-scale fleet operations [11].

This study addresses these gaps by integrating time series forecasting (LSTM, Prophet, ARIMA) with classification models (XGBoost, random forest) to develop an optimised predictive maintenance framework. The research aims to improve vehicle reliability, reduce downtime, and enhance fuel efficiency through intelligent predictive modelling.

Methodology

This section outlines the data collection process, feature engineering techniques, ML model selection, and evaluation metrics used for predictive maintenance and fuel optimisation.

Data collection and pre-processing

The dataset used in this study consists of historical maintenance records, real-time vehicle telemetry data, and fuel efficiency logs. The data was collected from fleet vehicles equipped with on board diagnostics (OBD-II) sensors and telematics devices. The dataset includes:

- **Vehicle attributes**: Model, engine size, fuel type, transmission type.
- **Maintenance data**: Service history, reported issues, past breakdowns.
- **Operational metrics**: Odometer readings, last service date, warranty expiration.
- **Fuel efficiency indicators**: Mileage per litre, engine performance, tire condition.

Data pre-processing steps:

- **Handling missing values**: Missing entries were imputed using linear interpolation and mean substitution.

- **Feature scaling**: Standardisation was applied to numerical features, while categorical variables were **one-hot encoded**.
- **Outlier detection**: Cook's distance method was used to identify and remove influential outliers.
- **Feature engineering**: New features such as **Service_Per_Year** (number of services per year) and **Odometer_Per_Year** were created to enhance model performance.

Model selection and implementation

To ensure accurate predictive maintenance and fuel optimisation, both time series forecasting models and classification models were used. Time series forecasting models (for predicting maintenance needs).

i. **ARIMA (Autoregressive integrated moving average)**
    a. Used for short-term maintenance prediction.
    b. Captures linear trends but struggles with complex dependencies.
ii. **Prophet**
    a. Accounts for seasonal trends in maintenance schedules.
    b. Robust against missing data and irregular time series.
iii. **LSTM (Long short-term memory)**
    a. A deep learning model that learns long-term dependencies in sequential data.
    b. Best suited for complex, non-linear vehicle maintenance forecasting.

Classification models (for maintenance optimisation)

i. **Random forest**
    a. An ensemble learning algorithm that classifies vehicles based on maintenance history.
    b. Provides **high accuracy** but is computationally expensive.
ii. **XGBoost (Extreme gradient boosting)**
    a. Optimised for large datasets, providing faster training and better generalisation.
    b. Achieves **higher precision in maintenance classification** than other models.
iii. **TPOT AutoML**
    a. An automated machine learning pipeline that optimises model selection and hyperparameters.
    b. Ensures the best possible model configuration without manual tuning.

Model training and hyperparameter tuning

Each model was trained using 80% of the dataset, while 20% was used for testing. To enhance performance, grid search and random search techniques were used for hyperparameter tuning. The best configurations were:

- **LSTM:** 3 hidden layers, 128 neurons, ReLU activation, dropout of 0.2.
- **XGBoost:** Learning rate of 0.05, maximum depth of 6, 500 estimators.
- **ARIMA:** (p,d,q) values optimised using AIC (Akaike Information Criterion).

Cross-validation with **k-fold (k=5)** was applied to avoid over fitting.

Model evaluation metrics

To assess model performance, the following metrics were used:

For time series forecasting models:

- **Mean squared error (MSE):** Measures the average squared differences between actual and predicted values.
- **Root mean squared error (RMSE):** Provides an interpretable error measure.
- **Mean absolute percentage error (MAPE):** Expresses errors as a percentage for easy comparison.

For classification models:

- **Accuracy:** Measures overall classification correctness.
- **Precision & recall:** Evaluates false positive and false negative rates.
- **F1-score:** Ensures a balance between precision and recall.

Results and discussion

This section presents the performance evaluation of predictive maintenance and fuel optimisation models. The results are analysed based on MSE, RMSE, MAPE, accuracy, and F1-score. The findings demonstrate that LSTM performs best for time series forecasting, while TPOT AutoML outperforms traditional classification models in maintenance prediction.

Model performance evaluation (Table 47.1)

Table 47.1 Performance comparison of predictive models

Model	MSE	RMSE	MAPE (%)	Accuracy (%)	F1-score
ARIMA	0.964	0.982	18.2	-	-
Prophet	0.076	0.275	12.5	-	-
LSTM	0.0758	0.0557	**7.8**	-	-
Random forest	-	-	-	94.5	0.92
XGBoost	-	-	-	97.6	0.96
TPOT AutoML	-	-	-	**98.2**	**0.98**

Source: Author

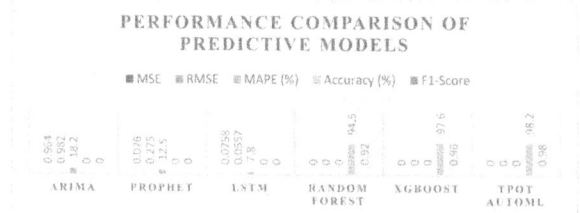

Figure 47.1 Represents the comparison of predictive models
Source: Author

a) LSTM achieves the lowest MSE (0.0758) and RMSE (0.0557), indicating its superior ability to forecast maintenance schedules accurately.
b) XGBoost and TPOT AutoML exhibit the highest accuracy (97.6% and 98.2%, respectively) for predictive maintenance classification.
c) MAPE results indicate that LSTM outperforms ARIMA and Prophet in long-term forecasting.
d) TPOT AutoML optimises hyperparameters automatically, achieving the best classification performance (F1-score = 0.98).

Model performance analysis

Time series forecasting for maintenance prediction

Time series forecasting models (ARIMA, Prophet, and LSTM) were evaluated for their ability to predict vehicle maintenance needs based on historical service records. LSTM consistently outperformed ARIMA and Prophet, capturing long-term dependencies and handling non-linear data patterns efficiently.

- ARIMA struggled with non-stationary data, leading to higher errors.
- Prophet performed better in short-term predictions but failed to capture long-term dependencies.
- LSTM provided the best forecast accuracy, reducing MAPE to 7.8%.

Predictive maintenance classification

Supervised learning models (Random Forest, XGBoost, and TPOT AutoML) were used for predictive maintenance classification. TPOT AutoML, which automates model selection and hyperparameter tuning, achieved the highest accuracy (98.2%) and best F1-score (0.98).

- Random forest showed good classification performance (accuracy: 94.5%) but required extensive parameter tuning.
- XGBoost outperformed random forest, reaching 97.6% accuracy with better recall and precision scores.
- TPOT AutoML provided the best results by optimising feature selection and model parameters automatically (Figure 47.2).

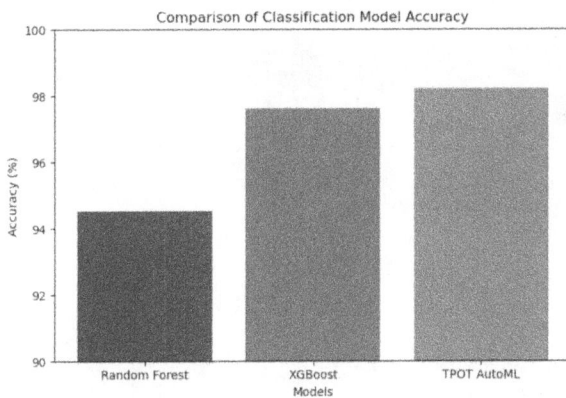

Figure 47.2 Comparison of model accuracy analysis of different algorithms
Source: Author

Impact on fuel efficiency

By integrating predictive maintenance models, fuel efficiency improved by 12–15%, as vehicles underwent servicing at optimal intervals, reducing fuel wastage. The key improvements observed were:

- **Reduced unplanned breakdowns** by 30%, minimising sudden maintenance costs.
- **Optimised fuel consumption**, leading to savings of up to 2.5 litres per 100 km.
- **Extended vehicle lifespan** through proactive servicing schedules.

Discussion

The results confirm that ML enhances predictive maintenance and fuel optimisation, offering significant operational advantages. The study highlights:

1. Deep learning models (LSTM) are highly effective for maintenance forecasting, outperforming traditional time series models.
2. Automated ML frameworks (TPOT AutoML) improve classification accuracy, minimising the need for manual tuning.
3. Predictive maintenance leads to cost savings, better fuel efficiency, and improved vehicle reliability.

Conclusion and future work

This study explores the effectiveness of ML-based predictive maintenance and fuel optimisation in the automotive industry. By leveraging time series forecasting models (ARIMA, Prophet, LSTM) and classification models (random forest, XGBoost, TPOT AutoML), the research demonstrates the potential of predictive analytics in reducing vehicle downtime, optimising fuel consumption, and minimising operational costs.

Key findings

The major findings from this research are:

1. LSTM models outperform ARIMA and Prophet in predicting vehicle maintenance schedules, achieving the lowest MSE (0.0758) and RMSE (0.0557).
2. TPOT AutoML outperforms traditional classification models by automating hyperparameter tuning and achieving the highest accuracy (98.2%) in maintenance classification.
3. Predictive maintenance models lead to a 12–15% improvement in fuel efficiency, reducing unnecessary servicing and fuel wastage.
4. Real-time predictive analytics can minimise unplanned breakdowns by 30%, leading to lower repair costs and extended vehicle lifespan.

References

[1] Moujahid, A., Dulva Hina, M, El Araki Tantaoui, M., El Khadimi, A., Ortalda, A., Ramdane-Cherif, A., Soukanec, A. Machine learning techniques in ADAS: A review. *Conf. Paper*, 2018.

[2] Mohammadnazar, A., Khattak, Z. H., Khattak, A. J. Assessing driving behavior influence on fuel efficiency using machine-learning and drive-cycle simulations. *Transport. Res. Part D*, 2024;126:104025.

[3] Campos-Ferreira, A. E., Lozoya-Santos, J. d. J., Tudon-Martinez, J. C., Ramirez Mendoza, R. A., Vargas-Mar-

tinez, A., Morales-Menendez, R., Lozano, D. Vehicle and driver monitoring system using on-board and remote sensors. *Sensors,* 2023;23:814.

[4] Rana, K., Khatri, N. Automotive intelligence: Unleashing the potential of AI beyond advance driver assisting system, a comprehensive review. *Comp. Elec. Engg.,* 2024;117:109237.

[5] Agarwal, S., Kavitha, R. Advanced driver assistance system using machine learning. *Internat. J. Res. Publ. Rev.,* 2024;5(3):1364–1370.

[6] Bathla, G., Bhadane, K., Singh, R. K., Kumar, R., Aluvalu, R., Rajalakshmi, K., Kumar, A., Thakur, R. N., Basheer, S. Autonomous vehicles and intelligent automation: Applications, challenges, and opportunities. *Mobile Inform. Sys.,* 2022;2022:36.

[7] Schoen, A., Byerly, A., Hendrix, B., Mahesh Bagwe, R., Euzeli Cipriano dos Santos Jr., Miled, Z. B. A machine learning model for average fuel consumption in heavy vehicles. 2019:68(7):6343–635.

[8] Aravind, R. Optimizing ADAS and autonomous driving systems with advanced ethernet protocols and machine learning. *Internat. J. Sci. Res. (IJSR),* 2023;12(10):2147–2155.

[9] Tunnell, J. A., Asher, Z. D., Pasricha, S., Bradley, T. H. Towards improving vehicle fuel economy with ADAS. *SAE Tec. Paper,* 2018.

[10] Bhagat, R., Singh, P., Srivastava, A. K., Khanna, S. A comprehensive review on advanced driver assistance systems (ADAS). *Internat. J. Adv. Engg. Manag. (IJAEM),* 2023;5(5):115–122.

[11] Qu, F., Dang, N., Furht, B., Nojoumian, M. Comprehensive study of driver behavior monitoring systems using computer vision and machine learning techniques. *J. Big Data,* 2024;11:32.

48 Performance and emission characteristics of diesel engine using blends of Calophyllum inophyllum biodiesel

Dodda Hanamesha[1], D. Madhu[2], Rudresh B. M.[3] and Lakshmikant Shivanayak[4,a]

[1]Department of Mechanical Engineering, Government Engineering College, Talakal, Koppal, Karnataka, India

[2]Government Engineering College, Chamarajanagara, Karnataka, India

[3]Department of Mechanical Engineering, Government Engineering College, K R Pete, Mandya, Karnataka, India

[4]Assistant Professor, Department of Mechanical Engineering, Government Engineering College, K R Pete, Mandya, Karnataka, India

Abstract

The energy requirement is growing at faster rate every day because of which the usage of fossil fuels is increasing which in turn causes the simultaneous depletion of fossil fuel energy resources. So, it encourages the researchers in every nation to find a renewable, environmentally friendly, alternate fuel source in replacement of petroleum diesel in diesel engines recently. The biodiesel, because of its favourable properties is found suitable as one of alternate renewable energy source in place of diesel in diesel engines. In this research work, because of non-edible, very high yield value per unit area of land and also very high biodiesel conversion rate, an effort is made the using of the blends of Calophyllum inophyllum biodiesel in a computerised, single cylinder, water cooled, direct injection, four stroke diesel engine. The determined fuel properties of Calophyllum inophyllum biodiesel were met with those for diesel fuel according to American Society for Testing and Materials (ASTM) standards. Using blends of Calophyllum inophyllum biodiesel, the performance and emission characteristics of diesel engine are found out and compared with diesel fuel considering it as a reference fuel. For the blends of Calophyllum inophyllum biodiesel and also diesel fuel, when engine load is increased, brake specific fuel consumption (BSFC) decreased whereas brake thermal efficiency (BTE) and exhaust gas temperature (EGT) increased. In the emission test, when engine load is increased, CO, CO_2, unburnt hydrocarbons (UBHC) and nitrogen oxide (NOx) emissions increased whereas at lower loads smoke opacity increased but it decreased at higher loads. All blends of biodiesel showed lower emissions of CO, UBHC and smoke opacity compared to those obtained for diesel fuel whereas CO_2 emissions obtained for all biodiesel blends are found higher than that obtained for diesel fuel. Overall, it can be concluded that the blends of Calophyllum inophyllum biodiesel can be used in place of diesel in diesel engine based on the obtained results by the diesel engine with respect to the performance and emission characteristics without any engine modifications.

Keywords: Diesel engine, Calophyllum inophyullum biodiesel, performance, emission

Abbreviations:

CO: carbon monoxide, CO_2: carbon dioxide, BTE: brake thermal efficiency, BP: brake power, NOx: nitrogen oxides, HC: hydrocarbons, η_{mech}: mechanical efficiency, EGT: exhaust gas temperature, ppm: parts per million, BSFC: brake specific fuel consumption, UBHC: unburnt hydrocarbons, ASTM: American Society for Testing and Materials

Introduction

The most basic requirement for human life is the energy. Fossil fuel consumption has significantly increased and using these energy sources has a significant adverse effect on the environment. A rapidly developing nation like India depends heavily on petroleum diesel for development and has higher energy needs overall. Due to their low fuel consumption and low emissions except NOx and particulate matter, diesel engines are primarily used for power generation, agricultural purposes and road transportation. This emission reduces the air quality and contributes to the acceleration of global warming.

The need for improved air quality as well as the rising cost of crude oil necessitates the development of alternative diesel engine fuels. Engine emissions are the primary reason of ozone layer depletion and environmental degradation. In order to meet the energy needs of the future, the world is now focusing on renewable energy sources like biomass and biofuels. In order to solve the energy crisis, biofuels have emerged as important as the energy required for industry. Renewable sources of energy such as solar, wind and biomass account for approximately 8% of the total energy sources with biofuels accounting for nearly 23%. It has been proven that biodiesel can be used efficiently in CI engines with a little or no

[a]lakshmikant.a.s@gmail.com

DOI: 10.1201/9781003685876-48

modifications of the engine. Additionally there are abundant agricultural areas of land available in all nations where inexpensive vegetable seeds can be grown [1]. The searching for useful energy as well as the need for a clean and green environment remains a vital focus of discussion for any researcher. The rapidly diminishing petroleum reserves have already sent a warning signal to the entire world to seek alternative energy sources to meet the rising energy demands. Furthermore, attention must be paid to the harmful emissions that come from fossil fuels [2]. Because of the depletion of fossil fuels and the serious concern about pollution from the combustion in internal combustion engines, particularly petrol and diesel engines, many researchers around the world have been working to develop promising substitute fuels for IC engines that can fully or partially replace fossil fuels. Biodiesel is a renewable alternative fuel and because its properties are similar to diesel, it can be used in compression ignition engines with no or minimal modifications, particularly when blended with diesel [3]. Diesel engines play an important role in the industrial and automotive sectors as a result of their greater efficiency and reduced fuel consumption. Since biodiesel is renewable, environmentally friendly, biodegradable, non-toxic and can be produced using resources that are readily available locally, it is used extensively in developing countries. Biodiesel derived from non-edible sources is preferred in India. Biodiesel made from pongamia, neem, jatropha, cotton seed, mahua seeds and other plants is commonly used to produce alternative fuels. Compared to conventional diesel fuel, biodiesels have a higher oxygen content, lubricity, lower volatility, lower sulphur and aromatic content and higher miscibility [4].

When compared to diesel fuel, biodiesel derived from animal fatty acids and vegetable oils appears to be the most environmentally friendly alternative fuel. In many countries, finding a suitable and sustainable alternative fuel has become a high priority. It will also be very important in many industries in the coming years. Biodiesel is a non-petroleum based fuel consisting of alkyl esters obtained from vegetable oil triglycerides or animal fat free fatty acids with short-chained alcohols [5]. Calophyllum inophyllum, is commonly found in eastern Africa, Papua, Madagascar, New Guinea, Northern Australia, India, tropical America. In India, they are found along the shorelines of Kerala, Maharashtra, Karnataka,

Tamil Nadu, Orissa, Andhra Pradesh, Andaman and Arunachal Pradesh. So, in this research work an attempt is made the using of Calophyllum inophyllum biodiesel in its blends in diesel engine to replace diesel fuel and in this research paper, it is presented the performance and emission characteristics of diesel engine using the blends of Calophyllum inophyllum biodiesel in comparison with diesel fuel as reference fuel.

Experimental procedure and specifications

A computerised water-cooled, single-cylinder, direct-injected, four-stroke diesel engine is used in the experiments. The engine was run from the no load to full load (at 3.5 kW, rated output). The diesel engine specifications are shown in the following Table 48.1.

Table 48.1 The specifications of the diesel engine

Particulars	Engine specification
Make	Kirloskar
Stroke	4 stroke
Cylinder	Single
Bore	87.5 mm
Stroke	110 mm
Cooling type	Water cooled
Cylinder volume	661 cc
Power	3.5 kW@ 1500rpm
EGR	Water cooled, SS, Range: 0–15%
Dynamometer	Type eddy current
Injection variation	0–25°bTDC
CR range	12–18

Source: Author

For determination of exhaust gas temperature, a type-K thermocouple was used and both thermocouple and combustion analyser were connected to a data acquisition system for collection of the data. The exhaust emissions CO, CO_2, HC, O_2 and NOx were measured using the AVL DIGAS 444N five-gas analyser. For measurement of smoke opacity, an AVL 437 smoke metre was used. The experimental setup is as shown in the above Figure 48.1.

The Calophyllum inophyllum vegetable oil was used to make biodiesel using the standard esterification and trans-estrification method American Society for Testing and Materials (ASTM) D6751 and the fuel met the standard specifications. The cost of raw oil is Rs. 800/litre and yield of biodiesel obtained is 88%. The properties of Calophyllum inophyllum

Figure 48.1 Experimental setup
Source: Author

Table 48.3 Operating conditions

Operating conditions	Units	Values
Injection time	Degrees	23bTDC
Compression ratio	--	17.5:1
Injection pressure	Bar	210

Source: Author

Table 48.4 Composition of biodiesel blend in volume percentage

Designation of biodiesel blend	Composition in percentage	
	Calophyllum inophyllum biodiesel	Diesel fuel
C-10	10	90
C-20	20	80
C-30	30	70
C-40	40	60

Source: Author

biodiesel and also diesel fuel are shown in Table 48.2. Table 48.3 shows the operating conditions used during experiments.

The Calophyllum inophyllum biodiesel blends are used along with diesel fuel as reference fuel. The composition of biodiesel blends is given in Table 48.4.

Results and discussion

By using blends of Calophyllum inophyllum biodiesel, the performance and emission characteristics of the diesel engine were found out and compared to standard diesel fuel as reference fuel.

Performance parameters

The following performance parameters were studied for blends of Calophyllum inophyllum biodiesel during experimentations.

Brake specific fuel consumption

The measurement of brake specific fuel consumption (BSFC) gives an overall efficiency of the engine. In Figure 48.2, it is shown that the variation of BSFC with BP when the engine is fuelled with Calophyllum inophyllum biodiesel blends. With increase in BP, BSFC of the engine decreased for all biodiesel blends. This is because at higher loads, the in-cylinder temperature is high, that leads to better combustion process which in turn minimises the fuel consumption. However, with increase in percentage

Table 48.2 Properties of Calophyllum inophyllum biodiesel and diesel fuel

Properties	C.I. biodiesel	Diesel	Biodiesel blends			
			C 10	C 20	C 30	C 40
Calorific value (kJ/kg)	38050	42995	42299	41906	41416	40859
Density (kg/m^3)	873	830	824	830	835	842
Flash point (°C)	185	53	50	65	71	75
Kinematic viscosity (m^2/sec) @ 40°C	5.79×10^{-6}	2.09×10^{-6}	3.35×10^{-6}	3.47×10^{-6}	3.77×10^{-6}	3.96×10^{-6}
Cetane number	50–55	52	-	-	-	-

Source: Author

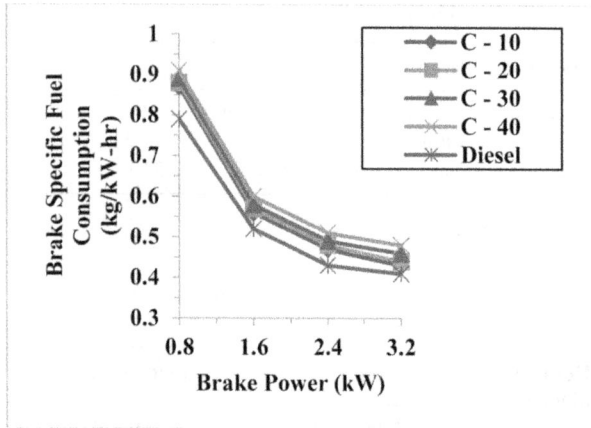

Figure 48.2 BSFC vs. BP for Calophyllum inophyllum biodiesel blends
Source: Author

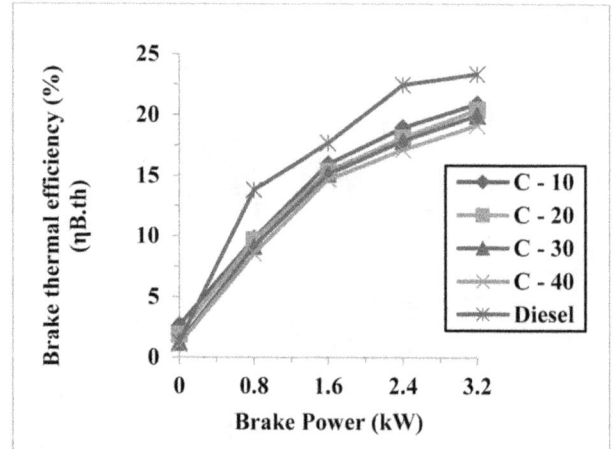

Figure 48.3 BTE vs. BP for Calophyllum inophyllum biodiesel blends
Source: Author

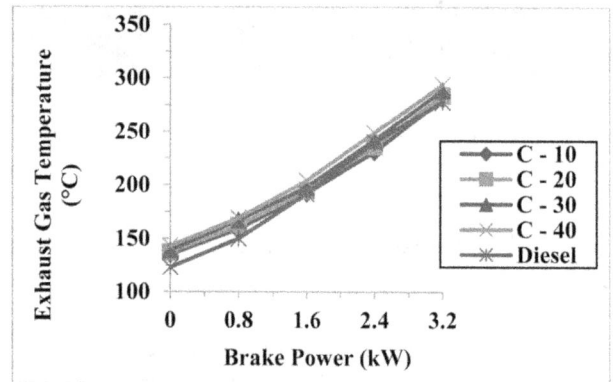

Figure 48.4 EGT vs. BP for Calophyllum inophyllum biodiesel blends
Source: Author

share of biodiesel in the blends, BSFC increased. This is because more biodiesel blend is required to produce the same amount of power due to its lower heating value and higher value of viscosity in comparison with conventional diesel fuel [6–8]. At full load, the BSFC for Calophyllum inophyllum biodiesel blends C-10, C-20, C-30 and C-40 are 0.43, 0.44, 0.46 and 0.48 kg/kW-hr, respectively in comparison with 0.41 kg/kW-hr for diesel fuel.

Brake thermal efficiency

Figure 48.3 shows that the variation of brake thermal efficiency (BTE) with BP when the engine is fuelled with Calophyllum inophyllum biodiesel blends. With increase in BP, BTE obtained by the engine increased for all biodiesel blends. This is because the in-cylinder temperature is high at higher loads that lead to better combustion process. However, with increase in percentage share of biodiesel in the blends, BTE decreased. This is because of the decreased calorific values of the blends which may the dominant factor and also higher viscosity of the blends that leads to poor atomisation and also improper mixing of the fuel with air [8–10]. At full load, the BTE for Calophyllum inophyllum blends C-10, C-20, C-30 and C-40 are 20.92%, 20.42%, 19.96% and 19.15%, respectively in comparison with 23.35% for diesel fuel.

Exhaust gas temperature

Figure 48.4 shows that the variation of exhaust gas temperature (EGT) with BP when the engine is fuelled with Calophyllum inophyllum biodiesel blends. With increase in BP, EGT for the engine increased for all biodiesel blends. This is because more fuel is required to be burnt at higher load condition. However, with increase in percentage share of biodiesel in the blends, EGT increased. This is because of enrichment of oxygen in biodiesel blends that in turn increases combustion temperature [6, 7]. At full load, the EGT for Calophyllum inophyllum biodiesel blends C-10, C-20, C-30 and C-40 are 281°C, 284°C, 289°C and 294°C, respectively in comparison with 278°C for diesel fuel.

Emission characteristics
Emissions of carbon monoxide

Figure 48.5 shows the variation of CO emissions with BP when the engine is fuelled with Calophyllum inophyllum biodiesel blends. With increase in BP, CO emissions from the engine increased for all biodiesel blends. This is because, even though in-cylinder temperature is high at higher load condition that leads to better combustion process by warmed-up condition, the higher quantity of fuel consumption takes place. However, with increase in percentage share of biodiesel in the blends, CO emissions decreased. This is because of presence of increased oxygen content in biodiesel blends that in turn increases the oxidation rate in leaner environment to convert CO into CO_2 emissions. From Figure 48.5, it is also noticed that the emissions of CO are low for blends of biodiesel at all loads compared to diesel fuel [6]. At full load, CO emissions for Calophyllum inophyllum biodiesel blends C-10, C-20, C-30 and C-40 are 0.22%, 0.20%, 0.19% and 0.17% vol, respectively in comparison with 0.53% vol for diesel fuel.

Emissions of carbon dioxide

Figure 48.6 shows the variation of CO_2 emissions with BP when the engine is fuelled with Calophyllum inophyllum biodiesel blends. With increase in BP, CO_2 emissions from the engine increased for all biodiesel blends. This is because of high combustion temperature at higher loads which can accelerate the oxidation process of CO into CO_2 emissions so that high formation of CO_2 emission in the engine's exhaust gases. The complete combustion of fuel is indicated by amount of CO_2 emissions in the exhaust gases. However, with increase in percentage share of biodiesel in the blends, CO_2 emissions increased. This is because of presence of increased oxygen content in the biodiesel blends that in turn increases the oxidation rate in the leaner environment to convert

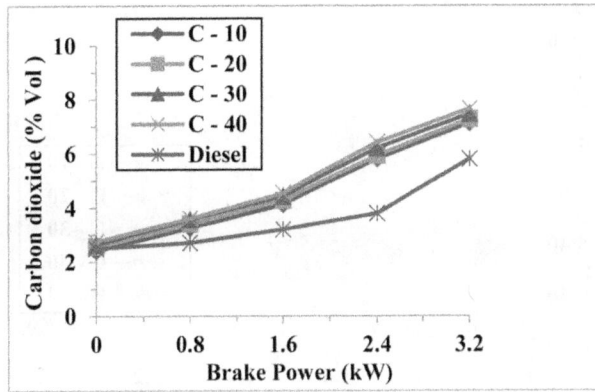

Figure 48.6 CO_2 emissions vs. BP for Calophyllum inophyllum biodiesel blends
Source: Author

CO emissions into CO_2 emissions. From Figure 48.6, it can also be noticed that for all biodiesel blends the CO_2 emissions are found higher compared to diesel fuel at all loads, [6, 7]. CO_2 emissions at full load for Calophyllum inophyllum biodiesel blends C-10, C-20, C-30 and C-40 are 7.12%, 7.27%, 7.46% and 7.65% vol, respectively in comparison with 5.8% vol for diesel fuel.

Unburnt hydrocarbon emissions

The lack of homogeneity and incomplete combustion are the main causes for the formation of unburnt hydrocarbons (UBHC) emissions in the diesel engine. Figure 48.7 shows that the variation of UBHC emissions with BP when the engine is fuelled with Calophyllum inophyllum biodiesel blends. With increase in BP, UBHC emissions from the engine increased for all biodiesel blends. This is due to incomplete combustion during combustion process which takes place because of supply of insufficient amount of oxygen and also in turn the formation of local rich mixtures in the combustion chamber. However, with increase in percentage share of biodiesel in the blends, UBHC emissions decreased. This is because of increased content of oxygen with addition of biodiesel proportion to diesel fuel that can compensate oxygen deficiency and therefore leads to a faster chemical reaction to improve the combustion process and thus results into more complete combustion process. From Figure 48.7, it is also noticed that compared to all blends of biodiesel, the UBHC emissions are found higher for diesel fuel at all loads [8, 11]. UBHC emissions at full load for Calophyllum inophyllum biodiesel blends C-10, C-20, C-30 and C-40

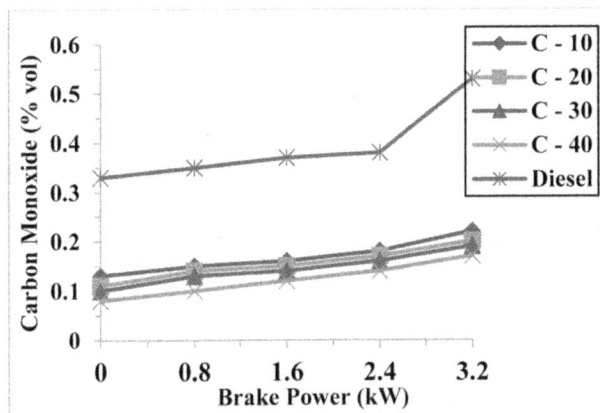

Figure 48.5 CO emissions vs. BP for Calophyllum inophyllum biodiesel blends
Source: Author

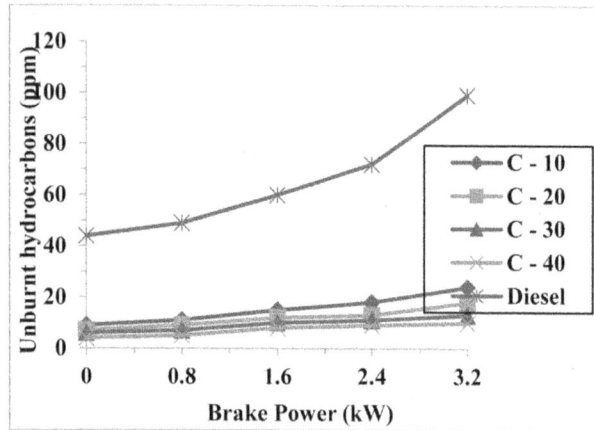

Figure 48.7 UBHC emissions vs. BP for Calophyllum ino-phyllum biodiesel blends.
Source: Author

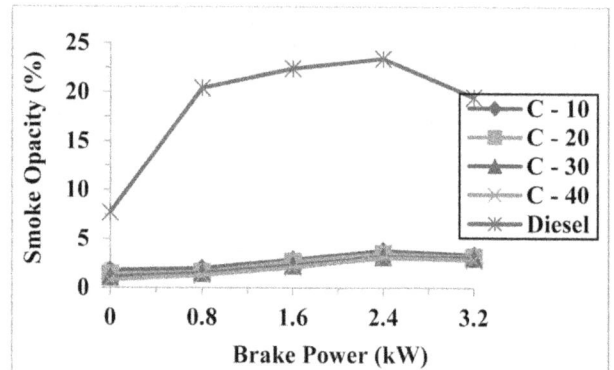

Figure 48.8 NOx emissions vs. BP for Calophyllum ino-phyllum biodiesel blends
Source: Author

are 24 ppm, 18 ppm, 13 ppm and 10 ppm, respectively in comparison with 99 ppm for diesel fuel.

Emissions of nitrogen oxides

In Figure 48.8, it is shown that variations of NOx emissions with BP when the engine is fuelled with Calophyllum inophyllum biodiesel blends, respectively. With increase in BP, NOx emissions from the engine increased for all biodiesel blends. This is because of maximum combustion temperature inside the cylinder at higher engine load. However, with increase in percentage share of biodiesel in the blends, NOx emissions increased. This is because of presence of inherent oxygen molecules in the blends of biodiesel that enhances the combustion process and thus increased the combustion temperature which in turn creates the favourable environment to form NOx emissions [6, 7, 10]. NOx emissions at full load for Calophyllum inophyllum biodiesel blends C-10, C-20, C-30 and C-40 are 1020 ppm, 1097 ppm, 1130 ppm and 1184 ppm, respectively in comparison with 1210 ppm for diesel fuel.

Smoke opacity

In Figure 48.9, it is shown that the variation of smoke opacity with BP when the engine is fuelled with Calophyllum inophyllum biodiesel blends. For all biodiesel blends, at lower BP, the smoke opacity from the engine increased and it decreased at higher BP. This is because of improper mixing of fuel droplets with air at lower loads that leads to poor combustion and it achieves proper mixing of fuel droplets with air at higher loads that leads to better

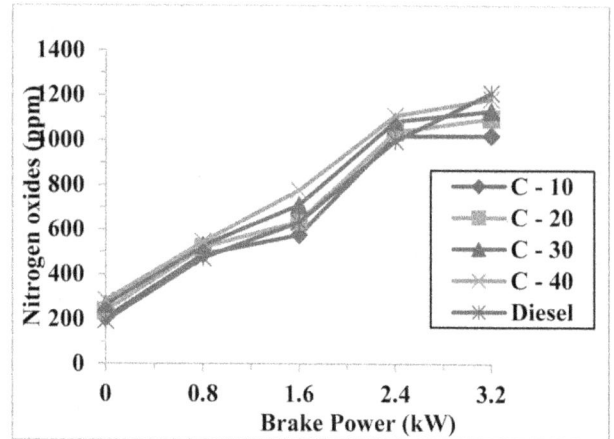

Figure 48.9 Smoke opacity vs. BP for Calophyllum ino-phyllum biodiesel blends
Source: Author

combustion and thus the smoke level is minimised consequently. However, with increase in percentage share of biodiesel in the blends, smoke opacity decreased. This is due to the higher oxygen content of biodiesel blends that results into full fuel oxidation which in turn reduces the tendency of formation of the smoke in the exhaust gases. From Figure 48.9, it is also noticed that at all loads, the smoke opacity of blends of biodiesel is lower than diesel. This is because of the presence of oxygen content in the blends of biodiesel that contributes for complete combustion process which is also considered as the indication of complete combustion process for blends of biodiesel in comparison with diesel [10]. The smoke opacity at full load for Calophyllum inophyllum biodiesel blends C-10, C-20, C-30 and C-40 are 3.4%, 3.2%, 3.1% and 2.9%, respectively in comparison with 19.5% for diesel fuel.

Conclusion

The diesel engine was operated using Calophyllum inophyllum biodiesel blends C-10, C-20, C-30 and C-40 along with diesel fuel as a reference fuel. During experiments, the same operating conditions were used and blends of biodiesel were compared with diesel fuel as reference fuel. Based on experimental results, the following conclusions are drawn.

1) As the engine load increased
   i) The BSFC decreased whereas BTE and EGT increased for all blends of biodiesel and also diesel fuel.
   ii) Emissions of CO, UBHC, NOx and CO_2 increased for all biodiesel blends and also diesel fuel.
   iii) At lower loads, the smoke opacity increased and it decreased at higher loads for all biodiesel blends and also diesel fuel.
2) When the percentage share of biodiesel is increased in the blends
   i) The BSFC and EGT increased whereas BTE decreased.
   ii) CO, UBHC and smoke opacity decreased whereas emissions of NOx and CO_2 increased for all biodiesel blends and diesel fuel.
3) CO emissions are found high for diesel fuel compared to all biodiesel blends.
4) CO_2 emissions are found low for diesel fuel compared to all biodiesel blends.
5) UBHC emissions from the engine are found high for diesel fuel compared to all biodiesel blends.
6) The smoke opacity of diesel fuel is found higher than all biodiesel blends at all loads.
7) All biodiesel blends showed better results with respect to emissions of CO, NOx, UBHC and smoke opacity compared to those obtained for diesel fuel whereas CO_2 emissions obtained for all biodiesel blends are found higher than that obtained for diesel fuel.

References

[1] Rajan, K., Senthil Kumar, K. R. Experimental study on diesel engine working characteristics using yellow oleander biodiesel with the effect of different injection timings. *Energy Sources Part A Recov. Utiliz. Environ. Effects*, 2024;46(1):4939–4952.

[2] Kathirvel, S., Layek, A., Muthuraman, S. Exploration of waste cooking oil methyl esters (WCOME) as fuel in compression ignition engines: A critical review. *Engg Sci. Technol. Internat. J.*, 2016;19:1018–1026.

[3] Bari, S. Performance, combustion and emission tests of a metro-bus running on biodiesel-ULSD blended (B20) fuel. *Appl. Energy*, 2014;124:35–43.

[4] Senthil Kumar, D., Thirumalini, S. Investigations on effect of split and retarded injection on the performance characteristics of engines with cashew nut shell biodiesel blends. *Internat. J. Ambient Energy*, 2022;43(1):2251–2259.

[5] Nantha Gopal, K., Thundil Karupparaj, R. Effect of pongamia biodiesel on emission and combustion characteristics of DI compression ignition engine. Ain Shams Engg. J. 2015;6:297–305.

[6] Borugadda, V. B., Paul, A. K., Chaudhari, A. J., Kulkarni, V., Sahoo, N., Goud, V. V. Influence of waste cooking oil methyl ester biodiesel blends on the performance and emissions of a diesel engine. *Waste Biomass Valor*, 2018;9:283–292.

[7] Kanwar, R., Sharma, P. K., Singh, A. N., Agrawal, Y. K. Performance and emission characteristics of a compression ignition engine operating on blends of castor oil biodiesel–diesel. *J. Inst. Eng. India Ser. C*, 2017;98:147–154.

[8] Anantha Kumar, S., Jayabal, S., Thirumal, P. Investigation on performance, emission and combustion characteristics of variable compression engine fuelled with diesel, waste plastics oil blends. *J. Braz. Soc. Mech. Sci. Eng.*, 2017;39:19–28.

[9] Gnanasekaran, S., Saravanan, N., Ilangkumaran, M. Influence of injection timing on performance, emission and combustion characteristics of a DI diesel engine running on fish oil biodiesel. *Energy*, 2016;116:1218–1229.

[10] Paul, G., Datta, A., Mandal, B. K. An experimental and numerical investigation of the performance, combustion and emission characteristics of a diesel engine fuelled with Jatropha biodiesel. *4th Internat. Conf. Adv. Energy Res. 2013, ICAER 2013, Energy Proc.*, 2014;54:455–467.

[11] Rajana, K., Senthil Kumar, K. R. Experimental study on diesel engine working characteristics using yellow oleander biodiesel with the effect of different injection timings. *Energy Sources Part A Recov. Utiliz. Environ. Effects*, 2024;46(1):4939–4952.

49 Rare earth element (Tb) doped calcium aluminate nanoparticles: Synthesis, structural optical and photoluminescence studies

B. S. Shashikala[1,a], H. B. Premkumar[2], G. P. Darshan[2], D. R. Lavanya[3] and H. Nagabhushana[4]

[1]Associate Professor, Department of Physics, K. S. Institute of Technology (VTU-affiliated), Bengaluru–560109, Karnataka, India

[2]Department of Physics, FMPS, MS Ramaiah University of Applied Sciences, Bengaluru–560054, Karnataka, India

[3]Department of Physics, School of Applied Sciences, REVA University, Bengaluru–560064, Karnataka, India

[4]Prof. C.N.R. Rao Centre for Advanced Materials, Tumkur University, Tumkur–572103, Karnataka, India

Abstract

This work emphasises the photoluminescence properties of Terbium doped calcium aluminate nanophosphor ($CaAl_2O_4$:Tb), which was fabricated using a bio-activated combustion method utilising lemon juice as a fuel. Regardless of the dopant, powder X-ray diffraction (PXRD) profiles support the monoclinic phase in every sample. The synthesised nanoparticles (NPs) were examined by scanning electron microscopy (SEM), diffuse reflectance spectroscopy (DRS) and photoluminescence (PL) studies. The PL emission spectra shows acute peaks at ~ 494, 544, 588 and 618 nm, which were ascribed to $^5D_4{}^7F_J$ (J= 6, 5, 4 and 3) transitions of Tb^{3+} ions, respectively. Concentration quenching was caused by the activator ions' dipole-dipole (d–d) interaction. Photometric characterisation verified the green colour emission of excellent purity. The aforementioned findings support the possibility that the developed nanophosphor might be a traditional material for use in white light emitting diodes (WLEDs) applications.

Keywords: Bio-inspired, combustion route, photoluminescence

Introduction

Combining nano science and nanotechnology, particularly with the introduction of engineered nanoparticles, is the most promising way to develop devices in a variety of fields, such as photonic materials, optical detectors, optoelectronic components, biomedicine, biochemical probes, laser and light emitting diodes, forensic and security applications [1–5]. Luminescent materials have many applications and attract the attention of researchers because of their measurable doping capacity, thermal stability, physical, chemical, structural, and spectroscopic properties [6–10]. The lighting industry is continuously researching LED lighting, which primarily depends on the development of reliable and bright phosphors. The upcoming generation of illumination sources, white light emitting diodes (WLEDs), will drastically change how people live their lives. The conventional WLED was made up of blue-emitting chips and colour-emitting phosphors that were triggered by blue light. Because of their elevated emission levels, WLED devices' performance was greatly impacted by the phosphor's quality. These gadgets can save over 70% on energy costs and don't use any harmful substances, in contrast to conservative light sources such as luminescent tubes and incandescent light bulbs. Incandescent and fluorescent lights are used instead of WLEDs because of their energy efficiency, compact size, quick decay period, and environmental friendliness [11–14]. Hence, the ability of luminescent nano crystals to be efficiently activated close to ultraviolet light and produce in the visible spectrum is crucial [15–18]. Aluminates ($CaAl_2O_4$) have drawn some attention among other host material types because of their remarkable physical and chemical stability, ease of synthesis, broad band gap, and affordability [19, 20]. When combined with such a host matrix, the rare earth ions produce good luminescent effects. This work established a bio-inspired combustion process for the Tb^{3+} activated $CaAl_2O_4$ phosphor utilising lemon juice as fuel. The synthesised samples were analysed using photoluminescence (PL), scanning electron microscopy (SEM), and PXRD characterisations.

[a]shashikala.yathish@gmail.com

DOI: 10.1201/9781003685876-49

Materials and methods

In this activity, substances of analytical reagent grade (AR) are used immediately without additional purification. aluminium nitrate [$Al(NO_3)_3.9H_2O$ (99.9%)], calcium nitrate [$Ca(NO_3)_2.4H_2O$ (99.9%)], terbium nitrate ($Tb(NO_3)_3 \cdot 5H_2O$) are utilised as oxidizers and used in appropriate stoichiometric amounts with fresh lemon juice as fuel. The pulp of the freshly bought lemons was extracted by peeling and squeezing them. A muslin cloth was then used to filter out the solid pulp. About 30 mL of the 1:4 diluted solution of filtered lemon juice was employed as fuel.

The stoichiometric concentrations of calcium, aluminium, and terbium nitrate were completely mixed in 10 mL of double-distilled water. The prepared lemon juice (15 mL) was completely dissolved in the above solution for approximately 15 minutes with a magnetic stirrer. The subsequent solution was put in a muffle furnace that had been warmed to between 450°C and 500°C. After boiling and going through auto-ignition and thermal dehydration, the reaction solution produces foam. The finished product was employed for additional research after being calcined for 3 hours at about 900°C.

Characterisation

Using the Shimadzu-made PXRD-7000 X-ray diffractometer, which emits CuKα radiation (λ=1.541 Å), the crystalline purity and phase of the produced samples were measured. The crystal structure has been refined using Full Prof Software suite by employing Rietveld study. 450 W Xenon lamps were used as the excitation source for the PL measurements, which were taken in a Horiba Flurolog-3 Spectro flourimeter.

Result and discussions

PXRD studies
PXRD profiles and Williamson–Hall's (W-H) plots of $CaAl_2O_4$:Tb^{3+} (1–11 mol%) NPs were revealed in Figure 49.1. Doped nano phosphors showed no discernible change in their XRD spectrums, hinting that the dopants were properly replaced in the host lattice. The strong and sharp diffraction profiles reveal the monoclinic phase and agree well with the reference material JCPDS No. 01-070-0134. The average crystallite size of the obtained samples was estimated by using Scherer's and W-H equations [21]

$$D = \frac{K\lambda}{\beta cos\theta} \quad (1)$$

$$\beta cos\theta = \frac{0.9\lambda}{D} + 4\varepsilon sin\theta \quad (2)$$

where, λ; wavelength of the X-rays, β; full-width at half maximum (FWHM) on 2θ scale and θ; Bragg's angle.

Table 49.1 shows the computed crystallite size (D) of the set samples. The crystallite size was found to be more in W-H method than Scherrer method since W-H method includes both crystallite size and lattice strain, while the Scherrer method includes only crystallite size.

Rietveld refinement studies
Rietveld refinement of the prepared samples was carried out using full prof suite, and refined structure and diffraction pattern are shown in Figure 49.2. The obtained data shows the crystal structure

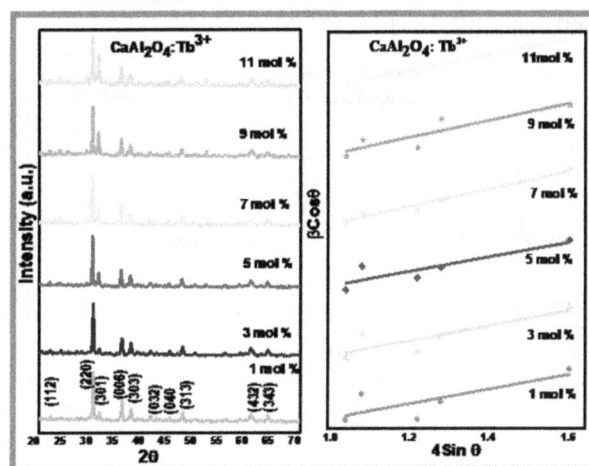

Figure 49.1 PXRD and W-H plots of $CaAl_2O_4$:Tb (1–11 mol%) NPs
Source: Author

Table 49.1 Crystallite size of $CaAl_2O_4$:Tb (1–11 mol%) NPs

Tb³⁺ conc. (mol%)	Crystallite size (nm) Scherer's method	W-H Method
1	20	33
3	21	35
5	26	28
7	23	31
9	17	36
11	33	

Source: Author

of $CaAl_2O_4$:Tb^{3+} (1–11 mol%) NPs effectively refine in monoclinic structure with space group P21/n. The refined lattice parameters are found to be a=8.7493 Å, b=8.0903 Å, c=15.1385 Å and refined cell volume was found to be 1071.581 Å^3. The low value of Chi-square (χ^2) =1.18 justifies the accuracy of refinement.

SEM studies

Scanning electron microscope (SEM) images of fabricated $CaAl_2O_4$:Tb^{3+} (1–11%) NPs were shown in Figure 49.3. Uneven particles with a broad size variation and aggregation were clearly seen.

Diffuse reflectance spectroscopy studies

The produced NPs' diffuse reflectance spectroscopy (DRS) spectrum was shown in Figure 49.4 (a). There is a distinct absorption band at 387 nm in the spectra that corresponds to the Tb^{3+} ions $^7F_6 \rightarrow {}^5D_3$ transitions. The following Kubelka-Munk theory equations were used to compute the optical energy gap (E_g) of the ready samples [22].

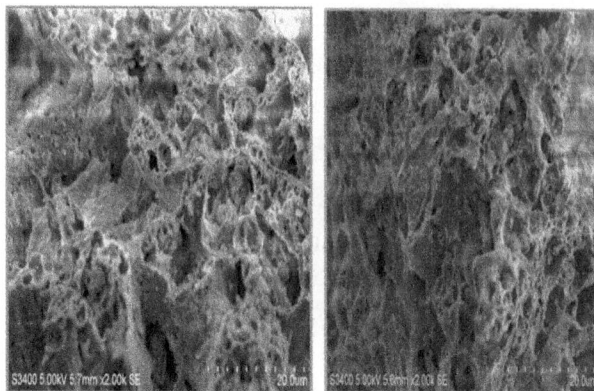

Figure 49.3 SEM images of optimised (a, b) $CaAl_2O_4$:Tb^{3+} (5 mol%)

Source: Author

$$F(R) = \frac{(1 - R^2)}{2R} \quad \text{---------(3)}$$

$$h\nu = \frac{1240}{\lambda} \quad \text{------------(4)}$$

where, R-reflection co-efficient, λ-absorption wavelength (nm).

The generated samples' energy gap (E_g) graphs were shown in Figure 49.4 (b). Table 49.2 presents the calculated E_g values.

Photoluminescence studies

Figure 49.5 (a) presents the PL excitation spectra of $CaAl_2O_4$:Tb^{3+} (1–11%) NPs at λ_{Emi} = 544 nm. The spectrum shows sharp peaks at ~314, 335, 357 and 377 nm, attributing to $^7F_0 \rightarrow {}^5H_3$, $^7F_0 \rightarrow {}^5D_4$ and $^7F_0 \rightarrow {}^5L_7$ transitions of Tb^{3+} ions, respectively.

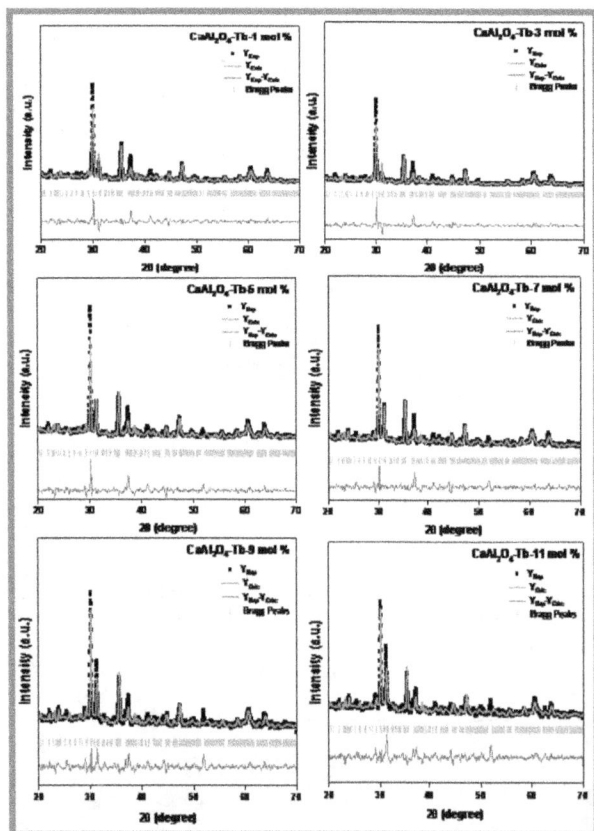

Figure 49.2 Rietveld refinement of $CaAl_2O_4$:Tb^{3+} (1–11 mol%) NPs.

Source: Author

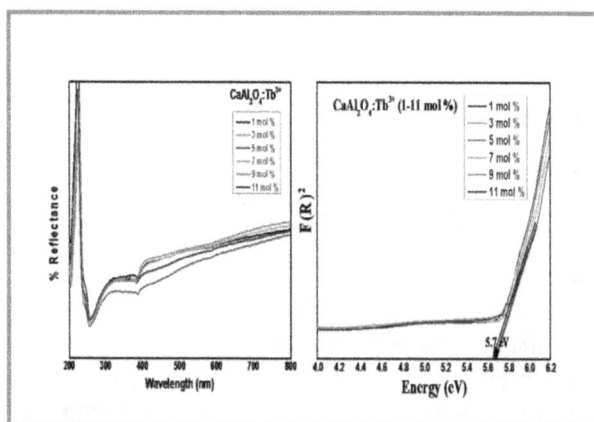

Figure 49.4 (a) DR spectra and (b) energy gap of the $CaAl_2O_4$:Tb^{3+} (1–11 mol%) NPs

Source: Author

Table 49.2 Energy gap of CaAl$_2$O$_4$:Tb^{3+} (1–11 mol%) NPs

CaAl$_2$O$_4$:Tb^{3+} (1–11 mol%)	Energy band gap (E_g) in eV
1	5.32
3	5.30
5	5.29
7	5.28
9	5.27
11	5.34

Source: Author

The PL emission spectra of CaAl$_2$O$_4$:Tb^{3+} (1–11%) NPs stimulated at 357 nm in the 450–650 nm range at room temperature are shown in Figure 49.5(b). Emission peaks were seen in the spectra at 494, 544, 588 and 618 nm ascribed to $^5D_4\rightarrow^7F_6$, $^5D_4\rightarrow^7F_5$, $^5D_4\rightarrow^7F_4$ and $^5D_4\rightarrow^7F_3$ of Tb^{3+} ion transitions [23]. The concentration quenching effect causes the PL intensity to increase until it reaches 5 mol%, after which it decreases. With higher activator concentrations, the inter-nuclear distance between the dopant usually decreases below the critical distance (R_c). When excitation energy passes through optical quenching centres, the emission intensity is decreased. Figure 49.6(a) depicts the variation in PL intensity with different Tb^{3+} ion concentrations.

In this study, Blasse's relation was used to calculate the R_c value between two nearby Tb^{3+} ions [24].

$$R_c = 2\left(\frac{3V}{4\pi N X_c}\right)^{1/3} \qquad (5)$$

where X_c is the ideal Tb^{3+} concentration (0.05), N is the number of cations/unit cell ($N = 4$), and V is the unit cell volume (97.435 Å^3). The R_c value was found to be ~ 11.6 Å and was greater than 5 Å. It was clear that the energy transfer between Tb^{3+} ions was caused by the non-radioactive energy transfer process, which could have resulted from multipole–multipole interaction. Multipolar interaction among the activator ions was found using a relation [25];

$$\frac{I}{x} = \frac{K}{\beta(x)^{\frac{Q}{3}}} \qquad (6)$$

where K and β are constants, I/x is the ratio of emission intensity (I) to activator ion concentration (x), and Q is the type of interaction ($Q = 6$, 8, and 10). The types of interactions, such as dipole-dipole (d–d), dipole-quadrupole (d–q), and quadrupole-quadrupole (q–q), are indicated by the measured value of Q. Figure 49.6(c) shows the log (I/x) vs. log (x) plot. The slope and Q value of the well-fitted logarithmic plot were found to be -1.365 and 6.43, respectively. The value of Q that was achieved was quite close to the theoretically expected value of 6. The concentration quenching was caused by the activator ions' endorse (d–d) contact.

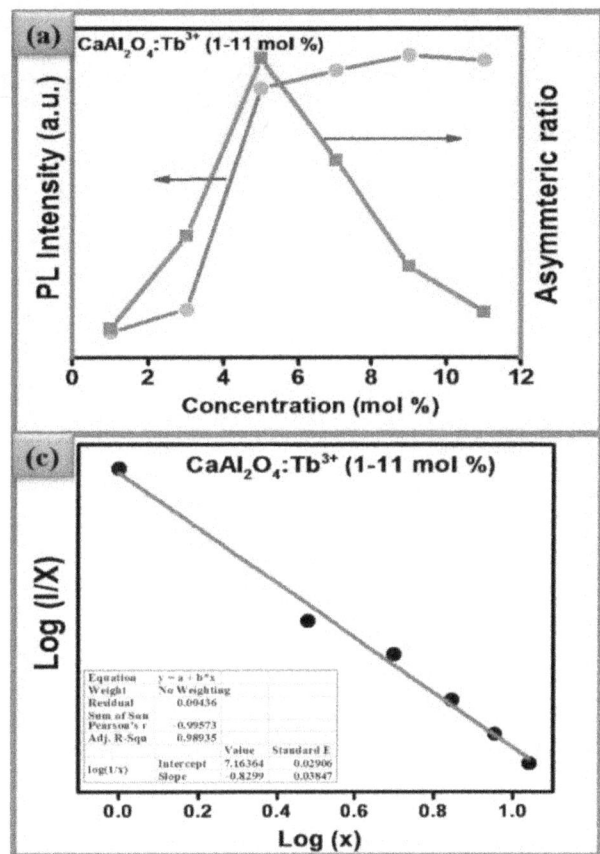

Figure 49.6 (a) Deviation of PL intensity and asymmetric ratio with concentration of Tb^{3+} ions, (c) Logarithmic plot of (x) and (I/x) in CaAl$_2$O$_4$:Tb^{3+} NPs
Source: Author

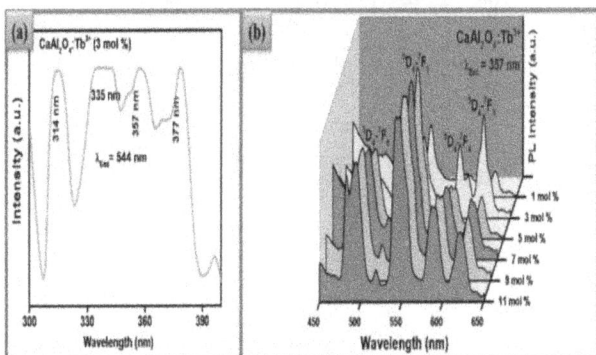

Figure 49.5 (a) PL excitation (b) PLE spectra of the CaAl$_2$O$_4$:Tb^{3+} (1–11 mol%) NPs
Source: Author

Photometric studies (CIE and CCT study)

Figure 49.7 displays the $CaAl_2O_4$:Tb (1–11 mol%) NPs' Commission Internationale de I'Eclairage (CIE) chromaticity diagram and correlated colour temperature (CCT) diagram obtained using 357 nm excitation. The green area of the CIE diagrams contained all of the synthetic chemicals with colour coordinates. The CCT is computed using CIE coordinates in order to establish the applicability of this green emission. A crucial metric for understanding how a light source's colour changes when heated to a specific temperature in comparison to a reference light source is CCT, which is expressed in Kelvin (K).

To obtain the CCT values, the light source's colour coordinates (x, y) were altered to (U_0, V_0) by using the subsequent Equations (7) and (8) [26];

$$U' = \frac{4x}{-2x + 12y + 3} \quad ------(7)$$

$$V' = \frac{9y}{-2x + 12y + 3} \quad ------(8)$$

Table 49.3 shows the estimated and tabulated photometric properties of $CaAl_2O_4$:Tb (1-11 mol%) and NPs.

Table 49.3 Photometric parameters of $CaAl_2O_4$:Tb (1–11 mol%) NPs

Tb conc. (mol%)	CIE		CT (K)
	X	Y	
1	0.3056	0.3081	7164
3	0.3225	0.3417	6015
5	0.3289	0.3479	6605
7	0.3087	0.3209	6804
9	0.3180	0.3413	6162
11	0.3215	0.3448	5984

Source: Author

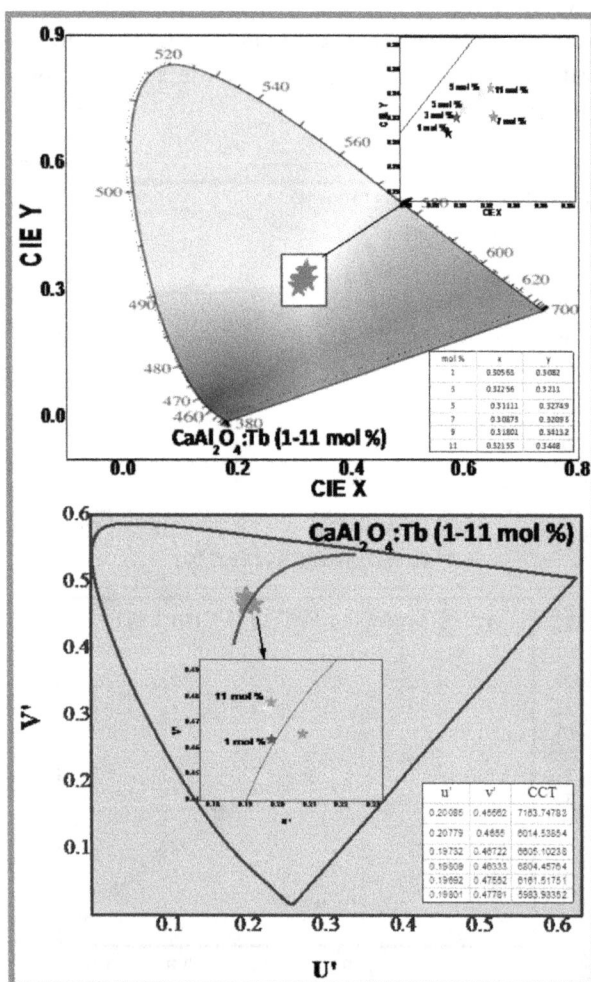

Figure 49.7 $CaAl_2O_4$ CIE and CCT graph: Tb (1–11 mol%) NPs
Source: Author

Conclusion

To summarise, lemon juice was used as fuel in combustion procedure to create unique $CaAl_2O_4$:Tb^{3+} (1–11 mol%) nano phosphors. The structural analysis was done using XRD and Rietveld refinement. The average crystallite size was found to be 33 nm. Photoluminescence emission spectra showed peak wavelengths of 494, 544, 588, and 618 nm attributing to $^5D_4 \rightarrow {^7F_J}$ (J = 6, 5, 4 and 3) transitions of Tb^{3+} ions. There was a visible rise in photoluminescence intensity up to 5 mol%, beyond which it decreased as the Tb^{3+} concentration increased due to concentration quenching (CQ). The results show that the observed CQ is mostly caused by the d–d interaction between Tb^{3+} ions. The predicted CIE values are found in the chromaticity's pure green zone and have a CCT value of 6455 K. These results disclose that the enhanced NPs are largely applicable for solid state lighting and WLEDs applications.

References

[1] Wanga, H., Liua, Y., Zhua, X., Weia, L., Jianga, X., Chenb, Y., Li, L. Preparation of $CaAlSiN_3$:Eu^{2+} red-emitting phosphor by a two-step method for solid-state lighting applications. *Ceram. Int.*, 2020;46:23035–23040.

[2] Fang, X., Hu, L., Ye, C., Zhang, L. One-dimensional inorganic semiconductor nanostructures: A new carrier for nanosensors. *Pure Appl. Chem.*, 2010;82:2185–2198.

[3] Wu, C., Wang, K., Batmunkh, M., Bati, A. S. R., Yang, D., Jiang, Y., Hou, Y., Shapter, J. G., Priya, S. Multifunctional nanostructured materials for next generation photovoltaics. *Nano Energy*, 2019;70:104480.

[4] Stanisz, M., Klapiszewski, Ł., Jesionowski, T. Recent advances in the fabrication and application of biopolymer-based micro- and nanostructures: A comprehensive review. *Chem. Eng. J.*, 2020;397:125409.

[5] Zhang, Q., Li, H., Ma, Y., Zhai, T. ZnSe nanostructures: Synthesis, properties and applications. *Prog. Mater. Sci.*, 2016;83:472–535.

[6] Wang, Y., Lin, C., Zheng, H., Sun, D., Li, L., Chen, B. Fluorescent and chromatic properties of visible-emitting phosphor KLa(MoO$_4$)$_2$:Sm^{3+}. *J. Alloys Comp.*, 2013;559:123–128.

[7] Li, J., Lu, G., Xie, B., Wang, Y., Guo, Y. Nano-preparation promoting effectively luminescent properties of one-dimensional rare earth oxides. *J. Rare Earths.*, 2012;30:1096–1101.

[8] Wen, H. L., Duan, C. K., Jia, G. H., Tanner, P. A., Brik, M. G. Glass composition and excitation wavelength dependence of the luminescence of Eu3+ doped lead borate glass *J. Appl. Phys.*, 2011;110:033536–033544.

[9] Wang, Y., Wang, L. Concentration-dependent effects of optical storage properties in CSSO: Dy. *J. Appl. Phys.*, 2007;101:053108–053108-5.

[10] Davtyan, S. P., Avetisyan, A. S., Berlin, A. A., Tonoyan, A. O. Synthesis and properties of particle-filled and intercalated polymer nanocomposites. Revi. J. Chem., 2013;3: 1–51.

[11] Li, J., Yan, J., Wen, D., Khan, W., Shi, J., Wu, M., Sua, Q., Tanner, P. A. Advanced red phosphors for white light-emitting diodes. *J. Mater. Chem. C Mater.*, 2016;4:8611–8623.

[12] Xie, R., Hirosaki, J. Silicon-based oxynitride and nitride phosphors for white LEDs—A review. *Sci. Technol. Adv. Mater.*, 2007;8:588–600.

[13] Liu, C., Xia, Z., Lian, Z., Zhou, J., Yan, Q. Structure and luminescence properties of green-emitting NaBaScSi2O7: Eu2+ phosphors for near-UV-pumped light emitting diodes. *J. Mater. Chem. C Mater.*, 2013;1:7139–7147.

[14] Naik, R., Prashantha, S. C., Nagabhushana, H., Sharma, S. C., Nagaswarupa, H. P., Anantharaju, K. S., Nagabhushana, B. M., Premakumar, H. B., Girish, K. M. A single phase, red emissive Mg2SiO4:Sm3+ nanophosphor prepared via rapid propellant combustion route. *Spectrochim. Acta A.*, 2015;40:516–523.

[15] Zhang, C., Lin, J. Defect-related luminescent materials: synthesis, emission properties and applications. *Chem. Soc. Rev.*, 2012;41:7938–7961.

[16] Feldmann, C., Justel, T., Ronda, C. R., Schmidt, P. J. Inorganic luminescent materials: 100 years of research and application. *Adv. Funct. Mater.*, 2003;13:511–516.

[17] Wang, L., Wang, X., Takeda, T., Hirosaki, N., Tsai, Y., Liu, R., Xie, R. Structure, luminescence, and application of a robust carbidonitride blue phosphor (Al1–xSix-CxN1–x:Eu2+) for near UV-LED driven solid state lighting. *Chem. Mater.*, 2015;27:8457–8466.

[18] Premkumar, H. B., Ravikumar, B. S., Sunitha, D. V., Nagabhushana, H., Sharma, S. C., Savitha, M. B., Mohandas Bhat, S., Nagabhushana, B. M., Chakradhar, R. P. S. Investigation of structural and luminescence properties of Ho3+ doped YAlO3 nanophosphors synthesized through solution combustion route. *Spectrochim. Acta A.*, 2013;115:234–243.

[19] Ryu, H., Bartwal, K. S. Operation by Eu luminescence in CaAl$_2$O$_4$:Eu^{2+} alloys by Zn substitution. *J. Alloy Compd.*, 2008;461:395–98.

[20] Zhang, Y., Chen, J., Xu, C., Li, Y., Jin Seo, H. Photoluminescence and abnormal reduction of Eu^{3+} ions in CaAl$_2$O$_4$:Eu nanophosphors calcined in air atmosphere. *Phys. B Condens. Matt.*, 2015; 472:6–10.

[21] He, K., Chen, N., Wang, C., Wei, L., Chen, J. Method for determining crystal grain size by X-ray diffraction. *Cryst. Res. Technol.*, 2018;53:1–6.

[22] Jamil, H., Dildar, I. M., Ilyas, U., Hashmi, J. Z., Shaukat, S., Sarwar, M. N., Khaleeq-ur-Rahman, M. Microstructural and optical study of polycrystalline manganese oxide films using Kubelka-Munk function. *Thin Solid. Film.*, 2021;732:138796.

[23] Basavaraj, R. B., Nagabhushana, H., Darshan, G. P., Daruka Prasad, B., Rahul, M., Sharma, S. C., Sudaramani, R., Archana, K. V. Red and green emitting CTAB assisted CdSiO3: Tb3+/Eu3+nanopowders as fluorescent labeling agents used in forensic and display applications. *Dyes Pigm.*, 2017;147:364–377.

[24] Blasse, G. *Philips Res. Report* 1969;24:131–144.

[25] Darshan, G. P., Premkumar, H. B., Nagabhushana, H., Sharma, S. C., Daruka Prasad, B., Prashantha, S. C., Basavaraj, R. B. Superstructures of doped yttrium aluminates for luminescent and advanced forensic investigations. *J. Alloy. Compd.*, 2016;686:577–587.

[26] Judd, B., Deane, J. Estimation of chromaticity differences and nearest color temperature on the standard 1931 ICI colorimetric coordinate system. *Opt. Soc. Am.*, 1936;26:421–426.

50 Azo-benzothiazole-based copper (II) complexes: Synthesis, spectral investigation, thermal kinetics and antibacterial investigations

Nagaraj Basavegowda[1], Harisha S.[2,a], Manjunath[2], Mohan Reddy R.[3], S. R. Kiran Kumar[2] and Shoukat Ali R. A.[4]

[1]Department of Biotechnology, Yeungnam University, Gyeongsan, Gyeongbuk, 38541, Republic of Korea

[2]Assistant Professor, Centre for Nanosciences, Department of Chemistry, K.S. Institute of Technology, Bengaluru–560062, Karnataka, India

[3]Department of Chemistry, BMS College of Engineering, Bengaluru–560019, Karnataka, India

[4]Department of PG Studies and Research in Chemistry, Davanagere University, Davanagere–577007, Karnataka, India

Abstract

Novel Cu(II) complexes with azo ligands derived from benzothiazole were synthesised and their molecular structures were elucidated using IR, Powder XRD, VSM, ESR, TGA, and UV spectroscopy. Thermo gravimetric analysis determined weight loss at various temperatures for both ligands and complexes. Broido's method was applied to calculate the activation energies of thermal decomposition, revealing first-order kinetics through $\log[\ln(1/y)]$ vs. 1000/K plots. Stability factors of the Cu azo complex of [3]L were tested against bacterial and fungal strains. This research enhances understanding of the thermal behaviour and decomposition kinetics of benzothiazole-based metal-azo complexes, with potential applications in catalysis, materials science, and biomedical fields. The methodology serves as a template for evaluating similar metal-organic complexes.

Keywords: Azo dye, benzothiazole, pyridone, Broido's plots, magnetic susceptibility, energy of activation, antimicrobial activity

Introduction

Azo dyes derived from heterocyclic compounds are promising ligands due to their strong affinity for various metal ions, forming stable and vividly coloured metal complexes with enhanced solubility in different solvents [1–6]. Donor sites from heteroatoms such as oxygen, nitrogen, and sulphur enable stable coordination, making azo dyes valuable in synthesising metal complexes and chelates [3–6]. Azo metal complexes have diverse applications in tissue staining, pharmaceuticals, and the food industry due to their stability and vibrant colours [7–12].

Transition metal complexes of heterocyclic azo dyes exhibit potent biological activity, particularly antimicrobial properties [13]. They are widely used in medical treatments, including silver bandages for burn care [14, 15], zinc-based disinfectants [16], and bismuth drugs for ulcers [17]. The increasing threat of drug-resistant bacteria and viral infections highlights the importance of metal-based drugs and treatments [18–22].

Azo dyes containing heterocyclic diazo-components also participate in significant biological reactions, including DNA, RNA, and protein synthesis inhibition, anticancer activity, and nitrogen fixation [23–26]. Azo rhodanine complexes exhibited significant antibacterial and antifungal properties [27]. Additionally, azo metal complexes have advanced technological applications, including use in inkjet printers, lasers, textile dyeing, and biomedical imaging [28]. Benzothiazole-based azo metal complexes serve as versatile materials in metal signalling, sensor molecules, optical devices, and trace metal analysis, with remarkable stability under intense infrared irradiation [29–31].

Based on these insights, copper-azo complexes were synthesised using copper acetate as a complexing agent. The complexes were characterised using IR, powder XRD, VSM, ESR, TGA, and UV-Vis techniques. Thermo gravimetric analysis evaluated their thermal stability, kinetics, and thermodynamic parameters (ΔS, ΔH, and ΔG) using Broido's method. The Cu-azo complex of 3L was tested for antimicrobial activity, confirming its stability and biological potential. These findings suggest copper-azo complexes have promising applications in

[a]harishas@ksit.edu.in

DOI: 10.1201/9781003685876-50

pharmaceuticals and material sciences, offering new avenues for developing metal-based drugs and antimicrobial agents.

Methods and materials

General

All chemicals and reagents used were purchased from SD Fine, Sigma-Aldrich, and Spectrochem, and utilised without any further purification. Copper(II) complexes were synthesised by reacting ligands 1L, 2L, and 3L with copper(II) acetate monohydrate in an ethanolic medium. The melting points of the ligands and their corresponding copper complexes were determined using a thermo electrical melting point apparatus with open capillary tubes, and the values obtained are uncorrected.

The electronic absorption spectra of the synthesised copper complexes were recorded in the 400–800 nm range using a SHIMADZU UV-Visible 1800 spectrophotometer. FTIR spectra were collected in the region of 400–4000 cm^{-1} with the aid of a Bruker Alpha-T FTIR spectrometer using KBr pellets. Powder X-ray diffraction (XRD) patterns were obtained with a Rigaku Miniflex 600 diffractometer operating at 40 kV and 15 mA. Magnetic susceptibility measurements were carried out using a vibrating sample magnetometer (VSM). Thermo gravimetric and derivative thermo gravimetric (TGA/DTG) analyses were performed on a TA Universal analyser under a nitrogen atmosphere. Electron spin resonance (ESR) spectra were recorded at room temperature using a Bruker EMX Plus series spectrometer.

Microbial strains

The in vitro antimicrobial activity was evaluated against six microbial strains, including two Gram-positive bacteria (*Staphylococcus aureus* and *Bacillus subtilis*), two Gram-negative bacteria (*Escherichia coli* and *Pseudomonas aeruginosa*), and two fungal species (*Aspergillus flavus* and *Candida albicans*). These microbial cultures were obtained from NCL, Pune, India. The strains were incubated overnight at 37 °C on nutrient agar and potato dextrose agar media for antimicrobial testing.

Experimental

General experimental procedure

A hot solution of recrystallised azo dye ¹L (2 mmol) in 30 mL of ethanol was prepared in a two-necked round-bottom flask. A hot ethanolic solution of copper acetate (1 mmol) was added, and the mixture was refluxed for 8 hours. After cooling, the precipitate was filtered, washed with ethanol, water, and petroleum ether, and dried under vacuum using anhydrous $CaCl_2$. A similar procedure was followed for preparing other azo metal complexes with different ligands and metal ions, as described in Schemes 1–3.

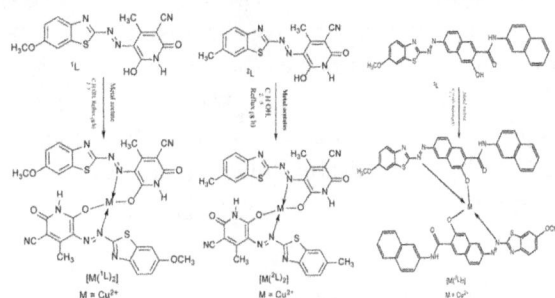

Scheme 1 – Synthesis of Copper complex of ¹L   Scheme 2 – Synthesis of Copper complex of ²L   Scheme 3 – Synthesis of Copper complex of ³L

Antimicrobial activity

The antimicrobial activity of the Cu azo complex of ³L was tested using the agar disc diffusion method. Streptomycin and Fluconazole served as reference drugs for bacteria and fungi, respectively. Pre-sterilised filter paper discs (6 mm) impregnated with the Cu azo complex (100 μg/mL) were placed on inoculated agar plates and incubated at 37°C for 24 hours (bacteria) and 28°C for 48 hours (fungi).

Minimum inhibitory concentration (MIC)

The MIC of the Cu–azo complex derived from ligand ³L was evaluated using the standard dilution technique. The complex was initially dissolved and then subjected to a two-fold serial dilution to obtain a range of concentrations from 5 to 75 μg/mL. These concentrations were employed to assess the inhibitory effect of the compound. The MIC was defined as the lowest concentration at which no visible microbial growth was observed on the agar surface.

Results and discussions

The metal complexes were synthesised using a ratio of 1:2 of metal salts to ligands which were shown in Schemes 1–3. Table 50.1 provides data on yield, molecular formula, solubility, and molecular weights. The coloured, non-hygroscopic complexes are stable at room temperature and soluble in DMF, dichloromethane, and DMSO.

UV-visible spectra of metal complexes

The UV-visible spectra of copper complexes from ligands ¹L and ²L were measured in DMSO and DMF, while those from ligand 3L were recorded in MDC at 300–700 nm (10^{-6}–10^{-8} M). Ligand peaks at 330–450 nm (π–π^* transitions) shifted to lower frequencies upon copper coordination, indicating ligand-to-metal charge transfer (LMCT) from nitrogen and oxygen donors [32]. The azo group (-N=N-) forms copper-nitrogen bonds, stabilising the complex [33]. Copper complexes showed two transitions at 19230–30303 cm^{-1} $^2E_g \rightarrow {^2}T_{2g}$ transitions, suggesting square planar geometry, with an LMCT band at 22222 cm^{-1} [34] suggesting square planar geometry, with an LMCT band at 22222 cm^{-1} [34].

Table 50.1 Electronic spectral data of synthesised azo metal complexes

Complexes	λ_{max} (in nm)	Wave number (in cm^{-1})	Assignment
¹LCu	330–520	19230–30303	$19230 - {^2}E_g \rightarrow {^2}T_{2g}$ 22222 – charge transfer
²LCu			
³LCu			

Source: Author

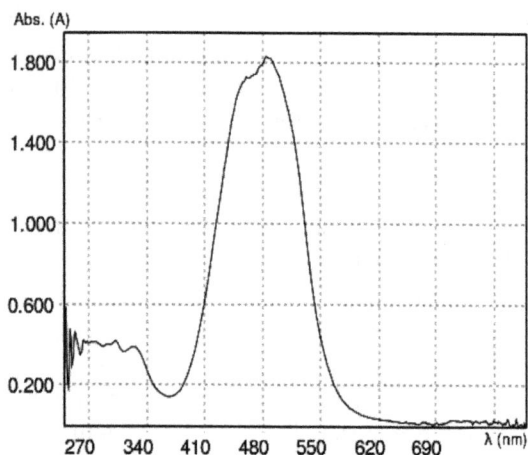
Figure S2 Electronic spectum of ¹LCu in in DMSO
Source: Author

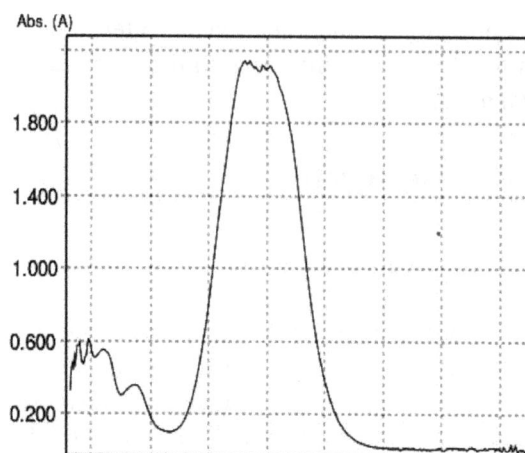
Figure S3 S3 Electronic spectum of ²LCu in DMF
Source: Author

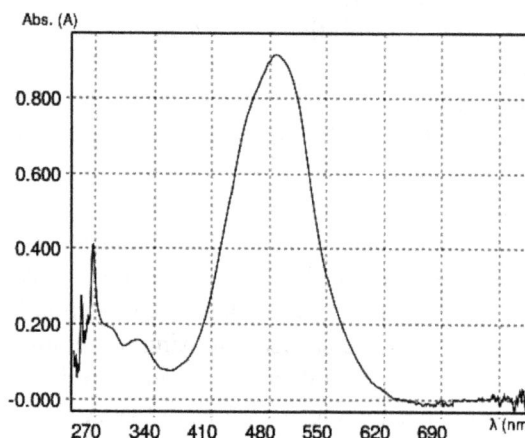
Figure S1 Electronic spectum of ¹LCu in DMF
Source: Author

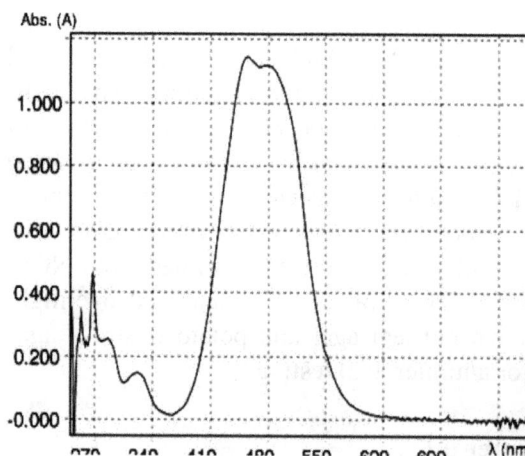
Figure S3 S3 Electronic spectum of ²LCu in DMF
Source: Author

The IR spectra of the synthesised copper complexes exhibit distinct bands corresponding to both ligand functionalities and metal–ligand bonding. Broad bands observed at 3412–3440 cm^{-1} in the free ligands, assigned to –OH stretching, are absent in the complexes, suggesting coordination through

Table S1 Physical characteristics of synthesized metal complexes

Sl No.	Complex	Molecular formula	Mol. Wt	Yield	Solubility
1	1LCu	$C_{30}H_{20}CuN_{10}O_6S_2$	744.21	65%	DMF/DSMO
2	2LCu	$C_{30}H_{20}CuN_{10}O_4S_2$	712.22	63%	DMF/DSMO
3	3LCu	$C_{58}H_{42}CuN_8O_6S_2$	1074.68	70%	MDC

Source: Author

hydroxyl oxygen atoms. Additionally, a broad band at lower frequencies indicates the presence of –NH groups. The azo (–N=N–) group shows a shift from 1507–1551 cm^{-1} in the ligands to 1455–1469 cm^{-1} in the complexes, indicating coordination through azomethine nitrogen (M←N). This shift (30–63 cm^{-1}) confirms metal ion involvement. Corresponding spectra are shown in Figures S5–S7 (Table 50.2).

Powder XRD studies
The diffraction patterns of the synthesised complexes were recorded across a 2θ range spanning from 10° to 80°, indicating the non-crystalline nature of the complexes. However, it appears that the ligands exhibit relatively higher crystallinity compared to their corresponding metal complexes. The crystallite size of the samples was determined using Scherrer's formula, $D=k\lambda/\beta\cos\theta$, where k is a constant (taken as 0.94), λ represents the wavelength of X-ray used (λ=0.154 nm), β is the full width at half maximum, and θ denotes Bragg's angle. The diffraction has been compiled in Table 50.3.

Magnetic susceptibility studies
The magnetic properties of Cu(II) azo complexes were analysed using a vibrating sample magnetometer (VSM). The Cu(II) complexes of ligands ^1L, ^2L, and ^3L showed magnetic moments of 1.84, 1.91, and 1.82 BM, respectively, suggesting a square planar

Table 50.2 IR spectral data of synthesised azo metal complexes dyes

Name of the complex	IR (v_{max} in cm^{-1})
^{1}**LCu**	3469 cm^{-1}(OH)
	2225 cm^{-1}(CN)
	1469 cm^{-1}(N=N)
	524 cm^{-1}(M-N)
	412 cm^{-1}(M-O)
^{2}**LCu**	3447 cm^{-1}(OH)
	2216 cm^{-1}(CN)
	1449 cm^{-1}(N=N)
	562 cm^{-1}(M-N)
	474 cm^{-1}(M-O)
^{3}**LCu**	3425 cm^{-1}(OH)
	2311 cm^{-1}(CN)
	1461 cm^{-1}(N=N)
	562 cm^{-1}(M-N)
	476 cm^{-1}(M-O)

Source: Author

geometry. The values exceed the expected 1.73 BM for one unpaired electron, indicating paramagnetic behaviour and monomeric nature without metal-metal interaction. The geometry prediction was supported by the measured magnetic moments. VSM plots of the Cu(II) complexes are shown in Figures S8–S10.

ESR spectral study of Cu(II) complexes
ESR analysis of paramagnetic metal (II) complexes offers insights into the distribution of unpaired electrons within the d-orbitals and the characteristics of the metal-ligand bonding. ESR spectra of Cu(II) complexes were recorded at room temperature using polycrystalline samples in DMSO solution (Figures 50.1–50.3). The spin Hamiltonian parameters of the complexes have been calculated and are provided in Table 50.4. Analysis of ESR spectra yielded calculated g values, with values of $g_{||}$ = 2.16 and $g_{\perp}$ = 2.02, where $g_{||} > g_{\perp} > g_e$. These results indicate that the unpaired electron is likely located in the $d_{x^2-y^2}$ orbital of copper in the metal complexes [38].

Table 50.3 Particle size, d-spacing and strain

Sample	2 Theta	whom	Particle size (nm)	d-Spacing	Strain
^{1}L	26.715	4.985	1.619	3.334	0.091
^{3}L	25.905	3.497	2.304	3.436	0.066
^{1}LCu	22.418	1.355	5.910	3.962	0.029
	27.403	0.741	10.904	3.251	0.013
^{3}LCu	24.072	1.228	6.537	3.693	0.025

Source: Author

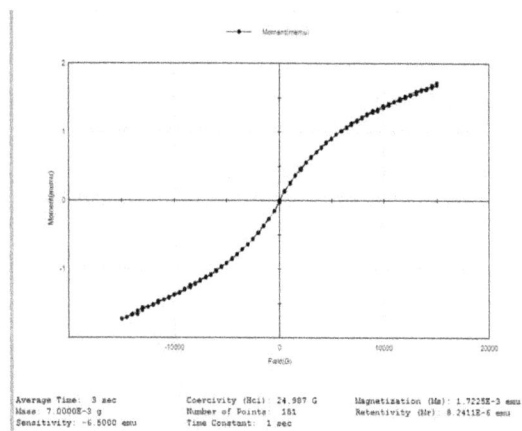

Figure S8 VSM spectrum of ^{1}LCu (II) Complex
Source: Author

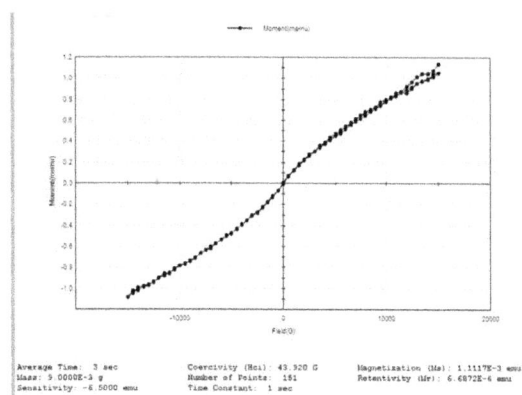

Figure S9 VSM spectrum of ^{2}LCu (II) Complex
Source: Author

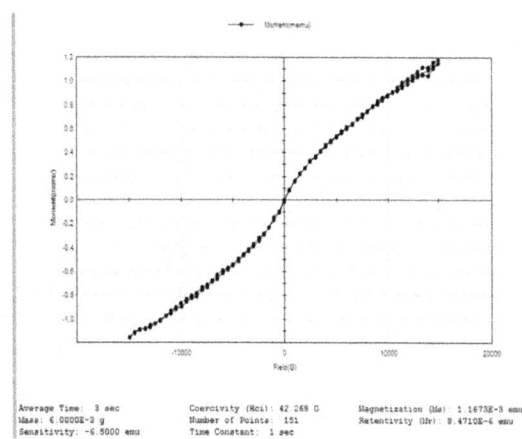

Figure S10 VSM spectrum of ^{3}LCu (II) Complex
Source: Author

The assessment of the covalent character of the metal-ligand bond relied on the value of $g_\parallel$. Literature suggests that $g_\parallel$ values above 2.3 indicate ionic character, while lower values indicate covalent character [39]. In this study, the $g_\parallel$ value was determined to

Table 50.4 ESR spectral data of Cu(II) complex of azo dye ligand (^{1}L, ^{2}L, ^{3}L)

Complex	$g_\parallel$	$g_\perp$	g_{av}	G
^{1}LCu	2.108	2.03	2.07	3.62
^{2}LCu	2.114	2.039	2.09	2.86
^{3}LCu	2.115	2.005	2.07	3.23

Source: Author

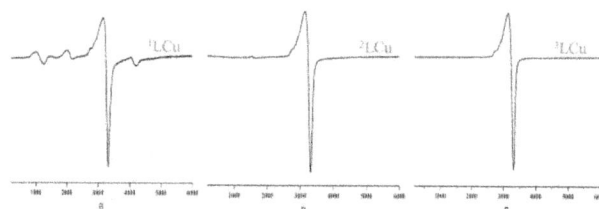

Figure 50.1 ESR spectra of Cu complexes of (^{1}L, ^{2}L and ^{3}L)
Source: Author

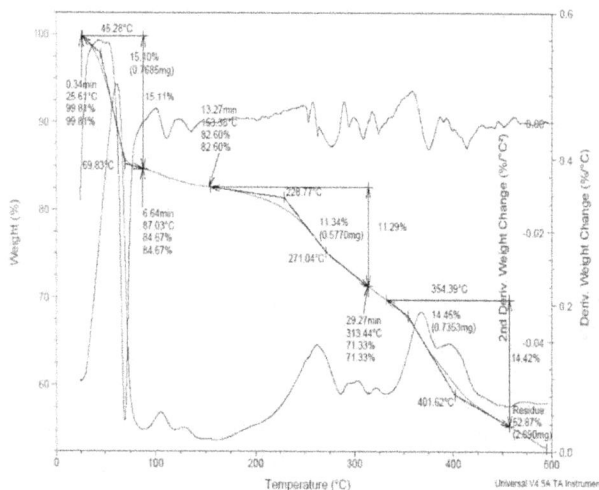

Figure 50.2 TG–DTG curve of ^{1}L (ligand)
Source: Author

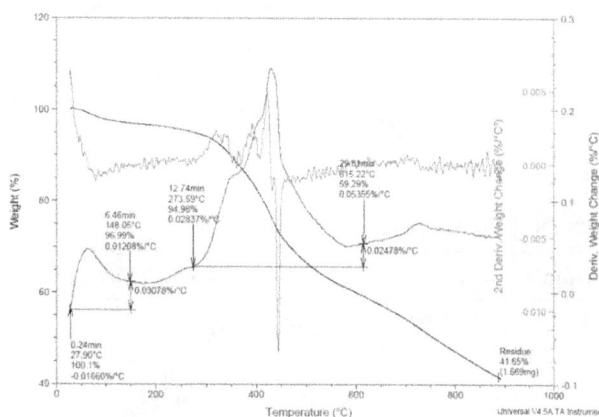

Figure 50.3 TG–DTG curve of ^{1}LCu complex
Source: Author

be 2.16, which is below 2.3, suggesting a substantial covalent character in the metal-ligand bond.

The exchange interaction parameter between two Cu(II) ions can be accessed using the Hathaway expression $G = (g_\parallel - 2)/(g_\perp - 2)$. According to this expression, a G value greater than 4 implies minimal exchange interaction between the Cu(II) centres within the solid-state complex. Conversely, a G value less than 4 is indicative of notable exchange interactions. In the present investigation, the calculated G value was found to be below 4, pointing to significant exchange interaction between Cu(II) ions in the complex [39]. The ESR spectrum of the Cu(II) complex suggests that the copper ion adopts a square planar geometry. Additionally, the extent of distortion in the geometry can be estimated by calculating g_{av} using the relation $g_{av}^2 = 1/3 \, (2 \, g_\parallel^2 + g_\perp^2)$.

Thermal properties of complexes

In this study, we utilised dynamic TGA technique to investigate the thermal behaviour of both ligands their corresponding metal complexes, enabling assessment of their thermal stability. Thermal analysis is instrumental in discerning the relative thermal stabilities of these materials, as well as in detecting the presence or absence of lattice/coordinated water molecules associated with the metal complexes, if present [40]

The focus of this experiment is the thermal behaviour analysis of three ligands and their corresponding metal complexes, accompanied by thermo grams depicted in Figures 50.2–50.7. Table 50.5 presents the proposed mode of fragmentation for both ligands and their respective metal complexes; along with the temperature range during mass loss occurs.

The thermal data suggests that the decomposition of the complexes proceeded in two or three distinct steps.

Figures 50.3, 50.5, and 50.7 depict the thermal behaviour of Cu(II) complexes derived from ^1L and ^3L, respectively, revealing three degradation steps for complexes derived from ^1L and ^3L, and two degradation steps for the complex derived from ^2L.

The Cu(II) complex of ^1L experienced degradation in two steps. During the first step, a weight loss of 3.29–3.85% was observed in the temperature range of 25–145°C, likely due to the removal of methoxy groups from the benzothiazole ring. DTG peaks with T_{max} ranging from 77°C to 95°C were observed

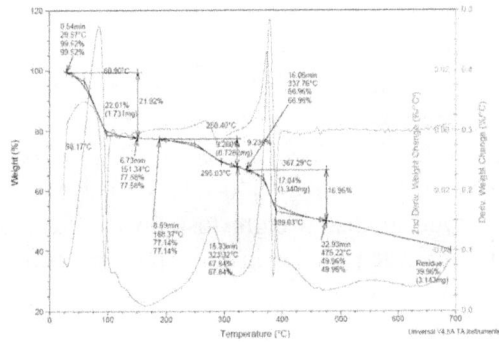

Figure 50.4 TG–DTG curve of ^2L (ligand)
Source: Author

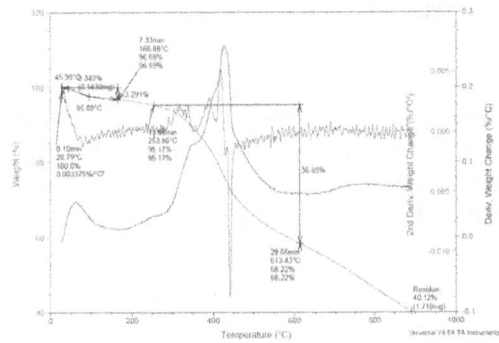

Figure 50.5 TG–DTG curve of ^2LCu complex
Source: Author

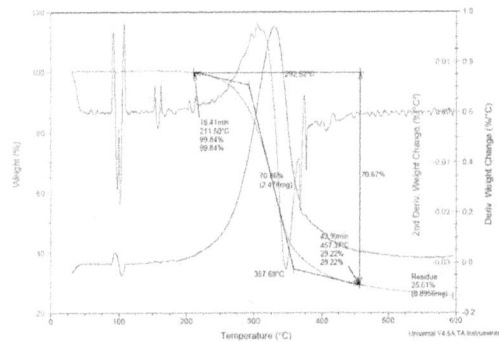

Figure 50.6 TG–DTG curve of ^3L (ligand)
Source: Author

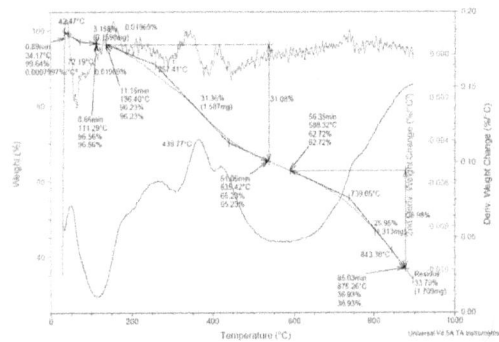

Figure 50.7 TG–DTG curve of ^3LCu complex
Source: Author

for these complexes. The second degradation step occurred within the temperature range of 255–635°C, leading to a weight loss of 35.42–36.89%, likely attributed to the loss of the pyridone moiety of the molecule. The DTG peaks observed for this degradation step ranged from T_{max}=440–445°C. The complexes displayed gradual weight loss up to 752°C, presumably attributed to the loss of the remaining organic moiety.

Determination of kinetic parameters

The kinetic and thermodynamic parameters of the complexes were determined using Broido's method [41, 42]. Degradation was monitored by weighing the sample during heating, assuming the formation of volatile degradation products. Activation energy (Ea) was used to calculate enthalpy change (ΔH), entropy change (ΔS), and free energy change (ΔG). Positive ΔH indicates endothermic decomposition, while positive ΔG reflects non-spontaneous processes. Higher

Ea causes rapid degradation, whereas lower Ea results in gradual decomposition [43, 44]. Broido's plots of log[ln(1/y)] versus 1000/K and ln(a/a–x) versus time, shown in Figures 50.8–50.19, were straight lines, confirming first-order kinetics. The value of "y" represents the remaining compound at temperature T (in Kelvin). The slopes of these plots were used to calculate activation energies at different decomposition stages. The obtained kinetic parameters (ΔH, ΔS, and ΔG) are summarised in Table 50.6. The stability of the complexes is strongly influenced by the activation energy values [45].

Antimicrobial activity

The antimicrobial activity of the Cu azo complex of ^3L was evaluated using the agar disc diffusion method, and the minimum inhibitory concentration (MIC) was determined using a dilution method. The complex demonstrated *in vitro* antimicrobial

Table 50.5 Thermal data of the complexes of ^1LCu, ^2LCu, and ^3LCu

Entry	Mass loss temp (°C)	DTG$_{max}$ (°C)	% Weight loss		Proposed mode of degradation and fragments loss
			Obsd.	Calcd.	
^1LCu	25–145	77	3.88	3.85	Loss due to two methoxy groups
	395–635	445	35.42	35.22	Loss due to two molecules of pyridone moiety
^2LCu	30–90	95	3.29	3.53	Loss due to two methyl groups
	255–615	442	36.89	35.72	Loss due to two molecules of pyridone moiety
	570–810	715	14.64	14.52	Loss of naphthol moiety
^3LCu	35–95	72	3.15	3.18	Loss of methoxy groups
	145–595	439	31.36	31.23	Loss of benzothiazole moiety
	620–820	739	25.95	25.80	Loss of naphthol moiety

Source: Author

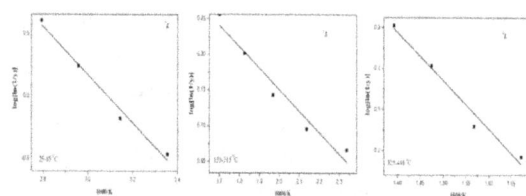

Figure 50.8 log [ln(1/y)] vs. 1000/K plots for thermal decomposition of azo ligand ^1L

Source: Author

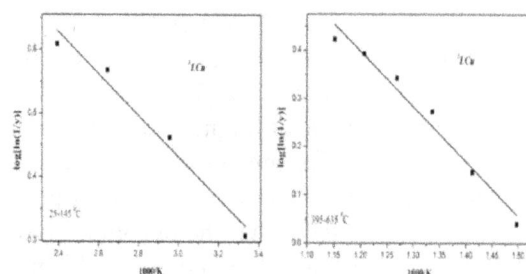

Figure 50.9 log [ln(1/y)] vs. 1000/K plots for thermal decomposition of azo complex ^1LCu

Source: Author

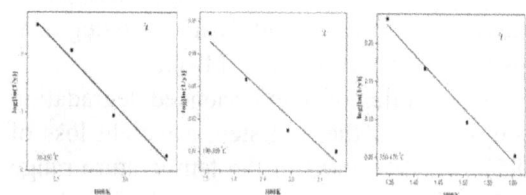

Figure 50.10 log [ln(1/y)] vs. 1000/K plots for thermal decomposition of azo ligand ^2L

Source: Author

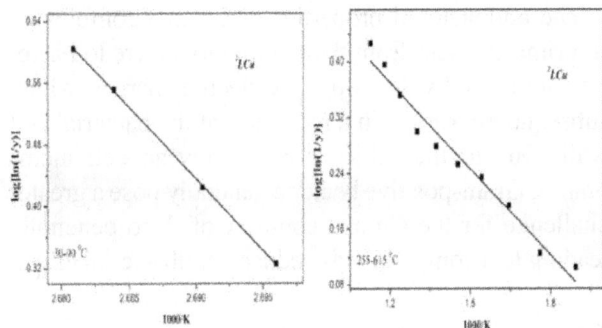

Figure 50.11 log [ln(1/y) vs. 1000/K plots for thermal decomposition of azo complex ^{2}LCu

Source: Author

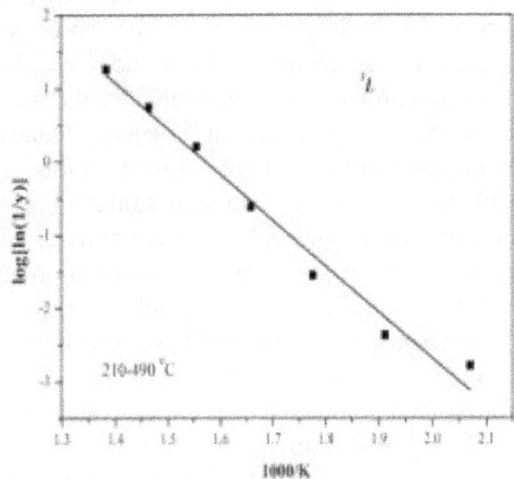

Figure 50.12 log [ln(1/y) vs. 1000/K plots for thermal decomposition of azo ligand ^{3}L

Source: Author

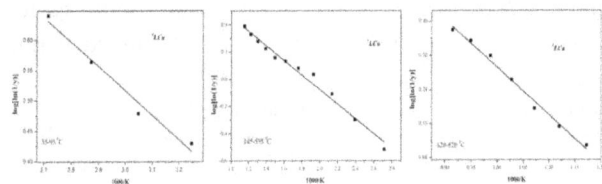

Figure 50.13 log [ln(1/y) vs. 1000/K plots for thermal decomposition of azo complex ^{3}LCu

Source: Author

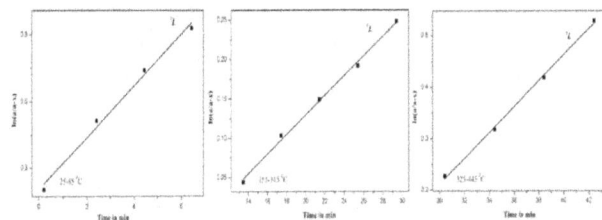

Figure 50.14 Plots of ln(a/a-x) vs. Time for thermal decomposition of azo ligand ^{1}L

Source: Author

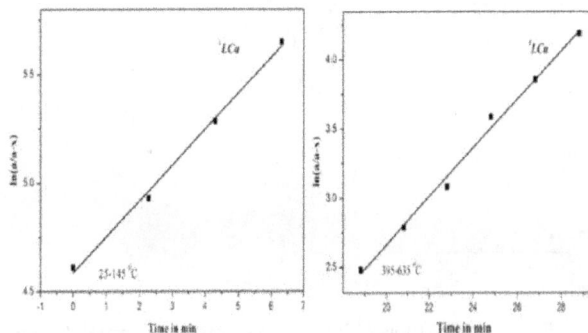

Figure 50.15 Plots of ln(a/a-x) vs. time for thermal decomposition of azo complex ^{1}LCu

Source: Author

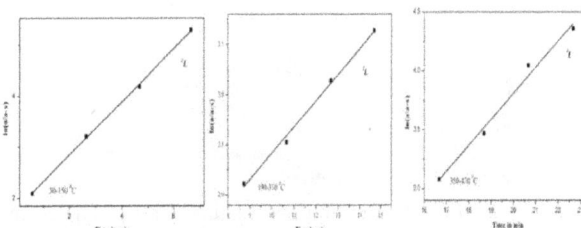

Figure 50.16 Plots of ln(a/a-x) vs. Time for thermal decomposition of azo ligand ^{2}L

Source: Author

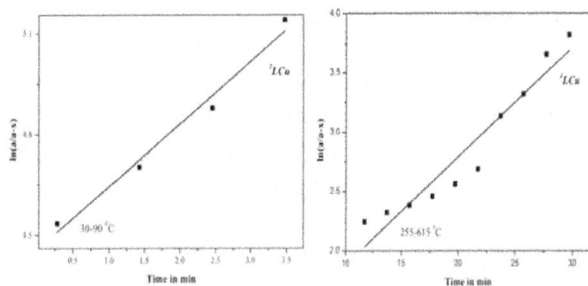

Figure 50.17 Plots of ln(a/a-x) vs. time for thermal decomposition of azo complex ^{2}LCu

Source: Author

activity against four bacterial strains, comprising two Gram-positive strains (*S. aureus, Bacillus subtilis*) and two Gram-negative strains (*Escherichia coli, Pseudomonas aeruginosa*), as well as two fungal strains (*Aspergillus flavus, Candida albicans*).

Table 50.7 displays the outcomes of the antibacterial activity testing of the Cu azo complex of ^{3}L. The MIC is described as the lowest concentration of the Cu azo complex of ^{3}L that hinders the growth of a microorganism. Cu azo complex of ^{3}L demonstrated MIC values of 28 μg/mL for *E. coli* and 31 μg/mL for *P. aeruginosa*.

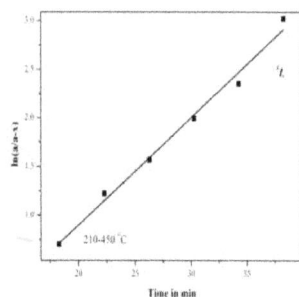

Figure 50.18 Plots of ln(a/a-x) vs. time for thermal decomposition of azo ligand ³L

Source: Author

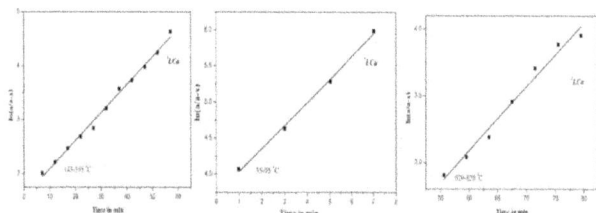

Figure 50.19 Plots of ln(a/a-x) vs. time for thermal decomposition of azo complex ³LCu

Source: Author

Considering the MIC values, *E. coli* and *P. aeruginosa* displayed the highest sensitivity to the Cu azo complex of ³L, whereas *B. subtilis, C. albicans,* and *A. flavus* exhibited the lowest sensitivity among the tested microbes. The antimicrobial efficacy of the tested Cu azo complex of ³L was compared to that of the positive control drugs *streptomycin* and *fluconazole.*

The antibacterial properties of Cu azo complex of ³L primarily stem from their ability to adhere to bacteria, facilitated by their opposite electric charges, which subsequently results in a reduction of the bacterial cell wall. Due to their thicker peptidoglycan cell membranes, Gram-positive bacteria generally pose a greater challenge for the Cu azo complex of ³L to penetrate, leading to a comparatively reduced antibacterial effect.

Conclusion

We synthesised metal complexes of benzothiazole-based azo dyes with copper. The proposed structures were confirmed through IR, Powder XRD, VSM, ESR, TGA, and UV spectroscopy. Thermo gravimetric analysis determined weight loss at various temperatures and evaluated thermodynamic parameters. High activation energy values indicated strong thermal stability. Broido's plots of log[ln(1/y)] versus 1000/K exhibited straight lines, confirming first-order kinetics. Negative entropy of activation values suggests increased structural order in the complexes compared to the ligands.

The synthesised azo metal complexes show potential in memory storage, electronic devices, and non-linear optical materials due to their stability. Their catalytic potential in organic transformations, such as cross-coupling reactions, was also noted. Integration with polymers could enhance optoelectronic and sensor properties. The Cu-azo complex of 3L exhibited significant antimicrobial activity against *E. coli* and *P. aeruginosa*, highlighting its biomedical potential. Stability was evaluated using

Table 50.6 Thermodynamic parameters of the thermal decomposition of azo metal complexes using Broido's methods

Entry	Decomposition range (°C)	DTG$_{max}$ (°C)	E$_a$ kJ mol^{-1}	ΔH kJ mol^{-1}	ΔS J mol^{-1} K^{-1}	ΔG kJ mol^{-1}
¹L	First (25–85)	45	36.86	36.48	-103.72	4.73
	Second (155–315)	228	6.08	4.17	-104.90	23.99
	Third (325–460)	354	22.33	19.38	-104.51	37.05
¹LCu	First (25–145)	77	6.23	5.59	-103.06	7.98
	Second (395–635)	445	21.89	18.19	-104.77	46.64
²L	First (30–150)	60	48.35	47.84	-102.86	6.26
	Second (190–310)	250	1.59	-0.48	-104.60	26.19
	Third (350–470)	367	13.36	10.30	-104.27	38.30
²LCu	First (30–90)	95	356.7	355.9	-100.73	9.93
	Second (255–615)	442	7.89	4.22	-102.84	45.50
³L	First (210–450)	357	120.02	117.05	-101.04	36.25
³LCu	First (35–95)	72	7.75	7.15	-103.77	7.49
	Second (145–595)	439	10.32	6.66	-102.72	45.18
	Third (620–820)	739	21.44	15.30	-102.77	75.97
	Third (740–840)	792	26.86	20.26	-101.78	80.72

Source: Author

Table 50.7 Antimicrobial activity of the Cu azo complex of ^{3}L

| Tested Compounds | In vitro activity zone of inhibition in mm (MIC in μg/ml)[a] | | | | | |
| | Gram positive | | Gram negative | | Fungi | |
	S. aureus	*B. subtilis*	*E. coli*	*P. aeruginosa*	*C. albicans*	*A. flavus*
Cu azo complex of ^{3}L	13(50)	12(70)	13(28)	12(31)	11(75)	14(75)
Streptomycin	11.6(05)	10.4(05)	14.5(05)	13(05)	Nt	Nt
Fluconazole	Nt	Nt	Nt	Nt	16(05)	18(05)

[a]The values given are means of three experiments.

Nt – denotes not tested.

Source: Author

thermo gravimetric analysis, confirming their promising properties for diverse applications.

Declaration of competing interest

There are no conflicts to declare.

Supplementary information (SI)

The physical characteristics of the synthesised metal complexes data are provided in Table S1, while the characterisation of the synthesised azo metal complexes by UV, FTIR, and magnetic susceptibility is presented in the supporting information through Figures S1–S10.

Acknowledgement

The authors, thankful to K.S. Institute of Technology, Bangalore for providing lab facility to carry out the research work.

References

[1] AbouEl-Enein, S. A., Emam, S. M., Polis, M. W., Emara, E. M. Synthesis and characterization of some metal complexes derived from azo compound of 4,40-methylenedianiline and antipyrine: Evaluation of their biological activity on some land snail species. *J. Mol. Struc.*, 2015;1099:567–578.

[2] El-Bindary, A. A., Diab, M. A., El-Sonbati, A. Z., Salem, O. L. Geometrical, molecular docking and potentiometric studies of sulfoxine and its new azo derivative with their metal complexes. *J. Mol. Liquids*, 2016;219:737–747.

[3] Anuj Kumar, S., Saikat, B., Suman Kumar, B., Rabindranath, M. Azo-containing pyridine amide ligand. A six-coordinate nickel (II) complex and its one-electron oxidized species: Structure and properties. *Inorg. Chim. Acta*, 2010;363:2720–2727.

[4] Poulami, P., Sankar Prasad, P., Debprasad, P., Lai, C. K., Paula, B., Vitor, F., Surajit, C. Ruthenium and palladium complexes incorporating amino-azo-phenol ligands: Synthesis, characterization, structure and reactivity. *InorganicaChimica Acta*, 2015;429:122–131.

[5] El-Sonbati, A. Z., Diab, M. A., El-Bindary, A. A., Mohamed, G. G., Morgan, Sh. M. Thermal, spectroscopic studies and hydrogen bonding in supramolecular assembly of azo rhodanine complexes. *InorganicaChimica Acta*, 2015;430:96–107.

[6] Tugba, E., Muhammet, K., Nurcan, K., Gokhan, C., McKee, V., Mukerrem, K. An azo-azomethine ligand and its copper (II) complex: Synthesis, X-ray crystal structure, spectral, thermal, electrochemical and photoluminescence properties. *InorganicaChimica Acta*, 2015;430:268–279.

[7] Masoud, M. S., Hagagg, S. S., Ali, A. E., Nasr, N. M. Synthesis and spectroscopic characterization of gallic acid and some of its azo complexes. *J. Mol. Struc.*, 2012;1014:17–25.

[8] Mohamed, G. G. Structural chemistry of some new azo complexes. *Spectrochimica Acta Part A.* 2001;57:411–417.

[9] Bouhdada, M., El. Amane, M., El. Hamzaoui, N. Synthesis, spectroscopic studies, X-ray powder diffraction data and antibacterial activity of mixed transition metal complexes with sulfonate azo dye, sulfamate and caffeine ligands. *Inorg. Chem. Comm.*, 2019;101:32–39.

[10] Abdel-Nasser, M. A., Alaghaz, Ammar, Y. A., Bayoumi, H. A., Aldhlmani, S. A. Synthesis, spectral characterization, thermal analysis, molecular modeling and antimicrobial activity of new potentially N_2O_2 azo-dye Schiff base complexes. *J. Mol. Struc.*, 2014;1074:359–375.

[11] Sirin, B., Koray, S., Baris, A., Muhammet, K., Ahmet, K., Mukerrem, K. Preparation, spectral, X-ray powder diffraction and computational studies and genotoxic properties of new azo–azomethine metal chelates. *J. Mol. Struc.*, 2014;1076:213–226.

[12] Abdel-Nasser, M. A., Alaghaz, Zayed, M. A., Alharbi, S. A. Synthesis, spectral characterization, molecular modelling and antimicrobial studies of tridentate azo-dye Schiff base metal complexes. *J. Mol. Struc.*, 2015;1084:36–45.

[13] Munire, S., Pervin, D., Muhammet, K., Ugur, A., Hatice, T. D., Mukerrem, K. New tridentate azo–azomethines and their copper (II) complexes: Synthesis, solvent effect on tautomerism, electrochemical and biological studies. *J. Mol. Struc.*, 2015;1096:64–73.

[14] AL-Adilee, K. J., Abass, A. K., Taher, A. M. Synthesis of some transition metal complexes with new heterocyclic thiazolyl azo dye and their uses as sensitizers in photo reactions. *J. Mol. Struc.*, 2016;1108:378–397.

[15] Emine, B., Eylem, A., Ulku, S. Synthesis and structural characterization of new oxovanadium (IV) complexes derived from azo-5-pyrazolone with prospective medical importance. *J. Mol. Struc.*, 2017;1127:653–661.

[16] Al-Adilee, K., Kyhoiesh, H. A. K. Preparation and identification of some metal complexes with new heterocyclic azo dye ligand 2-[2- (1- Hydroxy -4- Chloro phenyl) azo]-imidazole and their spectral and thermal studies. *J. Mol. Struc.*, 2017;1137:160–178.

[17] El-Sonbati, A. Z., Diab, M. A., El-Bindary, A. A., Shoair, A. F., Hussein, M. A., El-Boz, R. A. Spectroscopic, thermal, catalytic and biological studies of Cu(II) azo dye complexes. *J. Mol. Struc.*, 2017;1141:186–203.

[18] Ferreira, G. R., de Oliveira, L. F. C. Synthesis, spectroscopic and structural studies of new azo dyes metal chelatesderivated from 1-phenil-azo-2-naphthol. *J. Mol. Struc.*, 2017;1146:50–56.

[19] Bouhdada, M., EL Amane, M. Synthesis, characterization and spectroscopic properties of the hydrazo dye and new hydrazo dye-metal complexes. *J. Mol. Struc.*, 2017;1150:419–426.

[20] El-Sonbati, A. Z., El-Bindary, A. A., Abd El-Meksoud, S. A., Belal, A. A. M., El-Boz, R. A. Spectroscopic, potentiometric and thermodynamic studies of azo rhodanines and their metal complexes. *J. Mol. Liquids*, 2014;199:538–544.

[21] Gulbahar, K., Barıs, A., Huseyin, Z., Muhammet, K., Koray, S., Mukerrem, K. A novel azo-azomethine based fluorescent dye and its Co (II) and Cu (II) metal chelates. *J. Mol. Liquids*, 2014;200:105–114.

[22] El-Bindary, A. A., El-Sonbati, A. Z., Diab, M. A., Morgan, S. M. Geometrical structure, potentiometric and thermodynamic studies of rhodanine azo dye and its metal complexes. *J. Mol. Liquids*, 2015;201:36–42.

[23] Shoair, A. F., El-Shobaky, A. R., Azab, E. A. Geometrical structure and potentiometric studies of 5-chloro-2,3-dihydroxy pyridine azo derivatives and their metal complexes. *J. Mol. Liquids*, 2015;203:59–65.

[24] Van Anh, N., Anh Vu, T. N., Polyanskaya, N., Utenyshev, A., Shilov, G., Vasi, M. Structure and properties of some S-containing azo-derivatives of 5-pyrazolone and their Cu(II), Co(II), and Ni(II) metal complexes. *Inorg. Chem. Comm.*, 2023;158(2):111648.

[25] El-Sonbati, A. Z., Mohamed, G. G., El-Bindary, A. A., Hassan, W. M. I., Diab, M. A., Morgan, Sh. M., Elkholy, A. K. Supramolecular structure, molecular docking and thermal properties of azo dye complexes. *J. Mol. Liquids*, 2015;212:487–502.

[26] El-Sonbati, A. Z., El-Deen, I. M., El-Bindary, M. A. Spectroscopic, thermal and geometrical structures of Cu(II) azo rhodanine complexes. *J. Mol. Liquids*, 2016;215:612–624.

[27] El-Sonbati, A. Z., Diab, M. A., El-Bindary, A. A., Morgan, Sh. M., Barakat, A. M. Spectroscopic, geometrical structures, DNA and biological activity studies of azo rhodanine complexes. *J. Mol. Liquids*, 2016;224:105–124.

[28] El-Bindary, A. A., Hassan, N., El-Afify, M. A. Synthesis and structural characterization of some divalent metal complexes: DNA binding and antitumor activity. *J. Mol. Liquids*, 2017;242:213–228.

[29] Hassan, N., El-Sonbati, A. Z., El-Desouky, M. G. Synthesis, characterization, molecular docking and DNA binding studies of Cu(II), Ni(II), and Mn(II) complexes. *J. Mol. Liquids*, 2017;242:293–307.

[30] El-Deen, I. M., Shoair, A. F., El-Bindary, M. A. Synthesis, structural characterization, molecular docking and DNA binding studies of copper complexes. *J. Mol. Liquids*, 2018;249:533–545.

[31] Bal, S., Connolly, J. D. Synthesis, characterization, thermal and catalytic properties of a novel carbazole derived Azo ligand and its metal complexes. *Arab. J. Chem.*, 2017;10:761–768.

[32] Kakanejadifard, A., Esna-ashari, F., Hashemi, P., Zabardasti, A. Synthesis and characterization of an azo dibenzoic acid Schiff base and its Ni(II), Pb(II), Zn(II) and Cd(II)

complexes. *Spectrochimica Acta Part A Mol. Biomol. Spectros.*, 2013;106:80–85.

[33] Abdallah, S. M. Metal complexes of azo compounds derived from 4-acetamidophenol and substituted aniline. *Arab. J. Chem.*, 2012;5:251–256.

[34] Gaber, M., El-Sayed, Y. S., El-Baradie, K. Y., Fahmy, R. M. Complex formation, thermal behaviour and stability competition between Cu (II) ion and Cu0 nanoparticles with some new azo dyes. Antioxidant and in vitro cytotoxic activity. *Spectrochimica Acta Part A Mol. Biomol. Spectros.*, 2013;107:359–370.

[35] Zhang, G., Wang, S., Ma, J. S., Yang, G. Syntheses, characterization and third-order nonlinear optical properties of a class of thiazolylazo-based metal complexes. *InorganicaChimica Acta*, 2012;384:97–104.

[36] Gaber, M., Hassanein, A. M., Lotfalla, A. A. Synthesis and characterization of Co(II), Ni(II) and Cu(II) complexes involving hydroxy antipyrine azo dyes. *J. Mol. Struc.*, 2008;875:322–328.

[37] Gaber, M., Fayed, T. A., El-Daly, S., El-Sayed, Y. S. Y. Spectroscopic studies of 4-(4,6-dimethylpyrimidin-2-ylazo) benzene-1,3-diol and its Cu(II) complexes. *Spectrochimica Acta Part A*, 2007;68:169–175.

[38] Selwin Joseyphus, R., Sivasankaran Nair, M. Synthesis, characterization and biological studies of some Co(II), Ni(II) and Cu(II) complexes derived from indole-3-carboxaldehyde and glycylglycine as Schiff base ligand. *Arab. J. Chem.*, 2010;3:195–204.

[39] Bagdatli, E., Gunkara, O. T., Ocal, N. Synthesis and characterization of new copper (II) and palladium (II) complexes with azo-, bisazo-5-pyrazolones. *J. Organometal. Chem.*, 2013;740:33–40.

[40] Monika, R., Cho, H.-J., Mirica, L. M., Sharma, A. K. Azodyes based small bifunctional molecules for metal chelation and controlling amyloid formation. *Inorganic Chimica Acta*, 2018;471:419–429.

[41] Abdel-Nasser, M. A., Alaghaz, Ammar, Y. A., Bayoumi, H. A., Aldhlmani, S. A. Synthesis, spectral characterization, thermal analysis, molecular modeling and antimicrobial activity of new potentially N_2O_2 azo-dye Schiff base complexes. *J. Mol. Struc.*, 2014;1074:359–375.

[42] El-Sonbati, A. Z., Diab, M. A., Morgan, Sh. M. Thermal properties, antimicrobial activity and DNA binding of Ni(II) complexes of azo dye compounds. *J. Mol. Liquids*, 2017;225:195–206.

[43] Mahapatra, B. B., Mishra, R. R., Sarangi, A. K. Synthesis, spectral, thermogravimetric, XRD, molecular modelling and potential antibacterial studies of dimeric complexes with bis bidentate ON–NO donor azo dye ligands. *J. Chem.*, 2013:1–11.

[44] Alghool, S., Abd El-Halim, H. F., Dahshan, A. Synthesis, spectroscopic, thermal and biological activity studies on azo-containing Schiff base dye and its Cobalt(II), Chromium(III) and Strontium(II) complexes. *J. Mol. Struc.*, 2010;983:32–38.

[45] Marchetti, F., Pettinari, C., Di Nicola, C., Tombesi, A., Pettinari, R. Coordination chemistry of pyrazolone-based ligands and applications of their metal complexes. *Coord. Chem. Rev.*, 2019;401:213069.

51 Synergistic catalysis of neem oil for biodiesel production using a K$_2$CO$_3$–NH$_4$OH system

Swarna S.[1,a], Puneetha J.[2], Shruthi K. S.[3], Manjunatha N. K.[4] and Shashidhar S.[5]

[1]Department of Chemistry, K. S. School of Engineering and Management, Bangalore, Karnataka, India

[2]Department of Chemistry, J S S Academy of Technical Education, Bangalore, Karnataka, India

[3]Amity University of Applied Sciences, Amity University, Bangalore, Karnataka, India

[4]Department of Chemistry, Sri Venkateshwara College of Engineering, Bangalore, Karnataka, India

[5]Department of Mechanical Engineering, UVCE, Bangalore, Karnataka, India

Abstract

In the current study, biodiesel was produced from high free fatty acid content neem oil using a catalyst mixture of K$_2$CO$_3$ in NH$_4$OH via the trans-esterification process. The inclusion of NH$_4$OH enhanced the basicity of K$_2$CO$_3$ by promoting the in-situ formation of KOH within an ammonium carbonate ((NH$_4$)$_2$CO$_3$) medium. This ammonium carbonate environment effectively moderated the formation of intermediate water during methoxide generation, leading to improved biodiesel yield. A maximum yield of 98.0% and a fatty acid methyl ester (FAME) content of 97.10% were achieved under optimised conditions: 1 g of K$_2$CO$_3$ mixed with 0.5 g of NH$_4$OH, an oil-to-methanol molar ratio of 1:6, a reaction temperature of 55°C, and a reaction time of 60 minutes. The resulting biodiesel was characterised using ^{1}H NMR and FTIR techniques. Furthermore, the physicochemical properties of both neem oil and the synthesised biodiesel were evaluated according to ASTM D6751 standards and found to be within acceptable limits.

Keywords: Neem oil, K$_2$CO$_3$–NH$_4$OH system, in-situ potassium hydroxide formation, catalytic blend, biodiesel production

Introduction

The growing global demand for energy, combined with the environmental consequences of fossil fuel combustion, has intensified the search for sustainable and eco-friendly fuel alternatives. Among these, biodiesel has garnered considerable attention due to its renewable origin, biodegradability, lower emissions, and compatibility with existing diesel engines [1]. It can be synthesised from various feed stocks through trans-esterification, typically involving vegetable oils and alcohol in the presence of a catalyst [2].

Commercial biodiesel production has traditionally depended on first-generation edible oils such as soybean, sunflower, rapeseed, and palm. However, reliance on these feed stocks has raised significant challenges, including elevated costs, competition with food resources, and environmental issues associated with intensive farming methods [3]. To address these drawbacks, attention has increasingly shifted toward second-generation non-edible oil sources such as neem (*Azadirachta indica*), jatropha, pongamia, and castor, which offer more sustainable and eco-friendly alternatives without threatening food security [4, 5].

Among these, neem oil presents several advantages. It is abundantly available in dry and semi-arid regions, requires minimal agricultural inputs, and can yield 25–45% oil from its seeds. Neem trees flourish in a wide range of environmental conditions, tolerating poor soils, limited rainfall, and high temperatures. Beyond its known medicinal and pesticidal properties, neem oil holds significant potential for biodiesel production, particularly in low-resource or rural settings [6].

Base-catalysed trans-esterification remains the most commonly adopted method for biodiesel synthesis, primarily due to its high conversion efficiency and operational simplicity. Potassium hydroxide (KOH) is widely employed in this process; however, its use is challenged by the high free fatty acid (FFA) content in certain oils, which leads to soap formation and consequently lowers biodiesel yield [7]. To address these limitations, alternative catalytic systems have been explored. Enzymatic catalysts, particularly lipases, offer high selectivity under mild reaction conditions, yet they are cost-intensive and sensitive to operational parameters [8].

[a]Swarna.s@kssem.edu.in

DOI: 10.1201/9781003685876-51

Heterogeneous catalysts, on the other hand, facilitate catalyst recovery and reuse but are often limited by low reaction rates and mass transfer issues [9]. Ionic liquids provide tenable catalytic properties and enhanced solubility characteristics, although their application is restricted by concerns regarding cost, recyclability, and environmental impact [10].

Potassium carbonate (K_2CO_3) has emerged as a promising substitute for KOH, as it produces less water during methoxide formation and tolerates higher FFA levels. Despite these advantages, K_2CO_3 exhibits limited solubility in methanol and comparatively lower catalytic activity, necessitating larger dosages that can lead to saponification [11]. In this context, a synergistic catalyst system combining K_2CO_3 and ammonium hydroxide (NH_4OH) has shown superior performance. The K_2CO_3–NH_4OH system enhances solubility and promotes in-situ KOH generation within an ammonium carbonate medium, effectively reducing soap formation and improving catalytic efficiency. This composite catalyst offers a cost-effective and efficient route for biodiesel production from high-FFA feed stocks [12, 13].

This study aims to fill that gap by evaluating the trans-esterification of neem oil using the K_2CO_3–NH_4OH catalyst system. Process parameters were optimised to enhance biodiesel yield, and the final product was analysed using ^{1}H NMR and FTIR spectroscopy. Additionally, its fuel properties were assessed according to ASTM D6751 and EN 14214 standards to verify its suitability as a renewable diesel substitute.

Materials and methods

In this research study, neem oil was procured from a local supplier in Gauribidanur, situated in Karnataka, India. The reagents employed for the biodiesel synthesis process included high-purity anhydrous methanol (99.9%), concentrated sulphuric acid (98.9%), potassium hydroxide (KOH, 85% purity), and potassium carbonate (K_2CO_3, 99.9% purity), all of which were of analytical grade. Additionally, 25% aqueous ammonium hydroxide solution (liquor ammonia) was utilised. All chemicals were sourced from Sigma Aldrich, India, and used without further purification.

Fatty acid profile of neem oil
Understanding the fatty acid profile of neem oil is crucial for applications in biodiesel production, as it

influences fuel properties like cetane number, oxidative stability, and cold flow behaviour. The fatty acid profile of neem oil, compiled from various literature sources, is provided in Table 51.1.

Biodiesel production

Acid esterification pre-treatment
The crude neem oil exhibited an initial acid value of approximately 10 mg KOH/g, corresponding to a FFA content of around 5%. Since such a high FFA content is unsuitable for direct base-catalysed trans-esterification, an acid esterification step was employed to lower the acid value. In this procedure, 100 g of crude neem oil were combined with methanol at 25% of the oil's weight and sulphuric acid at 1% by weight. The reaction mixture was refluxed at 55°C and continuously stirred at 600 rpm for 1 hour. After completion, the mixture was allowed to settle for 8 hours to facilitate phase separation. The upper layer comprising excess methanol, water, and sulphuric acid was carefully separated using a separating funnel. The bottom layer, now containing pre-treated oil with a reduced acid value of 1.8 mg KOH/g, was used for the subsequent base-catalysed trans-esterification.

K_2CO_3-NH_4OH catalysed trans-esterification
For the base-catalysed trans-esterification, 1 g of potassium carbonate (K_2CO_3) was initially reacted with 0.5 g (approximately 0.6 mL) of ammonium hydroxide (NH_4OH), maintaining a molar ratio of 1:2. Following this, 27 mL of methanol was incorporated into the mixture, ensuring a molar ratio of 1:6 between neem oil and methanol. The

Table 51.1 Fatty acid composition of neem oil (compiled from various studies)

Fatty acid	Common acronym	Neem oil composition (Wt. %)	References
Palmitic acid	C16:0	16.0–18.2	[14, 15]
Stearic acid	C18:0	14.0–18.0	[14–16]
Oleic acid	C18:1	25.0–54.0	[14, 16, 17]
Linoleic acid	C18:2	6.0–16.0	[14, 15, 17]
α-Linolenic acid	C18:3	0.5–1.2	[15, 17]
Arachidic acid	C20:0	0.5–1.2	[16, 17]
Behenic acid	C22:0	0.3–1.1	[15, 17]
Other saturated FAs	—	~5.0	[10, 17]

Source: Author

catalyst-methanol solution was stirred for 5 minutes to facilitate in-situ generation of potassium hydroxide (KOH). The pre-treated neem oil was then added, and the trans-esterification reaction was performed under reflux at 55°C for 1 hour, with continuous stirring at 600 rpm. Upon completion, the mixture was allowed to settle, enabling the separation of the biodiesel phase from the glycerol by-product. The upper biodiesel layer was subjected to several water washes to remove any residual water-soluble impurities. Finally, the purified biodiesel was dried in a hot air oven at 100°C. Under optimised reaction conditions, a maximum biodiesel yield of 98.0% by weight was obtained. A schematic representation of the entire biodiesel production process is shown in Figure 51.1.

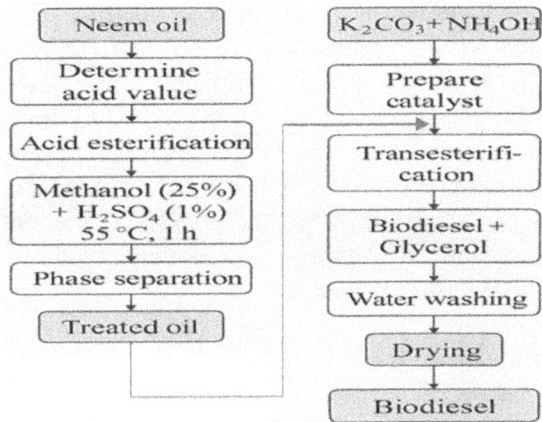

Figure 51.1 Preparation of neem oil biodiesel using K2CO3-NH4OH catalyst
Source: Author

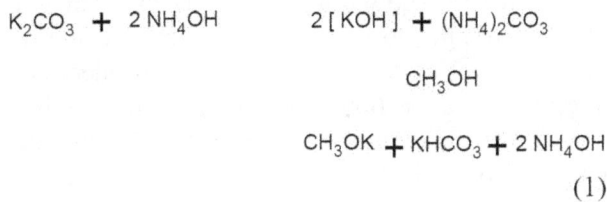

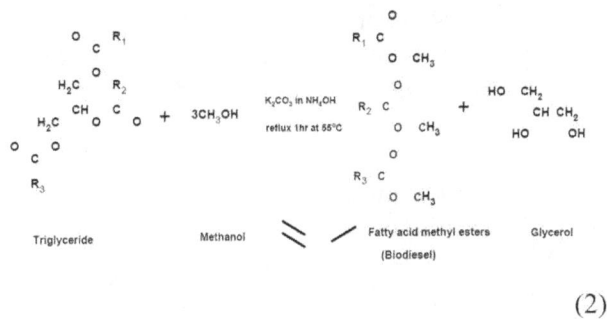

$$K_2CO_3 + 2\,NH_4OH \quad\longrightarrow\quad 2\,[KOH] + (NH_4)_2CO_3$$

$$\xrightarrow{CH_3OH}$$

$$CH_3OK + KHCO_3 + 2\,NH_4OH \tag{1}$$

$$(2)$$

$$Biodiesel\ Yield\ in\ weight\ \% = \frac{weight\ of\ Biodiesel\ obtained}{weight\ of\ oil\ taken} \times 100 \tag{3}$$

Results and discussion

The acid-catalysed pre-treatment conditions were selected based on established protocols in the literature for reducing the free fatty acid content of high-FFA feed stocks such as neem oil [18]. For the base-catalysed trans-esterification parameters specifically catalyst ratio, methanol-to-oil molar ratio, temperature, and reaction time—were optimised experimentally through a series of controlled trials aimed at maximising biodiesel yield and fatty acid methyl ester (FAME) content. The final conditions were determined by analysing the effect of each variable while keeping others constant, allowing for identification of the optimal combination.

Influence of NH_4OH in the trans-esterification process

The success of the methanolysis (trans-esterification) process largely depends on how efficiently methoxide ions are generated and sustained within the reaction environment. Potassium carbonate (K_2CO_3) is considered a promising catalyst since its reaction with methanol does not produce water, which can hinder ester formation (as shown in Equation 4). However, the interaction between K_2CO_3 and methanol is a reversible process that is sensitive to temperature. With increasing temperature, the equilibrium shifts backward, reducing the availability of active methoxide ions [19]. Previous studies indicate that higher quantities of K_2CO_3 are often needed to ensure sufficient methoxide generation, but an excess amount of catalyst can lead to unwanted saponification reactions, ultimately decreasing biodiesel yield [19, 20].

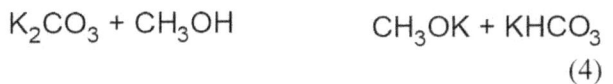

$$K_2CO_3 + CH_3OH \quad\longrightarrow\quad CH_3OK + KHCO_3 \tag{4}$$

To overcome this limitation, K_2CO_3 was initially reacted with ammonium hydroxide (NH_4OH) before its interaction with methanol. This pre-treatment leads to the in-situ formation of potassium hydroxide (KOH) and ammonium carbonate ((NH_4)$_2CO_3$). The generated KOH facilitates efficient methoxide production when reacted with methanol, while the hydrophilic ammonium carbonate captures water molecules present in the reaction, converting them into potassium bicarbonate ($KHCO_3$) and NH_4OH. This mechanism prevents the formation of free water, which is responsible for side reactions like saponification. Equation 1 represents this reaction pathway. The remarkable biodiesel yield of 98.0% from neem oil confirms that using a K_2CO_3 and NH_4OH mixture is a highly effective strategy for producing biodiesel from high-FFA feed stocks.

Optimisation of K_2CO_3 and NH_4OH ratios for maximum yield

The efficiency of in-situ KOH generation is influenced by the amounts of both K_2CO_3 and NH_4OH used. To determine the optimal catalyst loading, various combinations were tested by adjusting K_2CO_3 from 0.5 g to 1.5 g and altering the molar ratio of K_2CO_3 to NH_4OH between 1:1 and 1:4, while keeping other variables constant. In Figure 51.2, the highest yield of 98.0% was achieved using 1 g of K_2CO_3 and a 1:2 molar ratio with NH_4OH (0.5 g). Increasing the catalyst amount beyond this point resulted in a decline in yield, likely due to excessive early formation of KOH, which may lead to side reactions. Conversely, when lower amounts of catalyst were used, the yield remained suboptimal even with more NH_4OH, indicating that insufficient K_2CO_3 hindered the reaction.

At the optimal 1:2 molar ratio (1 g K_2CO_3 with 0.5 g NH_4OH), only the necessary amount of KOH was formed in the ammonium carbonate medium, supporting methoxide production without releasing free water. This approach resulted in a superior yield compared to other ratios. For instance, Lesmes Sanchez (2013) reported a 90.13% yield using 3 g of K_2CO_3 with canola oil [18], while Sukumar Puhan et al. (2005) achieved a 92% yield using 1 g of NaOH with mahua oil [19]. The improvement in yield in the current method can be attributed to the minimised water formation during the reaction, reducing ester hydrolysis and suppressing saponification side reactions [20, 21].

Influence of oil-to-methanol molar ratio on biodiesel production yield

The efficiency of the trans-esterification reaction largely depends on the molar ratio of methanol to oil (triglycerides). Although the stoichiometric requirement is three moles of methanol for every mole of triglyceride, this proportion is typically insufficient due to the reversible nature of the reaction. To drive the

equilibrium toward biodiesel formation and shorten the reaction time, an excess of methanol is commonly used [22, 23]. In the present study, the biodiesel yield from neem oil improved progressively with an increase in the methanol-to-oil molar ratio, reaching an optimum at 1:6. Beyond this point, specifically at 1:7, the increase in yield was marginal, and a further rise resulted in a slight decline, as shown in Figure 51.3.

Influence of reaction duration and temperature on biodiesel yield

Reaction time and temperature are key parameters influencing the efficiency of biodiesel synthesis. In this study, experiments were carried out across a temperature range of 45–65°C and reaction durations between 30 and 120 minutes. The catalyst mixture (1 g of K_2CO_3 and 0.5 g of NH_4OH) and methanol (1:6 molar ratio to oil) were held constant during these trials. The optimal conditions were identified at 55°C with a reaction time of 60 minutes, yielding a peak conversion efficiency of 98.0%. Beyond this temperature, a slight decline in yield was observed, likely due to enhanced saponification competing with the trans-esterification process. Prolonging the reaction time beyond one hour did not produce significant improvements in biodiesel yield, as illustrated in Figure 51.4.

Spectral characterisation of neem oil biodiesel
^1H NMR spectral analysis

^1H NMR spectroscopy serves as a powerful, non-destructive method to verify the conversion of triglycerides in vegetable oils into methyl esters (biodiesel) [24]. Figure 51.6 displays the ^1H NMR spectra of crude neem oil (a) and the resulting biodiesel (b). In the spectrum of neem oil (Figure 51.5a), the multiplets signals in the δ range of approximately 4.10–4.297 ppm represents the protons associated with the glycerol backbone of triglycerides. After trans-esterification, these peaks vanish in the spectrum of neem oil biodiesel (Figure 51.6b), while a distinct singlet appears at 3.627 ppm, indicating the presence of methoxy (-OCH_3) protons a clear marker of FAME formation [25, 26]. Additionally, the presence of a triplet at around 2.284 ppm corresponds to α-CH_2 protons, another characteristic feature of methyl esters. The emergence of these two distinct peaks confirms the successful conversion of neem oil into biodiesel. The conversion efficiency from triglycerides to methyl esters was calculated using

Equation (5) [27], and a high conversion rate of 97.10% was achieved. The chemical shifts in the **¹H NMR** spectra of both neem oil and its biodiesel are tabulated in Table 51.2.

$$C = 100 \times \frac{2A_{ME}}{3A_{\alpha-CH2}}$$

$$\text{(5)}$$

Where:

- **C** = Percentage conversion of oil to methyl esters
- **A**$_{ME}$ = Integral area of methoxy (-OCH₃) protons
- **Aα-CH₂** = Integral area of α-methylene protons
- The integers 2 and 3 denote the number of protons in each respective group.

FTIR spectral analysis

Fourier transform infrared (FTIR) spectroscopy serves as an effective method to verify the conversion of triglycerides in oils into FAME. Figure 51.6 presents the FTIR spectra of raw neem oil (a) and the resulting biodiesel (b). A prominent absorption band between 1800 and 1700 cm⁻¹ in both spectra indicates the presence of carbonyl (C=O) functional groups, a key signature of ester compounds [28].

Significant spectral variations are evident in the fingerprint region (1500–900 cm⁻¹), which distinguishes the structural differences between the oil and its methyl ester. In the biodiesel spectrum, a notable peak at 1435 cm⁻¹ corresponds to the asymmetric stretching of methyl (–CH₃) groups, which is absent in the crude neem oil spectrum. Moreover, the disappearance of the 1378 cm⁻¹ peak—associated with O–CH₂ vibrations typical of triglycerides in the biodiesel confirms the breakdown of the original oil structure. A new peak at 1197 cm⁻¹ in the biodiesel, attributed to O–CH₃ stretching vibrations, further supports the formation of methyl esters and successful trans-esterification.

The FTIR characteristic peaks of both neem oil and its biodiesel are tabulated in Table 51.3.

Comparison of biodiesel yield from neem oil under different catalytic conditions

The performance of the K₂CO₃-NH₄OH catalytic system for neem oil trans-esterification was **evaluated**

Table 51.2 ¹H NMR chemical shift observed in the neem oil and its biodiesel

Type of proton (s)	Functional group	Chemical shift δ (ppm)	Chemical shift in ppm	
			Neem oil	Neem oil biodiesel
CH₃-C	Terminal methyl group	0.8–1.00	0.845–0.878	0.829–0.86
-(CH₂)n-	Backbone CH₂	1.22–1.42	1.236–1.283	1.222–1.270
-CH₂CH₂COOCH₃-	β-methylene proton	1.55–1.69	1.589	1.567–1.602
=CH-CH₂-	α – methylene group to one double bond	1.93–2.10	1.969–2.039	1.953–2.024
-CH₂OCOR	Glycosidic proton	4.1–4.3	4.104–4.297	Absent
-CO(CH₃)O	Methyl group to ester	3.67	Absent	3.627 (Singlet)
-CH₂COOR	α -methylene group to ester	2.3	2.273–2.316 (quintet)	2.246–2.284 (triplet)
-CH=CH-	Olefinic proton	5.2–5.3	5.247–5.354	5.292–5.334

Source: Author

Table 51.3 FTIR characteristic frequency observed in the neem oil and its biodiesel

Region (cm⁻¹)	Functional group	Frequency observed in the spectrum	
		Neem oil	Neem biodiesel
1425–1447	CH₃ asymmetric bending	Absent	1435
1188–1200	O-CH₃ stretching	Absent	1197
1370–1400	O-CH₂ groups in glycerol moiety	1378	Absent
1075–1111	O-CH₂-C asymmetric axial stretching	1112	Absent
1700–1800	C=O stretching	1744	1739
3000–3100	C-H stretching in alkene	3010	3003
2840–3000	C-H stretching in alkane	2853 (C-H in CH₂), 2918 (C-H in CH₃)	2852 (C-H in CH₂), 2918 (C-H in CH₃)
665–730	C=C bending in alkene	724	724

Source: Author

in comparison to traditional catalysts such as direct KOH, standalone K_2CO_3, and K_2CO_3 in water. As outlined in Table 51.2, the highest biodiesel yield of 98.0% with a FAME content of 97.10% was obtained under optimised conditions: 1 g of K_2CO_3 combined with 0.5 g of NH_4OH, an oil-to-methanol molar ratio of 1:6, a reaction temperature of 55°C, and a reaction time of 60 minutes. This superior yield is primarily attributed to the in-situ formation of KOH within the NH_4OH medium, which not only enhances methoxide production but also minimises water generation due to the buffering action of ammonium and carbonate ions. In contrast, when K_2CO_3 was used in water, the yield was lower due to limited solubility and increased formation of intermediate water, which hampers trans-esterification efficiency. The use of 1 g K_2CO_3 alone was also suboptimal, as it could not produce adequate methoxide for effective conversion. Meanwhile, the direct application of KOH yielded about 85%, but the formation of soap and ester hydrolysis due to water presence lowered its efficiency. These results are consistent with Singh & Singh, (2010) [27], who reported an 86.5% yield using KOH for a neem oil of high-FFA feed stock under similar conditions. Overall, the K_2CO_3-NH_4OH system proved to be a highly effective and FFA-tolerant catalytic route for biodiesel production from neem oil. A comparison of the biodiesel yields obtained from neem oil under different catalytic conditions is presented in Table 51.4 [29, 30].

Physiochemical properties

Physicochemical properties of crude neem oil and the resulting biodiesel were evaluated following ASTM D6751 and EN 14214 standards, as summarised in Table 51.5. Key fuel parameters analysed included density, kinematic viscosity, calorific value, flash point, cloud point, acid value, copper strip corrosion, oxidation stability, ester content, saponification value, iodine value, methanol content, and moisture content. The measured properties of the biodiesel were found to be in close compliance with standard specifications and comparable to conventional diesel fuel, indicating its suitability as an alternative fuel source. Although slight deviations were observed in certain parameters, such as density (876 kg/m³) and flash point (171°C), these values remain within acceptable operational margins and exceed the minimum thresholds required for safe and efficient combustion. Importantly, critical indicators like viscosity (4.7 mm²/s) and ester content (97.10%) were well within the specified ranges, ensuring fuel stability, engine compatibility, and performance reliability. Thus, the overall physicochemical profile confirms the viability of neem biodiesel for practical diesel engine applications.

Conclusion

Present study successfully demonstrated the production of biodiesel from non-edible neem oil using an

Table 51.4 Comparative study of neem oil biodiesel yield with different catalytic conditions

Initial FFA of a feed stock	FFA after pre-treatment	Catalyst type	Catalyst amount	Oil to methanol molar ratio	Temperature/ reaction time	Biodiesel yield (%)	Reference
		K_2CO_3 in NH_4OH (1:2 wt. ratio)	1 g of K_2CO_3 in 0.5 g of NH_4OH	1:6	55°C/60 min	98%	Present study
10%	1.8%	KOH	1 g	1:6	60°C/60 min	84% (Exp)	Present study
		KOH	0.75 wt%	1:6	65/60 min	86.5	[27]
		K_2CO_3	1 g	1:6	55°C/60 min	60%	Present study
		K_2CO_3 in H_2O (1:2 Wt. ratio)	1 g of K_2CO_3 in 0.5 g of H_2O	1:6	55°C/60 min	88%	Present study
		CaO (heterogeneous base)	5 wt%	1:12	65/180 min	83.5	[28]
		Sulphuric acid (acid cat.)	2 wt%	1:9	60/120 min	75.0	[29]
		Lipase (enzymatic)	10 wt%	1:3	45/360 min	70.2	[30]

Source: Author

Table 51.5 Physicochemical properties of neem oil and its corresponding biodiesel compared with standard diesel

Fuel property	Evaluation standard	Standard value (ASTM D6751-02) biodiesel	Neem oil	Neem biodiesel	Diesel
Appearance	-	-	Dark yellow	Light yellow	-
Density @ 15°C (kg/m³)	ASTM-D1298	880 Max.	920	876	840
Viscosity @ 40°C (mm²/s)	D445	1.9–6.0	42	4.7	3.72
Flash point (°C)	D93	130 min	225	171	50
Cloud point (°C)	D2500	-3–12	9	4	-7
Pour point (°C)	D97	-15–10	12	3	-15
Heating value (MJ/kg) (MJ/kg)	D240	-	38.45	39.320	42.50
Ester content (mass %)	Knothe method	96.5 min.	-	97.53%	
Acid value (mg KOH/g)	AOCS-Cd3d-63-oil D664-biodiesel	0.80 max.	10.0	0.501	0.051
Copper strip corrosion for 3 h at 100°C	D130	No.3 max.	<No.1	<No.1	1
Oxidation stability, at 110°C (h)	EN14112	Min. 3 h	-	8	26.5
Saponification value (mg KOH/g oil)	D1962	-	190.2	182	-
Iodine value (mg I$_2$/100 g)	EN 14111	Max. 120	75	66	-
Methanol content (% vol.)	EN14110	0.2 max.	-	Nil	-
Water content (%vol.)	D2709	Max. 0.05	0	Nil	0.003

Source: Author

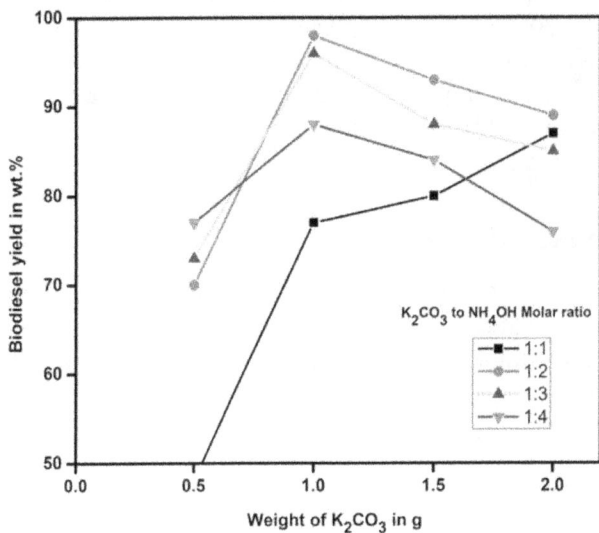

Figure 51.2 Effect of varying quantities of K$_2$CO$_3$ and NH$_4$OH
Source: Author

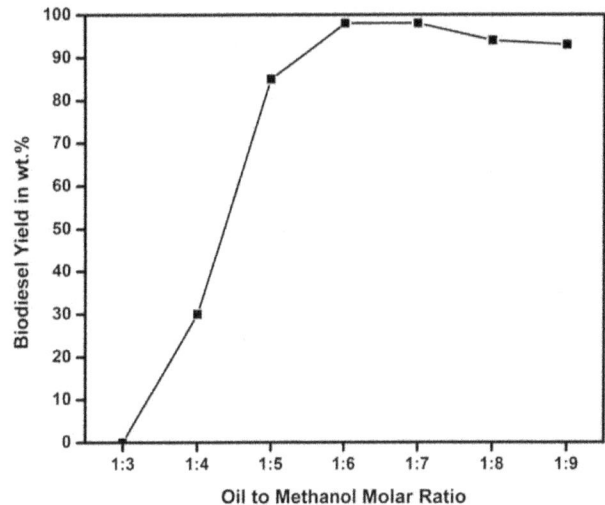

Figure 51.3 Influence of oil-to-methanol ratio on the trans-esterification efficiency of neem oil. On the biodiesel yield from neem oil
Source: Author

innovative catalytic system comprising potassium carbonate (K$_2$CO$_3$) dissolved in ammonium hydroxide (NH$_4$OH). The two-step process—acid-catalysed pre-treatment followed by base-catalysed trans-esterification—effectively reduced FFA content and enhanced conversion efficiency. Optimised reaction conditions, including a methanol-to-oil molar ratio

of 1:6, 1 g of K$_2$CO$_3$ mixed with 0.5 g of NH$_4$OH, a reaction temperature of 55°C, and a reaction time of 60 minutes, resulted in a maximum biodiesel yield of 98.0%.

Spectral analyses (FTIR and ^{1}H NMR) confirmed successful trans-esterification with the formation of FAME. The physicochemical properties of the

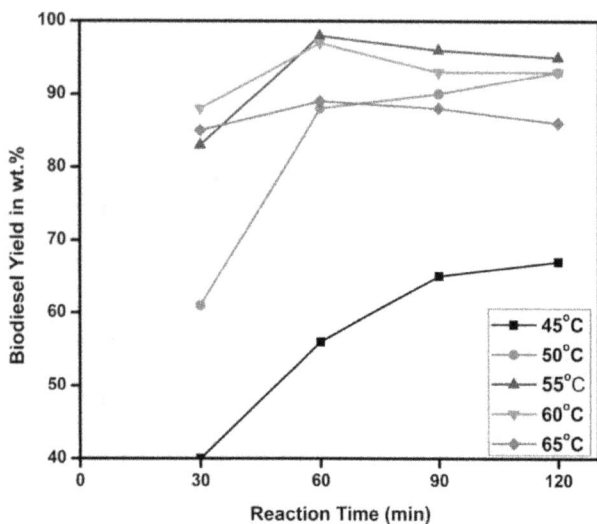

Figure 51.4 Effect of reaction duration and temperature on the biodiesel yield from neem oil
Source: Author

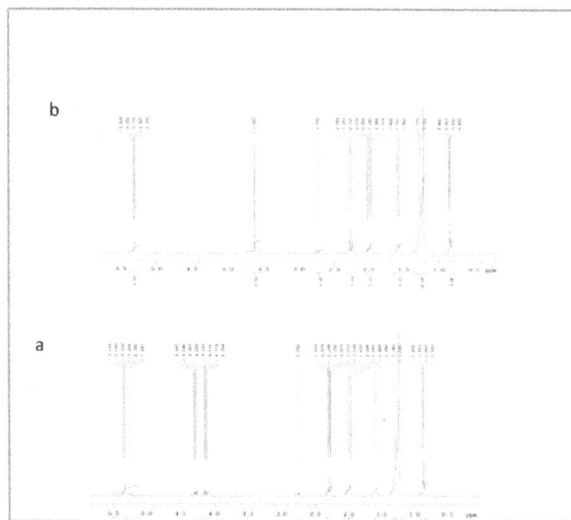

Figure 51.5 ¹H NMR spectra of neem oil (a) and its corresponding biodiesel (b), illustrating the transformation of triglycerides into methyl esters
Source: Author

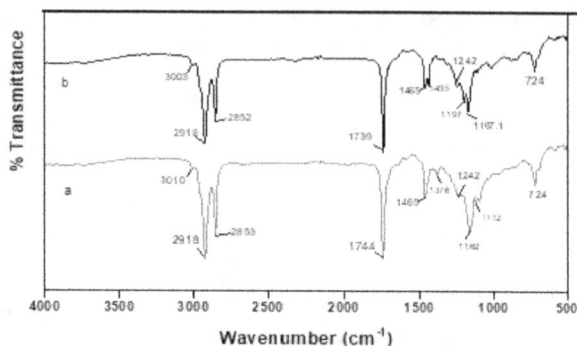

Figure 51.6 FTIR spectrum of crude neem oil (a) and its corresponding biodiesel (b)
Source: Author

reusability remains a limitation, as it is not recoverable for subsequent cycles. However, the method exhibited excellent reproducibility, with batch-to-batch variation within ±1% yield error, supporting its reliability for small-scale production. Further research is recommended to explore the feasibility of process scale-up, as well as to perform a detailed techno-economic analysis. This should include life-cycle assessments, feed stock logistics, and environmental impact evaluations to establish the industrial viability of this method for sustainable biodiesel production.

Acknowledgement

The author sincerely acknowledges **K. S. School of Engineering and Management (KSSEM), Bengaluru**, and **University of Mysore, India**, for providing the necessary infrastructure, equipment, and laboratory facilities essential for conducting this research work.

produced biodiesel were largely compliant with ASTM D6751 and EN 14214 standards. While minor deviations in parameters such as flash point and density were observed, these remained within acceptable performance margins and did not compromise fuel quality or applicability.

Although the K₂CO₃-NH₄OH system proved to be a highly efficient catalytic approach, catalyst

References

[1] Demirbas, A. Biofuels securing the planet's future energy needs. *Energy Conver. Manag.*, 2009;50(9):2239–2249.

[2] Atabani, A. E., Silitonga, A. S., Badruddin, I. A., Mahlia, T. M. I., Masjuki, H. H., Mekhilef, S. A comprehensive review on biodiesel as an alternative energy resource and its characteristics. *Renew. Sustain. Energy Rev.*, 2012;16(4):2070–2093.

[3] Pugazhvadivu, M., Jeyachandran, K. Investigations on the performance and exhaust emissions of a diesel engine using preheated waste frying oil as fuel. *Renew. Energy*, 2005;30(14):2189–2202.

[4] Ahmad, M., Khan, M. A., Iqbal, M., Khan, M. Z., Ali, H. Production of biodiesel from non-edible feedstocks: A

review of Neem (Azadirachta indica) oil as a sustainable source. *Renew. Energy Sustain. Dev.*, 2024;10(1):45–56.

[5] Patel, A., Arora, N., Mehtani, J., Pruthi, V., Pruthi, P. A. Assessment of fuel properties on the basis of fatty acid profiles of non-edible oil for the production of biodiesel. *Environ. Prog. Sustain. Energy*, 2016;35(2):470–475.

[6] Kumar, A., Sharma, S. Production of biodiesel from neem oil and its performance evaluation in a CI engine. *Renew. Energy*, 2015;76:362–368.

[7] Suleiman, K. K., Isah, M., Abdulfatai, A. S., Danyaro, Z. Kinetics and thermodynamics study of biodiesel production from neem oil using alumina as a catalyst. *FUDMA J. Sci.*, 2023;7(3): 65–71.

[8] Lam, M. K., Lee, K. T., Mohamed, A. R. Homogeneous, heterogeneous and enzymatic catalysis for trans esterification of high free fatty acid oil (waste cooking oil) to biodiesel: A review. *Biotechnol. Adv.*, 2010;28(4):500–518.

[9] Akpan, U. G., et al. Extraction and characterization of neem seed oil. *Leonardo J. Sci.*, 2006;8(9):43–52.

[10] Atabani, A. E., et al. Non-edible vegetable oils: A critical evaluation of oil extraction, fatty acid compositions, biodiesel production, characteristics, engine performance, and emissions production. *Renew. Sustain. Energy Rev.*, 2013;18:211–245.

[11] Adebayo, G. B., et al. GC-MS analysis of neem oil and its antibacterial properties. *FUDMA J. Sci. (FJS)*, 2021;5(3):32–37.

[12] Swarna, S., Swamy, M., Divakara, T., Krishnamurthy, K., Shashidhar, S., Srikanth, H. K_2CO_3 in NH_4OH as an effective catalyst mixture for the trans esterification of high acid value Mahua oil. *Asian J. Chem.*, 2021;33(11):2807–2812.

[13] Swarna, S., Srikanth, H. V., Anitha, R., Swamy, M. T., Divakara, T. R., Krishnamurthy, K. N., Shashidhar, S. Biodiesel synthesis from mixed non-edible oil feedstock using K_2CO_3 in NH_4OH catalyst mixture. *Energy Sources Part A Recov. Utiliz. Environ. Effects*, 2023;45(3):6901–6917.

[14] Ogbeide, S. O., et al. Characterization and biodiesel potential of neem seed oil. *FUW Trends Sci. Technol. J. (FT-STJ)*, 2022;7(1):413–417.

[15] Kaushik, N., and Vir, S. (2001). Variations in fatty acid composition of neem seeds collected from the Rajasthan state of India. Biochemical Society Transactions, 28(2), 880–882. https://doi.org/10.1042/0300-5127:0280880

[16] Sharma, A., Jain, S. Optimization of acid esterification for high-FFA non-edible oil feedstocks for biodiesel production. *Renew. Energy*, 2021;172:1285–1294.

[17] Baroi, C., Yanful, E. K., Bergougnou, M. A. Biodiesel production from Jatropha curcas oil using potassium carbonate as an unsupported catalyst. *Internat. J. Chem. Reactor Engg.*, 2009;7(A72):1–18.

[18] Lesmes Sanchez, L. N. Investigations of cost-effective biodiesel production from high FFA feedstock (Master's thesis). The University of Western Ontario, London, Ontario, Canada, 2013:71–721.

[19] Puhan, S., Vedaraman, N., Ram, B. V. B., Sankaranarayanan, G., Jeychandran, K. Mahua oil (Madhuca indica seed oil) methyl ester as biodiesel—Preparation and emission characteristics. *Biomass Bioenergy*, 2005;28(1): 87–93.

[20] Sharma, M., Jain, S., Rajvanshi, S. Effect of methanol to oil molar ratio on biodiesel production and process optimization: A review. *Biofuels*, 2022;13(6):675–684.

[21] Patel, A., Arora, N., Mehtani, J., Pruthi, V., Pruthi, P. A. Assessment of fuel properties and optimization of methanol to oil molar ratio for biodiesel production from microalgae and waste cooking oil. *Renew. Energy*, 2021;165:1130–1139.

[22] Knothe, G., Van Gerpen, J., and Krahl, J. (2005). The Biodiesel Handbook. Champaign, Illinois: AOCS Press, 1–272.

[23] Freedman, B., Butterfield, R. O., Pryde, E. H. Transesterification kinetics of soybean oil. *J. Am. Oil Chem. Soc.*, 1986;63(10):1375–1380.

[24] Ramadhas, A. S., Jayaraj, S., Muralitharan, C. Biodiesel production from high FFA rubber seed oil. *Fuel*, 2005;84(4):335–340.

[25] Knothe, G. Analytical methods used in the production and fuel quality assessment of biodiesel. *Fuel Proc. Technol.*, 2001;86(10):1059–1070.

[26] Knothe, G. Determining the degree of conversion of fatty acid methyl esters from vegetable oils by FTIR spectroscopy. *J. Am. Oil Chem. Soc.*, 2001;78(11):1025–1028.

[27] Singh, A., Singh, S. Production of biodiesel from non-edible neem oil and its performance evaluation. *Energy Sources Part A Recov. Utiliz. Environ. Effects*, 2010;32(20):1861–1869.

[28] Sharma, Y. C., Singh, B., Upadhyay, S. N. Advancements in development and characterization of biodiesel: A review. *Fuel*, 2016;163:187–194.

[29] Ali, B., Zafar, M. Production and analysis of biodiesel from neem oil using sulfuric acid. *Internat. J. Renew. Energy Res.*, 2015;5(3):623–628.

[30] Gupta, A., Roy, S. Enzymatic trans esterification of high-FFA Neem oil using immobilized lipase. *Biores. Technol.*, 2017;243:151–15.

52 The minimum pendant dominating Harary energy of a graph

Nataraj K.[1,2,a], Puttaswamy[2,3,b] and Purushothama S.[1,2,c]

[1]Department of Mathematics, Maharaja Institute of Technology, Mysore Belawadi, Srirangapatna Taluk, Mandya–571477, Karnataka, India

[2]Research Scholar, Visvesvaraya Technological University, Belagavi – 590018, ORCID Address: https://orcid.org/0009-0002-0099-7050

[3]Department of Mathematics, P.E.S. College of Engineering, Mandya–571401, Karnataka, India

Abstract

In this article, we present the idea of the minimum pendant dominating Harary energy of a graph, to as .We have computed the minimum pendant dominating Harary energy for various well-known families of graphs. Furthermore, we have established the limits for this energy measure.

Keywords: Minimum pendant dominating Harary set, minimum pendant dominating Harary matrix, minimum pendant dominating Harary energy

Introduction

Let G be a finite, non-empty, simple graph with no loops, multiple edges or directed edges. The number of vertices and edges in graph is denoted by n and m respectively. The symbols $\Delta(G)$ and $\delta(G)$ denotes the maximum and minimum degrees of the graph G, respectively.

If every vertex in $V - S$ is connected to at least one vertex in S, then the subset $S \subseteq G$ is referred to as a dominating set of G. The smallest size of a dominating set in G is the domination number, denoted by $\gamma(G)$.

If the induced sub-graph of S includes at least one pendant vertex, the dominating set S in G is calledpendant dominating set. The minimum size of a pendant dominating set in G is termed the pendant domination number of G, which is represented as $\gamma_{pe}(G)$.

In 1978, Gutman introduced the idea of graph energy [7]. Let G represent a graph, and let $A(G) = (a_{ij})$ denote its adjacency matrix. The Eigen values of $A(G)$ represented as $\lambda_1, \lambda_2, ..., \lambda_n$ are organised in non-increasing order. The energy of a graph refers to the total of the absolute values of the Eigen values found in its adjacency matrix.

$$E(G) = \sum_{i=1}^{n} |\lambda_i|$$

As the adjacency matrix $A(G)$ is real and symmetric and $\lambda_1, \lambda_2, ..., \lambda_n$ let be the distinct Eigen values of G with multiplicities $\alpha_1, \alpha_2, ..., \alpha_n$, respectively. The multi-set

$$spec(G) = \begin{pmatrix} \lambda_1 & \lambda_2 & ... & \lambda_n \\ \alpha_1 & \alpha_2 & ... & \alpha_n \end{pmatrix}$$

is called adjacency spec(G). The Eigen values of the adjacency matrix are indeed real, and their total corresponds to the number of elements present in the dominating set.

Refer to the reviews of the authors in ref [6], papers [3, 4, 7] and references cited therein for more information on the mathematical aspects of graph energy theory. In ref [10], the fundamental properties of a graph, including a variety of upper and lower bounds on its energy, were established. In addition, the molecular orbital theory of conjugated molecules has benefited greatly from the use of graph energy theory [5, 8].

Gungor and Cevik introduced the concept of Harary energy. The Harary matrix of G is an n*n square matrix where $(i, j)^{th}$ entry is $\frac{1}{d_{ij}}$, with d_{ij} representing the distance between the vertices v_i and v_j.Let $\lambda_1, \lambda_2, ..., \lambda_n$ be the Eigen values of Harary matrix of G. The Harary energy *HE(G)* is described as

[a]natraj.appu@gmail.com, [b]prof.puttaswamy@gmail.com, [c]psmandya@gmail.com.

DOI: 10.1201/9781003685876-52

$$HE(G) = \sum_{i=1}^{n} |\lambda_i|$$

Ref [13] contains additional research on Harary energy. Adiga et al. [1] defined the minimum covering energy, EC(G) of a graph G, which depends on its specific minimum covering set C. Building on this work, Rajesh Kanna et al. introduced the concepts of minimum dominating energy and minimum covering Harary energy of a graph [11, 12]. Now, we apply this idea to the graph's minimum pendant dominated by Harary energy. In this paper, we introduce the concept of the minimum pendant dominating Harary energy of a graph, denoted as $HE_{PD}(G)$. We have also calculated the minimum pendant Harary energy for several well-known families of graphs. The bounds for the minimum pendant dominating Harary energy have also been established.

The Harary energy reflects the "complexity" or "stability" of the graph, as the Eigen values are related to how the graph is connected and its overall structure. Higher Harary energy indicates that the graph has a more complex structure with a greater variety of connections between vertices. The Harary energy is used to study the structure of molecules where the vertices represent atoms and edges represent bonds between atoms. The energy of molecule can offer valuable insights into its stability and reactivity. Harary energy can be used to measure the robustness or connectivity of complex networks, such as social networks and computer networks, in the analysis process. Harary energy is a subset of the broader study of graph spectra, which uses the Eigen values of matrices associated with graphs to analyse their properties.

Results related to bounds on energy of graphs

Theorem 2.1. Let a_i and b_i where $1 \leq i \leq n$ are non-negative real numbers. Then

$$\left(\sum_{i=1}^{n} a_i^2\right)\left(\sum_{i=1}^{n} b_i^2\right) - \left(\sum_{i=1}^{n} a_i \ b_i\right)^2 \leq \frac{n^2}{4}(M_1 M_2 - m_1 m_2)^2$$

where $M_1 = \min_{1 \leq i \leq n}(a_i)$, $M2 = = \min_{1 \leq i \leq n}(b_i)$

$m_1 = \min_{1 \leq i \leq n}(a_i)$ and $m_2 = \min_{1 \leq i \leq n}(b_i)$

Theorem 2.2. Let a_i and b_i where $1 \leq i \leq n$ be positive real numbers then,

$$\left(\sum_{i=1}^{n} a_i^2\right)\left(\sum_{i=1}^{n} b_i^2\right) \leq \frac{1}{4}\left(\sqrt{\frac{M_1 M_2}{m_1 m_2}} + \sqrt{\frac{m_1 m_2}{M_1 M_2}}\right)^2 \left(\sum_{i=1}^{n} a_i \ b_i\right)^2$$

where M_i and m_i are similar to previous theorem.

Theorem 2.3. Let a_i and b_i where $1 \leq i \leq n$ be positive real numbers then,

$$\left| n \sum_{i=1}^{n} a_i \ b_i - \left(\sum_{i=1}^{n} a_i\right)\left(\sum_{i=1}^{n} b_i\right) \right| \leq \alpha(n)(A - a)(B - b)$$

where a, b, A and B are real constants and $1 \leq i \leq n$ for all i. Also, $a \leq a_i \leq A$, $b \leq b_i \leq B$, and

$$\alpha(n) = n \left\lfloor \frac{n}{2} \right\rfloor \left(1 - \frac{1}{n}\left\lfloor \frac{n}{2} \right\rfloor\right)$$

Theorem 2.4. Let $a_1, a_2, ..., a_n$ are non-negative numbers then,

$$n\left(\frac{1}{n}\sum_{i=1}^{n} a_i - \left(\prod_{i=1}^{n} a_i\right)^{\frac{1}{n}}\right) \leq n \sum_{i=1}^{n} a_i - \left(\sum_{i=1}^{n} \sqrt{a_i}\right)^2$$

$$\leq n(n-1)\left(\frac{1}{n}\sum_{i=1}^{n} a_i - \left(\prod_{i=1}^{n} a_i\right)^{\frac{1}{n}}\right)$$

Theorem 2.5. Let a_i and b_i where $1 \leq i \leq n$ be positive real numbers then,

$$\sum_{i=1}^{n} b_i^2 + tT \sum_{i=1}^{n} a_i^2 \leq (t + T)\left(\sum_{i=1}^{n} a_i \ b_i\right)$$

where t and T are real constants for all i and $1 \leq i \leq n$ holds for $ta_i \leq b_I \leq Ta_i$

The minimum pendant dominating Harary energy

Let G be a simple graph of order n with vertex set $V(G) = \{v_1, v_2, ..., v_n\}$ and edge set E(G). Let D be the minimum pendant dominating Harary set of the graph G. The minimum pendant dominating Harary matrix of G is the *n square matrix denoted by $H_{PD}(G) = h_{ij}$, where

$$h_{ij} = \begin{cases} 1, & \text{if } i = j \text{ and } v_i \in D \\ 0, & \text{if } i = j \text{ and } v_i \notin D \\ \dfrac{1}{d(v_i, v_j)}, & \text{otherwise} \end{cases}$$

The characteristic polynomial of $H_{PD}(G)$ denoted by $f_n(G,\lambda)$ is defined as $f_n(G,\lambda) = \det(\lambda I - H_{PD}(G))$. The Eigen values of $H_{PD}(G)$ are the minimum pendant dominating Harary Eigen values of the graph G.

Since $H_{PD}(G)$ is real and symmetric, its Eigen values are real and ordered in non-increasing fashion i.e., $\lambda_1 \geq \lambda_2 \geq ... \geq \lambda_n$. The minimum pendant dominating Harary energy of G is given by

$$HE_{PD}(G) = \sum_{i=1}^{n} |\lambda_i|$$

The trace of the minimum pendant dominating Harary matrix of graph G is same as the minimum pendant dominating set of $H_{PD}(G)$. In this paper, we explore mathematical aspects of the minimum pendant dominating Harary energy of a graph.

Example

Let G be a graph with vertex set $V(G) = \{v_1, v_2, ..., v_6\}$ as shown in Figure 52.1. The possible minimum pendant dominating Harary sets are $D_1 = \{v_1, v_2\}$ and $D_2 = \{v_2, v_5\}$.

(i)   Consider $\mathbf{D_1} = \{\mathbf{v_1}, \mathbf{v_2}\}$ then

Figure 52.1 Graph G showing Harary Energy
Source: Author

$$H_{PD_1}(G) = \begin{pmatrix} 1 & 1 & \frac{1}{2} & 1 & \frac{1}{2} \\ 1 & 1 & 1 & \frac{1}{2} & 1 \\ \frac{1}{2} & 1 & 0 & \frac{1}{3} & \frac{1}{2} \\ 1 & \frac{1}{2} & \frac{1}{3} & 0 & 1 \\ \frac{1}{2} & 1 & \frac{1}{2} & 1 & 0 \end{pmatrix}$$

The characteristic polynomial of $H_{PD_1}(G)$ is given by $f_n(G,\lambda) = \lambda^5 - 2\lambda^4 - \frac{46}{9}\lambda^3 - \frac{7}{9}\lambda^2 + \frac{131}{144}\lambda + \frac{1}{18}$ Minimum pendant dominating Eigen values are $\lambda_1 = -1.208033$,

$\lambda_2 = -0.602480$, $\lambda_3 = -0.059210$, $\lambda_4 = 0.368174$ and $\lambda_5 = 5.739444$. Hence the minimum pendant dominating Harary energy is $H_{PD_1}(G) = 5.739444$.

Consider $\mathbf{D_2} = \{\mathbf{v_2}, \mathbf{v_5}\}$ then

$$H_{PD_2}(G) = \begin{pmatrix} 0 & 1 & \frac{1}{2} & 1 & \frac{1}{2} \\ 1 & 1 & 1 & \frac{1}{2} & 1 \\ \frac{1}{2} & 1 & 0 & \frac{1}{3} & \frac{1}{2} \\ 1 & \frac{1}{2} & \frac{1}{3} & 0 & 1 \\ \frac{1}{2} & 1 & \frac{1}{2} & 1 & 1 \end{pmatrix}$$

The characteristic polynomial of $H_{PD_2}(G)$ is given by

$$f_n(G,\lambda) = \lambda^5 - 2\lambda^4 - \frac{46}{9}\lambda^3 - \frac{7}{9}\lambda^2 + \frac{131}{144}\lambda + \frac{1}{18}$$

The minimum pendant dominating Harary Eigen values and Harary energy are same as in the example. The above example shows that minimum pendant dominating Harary energy is influenced by choice of minimum pendant dominating Harary set.

The minimum dominating Harary energy of several standard graphs

In this part, we analyse the exact values of the minimum pendant dominating Harary energy for various standard graphs.

Theorem 4.1. The minimum pendant dominating Harary energy of complete graph K_n for $n \geq 2$ is

$$(n-3) + \sqrt{n^2 - 2n + 9}$$

Proof: Let K_n be a complete graph with vertex set $V(G) = \{v_1, v_2, v_3, ..., v_n\}$. The minimum pendant dominating Harary set is $D = \{v_1, v_2\}$ and the minimum pendant dominating Harary matrix is as follows:

$$H_{PD}(K_n) = \begin{pmatrix} 1 & 1 & 1 & 1 & \cdots & 1 & 1 \\ 1 & 1 & 1 & 1 & \cdots & 1 & 1 \\ 1 & 1 & 0 & 1 & \cdots & 1 & 1 \\ 1 & 1 & 1 & 0 & \cdots & 1 & 1 \\ \vdots & \vdots & \vdots & \vdots & \ddots & \vdots & \vdots \\ 1 & 1 & 1 & 1 & \cdots & 0 & 1 \\ 1 & 1 & 1 & 1 & \cdots & 1 & 0 \end{pmatrix}_{n \times n}$$

The characteristic polynomial of $H_{PD}(K_n)$ is given by

$$f_n(K_n, \lambda) = \begin{pmatrix} \lambda-1 & -1 & -1 & -1 & \cdots & -1 & -1 \\ -1 & \lambda-1 & -1 & -1 & \cdots & -1 & -1 \\ -1 & -1 & \lambda & -1 & \cdots & -1 & -1 \\ -1 & -1 & -1 & \lambda & \cdots & -1 & -1 \\ \vdots & \vdots & \vdots & \vdots & \ddots & \vdots & \vdots \\ -1 & -1 & -1 & -1 & \cdots & \lambda & -1 \\ -1 & -1 & -1 & -1 & \cdots & -1 & \lambda \end{pmatrix}$$

$$= (-1)^n \lambda (\lambda+1)^{n-3} (\lambda^2 - (n-1)\lambda - 2)$$

The Eigen values of K_n are $\lambda = 0$ with multiplicity 1, $\lambda = -1$ with multiplicity $(n-3)$ and $\lambda = \dfrac{(n-1)\pm\sqrt{n^2-2n+9}}{2}$ with multiplicity 1 each.

The spectrum of K_n is as follows:

$$spec(K_n) = \begin{pmatrix} \frac{(n-1)+\sqrt{n^2-2n+9}}{2} & 0 & -1 & \frac{(n-1)-\sqrt{n^2-2n+9}}{2} \\ 1 & 1 & (n-3) & 1 \end{pmatrix}$$

The minimum pendant dominating Harary energy of K_n is

$$HE_{PD}(K_n) = \left| \frac{(n-1)+\sqrt{n^2-2n+9}}{2} \right| + |-1|(n-3)$$
$$+ \left| \frac{(n-1)-\sqrt{n^2-2n+9}}{2} \right|$$

$$HE_{PD}(K_n)$$
$$= (n-3)$$
$$+ \frac{(n-1)+\sqrt{n^2-2n+9}-(n-1)+\sqrt{n^2-2n+9}}{2}$$

$$\therefore HE_{PD}(K_n) = (n-3) + \sqrt{n^2-2n+9}$$

Theorem 4.2. The minimum pendant dominating Harary energy of complete bipartite graph $K_{n,n}$ for $n \geq 2$ is $\dfrac{5n-4}{2} + \dfrac{\sqrt{4n^2+16n-16}}{4}$

Proof: Let $K_{n,n}$ be a complete bipartite graph with vertex set $V(G) = \{u_1, u_2, u_3, \ldots, u_n, v_1, v_2, v_3, \ldots, v_n\}$. The minimum pendant dominating Harary set is $D = \{u_1, v_n\}$ and the minimum pendant dominating Harary matrix is as follows:

$$H_{PD}(K_{n,n}) = \begin{pmatrix} 1 & \frac{1}{2} & \frac{1}{2} & \cdots & \frac{1}{2} & 1 & 1 & \cdots & 1 \\ \frac{1}{2} & 0 & \frac{1}{2} & \cdots & \frac{1}{2} & 1 & 1 & \cdots & 1 \\ \frac{1}{2} & \frac{1}{2} & 0 & \cdots & \frac{1}{2} & 1 & 1 & \cdots & 1 \\ \vdots & \vdots & \vdots & \ddots & \vdots & \vdots & \vdots & \ddots & \vdots \\ \frac{1}{2} & \frac{1}{2} & \frac{1}{2} & \cdots & 0 & 1 & 1 & \cdots & 1 \\ 1 & 1 & 1 & \cdots & 1 & 0 & \frac{1}{2} & \cdots & \frac{1}{2} \\ 1 & 1 & 1 & \cdots & 1 & \frac{1}{2} & 0 & \cdots & \frac{1}{2} \\ \vdots & \vdots & \vdots & \ddots & \vdots & \vdots & \vdots & \ddots & \vdots \\ 1 & 1 & 1 & \cdots & 1 & \frac{1}{2} & \frac{1}{2} & \cdots & 1 \end{pmatrix}_{2n \times 2n}$$

The characteristic polynomial of $H_{PD}(K_{n,n})$ is given by

$$f_n(K_{n,n}, \lambda) = \begin{pmatrix} \lambda-1 & \lambda-\frac{1}{2} & \cdots & \lambda-\frac{1}{2} & \lambda-1 & \lambda-1 & \cdots & \lambda-1 \\ \lambda-\frac{1}{2} & \lambda & \cdots & \lambda-\frac{1}{2} & \lambda-1 & \lambda-1 & \cdots & \lambda-1 \\ \vdots & \vdots & \ddots & \vdots & \vdots & \vdots & \ddots & \vdots \\ \lambda-\frac{1}{2} & \lambda-\frac{1}{2} & \cdots & \lambda & \lambda-1 & \lambda-1 & \cdots & \lambda-1 \\ \lambda-1 & \lambda-1 & \cdots & \lambda-1 & \lambda & \lambda-\frac{1}{2} & \cdots & \lambda-\frac{1}{2} \\ \lambda-1 & \lambda-1 & \cdots & \lambda-1 & \lambda-\frac{1}{2} & \lambda & \cdots & \lambda-\frac{1}{2} \\ \vdots & \vdots & \ddots & \vdots & \vdots & \vdots & \ddots & \vdots \\ \lambda-1 & \lambda-1 & \cdots & \lambda-1 & \lambda-\frac{1}{2} & \lambda-\frac{1}{2} & \cdots & \lambda-1 \end{pmatrix}$$

$$f_n(K_{n,n}, \lambda)$$
$$= \frac{(2\lambda+1)^{2n-4}(4\lambda^2-6n\lambda+(3n-7))(4\lambda^2+2n\lambda-(n-1))}{4^{n-1}}, \text{if n is odd}$$

$$f_n(K_{n,n}, \lambda)$$
$$= \frac{(2\lambda+1)^{2n-4}(4\lambda^2-6n\lambda+(2n-3))(4\lambda^2+2n\lambda-(n-1))}{4^n}, \text{if n is even}$$

The Eigen values of $K_{n,n}$ when n is odd is as follows:

$\lambda = -\frac{1}{2}$ with multiplicity $(2n-4)$, $\lambda = \frac{6n\pm\sqrt{36n^2-32n+48}}{8}$ and $\lambda = \frac{-2n\pm\sqrt{4n^2+16n-16}}{8}$ with multiplicity 1 each

The spectrum of $K_{n,n}$ is as follows:

$$spec(K_{n,n}) = \begin{pmatrix} \frac{6n+\sqrt{36n^2-48n+112}}{8} & \frac{-2n+\sqrt{4n^2+16n-16}}{8} \\ 1 & 1 \\ -\frac{1}{2} & \frac{6n-\sqrt{36n^2-48n+112}}{8} & \frac{-2n-\sqrt{4n^2+16n-16}}{8} \\ (2n-4) & 1 & 1 \end{pmatrix}$$

The minimum pendant dominating Harary energy of $K_{n,n}$ is

$$HE_{PD}(K_{n,n}) = \left| \frac{6n+\sqrt{36n^2-48n+112}}{8} \right|$$
$$+ \left| \frac{-2n+\sqrt{4n^2+16n-16}}{8} \right| + \left| -\frac{1}{2} \right|(2n-4)$$
$$+ \left| \frac{6n-\sqrt{36n^2-48n+112}}{8} \right|$$
$$+ \left| \frac{-2n-\sqrt{4n^2+16n-16}}{8} \right|$$

$$\therefore HE_{PD}(K_{n,n}) = \frac{5n-4}{2} + \frac{\sqrt{4n^2+16n-16}}{4}$$

The Eigen values of $K_{n,n}$ when n is even is as follows:

$\lambda = -\frac{1}{2}$ with multiplicity $(2n-4)$, $\lambda = \frac{6n\pm\sqrt{36n^2-32n+48}}{8}$ and $\lambda = \frac{-2n\pm\sqrt{4n^2+16n-16}}{8}$ with multiplicity 1 each

The spectrum of $K_{n,n}$ is as follows:

$$spec(K_{n,n}) = \begin{pmatrix} \frac{6n+\sqrt{36n^2-48n+112}}{8} & \frac{-2n+\sqrt{4n^2+16n-16}}{8} \\ 1 & 1 \\ -\frac{1}{2} & \frac{6n-\sqrt{36n^2-48n+112}}{8} & \frac{-2n-\sqrt{4n^2+16n-16}}{8} \\ (2n-4) & 1 & 1 \end{pmatrix}$$

The minimum pendant dominating Harary energy of $K_{n,n}$ is

$$HE_{PD}(K_{n,n}) = \left| \frac{6n + \sqrt{36n^2 - 32n + 48}}{8} \right|$$
$$+ \left| \frac{-2n + \sqrt{4n^2 + 16n - 16}}{8} \right| + \left| -\frac{1}{2} \right| (2n - 4)$$
$$+ \left| \frac{6n - \sqrt{36n^2 - 32n + 48}}{8} \right|$$
$$+ \left| \frac{-2n - \sqrt{4n^2 + 16n - 16}}{8} \right|$$

$$\therefore HE_{PD}(K_{n,n}) = \frac{5n - 4}{2} + \frac{\sqrt{4n^2 + 16n - 16}}{4}$$

Properties on minimum pendant dominating Harary energy

In this section, we explore several properties of the characteristic polynomials of the minimum pendant dominating Harary matrix.

Theorem 5.1. Let G be a simple graph with vertex set $V(G) = \{v_1, v_2, \ldots, v_n\}$, edge set $E(G)$ and let $D = \{u_1, u_2, \ldots, u_k\}$ be a minimum pendant dominating set. If the Eigen values of $HE_{PD}(G)$ are $\lambda_1, \lambda_2, \ldots, \lambda_n$ then

(i) $\sum_{i=1}^{n} |\lambda_i| = |D|$

(ii) $\sum_{i=1}^{n} \lambda_i^2 = |D| + 2 \sum_{i<j} \frac{1}{d(v_i, v_j)^2} = |D| + 2q'$

where $q' = \sum_{i<j} \frac{1}{d(v_i, v_j)^2}$

Proof: (i) By definition, the sum of Eigen values of $HE_{PD}(G)$ is identical to the trace of $HE_{PD}(G)$. Therefore, we have

$$\sum_{i=1}^{n} \lambda_i = \sum_{i=1}^{n} h_{ii} = |D| \text{ for all } i$$

(ii) Similarly, the sum of squares of the Eigen values of $HE_{PD}(G)$ is equal to the trace of $(HE_{PD}(G))^2$. Therefore, we have

$$\sum_{i=1}^{n} \lambda_i^2 = \sum_{i=1}^{n} \lambda_i \sum_{j=1}^{n} \lambda_j = \sum_{i=1}^{n} \sum_{j=1}^{n} h_{ij} h_{ji}$$
$$= \sum_{i=1}^{n} \sum_{j=1}^{n} (2h_{ij}^2 + h_{ii}^2) = \sum_{i=1}^{n} h_{ii}^2 + 2 \sum_{i<j}^{n} h_{ij}^2$$
$$= |D| + 2 \sum_{i<j} \frac{1}{d(v_i, v_j)^2}$$

Theorem 5.2: Let $G = (n, m)$ be a graph and let $\lambda_1(G)$ be the largest minimum pendant dominating Harary Eigen value of $HE_{PD}(G)$. Then,

$$\lambda_1(G) \geq \frac{2 \sum_{i<j} \frac{1}{d(v_i, v_j)^2} + |D|}{n}$$

Proof: Let G be a (n, m) simple graph and let $\lambda_1(G)$ be the minimum pendant dominating Harary Eigen value of $HE_{PD}(G)$. Then we have,

$\lambda_1 = \max_{X \neq 0} \left(\frac{X^T A X}{X^T X} \right)$ where X is any non-zero vector, X^T is its transpose and A is a matrix. If we choose $J = [1, 1, \ldots, 1]^T$ then we get,

$$\lambda_1(G) \geq \frac{J^T H_{PD}(G) J}{J^T J} = \frac{2 \sum_{i<j} \frac{1}{d(v_i, v_j)^2} + |D|}{n}$$

Corollary 5.1. Let G be a (n, m) simple graph with diameter 2 and $D = \{u_1, u_2, \ldots, u_k\}$ be the minimum pendant dominating Harary set. If $\lambda_1, \lambda_2, \ldots, \lambda_n$ are the Eigen values of $HE_{PD}(G)$ then

$$\sum_{i=1}^{n} \lambda_i^2 = \frac{4|D| + n^2 - n + 6m}{4}$$

Proof: Follows by noting that there are 2 m entries in $HE_{PD}(G)$ with 1 and $n(n-1) - 2m$ with $\frac{1}{2}$

Bounds on the minimum pendant dominating Harary energy

In this section, we will explore different bounds for minimum pendant dominating Harary energy of graphs.

Theorem 6.1: Let $G = (n,m)$ be connected graph and let $|D|$ be the minimum pendant dominating Harary set. Let $\Delta = |H_{PD}(G)|$ then

$$\sqrt{|D| + 2q' + n(n-1)\Delta^{\frac{2}{n}}} \leq HE_{PD}(G)$$

$$\leq \sqrt{(n-1)(|D| + 2q') + n\Delta^{\frac{2}{n}}}$$

Proof: By Theorem 2.4. consider

$$X = n\left(\frac{1}{n}\sum_{i=1}^{n} a_i - \left(\prod_{i=1}^{n} a_i\right)^{\frac{1}{n}}\right)$$

Put $a_i = |\lambda_i|$ in the above we get,

$$X = n\left(\frac{1}{n}(|D| + 2q') - \left(\prod_{i=1}^{n}|\lambda_i|\right)^{\frac{2}{n}}\right)$$

$$X = |D| + 2q' - n\Delta^{\frac{2}{n}}$$

Put $a_i = \lambda_i^2$ in the Theorem 2.4. we get,

$$X \leq n\sum_{i=1}^{n}\lambda_i^2 - \sum_{i=1}^{n}|\lambda_i|^2 \leq (n-1)X$$

$$\left(|D| + 2q' + n(n-1)\Delta^{\frac{2}{n}}\right) \leq \left(HE_{PD}(G)\right)^2$$

$$\leq (n-1)(|D| + 2q') + n\Delta^{\frac{2}{n}}$$

$$\therefore \sqrt{|D| + 2q' + n(n-1)\Delta^{\frac{2}{n}}} \leq HE_{PD}(G)$$

$$\leq \sqrt{(n-1)(|D| + 2q') + n\Delta^{\frac{2}{n}}}$$

Lemma 6.1. If $\lambda_1(G)$ is the largest minimum pendant dominating Harary Eigen value of $HE_{PD}(G)$ and $|D|$ is the cardinality of minimum pendant dominating Harary set. Then,

$$\lambda_1(G) \geq \frac{|D| + 2\sum_{i<j}\frac{1}{d(v_i,v_j)^2}}{n}$$

Proof: Let $V(G) = \{v_i/1 \leq i \leq n\}$ be vertex set and $E(G) = \{e_i/1 \leq i \leq n\}$ be edge set of $H_{PD}(G)$. Let $X = [1,1,...,1]$ be a non-zero vector. Then by Rayleigh's principle we have,

$$\lambda_1 = \max_{X \neq 0}\left(\frac{X^T H_{PD}(G)X}{X^TX}\right)$$

$$\geq \frac{J^T H_{PD}(G)J}{J^TJ}$$

$$= \frac{1}{n}\sum_{i=1}^{n}\lambda_i^2 = \frac{1}{n}\left(|D| + 2\sum_{i<j}\frac{1}{d(v_i,v_j)^2}\right)$$

$$\lambda_1(G) \geq \frac{|D| + 2\sum_{i<j}\frac{1}{d(v_i,v_j)^2}}{n}$$

where $J = [1,1,...,1]^T$ is a unit column matrix of order $n*1$

Lemma 6.2. Let G be a connected graph with diameter 2 and $\lambda_1(G)$ be the largest minimum pendant dominating Harary Eigen value of $HE_{PD}(G)$. Then

$$\lambda_1(G) \geq \frac{2m + n^2 - n + 2|D|}{2n}$$

Proof: Let G be a connected simple graph of diameter 2 and d_i denote the degree of vertex v_i. Clearly i^{th} row of $H_{PD}(G)$ consists of d_i number of $1's$ and $(n - d_i - 1)$ number of $\frac{1}{2}$ s. Then by Rayleigh's principle for $J = [1,1,...,1]^T$ is a column matrix we have,

$$\lambda_1(G) = \max_{X \neq 0}\left(\frac{J^T H_{PD}(G)J}{J^TJ}\right)$$

$$\lambda_1(G) \geq \frac{\sum_{i=1}^{n}\left[d_i * 1 + (n - d_i - 1) * \frac{1}{2}\right] + |D|}{n}$$

$$\therefore \lambda_1(G) \geq \frac{2m + n^2 - n + 2|D|}{2n}$$

Theorem 6.2. Let G be a (n,m) graph with diameter 2 and $\frac{2m + n^2 - n + 2|D|}{2n} \geq 1$ then

$$HE_{PD}(G) \leq \left(\frac{2m+n^2-n+2|D|}{2n}\right) + \sqrt{(n-1)\left[\left[\frac{4|D|+n^2-n+6m}{4}\right] - \left(\frac{2m+n^2-n+2|D|}{2n}\right)^2\right]}$$

Proof: Consider Cauchy's-Schwarz inequality

$$\left(\sum_{i=2}^{n} a_ib_i\right)^2 \leq \left(\sum_{i=2}^{n} a_i^2\right) \cdot \left(\sum_{i=2}^{n} b_i^2\right)$$

Put $a_i = 1$ and $b_i = |\lambda_i|$ in the above inequality

$$\left(\sum_{i=2}^{n}|\lambda_i|\right)^2 \le \left(\sum_{i=2}^{n}1\right)\cdot\left(\sum_{i=2}^{n}|\lambda_i|^2\right)$$

$$HE_{PD}(G) \le \lambda_1 + \sqrt{(n-1)\left(\frac{4|D|+n^2-n+6m}{4}-\lambda_1^2\right)}$$

Let $f(x) = x + \sqrt{(n-1)\left(\frac{4|D|+n^2-n+6m}{4}-x^2\right)}$

For decreasing function $f'(x) \le 0$ we have,

$$\Rightarrow 1 + \frac{(n-1)(-2x)}{2\sqrt{(n-1)\left(\frac{4|D|+n^2-n+6m}{4}-x^2\right)}} \le 0$$

$$\sqrt{\frac{4|D|+n^2-n+6m}{4n-4}} \le x$$

Clearly,

$$\sqrt{\frac{2m+n^2-n+2|D|}{2n}}$$

$$\in \left\{\sqrt{\frac{4|D|+n^2-n+6m}{4n}}, \sqrt{\frac{4|D|+n^2-n+6m}{4}}\right\}$$

Since $\frac{2m+n^2-n+2|D|}{2n} \ge 1$ by Lemma 7.2. we have,

$$\sqrt{\frac{2m+n^2-n+2|D|}{2n}} \le \frac{2m+n^2-n+2|D|}{2n} \le \lambda_1$$

Therefore,

$$f(\lambda_1) \le f\left(\frac{2m+n^2-n+2|D|}{2n}\right)$$

$$HE_{PD}(G) \le f\left(\frac{2m+n^2-n+2|D|}{2n}\right)$$

$$\therefore HE_{PD}(G)$$
$$\le \left(\frac{2m+n^2-n+2|D|}{2n}\right)$$
$$+ \sqrt{(n-1)\left[\left(\frac{4|D|+n^2-n+6m}{4}\right)-\left(\frac{2m+n^2-n+2|D|}{2n}\right)^2\right]}$$

Lemma 6.3. Let G be a graph with minimum pendant dominating set D. If the minimum pendant dominating Harary energy $HE_{PD}(G)$ is a rational number then $HE_{PD}(G) \equiv |D|(mod2)$

Proof: Let $\lambda_1, \lambda_2, ..., \lambda_n$ are the Eigen values of $HE_{PD}(G)$. Let $\lambda_1, \lambda_2, ..., \lambda_p$ be the positive and the remaining Eigen values are of negative sign then

$$\sum_{i=1}^{n}|\lambda_i|^2 = (\lambda_1+\lambda_2+\cdots+\lambda_p)-(\lambda_{p+1}+\lambda_{p+2}+\cdots+\lambda_n)$$

$$\sum_{i=1}^{n}|\lambda_i|^2 = 2(\lambda_1+\lambda_2+\cdots+\lambda_p)-(\lambda_1+\lambda_2+\cdots+\lambda_n)$$

$$HE_{PD}(G) = 2(\lambda_1+\lambda_2+\cdots+\lambda_p)-\sum_{i=1}^{n}|\lambda_i|$$

Hence $HE_{PD}(G) \equiv |D|(mod2)$

Theorem 6.3. Let G = (n,m) be a graph. If $|\lambda_1| \ge |\lambda_2| \ge \ldots \ge |\lambda_n|$ be a non-increasing order of Eigen values of $HE_{PD}(G)$ and $|D|$ be the minimum pendant dominating Harary set. Then

$$HE_{PD}(G) \ge \sqrt{n(|D|+2q')-\alpha(n)(|\lambda_1|-|\lambda_n|)^2}$$

where $\alpha(n) = n\lfloor\frac{n}{2}\rfloor\left(1-\frac{1}{n}\lfloor\frac{n}{2}\rfloor\right)$

Proof: Let a_i and b_i where $1 \le i \le n$ be positive real numbers then

$$\left|n\sum_{i=1}^{n}a_i b_i - \left(\sum_{i=1}^{n}a_i\right)\left(\sum_{i=1}^{n}b_i\right)\right| \le \alpha(n)(A-a)(B-b)$$

where a, b, A and B are real constants and $1 \le i \le n$ for all i. Also, $a \le a_i \le A$, $b \le b_i \le B$ and

$$\alpha(n) = n\lfloor\frac{n}{2}\rfloor\left(1-\frac{1}{n}\lfloor\frac{n}{2}\rfloor\right)$$

The above inequality is true if and only if $a_1 = a_2 = \ldots = a_n$ and $b_1 = b_2 = \ldots = b_n$. Let $a_i = |\lambda_i|$, $b_i = |\lambda_i|$, $a = b = |\lambda_n|$ and $A = B = |\lambda_1|$ then

$$\left|n\sum_{i=1}^{n}|\lambda_i|^2 - \sum_{i=1}^{n}|\lambda_i|^2\right| \le \alpha(n)(|\lambda_1|-|\lambda_n|)^2$$

Since $\sum_{i=1}^{n}|\lambda_i|^2 = |D|+2q'$ and $HE_{PD}(G) \le \sqrt{n(|D|+2q')}$ and the inequalities becomes

$$n(|D|+2q')-(HE_{PD}(G))^2 \le \alpha(n)(|\lambda_1|-|\lambda_n|)^2$$

$$\therefore HE_{PD}(G) \ge \sqrt{n(|D|+2q')-\alpha(n)(|\lambda_1|-|\lambda_n|)^2}$$

Theorem 6.4. Let G=(n,m) be a connected graph. Let $|\lambda_1| \ge |\lambda_2| \ge \ldots \ge |\lambda_n| > 0$ be the Eigen values of $HE_{PD}(G)$ then

$$HE_{PD}(G) \ge \frac{|D|+2q'+n|\lambda_1||\lambda_n|}{|\lambda_1|+|\lambda_n|}$$

Proof: Let a_i and b_i where $1 \le i \le n$ be positive real numbers then

$$\sum_{i=1}^{n} b_i^2 + tT \sum_{i=1}^{n} a_i^2 \leq (t+T) \left(\sum_{i=1}^{n} a_i \, b_i \right)$$

where t and T are real constants for all i and $1 \leq i \leq n$ holds for $ta_i \leq b_I \leq Ta_i$.

Put $a_i = 1$, $b_i = |\lambda_i|$, $t = |\lambda_n|$, and $T = |\lambda_1|$, in the above inequality we get,

$$\sum_{i=1}^{n} |\lambda_i|^2 + |\lambda_n||\lambda_1| \sum_{i=1}^{n} 1 \leq (|\lambda_n| + |\lambda_1|) \left(\sum_{i=1}^{n} |\lambda_i| \right)$$

$$\frac{|D| + 2q' + n|\lambda_n||\lambda_1|}{|\lambda_n| + |\lambda_1|} \leq HE_{PD}(G)$$

$$\therefore HE_{PD}(G) \geq \frac{|D| + 2q' + n|\lambda_n||\lambda_1|}{|\lambda_n| + |\lambda_1|}$$

Theorem 6.5. Let G be a simple connected graph then $\sqrt{|D| + 2q'} \leq HE_{PD}(G) \leq \sqrt{n(|D| + 2q')}$

Proof: The Cauchy's-Schwarz inequality is

$$\left(\sum_{i=1}^{n} a_i b_i \right)^2 \leq \left(\sum_{i=1}^{n} a_i^2 \right) \cdot \left(\sum_{i=1}^{n} b_i^2 \right)$$

Put $a_i = 1$ and $b_i = |\lambda_i|$ in the above expression

$$\left(\sum_{i=1}^{n} |\lambda_i| \right)^2 \leq \left(\sum_{i=1}^{n} 1 \right) \cdot \left(\sum_{i=1}^{n} |\lambda_i|^2 \right)$$

$$(HE_{PD}(G))^2 \leq n \cdot \left(\sum_{i=1}^{n} |\lambda_i|^2 \right)$$

$$(HE_{PD}(G))^2 \leq n(|D| + 2q')$$

$$HE_{PD}(G) \leq \sqrt{n(|D| + 2q')} \quad \text{---------- (1)}$$

Now, consider

$$(HE_{PD}(G))^2 = \sum_{i=1}^{n} |\lambda_i|^2$$

$$(HE_{PD}(G))^2 \geq |D| + 2q'$$

$$HE_{PD}(G) \geq \sqrt{|D| + 2q'} \quad \text{------------- (2)}$$

From (1) and (2) we conclude that

$$\sqrt{|D| + 2q'} \leq HE_{PD}(G) \leq \sqrt{n(|D| + 2q')}$$

Theorem 6.6. Let G = (n,m) be a simple connected graph then

$$HE_{PD}(G) \geq \sqrt{n(|D| + 2q') - \frac{n^2}{4}(|\lambda_1| - |\lambda_n|)^2}$$

where $|\lambda_1|$ and $|\lambda_n|$ are the maximum and minimum Eigen values of $|\lambda_i|$'s of $HE_{PD}(G)$.

Proof: Let $\lambda_1, \lambda_2, ..., \lambda_n$ be Eigen values of the minimum pendant dominating Harary matrix of G. Put $a_i = 1$ and $b_i = |\lambda_i|$ in the inequality of Theorem 2.1. we get,

$$\left(\sum_{i=1}^{n} 1^2 \right) \left(\sum_{i=1}^{n} |\lambda_i|^2 \right) - \left(\sum_{i=1}^{n} |\lambda_i| \right)^2 \leq \frac{n^2}{4}(|\lambda_1| - |\lambda_n|)^2$$

$$n(|D| + 2q') - (HE_{PD}(G))^2 \leq \frac{n^2}{4}(|\lambda_1| - |\lambda_n|)^2$$

$$\therefore HE_{PD}(G) \geq \sqrt{n(|D| + 2q') - \frac{n^2}{4}(|\lambda_1| - |\lambda_n|)^2}$$

Conclusion

Modern graph theory includes dominance theory and graph energy, which have numerous applications in network theory, computer science, chemistry, and even physics. In recent years many scholars are working in this area and also, they are introducing new domination parameters. In this paper we have initiated the pendant domination parameter for Harary energy. We have made significant progress in our research by calculating exact values of the minimum pendant dominating Harary energy for standard family graphs and establishing bounds for this parameter in terms of other graph properties such as the degree and order of the graph.

A mathematical chemistry concept that measures a graph's energy is called Harary energy. An understanding of a graph's chemical and structural properties can be gained from its energy. The study of molecular graphs and the properties of chemical compounds is primary application of the term, which was first used in context of graph theory. It helps in analysing the structural characteristics of molecules by representing them as graphs and examining the relationship between the graph's energy and the molecule's chemical properties. Here's an overview of its uses: Harary energy provides a numerical measure of the complexity of graph, based on the Eigen values of its Laplacian matrix. This measure has applications in various fields including chemistry, network analysis and the study of complex systems. It helps in understanding the underlying structure and stability of the system represented by the graph.

References

[1] Adiga, C., Bayad, A., Gutman, I., Srinivas, S. A. The minimum covering energy of a graph. *Kragujevac J. Sci.*, 2012;34:39–56.

[2] Bapatand, R. B., Pati, S. Energy of a graph is never an odd integer. *Bull. Kerala Mathem. Assoc.*, 2011:129–132.

[3] Cvetkovi´c, D., Gutman, I. (eds.). Applications of Graph Spectra. Belgrade: Mathematical Institution, 2009;13:7–32.

[4] Cvetkovi´c, D., Gutman, I. (eds.). Selected Topics on Applications of Graph Spectra. Belgrade: Mathematical Institute SANU in Belgrade, 2011;14(22):85–112.

[5] Graovac, A., Gutman, I., Trinajsti´c, N. Topological Approach to the Chemistry of Conjugated Molecules. Berlin New York, Springer-Verlag (1977). "Lecture Notes in Chemistry". 124

[6] Gutman, I., Li, X., Zhang, J. Graph Energy. Dehmer, M., Emmert, F. (eds.). Analysis of Complex Networks, From Biology to Linguistics. Weinheim: Wiley-VCH, 2009:145–174.

[7] Gutman, I. (2001). The energy of a graph: Old and new results. In A. Betten, A. Kohnert, R. Laue, & A. Wassermann (Eds.), Algebraic Combinatorics and Applications (pp. 196-211). Springer-Verlag, Berlin.

[8] Gutman, I., Polansky, O. E. Mathematical Concepts in Organic Chemistry. Berlin: Springer, 1986;214

[9] Koolen, J. H., Moulton, V. Maximal energy graphs. *Adv. Appl. Math.*, 2001:47–52.

[10] McClelland, B. J. Properties of the latent roots of a matrix: The estimation of π-electron energies. *J. Chem. Phys.*, 1971;54:640–643.

[11] Rajesh Kanna, M. R., Dharmendra, B. N., Pradeep Kumar, R. Minimum covering Harary energy of a graph. *Int. J. Pure Appl. Math.*, 2014;90(3):371–385.

[12] Rajesh Kanna, M. R., Raghavendra, R., Jagadeesh, R. C., Pradeep Kumar, R. Minimum dominating Harary energy of a graph. *Adv. Appl. Discre. Math.*, 2016;17(3):255–274.

[13] Cui, Z., Liu, B. On Harary matrix, Harary index and Harary energy. *Math. Commun. Math. Comput. Chem.*, 2012;68:815–823.

53 Minimum pendant dominating colour energy of a graph

Purushothama S.[1,a], Mamatha N.[2,b], Nayaka S. R.[3,c], Nataraj K.[4] and K. N. Prakasha[5,d]

[1]Associate Professor, Department of Mathematics, Maharaja Institute of Technology Mysore, Karnataka–571477, India

[2]Department of Mathematics, K. S. Institute of Technology, Bengaluru, Karnataka–560109, India

[3]Department of Mathematics, PES College of Engineering, Mandya, Karnataka–571401, India

[4]Department of Mathematics, Maharaja Institute of Technology Mysore, Karnataka–571477, India

[5]Department of Mathematics, Vidya Vidyavardhaka College of Engineering, Mysuru, Karnataka–570002, India

Abstract

In this article, we introduced the idea of minimum pendant dominating colour energy of a graph $E_c^{pe}(G)$ and computed the minimum pendant dominating colour energy $E_c^{pe}(G)$ of few families of graphs. Further, the properties and upper and lower bounds for $E_c^{pe}(G)$ are established.

Keywords: Colour energy, dominating set, minimum dominating set, minimum pendant dominating energy, pendant dominating set

Introduction

Let $G = (V, E)$ be a graph with n nodes and m edges. The degree of v_i written by $d(v_i)$ is the number of edges incident with v_i. The maximum node of degree is denoted by $\Delta(G)$ and minimum node of degree is denoted by $\delta(G)$. The adjacency matrix $A_D(G)$ of G is defined by its entries as $a_{ij} = 1$ if $v_i\, v_j \in E(G)$ or $v_i \in D$ if $(i = j)$ where D is a dominating set of G and 0 otherwise. The Eigen values of graph G are the Eigen values of its adjacency matrix $A_D(G)$, denoted by. A graph $\lambda_1 \geq \lambda_2 \geq \ldots \geq \lambda_n$ is considered singular if it has minimum one Eigen value equal to zero. In the case of singular graphs, it is clear that $det(A) = 0$. A graph is said to be nonsingular if all of its Eigen values are non-zero. A graph G is referred to be k-regular if every node in G has degree k.

The energy of a graph G was introduced by Gutman in 1978 [7]. Initially, the concept of graph energy did not capture significant attention from mathematicians. However, over time, its value was recognised, sparking global mathematical research in the field of graph energy. Nowadays, in connection with graph energy, energy like quantities was considered also for other matrices. The energy $E(G)$ of G is defined to be the sum of the absolute values of the Eigen values of G. i.e.,

$$E(G) = \sum_{i=1}^{n} |\lambda_i|.$$

Adiga et al. [2] have defined colour energy $E_c(G)$ of a graph G Siva Kota Reddy et al. [8] have defined the minimum dominating colour energy of a graph. Motivated by these two papers, we introduced the concept of minimum pendant dominating colour energy $E_c^{pe}(G)$ of a graph G and computed minimum pendant dominating chromatic energies of some standard graphs are computed. Upper and lower bounds for $E_c^{pe}(G)$ are also established.

Colouring and colour energy

A colouring of graph G is a coloring of its vertices such that no two adjacent vertices receive the same color. The minimum number of colours needed for colouring of a graph G is called chromatic number and is denoted by $\chi(G)$ [2].

Consider the vertex coloured graph. Then entries of the matrix $A_c(G)$ are as follows. If $c(v_i)$ is the colour of v_i, then

$$a_{ij} = \begin{cases} 1, & \text{if } v_i \text{ and } v_j \text{ are adjacent with } c(v_i) \neq c(v_j), \\ -1, & \text{if } v_i \text{ and } v_j \text{ are non-adjacent with } c(v_i) = c(v_j), \\ 0, & \text{otherwise.} \end{cases}$$

The characteristic polynomial of $A_c(G)$ is denoted by $f_n(G,\rho)$ and is defined by $f_n(G,\rho) = det(\rho I - A_c(G))$. The colour Eigen values of the graph G are the Eigen values of $A_c(G)$. Since $A_c(G)$ is real and symmetric, its Eigen values are real numbers and are labelled in non-increasing order $\rho_1 \geq \rho_2 \geq \ldots \geq \rho_n$. The colour energy of G is defined as

[a]psmandya@gmail.com, [b]mamathan@ksit.edu.in, [c]nayakaabhi11@gmail.com, [d]prakashamaths@gmail.com

DOI: 10.1201/9781003685876-53

$$E_c(G) = \sum_{i=1}^{n} |\rho_i|$$

Minimum dominating colour energy of a graph

Let G be a simple graph of order n with the vertex set $V = \{v_1, v_2, \ldots, v_n\}$ and edge set E. Let D be the minimum dominating set of a graph G. The minimum dominating colour matrix of G is the $n \times n$ matrix defined by $A_c^D(G) = (a_{ij})$. Where

a_{ij}

$$= \begin{cases} 1, & \text{if } v_i \text{ and } v_j \text{ are adjacent with } c(v_i) \neq c(v_j) \\ & \text{or if } i = j \text{ and } v_i \in D, \\ -1, & \text{if } v_i \text{ and } v_j \text{ are non} - \text{adjacent with } c(v_i) = c(v_j), \\ 0, & \text{otherwise.} \end{cases}$$

The characteristic polynomial of $A_c^D(G)$ is denoted by $f_n(G,\lambda)$ and is defined by $f_n(G,\lambda) = det(\lambda I - A_c^D(G))$. The minimum dominating Eigen values of the graph G are the Eigen values of the matrix $A_c^D(G)$. We note that these Eigen values are real numbers since $A_c^D(G)$ is real and symmetric. So, we can label them in the non-increasing order $\lambda_1 \geq \lambda_2 \geq \ldots \geq \lambda_n$. The minimum dominating energy of G is defined to be the sum of the absolute Eigen values of $A_c^D(G)$. In symbols, we write

$$E_c^D(G) = \sum_{i=1}^{n} |\lambda_i|$$

If the colour used is minimum then the energy is called minimum dominating chromatic energy and it is denoted by $E_\chi^D(G)$. For more information about minimum dominating colour energy of a graph is explained in ref [8].

Minimum pendant dominating colour energy of a graph

A dominating set S in G is called a pendant dominating set if $< S >$ contains at least one pendant vertex. The least cardinality of the pendant dominating set in G is called the pendant domination number of G, denoted by $\gamma_{pe}(G)$ [9, 10]. Let G be a simple graph of order n with the vertex set $V = \{v1, v_2, \ldots, v_n\}$ and edge set E. Let D be the $\gamma_{pe} -$ set of a graph G. The minimum pendant dominating colour matrix of G is the $n \times n$ matrix defined by $A_c^{pe}(G) = (a_{ij})$. Where

a_{ij}

$$= \begin{cases} 1, & \text{if } v_i \text{ and } v_j \text{ are adjacent with } c(v_i) \neq c(v_j) \\ & \text{or if } i = j \text{ and } v_i \in D, \\ -1, & \text{if } v_i \text{ and } v_j \text{ are non} - \text{adjacent with } c(v_i) = c(v_j), \\ 0, & \text{otherwise} \end{cases}$$

The characteristic polynomial of $A_c^{pe}(G)$ is denoted by $f_n(G,\lambda)$ and is defined by $f_n(G,\lambda) = det(\lambda I - A_c^{pe}(G))$. The minimum dominating Eigen values of the graph G are the Eigen values of the matrix $A_c^{pe}(G)$. We note that these Eigen values are real numbers since $A_c^{pe}(G)$ is real and symmetric. So, we can label them in the non-increasing order $\lambda_1 \geq \lambda_2 \geq \ldots \geq \lambda_n$. The minimum dominating energy of G is defined to be the sum of the absolute Eigen values of $A_c^{pe}(G)$. In symbols, we write

$$E_c^{pe}(G) = \sum_{i=1}^{n} |\lambda_i|$$

Example 1

Let G be a graph on 6 vertices as shown in the Figure 53.1. The chromatic number of the graph G is 3, i.e., $\chi(G) = 3$. The possible minimum pendant dominating set is $D = \{v_2, v_5\}$

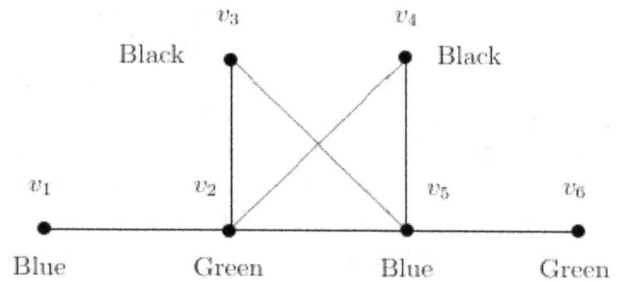

Figure 53.1 A simple graph with 6 vertices

Source: Author

The pendant dominating colour matrix of Figure 1 is given by

$$A_c^{pe}(G) = \begin{pmatrix} 0 & 1 & 0 & 0 & -1 & 0 \\ 1 & 1 & 1 & 1 & 1 & -1 \\ 0 & 1 & 0 & -1 & 1 & 0 \\ 0 & 1 & -1 & 0 & 1 & 0 \\ -1 & 1 & 1 & 1 & 1 & 1 \\ 0 & -1 & 0 & 0 & 1 & 0 \end{pmatrix}$$

The characteristic polynomial is given by

$$f_n(G,\lambda) = \lambda^6 - 2\lambda^5 - 9\lambda^4 + 14\lambda^3 + 20\lambda^2 - 24\lambda = 0$$

The minimum pendant dominating colour Eigen values are $\lambda_1 = -2$, $\quad = -$, $\lambda_3 = 0$, $\lambda_4 = 1$, $\lambda_5 = 2$, $\lambda_6 = 3$.

Hence, minimum pendant dominating colour energy, $LE_c^D \approx 10$.

Note that the minimum pendant dominating colour energy of the graph G depends on its minimum pendant dominating set.

Minimum pendant dominating colour energy of some standard graphs

Theorem 2.1. For $n > 2$, the minimum pendant dominating colour energy of a complete graph K_n is $(n-3)+\sqrt{n^2-2n+9}$.

Proof: Let K_n be the complete graph with the vertex set $=\{v1, v_2, \ldots, v_n\}$. Clearly, $D=\{v_1\}$ is a minimum dominating set of K_n. Then

$$A_c^{pe}(K_n) = \begin{pmatrix} 1 & 1 & 1 & \ldots & 1 & 1 \\ 1 & 1 & 1 & \ldots & 1 & 1 \\ 1 & 1 & 0 & \ldots & 1 & 1 \\ \cdot & \cdot & \cdot & \ldots & \cdot & \cdot \\ \cdot & \cdot & \cdot & \ldots & \cdot & \cdot \\ \cdot & \cdot & \cdot & \ldots & \cdot & \cdot \\ 1 & 1 & 1 & \ldots & 0 & 1 \\ 1 & 1 & 1 & \ldots & 1 & 0 \end{pmatrix}$$

The characteristic polynomial of $E_c^{pe}(K_n)$ is given by, $(\lambda)(\lambda+1)^{n-3}(\lambda^2-(n-1)\lambda-2)=0$.

The minimum pendant dominating colour Eigen values are
$\lambda = 0$ [1 time] $\lambda = 1$ [(n-3) times],

$$\lambda = \frac{(n-1)\pm\sqrt{n^2-2n+9}}{2} \text{ [one time each]}$$

Hence, the minimum pendant dominating colour energy of K_n is

$$E_c^{pe}(K_n) = |-1|(n-3) + \left|\frac{(n-1)+\sqrt{n^2-2n+9}}{2}\right| +$$

$$\left|\frac{(n-1)-\sqrt{n^2-2n+9}}{2}\right|$$

$$E_c^{pe}(K_n) = (n-3) + \left|\frac{(n-1)+\sqrt{n^2-2n+9}}{2}\right| + \left|\frac{(n-1)-\sqrt{n^2-2n+9}}{2}\right|$$

Therefore, $E_c^{pe}(K_n)=(n-3)+\sqrt{n^2-2n+9}$.

Theorem 2.2. For $n \geq 4$, the minimum pendant dominating colour energy of a star graph is $(n-1)+\sqrt{n^2+2n-7}$

Proof: Let $K_{1,n-1}$ be a star graph with the vertex set $V = \{v_1, v_2, \ldots, v_n\}$ having the vertex v_1 at the centre. The set $D = \{v_1, v_2\}$ is the minimum pendant dominating set of star graph. Then,

$$A_c^{pe}(K_{1,n-1}) = \begin{pmatrix} 1 & 1 & 1 & \ldots & 1 & 1 \\ 1 & 1 & -1 & \ldots & -1 & -1 \\ 1 & -1 & 0 & \ldots & -1 & -1 \\ \cdot & \cdot & \cdot & \ldots & \cdot & \cdot \\ \cdot & \cdot & \cdot & \ldots & \cdot & \cdot \\ \cdot & \cdot & \cdot & \ldots & \cdot & \cdot \\ 1 & -1 & -1 & \ldots & 0 & -1 \\ 1 & -1 & -1 & \ldots & -1 & 0 \end{pmatrix}$$

The characteristic polynomial of $E_c^{pe}(K_{1,n-1})$ is given by,

$$f_n(K_{1,n-1},\lambda)=(\lambda-1)^{n-3}(\lambda-2)(\lambda^2+(n-3)\lambda-(2n-4))$$

The minimum pendant dominating colour Eigen values are: $\lambda = 1$ [(n-3) times], $\lambda = 2$ [1 time],

$$\lambda = \frac{-(n-3)\pm\sqrt{n^2+2n-7}}{2} \text{ [one time each]}$$

Hence, the minimum pendant dominating colour energy is

$$E_c^{pe}(K_{1,n-1}) = |1|(n-3)+|2| + \left|\frac{(3-n)+\sqrt{n^2+2n-7}}{2}\right| + \left|\frac{(3-n)-\sqrt{n^2+2n-7}}{2}\right|$$

$$= (n-1) + \sqrt{n^2+2n-7}$$

Therefore, $E_c^{pe}(K_{1,n-1})=(n-1)+\sqrt{n^2+2n-7}$

Theorem 2.3. For $n \geq 3$, the minimum pendant dominating colour energy of a bi-star graph if $m = n$ is $(3n-4)+\sqrt{n^2+12n-12}$

Proof: Let $B(m, n)$ be a bistar graph with the vertex set $V = \{v_1, v_2, \ldots, v_{2n}\}$. The minimum pendant dominating set is $D = \{v_2, v_3\}$. Then

$$A_c^{pe}(B(n,n)) = \begin{pmatrix} 0 & 1 & -1 & \dots & -1 & -1 \\ 1 & 1 & 1 & \dots & 1 & 1 \\ -1 & 1 & 1 & \dots & -1 & -1 \\ \cdot & \cdot & \cdot & \cdots & \cdot & \cdot \\ \cdot & \cdot & \cdot & \cdots & \cdot & \cdot \\ \cdot & \cdot & \cdot & \cdots & \cdot & \cdot \\ -1 & 1 & -1 & \dots & 0 & -1 \\ -1 & 1 & -1 & \dots & -1 & 0 \end{pmatrix}$$

The characteristic polynomial of $E_c^{pe}(B(n,n))$ is given by

$$f_n(B(n,n),\lambda) = (\lambda-1)^{(2n-4)}(\lambda-2)(\lambda+(n-2))(\lambda^2+(n-2)\lambda-(4n-4)).$$

The minimum pendant dominating Eigen values are $\lambda = 1$, [(2n-4) times], $\lambda = 2$ [1 time], $\lambda = -(n-2)$ [1 time] $\lambda = \dfrac{-(n-2)\pm\sqrt{n^2+12n-12}}{2}$ [one time each] Hence, the minimum pendant dominating colour energy is

$$E_c^{pe}(B(n,n)) = |\ 1\ |\ (2n-4)+|\ 2\ |+|\ -(n-2)\ |$$
$$+\left|\frac{-(n-2)+\sqrt{n^2+12n-12}}{2}\right|+\left|\frac{-(n-2)-\sqrt{n^2+12n-12}}{2}\right|$$
$$E_c^{pe}(B(n,n)) = (3n-4)+\sqrt{n^2+12n-12}$$

Therefore, $E_c^{pe}(B(n,n)) = (3n-4)+\sqrt{n^2+12n-12}$

Theorem 2.4. For $n \geq 2$, the minimum pedant dominating colour energy of a complete bipartite graph $K_{n,n}$ is $(2n-1)+\sqrt{4n^2+4n-7}$

Proof: Let $K_{n,n}$ be a complete bi-partite graph with the vertex set $V = \{v_1, v_2, \dots, v_{2n}\}$ The minimum pendant dominating set of a complete bi-partite graph is $D = \{v_1, v_n\}$. Then

$$A_c^{pe}(K_{n,n}) = \begin{pmatrix} 1 & -1 & -1 & \dots & 1 & 1 \\ -1 & 0 & -1 & \dots & 1 & 1 \\ -1 & -1 & 0 & \dots & 1 & 1 \\ \cdot & \cdot & \cdot & \cdots & \cdot & \cdot \\ \cdot & \cdot & \cdot & \cdots & \cdot & \cdot \\ \cdot & \cdot & \cdot & \cdots & \cdot & \cdot \\ 1 & 1 & 1 & \dots & 0 & -1 \\ 1 & 1 & 1 & \dots & -1 & 1 \end{pmatrix}$$

The characteristic polynomial is given by,
$$f_n(K_{n,n},\lambda) = (\lambda-1)^{(2n-3)}(\lambda-2)(\lambda^2+(2n-3)\lambda - (4n-4)).$$

The minimum pendant dominating Eigen values are
$\lambda = 1$, [(2n-3) times], $\lambda = 2$ [1time], $\dfrac{-(2n-3)\pm\sqrt{4n^2+4n-7}}{2}$ [1 time each]

Hence, the minimum pendant dominating energy is

$$E_c^{pe}(K_{n,n}) = |\ 1\ |\ (2n-3)+|\ 2\ |\ (1)$$
$$+\left|\frac{-(2n-3)+\sqrt{4n^2+4n-7}}{2}\right|$$
$$+\left|\frac{-(2n-3)-\sqrt{4n^2+4n-7}}{2}\right|$$
$$E_c^{pe}(K_{n,n}) = (2n-1)+\sqrt{4n^2+4n-7}$$

Therefore, $E_c^{pe}(K_{n,n}) = (2n-1)+\sqrt{4n^2+4n-7}$

Properties of minimum pendant dominating colour Eigen values of a graph

Theorem 3.1. Let $|\lambda I - A_c^{pe}| = a_0\lambda^n + a_1\lambda^{n-1} + a_2\lambda^{n-2} + \dots + a_n$ be the characteristic polynomial of A_c^{pe}. Then

i. $a_0 = 1$

ii. $a_1 = -|D|$

iii. $a_2 = \binom{D}{2} - (m + m_c)$

where m is the number of edges and m_c is the number of pairs of non-adjacent vertices receiving the same colour in G.

Proof: i. It follows from the definition, $f_n(G,\lambda): det(\lambda I - A_c^{pe})$, that $a_0 = 1$.

ii. The sum of determinants of all 1×1 principal sub matrices of A_c^{pe} is equal to the trace of, which trace of A_c^{pe}, which $a_1 = (-1)^1$ trace of $\left[A_c^{pe}\right] = -|D|$.

iii. The sum of determinants of all the 2×2 principal sub matrices of $\left[A_c^{pe}\right]$ is

$$a_2 = \sum_{1 \leq i \leq j \leq n} (a_{ii}a_{jj} - a_{ij}a_{ji})$$

$$= \sum_{1 \leq i \leq j \leq n} a_{ii}a_{jj} - \sum_{1 \leq i \leq j \leq n} a_{ij}a_{ji}$$

$$= \binom{D}{2} - (m + \text{Number of pairs of non-adjacent})$$

vertices receiving the same colour in G)

Theorem 3.2. If D is a $\gamma_{pe}-$ set of a graph G and λ_1, $\lambda_2, ..., \lambda_n$ are the Eigen values of $A_c^{pe}(G)$ then

i. $\sum_{i=1}^n \lambda_i = |D|$

ii. $\sum_{i=1}^n \lambda_i = |D| + 2(m + m_c)$,

where m_c is the number of pairs of non-adjacent vertices receiving the same colour in G.

Proof: (i) We know that the sum of the Eigen values of $A_c^{pe}(G)$ is the trace of $A_c^{pe}(G)$.

$$\sum_{i=1}^n \lambda_i = \sum_{i=1}^n a_{ii} = |D|$$

(ii) The sum of squares of Eigen values of $A_c^{pe}(G)$ is the trace of $A_c^{pe}(G)^2$
Therefore,

$$\sum_{i=1}^n \lambda_i^2 = \sum_{i=1}^n \sum_{j=1}^n l_{ij} l_{ij} = \sum_{i=1}^n (l_{ii})^2 + \sum_{i \neq j} (l_{ji} l_{ji})$$

$$= \sum_{i=1}^n (l_{ii})^2 + 2 \sum_{i<j} (l_{ij})^2$$

$$= |D| + 2(m + m_c)$$

Upper and lower bounds

Theorem 4.1. Let G be any graph having n vertices and m edges and D is the minimum pendant dominating set of G. Then

$$E_c^{pe}(G) \geq \sqrt{2(m + m_c) + |D| + n(n-1)P^{\frac{2}{n}}}$$

where $P = |det A_c^{pe}(G)|$

Proof: By definition we have,

$$\left(E_c^{pe}(G) \right)^2 = \left(\sum_{i=1}^n |\lambda_i| \right) \cdot \left(\sum_{j=1}^n |\lambda_j| \right)$$

$$= \sum_{i=1}^n |\lambda_i|^2 + \sum_{i \neq j} |\lambda_i| \, |\lambda_j|$$

Now, by using arithmetic and geometric mean inequality, we get

$$\frac{\sum_{i \neq j} |\lambda_i| \, |\lambda_j|}{n(n-1)} \geq \left[\prod_{i \neq j} |\lambda_i| \, |\lambda_j| \right]^{\frac{1}{n(n-1)}}$$

$$E_c^{pe}(G) \geq \sum_{i=1}^n |\lambda_i|^2 \geq n(n-1) \left[\prod_{i \neq j} |\lambda_i| \, |\lambda_j| \right]^{\frac{1}{n(n-1)}}$$

$$\geq \sum_{i=1}^n |\lambda_i|^2 + n(n-1) \left[\prod_{i=1}^n |\lambda_i|^{2(n-1)} \right]^{\frac{1}{n(n-1)}}$$

$$= \sum_{i=1}^n |\lambda_i|^2 + n(n-1) \left[\prod_{i=1}^n |\lambda_i| \right]^{\frac{2}{n}}$$

$$= 2(m + m_c) + |D| + n(n-1)P^{\frac{2}{n}}$$

Therefore,

$$E_c^{pe}(G) \geq \sqrt{2(m + m_c) + |D| + n(n-1)P^{\frac{2}{n}}}$$

Theorem 4.2. Let G be any graph. Then
$$E_c^{pe}(G) \leq \sqrt{2n(m + m_c) + n|D|}$$

Proof: Let $\lambda_1 \geq \lambda_2 \geq ... \geq \lambda_n$ be the Eigen values of $A_c^{pe}(G)$ arranged in the non-increasing order. By using Cauchy's-Schwarz inequality we have,

$$\left(\sum_{i=1}^n a_i b_i \right)^2 \leq \left(\sum_{i=1}^n a_i^2 \right) \left(\sum_{i=1}^n b_i^2 \right)$$

Put $a_i = 1, b_i = |\lambda_i|$ then,

$$\left(\sum_{i=1}^n |\lambda_i| \right)^2 \leq \left(\sum_{i=1}^n 1 \right) \left(\sum_{i=1}^n |\lambda_i|^2 \right)$$

$$\left[E_c^{pe}(G) \right]^2 \leq n \left[\sum_{i=1}^n \lambda_i^2 \right] = n(2(m + m_c) + |D|)$$

i.e., $E_c^{pe}(G) \leq \sqrt{2n(m + m_c) + n|D|}$

Conclusion

The study of theory of domination and energy of graph is an important area in graph theory and also remarkable research is going on in this area.

Motivated by colour energy and dominating colour energy of graph. In this article we introduced the idea of minimum pendant dominating colour energy $E_c^{pe}(G)$ of a graph G and computed minimum pendant dominating chromatic energies of some standard graph. Properties and upper and lower bounds for $E_c^{pe}(G)$ are also established.

References

[1] Adiga, C., Bayad, A., Gutman, I., Srinivas, S. A. The minimum covering energy of a graph, *Kragujevac J. Sci.*, 2012;34:39–56.

[2] Adiga, C., Sampathkumar, E., Sriraj, M. A. Colour energy of a graph. *Proc. Jangjeon Math. Soc.*, 2013;16(3): 335–351.

[3] Adiga, C., Shivakumaraswamy, C. S. Bounds on the largest minimum degree Eigen values of graphs. *Int. Math. Forum.*, 2010;5(37):1823–1831.

[4] Jahanbani, A. Some new lower bounds for energy of graphs. *Appl. Math. Comput.*, 2017; 296:233–238.

[5] Sampathkumar, E., Sriraj, M. A. Vertex labelled/colored graphs, matrices and signed graphs. *J. Combin. Inform. Sys. Sci.*, 2013;38(1–2):113–120

[6] Bapat, R. B., Pati, S. Energy of a graph is never an odd integer. *Bull. Kerala Math. Assoc.*, 2011;1:129–132.

[7] Gutman, I. The energy of a graph. *Ber. Math-Statist. Sekt. Forschungsz.Graz,* 1978; 103:1–22.

[8] Siva Kota Reddy, P., Prakasha, K. N., Gavirangaiah, K. Minimum dominating colour energy of a graph. *Internat. J. Math. Combin.*, 2017;3:22–31.

[9] Nayaka, S., Puttaswamy, R., Purushothama, S. Pendant domination in some generalized graphs. *Internat. J. Sci. Engg. Sci.*, 2017;1(7):13–15.

[10] Nayaka, S., Puttaswamy, R., Purushothama, S. Pendant domination in graphs. *J. Combinat. Math. Combinat. Comput.*, 2020;112:219–229.

54 Structures of partially ordered ternary semi-rings

Dr. A. Rajeswari[1,a], Dr. Chandrakala H. K.[2] and Dr. Sheela[3,b]

[1]Department of Mathematics, East West Institute of Technology, Bangalore–560091, Karnataka, India

[2]Department of Mathematics, Bangalore Institute of Technology, Bangalore, Karnataka, India

[3]Data Analyst, FedEx Express, , Bangalore, Karnataka, India

Abstract

In this section we introduce the idea of anti-inverse ternary semi-groups with respect to ternary multiplication and usual addition. Studied some structural properties of ternary semi-rings and proved if $a\varrho b \Longleftarrow \Longrightarrow a^3 = a + b = ab^2$ partially ordered semi-rings.

Keywords: Ternary semi-ring, simple ternary semi-ring, partially ordered ternary semi-group

Introduction

Semi-ring hypothesis is one of the most creating part of Math with wide application in many areas, for example, Computer science, coding hypothesis, topological space and numerous analysts concentrates on various design of semi-rings like Boolean like semi-rings, ternary semi-rings, complemented ternary semi-rings, gamma semi-rings, complemented semi-rings and so on. In this paper, we talk about certain properties of complemented semi-rings. In 2003, Dutta and Kar [1] first introduced the concept of ternary semi-ring and further studied on properties of regular ternary semi-rings. Though the concept of ternary semi-ring generalises the concepts of semi-ring it is not closely a generalisation of semi-rings in there are several concepts like the lateral ideal semi-rings. Recently Jaya Lalitha et al., [2] proved few properties of regular ternary semi-groups. Several mathematicians have taken a view at ternary structures, we concentrate few structural properties of anti-inverse ternary semi-groups and characterised them by utilising the idea of singular ternary semi-rings and proved some structural properties. We studied the results of Madhusudana Rao et al. [3], which introduces the notion of anti-inverse ternary semi-groups and studied mono-ternary semi-rings by using lateral singular ternary semi-rings and characterise them.

Definition1.1: An empty set T_s together with a binary operation called usual addition and a ternary multiplication denoted by [] is defined as a ternary semi-ring if T_s is an additive commutative semi-group fulfilling the following conditions

(i) $[[abc]de]=[ab[cde]]=[a[bcd]e]$,

(ii) $[acd]+[bcd]=[(a+b)cd]$,

(iii) $[abd]+[acd]=[a(b+c)d]$,

(iv) $[abc]+[abd]=[ab(c+d)], \forall a,b,c,d,e \in T_s$.

Examples

The set of all rational numbers, Q with respect to usual addition and ternary multiplication.

Definition 1.2: A T.S T_g is in the case that an anti-inverse ternary semi-group with respect to ternary multiplication if for each element $a \in T_g$ there exists b, c $\in T_g$ such that $ca(bcb)=a$, $ab(cac)=b$, $bc(aba)=c$.

Definition 1.3: A semi-group $(T_g,+)$ is left singular ternary semi-group in the case that $a+b=a, \forall a,b \in T_g$.

Definition 1.4: A semi-group $(T_g,+)$ is right singular in the case that $a+b=b, \forall a,b \in T_g$.

Definition 1.5: A semi-group $(T_g,+)$ is singular ternary semi-group in the case that it is right as well as left singular.

Definition 1.6: A T.S $(T_g,[])$ is left singular in the case that $ab^2=a, \forall a,b \in T_g$.

Definition 1.7: A T.S$(T_g,[])$ is right singular in the case that $b^2a=a, \forall a,b \in T_g$.

Definition 1.8: A T.S$(T_g,[])$ is lateral singular in the case that $bab=a, \forall a,b \in T_g$.

Definition 1.9: A ternary semi-ring T_s is commutative ternary semi-ring in the case that $abc=cab=bac=bca=cba=acb, \forall a,b,c \in T_s$.

[a]rajeswari.bhat83@gmail.com, [b]sheelajsm@gmail.com

DOI: 10.1201/9781003685876-54

Partially ordered ternary semi-groups

Lemma 1.1: Let $(T_s,+,\cdot)$ be a ternary semi-ring. If T_s contains ternary multiplicative identity which is also additive identity, then $ab+ee=(a+e)(b+e), \forall a,b \in T_s$.

Proof: For all $a,b \in T_s$ there exists $e \in T_s$ such that $(a+e)(b+e)=a(b+e)+e(b+e)=ab+ae+eb+ee=a(b+e)+eb+ee=ab+eb+ee=(a+e)b+e(a+e)(b+e)=ab+ee$

Lemma 1.2: Let $(T_s,+,\cdot)$ be a simple ternary semi-ring then $ab+ee=(a+e)(b+e)$, $\forall a,b \in T_s$.

Proof: For all $a,b \in T_s$ there exists $e \in T_s$ such that $(a+e)(b+e)=a(b+e)+e(b+e)=ab+ae+ee=ab+ae+ee=ab+(a+e)=ab+ee=>(a+e)(b+e)=ab+ee$

Theorem 1.3: Let $(T_s,+,\cdot)$ be a simple ternary semi-ring in which T_s is idempotent such that a relation ϱ on T_s defined by $a\varrho b \Longleftrightarrow a^3=a+b=ab^2 \forall a, b \in T_s$. If $(T_s,\cdot)$ is singular then $(T_s,+,\cdot,\le)$ is a partially ordered ternary semi-group.

Proof: Given $(T_s,\cdot)$ is singular, thus, $ab^2=a$.
$a+b=ab^2+b=ab(b+e)=abb=ab^2=a$.
$a^3=a$ Thus $a^3=a+b=ab^2=a, \forall a,b \in T_s$

Reflexive:

$a\varrho b=\Longrightarrow a^3=a+b=ab^2=a$

Replace "b" by "a", $a^3=a+a=aa^2=a=\Longrightarrow a=a=a$ this$=\Longrightarrow a\varrho a$

$=\Longrightarrow \varrho$ is reflexive

Anti-symmetric:

$a\varrho b=\Longrightarrow a^3=a+b=ab^2=a$ and $b\varrho a$
$=\Longrightarrow b^3=b+a=ba^2=b$
Consider, $a^3=ab^2=(bab)b^2=bab^3=ba(b+a)=bab+baa$
$\quad=bab+ba(bab)=bab(ee+ab)=babab=b(aba)b$
$\quad=bbb=b^3=b$
Let $a\varrho b=\Longrightarrow a^3=a+b=ab^2=a$ and $b\varrho a=\Longrightarrow b^3=b+a=ba^2=b$
$=\Longrightarrow \varrho$ is anti-symmetric

Transitive:

Consider, $a^3=a+c...(1)$
$\quad a^3=a=ac^2$, [since T_s is singular] $\qquad (2)$
From (1) and (2) $a^3=a+c=ac^2=\Longrightarrow a\varrho c$.
$\quad$ Thus,ϱ is transitive.
$\quad$ Let
$\quad (a+c)^3=a^3=a=a+b$
$\quad =(a+c)+(b+c) \qquad (3)$
$\quad =a+(c+b+c)=a+c+c=a+c=ab^2+cb^2$ [Since T_s is singular]
$\quad =(a+c)b^2$
$\quad (a+c)^3=(a+c)(b+c)^2 \qquad (4)$
$\quad$ From (3) and (4), $(a+c)^3=(a+c)+(b+c)=(a+c)(b+c)^2$.
$\quad$ Thus, $(a+c)\varrho(b+c)$.
$\quad$ Similarly we can prove that $(c+a)\varrho(c+b)$.

Therefore, $(T_s,+,\cdot,\varrho)$ is a partially ordered ternary semi-group.

Pre-multiplied by "x" and post-multiplied by "y" on both sides,
$=\Longrightarrow xay +xab^2y=xay \Longrightarrow xay+ xabby=xay$
$\Longrightarrow xay+xa(yby)by=xay \Longrightarrow xay+(xay)(xbx)yby=xay$
$\Longrightarrow xay+(xay)(xb(yxy))yby=xay \Longrightarrow xay+(xay)(xby)$
$xy^2by=xay \Longrightarrow xay+(xay)(xby)^2=xay \Longrightarrow (xay)\rho(xby)$

Let $a+ab^2=a$

Pre-multiplied both sides by "xy" on both sides, we get,
$=\Longrightarrow xya+xyab^2=xya \Longrightarrow xya+xyabb=xya \Longrightarrow xya+xya(xbx)b=xya$
$\Longrightarrow xya+(xya)x(yby)xb=xya \Longrightarrow xya+(xya)(xyb)(xyxxb)=xya \Longrightarrow xya+(xya)(xyb)(xyb)=xya \Longrightarrow xya+(xya)(xyb)^2=xya \Longrightarrow (xya)\rho(xyb)$.

Let $a\rho b \Longrightarrow a+ab^2=a$
$ac^2+ac^2bb=a(a+c)+(a+c)bc^2bc^2=a$
$(a+c)+(a+c)(b+c)(b+c)=a(a+c)+(a+c)(b+c)^2=a=>(a+c)\rho(b+c)$

Similarly we can prove, $(c+a)\rho(c+b)$

Thus $(T_s,+,\cdot,\rho)$ is a partially ordered ternary semi-group

Theorem 1.5: Let $(T_s,+,\cdot)$ be a ternary semi-ring such that $(T_s,+)$ is idempotent left and right singular. Define a relation ψ on T_s by $a\psi b$ by $aba + b + a = a$ $\forall a, b \in T_s$. If $(T_s,\cdot)$ is lateral singular then $(T_s,+,\cdot,\le)$ is partially ordered ternary semi-group.

Proof:

Reflexive

Consider,
$\quad aba+b+a=ab+b+a=ab+a=a$ $a=a$
$\quad$ Thus, ψ is reflexive.

Anti-symmetric

Let $bab+a+b=b$ and $aba+b+a=a$
$\quad$ Consider, $bab+a+b=b$
$\quad =\Longrightarrow bab+aba+b+a+b=b \Longrightarrow a+b+b+a+b=b$
$\quad \Longrightarrow a+b+b=b \Longrightarrow a+b=b \Longrightarrow a=b$
$\quad$ Thus, ψ is anti-symmetric.

Transitive:

Let $aba+b+a=a$ and $bcb+c+b=b$.
Consider
$aba+b+a=a \Longrightarrow aba+bcb+c+b+a=a \Longrightarrow b+c+c+b=a$
$\quad \Longrightarrow c+c+a+b=a \Longrightarrow aca+c+a=a$ [Since T_s is L.S]
$\quad$ Thus, ψ is transitive.

Theorem 1.4: Let $(T_s,+,\cdot)$ be an idempotent mono-ternary semi-ring, such that a relation ρ on T_s defined by

$a\rho b \Longleftrightarrow a+ab^2=a \forall a, b \in T_s$. If $(T_s, \cdot)$ is singular then $(T_s, +, \cdot, \leq)$ is a partially ordered ternary semi-group.

Proof:

Given $(T_s, \cdot)$ is singular, we have $ab^2=a$ and $ba^2=b$.

Therefore, $a+ab^2=a$ and $b+ba^2=b$

Reflexive:

Consider, $a+ab^2=a$

$\Longrightarrow a+abb=a \Longrightarrow a+ab(aba)=a \Longrightarrow a+aaa \Longrightarrow a+a=a \Longrightarrow a=a$. Thus, $a\rho a$ ρ is reflexive.

Anti-symmetric:

Let $a\rho b \Longrightarrow a+ab^2=a$ and $b\rho a \Longrightarrow b+ba^2=b$

Consider

$a=a+ab^2=a+abb=a+aba^2b=a+ab(a+b)=a+aba+ab$ $b=a+b+ab^2=a^2b+ab^2=a^2b+a$

$=a^2b+a^3=a^2(b+a)=a^2ba^2=aabaa=aba=b$

Thus, ρ is anti-symmetric.

Transitive:

Let $a\rho b \Longleftrightarrow a+ab^2=a$ and $b\rho c \Longleftrightarrow b+bc^2=b$

Consider,

$a+ab^2=a \Longrightarrow a+ a(b+ bc^2)^2=a \Longrightarrow a+a(b+bc^2)(b+bc^2)=a$ $\Longrightarrow a+(ab+abc^2)(b+bc^2)=a \Longrightarrow a+ab(b+bc^2)+abc^2(b+bc^2)=a \Longrightarrow a+abb+abbc^2+abc^2b+abc^2bc^2=a \Longrightarrow a+ab^2+ab^2c^2+ab+abbc^2=a \Longrightarrow a+a+ac^2+ab^2+ab^2c^2=a \Longrightarrow a+a+a+ac^2=a \Longrightarrow a+a+ac^2=a \Longrightarrow a+ac^2=a$.

Thus, $a\rho c$, ρ is transitive.

Compatibility:

Let $a\rho b \Longrightarrow a+ab^2=a$

Post-multiplied by "xy" on both sides we get,

$\Longrightarrow axy+ab^2xy=axy \Longrightarrow axy+abbxy=axy \Longrightarrow axy+a(xbx)bxy=axy \Longrightarrow axy+ax(yby)xbxy=axy$

$\Longrightarrow axy+(axy)b(xyx)xbxy=a \Longrightarrow axy+(axy)b(xy)(xb)xy=axy$

$\Longrightarrow axy+(axy)(bxy)(bxy)=axy \Longrightarrow axy+(axy(bxy)=axy \Longrightarrow (axy)\rho(bxy)$

Let $a+ab^2=a$

Compatibility:

Let $aba+b+a=a$. Post-multiplied by "xy" on both sides, we get

$abaxy+bxy+axy=axy$ (1)

Consider,

$abaxy=a(xbx)axy=ax(yby)xaxy=(axy)b(xyx)xaxy=(axy)(bxy)xx(xax)xy=(axy)(bxy)(xxx)axy=(axy)(bxy)(xax)xy=(axy)(bxy)(axy)$.

Thus, from (1)$\Longrightarrow (axy)(bxy)(axy)+bxy+axy=axy=\Longrightarrow (axy)\psi(bxy)$

Similarly, we can prove that $(xay)\psi(xby)$ and $(xya)\psi(xyb)$.

Consider

$(a+c)(b+c)(a+c)+b+c+a+c=(ab+ac+cb+cc)(a+c)+c+a$

$=aba+abc+aca+acc+cbc+acc+cba+cbc+cca+ccc+c$

$=ab(a +c)+c+caccc+cba+b+cccac+c+c$

$=abc+c+cac+cba+b+cac+c=abc+c+a+c-ba+b+a+c=abc+a+cba+b+c$

$=abc+bab+cbcac+b+c=abc+bab+bac+b+c$

$=abc+ba(b+c)+b+c=abc+bab+b+c$

$=abaca+abaab+b+c= aba(ca+ab)+b+c$

$=aba(ca+aaba)+b+c=aba(c+aab)a+b+c$

$=aba(c+aaba)a+b+c=aba(c+aaaba)a+b+c$

$=aba(c+aba)a+b+c=aba(c+b)a+b+c=ababa+b+c$

$= aaa+b+c=a+b+c=b+c$

[Since T_s is R.S]

Thus, $(a+c)\psi(b+c)$. Similarly, we can prove that $(c+a)\psi(c+b)$

$(T_s, +, \cdot, \psi)$ is a partially ordered ternary semi-group.

Conclusion

We introduce the idea of anti-inverse ternary semi-groups with respect to ternary multiplication and usual addition. Studied some structural properties of ternary semi-rings and proved is $a\rho b \Longleftrightarrow a^3= a + b = ab^2$ partially ordered semi-ring.

Acknowledgement

The authors wish to express their deep gratitude to the learned referee for her valuable comments and suggestions.

References

[1] Dutta, T. K., Kar, S. A note on regular ternary semi-rings. *Kyung-pook Math. J.*, 2006;46:357–365.

[2] Jaya Lalitha, G., Sarala, Y., Madhusudhana, R. Regular ternary semi-groups. *Internat. J. Chem. Sci.*, 2017;15(4):1–6.

[3] Madhusudana Rao, D., et al. Concepts on Ternary semi-rings. *Internat. J. Modern Sci. Engg. Technol. (IJMSET)*, 2014;1(7):105–110.

[4] Rajeswari, A. "Structure of semi-rings satisfying identities", published in *Elsevair Publications* in the Proceedings of the International Conference on mathematical sciences 2014;7(10). ISBN – 978-93-5107-261-4.

[5] Sheela, N., Rajeswari, A. Anti-regular semi-rings. *Internat. J. Res. Sci. Engg.*, 2017;3(3). ISSN 2394-8299.

[6] Sheela, N., Rajeswari, A. Structures of anti-inverse semi-rings. *Ann. Pure Appl. Math.*, 2018;16(1):215–222.

55 Optimisation of electric vehicle battery thermal management system using computational fluid dynamics and thermal analysis

Dharmila Chowdary A.[1,a], *Jalaja P.*[2,b], *M. Krishna*[3,c] *and Shivakumar H. M.*[4,d]

[1]Department of Computer Science, University of Texas at Dallas, Richardson,Texas,75080, USA

[2]Department of Mathematics, K.S. Institute of Technology, Bengaluru–560109, Karnataka, India

[3]Department of Mechanical Engineering, R V College of Engineering, Bengaluru–560059, Karnataka, India

[4]Department of Mathematics, East West Institute of Technology, Bengaluru–560091, Karnataka, India

Abstract

The implementation of battery thermal management techniques in electric vehicles is vital for enhancing battery lifespan and mitigating thermal degradation, particularly for lithium-ion and post-lithium batteries. This study aimed to optimise the cooling efficiency of a battery thermal management system (BTMS) through simulation and experimental validation. A CAD-based model for computational fluid dynamics (CFD) and thermal analysis was developed in ANSYS software, utilising a turbulence model and an equivalent circuit battery model. The simulation, with 8,204,449 elements, showed temperature distributions from 30°C to 45°C. Experimental validation involved thermocouples on a lithium-ion cell, cooled by a water-glycol mixture. Simulated results showed an average cell temperature of 31°C, while experimental results recorded 35°C. Maximum heat transfer observed was 386.5 W in simulations and 355.4 W experimentally. A 16% reduction in cell temperature was achieved by increasing the cooling fluid's velocity and temperature, suggesting further optimisation potential for enhanced BTMS performance in electric vehicle applications.

Keywords: Thermal management system, CFD, EV battery, cooling effect

Introduction

The efficient thermal management of electric vehicle (EV) battery systems is critical to improving their performance, durability, and safety. Lithium-ion (Li-ion) batteries, the primary energy storage systems in most electric vehicles, face significant challenges related to temperature control during operation. Overheating can result in battery degradation, thermal runaway and reduced lifespan, making precise temperature regulation essential. Therefore, the development and optimisation of battery thermal management systems (BTMS) are crucial to mitigate these risks and ensure the reliable operation of EVs [1, 2]. Battery thermal management is particularly important for maintaining optimal temperature ranges for lithium-ion and post-lithium batteries. Typically, these batteries should operate within a temperature window of 15–35°C, with no more than a 5°C difference in temperature between individual cells to prevent performance degradation [3]. Active and passive cooling systems, including liquid cooling and phase change materials (PCMs), are employed to maintain consistent temperature levels and prevent excessive heating [4, 5]. The use of PCMs in combination with liquid cooling has shown promising results in improving cooling efficiency, particularly with materials such as paraffin wax enhanced with expanded graphite [6, 7]. To enhance the performance of lithium-ion batteries, several strategies are being explored. Research on hybrid cooling systems, such as PCM-liquid cooling combinations, has demonstrated substantial improvements in temperature regulation, reducing peak temperatures and enhancing overall thermal performance. For example, studies have highlighted that adjusting the coolant mass flow rate can effectively reduce battery pack temperatures, though this approach requires careful optimisation to balance temperature uniformity with cooling efficiency [8, 9]. In addition, methods like tapered air channels, coupled with computational fluid dynamics (CFD) simulations, have shown a reduction in peak temperatures by up to 20% while improving temperature uniformity across the battery pack [10].

[a]dharmilaandra@gmail.com / dxa230019@utdallas.edu, [b]pjmaths@gmail.com / jalajap@ksit.edu.in, [c]krishnam@rvce.edu.in, [d]hmsk78@gmail.com

DOI: 10.1201/9781003685876-55

Optimisation of BTMS is also integral to the development of efficient and safe EVs. Recent research has focused on optimising coolant channel designs and flow rates to improve heat transfer and reduce temperature gradients within the battery pack. Numerical simulations using CFD and experimental methods have been employed to refine the design of cooling plates and fluid paths, leading to a reduction in both temperature differences and pressure drops within the cooling system [11, 12]. Moreover, the integration of smart battery management systems (BMS) allows for better monitoring and control of battery temperature, contributing to safer and more efficient energy storage systems [13, 14]. As the demand for EVs continues to rise, the need for advanced and optimised thermal management solutions becomes increasingly important. This research aims to explore the application of CFD simulations and thermal analysis in optimising BTMS for lithium-ion batteries, focusing on improving temperature uniformity, reducing peak temperatures, and enhancing overall system efficiency. By investigating various cooling strategies and optimisation techniques, this study seeks to contribute to the development of more reliable and sustainable thermal management systems for electric vehicle applications [15, 16].

The literature highlights limited research on co-optimising battery pack architecture, coolant flow, and cold plate design for high-power-density applications like data centres and electric vehicles. A holistic approach integrating thermal performance, power efficiency, and system reliability is essential, addressing challenges in high heat flux scenarios through comprehensive optimisation strategies. The objective is to develop, optimise, and validate a BTMS through numerical and experimental analysis to minimise energy consumption and enhance vehicle performance.

Mathematical model development

Heat generation in the battery cell
The heat generated in the battery cell (Q_{cell}) is the combined effect of Joule heating and entropy change.

$$Q_{gen} = I^2 R + IT \frac{\partial V}{\partial T} \qquad (1)$$

where is I is current (A), R is Internal resistance (Ω), T is temperature (K) and V is voltage (V), T is temperature (K), V is voltage (V) and $\frac{\partial V}{\partial T}$ is entropy coefficient (V/K).

Energy conservation equation (heat transfer)
Using the general heat transfer equation, the thermal energy balance in the battery is:

$$\rho C_p \frac{\partial T}{\partial t} = \nabla \cdot (k \nabla T) + Q_{gen} \qquad (2)$$

where ρ is density of the material (kg/m³), C_p is a specific heat capacity (J/kg·K), k is thermal conductivity (W/mK), $\nabla (k \nabla T)$ is heat conduction term and Q_{gen} heat generation rate per unit volume (W/m³)

Convective heat transfer in the cooling system
For a liquid or air-cooled system, the heat flux at the battery surface due to convection is:

$$Q_{conv} = h (T_{battery} - T_{Coolant}) \qquad (3)$$

where h is convective heat transfer coefficient (W/m²·K), $T_{battery}$ is battery surface temperature (K) and $T_{coolant}$ is coolant temperature (K).

CFD governing equations
To analyse fluid flow and heat transfer in the cooling system, the **Navier-Stokes equations** are solved alongside the energy equation.

$$\nabla \vec{v} = 0 \qquad (4)$$

Momentum equation:

$$\rho \left(\frac{\partial \vec{v}}{\partial t} + \vec{v} \cdot \nabla \vec{v} \right) = -\nabla P + \mu \nabla^2 \vec{v} + \vec{F} \qquad (5)$$

Energy equation:

$$\rho C_p \left(\frac{\partial T}{\partial t} + \vec{v} \cdot \nabla T \right) = k \nabla^2 T + Q_{source} \qquad (6)$$

where v is velocity field (m/s), P is pressure (Pa), μ is dynamic viscosity (Pa s), F is body force (N) and Q_{source} heat source due to battery or external conditions.

Cooling channel analysis: For liquid-cooled systems, consider heat exchange in the channels:

$$m C_p (T_{in} - T_{out}) = \dot{Q} \qquad (7)$$

where is m is mass flow rate of coolant (kg/s), Q is heat removed by coolant (W) and T_{in} and T_{out} are coolant temperatures at inlet and outlet (K).

FEM modelling and simulation

Before proceeding with the cell module analysis involving a cold plate, it was necessary to determine the heat generated by a single cell. For this purpose, the heat pulse power characterisation (HPPC) test was conducted. This test provided data on the changes in voltage over time and variations in current corresponding to the pulse applied during the charge cycle. The resulting test data was subsequently used as input for the equivalent circuit model (ECM) in ANSYS Fluent. The ECM model, coupled with Fluent meshing, was implemented after ensuring the completion of a high-quality mesh. Unlike conventional hexahedral or tetrahedral cells, polyhedral-shaped cells were chosen for their ability to accommodate complex geometries more effectively due to their arbitrary number of planar faces, resembling a honeycomb arrangement with hexagonal patterns.

The meshing process utilised polyhedral elements, which have more than six faces and are highly suitable for intricate geometries. A minimum element size of 0.1 mm and a maximum size of 14 mm were employed, striking a balance between computational efficiency and accuracy. A total of 82.04 million polyhedral cells were created, resulting in a dense and well-defined mesh. The quality of the mesh was evaluated based on key metrics. Skewness, which measures the deviation of cells from an ideal shape, was recorded at 0.55, remaining within the acceptable range of less than 0.6, with a preferred value below 0.3. This ensured minimal numerical diffusion and reduced the likelihood of instability. Orthogonality, which assesses the deviation of cell faces from perpendicular, was measured at 0.94, exceeding the acceptable threshold of 0.7 and ensuring proper alignment for accurate solutions.

Following the volume meshing process, the 3D model was divided into three distinct regions: a solid region representing the cold plate, solid cells, and a fluid region for the coolant. Figure 55.1 illustrates the overall 3D model, and also provides a sectional view of the meshed model, showcasing the application of polyhedral elements. This approach facilitated the effective representation of the regions and ensured the model's readiness for subsequent simulation and analysis.

The material properties of the cold plate and lithium-ion cell were utilised in the analysis and are outlined in Table 55.1. These properties play a vital role in defining the behaviour of the materials under the given thermal and flow conditions. Accurate representation of material properties ensures realistic

Figure 55.1 Meshed model of a battery pack with air cooling
Source: Author

Table 55.1 Material properties for the cold plate and lithium ion cell

Property	Cell material	Cold plate material	
Thermal conductivity (W/mK)	1.5	205	
Specific heat capacity (j/kg K)	850	897	
Density kg/m³⁺		2200	2700
Young's Modulus (GPa)		15	70
Poisson's ratio		0.25	0.33
Electric conductivity		0.05	

Source: Author

simulation results, which are crucial for evaluating system performance under operational conditions.

The boundary conditions applied to the system were designed to capture both thermal and velocity fields. For the temperature field, a combination of boundary conditions was implemented, including the Dirichlet condition, which specifies a fixed temperature on a boundary, the Neumann condition, which imposes a prescribed heat flux, and the Robin condition, which accounts for convective heat transfer with an ambient temperature and a heat transfer coefficient. For the velocity field, the no-slip condition was used at solid walls to enforce zero velocity, while specific velocity values were applied at inlet and outlet boundaries. Depending on the flow geometry and physics, symmetry or periodic boundary conditions were also employed. The test conditions, such as the flow rate (2.5 LPM), inlet velocity 0.734 (m/s) inlet temperature 18°C and ambient temperature 25°C.

The convergence behaviour of key quantities, including continuity, x-velocity, y-velocity, z-velocity, energy, k, and omega, was tracked over 60 iterations. The logarithmic y-axis of the convergence plot, shown in Figure 55.2, demonstrated a general decrease in residuals, indicating that the solution was converging with each iteration. Continuity and velocity components displayed faster convergence rates compared to other quantities, confirming the stability and reliability of the numerical simulation. This convergence plot is essential for monitoring residuals and validating the accuracy of computational simulations, such as CFD or finite element analysis, ensuring that the solution approaches the desired level of precision.

Experimental study

To assess the thermal performance of a battery pack with a cold plate cooling system, a comprehensive experimental setup comprising hardware and software components is essential. Temperature sensors, specifically thermocouples, are strategically positioned at the cold plate inlet, cold plate outlet, the battery cell surfaces, and potentially in the surrounding air, as illustrated in Figure 55.3. These thermocouples are critical for capturing temperature data and were chosen based on their fast response time and suitable temperature range. The core of the data collection system is a data acquisition system (DAQ), which converts analogue signals from the sensors into digital data for analysis. A Tektronix data logger is employed, along with software capable of recording data from up to 20 sensors at a time interval of one minute. This ensures accurate and continuous monitoring of the thermal behaviour of the battery pack. For the experimental analysis, the locations of

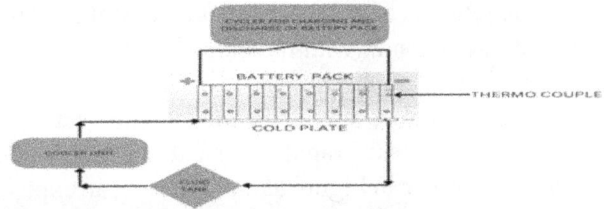

Figure 55.3 Schematic diagram of the battery thermal management experimental setup
Source: Author

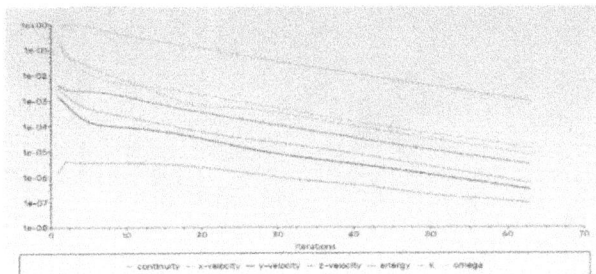

the thermocouples are depicted in Figure 55.3, where the blue small circles represent the positions of the thermocouples placed along the cell module. Due to the limited logging capacity, the sensors are positioned on alternate cells.

Results and discussion

Mathematical model results and discussion
This code models the cooling process in a BTMS by simulating heat transfer and temperature dynamics. It utilises key parameters such as the convective heat transfer coefficient, battery surface area, specific heat capacity, initial temperature, and coolant temperature (Figure 55.4).

The program computes the rate of heat dissipation and predicts the cooling behaviour of the battery over time, offering valuable insights into the system's thermal management efficiency. The Python

```
In [2]:   1  import numpy as np
          2  import pandas as pd
          3
          4  # Constants and Input Parameters
          5  R = 0.01  # Internal resistance (Ohms)
          6  T = 300  # Temperature (K)
          7  dV_dT = 0.0001  # Entropy coefficient (V/K)
          8
          9  rho = 2700  # Density of the material (kg/m³) - example for aluminum
         10  Cp = 900  # Specific heat capacity (J/kg·K)
         11  k = 200  # Thermal conductivity (W/mK)
         12  h = 50  # Convective heat transfer coefficient (W/m²·K)
         13  Tbattery = 320  # Battery surface temperature (K)
         14  Tcoolant = 300  # Coolant temperature (K)
         ...
         76  # Plot results
         71  plt.figure(figsize=(10, 6))
         72  plt.plot(time, Tbattery, label="Battery Temperature (K)", color='blue', linewidth=2)
         73  plt.axhline(y=Tcoolant, color='red', linestyle='--', label="Coolant Temperature (K)")
         74  plt.title("Battery Cooling vs. Time", fontsize=14)
         75  plt.xlabel("Time (s)", fontsize=12)
         76  plt.ylabel("Temperature (K)", fontsize=12)
         77  plt.legend(fontsize=10)
         78  plt.grid()
         79  plt.show()
```

Figure 55.2 Convergence testing of the battery thermal management system in CFD
Source: Author

Figure 55.4 A representative Python code for the CFD analysis of a battery thermal management system
Source: Author

implementation is based on the mathematical model outlined in Section 2, with the output presented as a cooling vs. time graph.

The simulation results indicate that the battery temperature decreases rapidly from its initial value of 360 K as shown in Figure 55.5, following an exponential cooling pattern as it approaches the coolant temperature of 300 K. The cooling behaviour is driven by convective heat transfer, where the temperature difference between the battery and the coolant facilitates heat dissipation. The cooling rate slows down as the battery temperature stabilises, and after approximately 600 seconds, the system reaches a steady-state temperature, indicating effective heat dissipation. This behaviour suggests that the thermal management system is functioning well, maintaining the battery temperature within safe limits to prevent thermal runaway.

The results highlight the key factors influencing the cooling process, including the convective heat transfer coefficient, the battery's surface area, and the material properties such as density and specific heat capacity.

Optimising these parameters, such as enhancing the coolant flow rate or increasing the surface area of the battery, could improve the cooling efficiency. While the simulation assumes uniform cooling, real-world systems may experience temperature gradients that could affect the accuracy of the results. Nonetheless, the insights from this study provide valuable guidance for designing and optimising thermal management systems in battery applications, ensuring their safe and efficient operation.

FEM results and discussion

The test conditions, including the velocity at the cold plate inlet and the inlet temperature, are summarised in Table (Figure 55.6). After solving the problem under steady-state conditions, the resulting temperature distribution of the battery pack is presented in Figure 55.6(a). The temperature variation along the battery cells and the cold plate is illustrated in Figure 55.6(b), while Figure 55.6(c) displays the temperature along the cold plate and cells. A contour plot of the temperature distribution is shown in Figure 55.6b, highlighting the total temperature field, with the highest temperatures depicted in bright red regions. The contour range spans from approximately 18–34°C, indicating a notable temperature gradient within the domain. This analysis confirms the presence of significant thermal variations, characteristic of a heat transfer problem. Figure 55.7 presents the average temperature of cell-1 during charging at a 1C rate, where the temperature is observed to increase steadily with iterations. The maximum temperature for cell-1 is approximately 32°C. Lastly, the outlet temperature of the cold plate, observed at the end of the charging cycle, is approximately 23°C. The results provide details on the maximum and minimum cell temperatures, along with the average temperature of the cells during the test.

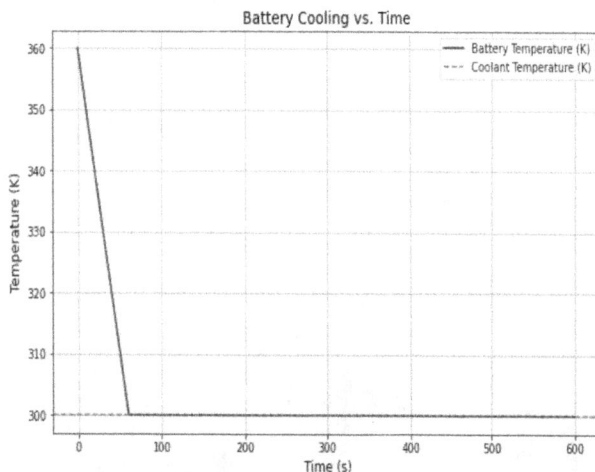

Figure 55.5 Python-simulated output of CFD analysis for the battery thermal management system
Source: Author

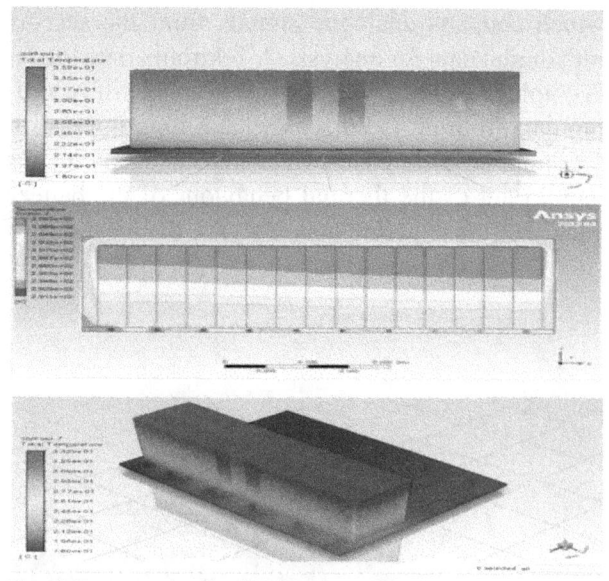

Figure 55.6 Temperature distribution across the battery cells during charging, incorporating cooling through the thermal management system
Source: Author

Table 55.2

TEST CONDITION	VALUE	UNIT
FLOW RATE	7	LPM
INLET VELOCITY	2.056	M/SEC
INLET TEMPERATURE	18	C
AMBIENT TEMPERATURE	25	C

Source: Author

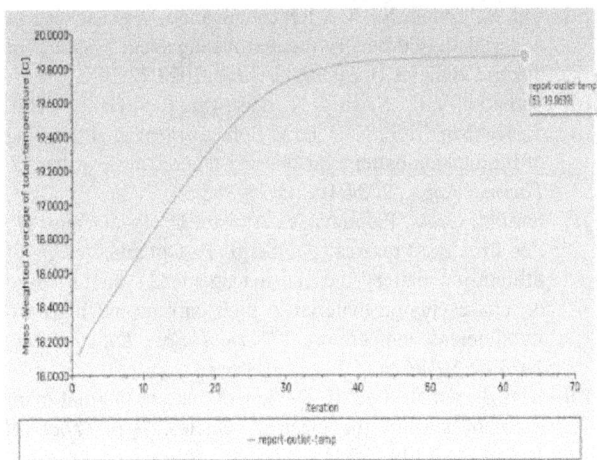

Figure 55.7 Temperature distribution across the battery cells during charging, incorporating cooling through the thermal management system
Source: Author

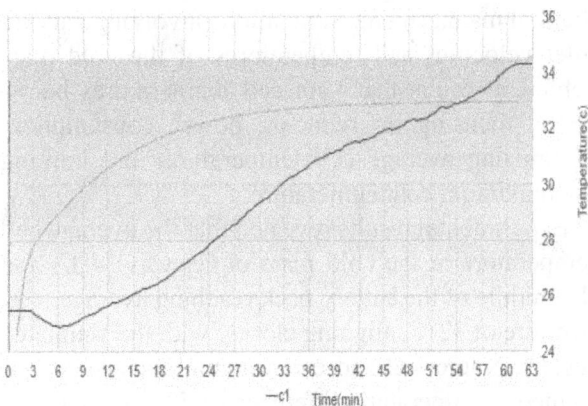

Figure 55.8 Experimental (red) and theoretical (green) analysis of the impact of charging and discharging during cooling on the temperature of a typical battery cell
Source: Author

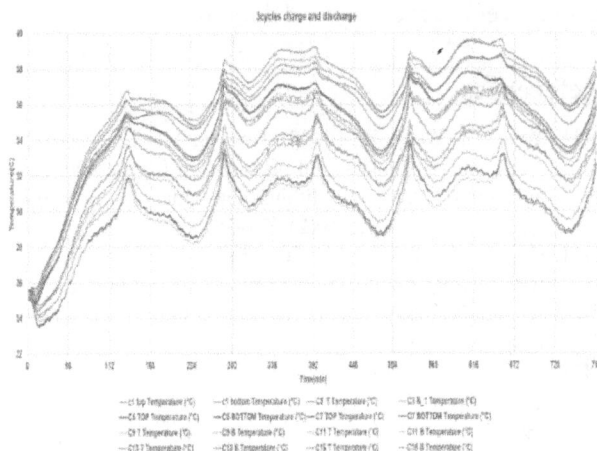

Figure 55.9 Experimental analysis of the effect of charging and discharging during cooling on battery cell temperature
Source: Author

Experimental and CFD comparison results and discussion

Examining the Figure 55.8 reveals a consistent trend of rising temperature over time for both the experimental (brown line) and simulated (green line) data of cell 1, as shown in Figure 55.8. This indicates on-going heat generation within the cell. A notable observation is the persistent temperature difference between the two lines.

Further analysis of the graph in Figure 55.9 reveals the thermal behaviour of a battery module undergoing three charge and discharge cycles. The x-axis represents time in minutes, while the y-axis indicates temperature in degrees Celsius (°C). Multiple lines in the graph correspond to temperatures recorded at different locations within the battery module, such as the top and bottom sides of the cells.

Key observations show that temperatures generally increase during charging and decrease during discharging, reflecting the internal heat generation and dissipation processes within the battery. The highest temperatures, or peaks, occur during charging cycles, likely due to endothermic (heat-absorbing) reactions at the electrodes. As the charging current reaches its maximum toward the end of the cycle, heat generation also peaks, resulting in these temperature spikes.

Conclusion

The CFD and thermal simulations of the battery pack, conducted using Python and ANSYS Fluent, were successfully completed with a residual error of 1×10^{-6}, and the results were documented during the battery's charge cycle. The average maximum cell temperature was observed to be 45°C at the end of the charge cycle, with intracell temperature differences maintained below 10°C, ensuring effective thermal management. Simulations involving varying inlet velocities and temperatures of the cold plate achieved results that kept cell temperatures below 45°C, focusing on reducing power consumption, minimising average cell temperature, and limiting the maximum cell temperature.

Experimental results revealed that the average cell temperature at the cold plate outlet was 34°C, and the middle of the battery pack reaching average temperature of 32°C, aligning closely with the simulated results. Numerical and experimental analyses highlighted a temperature difference of 5°C, with a 16% reduction in cell temperature compared to passive cooling, demonstrating the enhanced efficiency of the thermal management system. The improved thermal performance allows the battery pack to sustain higher charge and discharge cycles, thereby enhancing its overall efficiency and durability.

References

[1] Yang, H., Wang, Z., Li, M., Ren, F., Ma, B. Numerical study on cross-linked cold plate design for thermal management of high-power lithium-ion battery. *Energy Storage and Saving*, 2022;10(2):375–385.

[2] Gungor, S., Göçmen, S., Cetkin, E. A review on battery thermal management strategies in lithium-ion and post-lithium batteries for electric vehicles. *J. Thermal Engg.*, 2023;10(2):1071–1092.

[3] Lazim, A., Ismael, M. Cooling strategies of lithium-ion battery pack: A review. *Basrah J. Engg. Sci.*, 2024;24:39–47.

[4] Sun, J., Wei, M., Cai, S., Zhao, Y., Wright, E. Pack-level modelling and thermal analysis of a battery thermal management system with phase change materials and liquid cooling. *Energies*, 2023;16:5815.

[5] Wu, W. Optimization of an air-cooling thermal management system for lithium-ion battery packs via particle swarm algorithm. *J. Physics Conf. Series*, 2023; 2636:012006.

[6] Carlucci, A. P., Darvish, H., Laforgia, D. Thermal performance of a 48V prismatic lithium-ion battery pack under WLTC driving cycles with a liquid cooling system. *SAE Technical Paper Series*, 2023 24 0152 (2023).

[7] MasthanVali, P. S. N., Murali, G., Chinnasamy, S. Comparison study on cooling management of composite phase change material battery pack: Two different cases. *Thermal Sci.*, 2023;27:218.

[8] Murali, G., Nagavamsi, V., Srinath, A., Marianthiran, A. P. Battery thermal management system using phase change material on trapezoidal battery pack with liquid cooling system. 2020:5288–5300.

[9] Cai, S., Zhang, X., Ji, J. Recent advances in phase change materials-based battery thermal management systems for electric vehicles. *J. Energy Storage*, 2023;72:108.

[10] Satheesh, V. K., Krishna, N., Kushwah, P. S., Garg, I., Rai, S., Hebbar, G. S., Nair, D. V. Enhancement in air-cooling of lithium-ion battery packs using tapered airflow duct. *J. Thermal Engg.*, 2024;10(2):375–385.

[11] Farulla, G. A., Palomba, V., Aloisio, D., Brunaccini, G., Ferraro, M., Frazzica, A., Sergi, F. Optimal design of lithium-ion battery thermal management systems based on phase change material at high current and high environmental temperature. *Thermal Sci. Engg. Prog.*, 2023;42:101862.

[12] Kim, J., Oh, J., Lee, H. Review on battery thermal management system for electric vehicles. *Appl. Thermal Engg.*, 2019;149:192–212.

[13] Mathewson, S. Experimental measurements of LIFEPO4 battery thermal characteristics. 2014. [Online]. Available: https://uwspace.uwaterloo.ca/handle/10012/8378.

[14] Alihosseini, A., Shafaee, M. Experimental study and numerical simulation of a Lithium-ion battery thermal management system using a heat pipe. *J. Energy Storage*, 2021;39:102616.

[15] Ye, B., Rubel, , Li, H. Design and optimization of cooling plate for battery module of an electric vehicle. *Appl. Sci.*, 2019;9(4):754.

[16] Lipu, , Miah, , Ansari, S., Wali, , Jamal, T., Elavarasan, , Kumar, S., Ali, , Sarker, , Aljanad, A., Tan, Smart battery management technology in electric vehicle applications: Analytical and technical assessment toward emerging future directions. *Batteries*, 2022;8(11).

56 Enhancing waste management efficiency using the transportation problem

Jyothi P.[1,a], Vatsala G. A.[2,b], Radha Gupta[3,c] and Chaitra M.[4,d]

[1]Professor, Department of Management, Global Institute of Management Sciences, Bangalore University, Karnataka, India

[2]Department of Mathematics, Dayananda Sagar Academy of Technology and Management, Vishveswarayya Technological University, Karnataka, India

[3]Department of Mathematics, Dayananda Sagar College of Engineering, Vishveswarayya Technological University, Karnataka, India

[4]Department of Mathematics, BGS College of Engineering and Technology, Visvesvarayya Technological University, Karnataka, India

Abstract

Finding the most efficacious method to "transport" waste from incompatible sources (such as neighbourhoods, collection locations, or regions) to various disposal or processing sites (such as incinerators, landfills, recycling facilities, or) is the primary objective of waste management problems, which are customarily framed as transportation problems. In a transportation problem, the aim is to protecting the transportation costs while attaining the supply and demand constraints. Similarly, in a circumstances of waste management, the objective is to cut back on the cost (or effort) of transporting waste from various sources to the disposal sites, while safeguarding that the capacity of each site is not exceeded and all the waste is collected and processed. The intention of the paper is to suppress the total transportation cost, which is the sum of the costs of transporting waste from each neighbourhood to each disposal site, while satisfying the supply and demand constraints which helps the system of waste management works effectively. This paper helps in environmental sustainability and move towards societal benefit.

Keywords: Neighbourhood, landfill, non-negativity constraints, logistics

Introduction

This paper intended to minimise transportation costs in waste management by optimising the allocation of waste from different neighbourhoods to disposal sites. The purpose is to develop an efficient distribution strategy that meets supply and demand constraints while ensuring resources are utilised effectively and disposal capacities are not exceeded.

Vitally important components of the waste management

- **Demand (waste disposal capacity)**: The capacity or requirement of each disposal site to receive waste refers to demand (e.g., landfill, recycling centre, etc.).
- **Supply (waste to be collected)**: The amount of waste generated in each neighbourhood, region, or source point refers to supply.
- **Transportation costs**: The cost (in terms of money, time, or resources) to transport waste from a particular neighbourhood to a specific disposal site refers to transportation cost.
- **Objective**: The purpose is to reduce the total transportation cost, which is the sum of the costs of transporting waste from each neighbourhood to each disposal site, while fulfilling the supply and demand constraints.
- **Constraints related to supply**: The total waste transported from a neighbourhood cannot exceed the amount of waste generated in that neighbourhood.
- **Constraints related to demand**: The total waste transported to a disposal site must not exceed its capacity.
- **On-negativity constraints**: The amount of waste transported cannot be negative, meaning every transportation amount must be ≥ 0.

Objective of this study

When transportation problem is applied to system of **waste management**, it helps in optimising the

[a]hellojyothi95@gmail.com, [b]dr.vatsala.ga@gmail.com, [c]radha.gaurav.gupta@gmail.com, [d]chaitramg16@gmail.com

DOI: 10.1201/9781003685876-56

collection, transportation, and disposal of waste while reducing operational costs and environmental impact. The **transportation problem** is one of the types of optimisation problems used to minimise costs and improve efficiency in logistics. Controlling/suppressing the total transportation cost while adhering to both supply and demand constraints is the primary objective. This signifies ensuring that all waste from each neighbourhood is delivered to disposal sites without exceeding the capacity of the sites.

Primary targets

Capacity optimisation, cost minimisation, environmental sustainability, efficient waste collection & distribution, compliance with regulations, service level improvement, resource optimisation, integration with smart waste management systems. Industries and cities can accomplish **cost-effective, efficient, and environmentally friendly waste collection and disposal processes** by applying **the transportation problem model** in waste management.

Formulation of a problem

Consider a city where waste is collected from 6 neighbourhoods (A, B, C, D, E, and F) and needs to be transported to 4 waste disposal sites (W, X, Y, Z). Every garbage disposal location has a specified capacity to receive waste, and each neighbourhoods has a specific amount of waste that needs to be collected.

1. Restrictions on demand: Every disposal location can only manage a specific amount of demand. A site's disposal capacity cannot be exceeded by the amount of waste that is brought there.
2. Restrictions on supply: A specific quantity of waste must be transported to disposal locations in every community. The allotment has to meet these amounts.
3. Transportation cost: The cost of each transportation route is found by the quantity of garbage transported as well as the distance between a neighbourhoods and a disposal location. The average of these separate expenses for each route makes up the overall cost of transportation.

Formulation of the transportation model

In mathematical terms, the total transportation cost is often represented as a sum of products of transportation costs and the amounts of waste transported along various routes:

$$Total\ transportation\ Cost = \sum_{i,j} C_{ij}X_{ij}$$

where C_{ij} the transportation cost per unit of waste between neighbourhoods i and disposal site j,

X_{ij} is the amount of waste transported from neighbourhoods i to disposal site j

Optimisation problem minimise

$$\sum_{i,j} C_{ij}X_{ij}$$

Subject to:

- **Supply constraint** for each neighbourhood:
- $\sum_j X_{ij} = S_i$ for all neighbourhoods i

where S_i is the waste generated by neighbourhood i.

- **Demand constraint** for each disposal site: $\sum_j X_{ij} = D_j$ for all disposal sites j
  where D_j is the capacity (or demand) of disposal site j
- **Non-negativity**: $X_{ij} \geq 0$ *for all i,j*

The aim is to find values for X_{ij} (the allocation of waste) that controlling the total transportation cost while satisfying the supply and demand constraints.

This is a classical transportation problem in optimisation, and solving it efficiently ensures that waste management is done cost-effectively, reducing the overall expenses for the system.

Real time data collected for the problem

Waste collection points (supply):

- Neighbourhood A: 80 tons of waste
- Neighbourhood B: 90 tons of waste
- Neighbourhood C: 70 tons of waste
- Neighbourhood D: 60 tons of waste
- Neighbourhood E: 50 tons of waste
- Neighbourhood F: 40 tons of waste

Waste disposal sites (demand):

- Site W: 120 tons capacity
- Site X: 100 tons capacity

- Site Y: 80 tons capacity
- Site Z: 70 tons capacity

Minimise $Z = 5x_{11} + 8x_{12} + 6x_{13} + 7x_{14} + 6x_{21} + 9x_{22} + 7x_{23} + 8x_{24} + 7x_{31} + 6x_{32} + 8x_{33} + 9x_{34} + 5x_{41} + 8x_{42} + 6x_{43} + 7x_{44} + 7x_{51} + 6x_{52} + 5x_{53} + 8x_{54} + 9x_{61} + 6x_{62} + 7x_{63} + 4x_{64}$ \qquad (3)

Supply constraints (total waste transported from each neighbourhood):

$x_{11} + x_{12} + x_{13} + x_{14} = 80$ (Neighbourhood A)
$x_{21} + x_{22} + x_{23} + x_{24} = 90$ (Neighbourhood B)
$x_{31} + x_{32} + x_{33} + x_{34} = 70$ (Neighbourhood C)
$x_{41} + x_{42} + x_{43} + x_{44} = 60$ (Neighbourhood D)
$x_{51} + x_{52} + x_{53} + x_{54} = 50$ (Neighbourhood E)
$x_{61} + x_{62} + x_{63} + x_{64} = 40$ (Neighbourhood F)

Demand constraints (total waste transported to each site):

$x_{11} + x_{21} + x_{31} + x_{41} + x_{51} + x_{61} = 120$ (site W)
$x_{12} + x_{22} + x_{32} + x_{42} + x_{52} + x_{62} = 100$ (site X)
$x_{13} + x_{23} + x_{33} + x_{43} + x_{53} + x_{63} = 80$ (site Y)
$x_{14} + x_{24} + x_{34} + x_{44} + x_{54} + x_{64} = 70$ (site Z)

Non-negativity constraints $x_{ij} \geq 0$ for all i,j

Decision variables

Let x_{ij} represent the amount of waste (in tons) transported from neighbourhood i to site j

x_{11} amount of waste (in tons) transported from neighbourhood A to site W

x_{12} amount of waste (in tons) transported from neighbourhood A to site X

x_{13} amount of waste (in tons) transported from neighbourhood A to site Y

x_{14} amount of waste (in tons) transported from neighbourhood A to site Z

x_{21} amount of waste (in tons) transported from neighbourhood B to site W

x_{22} amount of waste (in tons) transported from neighbourhood B to site X

x_{23} amount of waste (in tons) transported from neighbourhood B to site Y

x_{24} amount of waste (in tons) transported from neighbourhood B to site Z

Similarly,

x_{31} amount of waste (in tons) transported from neighbourhood C to site W

x_{32} amount of waste (in tons) transported from neighbourhood C to site X

x_{33} amount of waste (in tons) transported from neighbourhood C to site Y

x_{34} amount of waste (in tons) transported from neighbourhood C to site Z

x_{41} amount of waste (in tons) transported from neighbourhood D to site W

x_{42} amount of waste (in tons) transported from neighbourhood D to site X

x_{43} amount of waste (in tons) transported from neighbourhood D to site Y

x_{44} amount of waste (in tons) transported from neighbourhood D to site Z

x_{51} amount of waste (in tons) transported from neighbourhood E to site W

x_{52} amount of waste (in tons) transported from neighbourhood E to site X

x_{53} amount of waste (in tons) transported from neighbourhood E to site Y

x_{54} amount of waste (in tons) transported from neighbourhood E to site Z

x_{61} amount of waste (in tons) transported from neighbourhood F to site W

x_{62} amount of waste (in tons) transported from neighbourhood F to site X

x_{63} amount of waste (in tons) transported from neighbourhood F to site Y

x_{64} amount of waste (in tons) transported from neighbourhood F to site Z

The problem is solved by converting into transportation problem

It is unbalanced transportation problem. It can be converted into balanced and can be solved with the different varieties of methods, Northwest corner rule, Vogel's approximation method, MODI method, etc. The problem is solved and the solution is as below. Best possible allocation are shown in the Table 56.1 and total transportation cost is found

Table 56.1 Initial basic table

	W	X	Y	Z	Supply
A	5	8	6	7	80
B	6	9	7	8	90
C	7	6	8	9	70
D	8	7	5	6	60
E	7	6	5	8	50
F	9	6	7	4	40
Demand	120	100	80	70	

Source: Author [Dr.Jyothi.P]

Table 56.2 Best possible allocations

	W	X	Y	Z		Supply
A	80					80 to W
B	40			50		40 to W, 50 to dummy
C		70				70 to X
D			60			60 to Y
E		30	20			(30 to X, 20 to Y)
F			40			(40 to Z)
Demand	120	100	80	70	20	

Source: Author [Dr.Jyothi.P]

Total transportation cost
$80 \times 5+40 \times 6 +70 \times 6 +60 \times 5 +30 \times 6 +20 \times 5+40 \times 4=1800$

Interpretation of the result

Best possible allocations areas mentioned in the Table 56.2.

A to W, B to W, C to X, D to Y, E to X and Y, F to Y.

The above allocations help in optimised waste collection and distribution with most cost efficient. By imposing streaming data and optimising transportation routes, purposes are met in our study are:

1. Optimised waste collection & distribution – Establishing that trash from each neighbourhood ends up in the right dumping or recycling station without encumbering any of them.
2. Benefits related to Environmental – Cut back on carbon footprints through the scaling down of redundant travel distances and downtime.
3. Outlay Savings – The most economical routes, controlling fuel consumption, and decreasing the functioning expenses is acknowledged in the process.
4. Enhanced supply chain management – Streamlining waste collection slots according to actual demand and disposal capacity.

Our study involves minimisation of comprehensive transportation costs while meeting supply and demand constraints. Concurrent is collected and following are achieved by ensuring that all waste from each neighbourhood is transported to depute disposal sites without exceeding their capacity and selecting the most profitable routes to optimise transportation charges.

With the integration of con-current data and optimisation techniques, waste management firms are able to develop a cost-efficient, sustainable, and effective recycling and waste disposal system. The positives of using the transportation problem in waste management are evident – our discussion demonstrates that using the problem helps in optimising the transportation logistics of moving commodities, e.g., waste, from quite a few sources to multiple destinations at reduced costs. In system of waste management, the transportation problem is significant in improving trash gathering routes, determines the optimal routes for waste collection trucks to control fuel usage and travel time, assists in slotting the collection of waste to avoid overflow and delay, economical disposal & recycling, assign waste to nearest or cheapest disposal points (e.g., landfill, recycling plant, incineration plant), minimising environmental footprint, reduces emissions by streamlining transport routes and eliminating irrelevant travel, saves transportation costs while optimising recycling, expedite green waste disposal by ensuring fair disposal in different facilities, effective hazardous waste management, enables safe and regulatory-approved transport of hazardous waste to approved treatment facilities, reduces the risk of environmental pollution by proper disposal of wastes, building reverse logistics, facilitates waste-to-energy plants by effectively moving organic or recyclable waste to respective plants, enables improved sorting and redistribution of recyclable resources. The industries of waste management and sustainable energy are meticulously associated, as effective waste treatment can be used for energy generation and put a stop to environmental degradation. By equalising waste management with sustainable energy initiatives, communities can control their environmental impact, improve energy security, and promote a more circular economic system. This association is based on a range of foundational beliefs such as landfill gas recovery, recycling and energy conservation, energy generated from waste biomass, waste-to-energy (WTE) conversion, circular economy measures, utilisation of renewable energy sources and the resource efficiency

Conclusion

Conclusion: The Transportation Problem to Improve Waste Management Efficiency. A classical transportation problem proves to be effective in waste management operations by optimizing its effects in urban settings (successfully demonstrated in this study). The authors developed an opportunity to achieve it by combining real-time data and optimization techniques:

Minimization of costs: The overall cost of transport was reduced to minimum of 1800 since the routes were allocated efficiently, and therefore, the economic worth of the method was achieved.

Capacity compliance: The number of waste generated by six neighbourhoods were spread among four disposal sites without overloading the sites according to the maximum capacity.

Environmental sustainability: The model decreased excessive travel as transport routes have been optimized and therefore the amount of fuel used to travel was minimized with a subsequent reduction of associated carbon emissions.

Operational improvement: The solution advanced planning and distribution of waste collection and improved logistics and service delivery.

Scalability and relevance: The framework can be scaled to different situations of urban and industrial waste and to be integrated into smart waste management systems.

Finally, the transportation model is a reasonable approach to controlling solid waste, which is strategically, affordable, and environmentally sensible, thus complying with contemporary sustainability and urban planning concepts.

References

[1] Sharma, J. K., Swarup, K. Time minimizing transportation problem. *Proc. Ind. Acad. Sci. (Math. Sci.)*, 1977;86: 513–518.

[2] Veena, A., Kowalski, K., Lev, B. Solving transportation problem with mixed constraints. *Internat. J. Manag. Sci. Engg. Manag.*, 2006;1:47–52.

[3] Okunbor, D. Management decision-making for transportation problems through goal programming. *J. Acad. Busin. Econ.*, 2004;4:109–117.

[4] Shalini, K., Polasi, S., Lakshmi, M. V. Selection of organic food farming by analytic hierarchy process (AHP). *Tuijin Jishu/J. Propuls Technol.*, 2023;44(3):3919–3939.

[5] Sridevi, P., Sammeta, V. M., Harish Babu, G. A., TJPRC. Optimum allocation of working hours for plant tissue culture. *IJMPERD*, 2020;10(3):13425–13434.

[6] Bhatia, M., Rana, A. A mathematical approach to optimize crop allocation—a linear programming model. *Int. J. Des. Nat. Ecodyn.*, 2020;15:245–252.

[7] Heydari, M., Othman, F., Salarijazi, M., Ahmadianfar, I., Sadeghian, M. Predicting the amount of fertilizers using linear programming for agricultural products from optimum cropping pattern. *J. Geogr. Stud.*, 2018;2:22–9.

[8] Srilatha, C., Supriya, S., Suhasini, K., Supriya, K. Optimal crop plans by using linear programming technique in Jayashankar Bhupal pally district of Telangana: A case study. 2022;11:2142–2145.

[9] Shalini, K., Polasi, S. Models of goal programming and R program ming to earmark acreage. In: Kamalov, F., Sivaraj, R., Leung, H.-H., editors. Advances in mathematical modeling and scientific com puting: international conference on recent developments in math ematics, Dubai, 2022—Volume 2. Cham: Springer International Publishing; 2024:545–545.

[10] Parekodi, J., Ga, V., Gupta, R. Goal programming model to budget ary allocation in garbage disposal plant: Performance and safety management. Singapore: Springer; 2019:77–90.

[11] Ghanashyam, K. J., Ga, V. A new approach to optimize the produc tion of crop using compost from solid waste. *Int. J. Adv. Sci. Engg. Technol.*, 2018;6(1):81–85.

[12] Guo, G., Obłój, J. Computational methods for martingale optimal transport problems. *Ann. Appl. Probab.*, 2019;29(6):3311–3347.

[13] Singh, S., Tuli, R., Sarode, D. A review on fuzzy and stochastic extensions of the multi index transportation problem. *Yugoslav. J. Oper. Res.*, 2017;27(1):3–29.

[14] Latpate, R., Kurade, S. S. Multi-objective multi-index transportation model for crude oil using fuzzy NSGA-II. *IEEE Trans. Intell. Transp. Syst.*, 2022;23(2):1347–1356.

[15] Jagtap, K., Kawale, S. Multi-dimensional-multi-objective-transportation-problem-by-goal-programming. *Int. J. Sci. Engg. Res.*, 2017;8(6):568–573.

[16] Kaur, L., Rakshit, M., Singh, S. A new approach to solve multi-objective transportation problem. *Appl. Appl. Math. Int. J. (AAM)*, 2018;13(1):10.

[17] Carlier, G. On a class of multidimensional optimal transportation problems. *J. Convex Anal.*, 2003;10(2):517–530.

[18] Arora, S. R., Ahuja, A. A paradox in a fixed charge transportation problem. *Ind. J. Pure Appl. Math.*, 2000;31: 809–822.

[19] Tkacenko, A., Alhazov, A. The multiobjective bottleneck transportation problem. *Comp. Sci. J. Moldova Kishinev*, 2001;9:321–335.

[20] Tkacenko, A. The generalized algorithm for solving the fractional multi-objective transportation problem. *ROMAI J.*, 2006;2:197–200.

[21] Pandian, P., Natarajan, G. Fourier method for solving transportation problems with mixed constraints. *Int. J. Contemp. Math Sci.*, 2010;5:1385–1395.

[22] Hassan, I., Raza, M. A., Khalil, M., Ilahi, R. Determination of optimum cropping pattern in the Faisalabad division (Pakistan). *Int. J. Agri. Biol.*, 2004;6(5):901–903.

[23] Hassan, I., Raza, M. A., Khan, I. A., Ilahi, R. Use of linear programming model to determine the optimum cropping pattern, production and income level: A case study from Dera Ghazi Khan Division. *J. Agri. Soc. Sci.*, 2005;1(1):32–34.

[24] Singh, D. K., Jaiswal, C. S., Reddy, K. S., Singh, R. M., Bhandarkar, D. M. Optimal cropping pattern in a canal command area. *Agri. Water Manag.*, 2001;50(1):1–8.

57 The properties of complementary centroidal mean

Venkataramana B. S.[1,a], K. M. Nagaraja[2,b], Sampathkumar R.[3,c], Harish A.[4,d] and Lakshmi Janardhan R. C.[5,e]

[1]Department of Mathematics, K. S. Institute of Technology, Raghuvanahalli, Kanakapura Road, Bangalore–560109, Karnataka, India

[2]Department of Mathematics, J. S. S. Academy of Technical Education, Uttarahalli-Kengeri Main Road, Bangalore–560060, Karnataka, India

[3]Department of Mathematics, RNS Institute of Technology, Uttarahalli-Kengeri Main Road, Bangalore–560098, Karnataka, India

[4]Department of Mathematics, Government First Grade College of Arts, Science and Commerce, Sira Tumkur–560098, Karnataka, India

[5]Department of Mathematics, Government First Grade College, Vijayanagar, Bangalore–560040, Karnataka, India

Abstract

In this study, we thoroughly examined the various types of Schur convexities linked with the complementary centroidal mean and rigorously confirmed their validity. Schur convexity is an important mathematical characteristic in the study of inequalities because it ensures that certain mean functions maintain the majorization-based ordering of input sequences. Additionally, a chain of inequity incorporating well-known literary approaches was formed.

2010 Subject classification: Primary 26 D10, secondary 26 D15

Keywords: Properties, Schur convexities, centroidal mean, complementary and invariant

Introduction

Based on proportions and their significance, the Pythagorean School of mathematics in the fourth century A. D. created and researched the mathematical means notion [1]. Subsequently, a number of writers advanced this topic in light of applications to numerous branches of technology and science [25]. The relationship between series and significant means [9], Greek and functional means [6], and various inequalities involving means were established by Lokesh et al. in recent years through the study of homogeneous functions.[7, 27], examined oscillatory mean, oscillatory type mean in Greek means, new means, and its generalisations [3–5, 8, 10] revealed a number of inequality results. The features of means and the good number of Schur convexity results are covered in the following ref [11, 16, 18, 19, 23, 24]. References to the work on mean inequality refinement can be found in [12–15, 17, 20–22, 26]. In literature, it was well known that for all real numbers a and b $A = \dfrac{a+b}{2}$, $G = \sqrt{ab}$, $H = \dfrac{2ab}{a+b}$,

$H_e = \dfrac{a + \sqrt{ab} + b}{3}$ and $C = \dfrac{a^2 + b^2}{a+b}$, respectively called as arithmetic, geometric, harmonic, heron and contra harmonic means.

Definition 1.1: [25] A mean is a mapping $M : R_+^2 \to R_+$, which has the property; $p \wedge q \leq M(p,q) \leq p \vee q$,for all $p,q > 0$. *where* $p \vee q = Max(p,q)$ and $p \wedge q = Min(p,q)$
Note: Regarding the properties of means states that each mean is reflexive, that is, $M(p,p) = p$, for all $p > 0$.

Definition 1.2: [2, 26] The mean of p and q are said to be complimentary with respect to each other if $p + q = 2A$ and invariant if $p \times q = G^2$.

Definition 1.3: [1] For all real numbers a and b the centroidal mean is defined as $C = \dfrac{2(a^2 + ab + b^2)}{3(a+b)}$

[a]venkynvs@gmail.com, [b]nagkmn@gmail.com, [c]r.sampathkumar1967@gmail.com, [d]harishharshi@gmail.com, [e]ljrcmaths@gmail.com

DOI: 10.1201/9781003685876-57

Definition 1.4: [25] For all real numbers a and b the complimentary centroidal mean is denoted by ^cC and is defined as;

$$^cC = 2A - C = 2\left(\frac{a+b}{2}\right) - \left(\frac{2(a^2+ab+b^2)}{3(a+b)}\right)$$

$$= \frac{1}{3}\left(\frac{a^2+b^2}{a+b}\right) + \frac{2}{3}\left(\frac{2ab}{a+b}\right)$$

Definition 1.5: [25] For all real numbers a and b the invariant centroidal mean is denoted by ^iC and is defined as;

$$^iC = \frac{G^2}{C} = \frac{ab}{\dfrac{2(a^2+ab+b^2)}{3(a+b)}} = \frac{3}{2}\left(\frac{ab(a+b)}{a^2+ab+b^2}\right)$$

Lemma 1.1: [1] Let $\Omega \subseteq R^n$ Ω be symmetric with non-empty geometrically convex set and let $\varphi : \Omega \to R_+$ be continuous on Ω and differentiable in Ω^0. If φ is symmetric on Ω and

i. $(x_1 - x_2)\left(\dfrac{\partial \varphi}{\partial x_1} - \dfrac{\partial \varphi}{\partial x_2}\right) \geq 0 (\leq 0)$

ii. $(\ln x_1 - \ln x_2)\left(x_1\dfrac{\partial \varphi}{\partial x_1} - x_2\dfrac{\partial \varphi}{\partial x_2}\right) \geq 0 (\leq 0)$

iii. $(x_1 - x_2)\left(x_1^2\dfrac{\partial \varphi}{\partial x_1} - x_2^2\dfrac{\partial \varphi}{\partial x_2}\right) \geq 0 (\leq 0)$

is Schur convex (concave), Schur-geometrically convex (concave) and Schur-harmonically convex (concave) function, respectively.

Schur convexity results

In this section, the results of Schur convexity of the complimentary centroidal mean are discussed.

Theorem 2.1. For $b \geq a \geq 0$, the complimentary centroidal mean is Schur concave.
Proof: From definition 1.4 the complimentary centroidal mean is given by;

$$^cC = \frac{1}{3}\left(\frac{a^2+b^2}{a+b}\right) + \frac{2}{3}\left(\frac{2ab}{a+b}\right) \qquad (2.1)$$

Differentiating Eqn (2.1) partially with respect to a and b, on subtracting gives;

$$\frac{\partial^cC}{\partial a} - \frac{\partial^cC}{\partial b} = -\frac{2}{3}\left(\frac{a-b}{a+b}\right) > 0$$

Therefore, $(a-b)\dfrac{\partial^cC}{\partial a} - \dfrac{\partial^cC}{\partial b} = -\dfrac{2}{3}\left(\dfrac{(a-b)^2}{a+b}\right) < 0.$

Thus complimentary centroidal mean is $^cC = \dfrac{1}{3}\left(\dfrac{a^2+b^2}{a+b}\right) + \dfrac{2}{3}\left(\dfrac{2ab}{a+b}\right)$ is Schur concave.

Theorem 2.1 graphical representation is given in Figure 57.1.

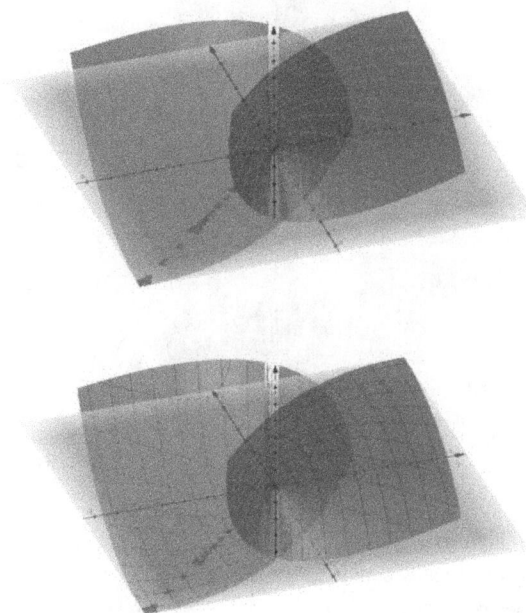

Figure 57.1 The complimentary centroidal mean is Schur concave
Source: Author

Theorem 2.2. For $b \geq a \geq 0$, the complimentary centroidal mean is Schur geometrically convex.
Proof: From definition 1.4 the complimentary centroidal mean is given by;

$$^cC = \frac{1}{3}\left(\frac{a^2+b^2}{a+b}\right) + \frac{2}{3}\left(\frac{2ab}{a+b}\right) \qquad (2.2)$$

Differentiating Eqn (2.2) partially with respect to a and b, on subtracting gives;

$$a\frac{\partial^cC}{\partial a} - b\frac{\partial^cC}{\partial b} = \left(\frac{(a-b)(a^2+b^2)}{3(a+b)^2}\right) < 0$$

Therefore,

$$(\ln a - \ln b)\left(a\frac{\partial^c C}{\partial a} - b\frac{\partial^c C}{\partial b}\right) = (\ln a - \ln b)\left(\frac{(a-b)(a^2+b^2)}{3(a+b)^2}\right)$$
$$> 0$$

Thus complimentary centroidal mean is $^cC = \frac{1}{3}\left(\frac{a^2+b^2}{a+b}\right) + \frac{2}{3}\left(\frac{2ab}{a+b}\right)$ is Schur geometrically convex.

Theorem 2.2 graphical representation is given Figure 57.2.

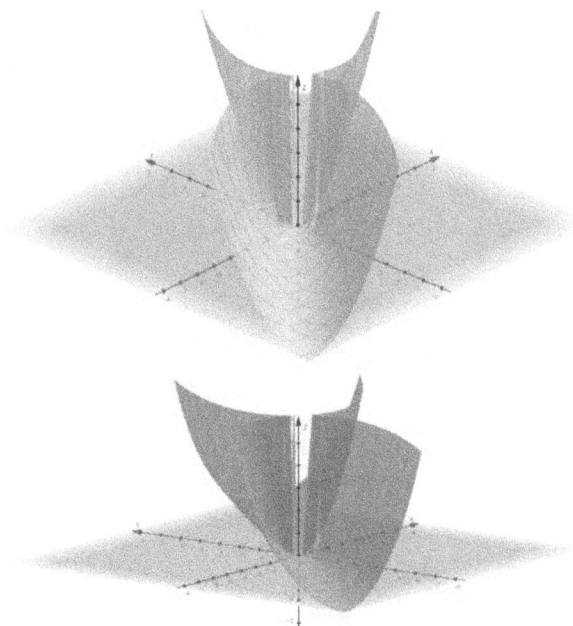

Figure 57.2 The complimentary centroidal mean is Schur geometrically convex
Source: Author

Theorem 2.3. *For $b \geq a \geq 0$, the complimentary centroidal mean is Schur harmonically convex.*
Proof: From definition 1.4 the complimentary centroidal mean is given by

$$^cC = \frac{1}{3}\left(\frac{a^2+b^2}{a+b}\right) + \frac{2}{3}\left(\frac{2ab}{a+b}\right) \quad (2.3)$$

Differentiating Eqn (2.3) partially with respect to a and b, on subtracting gives;

$$a^2\frac{\partial^c C}{\partial a} - b^2\frac{\partial^c C}{\partial b} = \left(\frac{(a^2-b^2)}{3}\right) < 0$$

Therefore,

$$(a-b)\left(a^2\frac{\partial^c C}{\partial a} - b^2\frac{\partial^c C}{\partial b}\right) = \left(\frac{(a-b)(a^2-b^2)}{3}\right)$$
$$> 0$$

Thus complimentary centroidal mean is

$^cC = \frac{1}{3}\left(\frac{a^2+b^2}{a+b}\right) + \frac{2}{3}\left(\frac{2ab}{a+b}\right)$ is Schur harmonically convex.

Theorem 2.3 graphical representation is given in Figure 57.3.

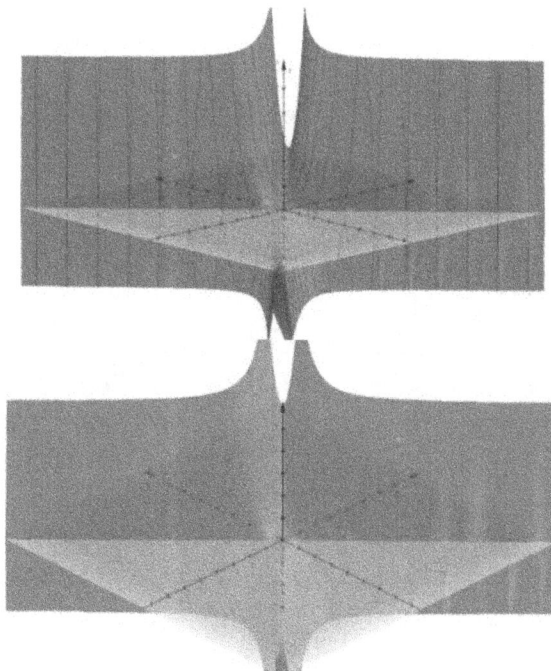

Figure 57.3 The complimentary centroidal mean is Schur harmonically convex
Source: Author

Theorem 2.4. *For $b \geq a \geq 0$, the complimentary centroidal mean is M-power convex for $M \geq 3$.*
Proof: From definition 1.4 the complimentary centroidal mean is given by;

$$^cC = \frac{1}{3}\left(\frac{a^2+b^2}{a+b}\right) + \frac{2}{3}\left(\frac{2ab}{a+b}\right) \quad (2.4)$$

Differentiating Eqn (2.4) partially with respect to a and b, on subtracting gives;

$$a^{1-m}\frac{\partial^c C}{\partial a} - b^{1-m}\frac{\partial^c C}{\partial b} =$$
$$\left(\frac{3ab(a^{m+1}-b^{m+1}) + 2a^2b^2(a^{m-1}-b^{m-1}) + b^3a^3(a^{m-3}-b^{m-3})}{3(a+b)^2}\right)$$

Therefore,

$$(a^m - b^m)\left(a^{1-m}\frac{\partial^c C}{\partial a} - b^{1-m}\frac{\partial^c C}{\partial b}\right) =$$
$$(a^m - b^m)\left(\frac{3ab(a^{m+1}-b^{m+1}) + 2a^2b^2(a^{m-1}-b^{m-1}) + b^3a^3(a^{m-3}-b^{m-3})}{3(a+b)^2}\right)$$

For $M \geq 3$.

$$(a^m - b^m)\left(a^{1-m}\frac{\partial^c C}{\partial a} - b^{1-m}\frac{\partial^c C}{\partial b}\right) =$$

$$(a^m - b^m)\left(\frac{3ab(a^{m+1} - b^{m+1}) + 2a^2b^2(a^{m-1} - b^{m-1}) + b^3 a^3(a^{m-3} - b^{m-3})}{3(a+b)^2}\right)$$

which is > 0.

Thus, complimentary centroidal mean $^c C = \frac{1}{3}\left(\frac{a^2 + b^2}{a+b}\right) + \frac{2}{3}\left(\frac{2ab}{a+b}\right)$ is M-power convex for $M \geq 3$.

For $M < 3$, Put $a = e^{-t}$ and $b = e^t$ such that $a < b$.

The right hand side of Theorem 2.4 is

$$^c C = 2\sinh m \; [6\sinh(m+1)t + 4\sinh(m-1)t + 2\sinh(m-3)t]$$

The Taylor's series expansion gives,

$$^c C = m[(6m - 2)t^2 + \frac{1}{3}(6m^3 - 4m^2 + 21m - 13)t^4 + \cdots - -]$$

By considering terms up to 2nd degree, the following conclusion is drawn.

(i) *For* $0 \leq M \leq \frac{1}{3}$; Theorem 2.4 is $+ve$
Thus, complimentary centroidal mean is $^c C = \frac{1}{3}\left(\frac{a^2 + b^2}{a+b}\right) + \frac{2}{3}\left(\frac{2ab}{a+b}\right)$ is M-power convex for $0 \leq M \leq \frac{1}{3}$.

(ii) *For* $\frac{1}{3} < M \leq 3$; Theorem 2.4 is $-ve$
Thus, complimentary centroidal mean is $^c C = \frac{1}{3}\left(\frac{a^2 + b^2}{a+b}\right) + \frac{2}{3}\left(\frac{2ab}{a+b}\right)$ is M-power concave for $\frac{1}{3} < M \leq 3$.

(iii) *For* $M < 0$; Theorem 2.4 is $+ve$
Thus, complimentary centroidal mean is $^c C = \frac{1}{3}\left(\frac{a^2 + b^2}{a+b}\right) + \frac{2}{3}\left(\frac{2ab}{a+b}\right)$ is M-power convex for $M < 0$.

Inequality chain

In this section, an inequality chain involving the complimentary centroidal means is established.
Case 1: Consider

$$^c C - A = \left(\frac{a^2 + 4ab + b^2}{3(a+b)}\right) - \frac{a+b}{2}$$

$$= \frac{-(a-b)^2}{6(a+b)} < 0$$

$^c C < A$

Case 2: Consider

$$C - A = \left(\frac{2(a^2 + ab + b^2)}{3(a+b)}\right) - \frac{a+b}{2} = \frac{(a-b)^2}{6(a+b)} > 0$$

$A < C$

Case 3: Consider

$$^c C - {}^i C = \left(\frac{a^2 + 4ab + b^2}{3(a+b)}\right) - \frac{3ab(a+b)}{2(a^2 + ab + b^2)} = \frac{(a-b)^2(2a+b)(a+2b)}{6(a+b)(a^2 + ab + b^2)} > 0$$

$^c C > {}^i C$

Case 4: *Consider*

$$(^c C)^2 - G^2 = \left(\frac{a^2 + 4ab + b^2}{3(a+b)}\right)^2 - ab = \frac{(a-b)^2(a^2 + ab + b^2)}{9(a+b)^2} > 0$$

$^c C > G$

Case 5: Consider

$$^i C - H = \frac{3ab(a+b)}{2(a^2 + ab + b^2)} - \frac{2ab}{(a+b)} = \frac{-ab(a-b)^2}{2(a+b)(a^2 + ab + b^2)} < 0$$

$^i C < H$

Case 6: Consider

$$^c C - He = \left(\frac{a^2 + 4ab + b^2}{3(a+b)}\right) - \frac{a + \sqrt{ab} + b}{3} = \frac{-\sqrt{ab}(\sqrt{a} - \sqrt{b})^2}{3(a+b)} < 0$$

$^c C < H$

Combining Case 1 to Case 6, provides the following inequality chain involving the complimentary centroidal mean.

$$^iC < H < G <^cC < H_e < A < C$$

Conclusion

This paper investigated the mathematical properties of the complementary centroidal mean, including its monotonicity, symmetry, and Schur-convexity. Several inequalities were established to compare it with other well-known means, highlighting its distinct characteristics. These results contribute to the theoretical development of mean values and provide a basis for future research in inequality theory and mathematical analysis.

References

[1] Bullen, P. S. Handbook of means and their inequalities. Dordrecht: Kluwer Academy Publication, 2003;1–59.

[2] Lakshmi Janardhana, R. C., Nagaraja, K. M., Narasimhan, G. A note on G–Complementary to Heron Mean. *Internat. J. Res. Anal. Rev.*, 2018;5(3):369–373.

[3] Lokesha, V., Nagaraja, K. M., Padmanabhan, S., Naveen Kumar, B. Oscillatory type mean in Greek means. *Int. e-J. Engg. Maths. Theory Appl.*, 2010;9:18–26.

[4] Lokesha, V., Naveen Kumar, B., Nagaraja, K. M., Bayad, A., Saraj, M. New means and its properties. *Proc. Jangjeon Math. Soc.*, 2010;14(3):243–254.

[5] Lokesha, V., Nagaraja, K. M., Naveen Kumar, B., Padmanabhan, S. Extension of homogeneous function. *Tamsui Oxford J. Math. Sci.*, 2010;26(4):443–450.

[6] Lokesha, V., Padmanabhan, S., Nagaraja, K. M., Simsek, Y. Relation between Greek means and various means. *Gen. Math.*, 2009;17(3):3–13.

[7] Lokesha, V., Nagaraja, K. M., Simsek, Y. New inequalities on the homogeneous functions. *J. Indone. Math. Soc.*, 2009;15(1):49–59.

[8] Lokesha, V., Zhang, Z.-H., Nagaraja, K. M. Gnan mean for two variables. *Far East J. Appl. Math.*, 2008;31(2):263–272.

[9] Lokesha, V., Nagaraja, K. M. Relation between series and important means. *Adv. Theoret. Appl. Math.*, 2007;2(1):31–36.

[10] Lokesha, V., Zhang, Z.-H., Nagaraja, K. M. rth Oscillatory mean for several positive arguments. *Ultra Scientist*, 2006;18(3):519–522.

[11] Murali, K., Nagaraja, K. M. Schur convexity of Stolarsky's extended mean values. *J. Math. Inequal.*, 2016;10(3):725–735.

[12] Nagaraja, K. M., Lokesha, V., Padmanabhan, S. A simple proof on strengthening and extension of inequalities. *Adv. Stud. Contemp. Math.*, 2008;17(1):97–103.

[13] Nagaraja, K. M., Araci, S., Lokesha, V., Sampathkumar, R., Vimala, T. Inequalities for the arguments lying on linear and curved path. *Honam Math. J.*, 2020;42(4):747–755.

[14] Nagaraja, K. M., Murali, K., Lakshmi Janardhana, R. C. Improvement of harmonic and contra harmonic mean inequality chain. *Int. J. Pure Appl. Math.*, 2017;114(4):771–776.

[15] Nagaraja, K. M., Reddy, P. S. K., Naveen Kumar, B. Refinement of inequality involving ratio of means for four positive arguments. *Bull. Internat. Math. Virt. Inst.*, 2013;3:135–138.

[16] Nagaraja, K. M., Reddy, P. S. K., Sudhir Kumar, S. Generalization of alpha-centroidal mean and its dual. *Iranian J. Math. Sci. Inform.*, 2013;8(2):39–47.

[17] Nagaraja, K. M., Reddy, P. S. K. Double inequalities on means via quadrature formula. *Notes Numb. Theory Dis. Math.*, 2012;18(1):22–28.

[18] Nagaraja, K. M., Reddy, P. S. K. Alpha-centroidal mean and its dual. *Proc. Jangjeon Math. Soc.*, 2012;15(2):163–170.

[19] Nagaraja, K. M., Reddy, P. S. K. A note on power mean and generalized contra-harmonic mean. *Scientia Magna*, 2012;8(3):60–62.

[20] Narasimhan, G., Nagaraja, K. M., Sampath Kumar, R., Murali, K. Establishment of a new inequality using extended Heinz type mean. *IOP Conf. Series J. Phy. Conf. Series*, 2018;1139:012034.

[21] Naveen Kumar, B., Sandeep Kumar, Lokesha, V., Nagaraja, K. M. Ratio of difference of means and its convexity. *Internat. e J. Math. Engg.*, 2011;2(2):932–936.

[22] Sampath Kumar, R., Nagaraja, K. M., Narasimhan, G. New family of Heinz type means. *Internat. J. Pure Appl. Math.*, 2018;118(18):3427–3433.

[23] Sandeep Kumar, Lokesha, V., Misra, U. K., Nagaraja, K. M. Power type α-centroidal mean and its dual. *Gen. Math. Notes*, 2014;25(1):33–42.

[24] Toader, S., Toader, G. Greek means and arithmetic and geometric means. *RGMIA Monograph*, 2005;1–95. http://rgmia.vu.edu.au/html.

[25] Toader, S., Toader, G. Complementary of a Greek mean with respect to another. *Automat. Comput. Appl. Math.*, 2004;13(1):203–208.

[26] Vimala, T., Nagaraja, K. M., Sampath Kumar, R. A harmonic and Heron mean inequalities for arguments in different intervals. *Internat. J. Math. Arch.*, 2019;10(7):47–52.

[27] Yang, Z.-H. On the homogeneous functions with two parameters and its monotonicity. *J. Inequal. Pure Appl. Math.*, 2005;6(4), article 101.

58 Soft expert graph and its applications

Supriya M. D.[1,a] and P. Usha[2,b]

[1]Assistant ProfessorDepartment of Mathematics, KSIT (Affiliated to Visvesvaraya Technological University, Belagavi), Bengaluru–560109, Karnataka, India

[2]Department of Mathematics, Siddaganga Institute of Technology (Affiliated to Visvesvaraya Technological University, Belagavi), Tumakuru–572103, Karnataka, India

Abstract

A new concept, soft expert graph is introduced. Let the sets U, E, X and O, respectively denotes universal, parameter, expert and opinion sets. Let $Z = E \times X \times O$ and $A \subseteq Z$. A bi-partite graph $G_{SE} = (V, E)$ is called a soft expert graph, for any F:A→P(U) if (F, A) is a soft expert set over $\cup$ and the vertex set $V(G_{SE}) = U \cup \{E \times X\}$ is partitioned into two subsets U and $E \times X$ such that a vertex u_i is adjacent to a vertex (l_j, x_i), if the expert agrees with parameter l_j. The soft expert graph is useful and efficient for studying the opinions of all experts. This novel concept not only advances the theoretical underpinnings of decision-making under uncertainty but also provides a practical and efficient tool for decision-makers across various domains.

Keywords: Soft set, soft expert set, soft expert graph

Introduction

In almost every field, uncertainty is prevalent. To address these uncertainties, Molodtsov [8] introduced of soft set theory, a mathematical framework designed specifically to handle uncertainties. Building on Molodtsov's seminal work, Chen et al. [4] and Maji et al. [6, 7] expanded the theory with various soft set operations and applications, while Alkhazaleh et al. [2] extended it to soft multisets, a generalisation of soft sets. These developments have spurred numerous studies and models, particularly in decision-making and medical assessment. Despite these advancements, most existing models are limited to single-expert scenarios. To address this limitation, Shawkat et al. [1] proposed soft expert sets, a framework that aggregates multiple expert opinions into a unified model. Inspired by this approach, the paper introduces the soft expert graph, a novel concept aimed at enhancing the utility of expert knowledge in decision-making processes.

The primary objective of this paper is to present the soft expert graph as a new conceptual tool, demonstrating its advantages over current methods. We showcase its applicability through case studies that highlight its practical benefits in real-world decision-making scenarios.

Preliminaries

We recall some fundamental definitions of the soft set theory.

Definition 2.1. [8] Let E be a set of parameters and U be a universe. Let $P(U)$ be a power set of U and $A \subseteq E$. A pair (F, A) is referred to as a soft set over U, where F is a mapping $F : A \to P(U)$.

Definition 2.2. [8] Consider a set of parameters, $E = \{l_1, l_2, ..., l_n\}$. Then the *NOT* set of E is denoted and defined by $\sim E = \{\sim l_1, \sim l_2, ..., \sim l_n\}$ where $\sim l_n = not\ l_i$, for all i.

Definition 2.3. [8] A soft set (F, A)'s complement is denoted as $(F, A)^c$ and it is defined as $(F, A)^c = (F^c, \sim A)$, where $F^c: \sim A \to P(U)$ is a mapping determined by $F^c(a) = U - F(\sim a)$, for every $a \in \sim A$.

Definition 2.4. [1] Let the sets U, E, X and O, respectively denotes universal, parameter, expert and opinion sets. Let $Z = E \times X \times O$ and $A \subseteq Z$. A pair (F, A) is referred to as a soft expert set over U, if F is a mapping $F : A \to P(U)$, where $P(U)$ stands for the power set of U.

Definition 2.5. [1] A agree-soft expert set $(F, A)_1$ over U is a soft expert subset of (F, A) which is defined as, $(F, A)_1 = \{F_1(a) : a \in E \times X \times \{1\}\}$.

[a]mdsupriya.5@gmail.com, [b]pushamurthy@gmail.com

DOI: 10.1201/9781003685876-58

Definition 2.6. [1] A disagree-soft expert set $(\boldsymbol{F},\mathsf{A})_0$ over U is a soft expert subset of $(\boldsymbol{F},\mathsf{A})$ defined as, $(\boldsymbol{F},\mathsf{A})_0 = \{\boldsymbol{F}(\mathbf{a}) : \mathbf{a} \in \mathsf{E} \times \mathsf{X} \times \{0\}\}$.

Definition 2.7. [5] A graph G is a pair (V,E), where V and E are finite set of vertices and edges respectively, where an edge is an unordered pair of distinct vertices.

Definition 2.8. [5] A complement or inverse of a graph G is represented by the notation G^c on the same vertices, where two distinct vertices of G^c are adjacent if and only if they are not adjacent in G.

Definition 2.9. [5] A graph $G = (V,E)$ is said to be a bipartite graph, if the vertex set $V(G)$ is partitioned into two sets, V_1 and V_2 and each edge connects a vertex in V_1 to a vertex in V_2,

Definition 2.10. [3] A soft graph is a quadruple $(G, J, \mathbb{K}, A)$, where

(i) $G = (V, E)$ is a simple graph
(ii) (J, A) is a soft set over V
(iii) $(\mathbb{K}, A)$ is a soft set over E
(iv) $(J(a), \mathbb{K}(a))$ is a subgraph of G for all $a \in A$.

Soft expert graph

This section presents the idea of soft expert graphs and lists their fundamental characteristics. Let the sets $\mathsf{U}, \mathsf{E}, \mathsf{X}$ and O, respectively denotes universal, parameter, expert and opinion sets. Let $\mathsf{Z} = \mathsf{E} \times \mathsf{X} \times \mathsf{O}$ and $\mathsf{A} \subseteq \mathsf{Z}$.

Definition 3.1 A simple bi-partite graph $G_{SE} = (V, E)$ is called a **soft expert graph**, if

(i) $(\boldsymbol{F}, \mathsf{A})$ is a soft expert set over U where $\boldsymbol{F}{:}\mathsf{A} \rightarrow \mathsf{P}(\mathsf{U})$
(ii) the vertex set $V(G_{SE}) = \mathsf{U} \cup \{\mathsf{E} \times \mathsf{X}\}$ is partitioned into two subsets U and $\{\mathsf{E} \times \mathsf{X}\}$ such that a vertex u_i is adjacent to a vertex (l_i, x_i) if the expert x_i agrees with parameter l_i.

We list the following properties/observations of soft expert graph G_{SE}:

1. G_{SE} is a bi-partite graph.
2.   Degree of (l_i, x_i) denotes the number of accepted opinions of the expert x_i with respect to the parameter l_i.

3.   Degree of u_i denotes the total number of experts accepted opinions for u_i which satisfies parameters.
4.   Since G_{SE} represents agree-soft expert set, G_{SE} is also called agree-soft expert graph.
5.   The complement graph G_{SE}^c is called disagree-soft expert graph.

NOTE: In this paper for convenience, we assume two-valued opinions in the O, that is, $\mathsf{O} = \{0 = disagree, 1 = agree\}$.

Example 1: Let us assume that a business that produces new products and seeks feedback from certain experts on those products. Let $\mathsf{U} = \{u_1, u_2, u_3, u_4\}$ represents the set of products, $\{l_1, l_2, l_3\}$ = {easy to use, quality, affordable} represents the set of evaluation parameters and let $\mathsf{X} = \{a, b, c\}$ be a set of experts.

Let the three experts decide to agree on the company's product is,

$$(\boldsymbol{F}, \mathsf{Z})_1 = \{((l_1, a, 1), \{u_1, u_2, u_4\}), ((l_1, b, 1), \{u_1, u_4\}),$$

$$((l_1, c, 1), \{u_3, u_4\}), ((l_2, a, 1), \{u_4\}), ((l_2, b, 1), \{u_1, u_3\}),$$

$$((l_2, c, 1), \{u_1, u_2, u_4\}), ((l_3, a, 1), \{u_3, u_4\}),$$

$$((l_3, b, 1), \{u_1, u_2\}), ((l_3, c, 1), \{u_4\})\}.$$

The soft expert graph $_{SE}$ for the above is:

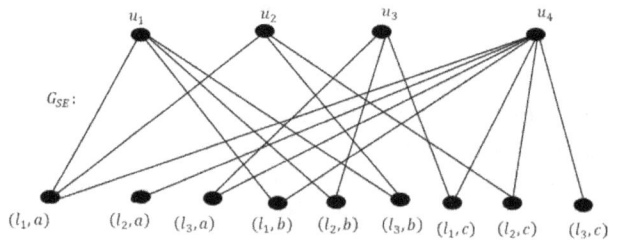

The complement of the above graph G_{SE} is:

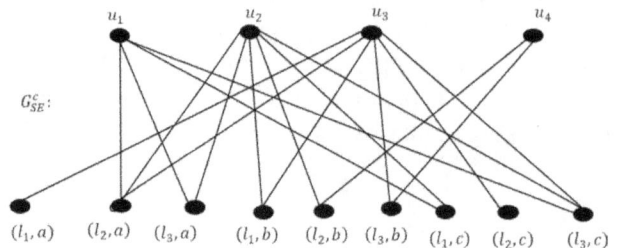

Using the above graph one can easily construct disagree soft expert set as

$$\{((l_1,a,0),\{u_3\}),((l_1,b,0),\{u_2,u_3\}),((l_1,c,0),\{u_1,u_2\}),$$

$$((l_2,a,0),\{u_1,u_2,u_3\}),((l_2,b,0),\{u_2,u_4\}),((l_2,c,0),\{u_3\}),$$

$$((l_3,a,0),\{u_1,u_2\}),(l_3,b,0),\{u_3,u_4\}),(l_3,c,0),\{u_1,u_2,u_3\})\}.$$

The soft expert graph G_{SE} is constructed to visualise satisfaction between products and parameters. Each vertex in V_1 represents a product and a vertex in V_2 represent a parameter, and edges connect products to parameters based on expert agreements.

The complement graph G_{SE}^c shows disagreements between products and parameters. It helps in understanding which products do not meet certain criteria according to experts.

Matrix representation of soft expert graph
This section describes representation of soft expert graph using a matrix.

Definition:
The soft expert adjacency matrix of G_{SE} is a rectangular $(0,1)$ matrix and is defined as

$$M_{SE} = (m_{ij}) = \begin{cases} 1 & if\ u_i \in F(a), for\ a \in Z \\ 0 & otherwise. \end{cases}$$

The matrix representation of the graph G_{SE} in example 1 is

$$M_{SE}(m_{ij}) = \begin{array}{c} \\ (l_1,a) \\ (l_2,a) \\ (l_3,a) \\ (l_1,b) \\ (l_2,b) \\ (l_3,b) \\ (l_1,c) \\ (l_2,c) \\ (l_3,c) \end{array} \begin{array}{cccc} u_1 & u_2 & u_3 & u_4 \\ \begin{pmatrix} 1 & 1 & 0 & 1 \\ 0 & 0 & 0 & 1 \\ 0 & 0 & 1 & 1 \\ 1 & 0 & 0 & 1 \\ 1 & 0 & 1 & 0 \\ 1 & 1 & 0 & 0 \\ 0 & 0 & 1 & 1 \\ 1 & 1 & 0 & 1 \\ 0 & 0 & 0 & 1 \end{pmatrix} \end{array}$$

This adjacency matrix represents the soft expert graph, where each element indicates whether a product meets a parameter according to an expert's opinion.

This adjacency matrix represents the soft expert graph, with each element denoting whether a product satisfies a given parameter based on an expert's evaluation.

Algorithm for best choice/decision
To make the best choice, we can follow the algorithm below:

STEP 1: Find the agree-soft expert set.
STEP 2: Obtain the soft expert matrix.
STEP 3: Find the column sum $\sum C_i$ which represents the degree of each vertex.
STEP 4: Find "i" for which $u_i = max\{\sum C_i\}$ which gives a vertex with maximum degree.
STEP 5: Conclude that i^{th} element product in the sample satisfies maximum number of parameters. Hence i^{th} element is the best choice.

In the above matrix,
$$\sum C_1 = 5, \sum C_2 = 3, \sum C_3 = 3, \sum C_4 = 7, u_4 = \sum C_4 = 7.$$

From this $u_i = max(\sum_{i=1}^{4} C_i) = 7$ for $i = 4$. Therefore, the product, that is u_4, is the best product produced by 4th firm among the four similar items produced by different company.

To find the best product, the algorithm calculates the sum of agreements for each product (vertex) and selects the product with the highest sum.

This example demonstrates how soft expert graphs can be used to make informed decisions based on expert opinions and product criteria.

Soft expert graph applications

We present soft expert graph applications in decision-making problems in this part.

Application
Let's say a business wishes to hire an employee for a position. There are eight candidates who appear for the interview which forms the universe $U = \{u_1,u_2,u_3,u_4,...,u_8\}$. A set of criteria are taken into account for hiring i.e., $E = \{l_1,l_2,l_3,l_4,l_5\} = $ {Qualification, experience, knowledge, communication skills, ability to handle challenges}. Let $X = \{a,b,c\}$ be set of experts (panel members).
Consider the agree-soft expert set,

$$(F,Z) = \{\{((l_1,a,1),\{u_1,u_2,u_4,u_7,u_8\}),((l_1,b,1),\{u_1,u_4,u_5,u_8\}),$$

$$((l_1,c,1),\ \{u_1,u_3,u_4,u_6,u_8\}),((l_2,a,1),\{u_3,u_5,u_8\}),$$

$$((l_2,b,1),\{u_1,u_3,u_4,u_5,u_6,u_8\}),$$

$$((l_2,c,1),\{u_1,u_2,u_4,u_7,u_8\}),((l_3,a,1),\{u_3,u_4,u_5,u_7\}),$$

$$((l_3,b,1),\{u_1,u_2,u_5,u_8\}),\ ((l_3,c,1),\{u_1,u_7,u_8\}),$$

$$\left(\left(l_4,a,1\right),\left\{u_1,u_7,u_8\right\}\right),\left(\left(l_4,b,1\right),\left\{u_1,u_4,u_5,u_6\right\}\right),$$
$$\left(\left(l_4,c,1\right),\left\{u_1,u_6,u_7,u_8\right\}\right),\left(\left(l_5,a,1\right),\left\{u_1,u_2,u_3,u_5,u_8\right\}\right),$$
$$\left(\left(l_5,b,1\right),\left\{u_1,u_4,u_5,u_8\right\}\right),\left(\left(l_5,c,1\right),\left\{u_1,u_3,u_5,u_7,u_8\right\}\right)\}.$$

The adjacency matrix of the soft expert graph corresponding to the above soft expert set is,

$M_{SE}(m_{ij}) =$

	u_1	u_2	u_3	u_4	u_5	u_6	u_7	u_8
(l_1,a)	1	1	0	1	0	0	1	1
(l_2,a)	0	0	1	0	1	0	0	1
(l_3,a)	0	0	1	1	1	0	1	0
(l_4,a)	1	0	0	0	0	0	1	1
(l_5,a)	1	1	1	0	1	0	0	1
(l_1,b)	1	0	0	1	1	0	0	1
(l_2,b)	1	0	1	1	1	1	0	1
(l_3,b)	0	0	1	1	1	0	1	0
(l_4,b)	1	0	0	1	1	0	0	1
(l_5,b)	1	0	0	1	1	0	0	1
(l_1,c)	1	0	1	1	0	1	1	1
(l_2,c)	1	1	0	1	0	0	1	1
(l_3,c)	1	0	0	0	0	0	1	1
(l_4,c)	1	0	0	0	0	1	1	1
(l_5,c)	1	0	1	0	1	0	1	1
$\sum C_i$	12	3	7	9	9	3	9	13

To find the best choice, we now employ algorithm 3.2.

$$\sum C_1 = 12, \sum C_2 = 3, \sum C_3 = 7, \sum C_4 = 9, \sum C_5 = 9,$$
$$\sum C_6 = 3, \sum C_7 = 9, \sum C_8 = 13.$$

Max $\sum C_i = \sum C_8 = 13 = u_8$.

So, the committee has to select 8th candidate (u_8) as the best candidate for the position.

Committee can decide the order of preferences based on the sum of C_i values ($\sum C_i$).

Note that if two or more values are equal then all of them lie in the same preference and it is left to the decision of the company to choose the best among those candidates (Table 58.1).

Table 58.1 Prioritized List of Candidates for Hiring

Preference	Candidate
1	u_8
2	u_1
3	u_4, u_5, u_7
4	u_3
5	u_2, u_6 (Reject)

Source: Author

Application

As per the pathology department at AIIMS, Delhi, the following details are observed:

1. Symptoms of sickness due to air pollution include dry cough and sneezing.
2. Symptoms of the common cold include cough, sneezing, and a runny nose.
3. Symptoms of the flu include body aches, fever, a runny nose, cough, and sneezing.
4. Symptoms of COVID include cough, sneezing, body soreness, high fever, and severe breathing problems (seek medical advice immediately).

Assume that there are eight patients who have undergone medical investigation for COVID infection which forms the universe $U = \{p_1, p_2, p_3, p_4, p_5, p_6, p_7, p_8\}$. Medical experts consider a set of parameters $E = \{l_1, l_2, l_3, l_4, l_5, l_6\}$ = {cough, sneeze, runny nose, body ache, fever, difficulty in breathing}.

Let $X = \{a, b\}$ be set of experts (lab reports).

Suppose the agree-soft expert set is,

$$(F, Z) = \{\left(\left(l_1,a,1\right),\left\{p_1,p_2,p_4,p_7,p_8\right\}\right),\left(\left(l_1,b,1\right),\right.$$
$$\left\{p_1,p_3,p_4,p_6,p_7,p_8\right\}),$$
$$\left(\left(l_2,a,1\right),\left\{p_1,p_3,p_4,p_5,p_6,p_8\right\}\right),$$

$$\left(\left(l_2,b,1\right),\left\{p_1,p_2,p_4,p_5,p_8\right\}\right),\left(\left(l_3,a,1\right),\left\{p_3,p_5,p_8\right\}\right),$$

$$\left(\left(l_3,b,1\right),\ \left\{p_3,p_5,p_8\right\}\right),\left(\left(l_4,a,1\right),\left\{p_1,p_4,p_5,p_6\right\}\right),$$

$$\left(\left(l_4,b,1\right),\left\{p_1,p_5,p_6,p_8\right\}\right),\left(\left(l_5,a,1\right),\left\{p_1,p_2,p_3,p_5,p_8\right\}\right),$$

$$\left(\left(l_5,b,1\right),\left\{p_1,p_3,p_5,p_8\right\}\right),\left(\left(l_6,a,1\right),\left\{p_1,p_4,p_5,p_8\right\}\right),$$

$$\left(\left(l_6,b,1\right),\left\{p_1,p_7,p_8\right\}\right)\}.$$

The matrix representation of the soft expert graph corresponding to the above data is,

$$M_{SE}(m_{ij}) = \begin{array}{c} \\ (l_1,a) \\ (l_2,a) \\ (l_3,a) \\ (l_4,a) \\ (l_5,a) \\ (l_6,a) \\ (l_1,b) \\ (l_2,b) \\ (l_3,b) \\ (l_4,b) \\ (l_5,b) \\ (l_6,b) \\ \sum C_i \end{array} \begin{array}{cccccccc} p_1 & p_2 & p_3 & p_4 & p_5 & p_6 & p_7 & p_8 \\ 1 & 1 & 0 & 1 & 0 & 0 & 1 & 1 \\ 1 & 0 & 1 & 1 & 1 & 1 & 0 & 1 \\ 0 & 0 & 1 & 0 & 1 & 0 & 0 & 1 \\ 1 & 0 & 0 & 1 & 1 & 1 & 0 & 0 \\ 1 & 1 & 1 & 0 & 1 & 0 & 0 & 1 \\ 1 & 0 & 0 & 1 & 1 & 0 & 0 & 1 \\ 1 & 0 & 1 & 1 & 0 & 1 & 1 & 1 \\ 1 & 1 & 0 & 1 & 1 & 0 & 0 & 1 \\ 0 & 0 & 1 & 0 & 1 & 0 & 0 & 1 \\ 1 & 0 & 0 & 0 & 1 & 1 & 0 & 1 \\ 1 & 0 & 1 & 0 & 1 & 0 & 0 & 1 \\ 1 & 0 & 0 & 0 & 0 & 0 & 1 & 1 \\ 10 & 3 & 6 & 6 & 9 & 4 & 3 & 11 \end{array}$$

Using Algorithm 3.2, one can easily identify with the help of soft expert graph that out of 12 observations made on these patients, majority of the symptoms are found in p_8, p_1, p_5. Therefore, it may be concluded that u_1, u_5, u_8 are affected by Corona virus and need medical treatment immediately and should be kept under isolation (Table 58.2).

Table 58.2 Prioritized Patient List for Immediate Medical Attention

Possible detection	Patients
Air pollution	p_2, p_6, p_7
Common cold	p_4
Flu	p_3
COVID 19	p_1, p_5, p_8

Source: Author

In various fields, expert opinions play a crucial role in decision-making processes. For example, in project selection, committees evaluate proposals based on criteria such as feasibility, impact, cost, and innovation. Similarly, in product development, teams consider factors like functionality, aesthetics, cost, and market demand to determine the best design. In medical diagnosis, doctors assess treatment options by considering symptoms, medical history, test results, and input from specialists. Likewise, in policy decision-making, government policymakers evaluate options based on economic impact, social implications, feasibility, and expert opinions from relevant fields.

Conclusion

Graph theory is a powerful mathematical tool used across various domains for solving challenging problems. The concept of soft expert graphs, proposed in this paper, offers a versatile approach to decision-making, especially in scenarios involving multiple criteria and expert opinions. We believe that soft expert graphs have the potential to transform how expertise is utilised in decision-making processes, making them more efficient and effective than existing methods. Additionally, an algorithm/technique has been provided, along with case studies to demonstrate its practical application.

References

[1] Alkhazaleh, S., Salleh, A. R. Soft expert sets. *Adv. Dec. Sci.*, 2011, Article ID: 757868, 1–12. https://doi.org/10.1155/2011/757868.

[2] Alkhazaleh, S., Salleh, A., Hassan, N. Soft multisets theory. *Appl. Math. Sci.*, 2011;5:3561–3573.

[3] Akram, M., Nawaz, S. Operations on soft graphs. *Fuzzy Inform. Engg.*, 2015;7:423–449.

[4] Chen, D., Tsang, E. C. C., Yeung, D. S., Wang, X. The parameterization reduction of soft sets and its applications. *Comp. Math. Appl.*, 2004;49(5–6):757–763.

[5] Harary, F. Graph theory. Narosa publishing House, Reprint, 2013: 298.

[6] Maji, P. K., Biswas, R., Roy, A. R. Soft set theory. *Comp. Math. Appl.*, 2003;45(4–5):555–562.

[7] Maji, P. K., Roy, A. R., Biswas, R. An application of soft sets in a decision-making problem. *Comp. Math. Appl.*, 2002;44:1077–1083.

[8] Molodtsov, D. Soft set theory-first results. *Comp. Math. Appl.*, 1999;37(4–5):19–31.

59 Convection dynamics of viscoelastic fluids under temperature-dependent viscosity and heat sources

Jayalatha G.[1,a], Sakshath T. N.[1,b], P. G. Siddheshwar[2,c] and Sekhar G. N.[3,d]

[1]Department of Mathematics, RV College of Engineering, Mysore road, Bengaluru-560059, Karnataka, India

[2]Department of Mathematics, Christ University, Bengaluru-560029, Karnataka, India

[3]Department of Mathematics, BMS College of Engineering, Bengaluru-5600019, Karnataka, India

Abstract

The impact of thermal sources on the instability behaviour of viscoelastic fluids exhibiting variable viscosity has been examined. A linear stability analysis is carried out using the normal mode technique. Numerical computations are performed under stress-free, isothermal boundary conditions, revealing that convection arises through an oscillatory mode in this configuration. Threshold values for stationary as well as oscillatory instabilities are evaluated. The findings demonstrate that both strain retardation and the heat source individually contribute to stabilising the system under free isothermal boundary conditions. In the limiting cases, the results for Jeffreys fluids reduce to those for Maxwell, Newtonian, and Rivlin-Ericksen fluids. This study is significant due to the potential to enhance or suppress convection through the application of a heat source or sink. The study of variable viscosity viscoelastic fluids in climate modelling provides more accurate predictions of oceanic and atmospheric circulations. Incorporating heat sources and sinks into models helps explain global climate shifts, extreme weather events and long-term ocean current changes.

Keywords: convection, viscoelastic fluid, linear stability, variable viscosity, heat source

Introduction

Rayleigh-Bénard convection (RBC) is a classical fluid dynamics phenomenon where a fluid layer heated from below and cooled from above develops characteristic convective patterns due to thermal buoyancy forces. RBC in viscoelastic fluids with variable viscosity reveals complex dynamics not seen in simple Newtonian fluids. These complexities offer deeper insights into natural and industrial processes where heat transfer, fluid flow and material properties interact under varying thermal conditions. Further exploration of this field can lead to advancements in predictive modelling and optimisation in engineering applications. Viscoelastic fluids, characterised by memory effects and time-dependent strain, add another layer of complexity. Elasticity allows fluids to store and release energy, affecting the stability and structure of convection cells. Stress relaxation and creep behaviour modify the convection on set and pattern formation. In Geophysics, Magma dynamics and mantle convection temperature-dependent viscosity and viscoelasticity plays a crucial role. The study of variable viscosity viscoelastic fluids in climate modelling plays a crucial role in understanding large-scale oceanic and atmospheric circulations. Since these circulations drive global weather patterns and climate changes, accounting for the effects of viscoelastic behaviour and viscosity which is temperature dependent is essential for accurate predictions.

The investigation of RBC in Newtonian fluids, which are governed by Newton's law of viscosity, has been well-established for many years. Extensive research has been conducted on the instability in RBC for constant viscosity Newtonian fluids, driven by its wide-ranging applications in engineering and heat transfer [1, 2]. In many industrial processes, fluid involved is viscoelastic in nature. Viscoelastic fluids are characterised by their ability to exhibit both viscous and elastic behaviours, leading to unique flow properties and responses under different conditions [3].

Unlike Newtonian fluids, viscoelastic fluids exhibit stress relaxation and creep, influencing the on-set and dynamics of convection. The presence of internal heat generation in a horizontal layer leads to a non-linear temperature distribution. The initiation of thermal instability within a horizontal viscoelastic fluid layer characterised by a non-uniform

[a]jayalathag@rvce.edu.in, [b]tnsakshath@rvce.edu.in, [c]pg.siddheshwar@christuniversity.in, [d]gnsbms@gmail.com

DOI: 10.1201/9781003685876-59

baseline temperature distribution has been studied by Hamabata, Othman & Sweilam, Siddheshwar, et al., Jayalatha & Suma [4, 5, 7, 8]. It is well-established that viscosity is influenced by temperature. There are few works on thermal convection with viscosity dependent on temperature in viscoelastic fluid with external mechanisms controlling convection [9–13] and references therein. The literature highlights a gap in considering the combined effects of viscosity variation with temperature and with external heat source to control convection. The influence of temperature-dependent volumetric heat source (sink) on convection under 1 g (terrestrial gravity) or μg (micro-gravity) in variable viscosity viscoelastic fluids is studied considering FIFI boundary conditions.

The problem under consideration results in a non-uniform basic temperature gradient, leading to governing equations with variable coefficients. As demonstrated in this study, the temperature distribution in the basic state plays a significant role in determining the onset of convective instabilities.

Mathematical formulation

We examine an unbounded horizontal layer of Jeffreys fluid characterised by a uniform thickness d (refer to Figure 59.1). The lower boundary at $z = 0$ is maintained at a constant temperature $T_0 + \Delta T$, while the upper boundary at $z = d$ is kept at temperature T_0. The viscosity μ of the Jeffreys fluid is assumed to vary with temperature T.

The governing equations describing the behaviour of the Jeffreys fluid with temperature-dependent viscosity are given below:

$$q_{i,i} = 0 \qquad (1)$$

$$\rho_0 \left[\frac{\partial q_i}{\partial t} + q_j q_{i,j} \right] = -p_{,i} + \rho g_i + \tau'_{ij,j} \qquad (2)$$

$$\left(1 + \lambda_1 \frac{\partial}{\partial t}\right) \tau'_{ij,j}$$
$$= \left(1 + \lambda_2 \frac{\partial}{\partial t}\right) [\mu(T)] \left[\frac{\partial \vec{q}_i}{\partial x_j} + \frac{\partial \vec{q}_j}{\partial x_i}\right] \qquad (3)$$

$$\frac{\partial T}{\partial t} + q_j T_{,j} = k[T_{,jj}] + Q_1(T - T_0) \qquad (4)$$

The density equation of state is

$$\rho = \rho_0 [1 - (T - T_0)\alpha] \qquad (5)$$

The thermo rheological equation describing shear viscosity as a function of temperature is represented by the Nield model [6] as follows:

$$\mu(T) = \mu_0 \left[\frac{1}{(T-T_0)\delta+1}\right] \qquad (6)$$

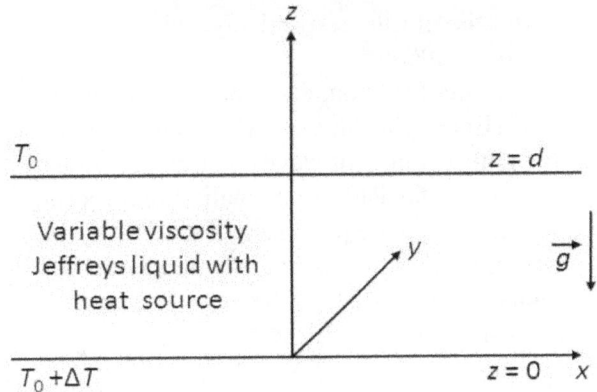

Figure 59.1 Physical configuration under investigation
Source: Author

To examine the stability characteristics, small perturbations are superimposed on the basic state. Utilising classical linear stability analysis, the corresponding linearized, dimensionless equations are derived as follows:

$$\left(1 + \Lambda\Lambda_1 \frac{\partial}{\partial t}\right) \left[h(z)\nabla^4 w + 2\frac{\partial h(z)}{\partial z}\nabla^2\left(\frac{\partial w}{\partial z}\right) + \frac{\partial^2 h(z)}{\partial z^2}\left(\frac{\partial^2 w}{\partial z^2} - \nabla_1^2 w\right)\right] - \frac{1}{Pr}\left(1 + \Lambda_1 \frac{\partial}{\partial t}\right)\frac{\partial}{\partial t}(\nabla^2 w) + \left(1 + \Lambda_1 \frac{\partial}{\partial t}\right) R\nabla_1^2\theta = 0 \qquad (7)$$

$$\frac{\partial\theta}{\partial t} - \nabla^2\theta - R_I\theta + wf(z) = 0 \qquad (8)$$

where

$$h(z) = \left[1 + \frac{V\sin(\sqrt{R_I} - \sqrt{R_I}z)}{\sin\sqrt{R_I}}\right]^{-1},$$

$$f(z) = \frac{-\sqrt{R_I}\cos(\sqrt{R_I} - \sqrt{R_I}z)}{\sin\sqrt{R_I}},$$

$$\theta = \frac{T-T_0}{\Delta T}, \nabla^2 = \frac{\partial^2}{\partial x^2} + \frac{\partial^2}{\partial y^2} + \frac{\partial^2}{\partial z^2} = \nabla_1^2 + \frac{\partial^2}{\partial z^2}.$$

To continue the analysis, perturbations are assumed to be periodic wave disturbances, enabling the solutions to Equations (7) and (8) to be represented in the following separable form:

$$w = e^{i\omega t}w(z)e^{i(lx+my)}, \theta = e^{i\omega t}\theta(z)e^{i(lx+my)}. \quad (9)$$

Here, l and m represent the horizontal components of the wave number, a and $a^2 = l^2 + m^2$. Here $\theta(z)$, $w(z)$ denote the amplitudes of the temperature and velocity, respectively and ω is the frequency of oscillation.

We consider FIFI boundary combination in studying RBC. Employing the Galerkin method in computing the critical values of wave and Rayleigh number, following the standard orthogonalisation procedure, we get homogenous equations in A and B.

For a solution which is non-trivial determinant of coefficient matrix needs to be zero. Expanding the determinant, we get

$$R = X + i\omega Y \quad (10)$$

Since R is real, imaginary part is equated to zero and an expression for ω^2 is obtained.

Results and discussions

In this study, a linear stability analysis is conducted on Walters B, Maxwell, Jeffreys, and Rivlin-Ericksen type fluids by adopting the Nield model [6] to describe viscosity variation with temperature, incorporating temperature-dependent internal heat generation. Here, the Prandtl number considered is larger compared to that typically assumed for

Newtonian fluids. Furthermore, the study explores Rayleigh–Bénard convection (RBC) with internal heat generation through an oscillatory mode. As reported by Siddheshwar et al. [7], when the elastic ratio exceeds or is less than unity, direct bifurcation is achievable for certain parameter ranges, whereas Hopf bifurcation occurs exclusively when the elastic ratio is below unity. Thus, the previously established necessary condition "elastic ratio less than unity" for oscillatory instability in the absence of a heat source or sink remains valid even in the presence of internal heat generation and temperature-dependent viscosity, as indicated explicitly by equation [10]. The frequency ω_c and critical wave number a_c, which govern the critical oscillatory Rayleigh number R_{oc} marking the onset of convection in a Jeffreys fluid, have been calculated for a range of parameter values and the obtained results are presented in Table 59.1. The data reveal that, in the case of oscillatory convection, stress relaxation and strain retardation have counteracting effects on the bifurcation behaviour, mirroring the trends observed in systems without internal heat generation. The presence of a heat source tends to reinforce the impact of stress relaxation, while a heat sink supports the influence of strain retardation in controlling the onset of convection.

For Jeffreys fluids, an increase in the stress relaxation parameter Λ_1 leads to a decrease in a_c and a rise in ω_c whereas increasing the strain retardation parameter Λ results in reductions in both a_c and ω_c,

Table 59.1 Oscillatory RBC – Critical values of *Roc*, a_c and ω_c for FIFI boundary for differentvalues of R_I and for the viscoelastic fluid with variable viscosity

R_I	Λ_1	Λ	Pr	$V=0$			$V=0.3$			$V=0.5$		
				R_{oc}	a_c	ω_c	R_{oc}	a_c	ω_c	R_{oc}	a_c	ω_c
-1	0.2	0.5	10	594.02	2.52	5.03	526.07	2.52	4.92	487.67	2.52	4.85
	0.3	0.33	10	384.95	2.50	7.66	340.03	2.50	7.52	314.67	2.51	7.43
	0.3	0.4	10	437.57	2.48	6.55	386.74	2.48	6.44	358.03	2.48	6.36
	0.3	0.4	15	435.61	2.47	6.84	384.59	2.47	6.76	355.76	2.47	6.71
0	0.2	0.5	10	558.20	2.50	4.62	491.88	2.50	4.51	459.28	2.50	4.43
	0.3	0.33	10	361.77	2.49	7.36	317.88	2.49	7.22	296.34	2.48	7.13
	0.3	0.4	10	410.02	2.46	6.26	360.49	2.46	6.15	336.16	2.45	6.08
	0.3	0.4	15	408.17	2.45	6.53	358.47	2.45	6.45	334.03	2.45	6.40
1	0.2	0.5	10	523.49	2.48	4.15	456.45	2.48	4.03	418.34	2.48	3.95
	0.3	0.33	10	339.28	2.47	7.032	294.92	2.47	6.89	269.73	2.47	6.79
	0.3	0.4	10	383.28	2.44	5.950	333.38	2.44	5.83	305.03	2.43	5.76
	0.3	0.4	15	381.53	2.44	6.20	331.46	2.43	6.11	302.99	2.43	6.05

Source: Author

regardless of the value of the internal heat generation parameter R_I. As R_I increases, the critical Rayleigh number R_{oc}, wave number a_c, and frequency ω_c all show a decreasing trend (Figure 59.2). In contrast, for Maxwell fluids, a_c decreases for negative values of R_I but increases when R_I is positive as seen from Table 59.2. For Newtonian fluids, a_c consistently decreases with an increase in R_I, as seen in Table 59.3.

Additionally, the results demonstrate that both R_{oc} and ω_c decrease as the viscosity variation parameter V increases (Figure 59.2). A rise in the Prandtl number Pr is found to lower R_{oc} while increasing ω_c. These effects of Pr and V are more significant on R_{oc} and ω_c than on the critical wave number a_c, and this observation holds true whether or not an endogenous thermal sources and sinks are present.

For the Rivlin–Ericksen fluid, oscillatory convection is absent as given by Dunn and Rajagopal [14], reinforcing the applicability of the principle of exchange of stabilities for this fluid class. Additionally, both Rivlin-Ericksen and Walters B fluids demonstrate behaviour akin to that of Newtonian fluids in the context of RBC. In prior studies involving Newtonian fluids with internal heat generation, [15] established that the behaviour observed under a heat sink could be deduced from that under a heat source. In contrast, the current analysis reveals that

Table 59.2 Oscillatory RBC – Critical values of R_{oc}, a_c and ω_c for FIFI boundary for different values of R_I and for the Maxwell fluid of constant/variable viscosity with $Pr=10$, $\Lambda_1=1$

R_I	V=0			V=0.3		
	R_{oc}	a_c	ω_c	R_{oc}	a_c	ω_c
-1.0	6.19929	3.79064	15.2244	6.00218	3.74125	14.1782
-0.9	6.17331	3.79017	15.2220	5.97489	3.74009	14.1696
-0.8	6.14744	3.78975	15.2196	5.94768	3.73800	14.1587
-0.7	6.12167	3.78938	15.2175	5.92056	3.73790	14.1524
-0.6	6.09601	3.78907	15.2154	5.89352	3.73687	14.1438
-0.5	6.07046	3.78881	15.2135	5.86657	3.73587	14.1352
-0.4	6.04501	3.78861	15.2117	5.83970	3.73398	14.1244
-0.3	6.01967	3.78847	15.2100	5.81291	3.73308	14.1158
-0.2	5.99443	3.78838	15.2084	5.78620	3.73316	14.1094
-0.1	5.96930	3.78835	15.2070	5.75958	3.73235	14.1008
0.0	**5.94428**	**3.78838**	**15.2057**	5.73611	3.73263	14.1091
0.1	5.91937	3.78847	15.2046	5.70659	3.73085	14.0836
0.2	5.89457	3.78862	15.2036	5.68021	3.73017	14.0749
0.3	5.86987	3.78884	15.2027	5.65391	3.72953	14.0663
0.4	5.84528	3.78911	15.2020	5.62770	3.72895	14.0577
0.5	5.82081	3.78946	15.2015	5.60156	3.72841	14.0490
0.6	5.79644	3.78986	15.2011	5.57551	3.72793	14.0403
0.7	5.77218	3.79034	15.2008	5.54953	3.72749	14.0316
0.8	5.74802	3.79088	15.2007	5.52363	3.72710	14.0229
0.9	5.72398	3.79149	15.2008	5.49781	3.72677	14.0142
1.0	5.70005	3.79217	15.2010	5.47206	3.72649	14.0054

Source: Author

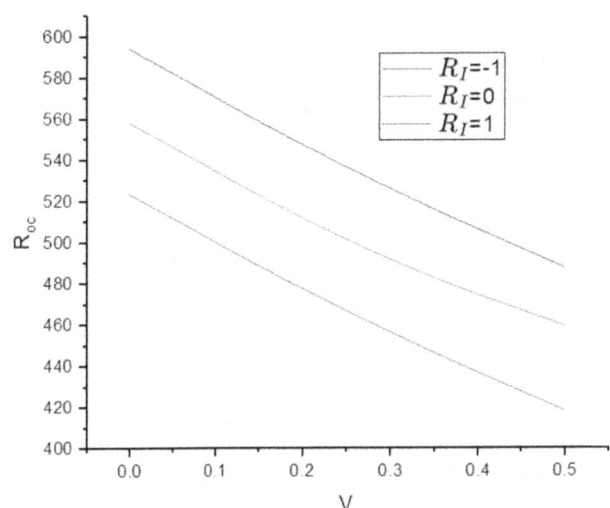

Figure 59.2 R_{oc} vs. V with $\Lambda_1=0.2$, $\Lambda=0.5$ for different values of R_I

Source: Author

due to asymmetry in the governing equations, the impacts of a thermal source and a thermal sink must be regarded as inherently different.

In the limiting case for V=0, present findings coincide with those of Siddheshwar et al. [7] and when $R_I=0$, findings of the present study coincide with those of Sekhar and Jayalatha [10] for constant/variable viscosity Jeffreys fluid. Our results are also validated for Newtonian liquid by comparing with Chandrasekhar [1].

IV. General results

1. Following are true for all V and R_I in findings of Walters B, Rivlin-Ericksen (and Newtonian), Jeffreys and Maxwell type fluids:

Table 59.3 Stationary RBC – Critical values of R_oc and ac for FIFI boundary for different values of R_I and for the Newtonian/Walters B/Rivlin-Ericksen fluid with constant/variable viscosity

R_I	V =0		V=0.3	
	R_{oc}	a_c	R_{oc}	a_c
-1.0	719.463	2.25707	634.530	2.25724
-0.9	713.187	2.25367	628.554	2.25262
-0.8	706.930	2.25023	622.593	2.24894
-0.7	700.691	2.24676	616.647	2.24522
-0.6	694.469	2.24325	610.717	2.24145
-0.5	688.265	2.23971	604.802	2.23764
-0.4	682.079	2.23613	598.902	2.23379
-0.3	675.911	2.23252	593.052	2.21588
-0.2	669.760	2.22886	587.148	2.22668
-0.1	663.627	2.22517	581.293	2.22266
0.0	**657.511**	**2.22144**	576.577	2.21801
0.1	651.413	2.21767	569.628	2.21449
0.2	645.333	2.21386	563.817	2.21033
0.3	639.269	2.21000	558.021	2.20612
0.4	633.224	2.20611	552.239	2.20186
0.5	627.195	2.20217	546.472	2.19754
0.6	621.183	2.19818	540.718	2.19318
0.7	615.189	2.19415	534.979	2.18875
0.8	609.212	2.19008	529.254	2.18427
0.9	603.252	2.18595	523.542	2.17974
1.0	597.309	2.18178	517.843	2.17514

Source: Author

$$R_{sc}^{Newtonian} = R_{sc}^{Rivlin-Ericksen}$$
$$= R_{sc}^{WaltersB}$$
$$> R_{sc}^{Jeffreys}$$
$$> R_{sc}^{Maxwell}$$

$$a_c^{Newtonian} = a_c^{Rivlin-Ericksen}$$
$$= a_c^{WaltersB}$$
$$< a_c^{Jeffreys}$$
$$< a_c^{Maxwell}$$

$$\omega_c^{Jeffreys} < \omega_c^{Maxwell}$$

2. Following is true for all V with four viscoelastic and Newtonian fluids considered:

$$R_c(R_I < 0) > R_c(R_I = 0) > R_c(R_I > 0)$$

3. Following is true for all V with four viscoelastic and Newtonian fluids considered:

$$a_c(R_I < 0) > a_c(R_I = 0) > a_c(R_I > 0)$$

4. Following observation is true for all V with Jeffreys and Maxwell fluids:

$$\omega_c(R_I < 0) > \omega_c(R_I = 0) > \omega_c(R_I > 0)$$

Conclusion

The findings of this study reveal that the variable viscosity parameter tends to enhance the system's instability, even when internal heat generation is present. Through appropriate limiting cases, the behaviour of Jeffreys fluids transitions into that of Maxwell, Newtonian, Walters B, and Rivlin-Ericksen fluids. Among these, Maxwell fluids exhibit greater susceptibility to instability compared to Jeffreys fluids, while Rivlin–Ericksen, Newtonian and Walters B fluids demonstrate relatively higher stability.

The analysis further confirms that in Rayleigh–Bénard convection, Rivlin–Ericksen and Walters B fluids exhibit dynamic characteristics closely resembling those of Newtonian fluids. This behaviour persists irrespective of the inclusion of temperature-dependent heat sources or sinks, as well as the temperature dependence of viscosity.

Additionally, the results suggest that for viscoelastic fluids with variable viscosity, the presence of a heat source promotes the onset of convection, acting as a destabilising influence, whereas a heat sink has a stabilising effect on the system.

Acknowledgement

The authors express their gratitude to their respective Institutions for their support.

References

[1] Chandrasekhar, S. *Hydrodynamic and hydromagnetic stability.* Oxford: Oxford University Press, 1961.
[2] Platten, J. K., Legros, J. C. *Convection in liquids.* Berlin: Springer, 1984.
[3] Joseph, D. D. *Fluid dynamics of viscoelastic liquids.* NewYork: Springer, 1990.

[4]   Hamabata, H. Over stability of a viscoelastic liquid layer with internal heat generation. *Int. J. Heat Mass Transfer*, 1986;29:645–647.

[5]   Othman, M. I. A., Sweilam, N. H. Electrohydrodynamic instability in a horizontal viscoelastic fluid layer in the presence of internal heat generation. *Can. J. Phys.*, 2002;80:697–705.

[6]   Nield, D. A. The effect of temperature-dependent viscosity on the onset of convection in a saturated porous medium. *J. Heat Transf.*, 1996;118:803–805.

[7]   Siddheshwar, P. G., Sekhar, G. N., Jayalatha G. Analytical study of convection in Jeffreys liquid with a heat source. *4th Internat. Conf. Fluid Mec. Fluid Power*, 2010;481:1–10.

[8]   Jayalatha, G., Suma, N. Effect of heat source on Rayleigh–Bénard convection in rotating viscoelastic liquids. *Heat Transfer*, 2021;50(8):7672–7690.

[9]   Sekhar, G. N., Jayalatha, G. Elastic effects on Rayleigh-Bénard-Marangoni convection in liquids with temperature-dependent viscosity. *Proc. ASME*, 2009:925–934.

[10]   Sekhar, G. N., Jayalatha, G. Elastic effects on Rayleigh–Bénard convection in liquids with temperature dependent viscosity. *Int. J. Therm. Sci.*, 2010;49:67–79.

[11]   Siddheshwar, P. G., Sekhar, G. N., Jayalatha, G. Surface tension driven convection in viscoelastic liquids with thermorheological effect. *Int. Commun. Heat Mass Transf.*, 2011;38:468–473.

[12]   Jayalatha, G., Sakshath, T. N., Siddheshwar, P. G., Sekhar, G. N. Impact of variable viscosity and gravity variations on Rayleigh-Bénard instabilities of viscoelastic liquids in energy sustainable system. *2024 8th Internat. Conf. Comput. Sys. Inform. Technol. Sustain. Sol. (CSITSS)*, 2024:1–6.

[13]   Sujatha, P. R., Jayalatha, G., Nivya, M., Manjunath, S. N. Investigation of variable viscosity impact on convection dynamics in viscoelastic ferromagnetic fluids. *2024 8th Internat. Conf. Comput. Sys. Inform. Technol. Sustain. Sol. (CSITSS)*, 2024:1–8.

[14]   Dunn, J. E., Rajagopal, K. R. Fluids of differential type, critical review andthermodynamic analysis. *Int. J. of Engg. Sci.*, 1995;33:689–729.

[15]   Watson, P. M. Classical cellular convection with a spatial heat source. *J. Fluid Mech.*, 1968;32:399–411.

For Product Safety Concerns and Information please contact our EU
representative GPSR@taylorandfrancis.com
Taylor & Francis Verlag GmbH, Kaufingerstraße 24, 80331 München, Germany

www.ingramcontent.com/pod-product-compliance
Lightning Source LLC
Chambersburg PA
CBHW081038220326
41598CB00038B/6913